The Science of Hi-fidelity

Second Edition

Kenneth W. Johnson
Willard C. Walker
and
John D. Cutnell

Department of Physics and Astronomy
Southern Illinois University
Carbondale, Illinois 62901

KENDALL/HUNT
PUBLISHING COMPANY
Dubuque, Iowa

Library of Congress Catalog Card Number: 81–81012

ISBN 0–8403–2297–6

Third Printing, 1984

Printed in the United States of America

C 402297 03

Contents

Preface to the Second Edition

The Science of Hi-fidelity is based on a very exciting physics course that we have been teaching to enrollments of 1,000 students per semester. Educators have discovered the effectiveness of teaching subjects that are of intrinsic student interest and relevance, and hi-fidelity possesses both of these ingredients to an unparalleled extent. Certainly hi-fi ranks as one of the all-time favorites among college students, and a large fraction of their time is devoted to the listening and enjoyment of music. In addition, hi-fi also represents a vehicle through which most of the basic laws of physics can be taught in a unified and coherent manner. A preliminary survey of the contents of this book will quickly establish that the scientific scope of hi-fi is so comprehensive that most of the traditional areas of physics must be thoroughly understood and integrated in order to elucidate its principles of operation.

Consumer education is another area that is of great importance to students today, and the attractiveness of physics would be greatly enhanced if a bridge could be established between learning scientific principles and how they ultimately relate to the quality, and hence price, of consumer products. Stated in more basic terms the following question summarizes students' feelings: "How can I use physics to get a good bargain when I buy hi-fi equipment?" This is a very fundamental question because, from the point of view of the student, an affirmative answer will immediately elevate physics from a subject as being useful only to physicists, and their like, to one of direct value and utility in their own lives. Hi-fidelity has all of its roots deeply ingrained in physical laws and it follows that the study of such principles can automatically lead to an understanding of hi-fi specifications and, hence, consumer education. The symbiotic blend of physics and consumer education opens up an exciting new twist in physics teaching which leads to substantial increases in interest and participation.

In writing this book we have chosen to interweave the subjects of science and hi-fidelity as closely as possible, because an alternating sequence allows the students to recognize immediately the applicability of physics to hi-fidelity. For example, Chapters 3, 4, 5, 7, 9, 11, and 14 are the "science" chapters which are mainly concerned with elucidating the physical concepts and laws which lie at the root of all audio systems. The "hi-fi" chapters, 2, 6, 8, 10, 12, and 13 are devoted to the various aspects of hi-fi components themselves, such as speakers, amplifiers, turntables, and the like; it is also in these chapters where all of the consumer education information is found. We have discovered that the alternating sequence between science and hi-fi brings to the course a dynamic and versatile balance between the two.

We have chosen not to use a significant amount of mathematical formalism although a limited number of useful formulas are given. We strongly believe that the overwhelming majority of science concepts can be easily presented with good illustrations and classroom demonstrations. Minimizing math, however, does **not** imply becoming less technical. Quite the contrary, the study of physics through hi-fidelity is an extremely technical endeavor—and it is one which students thoroughly enjoy.

Since the 1st edition of **The Science of Hi-fidelity** appeared in 1977, we have received a considerable amount of correspondence from colleagues who are also teaching undergraduate courses that combine physics and hi-fidelity. In addition, we have received a significant amount of feedback from students who have taken our own course. The ideas and comments of these interested people have played an important role in the development of the 2nd edition.

The most noticeable improvement is the inclusion of all new artwork with the use of blue color to highlight important features. In addition to the new figures, the text of the 2nd edition has been considerably updated and expanded to include new material. Some of the changes are:

1. A new chapter (Chapter 14) on mechanics has been added. The chapter entitled '4-Channel Sound,' which appeared in the 1st edition, was removed because 4-channel sound systems have all but disappeared from the marketplace.
2. Chapter 9 (Heat in Hi-fidelity) has been expanded to include topics such as thermal expansion, specific heat, and thermal conductivity.
3. The technical specifications for amplifiers (Chapter 8) have been completely rewritten to conform to the new standards issued by The Institute of High Fidelity.
4. Many more examples have been added throughout the text to improve the teaching of both physics and hi-fi principles.
5. The number of homework problems located at the end of each chapter has been increased substantially. In addition, the sections entitled "Summary of Terms," as well as the Glossary, have been expanded.

We hope that the improvements in the 2nd edition will make it more enjoyable to learn the interrelationship between physics and hi-fidelity.

Kenneth W. Johnson
Willard C. Walker
John D. Cutnell

Carbondale, Illinois
September, 1980

Acknowledgments

The most enjoyable part of writing a book is the opportunity to acknowledge the people who have so generously contributed their ideas, expertise, and time. The largest gratitude goes to our wives and to our children. Besides continually supporting and encouraging this work, they cheerfully picked up the extra burdens which arose because of the long "working" evenings and weekends.

To Dr. Lee Grismore we extend our appreciation for enlightened administrative leadership by providing the resources so that new ideas could be readily translated into practice. As chairman of the physics department, Lee was instrumental in helping us to establish the course, and it was indeed a pleasure to have leadership of his calibre.

Dr. Richard McKenzie has been of immense aid to us during all phases of the program. First, as a lab instructor he provided many ideas during the development of the labs. Second, he critically read the entire 1st edition and offered numerous technical and pedagogical suggestions. In our opinion many of the fine qualities of this book are directly attributable to him.

Mr. Ralph Harper and Mr. Bob Adams were instrumental in the course development during its first two years, both as lab instructors and lecturers. They introduced many new ideas and the present course is a direct reflection of their outstanding contributions.

Our appreciation also extends to Mr. Jack Trux, President, and Mr. Harry Horning, Director of Training, of Bang and Olufsen of America. They provided an enormous amount of technical assistance throughout all stages of the program. Jack Trux and Harry Horning represent the epitome of professionalism and competence within the audio industry, and we are most grateful for their continued support.

We are also indebted to Mr. Len Feldman who gave considerable assistance during all stages of the course development.

The figures were drawn by Mr. John L. Yack, Coordinator of Southern Illinois University's Commercial Graphics-Design Department, and his assistants, Mr. Dan Ford, Mr.

Mark Green, and Mr. Kent Robbins. The high quality of the figures is a tribute to their outstanding abilities as designers and artists. It is also a pleasure to thank Mr. Greg Sarber who designed the cover.

Kenneth W. Johnson
Willard C. Walker
John D. Cutnell

Chapter 1

THE FASCINATING SCIENCE OF HI-FIDELITY

The sound of music is so appealing and universally enjoyed that an enormous amount of inventive genius has been devoted to all of its facets, from recording to playback. For many of us it has always been a source of amazement to look at the myriad of clever devices provided by technology which offer convenience, flexibility, and enjoyment to the listener. Radios, record players, and tape recorders have become so commonplace that few people stop and reflect upon the basic understanding of nature that scientists have acquired in order to construct this hardware. Everyone is familiar with the concept of radio, yet it seems incredible to presume that sound can be carried by a radio wave which, in itself, defies all of our senses of sight, sound, smell, and touch. At the broadcasting station music is picked up from a disc, sent into some mysterious electronic boxes, and then finally transmitted to the antenna where it seems to disappear magically into thin air. Then at the receiving end, the radio "pulls the rabbit from the hat" by capturing the invisible waves and converting them back to a faithful reproduction of the original sounds. Quiet, efficient, fast, and beautiful!

Even on a less abstract level there are countless examples of such happenings in the audio world. Record manufacturers have no trouble in storing Beethoven's Ninth Symphony in the ultra small grooves cut into a plastic disc called a record. Stop and think for a moment. . .what an incredible transformation has occurred! A very complex and involved piece, such as the Ninth, has literally been captured and frozen in time onto a vinyl medium. It's not at all obvious just how this is accomplished and, indeed, that it can be done at all! Imagine yourself living at a point in time when all of our technology was unknown and words like record, turntable, and stylus were nonexistent. Would the idea of catching sight or sound have been apparent to you? Photography—perhaps; sound recording—no way! Photography is easy for most of us to grasp conceptually because the camera catches a scene and transcribes it into another visual picture, sight-to-sight, so to speak. Recording is a horse of a totally different color because of one very fundamental point: Just how does one take a "picture" of sound? Unlike photography, the complex audio tones must be captured and filed away in a manner which is totally different from the original sensation. Listening to sounds involves our aural senses while the storage of it on records implicates, so to speak, the touch sense. We can touch the groves and feel the mechanical undulations which cause the stylus to vibrate up and down, back and forth.

Proceeding one step further, audio engineers have also learned to utilize the magnetic sense for accumulating sound (not normally found in humans—but it seems to be a virtue of tape decks). Examine a piece of magnetic tape; it doesn't look or feel like the "Ninth," and you certainly can't hear it by putting it up to your ear, yet it's there and requires only the appropriate sensor to play it back.

Most of us will readily admit that the topic of hi-fidelity, while extremely interesting, can be very technical and is often disguised in a mathematical jargon that mystifies the average person.

Is it that nature is so fundamentally difficult to understand?

Or is it that the "learned folk" simply don't take the time and effort to explain these ideas in a way that's understandable to all of us?

The answer is, we believe, very little of the former and a whole lot of the latter. Despite the fact that most people will readily concede to a total ignorance of advanced engineering, it does seem that the language of modern technology holds a certain fascination which readily captures their interest—a fact that is widely recognized by Madison Avenue in promoting sales. When was the last time that you opened a magazine to see an ad which blasts out something like:

Most people haven't the foggiest idea of what "Wow" is, much less "Flutter," and "Rumble" has more of a street fight connotation than a legitimate technical specification. Yet we accept this type of advertising and assume that there is indeed something fine about −40 dB Rumble and, hopefully, we will look for this feature the next time we are buying a turntable.

The real question, then, is this: Admitting the fact that most of us know very little science, is there any way that audio concepts can be made understandable (and even enjoyable)?

The answer is an emphatic YES!!!

Given an exciting and informative subject, such as hi-fi, it is easy to show how the basic laws of nature can be used to understand these systems completely—not by using an arsenal of mathematical formulas, but with simple explanations and common sense.

Consumer Education in Hi-fidelity

If one comprehends the basic principles behind a hi-fi system, then there are two immediate rewards. First, there is a great satisfaction in being able to understand a highly technological subject, and it is comforting to know that the many seemingly diverse devices such as tape recorders, record players, microphones, speakers, etc., have a great deal in common. Just by simply grasping a few basic laws a whole range of audio components can be explained. Secondly, it is possible to use this information to provide a sound basis for consumer education in the field of home entertainment. At the end of this book you'll be able to stand eyeball-to-eyeball with any salesperson and possess confidence in your ability to comprehend and judge audio equipment.

'Consumer Education' is a worthy goal but its attainment is not as easy as it sounds. To be a good consumer one must know how to specify products and this means knowing the technical terms and their implications. Therefore, a lot of time will be spent in the upcoming chapters learning how to translate general knowledge into specific details. For sure this is hard work, but it's at the heart of any consumer education program.

Here is a summary of the three goals that we have set out to attain in this book:

1. Examine, in detail, all of the features of hi-fidelity sound systems to show their diversity, flexibility, and utility.
2. Investigate the basic laws of nature in order to explain and understand how audio works. These concepts will come from the areas of sound, electricity, magnetism, heat, radiation, energy, and mechanics.
3. Use this basic scientific information as a foundation for consumer education in the field of hi-fidelity and audio.

The attainment of these goals should provide you with an excellent understanding of audio systems—a subject that is exciting, dynamic, and rapidly growing. We are sure that you will find the study of scientific technology quite fascinating and well worth your while.

Chapter 2

THE ANATOMY OF HI-FIDELITY SYSTEMS

2.1 The Building Blocks of a Hi-fi System

The popularity of audio has grown to such an extent that terms like "hi-fi" and "stereo" have become household words. They describe an immense number of systems whose commercial availability reflects the needs of almost every user. If you have ever browsed around a hi-fi store you've probably noticed the large variety of audio components on the shelves; radios, tape decks, receivers, preamplifiers, integrated amplifiers, etc. are quite common. Upon seeing this array of equipment you might well ask: Why are there so many different types of audio components? What does each type do? And, how do I choose a hi-fi system that fits my needs? The answer to each of these questions can be made clear by utilizing the *building block* approach, which illustrates how individual audio components are combined in order to produce a complete audio system.

The usefulness of the building block approach may be unfamiliar to some, but it is readily apparent in the way motor vehicles are assembled. Cars, trucks, buses, and earth movers certainly look different, but everyone knows that each must have a motor, a steering system, brakes, a transmission, and the like. These, then, represent the building blocks which must be integrated to produce an operational vehicle, as shown in Figure 2.1.

Hi-fi systems are certainly no different because they are also built-up from more elementary units, called the *sound sources,* the *preamplifier/control center* (or *"preamp"* for short) the *power amplifier,* and the *speaker(s).* Figure 2.2 illustrates how these four building blocks are assembled to produce a complete sound system. The sound source—be it a turntable, tape deck, tuner, or microphone—produces the tiny amounts of electricity which contain the musical information. These minute electrical "signals" are directed to the preamplifier/control center. After being processed by this

Figure 2.1. Even though motor vehicles may look different, they are all constructed from the same building blocks. These building blocks are the motor, steering system, brakes, etc.

Figure 2.2. The four basic building blocks of any sound system. Notice that the (1) sound source (turntable, tape deck, tuner, or microphone) directs its output of weak electrical signals to the (2) preamplifier/control center which, after the appropriate processing, passes them along to the (3) power amplifier, the unit which powers the (4) speakers.

unit, the signals are then sent to the power amplifier where they are substantially strengthened. Finally, the strengthened signals are routed to the speakers where they are converted into sound. It should be emphasized that all sound systems, no matter how simple or complicated they appear, contain these four building blocks.

In order to help you gain a better understanding of the building blocks themselves, the next four sections will provide brief introductions to the sound sources, the preamplifier/control center, the power amplifier, and the speakers. Section 2.6 will illustrate how audio manufacturers assemble the building blocks to produce the great variety of hi-fi systems which appear on the market. Finally, sections 2.7 and 2.8 will discuss the more important operational features of hi-fi systems.

Summary

All hi-fi systems are built up from four elementary units called the "building blocks." These blocks are:

1. the sound sources (turntable, tape deck, tuner, microphone)
2. the preamplifier/control center
3. the power amplifier
4. the speakers.

The minute electrical signals, which contain the musical information, originate at the sound sources. The signals are first sent to the preamplifier/control center and then to the power amplifier. After leaving the power amplifier the electrical signals are directed to the speakers, where the signals are converted into sound.

2.2 The Sound Sources (The First Building Block)

While the sound sources portrayed in Figure 2.2 are familiar to most people, the terminology associated with them may not be. Since this terminology will prove useful throughout subsequent discussions, it is summarized below.

A **turntable** is a very popular type of sound source, and it will be discussed in greater detail in Chapter 12. A turntable consists of three major parts:

Garrard Model GT 350AP Turntable

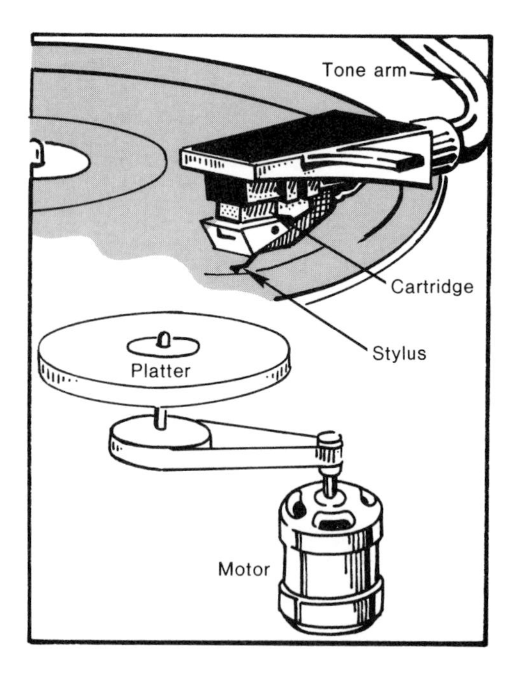

A. **Tone Arm**
This arm holds the cartridge and guides it across the record.

B. **Cartridge** (or **Pickup**)
This small, and usually expensive, unit is located at the end of the tone arm and contains the stylus (needle). The job of the cartridge is to produce an electrical signal from the movement of the stylus as it follows the record groove. The electrical signal generated by a magnetic phono cartridge is extremely weak (about 5 thousandths of a volt*), and it must be strengthened considerably by the preamplifier/ control center before it can cause the speakers to create audible sound.

C. **Platter and Motor**
The circular platter, which is rotated by the motor, holds the record and rotates at speeds of either 78, 45, or 33 ⅓ rpm (revolutions per minute).

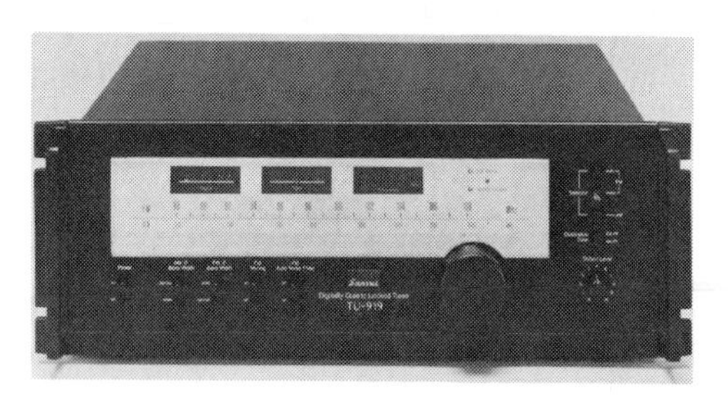

Sansui Model TU-919 Digital Controlled, Quartz-Locked AM/FM Tuner

A **tuner** is not a radio, but it is an electronic device that permits one to tune in radio stations. A radio involves a lot more than just tuning in a station, as will be discussed in section 2.6. A tuner's job is to process the radio waves which are picked up by the antenna and convert them into electrical signals containing the audio information. There are two general types:

A. AM tuners for tuning in AM stations. (What else?)
B. FM tuners which select FM stations. In addition, an FM tuner may contain a "stereo multiplexer" which allows the tuner to receive stereo as well as monaural broadcasts.

Most hi-fi units include both FM and AM tuners, because the addition of AM circuitry involves little further cost over that required by an FM tuner alone. The tuner itself does contain some amplifiers, so its output voltage is around 0.5 volts, which is considerable higher (by a factor of about 100) than that produced by a magnetic phono cartridge. However, the 0.5 volts is still not strong enough to actuate loudspeakers and, consequently, the electrical signal must be routed to the preamp-power amp combination, as shown in Figure 2.2. Want to know more about tuners? See Chapter 10.

A **tape deck** is a unit which records onto and plays back from a magnetic tape. Currently it is the only sound source that allows the user to store as well as retrieve audio information. There are three types of tape formats that are used

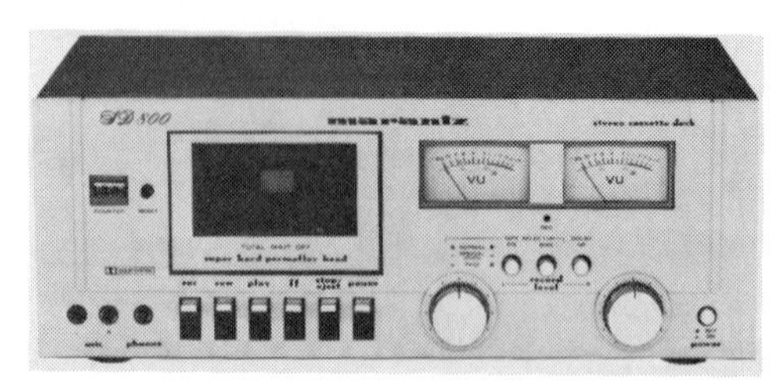

Marantz Model SD-800 Stereo Cassette Deck

*Voltage is the amount of electrical energy which is given to each unit, or "Coulomb," of moving electrical charge. The voltage ratings of some common devices are: flashlight batteries (1.5 volts), car batteries (12 volts), household electrical outlets (120 volts), and magnetic phono cartridges (about 0.005 volts).

today: open reel, cassette, and 8-track cartridge. The comparative advantages of each of these types will be discussed in Chapter 13.

In general, a tape deck has facilities for recording, playback, and erasure. It also provides some electrical amplification of the weak signals detected by the playback circuitry, and the resulting output voltage is approximately 0.5 volts. As is the case with any sound source, this voltage level is not strong enough to drive large speakers, although one can use headphones for listening. Therefore, the electrical output of the tape deck must follow the usual route through the preamp and power amplifier to the speakers.

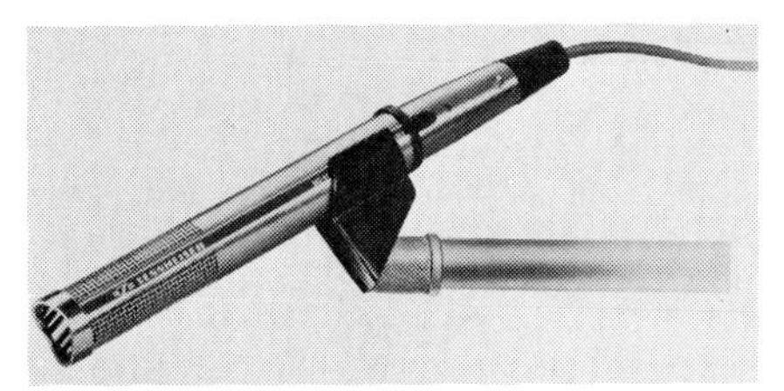

Sennheiser Electronic Corporation Model MD 402 11 Microphone

A **microphone** is an essential addition to any home sound system for those who are involved with live performances. With the connection of a microphone to the preamplifier/control center your hi-fi system becomes a public-address system suitable for home use. If you own a tape deck the microphone also can be used for recording live material. The recording may be as simple as someone dictating notes, or as complicated as a rock group utilizing a very sophisticated multiple-miking procedure with special microphones, microphone placement, and mixing units. A microphone is not needed (nor should it be used) when tape recording either records, AM/FM broadcasts, or other sources. The correct procedure for taping such material will be discussed in sections 2.7 and 2.8.

2.3 The Preamplifier/Control Center (The Second Building Block)

Rotel Preamplifier Model RC-5000

The preamplifier/control center, or"preamp" for short, is really the control center of a hi-fi system, because the majority of the adjustments on a hi-fi system are made with the knobs and buttons which are involved with the preamp's operation. The VOLUME, BASS and TREBLE tone controls, BALANCE, LOUDNESS, STEREO/MONO switch, and the HIGH and LOW frequency filters, if present, are representative of its versatility. (These controls are discussed in Chapter 8.) All of the above functions are located in the preamp, because designers have found it very easy and economical to manipulate the electrical signal at the lower power levels in the preamp rather than at the higher power levels which exist in the power amp.

In addition to controlling the tonal characteristics of the sound, the preamp is capable of routing the signal into any one of a variety of auxiliary devices selected by the user. This latter function is more fully explored in section 2.8.

2.4 The Power Amplifier (The Third Building Block)

Onkyo Stereo Power Amplifier Model M-505

Unlike the preamp, the power amplifier is reasonably simple looking since it has few controls accessible to the user. The power amplifier receives the relatively weak electrical signals from the preamp and produces signals powerful enough to drive speakers at loud listening levels. The ability of a power amplifier to produce sufficiently strong electrical signals is measured by its output power, which is rated in *watts*. High power ratings imply that the power amplifier can send large amounts of electrical power to the speakers which produce correspondingly loud sounds. This situation is similar to the manner in which a 100 watt light bulb produces more light than a 50 watt bulb.

From a consumer point of view, the output power is a most important consideration because the cost of an amplifier largely depends on the number of watts that it can produce. Other things being equal, more watts mean more dollars! Chapter 8 contains a discussion on power amplifiers.

2.5 The Speakers (The Fourth Building Block)

Bose Corporation

The speakers are the part of a hi-fi system which actually convert electricity into sound, and the fascinating way in which they accomplish this job will be discussed in Chapters 6 and 11. The conversion of electricity into sound is perhaps the most difficult task that a hi-fi component performs, and, for this reason, it has often been said that the speakers are the weakest link in the audio chain.

Choosing the right speakers involves the very personal judgment of what constitutes appealing sound quality. This is one reason why speakers are generally sold separately, and the fact is that there are more manufacturers of speakers than of any other component.

In addition to the consumer oriented matter of catering to personal tastes in sound quality, there is a good technical reason why speakers should be sold separately: speakers need to be physically set apart from the remainder of the system, especially the turntable. Because of its nature, the phono cartridge is quite sensitive to extraneous vibrations of the turntable platter. Speakers, on the other hand, are excellent vibrators since that is precisely how they produce the sound. Place these two close together on the same stand and disaster can result! Low-frequency vibrations from the speaker (which can easily be felt by placing your finger on the speaker cabinet) are transmitted right through the stand and force the record on the platter into unwanted vibration. A very unpleasant rumble or chattering noise may be produced. With loud sound levels these low-frequency vibrations can even cause tone arm jumping in which the stylus actually leaves the groove and skates across the record. This undesirable acoustic interaction between a loudspeaker and the turntable is termed *acoustic feedback,* and can be reduced by simply separating these two components.

A Cardinal Rule Of High Fidelity:
Always separate the speakers from the turntable. Never place the turntable on top of the speakers, or on the same shelf or table.

2.6 What You Can Buy At the Store

Using the building block concept, let us now look at the possibilities which exist when purchasing a sound system. We have emphasized that any working sound system must contain all four of the building blocks shown in Figure 2.2. However, it should be recognized that the buyer has the option of purchasing the blocks separately, partially combined, or completely assembled in a self-contained unit.

Generally speaking, hi-fi systems can be divided into two groups: The *component* systems and the *preassembled* systems. If you wish to assemble a hi-fi system by purchasing the individual blocks, either separately or in certain combinations, then the component system is best suited to your needs. Alternately, if you wish to buy a hi-fi system in which the four blocks have been already put together by the manufacturer, then the preassembled or "packaged" system is for you. We shall now discuss the merits of each of these systems, and show how they are both related to the building block approach.

Component Hi-Fi Systems

The component approach to buying an audio system is very popular among hi-fi enthusiasts. The most straightforward method is to purchase separate units for each of the four building blocks. While buying four separate units can be relatively expensive, this method allows the buyer great flexibility in choosing favorite makes and models for each of the building blocks. In addition, there are some listeners who require extremely large amounts of power for full audio enjoyment. By large power levels we mean over 200 watts per channel, and such monstrous amounts of power are usually available only from separate power amplifiers. Figure 2.3 shows a hi-fi system which results from assembling separate components for each of the four building blocks.

Figure 2.3. The zenith of high-fidelity systems consists of a four-part system where the sound sources, preamplifier/control center, power amplifier, and speakers are purchased as separate units. This approach is preferred by those who require extremely large powers, and is usually the most expensive of the possible component systems.

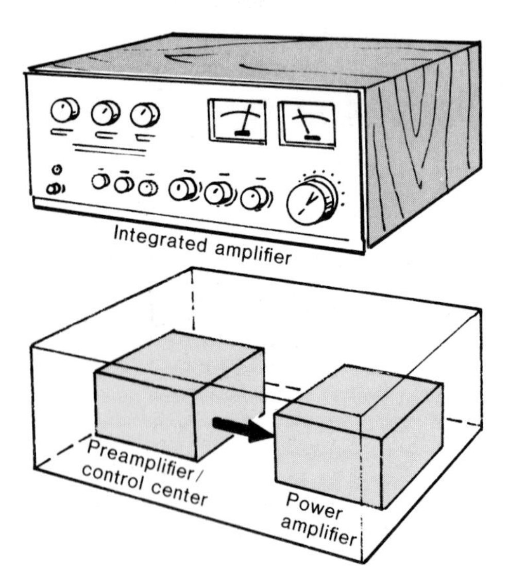

Figure 2.4. An integrated amplifier is a single unit which contains the preamplifier/control center and the power amplifier.

An **integrated amplifier** can also be used to assemble a component system. An integrated amplifier is a preamplifier/control center and a power amplifier contained in one unit, as shown in Figure 2.4. It is somewhat less expensive than buying two separate units because of the economy that results in having one enclosure rather than two, and the elimination of certain duplicated features, such as electrical plugs and controls. Furthermore, a separate preamp and power amp must each contain rather expensive "power supplies" within them. These power supplies provide the transistors with the necessary voltages and currents for proper operation. In an integrated amplifier a common power supply can be shared by both the preamp and power amp with a resulting savings in cost. Figure 2.5 shows the appearance of a component system based on the integrated amplifier, and it should be stressed once again that the buyer has the option of purchasing each of the units from the same, or from a different, manufacturer.

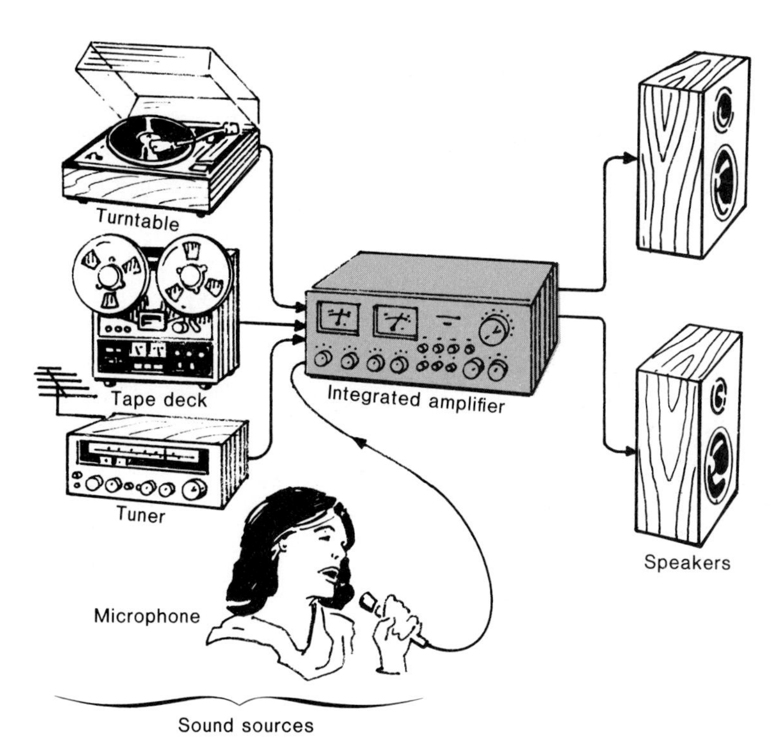

Figure 2.5. An integrated amplifier contains both a preamplifier/control center and a power amplifier. Only the sound sources and speakers must be added to complete the system.

Using a **receiver** is another alternative for assembling a component system. As shown in Figure 2.6, a receiver combines a tuner, a preamplifier/control center, and a power amplifier into one unit, with the concomitant cost savings over buying each component separately. Receiver based systems are, without a doubt, the most popular and least expensive systems for those people who favor the component approach. When buying a receiver, all you have to do is add the speakers and you can listen to your favorite radio station. Of course, as illustrated in Figure 2.7, a turntable, tape deck, or microphone can also be easily attached to the system.

Component systems are very popular because they offer flexibility, easy upgrading, and the assurance of quality performance. For example, the flexibility provided by the component approach becomes evident when a turntable, a tape deck, or a microphone, perhaps from different companies, can easily be added to arrive at a system which best suits your needs. Why would anyone want to purchase components from different manufacturers? We do it all the time when buying clothes: pants from one manufacturer, shirts from another, and shoes from a third. No problems. In fact, it is this diversity that allows the buyer to bargain hunt, fashion coordinate, and just plain have fun when shopping around.

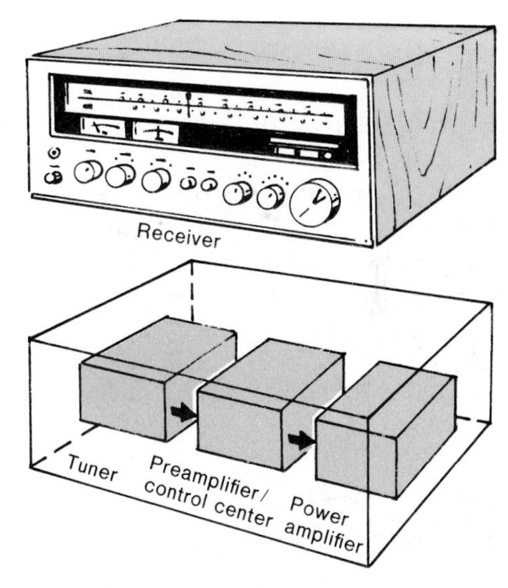

Figure 2.6. A receiver is a unit which contains three important hi-fi units: the tuner, the preamplifier/control center, and the power amplifier.

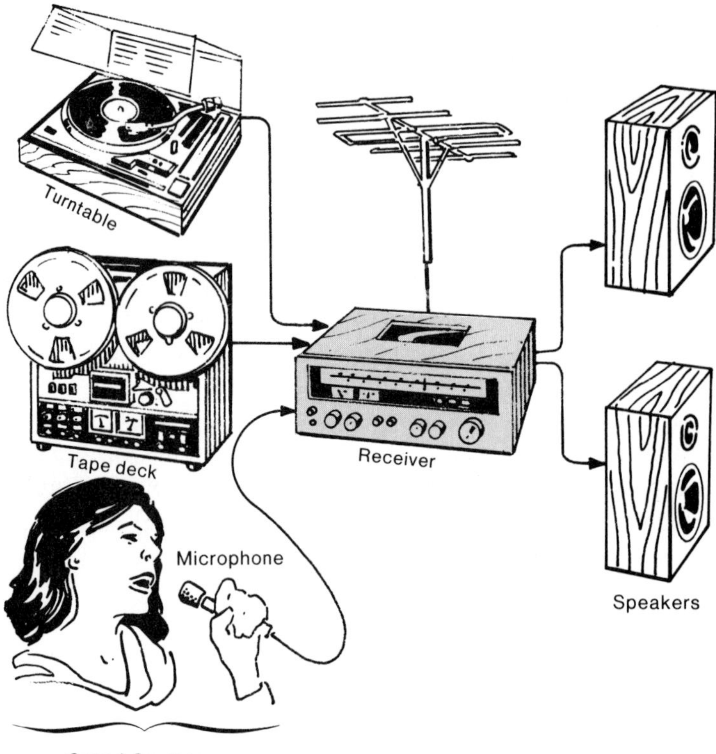

Figure 2.7. A receiver is, by far, the most popular of the audio components. Combining the tuner, preamplifier/control center and the power amplifier into one unit provides economy that lowers the cost, and yet permits the user substantial flexibility in selecting other sound sources and speakers for completing the system.

With components one can update part of a hi-fi system conveniently, because whenever new technology or wear-and-tear dictate a new component, the whole system does not have to be thrown out. If, for example, you should decide to trade-in a turntable for a newer model, it is a simple matter to disconnect the old unit from the remainder of the system and hook up the new one. In addition, other components such as reverberation units, equalizers, and noise reduction systems, can be easily added to an existing component system as your budget permits.

Probably the most compelling reason to buy a component system is that the most reputable manufacturers include a complete list of specifications with each component. These specifications can give the prospective buyer a rather complete picture of the component's performance capabilities, and they are an immeasurable aid in selecting the best quality sound system for the money.

Manufacturers of audio components are anxious to provide the customer with multiple choices, whether it is a receiver, integrated amplifier, or an elaborate four-part system. Within each group products can be purchased which meet the standards for hi-fidelity reproduction, and the final consideration as to which type best suits your needs should be based on convenience, flexibility, and economy. To give you some idea about the relative popularities of the various alternatives in component buying, the list below shows some approximate sales percentages: receivers 80%, integrated amps 15%, separate preamps and power amps 5%.

Preassembled Hi-Fi Systems

As the name suggests, the four building blocks in these systems have been selected and preassembled by the manufacturer. All you have to do is turn them on.

Portables are excellent examples of preassembled systems. Figure 2.8 illustrates a portable which combines a radio and a tape recorder; of course it is possible to buy portables which are either a radio or a tape recorder alone. Small radios which are often found around the home, AM/FM car radios, and CB radios, are also considered as portables, although they usually lack the convenient carrying handles. As you can see, portables may, or may not, use batteries as their only source of power, because many portables rely solely on household power (or the electrical power generated by a car) for their operation. In any case, portables are complete audio systems which contain all four building blocks. The source may be a turntable, tuner, tape deck, microphone, or some combination of the four.

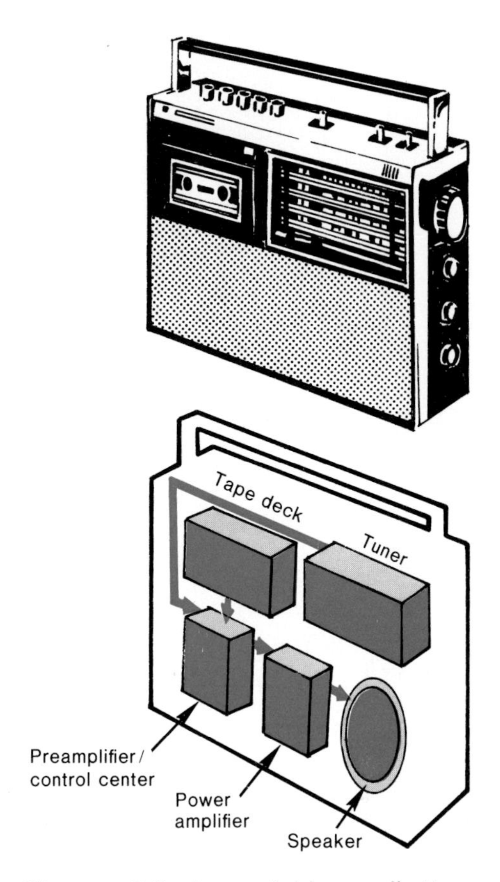

Figure 2.8. A portable radio/tape player and its four building blocks. Notice that there are two sound sources: a tape deck and a tuner.

Compacts are preassembled hi-fi systems which have become popular with many people. As shown in Figure 2.9, compacts are easily recognizable because the sound sources, preamp, and power amplifier are contained in a single unit, with the speakers usually attached by long wires, thus allowing the speakers to be located where ever you wish. (Remember, however, the Cardinal Rule of Hi-Fi; keep the speakers separate from the turntable!)

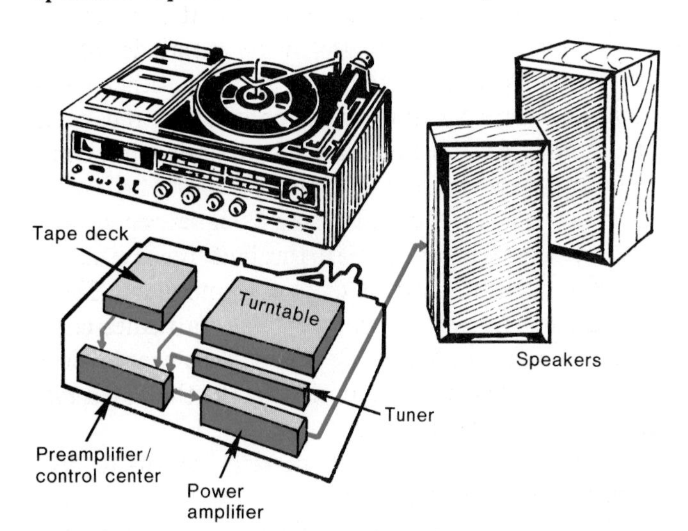

Figure 2.9. A typical compact system which contains three sound sources: a tape deck, a turntable, and a tuner.

Compacts almost always contain a turntable as a sound source, and the more expensive units may include a tuner, and perhaps either a cassette or a cartridge tape deck. Compacts offer the buyer the convenience of not having to select and assemble all the components—a task which the manufacturer has kindly done. Compacts can also be conveniently moved from room to room, although they are usually more bulky than the portables.

Although many of the more expensive compacts offer good sound, a valid complaint is that many compact manufacturers do not provide the technical specifications of the building blocks. In the absence of such data it becomes most difficult, if not impossible, for the consumer to make sensible decisions regarding the quality of the hi-fi equipment. The equipment could be very "low-fi!" Whenever purchasing audio components the buyer should look for these specification figures. Specifications often have a revealing, and many times surprising, story to tell, as we will discuss in later chapters. To understand the story, however, the buyer must know in advance how to interpret such information in order to make an intelligent decision; and that's what this book is all about!

Consoles also represent preassembled audio systems in which each building block, including the speakers, is contained in a large furniture-type enclosure, as shown in Figure 2.10. They are usually long, low cabinets with either a front door or top-lift access to the system. Quite clearly consoles are not very portable, and they are purchased by those whose interests are shared equally between music and furniture. Console manufacturers offer the consumer a variety of source combinations, with tuners and turntables being the most common. Some consoles can sound reasonably good in certain circumstances, although it is rare when a console manufacturer details the specifications which indicate the quality of the building blocks; the buyer could be getting cheap equipment in a fancy cabinet.

Notice that some consoles violate the cardinal rule of hi-fi by housing the speakers in the same cabinet as the turntable. In order to prevent the severe vibrations which could result from acoustic feedback, console designers sometimes deliberately restrict the volume of the bass notes to levels which are less than natural. Of course, this reduces the severity of the speaker vibrations and helps to eliminate the unpleasant side effects of acoustic feedback; however, don't forget that musical performance has also been deliberately compromised.

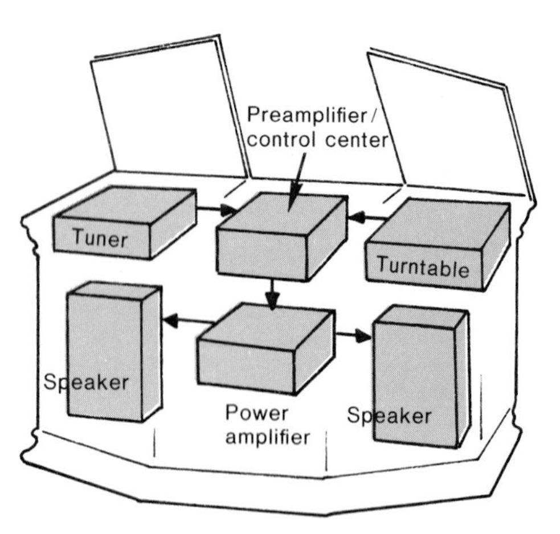

Figure 2.10. A console which contains a tuner and a turntable as the sound sources.

Summary

There are four building blocks which must be present in any hi-fi system: the sound source(s), the preamplifier/control center (the "preamp"), the power amplifier, and the speaker(s). With the component approach, the buyer selects each block, or certain combinations of the blocks, from the same, or different manufacturers. As indicated in Table 2.1, the receiver and the integrated amplifier are components which contain combinations of several building blocks; a receiver is the combination of a tuner, preamp, and power amplifier, while an integrated amplifier contains a preamp and a power amplifier. Preassembled hi-fi systems include such familiar units as portables, compacts and consoles. These units contain all four building blocks which have been preassembled for the convenience of the buyer.

Regardless of the type of audio system (component, portable, compact, or console), please do not confuse a tuner with a radio, a turntable with a phonograph (or a record player), and a tape deck with a tape recorder. Tuners, turntables and tape decks are only the sound sources. Radios, phonographs and tape recorders, respectively, are complete audio systems which are based on these sound sources.

The next time you walk into a hi-fi store and see the bewildering array of hardware on the shelves, don't be alarmed. By utilizing the building block approach you will be able to recognize immediately where each piece fits into the overall system. Table 2.1 also indicates how the preassembled hi-fi systems utilize the four building blocks.

	THE FOUR BUILDING BLOCKS OF A HI-FI SYSTEM						
	1. SOUND SOURCES				2. PREAMP/ CONTROL CENTER	3. POWER AMPLIFIER	4. SPEAKERS
	TUNER	TURNTABLE	TAPE DECK	MICROPHONE			
COMPONENTS							
RECEIVER	✓				✓	✓	
INTEGRATED AMPLIFIER					✓	✓	
PREASSEMBLED SYSTEMS							
PORTABLES	USUALLY INCLUDED	SOMETIMES INCLUDED	SOMETIMES INCLUDED	SOMETIMES INCLUDED	✓	✓	✓
COMPACTS	USUALLY INCLUDED	USUALLY INCLUDED	SOMETIMES INCLUDED	SOMETIMES INCLUDED	✓	✓	✓
CONSOLES	USUALLY INCLUDED	USUALLY INCLUDED	SOMETIMES INCLUDED	SOMETIMES INCLUDED	✓	✓	✓

Table 2.1 A listing of the building blocks which are contained in various component and preassembled hi-fi units. A checkmark indicates that the particular building block is found within the unit listed on the left. In addition, it is also possible to purchase each of the building blocks as a separate component.

2.7 How the Electrical Signal Is Routed Through a Receiver

Many people own hi-fi systems which utilize one or more sound sources. In addition to the familiar turntable and AM/FM tuner, tape decks are now widely used. With a tape deck it is possible to record favorite selections from other sound sources such as records, a tuner, or even another tape deck. A well designed hi-fi system allows such tape recording operations, because it permits the user to control how the electrical signals are routed through the system.

In order to visualize the various paths for the flow of electricity through a hi-fi system, consider Figure 2.11. This picture shows a turntable, a tape deck, an FM antenna and a speaker connected to the rear panel of a receiver. * Normally every component, except for the antenna, is attached to the receiver by two cables—one for each of the left and right stereo channels. For the sake of simplicity, only one channel, and hence only one cable, is drawn in the figure. The cables from the sources and speakers are attached to the appropriate connectors located on the back panel of the receiver. These connectors are called "jacks" and they are symbolically represented as small circles in the drawing.

Figure 2.11 illustrates an example where a turntable is connected to the PHONO jack, a tape deck is connected to the AUX (AUXiliary) jack, and an FM antenna is attached to the antenna connector. Most receivers also contain an AM tuner (not illustrated in the figure) with its own built-in antenna, although there is usually a provision for attaching an external AM antenna for better reception.

The set-up shown in Figure 2.11 will allow the user to select any one of the three sound sources. However, the tape deck which is attached to the AUX jack can be used only to play back pre-recorded tapes; it cannot be used to record music. In order to record, the tape deck must be connected differently, as will be discussed in the next section.

Note: The AUX input jack may also be used to connect another tuner (either AM or FM) to the receiver. However, the AUX jack may not be used to attach another turntable for reasons which will be discussed later.

As an example of how a hi-fi system operates, we will now follow the electrical signal as it originates at the turntable, travels through the receiver, and finally arrives at the speakers, where it is converted into sound.

*Because of its popularity, the receiver will be used throughout this section for illustrating how the signal flows through a hi-fi system. However, the concepts to be discussed apply equally well with either an integrated amplifier or a separate preamp and power amp combination.

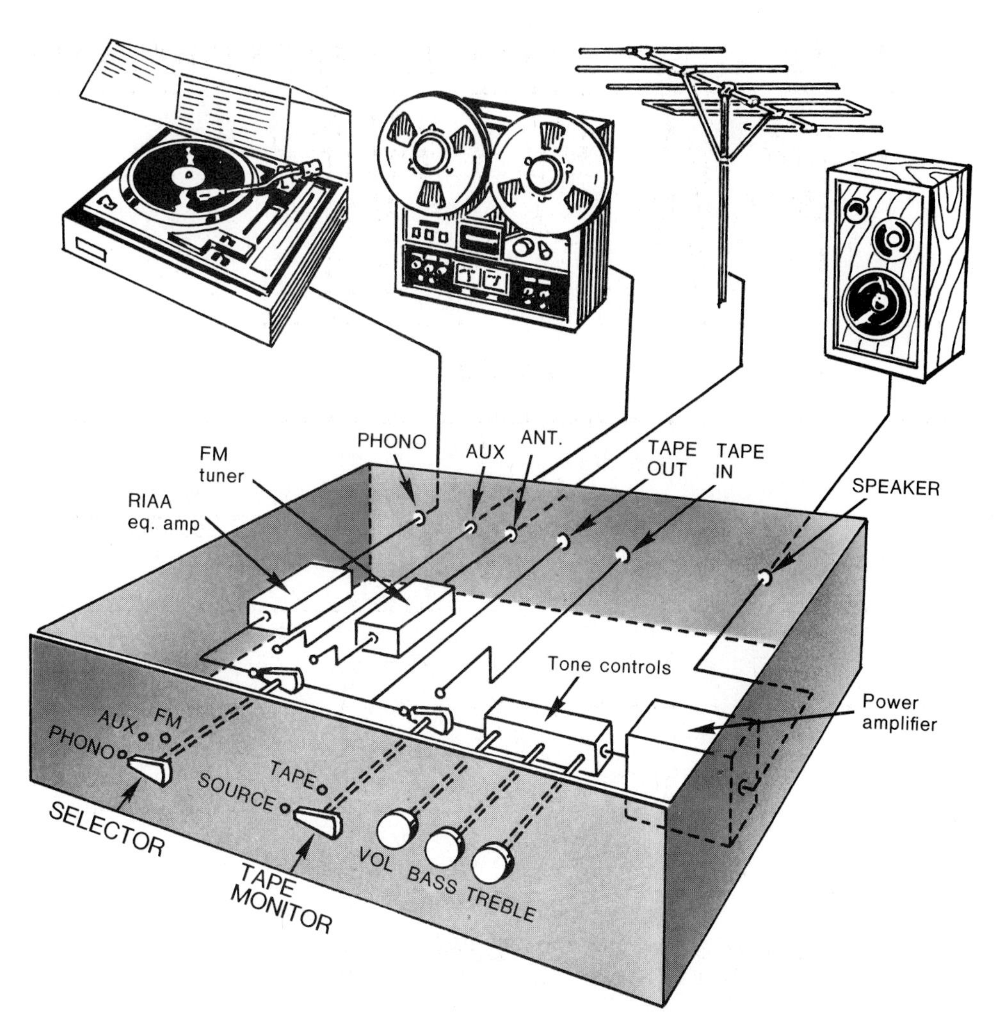

Figure 2.11. A component hi-fi system with a transparent view of the receiver. Only one of the two stereo channels is shown for simplicity.

The Phono Signal First Enters the Receiver and Passes Through the RIAA Equalization Amplifier

As illustrated in Figure 2.12, the signal from the turntable cartridge enters the receiver and first encounters the RIAA equalization amplifier. One of the tasks of this amplifier is to boost the relatively weak signal to a much stronger value. In addition to boosting the signal, the amplifier also "equalizes" the signal—a required process for phono signals which will be explained in more detail in Chapter 12. Notice that the RIAA equalization amplifier is unique to the PHONO input, and it is not present in either the AUX or the ANT inputs. Therefore, a turntable must be connected only to the PHONO jack and not, for instance, to the AUX jack.

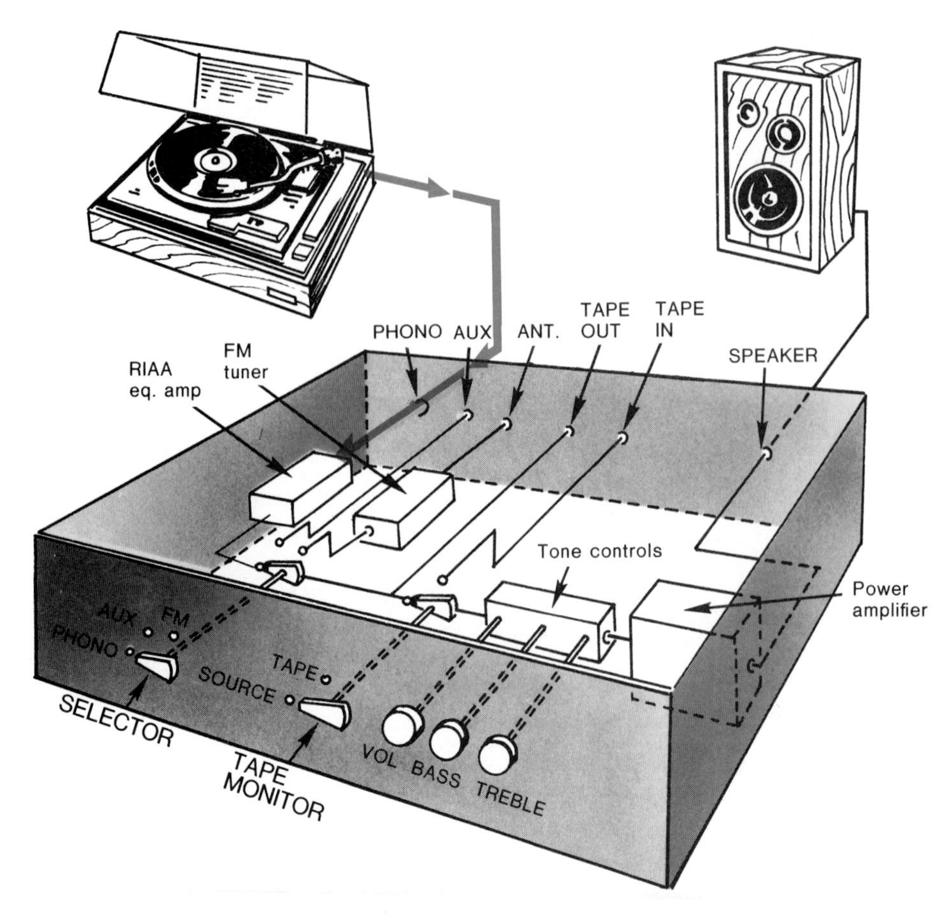

Figure 2.12. The signal from the turntable first encounters the RIAA equalization amplifier after it enters the receiver.

Most turntables which are used in component hi-fi systems are equipped with magnetic cartridges. Occasionally, however, some turntables utilize ceramic cartridges which do not work according to the same principle as magnetic cartridges. The electrical signal from a ceramic cartridge is much stronger than that from a magnetic cartridge (typically 0.1 volt versus 0.005 volt). If you own a turntable which has a ceramic cartridge do not attempt to connect it to the standard PHONO jack; the receiver must have a special ceramic phono input jack that is needed to accept this signal.

The Signal Next Arrives at the SELECTOR Switch

Upon leaving the RIAA equalization amplifier the boosted and equalized phono signal encounters the SOURCE SELECTOR (or SELECTOR, for short) switch. As shown in Figure 2.13, the SELECTOR switch is mechanically linked to a front-panel knob. This knob allows the user to select which source (the turntable, in this case) is to be sent to the remainder of the hi-fi system. If the SELECTOR switch is rotated to either the AUX or FM

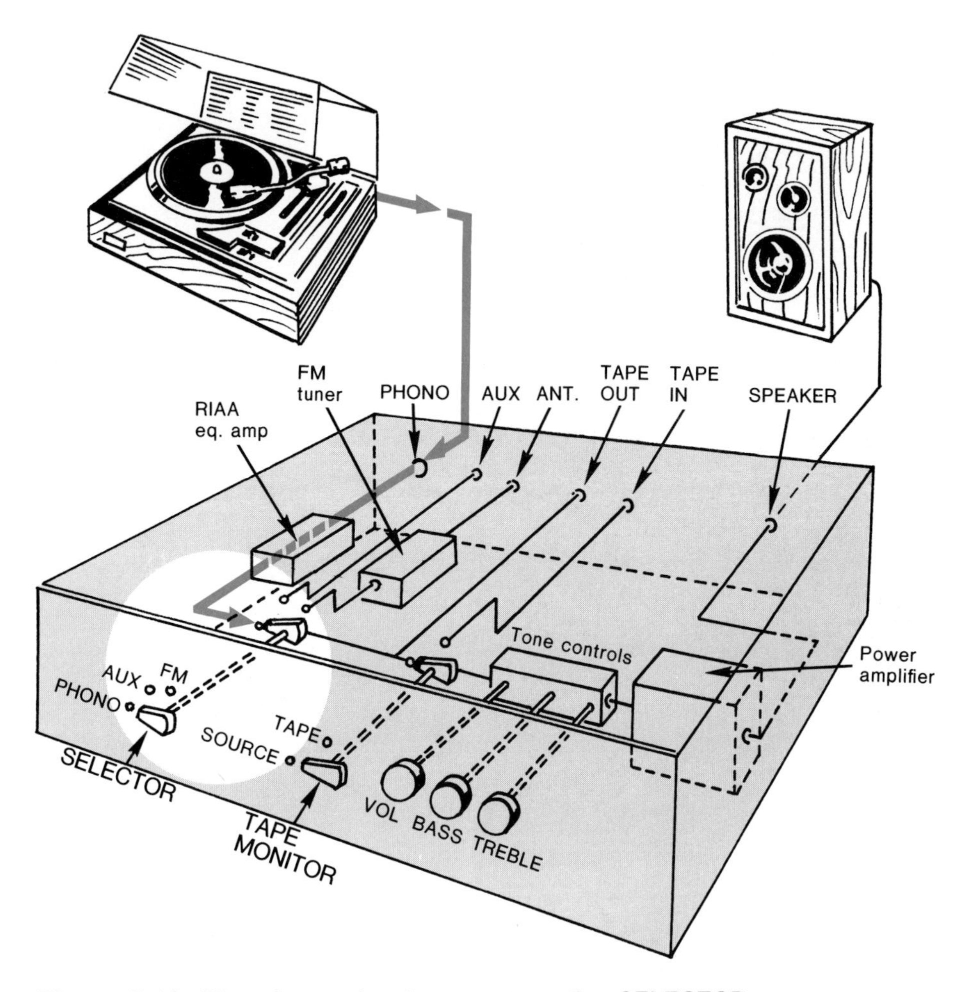

Figure 2.13. The phono signal encounters the SELECTOR switch after being processed by the RIAA equalization amplifier.

position, then the signal from the musical source attached to the corresponding jack will be sent to the speaker. It should be noted that the SELECTOR switch may have a variety of other names, depending on the manufacturer. For example, SOURCE, FUNCTION, and MODE are alternate labels that are sometimes used.

Please note that any source signal, regardless of its point of origin, travels the same path through the receiver once it passes beyond the SELECTOR switch.

The Signal Encounters a Fork in the Path and the TAPE MONITOR Switch

After passing through the SELECTOR switch, the signal arrives at a junction where the path splits into two parts, as shown in Figure 2.14. One branch of the path leads part of the signal directly to the TAPE OUT jack on the rear panel of the receiver. It is the TAPE OUT jack which allows the signal to be sent to a tape deck for recording purposes. The

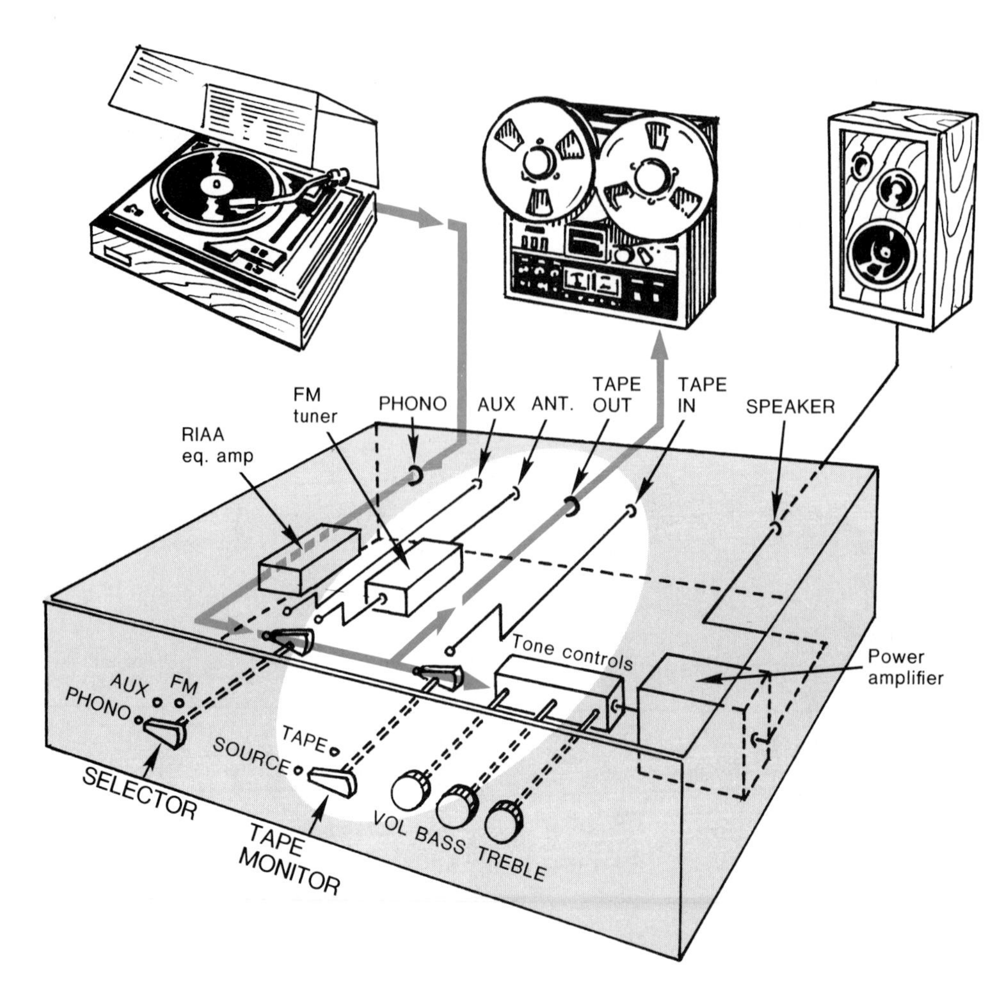

Figure 2.14. The signal from the source (e.g., turntable) is always present at the TAPE OUT jack for tape recording purposes. In addition, the signal is also sent to the tone control section when the TAPE MONITOR switch is set to its SOURCE position.

signal from the appropriate source (as determined by the setting of the SELECTOR switch) is always present at the TAPE OUT jack, and if you wish to make a tape recording simply connect a tape deck to this jack.

The second branch of the path directs the electrical signal to the TAPE MONITOR switch. Like the SELECTOR switch, the TAPE MONITOR switch is also controlled by a front-panel knob, as shown in Figure 2.14. When this knob is set to the SOURCE position, the signal from the source is allowed to proceed through the remainder of the receiver and out to the speakers. Thus, a well designed receiver makes it possible for the user to make a tape recording while simultaneously listening to the music coming directly from the source. The function of the TAPE MONITOR switch when it is placed in the TAPE position will be considered separately in section 2.8.

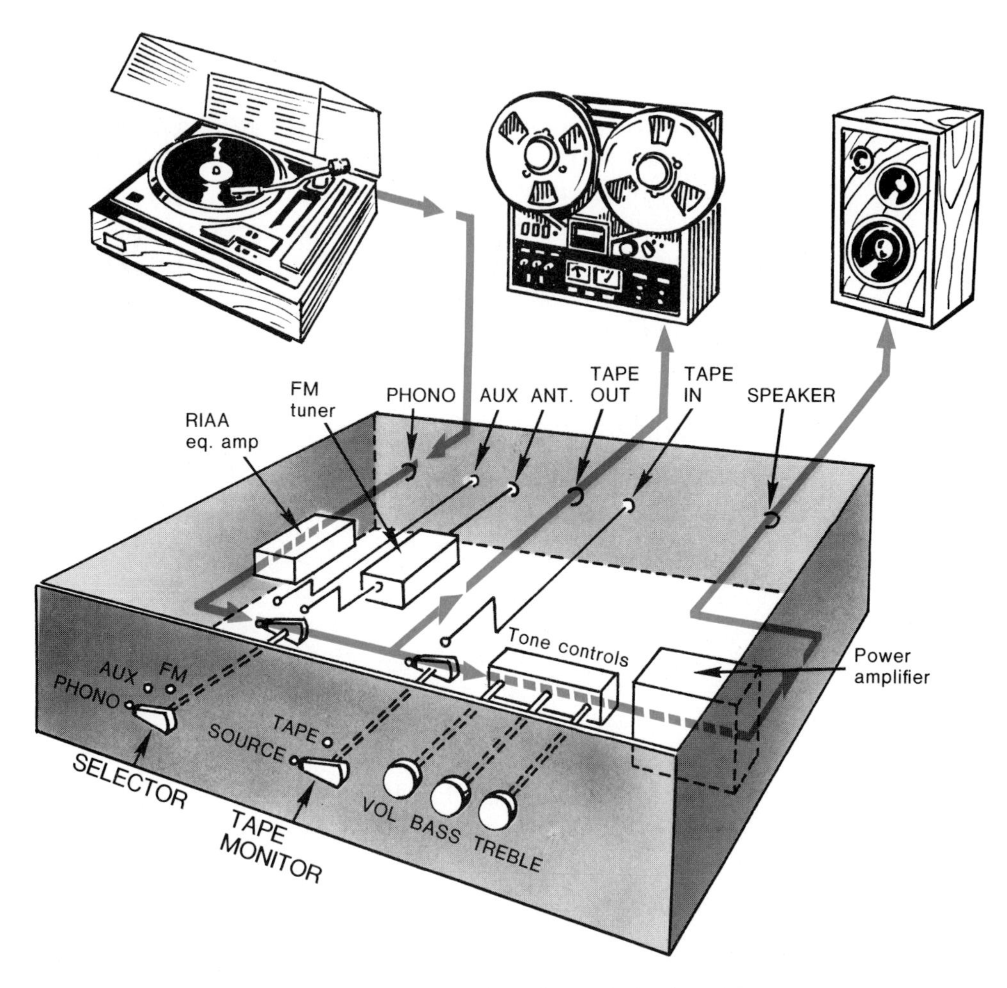

Figure 2.15. After leaving the TAPE MONITOR switch, the signal proceeds through the tone control section, the power amplifier, and finally arrives at the speaker.

The Signal Proceeds from the TAPE MONITOR Switch to the Tone Control Section, to the Power Amplifier, and Finally Arrives at the Speaker

After the signal passes through the TAPE MONITOR switch, it flows into the tone control section where the VOLUME, BASS, and TREBLE controls are located, as illustrated in Figure 2.15. Many receivers will also have a LOUDNESS control, as well as HIGH and LOW filters (not shown in the figure). These additional features will be explained in Chapter 8, where more information on the VOLUME, BASS, and TREBLE controls will also be provided.

Here is an important point. Notice in Figure 2.15 that the signal proceeds to the TAPE OUT jack before it arrives at the tone control section. This means that the controls, like VOLUME, BASS, TREBLE, etc., have absolutely no effect on what is being tape recorded! For example, suppose that you are listening to records and, at the same time, tape recording them. Suddenly someone walks into the room and turns down the VOLUME control. Disaster? Absolutely not, because the "sound" has been already recorded before it even reached the VOLUME control! Likewise, any other function located in the tone control section cannot affect your tape recording—a very nice safeguard feature.

After leaving the tone control section, the signal proceeds to the power amplifier where it acquires a final "build-up" and the ability to produce adequate sound levels from the loudspeaker.

Summary

With an understanding of how the signal flows through a receiver it is easy to operate any hi-fi system:

1. Using the SELECTOR switch, select which source you want to hear: AM or FM, PHONO, or AUX (this input can be used to play back a tape deck). Want to make a tape recording simultaneously? No problem. Simply connect a tape deck to the TAPE OUT jack located on the back of the receiver.
2. With the TAPE MONITOR switch set to the SOURCE position, the music coming directly from the source will be heard through the speaker.

2.8 More on the TAPE MONITOR Switch

Since the TAPE MONITOR switch is the key to understanding fully the capabilities of a hi-fi system, we will now take a closer look at this control. Suppose that you are taping a record by using a tape deck which has separate record and playback heads.* As a preliminary procedure for making a good recording, you would first like to compare the recorded music with the nonrecorded music coming directly from the record. This comparison would allow you to adjust the controls on the tape deck to make the tape recording exactly as you would like it, thus eliminating any guesswork. When the tape recording is OK, you can rewind the tape, restart the record and, without changing any controls, produce a "perfect" tape! The TAPE MONITOR switch will allow you to do this if the tape deck is properly connected to the receiver.

The connections between the receiver and the tape deck require two cables per stereo channel, one for recording and one for playback. When connecting cables between the receiver and tape deck it will help to remember the simple rule: OUTPUTS plug into INPUTS. Using this rule, the recording cable is connected from the TAPE OUT jack of the receiver to the INPUT jack of the tape deck. With this connecting cable the signal will travel out of the receiver and

*See footnote on next page.

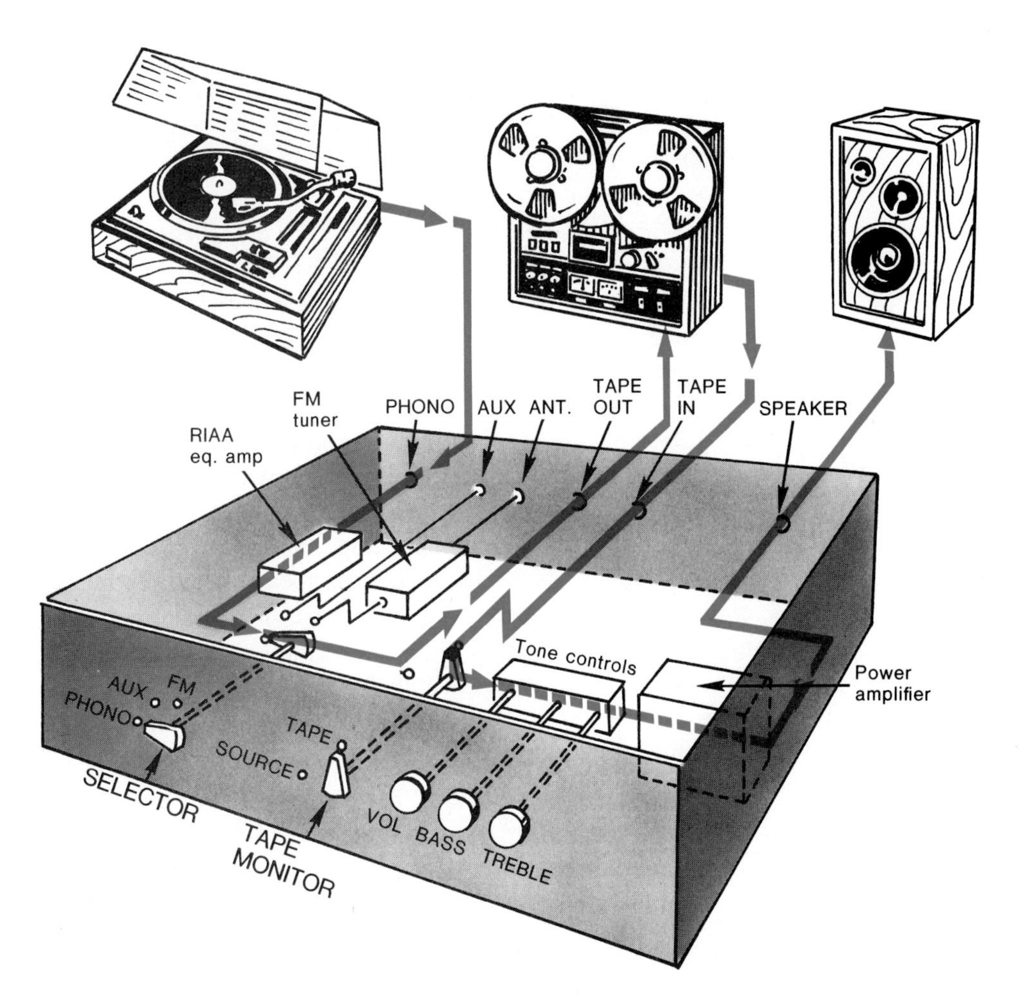

Figure 2.16. When the TAPE MONITOR switch is set to its TAPE position, the playback signal from the tape deck, rather than the signal coming directly from the turntable, is sent to the speaker.

into the tape deck. Once inside the tape deck, the signal is directed to the recording head* where it is imprinted on the magnetic tape. If the tape deck is also operating in the playback mode, the magnetized tape moves past the playback head* and a playback signal is produced. To direct this signal into the receiver a second cable must be connected between the OUTPUT jack of the tape deck and the TAPE IN jack of the receiver, in accordance with the above rule. To hear the recorded tape the TAPE MONITOR switch must be set to its TAPE position, as shown in Figure 2.16, so that the electrical signals coming from the tape deck can proceed through the remainder of the system. Thus, the

*A recording head on a tape deck is a small magnet, often called an electromagnet, which is in close contact with the moving tape. The recording head receives the audio signal and causes a magnetic copy of it to be placed on the tape. The playback head operates in an opposite manner. As the magnetic copy on the tape passes by the playback head, an audio signal is produced in the head which is then sent to the remainder of the hi-fi system for amplification and conversion into sound. These heads and their functions will be discussed in Chapter 13.

listener can compare the recorded music with the nonrecorded music coming directly from the sound source by switching the TAPE MONITOR switch between its TAPE and SOURCE positions.

Devices Other Than a Tape Deck Which Can Be Used with the TAPE MONITOR Switch

There are many devices on the market today which are designed to change the characteristics of sound being produced by an audio system. For example, there are noise reduction systems, such as Dolby, dbx, Autocorrelator, and DNL, which reduce the annoying hiss and noise often associated with tape recordings and records. Also, equalizers can be purchased which will either enhance or diminish the sound at selected frequencies, and thereby compensate for deficiencies in the room acoustics, speakers, or in the recording itself. Reverberation units are becoming very popular because they permit a hi-fi system to simulate the acoustics of large concert halls by adding electronic echoes, or reverberation, to the music. The user can insert many of these additional devices in place of the tape deck shown in Figure 2.16. The TAPE MONITOR facility then permits the audio signal to pass through these components before it reaches the remainder of the receiver.

> *A Hint to the Consumer.* Some amplifier and receiver manufacturers do not include the important TAPE MONITOR switch (and the associated TAPE OUT/TAPE IN jacks), and such products seriously lack versatility. Often these units will contain TAPE jacks which are source inputs like the FM, AUX, or PHONO jacks. But the presence of these TAPE jacks should not be confused with the valuable TAPE MONITOR feature. Purchasing an amplifier or a receiver which has a TAPE MONITOR facility will greatly enhance the versatility of any hi-fi system.

Other Names Given to the TAPE MONITOR Switch and the TAPE OUT/TAPE IN Jacks

Unfortunately, the hi-fi industry has not adopted standard names for the TAPE MONITOR switch and the TAPE OUT/TAPE IN jacks. So do not be too surprised if your receiver (or integrated amplifier, or preamplifier) has different names for these features. Figure 2.17 illustrates some of the common names that different manufacturers use on their products.

Summary

The TAPE MONITOR facility is a very important feature that allows the user to direct the audio signal out of the receiver, through another component such as a tape deck, and then send it back into the receiver. When the TAPE MONITOR switch is set to its TAPE position, the listener hears the signal which is coming from the tape deck. When the TAPE MONITOR switch is set to the SOURCE position, the listener hears the signal which is coming directly from one of the source jacks (PHONO, AUX, or FM), and not the tape recorded signal.

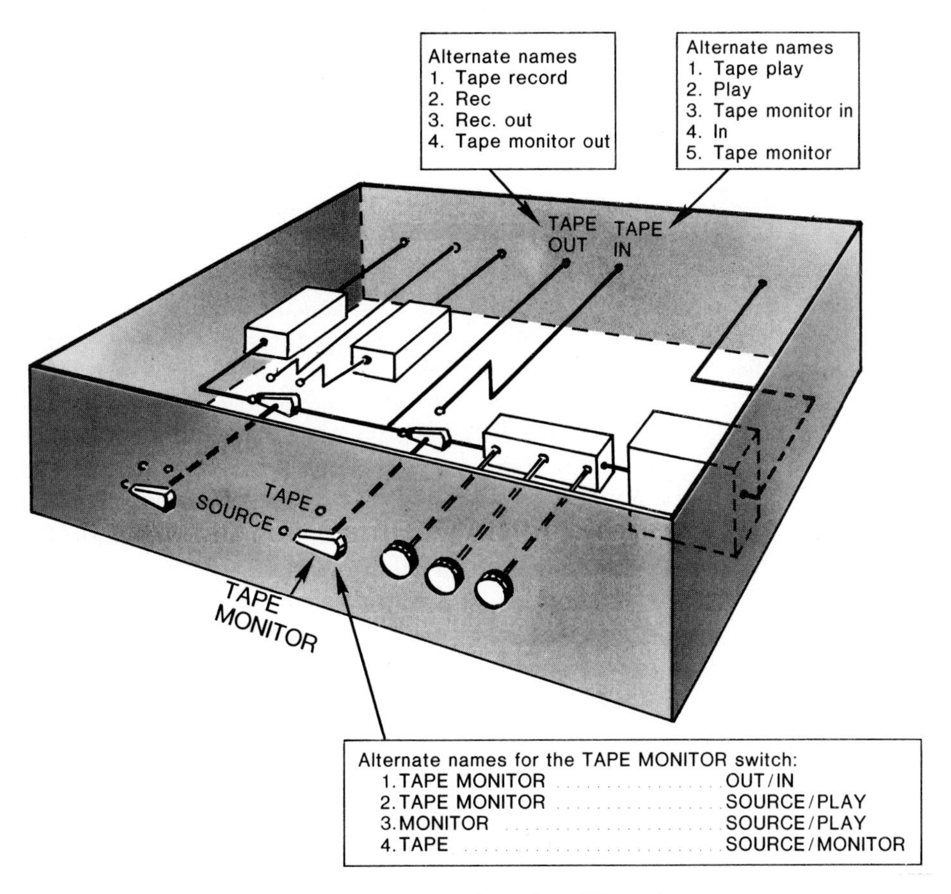

Figure 2.17. Some alternate names given by different manufacturers to the TAPE MONITOR switch and the TAPE OUT/TAPE IN jacks.

Summary of Terms

Acoustic Feedback—An undesirable noise (similar in sound quality to rumble) created when vibrations from the loudspeakers are picked up by the cartridge and amplified by the sound system. The problem of acoustic feedback usually can be reduced by physically separating the loudspeakers from the turntable. In public address systems the microphone/speaker combination also gives rise to acoustic feedback which is often heard as an annoying screech.

Amplification—The process of making electrical signals larger.

AUX (Auxiliary) Jacks—The source input jacks to a preamp where a tape deck or a tuner may be connected.

Building Blocks—The elementary units which comprise a hi-fi system. The four building blocks are: the sound sources (turntable, tape deck, tuner, microphone), the preamplifier/control center, the power amplifier, and the speaker(s).

Cartridge—(See **Phono Cartridge**)

Channel—A complete and separate "sound" path through an audio system. A stereophonic system has two channels designated as "left" and "right." A monophonic system has only one channel.

Compact—A complete audio system which usually contains all components, except for the speakers, in a single case. The compacts may contain any or all sound sources.

Complete System—A phrase describing an audio system which contains all four building blocks: a sound source, a preamp, a power amp, and speakers.

Component—Any element of an audio system which is designed to do a particular job. Typical components include receivers, tuners, speakers, tape decks, turntables, integrated amps, etc.

Console—A complete audio system, including the speakers, contained in a single furniture-type cabinet. Because the turntable and speakers share a single cabinet, consoles usually pose the threat of acoustic feedback at high sound levels.

Electrical Signal—Electricity which contains the sound information. All audio systems, beginning with the sound sources, process only electrical signals, and the sound itself is not recreated until the signals reach the speakers. In AM and FM broadcasting the radio wave is called the "signal."

Hi-fi (High-fidelity)—A phrase which is synonymous with "quality performance" in audio components.

Integrated Amplifier—A single audio component which combines a preamplifier control/center and a power amplifier. An integrated amplifier does not contain a sound source or speakers.

Jack—A small, round electrical connector found on all hi-fi units to which the wires, interconnecting the components, are attached.

Microphone—A device which converts sound into electrical signals.

Millivolt—One-thousandth of a volt.

Mono (Monophonic)—A word which is used to describe an audio system which has only one "sound" channel.

Phono Cartridge—A small unit which is located on a turntable at the end of the tone arm and holds the stylus (needle). It is the job of the cartridge to transcribe the intricate movements of the stylus, as it vibrates in the groove, into an electrical signal which is sent to the PHONO input of the preamplifier/control center.

Phonograph—(See **Record Player**).

Phono Jacks—The source input jacks to a preamp where a turntable is connected.

Pick-Up—(See **Phono Cartridge**).

Platter—A heavy, metal disc which is located on a turntable and which holds and turns the records during playback.

Portables—Complete audio systems which are small enough to be conveniently carried.

Power—The amount of energy being produced, or consumed, every second. The unit of power is called a watt. If the power rating of an amplifier is, for example, 20 watts, then the amplifier will produce and deliver to the speakers 20 units of electrical energy per second. Likewise, a speaker which is producing 3 watts of acoustical power creates 3 units of sound energy each second.

Power Amplifer—An electronic device which receives the relatively weak electrical signals from the preamplifier/control center and boosts them to such an extent that the signals are capable of driving speakers. The output power of a power amplifier is measured in watts. The power amp may be purchased as a separate audio component, or it may be incorporated into either an integrated amplifier or a receiver.

Power Supply—An electrical unit found inside amplifiers which provides the source of energy for their operation.

Preamplifier/Control Center (or Preamp)—A collection of electronic circuits which accepts inputs from the various sound sources (turntable, tuner, tape deck, and possibly a microphone), adjusts the signals in level and tonal characteristics according to front panel control settings, and finally passes the signals on to the power amplifier. A preamp may be purchased as a separate audio component or it may be incorporated as part of either an integrated amplifier or a receiver.

Preassembled Hi-Fi Systems—Complete audio systems in which the four building blocks have been selected and assembled by the manufacturer for the convenience of the consumer.

Radio—A complete audio system which has a tuner for its sound source.

Receiver—A single unit that combines a tuner, preamp, and a power amp. A receiver does not have speakers incorporated into it.

Record Player (or Phonograph)—A complete audio system, including a preamp, a power amp, and speaker(s), which has a turntable for its source of sound.

RIAA Equalization Amplifier—A collection of electronic circuits which boosts and "equalizes" the signal coming from magnetic phono cartridges. The RIAA (Recording Industry Association of America) equalization amplifier is found in the preamplifier/control center section of receivers and integrated amplifiers.

Rumble—A low frequency noise often resulting from turntable vibrations which are picked up by the cartridge and reproduced by the audio system.

Selector Switch—A switch which is located on the front panel of a preamp, integrated amp, or a receiver which directs the electrical signals from a sound source (turntable, tuner, tape deck, microphone) to the remainder of the audio system.

Signal—(See **Electrical Signal**).

Sound Source—A general phrase used to designate a turntable, tuner, tape deck, or a microphone. Sound sources are not complete audio systems because the sources require additional components (preamp, power amp, and speakers) to produce sound. However, headphones may be used directly with tuners or tape decks.

Source—(See **Sound Source**).

Source of Sound—(See **Sound Source**).

Speaker—An electromechanical device which converts the incoming electrical signals into sound.

Stereo—An audio system which has two channels called "left" and "right". "Stereo" does not necessarily imply hi-fidelity, or good quality.

Stylus—A carefully shaped and polished "needle" which moves in the record groove. All high-quality styli are made from diamonds.

Tape Deck—A magnetic tape machine which includes no power amplifiers or speakers, and is designed to be used as a sound source in a component system. Most tape decks, however, permit listening with headphones.

TAPE IN Jacks—Jacks which are located on the rear panel of a preamp, an integrated amp, or a receiver, to which the outputs from a tape deck, or other audio accessories, are connected.

Tape Monitoring—Listening to a tape recording a fraction of a second after the recording was made. This is possible only if the tape deck has separate record and playback heads. The tape monitoring feature is used in conjunction with the TAPE MONITOR switch found on most preamps.

Tape Monitor Switch—A circuit-interrupt switch which is located in the preamp. The TAPE MONITOR switch allows a tape deck to be inserted for either playback-only purposes, or for tape monitoring a recording which has just been made. In addition, the TAPE MONITOR switch also permits a variety of other audio accessories, such as noise reduction units, reverberation units, filters, and equalizers to be connected to the system.

TAPE OUT Jacks—Jacks which are located on the rear panel of a preamp, integrated amp, or a receiver, to which a tape deck is attached for recording purposes. The program material, as selected by the SOURCE switch, is present at the TAPE OUT Jacks.

Tape Player—Either a tape deck or a complete system which contains only playback facilities with no provisions for recording.

Tape Recorder—A tape machine which includes a tape deck, preamp, power amp, and speakers; it is a complete, self-contained audio system.

Tone Arm—The arm of a turntable that extends over the record and holds the phono cartridge in place.

Tone Control Section—The section of a receiver, integrated amplifier, or a preamplifier/control center which contains the BASS, TREBLE, VOLUME, and FILTER controls.

Transistor—A solid state device which is used for amplifying electrical signals.

Tuner—An electronic component which receives AM and/or FM radio broadcasts and converts them into electrical signals whose frequencies lie in the audio range. A tuner is only a sound source, and a preamp, a power amp, and speakers must be added to complete the audio system. A tuner is not a radio.

Turntable—A sound source which plays records. A turntable usually consists of a platter upon which the record rests, a driving motor, a tone arm, and a cartridge.

Volt—A basic unit of electrical energy. The strength of electrical signals is measured in volts with higher volts representing signals which carry more electrical energy.

Watt—The basic unit in which power is measured; e.g., "50 watts of power".

Review Questions

1. Describe the four basic building blocks of a hi-fi system and show the order in which they are connected.
2. What are the two main functions of a preamplifier?
3. What are the differences between a preamplifier and a power amplifier?
4. What combinations of audio components are available which allow one to purchase a radio?
5. Explain the advantages of having a TAPE MONITOR facility on the preamplifier.
6. Define an integrated amplifier and explain why one would choose it as a component rather than a receiver.
7. Explain the difference between a tuner and a radio.
8. Explain the difference between a turntable and a record player.
9. Explain the difference between a tape deck and a tape recorder.
10. Define a receiver and explain why it is so popular as a component.
11. Explain the purpose of the TAPE MONITOR facility.

Exercises

NOTE: The following questions have up to five possible answers. Please select the *one* response which best answers the question.

1. The controls for VOLUME, BASS, TREBLE, and BALANCE are usually found in the:
   1. preamplifier.
   2. power amplifier.
   3. tuner.
   4. sound source.
2. The combination of a preamplifier, power amplifier, and a tuner (all contained within one unit) is called:
   1. an integrated tuner.
   2. an integrated amplifier.
   3. a receiver.
   4. a radio.
   5. a sound source.
3. A unit which contains only a preamp and a power amp is called:
   1. a receiver.
   2. a tuner.
   3. an integrated amplifier.
   4. a 3-part system.
   5. a sound source.
4. A unit which contains a tape deck, a preamp, a power amp, and speakers is called:
   1. an integrated amplifier.
   2. a tape recorder.
   3. a phonograph.
   4. a radio.
   5. This combination has no name because it is not commercially available.
5. If you were attempting to listen to your hi-fi system and **no** sound was coming through the speakers, then a possible cause for this sorry situation might be (assume that only a turntable and tuner are connected and functioning):
   1. the SELECTOR switch was set to FM.
   2. the VOLUME control was set to its maximum value.
   3. the SELECTOR switch was set to PHONO.
   4. the TAPE MONITOR switch was in the TAPE position.
   5. the TAPE MONITOR switch was in the SOURCE position.
6. Referring to the drawing below, identify what is wrong.

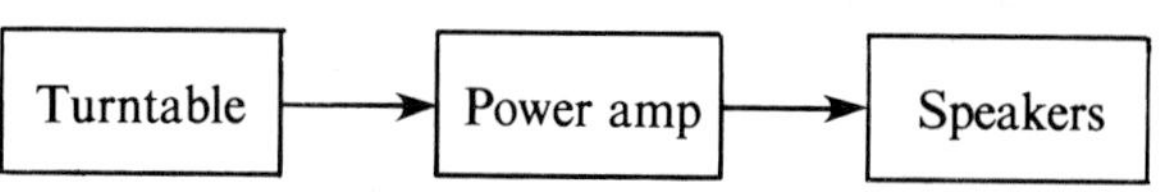

   1. The turntable should be replaced by a tuner, since a tuner does not require a preamp.
   2. A preamp should be added between the turntable and the power amp.
   3. A preamp should be added between the power amp and the speakers.
   4. Nothing is wrong with the above drawing.
   5. The power amp should be replaced by a preamp.
7. A tuner is:
   1. a special kind of radio.
   2. a radio without speakers.
   3. one building block of a radio.
   4. a radio without a power amplifier.
   5. a radio without a preamplifier.
8. To play a prerecorded tape through the TAPE MONITOR jacks of a hi-fi system involves the following steps. Connect the tape deck to the _______ jack, and set the TAPE MONITOR switch to the _______ position.
   1. TAPE IN, TAPE
   2. TAPE IN, SOURCE
   3. TAPE OUT, TAPE
   4. TAPE OUT, SOURCE
   5. AUX, TAPE
9. Any sound source applied to the SELECTOR switch of a preamplifier can be monitored without involving the tone control section from the:
   1. TAPE IN jack.
   2. TAPE OUT jack.
   3. PHONO jack.
   4. AUX jack.
   5. speaker terminal.
10. Which of the following is an advantage of having a component stereo system compared to a console stereo?
    1. Component systems use less electricity.
    2. The speakers are separated from the turntable.
    3. Component systems weigh more and, therefore, they are less vulnerable to vibrations.
    4. A component system is easier to assemble.

11. Which one of the following represents a correctly assembled hi-fi system?
    1. Preamp→tuner→power amp→speakers.
    2. Turntable→power amp→speakers.
    3. Tape deck →integrated amp→preamp→ speakers.
    4. Tape deck→receiver→speakers.
    5. Tuner→receiver→preamp→speakers.
12. The term high fidelity means:
    1. a total faithfulness to the original sound.
    2. a high-powered amplifier.
    3. stereo recording.
    4. the audio system contains four building blocks.
    5. the ability of a system to accept tape decks, turntables, and tuners as sources of sound.
13. If you wanted to listen to a record, the SELECTOR switch must be set to ____ and the TAPE MONITOR switch set to ____.
    1. PHONO, SOURCE
    2. AUX, SOURCE
    3. PHONO,TAPE
    4. FM, TAPE
    5. AUX, TAPE
14. In the middle of taping your favorite FM program you accidentally turn down the VOLUME control. How does this affect your tape?
    1. Since all signals coming out of your preamp are affected by the VOLUME control, your tape is ruined.
    2. Since the VOLUME control is on the power amp and the tape is made from the preamp, nothing happens to your tape.
    3. Since the tape is made from the power amp, nothing happens.
    4. Since the VOLUME control of the preamp occurs after the signal has been sent out of the preamp for taping, nothing happens to your tape.
    5. The tape deck ceases to record the signal because the speakers are no longer producing sufficient sound.
15. An electronic device which produces large currents that can drive speakers is called:
    1. a tuner.
    2. a tape deck.
    3. a multiplexer.
    4. a preamplifier.
    5. an integrated amplifier.
16. If you bought only a tape deck and a pair of loudspeakers, would you be able to listen to them?
    1. Yep! Sure would.
    2. No, because a preamp is needed.
    3. No, because a power amp is needed.
    4. No, because a sound source is needed.
    5. No, because an integrated amp is required.
17. Which one of the following units contains all four building blocks of a hi-fi system?
    1. An automatic turntable.
    2. A cassette deck.
    3. A tuner.
    4. A radio.
    5. A microphone.
18. Of the five units listed below, the *smallest* electrical signals come out of the:
    1. preamplifier.
    2. magnetic phono cartridge.
    3. power amplifier.
    4. integrated amplifier.
    5. receiver.
19. Electrical signals entering the ________ jack pass through the RIAA equalization amplifier.
    1. FM
    2. AUX
    3. PHONO
    4. TAPE IN
    5. TAPE OUT
20. In a hi-fi system the *largest* electrical signals are produced by the:
    1. power amplifier.
    2. preamplifier.
    3. tuner.
    4. magnetic phono cartridge.
    5. microphone.
21. If you were making a tape recording (from the TAPE OUT jack) of a record, what control on the preamp would cause the recorded tape to have more bass?
    1. The BASS control.
    2. The TREBLE control.
    3. The VOLUME control.
    4. The BALANCE control.
    5. No control on the preamp can cause the recorded tape to possess more bass.

22. Placing speakers on the same stand as the turntable:
    1. is an acceptable practice for hi-fi performance.
    2. results in a better sound quality due to the close proximity of the two components.
    3. reduces the power delivered to the speakers by the amplifier.
    4. can cause unwanted vibrations throughout the system.
    5. will cause the tuner to malfunction.
23. An integrated amplifier always:
    1. has less power than a receiver.
    2. consists of a preamp and a power amp.
    3. lacks input jacks for connecting a tuner.
    4. lacks a tone control section.
    5. consists of a turntable, a preamp, and a power amp.
24. If you were to buy a compact stereo system, which building block would *not* be included?
    1. A sound source.
    2. A preamp.
    3. A power amp.
    4. The speakers.
    5. All of the building blocks would be included in a compact.
25. Suppose that you wanted to copy a prerecorded tape onto a blank tape. Your receiver has only PHONO and AUX jacks, as well as the TAPE OUT/TAPE IN jacks for the TAPE MONITOR facility. The tape deck containing the prerecorded tape should be connected to the _____ jack of the receiver, while the deck with the blank tape should be connected to the _____ jack.
    1. TAPE IN, TAPE OUT.
    2. AUX, TAPE OUT.
    3. TAPE IN, AUX.
    4. AUX, TAPE IN.
    5. TAPE OUT, AUX.

Chapter 3

WAVES AND SOUND

3.1 The Importance of the Wave Concept in Hi-fi

Throughout our physical world the concept of waves plays such an important and pervasive role that phrases like "tidal waves," "microwaves," "radio waves," etc., are commonplace. In audio reproduction both the sound itself and the electricity which flows through hi-fi equipment are forms of waves. In fact, without waves there would be no hi-fi, since they actually carry the audio information. It is important, therefore, to develop a feeling for waves and to see how hi-fidelity systems are almost totally characterized by wave-related terminology. Pick up, for example, the specification sheet for an amplifier, as shown in Figure 3.1. This sheet indicates how well the amplifier will reproduce the original sound. At this point you may not understand the technical jargon in Figure 3.1, but don't worry. The terms are listed only to emphasize that they are based on the concept of waves, as indicated in the right hand column of the figure. Similarly, the specifications for all other audio components, such as turntables, tape decks, speakers, tuners, etc., are based largely on the concept of waves. With an understanding of waves, the educated consumer is able to interpret this data and know whether or not the component is a quality product.

3.2 Describing Waves

Water waves are a fine starting point for our study of waves, because everyone is familiar with them. When a wave moves along, the flat, undisturbed surface of the water is changed into the familiar shape which is shown in the margin on the next page. You may be already aware of several important characteristics which pertain to water waves: (1) they travel from place to place, (2) they disturb things as they go, and (3) they are made up of repeating patterns called "cycles." Several additional characteristics, however, are important to the way in which hi-fi uses waves, and these will now be discussed.

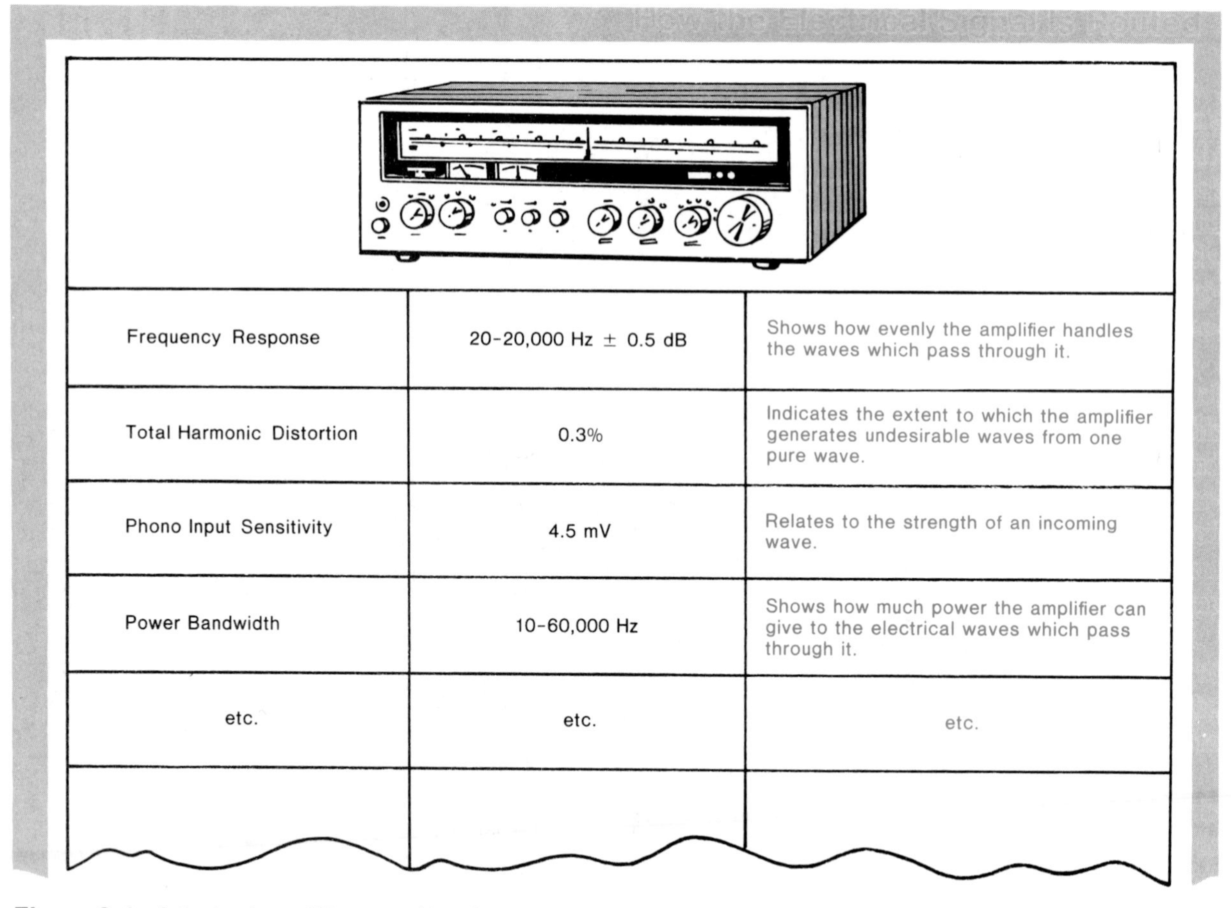

Frequency Response	20-20,000 Hz ± 0.5 dB	Shows how evenly the amplifier handles the waves which pass through it.
Total Harmonic Distortion	0.3%	Indicates the extent to which the amplifier generates undesirable waves from one pure wave.
Phono Input Sensitivity	4.5 mV	Relates to the strength of an incoming wave.
Power Bandwidth	10-60,000 Hz	Shows how much power the amplifier can give to the electrical waves which pass through it.
etc.	etc.	etc.

Figure 3.1. A typical amplifier specification sheet. Each of the terms has its origin in the concept of a "wave."

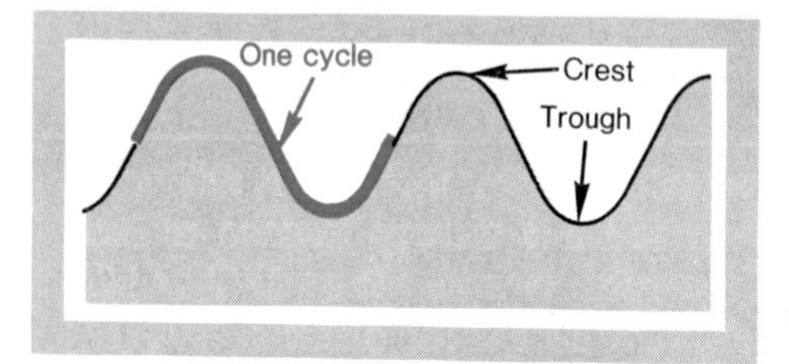

Suppose you were in the ocean with a camera on a day when the swells were moderately large. If you photographed the waves at water level, your picture would show an undulating surface like the one shown in the margin. Getting fancy, you could take the picture home and draw on it:

1. a dashed line that indicates the flat undisturbed water level for a calm sea.
2. a vertical scale which shows how much water at any point along the wave has been displaced above or below the flat level.
3. a horizontal scale which gives the horizontal distance along the wave.

The result would be as shown in Figure 3.2, and it has the advantage that the additional lines allow anyone to determine two very important properties of the wave: its amplitude and its wavelength.

Amplitude

The **amplitude** of a wave is the vertical distance from the flat position (i.e., the calm sea) to the crest of the wave. It is also the vertical distance from the flat position to the

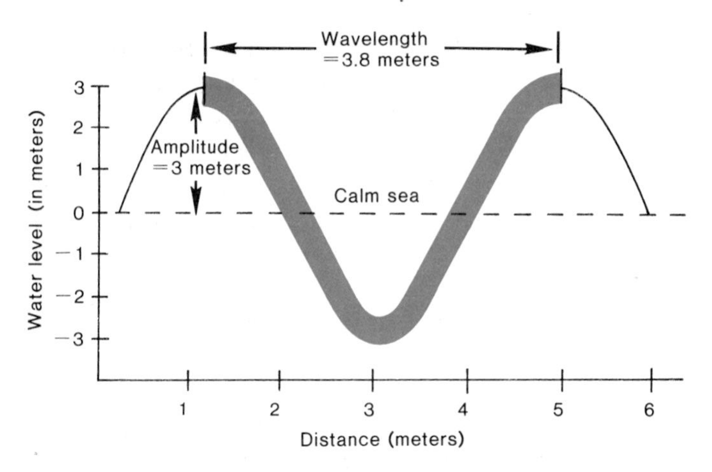

Figure 3.2. Two important parameters of a water wave are its amplitude and its wavelength.

bottom of the trough. In Figure 3.2 the amplitude is 3 meters. At this point it should not be too difficult to associate the amplitude of a wave with the amount of energy that the wave carries. A 10 meter tidal wave can do much more damage, and hence carries a lot more energy, than a 6 cm ripple! Therefore, we conclude that a large amplitude is associated with the ability of the wave to carry large amounts of energy.

A point of confusion often arises because many people think that the amplitude of a water wave is the vertical distance from trough to crest, which in this case would be 6 meters. In our example 6 meters is not the amplitude but, in fact, represents twice the amplitude.

Wavelength

The **wavelength** is the horizontal distance between two successive crests of the wave. According to the horizontal scale in Figure 3.2, the wavelength shown in the picture is about 3.8 meters. In other words, the wavelength represents how far the wave travels horizontally before its cyclic pattern is repeated. Actually, the wavelength is also the horizontal distance between two successive troughs or, for that matter, the distance between any two identical points on the wave.

Wave Period and Frequency

Water waves, as we all know, are not stationary but move across the ocean with a certain speed. While Figure 3.2 is adequate for many purposes, it does not convey this sense of motion, because, like a snapshot from a camera, it freezes the wave's movement in time. However, there is a useful way of graphically representing a wave's motion. Simply jump into the water with a ruler and a watch, and measure the height of the water as the wave passes by. At one instant of time the wave might reach 1 meter on the ruler, while at a later time the water would reach another level. As time goes on this type of measurement would record a water level which fluctuates between the limits established by the amplitude of the wave. A graphical representation of these measurements can be obtained by recording the reading from

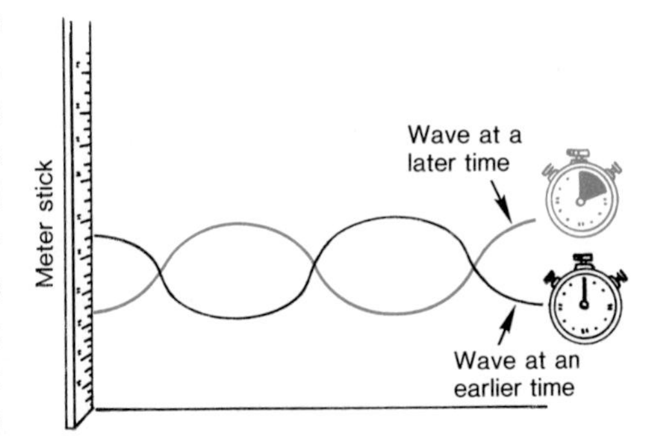

the ruler on the vertical axis and the corresponding time on the horizontal axis. The resulting graph is shown in Figure 3.3, and it gives the water level at any time as the wave passes by. Note that the only difference between Figure 3.2 and Figure 3.3 is that the horizontal axis in one case represents "distance" while in the other it represents "time." The advantage of Figure 3.3 is that two other important properties of the wave can be deduced—its period and its frequency.

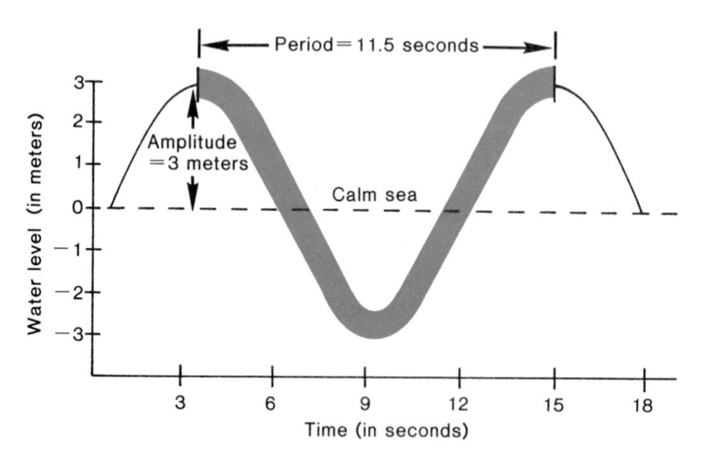

Figure 3.3. Another method of graphically displaying a wave is to measure its level at each instant of time as the wave passes.

The **period** of the wave is the time it takes for two successive crests to pass a particular point. This is easy to visualize if you are standing by the sea and timing the waves as they arrive one after the other. The period is the time that elapses between two successive waves which wash ashore. More precisely, the period is the time required for the wave to travel exactly one wavelength. Looking at the example in Figure 3.3, you would expect a complete wavelength to pass by every 11.5 seconds. Therefore, the period is 11.5 seconds.

The **frequency** of the water wave is simple to understand. It is merely the number of wavelengths, or crest-trough combinations, which pass by any particular point in one second. If you were watching from shore, the frequency of a wave would be the number of crest-trough combinations which break ashore each second. If three such combinations were to arrive each second, then the frequency would be three cycles per second, where we have followed the normal convention by using the word "cycle" to indicate a crest-trough combination. Figure 3.4 shows the appearance of two waves which have different frequencies.

The idea of frequency arises so often in hi-fidelity that engineers use the phrase "1 Hertz" in place of "1 cycle per second." This substitution is made in honor of Heinrich

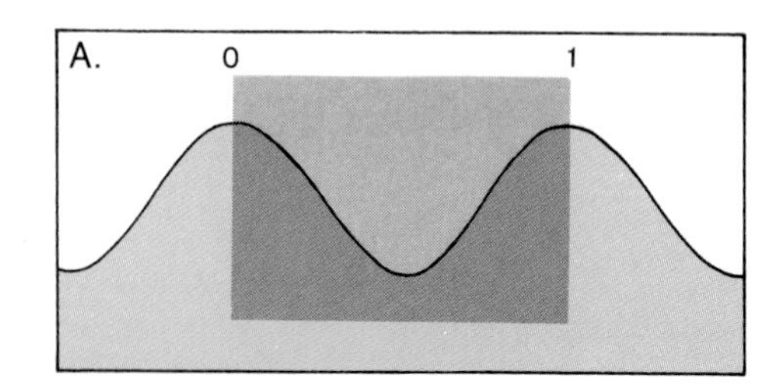

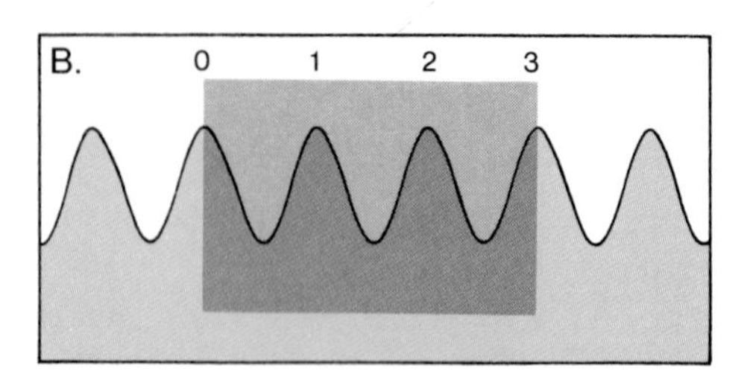

Figure 3.4. The frequency of a wave is the number of crests or troughs that pass a given point in one second. Each shaded box represents one second of time, and, therefore, the frequency of wave B is three times that of wave A.

Hertz who lived during the late 1800s and carried out many famous experiments with waves. Thus, a wave whose frequency is 5 cycles per second is said to have a frequency of 5 Hertz, or simply 5 Hz. Very often the frequencies encountered in hi-fi applications run into the thousands of Hertz, and most audio literature abbreviates the phrase "1,000 Hertz" as 1 kiloHertz, or 1 kHz. The prefix "kilo" comes from the metric system and means one-thousand. A frequency of 8,000 cycles per second would, then, be written as 8 kHz.

The Relationship between Period and Frequency

With a little thought the frequency can be calculated if the period is known, and vice versa. After all, if more crest-trough combinations pass each second (larger frequency), the time required for each combination to pass will be correspondingly less (smaller period). Indeed, period and frequency are related by the following important formula:

$$\text{Frequency} = \frac{1}{\text{Period}}. \qquad \text{(Equation 3.1)}$$

To help clarify Equation 3.1 imagine that you are watching a freight train move through a crossing as shown in Figure 3.5. The length of each car corresponds to one wavelength in wave terminology. The frequency is equal to the number of cars which pass by each second, while the period is the time required for one car to pass. If it takes 5 seconds for one boxcar to pass (the period) then 1/5 of a car will pass each second (the frequency), as predicted by Equation 3.1. Returning to the waves discussed earlier, if it takes 5 seconds for one complete crest-trough cycle of a water wave to break ashore, then it is only common sense that 1/5 of a cycle breaks ashore each second. This same result can be obtained by using Equation 3.1:

$$\text{Frequency} = \frac{1}{\text{Period}} = \frac{1}{5 \text{ seconds}} = \frac{1}{5} \text{ Hz}.$$

From a hi-fi point of view, the concept of frequency is present everywhere in the technical specifications of audio equipment, and it is more useful than the idea of a wave's period. In any case, Equation 3.1 allows the period to be calculated if the frequency is known and vice versa.

Letting f represent the frequency and T the period, Equation 3.1 is often written as

$$f = \frac{1}{T}.$$

Question: What is the period of a wave whose frequency is 0.5 Hz? Using the above equation one obtains

$$0.5 \text{ Hz} = \frac{1}{T}.$$

Solving for T yields T = 2.0 seconds.

The Speed of a Wave

A wave moving across the water, like a train in motion, is characterized by its speed. The speed of any object is just the distance it travels in a unit of time. For example, a speed of 60 kph means that a train will travel 60 kilometers in one

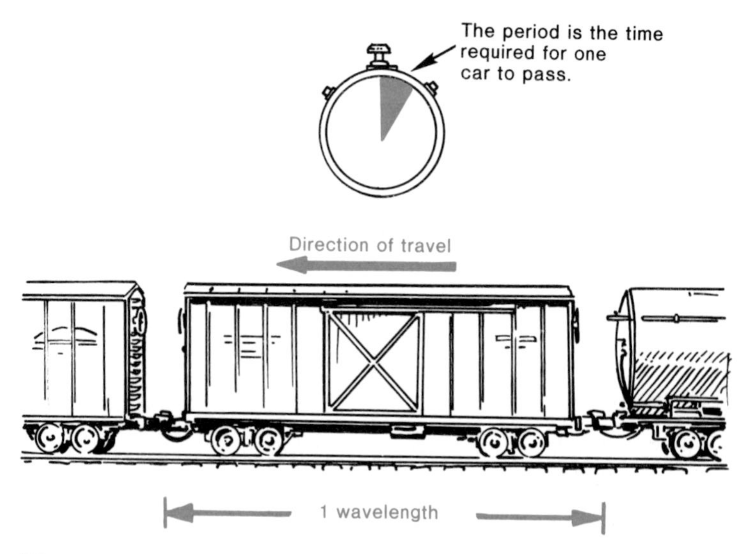

Figure 3.5. A moving train has the characteristics of a traveling wave.

hour. Sound travels through air at a speed of 344 meters per second, while light clips along at the astonishing rate of 300,000 kilometers per second!! This can be formulated in a very straight-forward manner:

$$\text{Speed} = \frac{\text{Distance traveled}}{\text{Elapsed time}}. \qquad \text{(Equation 3.2)}$$

Using v to stand for speed, d for distance, and t for time, Equation 3.2 can be written as

$$v = \frac{d}{t}.$$

Question: If the speed of a sound wave is 344 meters per second, how long does it take sound to travel from the stage of an outdoor rock concert to the last row of seats 150 meters away? Using the equation one obtains

$$344\ \frac{\text{meters}}{\text{second}} = \frac{150 \text{ meters}}{t},$$

or $t = 0.44$ seconds.

The Relationship between the Speed of a Wave and Its Wavelength and Frequency

An important relation exists between a wave's speed, its frequency and its wavelength. To see this, think about the train in Figure 3.5. If the length of each car is multiplied by the number of cars which pass by each second, the result will give the distance the train moves each second—which is its speed! Explicitly, this relation can be written as

$$\underbrace{(\text{Length of each car})}_{\text{Wavelength}} \times \underbrace{\begin{pmatrix}\text{Number of cars} \\ \text{passing every} \\ \text{second}\end{pmatrix}}_{\text{Frequency}} = \underbrace{\begin{pmatrix}\text{Distance the} \\ \text{train moves} \\ \text{each second}\end{pmatrix}}_{\text{Speed}}$$

Suppose, for example, that each car of the train is 30 meters long and that ½ a car passes by every second. It is fairly obvious that the train must be moving with a speed of 15 meters per second. Mathematically this result can be obtained by using the above formula:

$$\underbrace{(30 \text{ meters})}_{\text{Wavelength}} \times \underbrace{(1/2 \text{ car per second})}_{\text{Frequency}} = \underbrace{\begin{matrix}15 \text{ meters} \\ \text{per second}\end{matrix}}_{\text{Speed}}$$

Using the Greek letter lambda λ to stand for wavelength, f for frequency, and v for speed, Equation 3.3 can be written as

$$\lambda f = v.$$

Question: What is the wavelength of a sound wave which has a frequency of 100 Hz and travels with a speed of 344 meters/sec? Using the equation, we obtain

$$\lambda(100 \text{ Hz}) = 344 \text{ meter/second}$$

$$\lambda = \frac{344}{100} = 3.44 \text{ meters}.$$

In audio terms the length of each car corresponds to the wavelength and the number of cars passing each second represents the frequency. Therefore, the relation between wavelength, frequency, and speed of a wave is given by Equation 3.3. This is an important equation because it applies to any type of wave, and allows the calculation of any one of the three terms if the other two are known.

$$\text{Wavelength} \times \text{Frequency} = \text{Speed of wave} \qquad \text{(Equation 3.3)}$$

Summary

There are five parameters which characterize the properties of waves:

1. amplitude.
2. wavelength.
3. frequency.
4. period.
5. speed.

The five properties are not all independent, because relationships exist among several of them. Properties (3) and (4) are related, because

$$\text{Frequency} = \frac{1}{\text{Period}}.$$

In addition, (2), (3), and (5) are related via the formula

$$\text{Speed of wave} = \text{Wavelength} \times \text{Frequency}.$$

3.3 Transverse and Longitudinal Waves

Waves found in nature are generally classified as one of two types: transverse or longitudinal. Both of these kinds can be illustrated with the use of the remarkable toy called a "slinky," which is nothing more than a long, loosely coiled spring. Figure 3.6 shows how both types of waves can be generated by simply moving the slinky in different directions. The up-and-down hand motion in (A) produces a transverse wave. By contrast, the back-and-forth horizontal movement in (B) generates a longitudinal wave. Of the two types of waves the transverse wave is probably more familiar because it appears to be similar to a water wave.* However, both types are important in hi-fi, and we will now look more closely at each one.

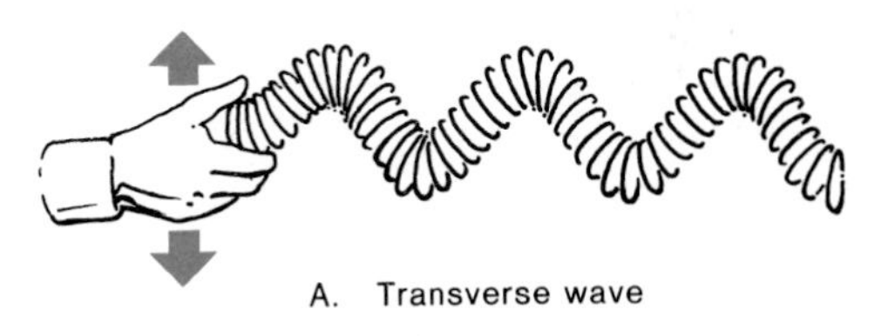

A. Transverse wave

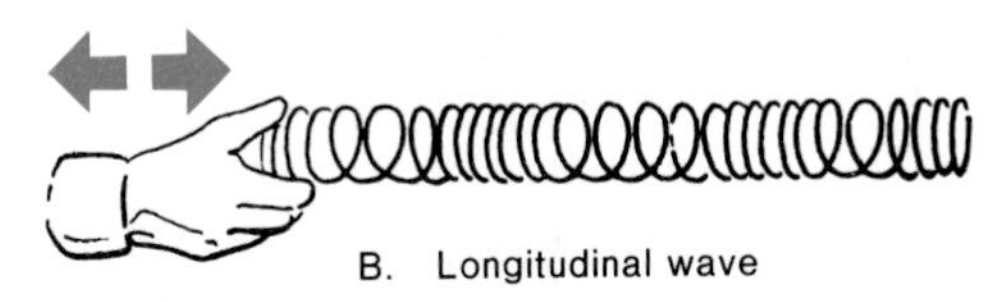

B. Longitudinal wave

Figure 3.6. Two types of waves, transverse and longitudinal, that can be generated by the use of a slinky.

*Strictly speaking, water waves are not purely transverse, but have both transverse and longitudinal components.

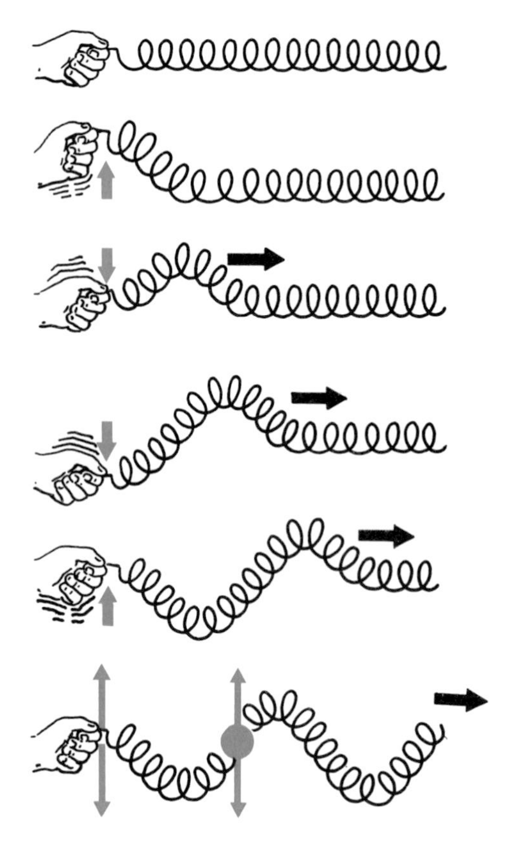

Figure 3.7. Generating a transverse wave on a slinky using an up and down motion of the hand.

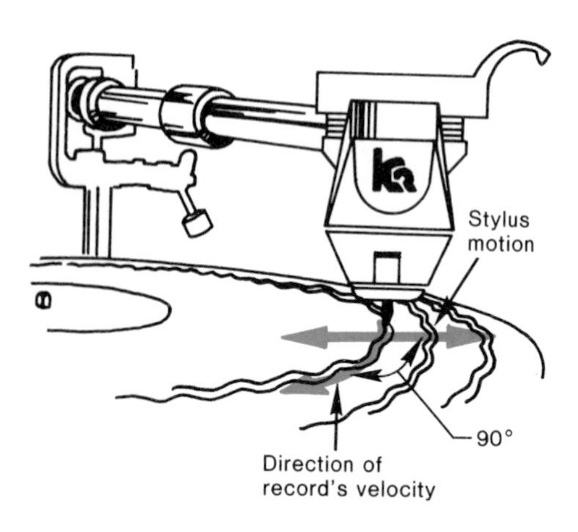

Figure 3.8. An example of a transverse wave. A monophonic record groove forces the stylus to vibrate horizontally in a direction which is transverse, or 90°, to the record's velocity.

Transverse Waves

A transverse wave is so-named because the motion of the medium (the slinky in Figure 3.6, for example) is 90°, or transverse, to the direction in which the wave travels. This is illustrated more clearly in Figure 3.7, which shows that an upward pointing pulse can be sent traveling to the right by jerking the slinky upward. How far the pulse moves in one second will be determined by its speed. Likewise, a downward pointing pulse can be sent traveling to the right by jerking the slinky downward. Notice that by moving the hand upward and then downward one full cycle of the wave, consisting of an upward pulse followed by a downward pulse, is sent traveling to the right. An entire wave, consisting of many cycles, can be produced by simply continuing the up-down motion of the hand. The important thing to note in Figure 3.7 is the motion of the colored dot, which has been attached to the slinky as a reference point. This dot moves only up and down (like the hand) as the wave advances to the right; the dot does *not* move in the horizontal direction in which the wave is traveling. Because the dot moves 90° to the direction of the wave travel, the wave is called a "transverse" wave.

The groove on a monophonic record is an excellent example of a transverse wave. A monophonic record groove represents the mechanical equivalent of sound which has been "frozen" into the vinyl plastic. The groove forces the stylus to vibrate back and forth in a direction which is transverse to the motion of the record, as shown in Figure 3.8.

Another example of a transverse wave occurs when a guitar string is plucked. A transverse wave is generated and travels down the length of the string, as illustrated in Figure 3.9. Notice, once again, that the vibratory motion of the string is 90°, or transverse, relative to the velocity of the wave as it moves down the neck of the guitar.

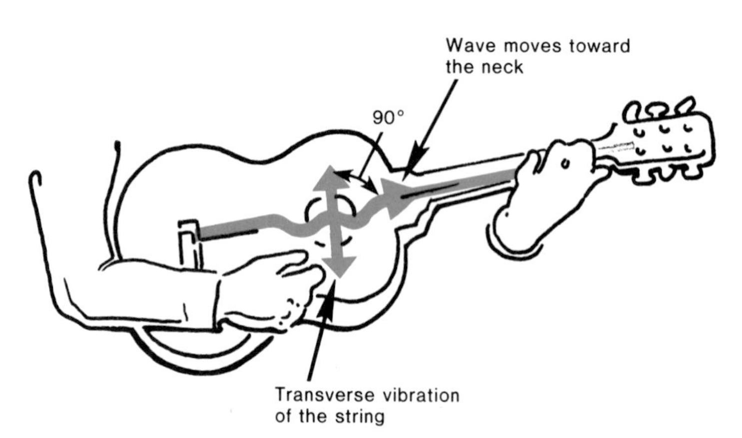

Figure 3.9. It is easy to generate a transverse wave on a guitar string.

The radio waves which bring those AM and FM stations to your home are also transverse waves. These waves, which are invisible, travel outward from the transmitting tower at the speed of light. When the radio waves strike the metal antenna, they force the electrons within it to oscillate back and forth much the same as a water wave pushes a rowboat up and down. Radio waves must be transverse because, as shown in Figure 3.10, they make the electrons vibrate in a direction which is transverse to the wave's velocity. The oscillating electrons in the antenna, in turn, cause the electrons within the hi-fi system to vibrate in synchronism and, with the aid of transistors, this oscillatory motion is converted into an audio signal.

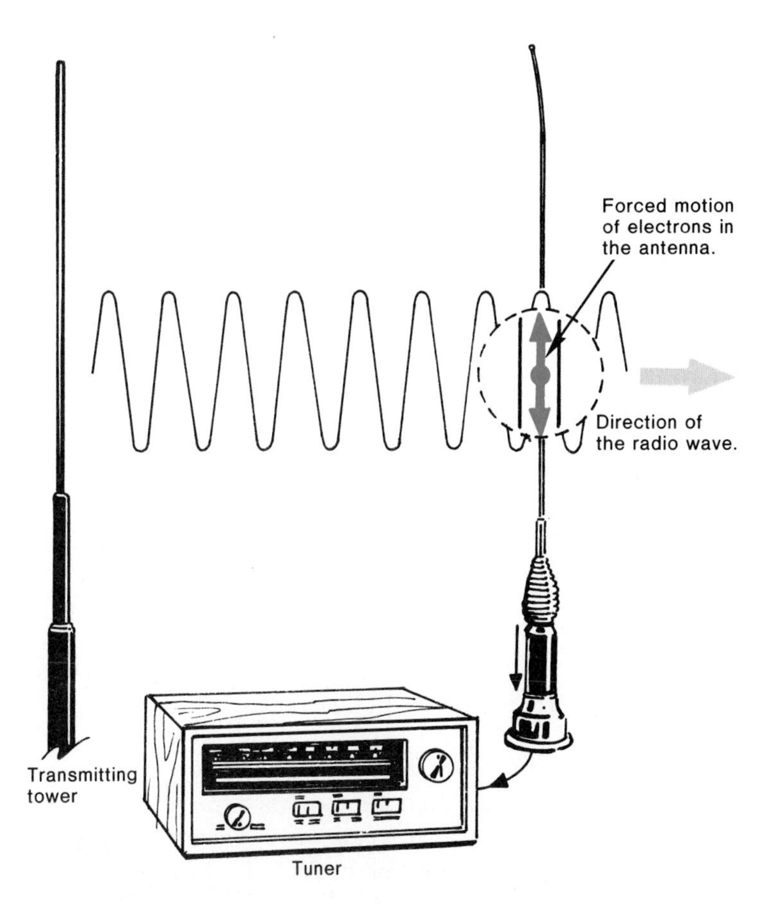

Figure 3.10. Radio waves force the electrons within the receiving antenna to oscillate transverse to the direction of the radio wave.

Longitudinal Waves

To see how a slinky can be used to generate a longitudinal wave, imagine that the slinky is resting on a table, as shown in Figure 3.11. A longitudinal wave is produced by sliding the hand back and forth parallel to the table and along the

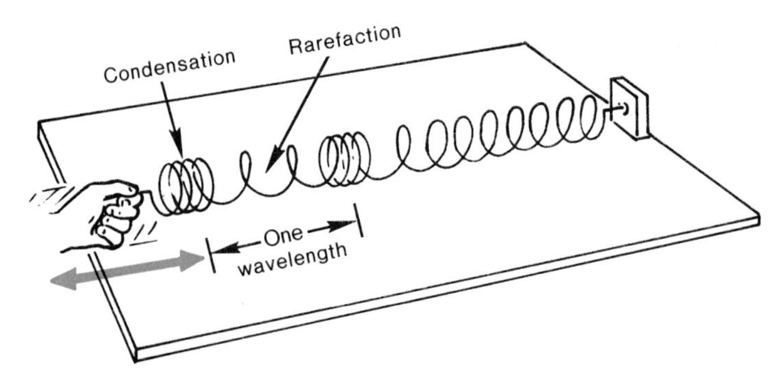

Figure 3.11. A longitudinal wave in a slinky showing the condensations and rarefactions.

direction of the slinky. Notice that this motion is different than that used in making the transverse wave in Figure 3.7. By pushing forward the slinky is compressed, as shown in Figure 3.12. The compression, or "condensation" as it is also called, is sent traveling to the right. In one second, it moves a distance which is determined by its speed. Likewise, by pulling the hand backward, the slinky is stretched. The stretched region is called a "rarefaction," and it is also sent traveling to the right. By moving the hand forward and then backward a compression, followed immediately by a rarefaction, can be generated, thus sending one complete cycle of the wave traveling to the right. The entire longitudinal wave can be produced by simply continuing the back-and-forth motion of the hand. The important thing to understand in Figure 3.12 is the motion of the colored reference dot which has been attached to the slinky. This dot moves back-and-forth horizontally (like the hand), parallel to the direction in which the wave travels; the dot does *not* move at all in the perpendicular direction. Because the coils of the spring vibrate back and forth parallel to the direction of wave travel, the wave is called a "longitudinal" wave.

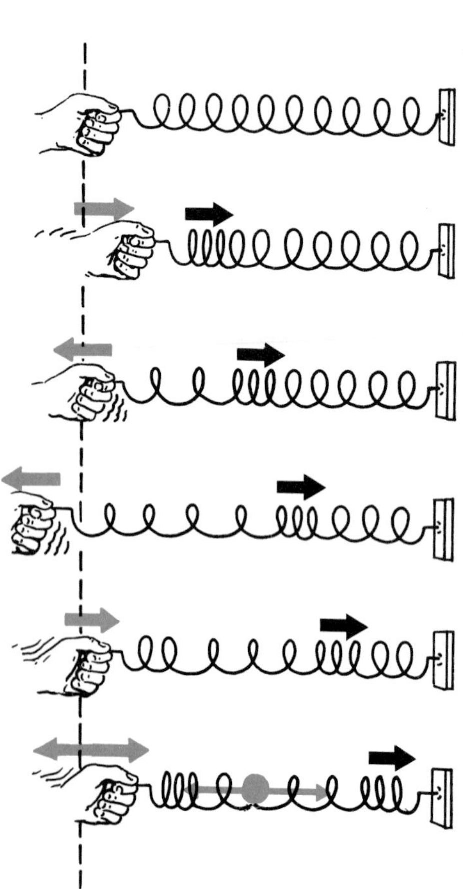

Figure 3.12. A longitudinal wave is generated by vibrating the hand in a direction which is parallel to the length of the slinky.

With a little thought, it should become clear that a longitudinal wave possesses all five wave characteristics mentioned in section 3.2, even though its appearance is far different from that of a transverse wave. Here is a listing of these traits and an explanation of how they apply to longitudinal waves:

1. **Wavelength**—the horizontal distance between two successive compressions (or rarefactions). See Figure 3.11.
2. **Period**—the time it takes for one complete wavelength (a compression followed by a rarefaction) to pass.
3. **Frequency**—the number of wavelengths which pass by every second.
4. **Speed**—how fast the compressions, or rarefactions, travel through the medium (down the slinky).

5. **Amplitude**—To answer the question of "What is meant by the amplitude of a longitudinal wave?" consider Figure 3.13. Both longitudinal slinky waves in Figure 3.13 have the same wavelength, but the compressions of (B) are much more tightly packed than those of (A). And, the rarefactions of (B) are more loosely packed than those of (A). Therefore (B) has a greater amplitude than wave (A). Remember, a large amplitude longitudinal wave is "compressed" and "rarefied" to a greater degree than is a small amplitude longitudinal wave.

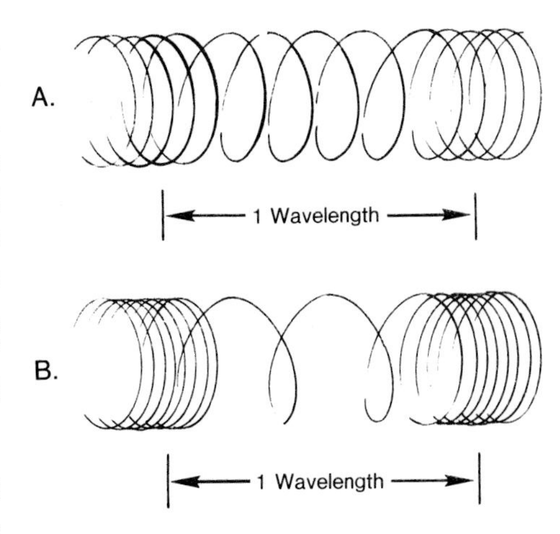

Figure 3.13. Two longitudinal waves with the same wavelength but with different amplitudes. Wave "B" has a larger amplitude than does "A".

It is also true that the longitudinal wave obeys all the same mathematical relations as does its transverse counterpart; namely,

$$\text{Frequency} = \frac{1}{\text{Period}},$$

$$\text{Wave Speed} = \text{Wavelength} \times \text{Frequency}.$$

Thus, a longitudinal wave has all five properties of a transverse wave (amplitude, frequency, etc.) and obeys the same two formulas given above. It only looks different!

The most important example of a longitudinal wave turns out to be sound itself. In fact sound is so important in audio that we will spend the next six sections discussing its properties. Another fine example of longitudinal waves are the electricity waves which carry the musical information from the sources, through the amplifiers, and on to the speakers. These electricity waves will be discussed in Chapter 7.

Summary

There are two main types of waves, transverse and longitudinal. In a transverse wave the vibration of the medium is perpendicular to the direction in which the wave travels. In a longitudinal wave the vibration of the medium is parallel to the direction in which the wave travels. Both types are characterized by amplitude, wavelength, frequency, period, and speed. Both obey the equations relating speed, wavelength, frequency, and period (Equations 3.1, 3.2, and 3.3).

3.4 Sound—How It Is Produced, Transmitted, and Detected

The two things which must always be present for the production of sound are (1) a vibrating object, such as a speaker, guitar string, or vocal cords, and (2) a medium in which the sound waves can propagate. The medium can be a gas, liquid, or solid—anything except a vacuum. A vacuum

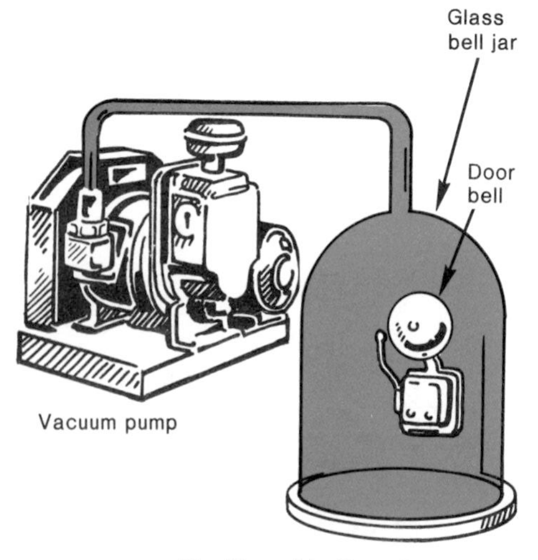

Figure 3.14. Removing the air inside the bell jar eliminates the sound even though the doorbell is still electrically "on."

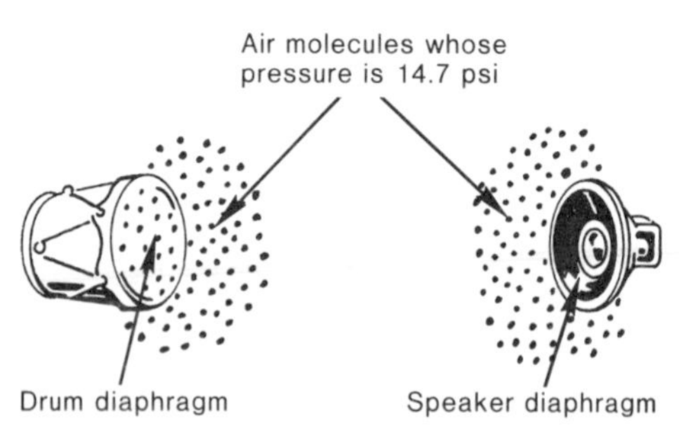

Figure 3.15. Both the drum and the speaker produce sound by a vibrating diaphragm. Normally, the air molecules surrounding these instruments are at a standard pressure of 14.7 pounds per square inch.

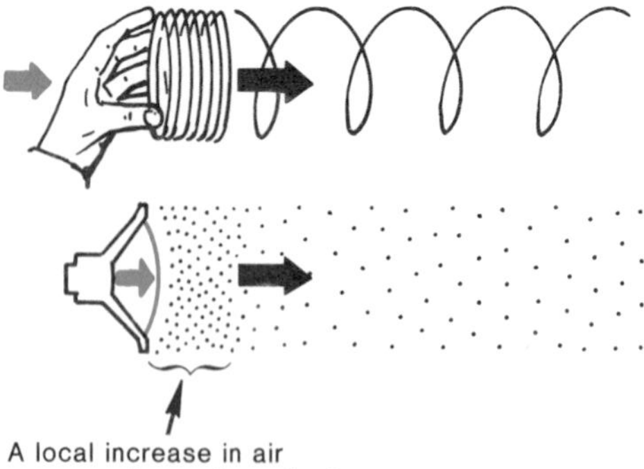

Figure 3.16. The creation of a condensation by a moving speaker diaphragm.

is excluded because sound requires the presence of matter for its existence. This requirement is not needed for either light or radio waves, which travel best in a vacuum. Witness the fact that light moves quite easily in the void of outer space during its journey from the sun to the earth.

Figure 3.14 illustrates an elegant experiment which demonstrates the necessity of matter for sound propagation. Place an electric door bell inside a bell jar which can be evacuated with a vacuum pump. When the jar is filled with air, the sound of the bell can be clearly heard. However, when the pump is turned on, air is removed from the jar, a vacuum results, and the sound disappears. Since the moon has no atmosphere, two astronauts can shout at each other all they want and never be heard!! That's why their space suits are equipped for radio communications.

In order to understand better how sound is produced let us follow the action of a hi-fi speaker. It was mentioned earlier that a vibrating object is required to produce sound, and a speaker, like a drum, has a diaphragm which is capable of moving back and forth. Of course the drum diaphragm is set into motion by a tap of the fingers, whereas the speaker diaphragm is forced to vibrate by the incoming electricity. As shown in Figure 3.15, when the speaker is not working it is surrounded by air molecules whose normal atmospheric pressure is 14.7 pounds per square inch or, in shorthand notation, 14.7 psi. As the speaker diaphragm moves outward, it compresses the air molecules directly ahead of it, which causes the air pressure in this region to increase slightly, as indicated in Figure 3.16. This area of increased pressure is called a "condensation," and it is analogous to the compressed region in the slinky which is also shown in the figure for comparison. The increase in air pressure which accompanies a condensation is somewhat like that resulting from the addition of air to a tire. As the pump squeezes more air into the tire, the air pressure in the tire increases.

Now, here's the interesting part. The air condensation, just like a compression in the slinky, moves away from the speaker at—you guessed it—the speed of sound! Although the compression in the slinky travels much slower than the speed of sound, we will assume for simplicity that both condensations move equally fast in Figure 3.16. In looking at the picture, it is important to keep in mind one important fact. The air molecules which comprise the condensation do not remain with it as the condensation propagates outward. The molecules pass the condensation on to their neighbors, so to speak, who in turn, pass it further on down the line. Remember Figure 3.12 in which a "dot" was firmly attached to the slinky? The "dot" simply oscillated about an equilibrium position and served to move the condensation (or rarefaction) down the slinky. This situation is like that at a football game when a hot dog is passed from hand to hand until it reaches the person at the end of the row. The hot dog can move a great distance even though each person only moves it a small amount. In the same manner a condensation can travel a large distance, although the individual molecules move over very small distances.

After creating a condensation the diaphragm reverses itself and begins to move inward, as shown in Figure 3.17. This motion produces a partial vacuum, called a "rarefaction," where the air pressure is below its normal value of 14.7 psi. Following behind the condensation, the rarefaction also travels away from the speaker at the speed of sound. Like the condensation, the rarefaction can travel a large distance as it is passed "hand-to-hand" by individual air molecules. A completely analogous situation is shown with the slinky in the figure.

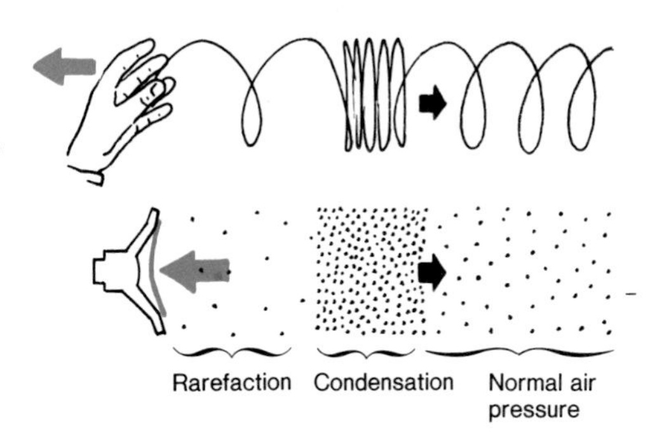

Figure 3.17. The creation of a rarefaction by a moving speaker diaphragm.

A sound wave, then, is comprised of condensations and rarefactions. Sound is not a mass movement of air, such as occurs on a windy day, but the cooperative venture of moving condensations and rarefactions from "hand-to-hand," from one air molecule to another. Sound is a series of alternating condensations and rarefactions which travel outward from a vibrating source. The condensations and rarefactions represent, respectively, regions where the air pressure is higher and lower than normal. Figure 3.18 depicts this concept by plotting a graph of the air pressure as it varies along a sound wave. Even though this graph looks like a transverse wave, be sure to remember that the sound wave itself is a longitudinal wave.

The incredible feature about these pressure variations is that they are so small. In the sound waves generated by a normal conversation, for example, the air pressure fluctuates between 14.7 + 0.000001 psi and 14.7 − 0.000001 psi. The ear is indeed a remarkable device in order to sense changes in the air pressure of only one-millionth of a psi! When these pressure changes occur rapidly enough (between 20 times per second and 20,000 times per second) the brain perceives them as sound. Pressure changes which occur more slowly than 20 times per second are not perceived as sound, but they are sensed as individual pressure changes like the ones which cause your ears to "pop" as you ascend or descend in an airplane. Pressure changes which take place over a period of hours or days are, of course, referred to as "weather" changes!

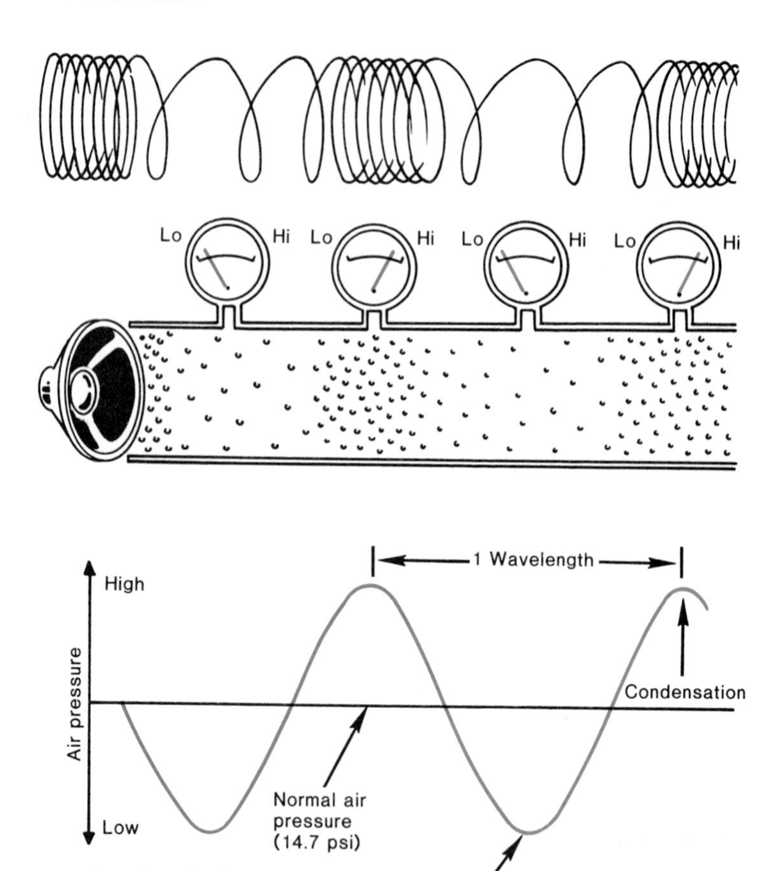

Figure 3.18. A sound wave is an alternating sequence of condensations and rarefactions. The graph shows that a condensation is a region of higher-than-normal air pressure, while a rarefaction is a region of lower-than-normal pressure.

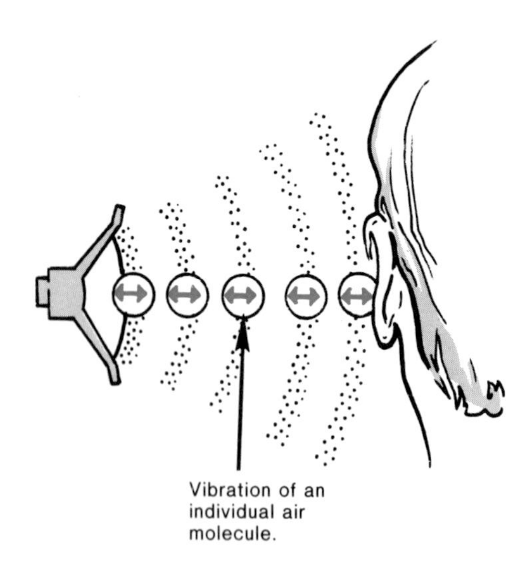

Figure 3.19. The pressure variations of the wave move the ear drum back and forth, producing the physiological sensation of a 1,000 Hz tone. Notice that the molecules on which the speaker diaphragm pushes are not the same ones which move the eardrum.

If a speaker diaphragm is vibrating back and forth at the rate of 1,000 Hz, then 1,000 condensations, each followed by a rarefaction, will be generated every second, thus forming a longitudinal sound wave whose frequency is precisely 1,000 Hz. These pressure fluctuations arrive at the eardrum and force it to vibrate in synchronism with the incoming waves 1,000 times each second. This oscillatory motion of the eardrum is shown in Figure 3.19 and is ultimately interpreted by the brain as a steady tone whose frequency is 1,000 Hz. Notice that the origin of the oscillatory motion is the speaker diaphragm. The air molecules between the speaker and the ear pass the oscillatory motion from "hand-to-hand" until it reaches the ear. Many hi-fi enthusiasts mistakenly believe that as the air pressure rises and falls at the eardrum because of the 1,000 Hz tone, the listener will also perceive the loudness of the sound to be rising and falling at a rate of 1,000 times each second. Not so! When the speaker is producing a sound of constant amplitude, the individual pressure fluctuations occur too rapidly to be sensed individually, and the listener only hears a tone of steady, constant loudness.

Summary

A sound wave is produced when an object vibrates in a medium, such as air or water. The sound wave consists of small pressure fluctuations which travel through the medium at the corresponding speed of sound. A pressure fluctuation whose pressure is greater than the normal average pressure level is called a condensation, while a fluctuation whose pressure is below the normal level is called a rarefaction. In a normal conversation these fluctuations are approximately one-millionth of a psi.

3.5 Power and the Pressure Amplitude of a Sound Wave

It should come as no surprise to learn that sound waves carry acoustical power. When two people are talking, it is the power contained within the sound waves which activates the hearing process. Moreover, common sense indicates that a loud sound carries more power than does a soft sound. In fact, the sonic booms generated by high speed aircraft carry sufficient power to cause considerable damage to buildings.

Sound power is measured in units called "watts," which are also the familiar units that are used to rate household electrical devices such as light bulbs, hair driers, power amplifiers, and the like. In a hi-fi system it is the speakers which convert the electrical power into sound power. For example, a 50 watt power amplifier is capable of delivering 50 watts of electrical power to each of the speakers. As we will see later in Chapter 6, speakers typically convert only a fraction (say, 1 watt) of this electrical power into sound power, and the remaining 49 watts of electrical power are wasted in the form of heat. When the newly-created sound waves enter the room, they carry the 1 watt of sound power with them, and eventually part of this power is intercepted by the listeners.

Recall, from section 3.4, that a sound wave is a series of condensations and rarefactions which are produced by a vibrating diaphragm. Figure 3.20 illustrates how the air pressure varies for two individual sound waves, one with a larger amplitude than the other. The power which these waves carry to the ear of a listener is related to the size of the pressure fluctuations, with more powerful waves having greater pressure amplitudes. It might seem that the sound power which is intercepted by the ear should be directly proportional to the pressure amplitude. Strangely enough, this is not the case! Doubling the pressure amplitude of a sound wave does *not*, for example, double the power which reaches the ear. Without going into all the scientific details, it turns out that the power reaching the ear is proportional

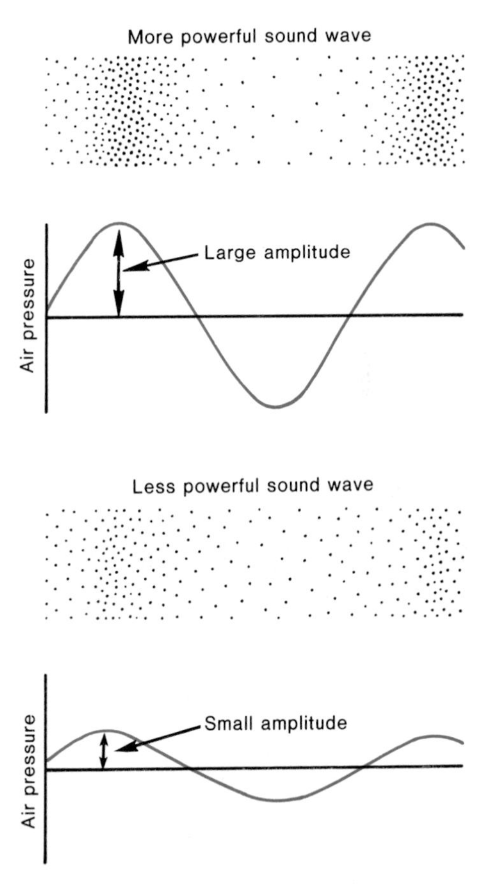

Figure 3.20. The power carried by a sound wave is associated with its amplitude. More powerful waves have larger amplitudes.

to the square of the pressure amplitude. This implies that it requires four times ($2^2 = 4$) more power to create a sound wave with twice the amplitude. Similarly, if the amplitude is tripled, the sound carries nine times ($3^2 = 9$) more power to the ear than it originally did. This fact, which relates the power of a wave (passing through an area such as the ear) to the square of its amplitude, is very important and we shall make use of it in Chapter 4.

3.6 The Intensity of a Sound Wave

Sound intensity is a concept that is commonly used by audio designers. Sound intensity is analogous to the "bushels-per-acre" concept used by farmers when comparing their crop yields. For example, if a farmer harvests a total of 2,000 bushels of wheat from 80 acres, the yield is 25 bushels-per-acre. Note that the yield is calculated by dividing the total wheat production (2,000 bushels) by the area of the land (80 acres). Expressing the yield in this manner makes it easy for different farmers to compare their productivities, even though each one may have a different size farm.

In a similar fashion sound intensity is calculated by dividing the total sound power by the area through which it passes perpendicularly. Figure 3.21 illustrates an example where a total of 5×10^{-6} watts of sound power is passing perpendicularly through a surface whose area is 2.0 m^2. Using the formula

$$\text{Sound Intensity} = \frac{\text{Total sound power which passes through a surface}}{\text{Area of the surface}}, \qquad \text{(Equation 3.4)}$$

we see that the sound intensity is 2.5×10^{-6} watts/m^2

$$\left(\frac{5 \times 10^{-6} \text{ watts}}{2 \text{ m}^2} = 2.5 \times 10^{-6} \text{ watts/m}^2\right).$$

The concept of sound intensity makes it easy to compare the relative strengths of two sound levels, even when two different areas are involved. For example, suppose that two different hi-fi systems are playing in two different rooms. In one of the rooms listener A measures that 12.5×10^{-9} watts of sound power cross an area of 2.5 m^2. In the other room listener B measures that 15×10^{-9} watts of sound power cross an area of 6 m^2. A's sound intensity is 5×10^{-9} watts/m^2 (12.5×10^{-9} watts/2.5 m^2), while B's sound intensity is 2.5×10^{-9} watts/m^2 (15×10^{-9} watts/6 m^2). Since the sound intensity in A's room is greater than the sound intensity in B's room, A's hi-fi system will sound louder to a listener who is comparing the two systems.

In section 3.5 we discussed the fact that the sound power intercepted by a small area such as the ear is proportional to the square of the pressure amplitude. The concept of intensity lends itself naturally to power on a per unit area basis, and, thus, intensity is proportional to the square of the pressure amplitude.

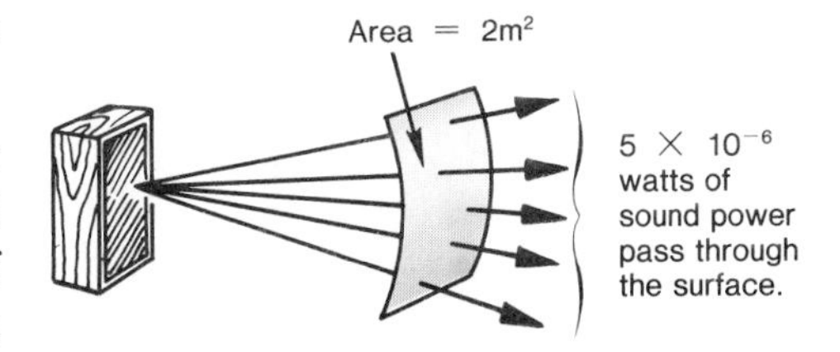

Figure 3.21. The intensity of a sound wave is obtained by dividing the total amount of sound power which passes perpendicularly through a surface by the area of the surface.

3.7 Properties of Sound That Are Important in Audio Systems

A substantial fraction of any audio system deals directly with sound waves, so it is worthwhile to spend some time understanding their important characteristics. In this section we will investigate such topics as loudness, pitch, the limits of human hearing, and the speed of sound.

Loudness

The loudness of the sound produced by a hi-fi system is a characteristic which is important to each of us. The discussions in sections 3.5 and 3.6 indicate that the intensity of a sound wave is associated with its loudness; larger intensities give rise to louder sounds, and vice versa. Intensity is an objective property of sound, because intensity can be accurately measured with electronic equipment. Loudness, on the other hand, is a subjective property of sound because each individual determines what is loud or what is soft. A person with good hearing may interpret a sound as being loud, but another person with a hearing disability is likely to interpret the same sound as being soft.

Strangely enough, even for normal hearing there is not a direct proportionality between the intensity of a sound wave and its perceived loudness. For example, a doubling of the intensity will not result in a doubling of the loudness. To be sure, the loudness will increase, but it will increase only slightly. The exact relationship between intensity changes and the associated loudness changes has many important consequences in hi-fi, and we shall discuss some of the more interesting ones in Chapter 5.

Frequency, the Limits of Human Hearing, and Pitch

As pointed out in section 3.4, sound is a series of alternating high and low pressure regions which travel outward from a vibrating source. The number of high/low pressure cycles which are produced each second is called the frequency of the sound wave. Numerous experiments, dealing with all types of listeners, have shown that the ear can detect all sound frequencies from 20 Hz to 20,000 Hz. Actually, these limits of hearing are only approximate. They only indicate the average hearing ability for people who are less than 25 years old, because the ability to hear high frequencies markedly decreases as one becomes older. Of course, sound can be generated outside the 20–20,000 Hz range. It is well-known, for example, that some dogs can detect sound waves whose frequencies are as high as 30 kHz, although this is certainly outside the human range. Sound waves whose frequencies lie below 20 Hz are called "infrasonic" and those above 20,000 Hz are labeled as "ultrasonic."

The frequency of a sound wave is commonly associated with what is called its musical pitch. Many listeners refer to a high-frequency sound wave as simply a "high-pitched" sound; the higher the frequency, the higher the pitch. Conversely, smaller sound frequencies are said to be "low-pitched" sounds. Very often in hi-fi one also hears phrases like "treble" and "bass." A treble sound is simply a high frequency sound whose frequency is typically greater than 5 kHz. A bass sound is one whose frequency typically lies below 300 Hz.

It should be noted that frequency, like intensity, is an objective property of sound because it can be accurately measured with an electric frequency counter. On the other hand, pitch, like loudness, is a subjective property because pitch is the listener's perception of a sound frequency. Different people might perceive the same sound frequency as having a different pitch. Because the frequency of a sound wave establishes the musical pitch that one hears, frequency is an important concept to designers of audio equipment. In principle, each designer strives for audio perfection whereby the sound heard through a hi-fi system would be identical to that heard at a live performance. Every audio component should be able to process correctly all the frequencies present in the music. Each component should neither add new frequencies to the music nor change the relative loudness of the frequencies already present. In practice, such "perfection" is only partially realized and reputable manufacturers provide technical data which specify how close their products come to perfection. Such terms as "total harmonic distortion," "intermodulation distortion," and "frequency response" are commonly used to describe the frequency-related virtues of audio components. Each of these terms will be carefully discussed in subsequent chapters.

The Speed of Sound

The speed of sound in air is about 344 meters per second. One can calculate the short time required for sound to travel from a speaker to a listener if the distance between the two is known; suppose this distance is 7 meters, which is an average room dimension. According to Equation 3.2, Speed = Distance/Time,

$$\frac{344 \text{ meters}}{\text{second}} = \frac{7 \text{ meters}}{\text{Time}}.$$

$$\text{Time} = \frac{7}{344} = 0.02 \text{ seconds}$$

—a mere two-hundredths of a second!

Our ears possess the remarkable ability of being able to localize sounds with surprising accuracy. A stereo system, with its two speakers allows one to distinguish instruments which are located on the left of the band from those on the right. This ability is partially due to the difference in time required for the sound to reach each ear. Take, for example, the situation in Figure 3.22 in which the sound from a bass guitar is originating from the left speaker. The sound reaches the left ear a short time before reaching the right ear. For low frequency tones it is primarily this time delay which the brain uses to place the source of the music on the left, as opposed to the right. A simple calculation shows how surprisingly short this time delay is.

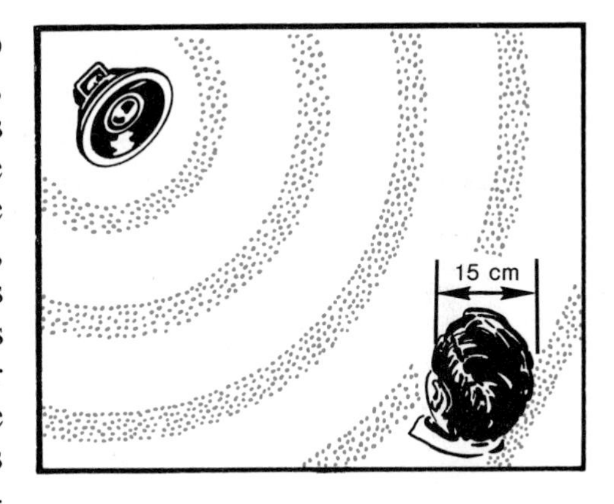

Figure 3.22. The slightly early arrival of sound at the left ear helps the listener to locate the speaker in the left corner of the room.

$$\text{Speed} = \frac{\text{Distance}}{\text{Time}}. \qquad \text{(Equation 3.2)}$$

$$344 \frac{\text{meters}}{\text{second}} = \frac{0.15 \text{ meters}}{\text{Time}}.$$

$$\text{Time} = \frac{0.15}{344} = 0.00044 \text{ second!}$$

Therefore, the sound arrives at the left ear 0.00044 seconds before it reaches the right ear. As short as this time may seem, it is sufficient to aid the brain in locating the bass guitar correctly.

It comes as a surprise to most people that sound travels even faster in liquids and solids than it does in air. Table 3.1 gives the speed of sound in some common liquids and solids.

		Speed of Sound		
		Miles Per Hour	Kilometers Per Hour	Meters Per Second
Gas	Air	769	1,238	344
Liquids	Alcohol	2,727	4,389	1,219
	Turpentine	2,975	4,788	1,330
	Water	3,204	5,156	1,432
Solids	Aluminum	11,386	18,324	5,090
	Glass	12,272	19,750	5,486
	Iron	11,454	18,433	5,120

Table 3.1. The speed of sound in air, liquids, and solids. Notice how sound travels faster as one changes from a gas, to a liquid, to a solid.

Note, for example, that sound in iron travels an astonishing 5,120 meters per second, which is about 15 times faster than in air!! The thing to remember here is that sound travels slowest in gases, faster in liquids, and fastest in solids.

Summary

The loudness, pitch and speed of a sound wave are important concepts when dealing with audio systems. Sound waves containing large amounts of power, and hence having large sound intensities, are interpreted as being loud sounds. A sound wave which has a high frequency is interpreted as a high-pitched sound. Humans can hear frequencies which lie in the range between 20 Hz and 20,000 Hz. The speed at which sound travels in air is approximately 344 meters/second. The speed of sound is least in gases, larger in liquids, and largest in solids.

3.8 Reflection of Sound

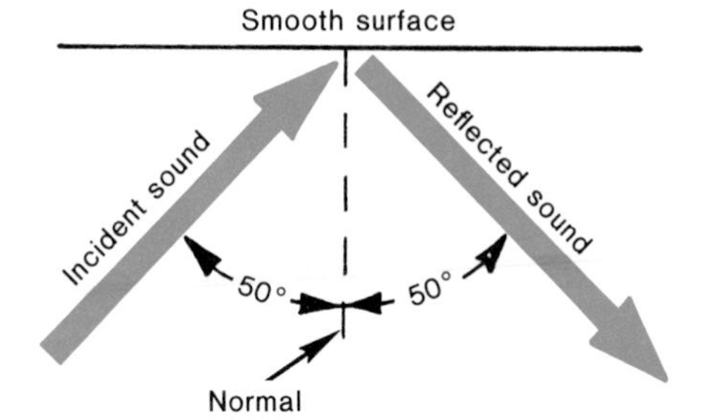

Figure 3.23. When sound is reflected from a smooth surface, the angle of incidence equals the angle of reflection.

Sound waves reflect from the walls and other surfaces of a room just like light reflects from a mirror. Figure 3.23 illustrates a sound wave being reflected from the smooth surface of a wall. In this example the incident wave strikes the surface at an angle of 50° with respect to the dotted line which is perpendicular to the surface; this dotted line is often called the "normal" line. Notice that the reflected sound leaves the surface at an angle of 50° with respect to the normal. Although 50° was chosen for the sake of illustration, the **law of reflection** states that, for any angle, the angle of reflection equals the angle of incidence. As we shall now discuss, the law of reflection plays an important role in determining how the room affects the way in which a hi-fi system sounds.

Reflection and the Intensity of Sound

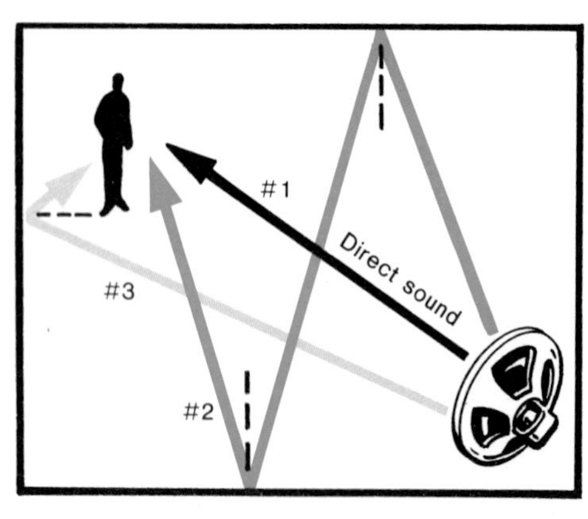

Figure 3.24. The sound that a listener hears is the sum of the direct sound plus the sound reflected from the walls. It is the latter which gives the sound its "presence" or "ambience."

One of the pronounced effects of reflection is that the sound intensity within a room is greater than it would have been without the benefit of reflection. Notice, as drawn in Figure 3.24, that sound spreads out into a wide area as it leaves the speaker. While some of the sound travels directly to the listener's ears, as shown by path #1, an appreciable amount arrives only after having been reflected from the walls and other surfaces (paths #2 and #3). The latter paths are, of course, drawn with equal angles of incidence and reflection, according to the law of reflection. The listener thus hears contributions from both the direct and the reflected sound. In many rooms the intensity of the reflected sound may actually exceed the intensity of the direct sound. Therefore, the total sound intensity, and hence loudness, is greatly enhanced by the presence of reflecting surfaces. Try singing in the shower. The highly reflective walls of the bathroom absorb little of the sound and reflect a great deal of it back to your ears. The increased loudness associated with the reflected sound boosts even the weakest voice.

Another interesting aspect of reflection is that, within a typical listening room, a hi-fi system sounds more or less equally loud anywhere within the room, except for places that are extremely close (less than 1 meter) to the speakers. Try walking around your room while listening for differences in the sound loudness. You will probably discover that any differences are rather small, and that the sound loudness is remarkably constant. The reason for this uniformity is that the sound is usually reflected from many surfaces before reaching your ears. Thus, the sound is "bounced" into all parts of the room, with the net result being that the loudness becomes more or less evenly distributed everywhere.

Anyone who has moved a hi-fi system outdoors for a party has noticed that the sound does not appear as loud outside as it did inside. Outside, the sound waves experience fewer reflections before reaching the listener because there are fewer surfaces to reflect the sound; thus, the sound intensity is less. In fact, if there were no reflection at all, the listener would hear only the sound coming directly from the speaker. This loss of intensity is most obvious for those highly valued bass notes because their intensity is greatly dependent on reflections. Due to the lack of reflecting surfaces, much more electrical power is needed in order to produce the same sound intensity outdoors than would be required in an enclosed room (assuming equal distances from the speaker). The cost of the power amplifiers needed to produce an outdoor rock concert, for example, is not trivial! In comparison, the reflective properties of a listening room allow one to purchase a significantly less powerful, and correspondingly less expensive, amplifier than would be needed if there were no reflections.

There is another effect which results when speakers are placed outdoors, in addition to the overall decrease in loudness discussed above. It is well-known that the sound becomes even less loud as one moves further from the speakers. This decrease in loudness comes about because the sound intensity decreases at larger distances from the speakers. Figure 3.25 illustrates how the sound power spreads out over larger and larger areas as it travels outward. Notice, for example, that the same amount of sound power passes through both the #1 and #2 surfaces. However, the area of the #2 surface is substantially larger than the area of #1, so that the sound intensity at #2 is considerably less. The sound power, so to speak, is "diluted" over larger and larger areas.

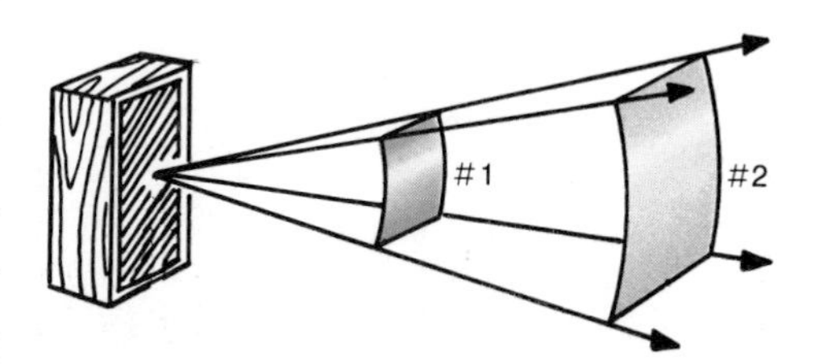

Figure 3.25. The intensity of the direct sound decreases as one moves further away from a loudspeaker, because the sound power spreads out over larger and larger areas.

Therefore, the further that one is from an outdoor speaker, the less will be the sound intensity and, hence, the less will be the loudness.

Reflection and the Reverberation of Sound

The reflection of sound also plays an important role in determining the "reverberation" characteristics of either a room or an auditorium. Upon examination of Figure 3.26 it should be apparent that the listener hears the reflected sound a little later than the direct sound, because the reflected sound travels a longer distance. The two travel times can be easily computed from the distances and the speed of sound by using Equation 3.2:

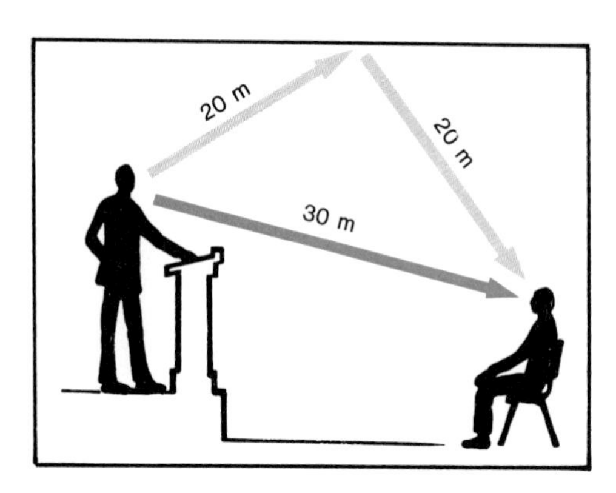

Figure 3.26. There is a delay in the time of arrival between the direct sound and the reflected sound. Direct sound follows the shortest path and arrives first.

Direct sound

$$\text{Time} = \frac{30 \text{ meters}}{344 \text{ meters/sec}} = 0.087 \text{ seconds.}$$

Reflected sound

$$\text{Time} = \frac{40 \text{ meters}}{344 \text{ meters/sec}} = 0.116 \text{ seconds.}$$

The reflected sound thus arrives 0.029 seconds after the direct sound. Although Figure 3.26 illustrates only one reflected sound wave, in reality many reflected waves reach the listener because there are many possible reflecting paths (see Figure 3.24, for example). Since each path length is different, each reflected sound arrives at a slightly different time. However, the arrival times of the reflected waves are so close together that each wave cannot be perceived individually. The listener hears only a "blur" of sound. This blur of reflected sound is called "reverberation."

As mentioned earlier, the amount of reflected, or reverberated, sound within a room may be substantial and contribute as much to what you hear as does the direct sound. Reverberated sound, however, produces an entirely different aural sensation. When the source of sound is turned off, the direct sound ceases abruptly. Reverberated sound, on the other hand, persists even after the direct sound stops, because the reflected sound bounces around until it has been absorbed by the walls, furnishings, and people.

Reverberation is responsible for providing the listener with the aural sensation of being either in a large concert hall or in a small room. The open, spacious sound of a large hall is due to the long time delays associated with the many long reflecting paths. On the other hand, the less-spacious dimensions of an ordinary living room produce an entirely different reverberated sound pattern that the listener typically associates with a small room. Each concert hall possesses unique reverberation qualities due to its size and furnishings, and music played in one hall sounds markedly different when played in another. In fact, a few very experienced listeners can identify the hall in which they are sitting by simply listening to the music!

Reverberation is such an important concept in acoustics that audio designers routinely use the concept of "reverberation time." Reverberation time is the time required for the reverberated sound to decay to one-millionth of its original power level after the sound source has been turned off. The reverberation time of any room or auditorium is affected greatly by the amount of absorbing material present. With a large amount of such material the reverberated sound is quickly absorbed, and a very short reverberation time results. People are extremely good sound absorbers. Carpets, overstuffed furniture, drapes, and sound absorbing materials on the walls also help to reduce the reverberation time. A room with too much sound absorption (an abnormally small reverberation time) is called a "dead" room and musical instruments, as well as voices, do not sound natural in such an environment.

On the other hand, brick or tile-lined rooms absorb relatively little of the reverberated sound, and so the sound "hangs-around" for a long time; bathrooms are good examples of such rooms. Such rooms are said to be "live," and in them words or musical notes tend to become blurred together and cannot be clearly distinguished. Figure 3.27 shows what happens in this case. Notice that when the second note begins, the first note has not yet died away because

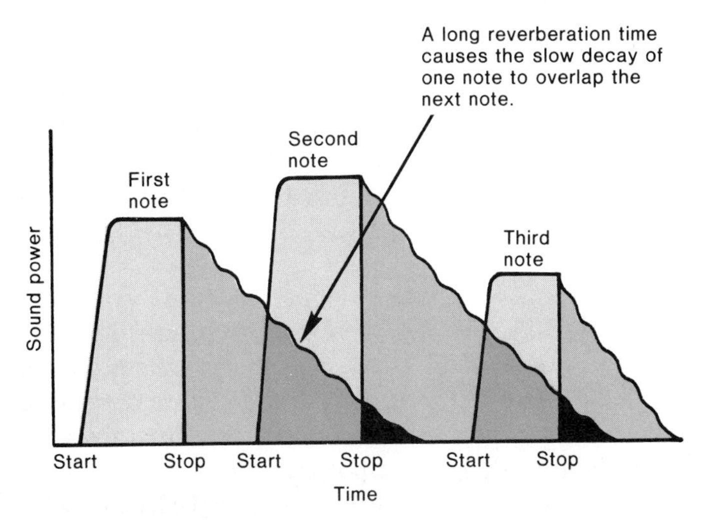

Figure 3.27. A long reverberation time is detrimental to good music reproduction, because there is too much overlapping and, hence, blurring of the notes.

of the long reverberation time. This obviously causes an excessive overlap of the two notes and results in a loss of clarity and definition. Excessively long reverberation times, often found in large auditoriums, require a public address system especially designed to reduce their ill effects. Often the reverberation time can be reduced by partially sound-proofing the room with acoustic tiles. These tiles are perforated with

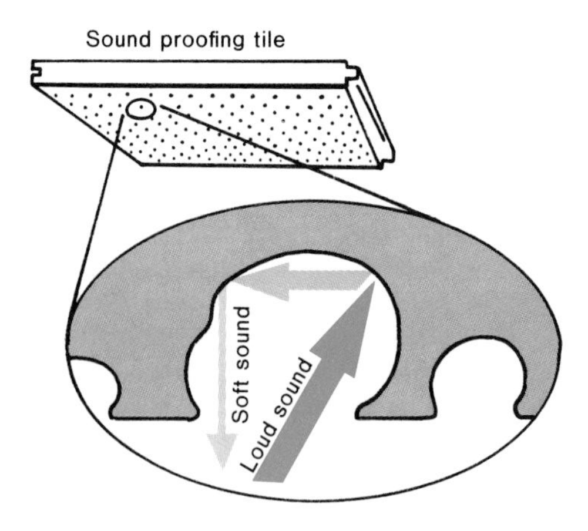

Figure 3.28. An acoustic tile traps the sound inside its cavities and reduces the sound loudness by multiple reflections and absorptions.

many tiny holes and utilize the principles of reflection and absorption in a clever way to reduce the overall loudness of the sound waves. When the sound strikes a tile it enters the holes and bounces around inside due to multiple reflections (Fig. 3.28). Each time a reflection occurs on an inside surface of the cavity, part of the sound is absorbed with the remainder being reflected. It is easy to see that the sound, when it finally emerges from the hole, is greatly reduced in loudness. Hence the sound proofing action is achieved.

In recording studios or concert halls very careful engineering is required to produce exactly the desired amount of reverberation. Depending on the type of music to be played, and the use for which the room is intended, typical reverberation times range from 0.5 to 2.2 seconds.

Reverberation Units

A reverberation unit is an audio component which electronically simulates the effects of reverberation. As mentioned in previous paragraphs, listening to a symphonic recording in your own room doesn't quite match the realism of a live performance, because the reverberation characteristics of the concert hall are missing. The reverberation unit attempts to simulate these characteristics and, hence, provide the listener with a more realistic illusion of hearing the music as it was actually performed.

Figure 3.29 illustrates a hi-fi system with a reverberation unit inserted into the TAPE MONITOR facility of the preamplifier. Although some reverberation units utilize four speakers (two in front of the room and two in the rear), only two speakers are illustrated for the sake of simplicity. The

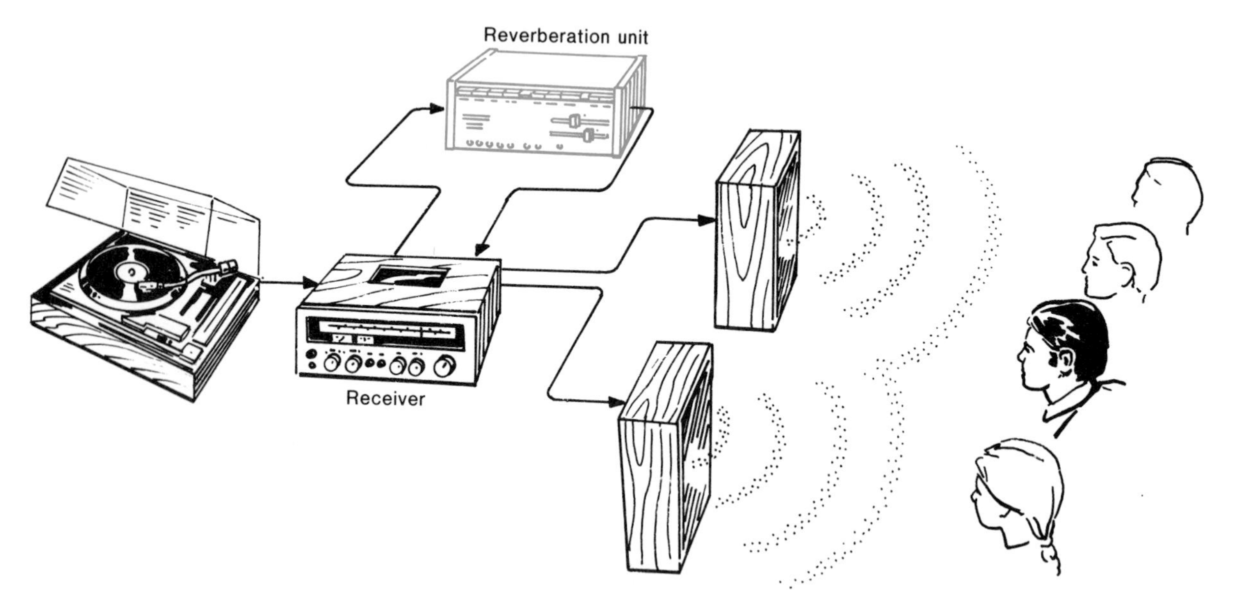

Figure 3.29. A reverberation unit can be inserted into a hi-fi system in order to create a "concert-hall" effect.

reverberation unit first splits-off a fraction of the audio signal and delays it in time by an amount which can be determined by the user. The delayed signal is then recombined with the undelayed portion of the signal, and the total is sent out of the reverberation unit and on through the remainder of the hi-fi system. The sound which emerges from the speakers has two components; the undelayed (or "direct") sound plus the delayed (or "reverberated") sound. Since the electronic time delay simulates the effect of reverberation, the listener can experience the exhilarating effect of hearing sound which appears to have been reflected from the surfaces of a much larger room.

Reflection and Loudspeaker Design

Some hi-fi loudspeakers directly exploit the law of reflection as the basis for their operation. These loudspeakers are designed so that a substantial amount of sound exits from both the sides and rear of the cabinet, as well as from the front. Figure 3.30 indicates how the walls of the room reflect the "side-and-rear" sound back to the listener, thereby increasing the amount of reflected sound that the listener hears. The additional reverberation adds an acoustic spaciousness to the music, which sounds like it has originated from within a much larger room. Thus the "reflection" loudspeakers attempt to give an open, spatial quality to the music, much the same as the reverberation units do. Obviously, these type of speakers should be placed near one, or more, walls in order to achieve their best results, and manufacturers usually suggest optimum distances for different types of listening rooms.

Figure 3.30. The reflection type loudspeakers deliberately reflect a large fraction of the sound from the walls before it reaches the listener.

Summary

The law of reflection determines how sound reflects from a surface. It states that the angle of incidence equals the angle of reflection. Reflections within a room produce several effects. Reflections increase the sound intensity and they distribute the intensity more-or-less uniformly throughout the room. Outdoors, the reduced number of reflecting surfaces causes the sound intensity to be less than it would be inside a room. In addition, the intensity is not uniform, and it decreases as one moves away from the speaker. Reverberation is the result of numerous reflected sound waves which arrive so close together that the listener hears only a continuous blur of sound, rather than the individual reflected waves. Reverberation gives each room, auditorium, or concert hall its unique sound qualities. The reverberation time is the time it takes for the reverberated sound to decay to one-millionth of its original power level. Typical reverberation times range from 0.5 to 2.2 seconds. Reverberation times which are either too short or too long can have an adverse effect on music and speech. The qualities of a large concert hall can be simulated in a small room by adding a reverberation unit to a hi-fi system. Some speakers attempt to create a more spacious, open effect by deliberately reflecting a large portion of their sound from the walls of a room.

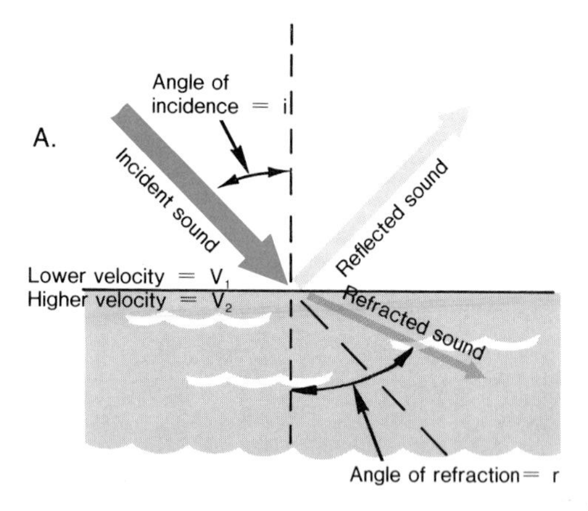

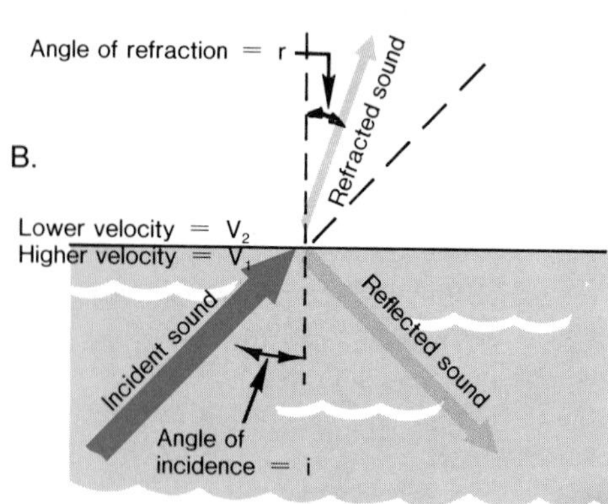

Figure 3.31. When a sharp boundary separates two regions (e.g., air and water) in which the speed of sound is different, the sound waves bend away from their original direction when crossing the boundary. (A) The sound wave is traveling from a lower velocity medium into a higher velocity medium. (B) The sound is traveling from a higher velocity medium into a lower velocity medium.

3.9 Refraction of Sound

We have just seen how the direction of sound can be changed by reflecting it from a surface. There is another way to bend a sound wave via a phenomenon known as "refraction." Refraction is the bending of a sound wave when it moves through a region of space where its speed changes.

To illustrate the concept of refraction, consider Figure 3.31 (A) which shows a sound wave traveling from air (speed of sound = 344 m/s) into a pool of water (speed of sound = 1,432 m/s). Part of the incident sound is reflected back into the air, according to the law of reflection. The remainder of the sound passes into the water. Notice how the sound abruptly bends and changes its direction of travel as it enters the water. This bending, or refraction, occurs whenever a sound wave moves from one medium into another. The amount of bending is given by the **law of refraction** which states:

1. If the sound travels from a low-velocity medium (e.g., air) into a higher-velocity medium (e.g., water), the refracted sound will bend *away* from the normal, as illustrated in Figure 3.31 (A).

2. Conversely, if the sound travels from a high-velocity medium into a low-velocity medium, the refracted sound will bend *toward* the normal, as shown in Figure 3.31 (B).
3. The amount of bending depends on the sound velocities in the two media; the greater the velocity difference, the more the sound wave will bend away from its original direction when it crosses the boundary.

It is worthwhile to note that the law of refraction is also obeyed by light, as well as by sound.

Generally speaking, the refracted sound in Figure 3.31 (A) becomes absorbed as it penetrates deeper into the water. There is a great similarity between this example and what actually happens to the sound in a listening room. Part of the sound power emitted by the speakers is reflected by the surfaces within the room. However, every time a reflection occurs some of the incident power also passes through the surface where it is refracted and then is partially absorbed. For this reason, the reflected power is always less than the incident power. Since each surface (be it due to a wall, furniture, rugs, etc.) absorbs a different amount of sound, the strength of reflected sound within a room depends strongly on the type of construction materials and furnishings of the room. Therefore, two rooms which have identical sizes can have markedly different reverberation characteristics.

It is not necessary to have an abrupt boundary between two media in order for the speed of sound to change. Scientists have known for some time that the speed of sound in air increases slightly as the temperature increases. If, for some reason, the sound should travel into regions of progressively warmer temperatures, the sound will not only speed up but it will also gradually bend due to the effects of refraction. Figure 3.32 shows that such a condition is likely to occur near a lake on a summer evening after the sun has set; the water cools the air adjacent to it while the air higher-up still retains some of the warmth of the day. Any sound which travels into the warmer air has its speed increased and, consequently, the sound gradually bends down towards a listener who is on the lake. Figure 3.33 shows the daytime situation in which the air near the water is actually warmer than the air higher up. This reverse-temperature situation causes any upward traveling sound to slow down and refract further into the upper atmosphere and away from the listener. This is the reason why sound from a radio may be heard across the lake at night but not during the day. We say that the sound "carries further" at night.

The law of refraction can be stated mathematically with the aid of the trigonometric "sine" function. In terms of the angles and velocities shown in Figure 3.31, the law of refraction is

$$\frac{\sin i}{\sin r} = \frac{v_1}{v_2}.$$

In this equation the angle of incidence is denoted by i, and the angle of refraction by r. v_1 and v_2 are, respectively, the velocities of sound in the incident and refracted media.

Example: Suppose that the sound wave is traveling from air (v_1 = 344 m/s) into water (v_2 = 1,432 m/s). If the angle of incidence i equals 10°, the angle of refraction r can be calculated using the above formula:

$$\frac{\sin 10°}{\sin r} = \frac{344 \text{ m/s}}{1{,}432 \text{ m/s}}$$

$$\sin r = 0.72$$

$$r = 46°.$$

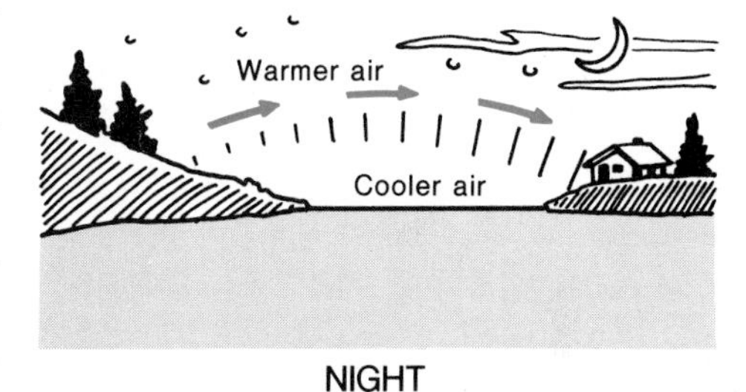

Figure 3.32. The higher velocity of sound waves in warmer air causes sound waves to be refracted down at night and travel back towards the ground.

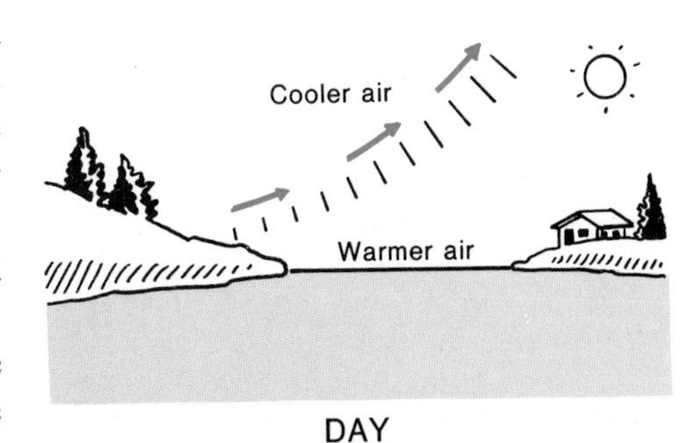

Figure 3.33. During the daytime the sound is refracted away from the earth's surface.

Actually, one does not have to be near a lake for refraction to take place. Very often an outdoor rock band playing late at night during the summer can often be heard from one mile away. In this case it is the ground itself which cools the air adjacent to it. The sound from the band, which would normally travel into space and become lost, is bent back toward the cooler earth and reaches the ears of sleeping (and complaining) neighbors.

Summary

If the speed of sound either increases or decreases the sound will bend (refract) away from its original direction of travel. In general, both reflection and refraction occur when sound impinges on a surface. The reflected wave is always less powerful than the incident wave because part of the sound power is refracted into the material and eventually absorbed. The amount of refraction, as stated by the law of refraction, depends on the velocities of sound in the two media, and whether the sound is traveling from a low-speed media into a high-speed media, or vice versa. If the changes in the speed of sound occur gradually, such as those which happen in the air over a lake at night, the sound waves gradually bend. This type of refraction leads to the commonly known fact that "sound carries further at night than during the day."

3.10 The Doppler Effect

At one time or another all of us have been standing on the sidewalk and listening to the siren on an approaching fire truck. No doubt you have noticed a distinct change in the pitch of the siren as the truck passes. When the truck is approaching, the siren has a high pitch and, as the truck passes and moves away, the pitch of the siren suddenly drops. This pitch change is due to the Doppler effect. The Doppler effect is a change in the pitch of the sound when there is relative motion between the sound source and an observer.

To understand the cause of the Doppler effect first consider the stationary fire truck in Figure 3.34. Its siren is emitting a sound of constant frequency in all directions. If this frequency were 1,000 Hz, then any listener, in front or behind, would receive 1,000 sound cycles every second. In Figure 3.35 the truck moves, however, and the sound cycles emitted in the forward direction are closer together and have a shorter wavelength than normal. This shortening of the wavelength occurs because as the truck moves forward it is "catching up" with the previously emitted wave crests. The person in front of the truck in Figure 3.35 will hear a higher pitch sound because the crests are closer together and, therefore, more of them arrive every second. A person standing behind the truck hears a lower pitch, because the siren is

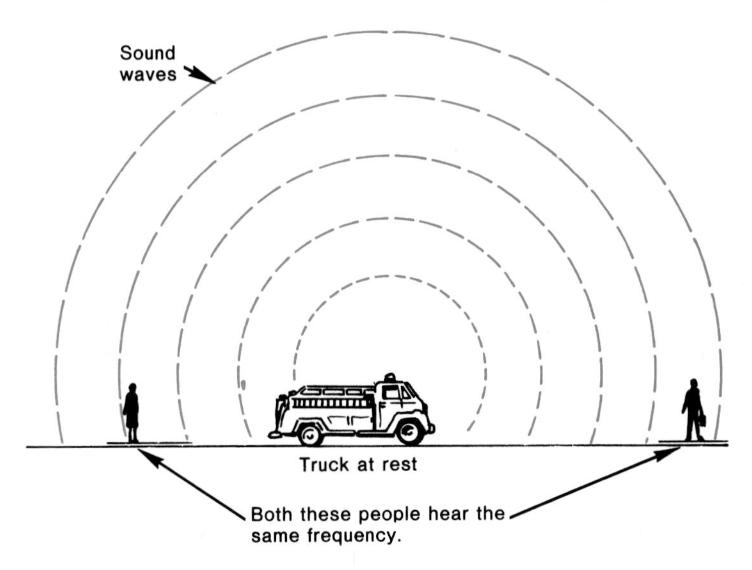

Figure 3.34. The sound wave emitted by the siren of a stationary fire truck. Each person hears the same frequency.

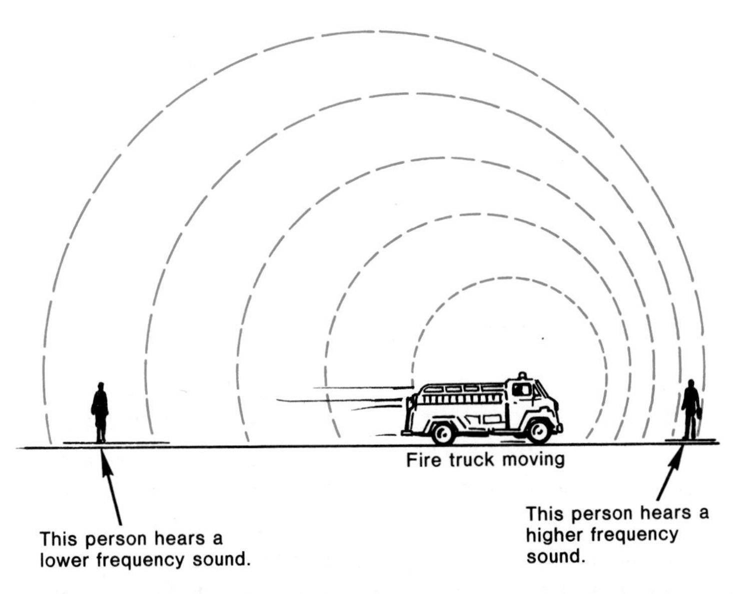

Figure 3.35. Due to the Doppler effect, the observer ahead of the moving truck hears a higher frequency sound, and the observer behind the truck hears a lower frequency sound.

moving away from the emitted wave crests and there is a greater distance between them. The result is a longer wavelength, and fewer crests arrive each second at the ear of a person standing behind the truck.*

How much does the pitch change due to the Doppler effect? It depends on the speed of the car—with higher speeds resulting in greater frequency changes. As an example, suppose you are sitting in the grandstand of a speedway, and two race cars are traveling with the speeds shown in Figure 3.36. Suppose, further, that each car is generating

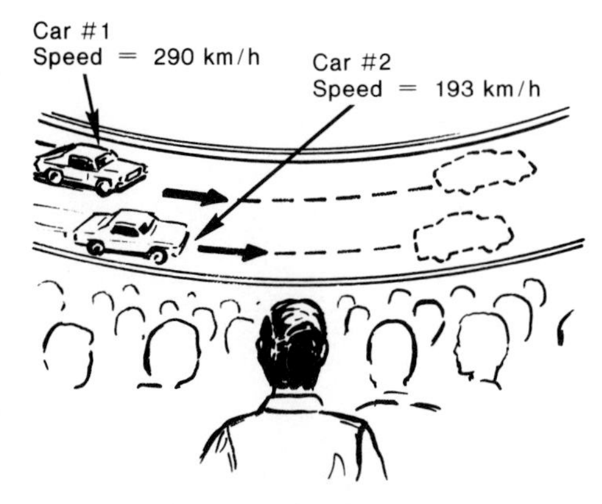

Figure 3.36. The Doppler effect for two race cars traveling at different speeds. Each car is producing a 1 kHz sound, when stationary, due to its exhaust.

*There is also a Doppler effect which arises when the source of sound is stationary and the observer moves. This is described in any standard physics text.

Using f′ to denote the frequency heard by the stationary observer, f to denote the frequency produced by the source, V to denote the speed of sound, and v to denote the velocity of the moving source, the Doppler effect is given by

$$f' = f\left(\frac{V}{V \pm v}\right).$$

The + sign is used when the source moves away from the observer, and the − sign is used when the source moves toward the observer.

Example: Calculate the frequency heard by a stationary observer f′ of a 1,000 Hz exhaust noise on a race car which is approaching the observer at a speed of 193 km/h. The speed of sound is 1,238 km/h.

$$f' = (1{,}000 \text{ Hz}) \times \left(\frac{1{,}238 \text{ km/h}}{1{,}238 \text{ km/h} - 193 \text{ km/h}}\right)$$

$$f' = 1{,}184 \text{ Hz}.$$

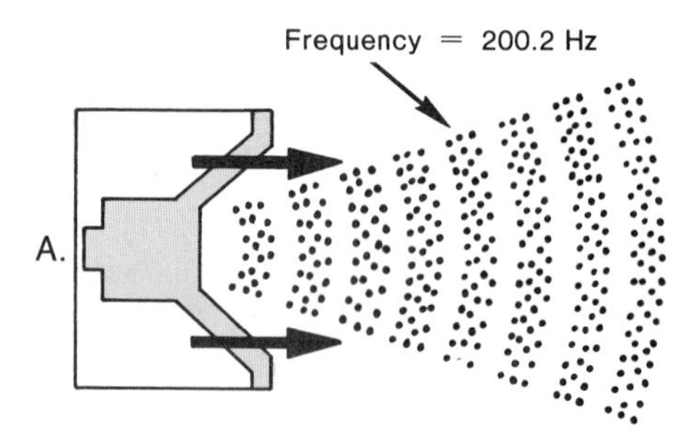

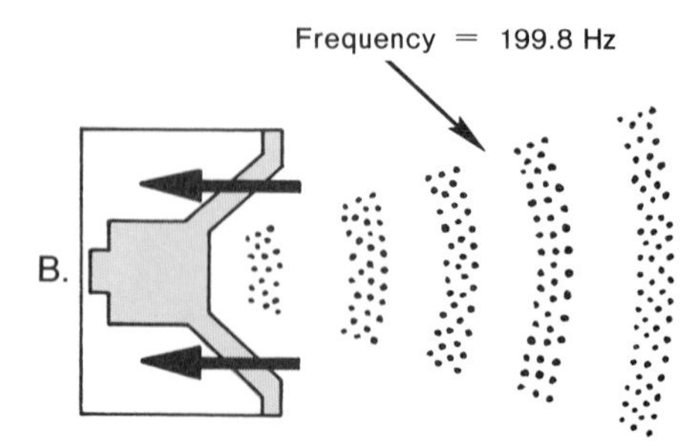

Figure 3.37. A speaker vibrating simultaneously at both 20 Hz and 200 Hz causes a Doppler frequency shift to occur in the 200 Hz tone.

an exhaust noise at 1,000 Hz when stationary. However, as indicated in Figure 3.35, you perceive a different pitch for each of the cars because they are traveling at different speeds. Car #1, traveling at a speed of 290 km/h, appears to have a frequency of 1,305 Hz when approaching you, while Car #2 appears to have a frequency of 1,184 Hz because it has a different speed of 193 km/h. In addition, you will hear the familiar frequency change as the cars pass by. For example, Car #1 appears to be emitting a sound of frequency 1,305 Hz while approaching, which then drops to 810 Hz as it passes, with the frequency change being an astounding 495 Hz!

The Doppler Shift in Audio

In audio the Doppler shift arises due to the back and forth motion of the speaker cone. The sounds that we hear from any musical selection are a complex mixture of many notes being played simultaneously. Suppose, for the sake of simplicity, there are only two notes being sent through the speaker, 20 Hz and 200 Hz. As the speaker oscillates at the slower rate of 20 Hz, it will simultaneously execute a much faster 200 Hz vibration. During one-half of each cycle of the 20 Hz sound, the speaker is moving toward the listener while simultaneously emitting the 200 Hz sound. Under this condition the 200 Hz tone will be perceived, due to the Doppler effect, as a higher pitched tone as shown in Figure 3.37 (A). For the second half of each 20 Hz cycle the speaker moves away from the listener and the 200 Hz tone appears lower in pitch as shown in Figure 3.37 (B). How much shift in frequency occurs? With reasonable numbers for the vibrational speed of the speaker cone at 20 Hz (0.3 meters/second) the 200 Hz note would appear to be 200.2 Hz as the cone moves forward and 199.8 Hz as it moves away. This constitutes a total frequency deviation of 0.4 Hz which occurs 20 times each second, and results in an audible "blurring" of the 200 Hz sound. Surprisingly enough, the Doppler effect, usually associated with everyday fire sirens and train whistles, has an audio hi-fi manifestation.

Summary

The Doppler effect arises because a listener perceives a sound source (such as a siren) to emit a different frequency when it is moving than when it is stationary. When a sound source is moving towards the observer, the observer hears a higher frequency. When the sound source moves away from an observer, the observer hears a lower frequency. The faster the sound source is moving, the greater will be the change in frequency.

Summary of Terms

Amplitude—The amplitude of a wave is the maximum excursion of a physical quantity from its equilibrium position. The physical quantity can be: the displacement of water for a water wave, the change in pressure for a sound wave, or the displacement of the coils for either a transverse or longitudinal slinky wave.

Compressions—(See **Condensations.**)

Condensations (or Compressions)—Regions along a sound wave where the pressure is largest. Also called crests. In a longitudinal slinky, compressions are regions where the coils are "bunched up."

Crest/Troughs—The locations along a wave where the displacement of the wave has either its greatest value (crest) or its smallest value (trough.).

Cycle—A term used with waves to designate a crest-trough combination.

Doppler Effect—An apparent change in the pitch of sound emitted by a source when either the source or the observer is moving relative to each other. When the source is approaching an observer the pitch sounds higher than normal, and when the source is receding from an observer the pitch appears lower than normal.

Energy—The ability of waves, or any object, to accomplish some type of work. Examples: sound waves forcing your eardrum to vibrate, electrical signals from a power amp which move the speaker diaphragm, etc.

Frequency—The number of wave cycles which pass a particular point each second. Frequency is measured in Hertz (Hz). The frequency of a wave is related to its period by:

$$\text{frequency} = \frac{1}{\text{period}}.$$

Hertz (Hz)—A unit of frequency which represents one complete wave cycle (crest-trough combination) passing by each second:

one Hertz = one cycle/second.

Infrasonic Sound—Sound whose frequency lies below the audible range: below 20 Hz.

Intensity—The amount of sound power which passes perpendicularly through a surface divided by the area of the surface.

KiloHertz (kHz)—1,000 Hz.

Law of Reflection—(See **Reflection, Law of**)

Law of Refraction—(See **Refraction, Law of**)

Limits of Human Hearing—The frequency range, from about 20 Hz→20 kHz, over which the normal young (25 years or less) person can hear sound.

Longitudinal Wave—A wave which causes the particles of the medium to vibrate back and forth parallel to the direction in which the wave travels. Sound waves are longitudinal waves.

Loudness—The subjective perception of the amount of power carried by a sound wave. A more powerful sound wave will produce the sensation of a greater loudness.

Period—The time required for one complete wave cycle (crest-trough combination) to pass a particular point. The period of a wave is related to its frequency via:

$$\text{period} = \frac{1}{\text{frequency}}.$$

Pitch—The "highness" or "lowness" of a musical tone as perceived by a listener. A high frequency vibrating source produces a sound of high pitch; a low-frequency vibrating source produces a sound of low pitch.

Power—The amount of energy being produced or consumed every second. The unit of power is called a watt. If the power rating of an amplifier is, for example, 20 watts, then the amplifier will produce and deliver 20 units of electrical energy per second. Likewise, a speaker which is producing 3 watts of acoustical power creates 3 units of sound energy each second.

Rarefactions—Regions along a sound wave where the pressure is smallest. Also called troughs. In a longitudinal slinky, rarefactions are regions where the coils are "stretched out."

Reflection—The ability of a sound wave to "bounce off" an object.

Reflection, Law of—When a sound wave impinges on a smooth surface, the angle of reflection equals the angle of incidence.

Refraction—The bending of a sound wave caused by differences in the wave's speed as it travels through nonuniform media.

Refraction, Law of—When sound travels from a low-velocity medium into a higher-velocity medium the refracted sound will bend away from the normal. If the sound travels from a higher-velocity medium into a lower-velocity medium, the refracted sound will bend toward the normal. The amount of bending depends on the sound velocities in the two media and the angle of incidence.

Reverberation—Numerous reflections of sound which arrive close together and cannot be perceived as individual echoes.

Reverberation Time—The time required for the reverberated sound to decay to one-millionth of its original level after the speakers have been turned off. Typical reverberation times are between 0.5 and 2.2 seconds.

Reverberation Unit—A component that will add reverberation to the sound when connected to an audio system.

Sound—A longitudinal wave consisting of an alternating series of condensations and rarefactions which must have a medium in which to travel.

Sound Intensity—(See **Intensity**)

Sound Power—(See **Power**)

Speed—A measure of how fast an object or a wave is moving. Speed is the distance which is traveled in some unit of time; e.g., 344 meters per second. Speed, unlike velocity, does not convey any information on the direction in which the object or wave is moving.

Tone—A musical note.

Transverse Wave—A wave which causes the particles of the medium to vibrate perpendicular (or transverse) to the direction in which the wave travels.

Troughs—(See **Crests/Troughs.**)

Ultrasonic—Sound whose frequency lies above the audible range: above 20 kHz.

Velocity—A term which indicates both the speed and direction that an object or wave is moving. Very often velocity and speed are used interchangeably although, technically, the two concepts are slightly different.

Wave—A process by which energy and information can be transmitted by the periodic motion of a physical quantity. The physical quantity can be: the displacement of water for a water wave, the change in pressure for a sound wave, or the displacement of the coils for either a transverse or longitudinal slinky wave.

Wavelength—The distance between successive crests, troughs, or identical parts of a wave, measured in the direction that the wave travels.

Review Questions

1. Explain the relationship between frequency, speed, and wavelength for transverse and longitudinal waves.
2. Why can sound, which is a longitudinal wave, be shown on a graph which looks like a transverse wave?
3. Explain how a sound wave travels through a medium in terms of condensations and rarefactions.
4. What is a Hertz? KiloHertz?
5. Explain the difference between reflection and refraction.
6. How is the power of a sound wave related to its pressure amplitude?
7. What effect does the Doppler effect have on the frequency of sound which is being produced by a moving speaker diaphragm?
8. How do the vibrations of a transverse wave differ from those of a longitudinal wave?
9. In terms of condensations and rarefactions, explain how a loud sound differs from a soft sound.
10. Explain what causes reverberation, and what is meant by the reverberation time.
11. What is the intensity of a sound wave, and how does it differ from the power of the wave?
12. Why does the sound intensity decrease when a loudspeaker is removed from a room and placed outdoors?
13. What is the difference between intensity and loudness?
14. What is the difference between frequency and pitch?

Exercises

NOTE: The following questions have up to 5 possible answers. Please select the **one** response which best answers the question.

1. A wave in which the individual particles of a medium vibrate back and forth parallel to the direction in which the wave travels is called a:
   1. transverse wave.
   2. longitudinal wave.
   3. light wave.
   4. directional wave.
2. Sound travels slowest in:
   1. tin.
   2. ethyl alcohol.
   3. water.
   4. air.
   5. wood.

3. Suppose that a cork is riding on water and is bobbing up and down at a rate of 6 times each second. Then the time required for each complete wave to pass is:
   1. 6 seconds.
   2. 3 seconds.
   3. 12 seconds.
   4. 1/6 second.
   5. 1/3 second.
4. Which one of the following characteristics of sound has most to do with the loudness of a sound wave?
   1. Amplitude.
   2. Pitch.
   3. Wavelength.
   4. Frequency.
   5. Period.
5. Sound is a longitudinal wave which requires ____________ for transmission.
   1. a vacuum
   2. empty space
   3. a power amplifier
   4. a medium
6. Ten crests and ten troughs of a wave pass a certain point every two seconds. The frequency of the wave is:
   1. 5 Hz.
   2. 10 Hz.
   3. 20 Hz.
   4. 8 Hz.
   5. 40 Hz.
7. The wavelength of a wave is:
   1. the distance a wave travels in 1 second.
   2. the distance a wave travels before it starts to repeat itself.
   3. the distance a wave travels before it dies out.
   4. the horizontal distance from a crest to the adjacent trough.
8. The horizontal distance from a crest to the next trough of a water wave is:
   1. one wavelength.
   2. 1/2 wavelength.
   3. 1 Hz.
   4. 1/2 Hz.
   5. one period.
9. Which is the correct formula that connects the wave's speed, frequency, and wavelength?
   1. Wave speed = frequency × wavelength.
   2. Frequency = (wave speed)/wavelength.
   3. Wavelength = (wave speed)/frequency.
   4. All of the above answers are correct.
10. The frequency of a sound wave tells you:
    1. how rapidly the pressure fluctuations are occurring.
    2. how large the pressure fluctuations are.
    3. the distance between successive condensations.
    4. how fast the pressure fluctuations are moving through space.
    5. how spread apart in space the pressure fluctuations are.
11. The low pressure regions in a sound wave are called:
    1. crests.
    2. rarefactions.
    3. condensations.
    4. refractions.
    5. reflections.
12. What is the wavelength of a typical sound wave whose frequency is 300 Hz? (You will have to recall the speed of sound in air to answer this one.)
    1. About 750 meters.
    2. About 300 meters.
    3. About 1.15 meters.
    4. About 0.5 meters.
    5. About 0.1 meters.
13. The distance between two successive, identical, parts of a wave is called the:
    1. frequency.
    2. intensity.
    3. amplitude.
    4. wavelength.
14. In a normal sound wave the pressure variations above and below atmospheric pressure are approximately:
    1. 14.7 pounds per square inch.
    2. 1 pound per square inch.
    3. one-thousandth of a pound per square inch.
    4. one-millionth of a pound per square inch.
15. If a transverse wave has a frequency of 20 Hz and a wavelength of 3 feet, calculate the wave's speed.
    1. 60 feet/second.
    2. 20/3 feet/second.
    3. 3/20 feet/second.
    4. (20 + 3) = 23 feet/second.
    5. A transverse wave has no speed.

16. A siren produces sound by blowing air through a series of holes drilled into a rotating disc. What will be the frequency of the sound if the disc contains 50 holes (spaced uniformly around its circumference), and is rotating at a speed of 30 revolutions each second?
    1. $30/50 = 0.6$ Hz.
    2. $50/30 = 1.7$ Hz.
    3. $50 \times 30 = 1{,}500$ Hz.
    4. $30 + 50 = 80$ Hz.
    5. 30 Hz.
17. Wavelength is the __________ it takes for a wave to repeat itself, and the period is the __________ it takes for a wave to repeat itself.
    1. time, time
    2. distance, distance
    3. time, distance
    4. distance, time
18. A transverse wave and a longitudinal wave each have a frequency of 100 Hz and a velocity of 400 m/s. The wavelength of the transverse wave is ____________ that of the longitudinal wave.
    1. the same as
    2. longer than
    3. shorter than
19. For a longitudinal wave, such as sound, the direction of vibration is:
    1. perpendicular to the direction in which the wave travels.
    2. independent of the direction of wave travel.
    3. parallel to the direction the wave travels.
    4. unknown because you have to know what material the wave is traveling through.
20. If the frequency of a sound wave increases, what happens to its wavelength?
    1. It remains the same.
    2. It increases.
    3. One cannot say because any change in amplitude must be taken into account.
    4. It decreases.
21. Suppose that the amplitude of a sound wave is decreased while the frequency of the wave remains the same. What subjective factor of the wave will also change?
    1. Its pitch.
    2. Its loudness.
    3. Its period.
    4. Its velocity.
    5. All of the above factors will change.
22. The human ear can hear the sound produced by a vibrating object when the object is vibrating in the range between:
    1. 20 Hz to 20 kHz.
    2. 200 Hz to 25 kHz.
    3. 12 Hz to 20 kHz.
    4. 12 Hz to 15 kHz.
    5. 20 kHz to 40 kHz.
23. A 1 kHz sound wave is traveling through air, and another 1 kHz sound wave is traveling through a bar of steel. The two sound waves will have:
    1. the same wavelengths even though the speed of sound is different in air than in steel.
    2. the same wavelengths since the speed of sound is the same in air and in steel.
    3. different wavelengths since the speed of sound is different in air than in steel.
    4. different wavelengths even though the speed of sound is the same in air and in steel.
    5. different periods.
24. A police car emits a sound whose frequency is 3 kHz when the car is at rest. When the car is approaching you will hear a frequency which is __________, and when the car is receding you will hear a frequency which is __________.
    1. greater than 3 kHz, greater than 3 kHz.
    2. less than 3 kHz, less than 3 kHz.
    3. less than 3 kHz, greater than 3 kHz.
    4. greater than 3 kHz, less than 3 kHz.
    5. equal to 3 kHz, equal to 3 kHz.
25. Reflection is a property of a wave in which:
    1. refraction occurs at the surface.
    2. the angle of reflection equals the angle of incidence.
    3. the wave reflects off a surface at an angle to the surface which is different than its incoming angle.
    4. the sound is absorbed by the surface.
26. A stereo is playing in a room and at a particular location, called A, 9×10^{-8} watts of sound power cross an area of 2 m^2. At location B the sound system produces 0.8×10^{-8} watts of sound power across an area of 0.1 m^2. In which location will the music sound louder? The music will sound louder in:
    1. location A because its sound power is larger.
    2. location A because its intensity is larger.
    3. location B because its sound power is smaller.
    4. location B because its intensity is larger.
    5. (The music will sound equally loud in both locations.)

27. A sound wave whose intensity is 6×10^{-9} watts/m^2 crosses an area of 3 m^2. What is the total sound power that crosses the area?
   1. 6×10^{-9} watts.
   2. $(6 \times 10^{-9}$ watts/m$^2)/(3$ m$^2) = 2 \times 10^{-9}$ watts/m^4.
   3. $(6 \times 10^{-9}$ watts/m$^2) \times (3$ m$^2) = 18 \times 10^{-9}$ watts.
   4. $(3$ m$^2)/(6 \times 10^{-9}$ watts/m$^2) = 0.5 \times 10^{+9}$ m^4/watt.
28. 2.5×10^{-8} watts of sound power cross an area of 5 m^2. What is the intensity of the sound wave?
   1. 2.5×10^{-8} watts
   2. $(2.5 \times 10^{-8}$ watts$) \times (5$ m$^2) = 12.5 \times 10^{-8}$ watts/m^2.
   3. $(2.5 \times 10^{-8}$ watts$)/(5$ m$^2) = 0.5 \times 10^{-8}$ watts/m^2.
   4. $(5$ m$^2)/(2.5 \times 10^{-8}$ watts$) = 2.0 \times 10^{+8}$ m^2/watt.
   5. $(2.5 \times 10^{-8}$ watts$)^2/(5$ m$^2) = 1.25 \times 10^{-16}$ watts^2/m^2.
29. A person is whistling a note at a certain pressure amplitude. How much more power is required in order to whistle a note that has a pressure amplitude that is four times greater?
   1. Four times more power is required.
   2. Eight times more power is required.
   3. Sixteen times more power is required.
   4. Two times more power is required.
30. Suppose that sound "A" is produced with nine times more power than sound "B". The pressure amplitude of "A" is ___________ than the pressure amplitude of "B".
   1. 9 times greater
   2. 9 times less
   3. 3 times greater
   4. 3 times less
   5. $9 \times 9 = 81$ times greater
31. What is the speed of the condensations and rarefactions, created by a vibrating speaker, which travel to your ear?
   1. They both travel at approximately 344 m/s.
   2. They both travel at approximately $(344 \times 2) = 688$ m/s.
   3. The condensations travel at the speed of sound, while the rarefactions travel at one-half the speed of sound.
   4. The condensations travel at the speed of sound, while the rarefactions have no speed.
   5. Condensations and rarefactions do not travel from a speaker to your ear.
32. The difference between pitch and frequency is that:
   1. pitch is measured in Hertz, while frequency is not.
   2. pitch is an objective concept, while frequency is a subjective concept.
   3. pitch is a subjective concept, while frequency is an objective concept.
   4. a high pitched note is produced by a low frequency sound wave.
   5. (There is no difference between pitch and frequency.)
33. Which of the following frequencies would be called *ultrasonic*?
   1. 10 Hz.
   2. 20 Hz.
   3. 1,000 Hz.
   4. 15 kHz.
   5. 30 kHz.
34. What is generally regarded as the *lowest* frequency that an average young person can hear?
   1. 1 kHz.
   2. 5 kHz.
   3. 10 kHz.
   4. 20 kHz.
   5. 20 Hz.
35. Which one of the following sound frequencies would be called *infrasonic*?
   1. 10 Hz.
   2. 100 Hz.
   3. 1 kHz.
   4. 10 kHz.
   5. 100 kHz.
36. Suppose that the sound intensity at the door of a listening room is 0.0002 watts/m^2. The door has a total area of 1.5 m^2. When the door is opened ___________ of sound power pass through the doorway.
   1. 0.0002 watts
   2. 0.0003 watts
   3. 0.0004 watts
   4. 1.5 watts
   5. 0.0015 watts
37. A loudspeaker is placed outdoors where there are almost no reflections. The sound intensity is measured to be 8×10^{-8} watts/m^2 at a distance of 5 meters from the loudspeaker. In general, what would happen to the sound intensity when the loudspeaker is moved indoors, assuming that the intensity is still measured at a distance of 5 meters from the loudspeaker? The intensity will:
   1. increase.
   2. decrease.
   3. remain the same.

38. When a sound wave travels to a wall of a listening room:
    1. all of it will be reflected.
    2. part of its power will be absorbed by the wall, and the remainder will be reflected.
    3. all of it will be refracted into the wall.
    4. the reflected sound wave will have more power than the incident wave.
    5. the reflected wave will have a different speed than the incident wave.

39. The Doppler effect is produced when:
    1. sound is reflected from a smooth surface.
    2. the sound wave changes its velocity.
    3. sound is refracted.
    4. there is relative motion between the sound source and the observer.
    5. two transverse waves interact.

40. Some reflection of sound is needed in a room for proper sound reproduction. However, excessive reflections:
    1. have no harmful effect on the sound.
    2. cause the reverberation time to decrease.
    3. cause a decrease in the sound intensity in the room.
    4. cause the individual notes to become blurred together, resulting in an unnatural sound.
    5. produce a "dead" room.

41. A live musical performance is recorded in a large auditorium. When this music is reproduced in a normal listening room, the reproduction:
    1. has the same reverberation time as the live performance.
    2. will sound exactly like the live performance, because the reverberation characteristics of the auditorium were also recorded.
    3. will be different than the live performance because the reverberation characteristics of the auditorium and the listening room are not the same.
    4. will not need any reflections from the listening room to sound natural.

Chapter 4

COMPLEX WAVES AND INTERFERENCE

4.1 Introduction

The subject of complex waves deals with some of the very exciting properties of sound which arise when more than one frequency is present at the same time. Very few people speak in a monotone, i.e., a single frequency, and virtually all speech and musical sounds are composed of many frequencies, each with a different relative loudness. It is precisely this tonal "brew" which gives musical sounds their quality or color and it underlies the reason why no two people or musical instruments sound exactly alike. Scientists have coined the phrase "complex sound waves" to describe sound waves which exhibit a complex pressure pattern caused by the coexistence of the many simpler waves which make it up. Figure 4.1 shows the pressure variations of a single frequency wave, often called a *sine* wave, and a complex sound wave. Voices and musical instruments routinely produce complex sound waves, and we shall be concerned with the technique by which such waves are constructed from the simpler sine waves. This important technique is called *The Principle of Linear Superposition* and it will be discussed in section 4.2. The remaining sections in this chapter will present some of the interesting and important consequences of this "principle" along with discussions on its applicability to audio.

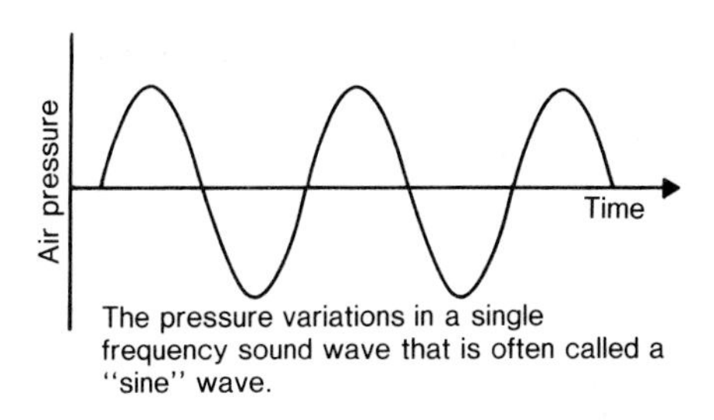

The pressure variations in a single frequency sound wave that is often called a "sine" wave.

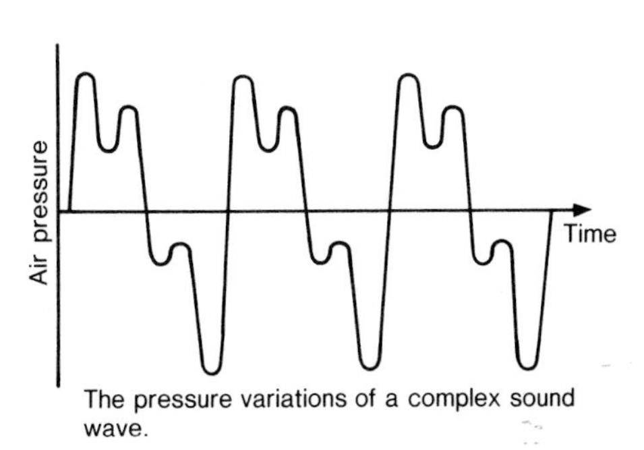

The pressure variations of a complex sound wave.

Figure 4.1. A comparison between the pressure variations of a simple sine wave and a more complex wave.

4.2 The Principle of Linear Superposition

The principle of linear superposition states that two or more waves, which occupy the same space at the same time, can be added together in order to obtain the total effect of all the waves. In order to visualize this process let us first examine the superposition of two transverse pulses moving

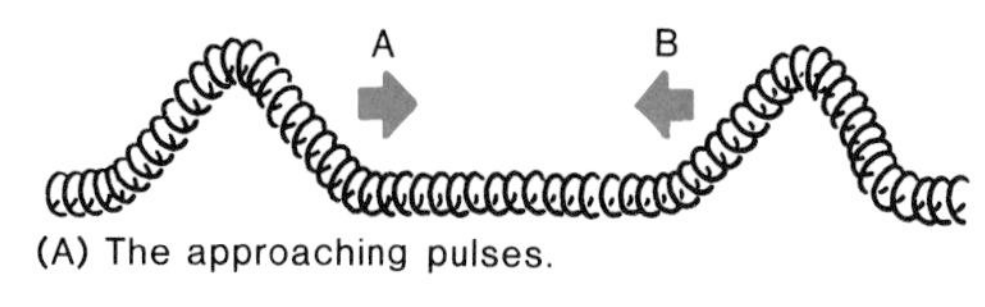

(A) The approaching pulses.

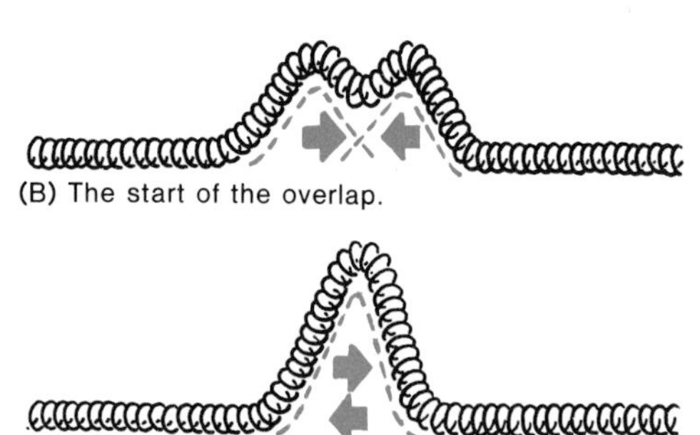

(B) The start of the overlap.

(C) Total overlap—notice that the slinky has twice the height of either pulse.

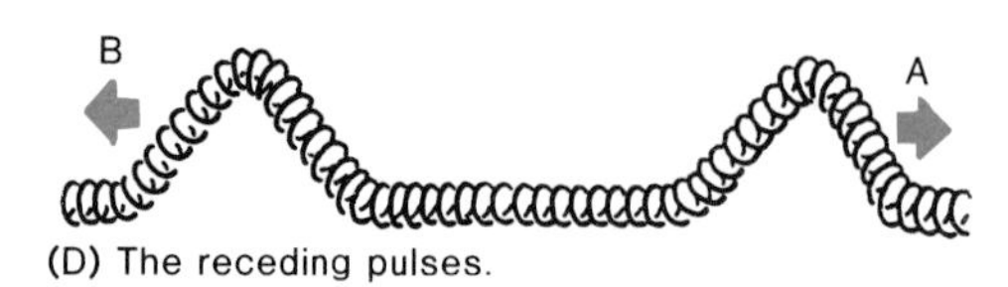

(D) The receding pulses.

Figure 4.2. Two transverse "up" pulses passing through each other and demonstrating the principle of linear superposition.

toward each other along a slinky. Figure 4.2 shows two approaching pulses which are both displaced in the "up" direction. Figure 4.3 portrays the case in which one of them is "up" and the other is "down."

Scanning down the sequence of drawings in both of these figures, two facts should become apparent:

1. The two pulses are able to pass easily through each other as if the other were not present. This observation is true even if the slinky pulses are replaced by sound waves, because we know from everyday experiences that when two people talk their "voices" pass right through each other with ease. Notice, in part (D) of each figure that the pulses retain their same shape (and speed, too) after they have passed through each other.
2. In the region where the two pulses are overlapping, parts (B) and (C) in each figure, the resulting shape of the slinky is simply the sum of the two pulses. *This is the principle of linear superposition.*

The second comment given above needs a little more elaboration. Figure 4.2 (B) shows an instant in time when the approaching pulses "A" and "B" just begin to overlap, as depicted by the dotted lines. Each pulse begins to lose its identity, and the slinky assumes a shape in the overlap region which results from simply adding together the individual dotted lines. In Figure 4.2 (C) the two "up" pulses completely overlap, and the slinky reaches a pulse height which is the sum of the two—or *twice the amplitude of each pulse.*

Figure 4.3 (C) shows the unusual result at the instant in time when an "up" pulse and a "down" pulse exactly overlap. Rather than adding together to yield a pulse with twice the amplitude, as shown in Figure 4.2 (C), the oppositely displaced pulses momentarily cancel each other and the slinky becomes perfectly straight. A moment later, however, the two pulses begin to move apart and the slinky once again becomes distorted as it conforms to the shape of the individual pulses.

Figures 4.2 and 4.3 show the principle of superposition as applied to transverse pulses, but the principle works equally well for all types of waves including sound, water waves, light waves, radio waves, electricity waves, microwaves, and so on. It is one of the most important laws in all of audio and has major consequences for sound reproduction. We shall devote the remainder of this chapter to showing examples of its effects.

Summary

The principle of linear superposition states that when two or more waves, transverse or longitudinal, are present in the same region of space at the same time, the resulting wave is the sum of the individual waves.

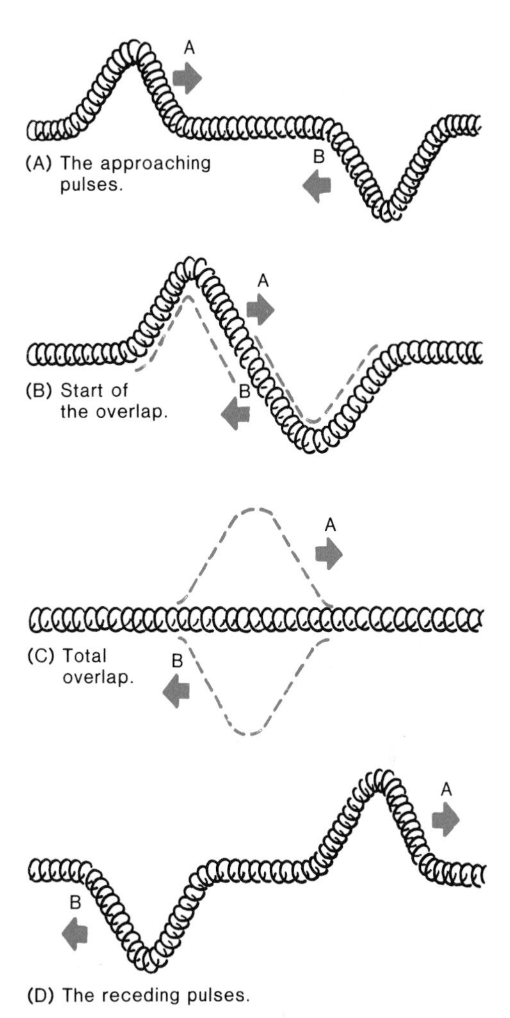

Figure 4.3. Two transverse pulses, one "up" and one "down," passing through each other and demonstrating the principle of linear superposition.

4.3 Constructive and Destructive Interference of Sound Waves

Constructive and destructive interference are terms used to describe the results of linear superposition. Let us set up a simple, but dramatic, experiment to show how the principle of superposition can have profound effects on the loudness of musical tones. Suppose that two stereo speakers are set up so that the sound which they produce intersects in the middle of the listening room, as shown in Figure 4.4. The source of sound is an audio generator which is producing a 344 Hz electrical signal of constant amplitude. The signal from the audio generator is fed into the AUX input for both channels of the receiver which, after amplification, sends it to the speakers where a 344 Hz sound is generated. As we shall now see, the relative position of the two speakers plays a vital role in determining what a listener in Figure 4.4 hears.

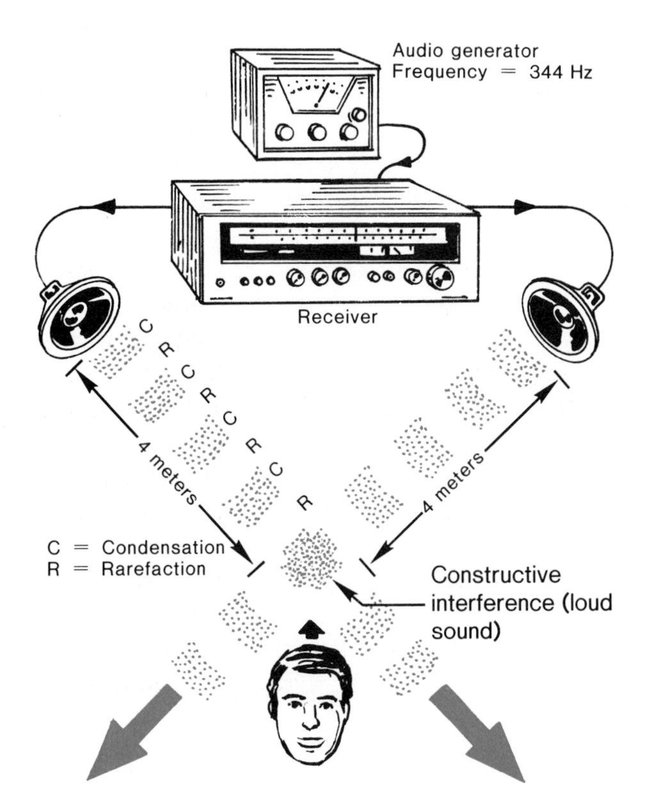

Figure 4.4. Two speakers, equally distant from the listener, produce a loud sound in the overlap region due to constructive interference.

Comment: The sound frequency was chosen for convenience to produce a wavelength of 1 meter. Remember how this is calculated?

Speed of sound = (frequency) × (wavelength) (Equation 3.3)

344 meters per second = (344 Hz) × (wavelength)
Therefore,

wavelength = 1 meter.

Constructive Interference of Two Sound Waves

If the distance from each speaker to the listener is exactly the same, as shown in Figure 4.4, then the crests from one sound wave will always meet the crests from the other as the two waves pass through each other; similarly, troughs will always meet troughs. Figure 4.5 depicts both the individual and combined pressure patterns in the crossover area. Notice that, according to the principle of linear superposition, the sum of the two waves results in a sound wave whose condensation crests are twice as large and whose rarefaction troughs are twice as deep as those of the individual waves. A listener sitting in the overlap region would hear a nice loud sound. When two waves always meet crest-to-crest and trough-to-trough they are said to be *in-phase,* or *interfering constructively,* with each other. So far so good. Now let us consider what happens if one of the speakers is moved; you will be quite surprised at the result!

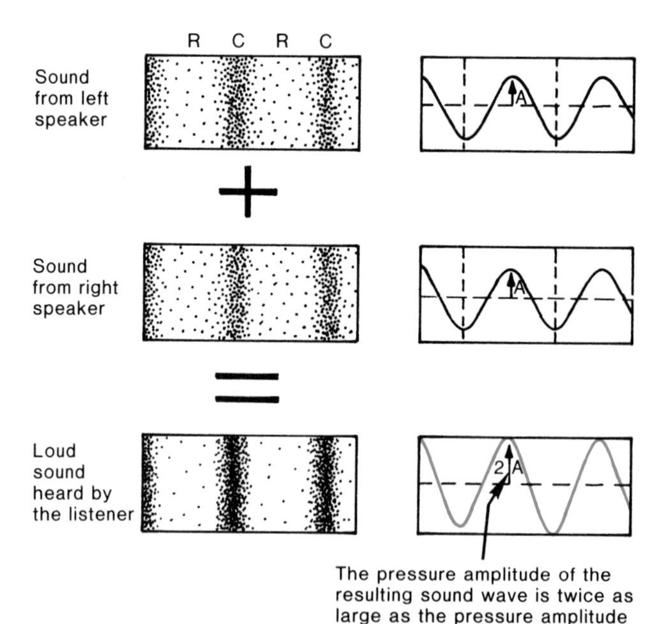

Figure 4.5. The sound waves from the left and right speakers arrive at the listening area in such a way that condensations always add together to produce condensations which are twice as strong; the same is true for the rarefactions. The result is that a loud sound will be heard in the overlap region.

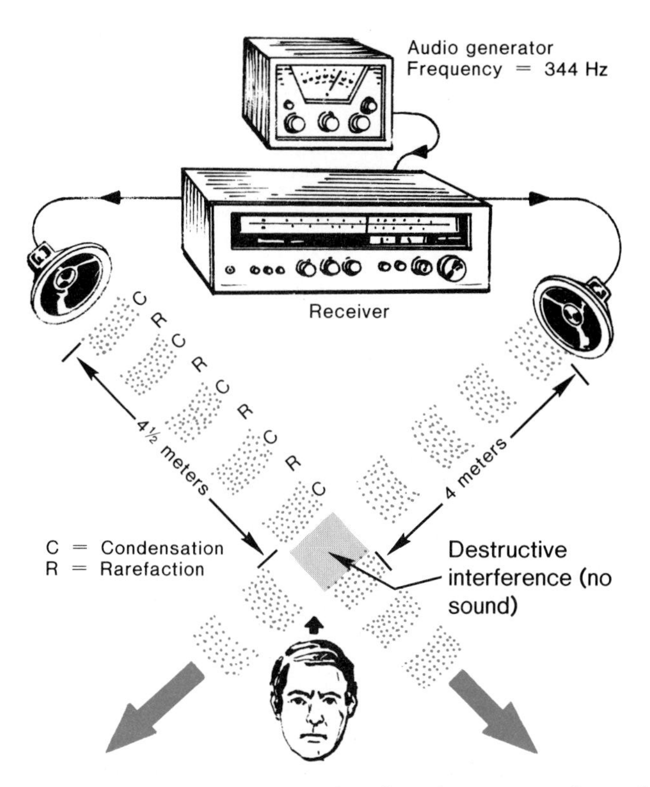

Figure 4.6. The left speaker has been moved one-half of a wavelength further away from the listener than the right speaker. No sound will be heard in the overlap area because of destructive interference.

Destructive Interference of Two Sound Waves

Figure 4.6 shows the situation when the left speaker, for example, is moved back by a distance of one-half of a wavelength, or 1/2 meter.* Now the whole story changes in the region where the two waves intersect. Since the sound from the left speaker must travel one-half of a wavelength further than the sound from the right speaker, a crest arriving from the left will meet a trough arriving from the right when they intersect. Similarly, a trough arriving from the left will meet a crest arriving from the right. According to the principle of linear superposition, the net effect is always a mutual cancellation of one wave by the other. Figure 4.7 shows that the condensations and rarefactions from the respective waves will exactly cancel each other, and the air pressure will always be a steady 14.7 psi. Of course a constant air pressure, devoid of any condensations and rarefactions, means that *there simply is no sound!!!* It is really incredible to think that one can be sitting in front of two speakers and, under certain conditions, hear absolutely no sound. The two waves

*When the left speaker is moved further back, its sound intensity (and hence, its amplitude) will decrease in the overlap region. To offset this effect we will assume that the power from the left channel of the amplifier is increased slightly, which will keep the amplitudes of the right and left sound waves equal at all times.

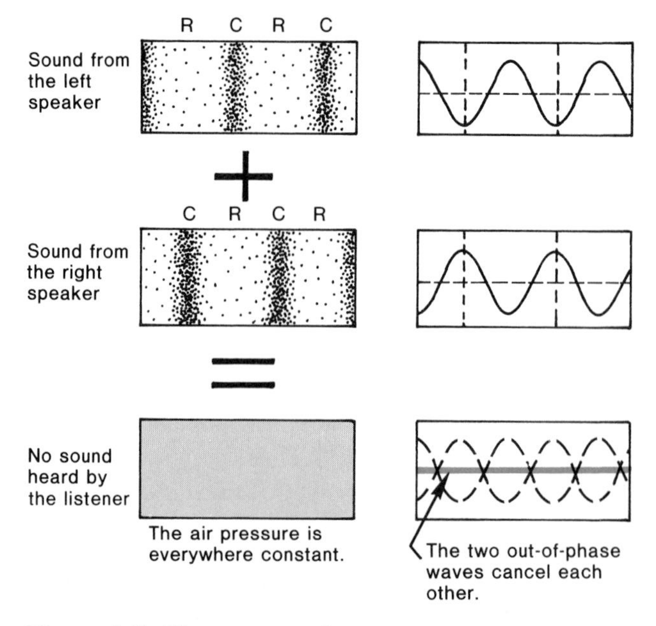

Figure 4.7. The two sound waves from the left and right speakers arrive at the listener in such a way that condensations from one speaker always meet rarefactions from the other. The pressure variations of the two waves exactly cancel out in the overlap region, creating a region of constant air pressure. Consequently, no sound will be heard in the overlap region!

depicted in Figure 4.7 are said to be *out-of-phase,* or *interfering destructively,* because the crests of one wave always meet the troughs of the other, and vice versa. Even though two sound waves can travel through each other apparently unaltered, we now see that it is a mistake to assume that nothing really happens. For in the overlap region one can actually experience a no-sound condition!

Interference, The General Picture

A little reasoning, and with the help of Figures 4.4 and 4.6, will show that the left speaker could be moved back *another* one-half wavelength ($4\ 1/2 + 1/2 = 5$ meters) and, once again, the waves would be in-phase and the listener would hear a loud sound. The condition of constructive interference results because the left sound wave travels exactly one whole wavelength (1 meter) farther than the right wave, and once again crest meets crest and trough meets trough in the overlap region. In general, every time the left speaker is moved backward (or forward) by one-half of a wavelength, the listener will experience a dramatic change in the sound loudness. Table 4.1 gives a summary of these phenomena. The variations in sound loudness exhibited in Table 4.1 are quite amazing, and it is always interesting to discover the intriguing ways that nature can operate—ways that very few people are aware of.

Sound Frequency = 344 Hz 1 Wavelength = 1 meter				
Distance of Left Speaker	**Distance of Right Speaker**	**The Two Sound Waves Are:**	**Type of Interference**	**Comment**
4 meters	4 meters	In-phase	Constructive	Loud sound
4 1/2 meters	4 meters	Out-of-phase	Destructive	No sound
5 meters	4 meters	In-phase	Constructive	Loud sound
5 1/2 meters	4 meters	Out-of-phase	Destructive	No sound
6 meters	4 meters	In-phase	Constructive	Loud sound

Table 4.1. A summary of the constructive and destructive interference phenomena as one speaker is moved farther from the listener.

Although the example shown above is very illustrative of the process of constructive and destructive interference, a more realistic situation is to keep the two speakers fixed and permit the listener to move about the room. This latter effect has direct applicability to stereo speaker systems, and it illustrates the effect of "losing" or "gaining" certain frequencies depending upon where the listener sits in the room. Consider Figure 4.8, which is essentially the same as Figure 4.4 except the sound emanating from each speaker is shown to be spreading outward as concentric circles. Each solid circular line represents a condensation in the sound wave and, of course, the distance between two solid circular lines is the wavelength of the sound. Similarly each dashed circular line represents a rarefaction. As shown in Figure 4.8, the sound patterns from the two speakers overlap, and there are places where constructive and destructive interference occur. Any place where two solid or two dashed circles intersect is a region of constructive interference, and a listener stationed at such a location will hear a loud sound from the speakers. Any place where a solid and a dashed circle intersect is a region of destructive interference, and a listener situated there will hear no sound. Since there are many places of each kind within the room, it is possible for the listener to walk about the room and hear dramatic variations in loudness.

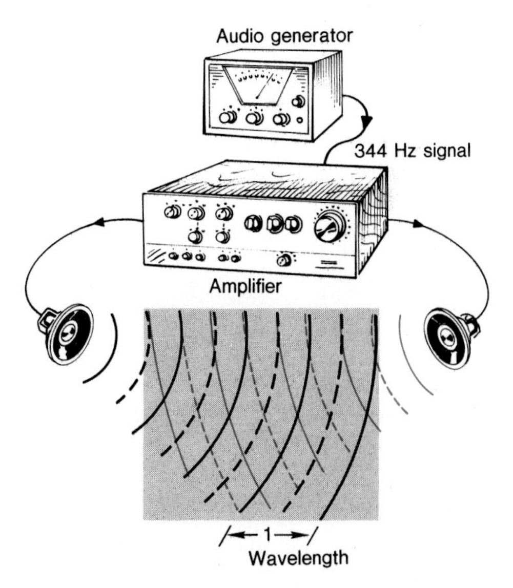

Figure 4.8. A more realistic view of the interference pattern set up by two speakers. The solid circles denote condensations, while the dashed circles denote rarefactions. Where two solid or two dashed circles intersect, constructive interference occurs. Where a solid and a dashed circle intersect, destructive interference occurs.

In the regions of constructive interference the in-phase sound waves from each speaker add to produce a resulting sound wave which has twice the amplitude of each individual wave. According to the discussion in sections 3.5 and 3.6, the resulting sound has *four* times the intensity of that produced by a single speaker alone. (Recall that the sound intensity is proportional to the amplitude squared, hence doubling the amplitude means four times more intensity.) It would be natural to believe that the sound intensity is twice that of one speaker because, after all, there are two speakers producing sound. But four times as much intensity?

Well, the extra intensity found in the regions of constructive interference has been "stolen" from the regions of destructive interference where there is no sound at all. Therefore interference is indeed a very interesting phenomena. According to Figure 4.8, interference redistributes the sound energy within a room by taking the energy away from certain regions (destructive interference) and giving that energy to other regions (constructive interference). Interference, so to speak, "robs Peter to pay Paul."

The effects of interference mean that certain frequencies may sound very quiet at one place in a room, while they sound abnormally loud at another. This conclusion is not limited to strictly pure tones, such as that used in Figure 4.8. When music, rather than a pure tone, is being played through the speakers, interference effects can and do occur. However, they are very complicated, because the loudness of various frequencies in the music is constantly increasing and decreasing as the music changes. Therefore, the intersecting sound patterns shown in Figure 4.8 are also constantly changing. The net effect is that some instruments are more difficult to hear than others at certain listener locations. In designing concert halls great effort is devoted to eliminating acoustically uneven spots, so that the music can be heard equally well no matter where the listener sits. Judicious use of reflected sound waves is the architect's main tool for "filling in" or "smoothing out" those spots where interference has diminished or exaggerated the sound level.

An Example of Microphone Placement

Another example of the interference idea, and one of practical value, is the question of the microphone placement when using a public address system. Consider a group of people sitting around a table with the microphone* placed somewhere in the center. Should the mike be placed on a stand or should it be laid directly on the table top itself as shown in Figure 4.9? Almost everyone, without even thinking, will answer that the stand is the more favorable setup because that is the way it is usually done. However, almost everyone is wrong! As we shall now show, the use of the table-plus-stand can lead to an undesirable partial sound cancellation caused by destructive interference at certain frequencies. To understand this we only need to remember that sound arrives at the mike via two paths: (a) directly from the speaker and (b) along a path which has been reflected from the table top, as indicated in Figure 4.10. The arrival of these two waves should arouse your suspicions

Figure 4.9. Should the microphone be placed directly on top of the table or should a microphone stand be used?

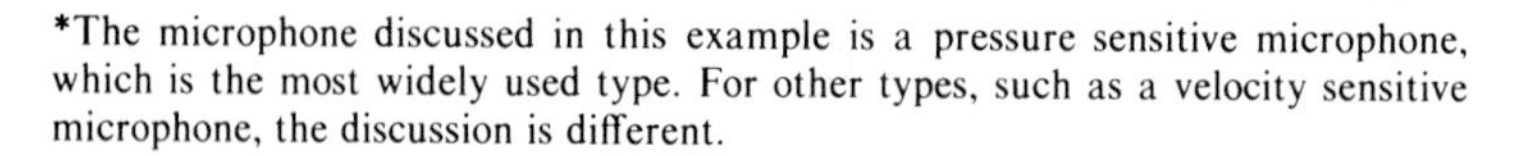

*The microphone discussed in this example is a pressure sensitive microphone, which is the most widely used type. For other types, such as a velocity sensitive microphone, the discussion is different.

because the two will overlap at the mike and the conditions for either constructive or destructive interference might be right. For example, if the path of the reflected wave is one-half of a wavelength longer than that of the direct wave, the two waves will arrive out-of-phase. A partial cancellation results in this case because the two waves, in general, may not be equally loud and, hence, their amplitudes would not be the same.

To illustrate this problem suppose the direct sound travels 2 meters and the reflected sound covers a greater distance of 2 1/2 meters. Now, as a person speaks, the voice emits different frequencies within the audio range, and each of these frequencies will give rise to direct and reflected sounds. For the sound wave whose frequency is 344 Hz the conditions are perfect for destructive interference since, like the speaker example, the reflected wave travels one-half a wavelength, or 1/2 meter, further. The sound intensity reaching the microphone might look something like Figure 4.11. The figure shows that the sound near 344 Hz is substantially quieter than normal because of the destructive interference which has occurred between the direct and reflected waves. Therefore, the voices, as amplified by the public address system, will sound unnatural due to the loudness "hole" near 344 Hz.

While the 344 Hz tone suffers a loss in intensity due to destructive interference, the 688 Hz tone will exhibit constructive interference and, hence, produce a louder than normal sound at the mike. The reason for the constructive interference is that the wavelength of the 688 Hz sound is 1/2 meter, which is precisely the path difference between the direct and reflected sounds. A path difference of one whole wavelength causes the waves to arrive in-phase at the mike. Figure 4.11 shows the "bump" in the frequency response curve at 688 Hz which implies that the sound at this frequency will be louder than normal and will, once again, appear unnatural.

What is the solution to the problem of destructive and constructive interference? Simple. Place the microphone directly on the tabletop thereby forcing the direct and reflected waves to travel the *same distance*. The two waves will always arrive at the microphone in-phase regardless of their frequencies. There will never be an out-of-phase condition eliminating part of the sound. By this manner the microphone is more efficiently picking up the sound. Therefore, either place a microphone extremely close to, or very far away from, a reflecting surface in order to avoid destructive interference problems.

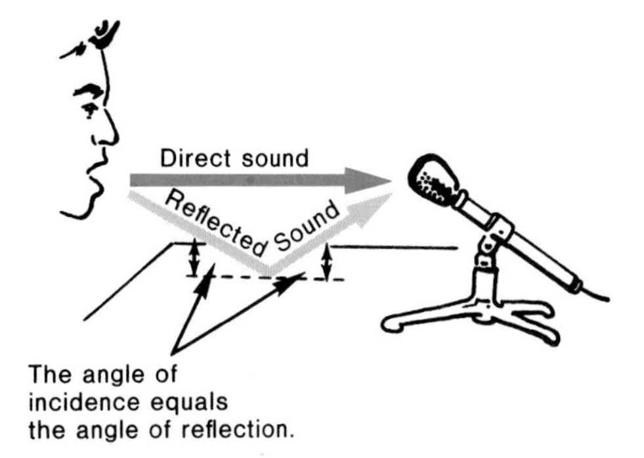

Figure 4.10. The direct and reflected sound waves arrive at the mike by two different paths. Interference between the two waves will occur.

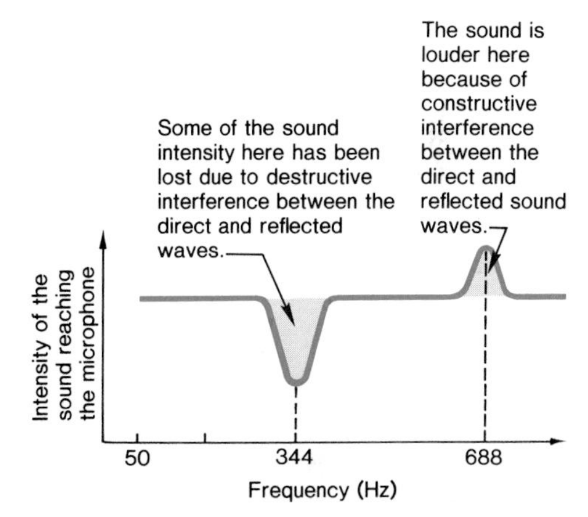

Figure 4.11. Interference can produce a partial cancellation of the sound at certain frequencies or an enhancement at other frequencies, depending on whether the waves are in-phase or out-of-phase when they reach the mike.

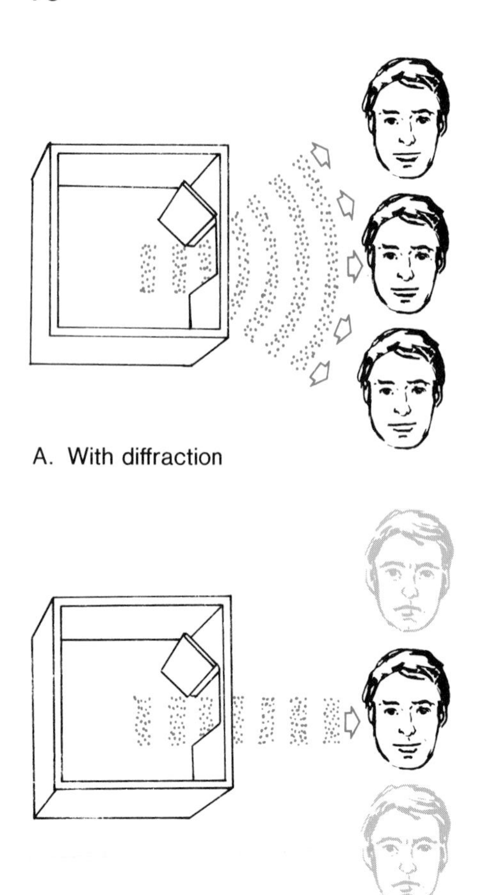

B. Without diffraction

Figure 4.12. Diffraction is the ability of a wave to bend around an obstacle, such as the edge of a doorway. The source of sound waves within the room may be due to a loudspeaker or a live performer.

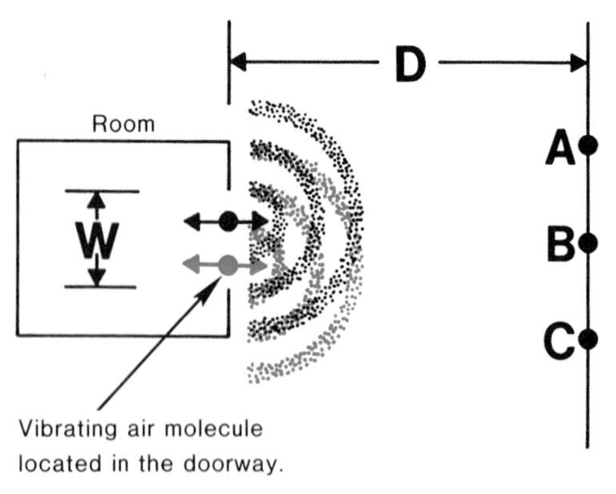

Figure 4.13. Each vibrating molecule in the doorway generates a spherical sound wave that expands outward and bends, or diffracts, around the edges of the doorway as it leaves the room. Because of interference effects, the total sound intensity is greatest at the point B, and it decreases gradually until it becomes zero at the points A and C.

Summary

1. Two waves are *out-of-phase* when the crests of one always fall in the troughs of the other. The result is a cancellation of the sound and is referred to as *destructive interference.*
2. Two waves are *in-phase* when crests always coincide with crests, and troughs with troughs. The result is an enhanced wave which is louder than either of the two. The result is called *constructive interference.*

4.4 Diffraction and the Dispersion of Sound

There is a very simple phenomenon which you can experience by simply turning on the stereo, walking outside the room, and going around the corner. It should come as no surprise that the sound can still be heard, because nature has been very kind to us by allowing sound waves to bend around corners, as illustrated in Figure 4.12 (A). If sound waves did not bend around corners, the only location outside the room from which sound could be heard would be directly in front of the doorway, as depicted in Figure 4.12 (B) (assuming, of course, that no sound is transmitted directly through the walls).

Diffraction is the ability of a wave to bend around an obstacle, such as the edges of an open doorway, and reach places that otherwise would be inaccessible.

Although all sound waves are capable of diffracting around the edges of a doorway to some extent, certain frequencies diffract more readily than others. Careful listening reveals that the diffracted sound contains louder low frequency bass notes, and relatively quieter higher frequency notes. It is a fact that low frequency sound bends more easily than does high frequency sound.

The diffraction of sound is another fine example of interference. To understand better the relationship between the two, consider the room drawn in Figure 4.13. Imagine that there exists a source of sound within the room, although it is not shown in the picture. Remember, from the discussion in section 3.4, that a sound wave is a collection of vibrating air molecules which pass along the condensations and rarefactions by "bumping into" their neighbors. If the sound wave is to pass through the doorway, each of the air molecules in the doorway must be set into vibration by those molecules vibrating immediately inside the room. Figure 4.13 illustrates two of the many air molecules located in the doorway which are vibrating back and forth in the direction in which the sound wave travels. In order for the sound to leave the room, the doorway molecules must, in turn, bump into the outside molecules in order to start them vibrating.

Therefore, as far as the outside is concerned, each molecule in the doorway is a source of sound waves in its own right. As illustrated in the picture, the sound waves originating from each doorway molecule expand spherically outward, much like the water waves that result when a stone is dropped into a pond. Since the sound waves originating from these two molecules expand spherically outward, some sound will reach locations off to either side of the doorway. The net effect is a "bending" around the door!

The idea of regarding each point on a wave, such as the sound wave in the doorway, as a miniature source of spherically expanding waves originated with the Dutch scientist Christian Huygens in the late 17th century, and is called *Huygen's principle.*

According to the principle of linear superposition, the sound waves generated by all the air molecules in the doorway must be added together to obtain the total sound at any location outside the room. The result of this adding process gives features which are similar to those discussed in section 4.3. There are places outside of the room where no sound can be heard because of destructive interference, and places where a loud sound results from constructive interference. For example, in Figure 4.13 the sound is loudest at B, directly opposite the middle of the doorway, with the intensity decreasing symmetrically on either side of the opening until nothing can be heard at points A and C.

The extent to which the sound spreads out once it leaves the room depends on the frequency of the sound wave. For low frequency (long wavelength) sound the distance between A and C turns out to be large. Since the low frequency sound spreads out, or disperses, over a wide area, it is said to have a "wide dispersion." Conversely, the distance between A and C turns out to be small for high frequency (short wavelength) sound. High frequency sound thus has a relatively narrow dispersion.

It is possible to calculate the distance d_{AC} between the points A and C in Figure 4.13 by using the formula:

$$d_{AC} = \frac{2\,D\,v}{W\,f},$$

where D and W are distances which are defined in Figure 4.13, v is the speed of sound and f is the sound frequency.

Question: How does the distance over which the sound disperses compare for a 100 Hz bass tone and a 5,000 Hz treble tone? Since d_{AC} is inversely proportional to frequency, the distance is 5,000/100 or 50 times greater for the bass tone than it is for the treble tone.

In the above discussion it has not been necessary to specify the manner in which the sound is produced within the room. Therefore, Figures 4.12 and 4.13 can also serve as illustrations of sound emerging from a hi-fi speaker. The walls of the room correspond to the speaker cabinet, while the doorway corresponds to the opening through which the moving part of the speaker (the diaphragm) generates the sound. Thus the sound from a loudspeaker spreads out into the listening room with the low-frequency bass tones naturally having a wider dispersion than the high-frequency treble tones.

When positioning a speaker it is very important that the listening chair be situated somewhere near the axis of the speaker since, as shown in Figure 4.14, the high frequency tones are dispersed in a narrow cone centered around this axis. Thus, you could locate yourself anywhere in the area encompassed by this cone, such as position (1), and enjoy the full frequency range of the music. Position (2), however, is far less desirable as a listening position, because the high-frequencies are not being diffracted into this area. Many

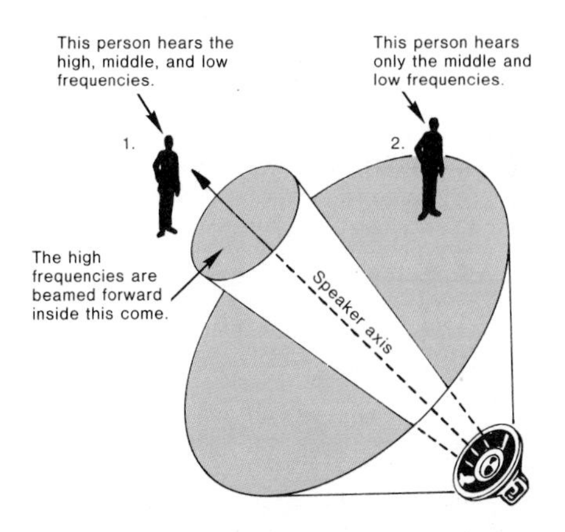

Figure 4.14. Because the high frequencies are not dispersed as well as the middle and low frequencies, you should sit as close as possible to the speaker's axis in order to hear the high, middle, and low frequencies.

speaker manufacturers will list in their specification sheet just how far the listener can be removed from the speaker axis and still enjoy the complete tonal range of the music. This point will be discussed more fully in Chapter 6, along with a simple test which you can perform to check the dispersion of your own speakers.

Summary

Diffraction is the ability of a wave to bend around an obstacle and reach otherwise inaccessible locations. It is responsible for the fact that you can "hear around the corners." Diffraction is the result of interference between sound waves generated by the vibrating molecules which are located near the obstacle. Low frequency tones diffract more than high frequency tones. Hi-fi speakers are designed to spread out, or disperse, their sound over a wide area, and audio engineers use diffraction as a major method for achieving wide dispersion.

4.5 Transverse Standing Waves

Transverse standing waves are another very interesting phenomena which are the result of the principle of linear superposition. Standing waves can be produced by tying one end of a rope to a wall and then shaking the rope up and down at one of the "right frequencies." Figure 4.15 shows some of the beautiful standing wave patterns which can be generated in this manner. The phrase "right frequency" is important because, as we shall later see, standing waves cannot be established at any arbitrary frequency of vibration. Suppose, for example, that the "right frequency" in Figure 4.15 happens to be 1 Hz. Then the standing wave pattern shown in part (A) will be produced if the rope is vibrated up and down once every second. If the rope is vibrated at *twice* its original rate, or 2 Hz, the simple standing wave of (A) disappears and it is replaced by the standing wave shown in part (B), which has two loops instead of one. Tripling the original frequency to 3 Hz produces the three loop pattern shown in (C), while (D) exhibits a standing wave generated by a 4 Hz movement of the hand. Therefore, all the patterns shown in these photos (and more!) can be generated by vibrating the rope up and down at any frequency which is an integer multiple of the first "right frequency."

Notice, also from Figure 4.15, that standing waves possess *nodes* and *antinodes*. The nodes are points on the rope which do not vibrate at all; a fly could land on a node and feel right at home. An antinode, on the other hand, represents a point along the rope where maximum vibration occurs. The picture shows that the number of nodes and antinodes which occur in a standing wave pattern increases as the frequency of vibration increases.

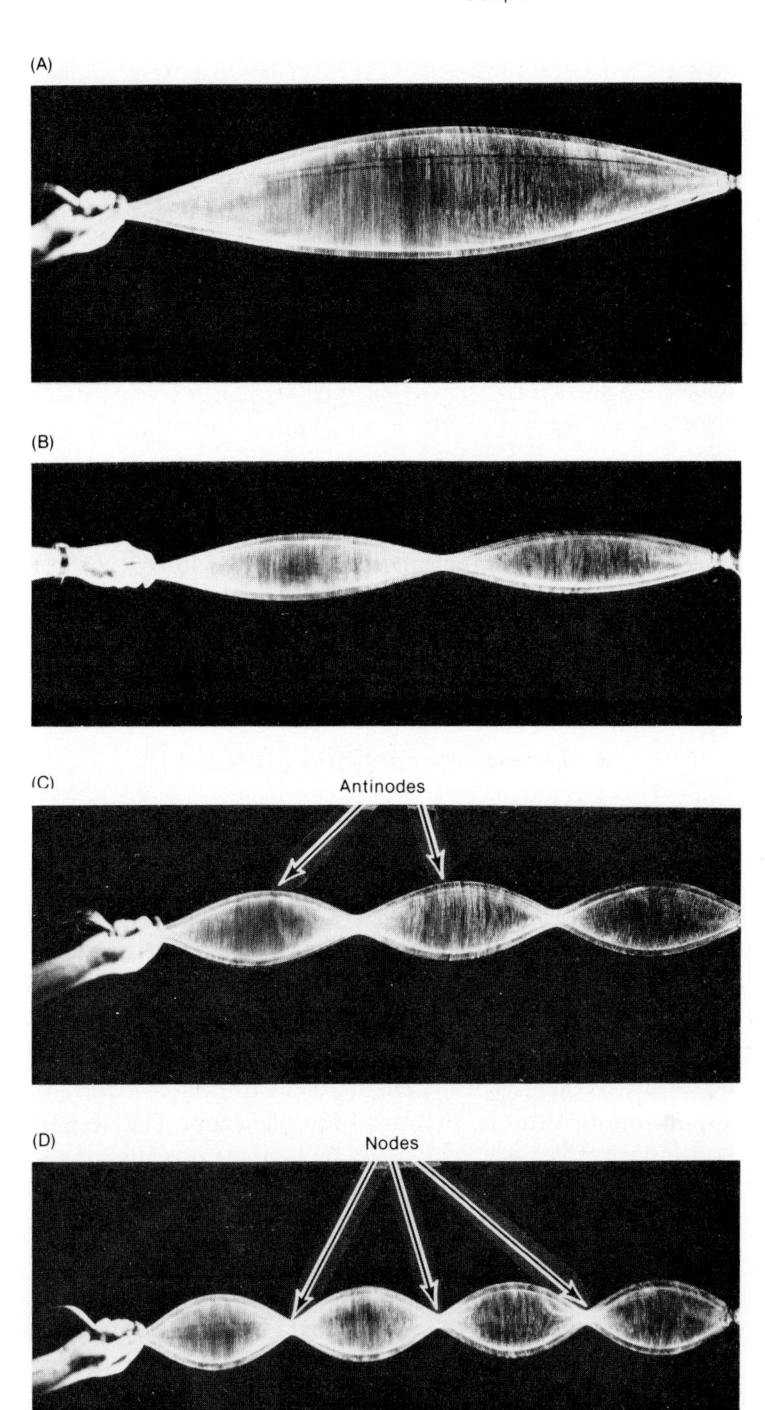

Figure 4.15. Shaking a rope at the "right frequencies" will set up standing waves. Pattern (B) results by vibrating the hand at twice the frequency as in (A). In (C) the frequency is three times that of (A), and in part (D) it is four times that of (A).

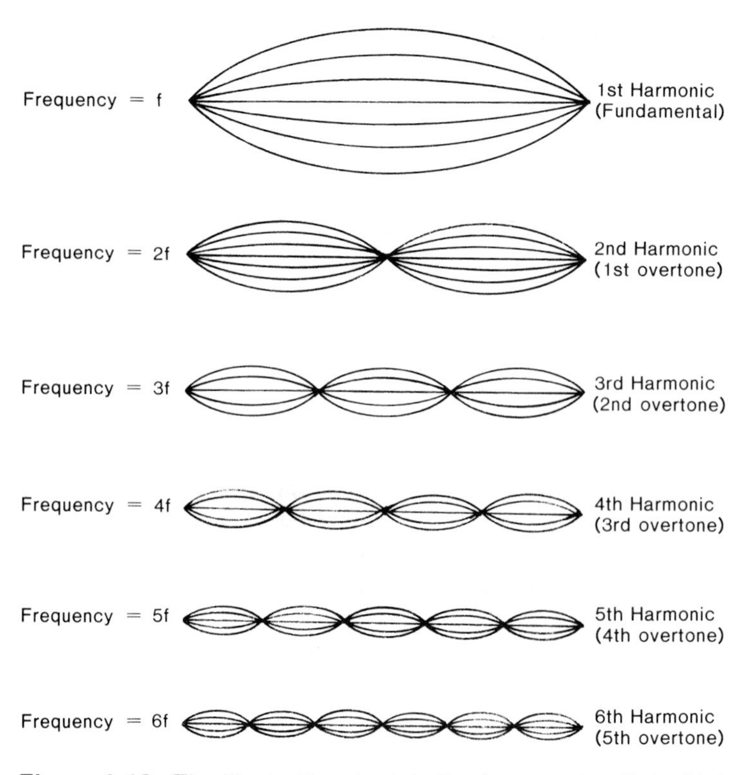

Figure 4.16. The illustration depicts the frequencies (left side) and standard names (right side) associated with standing wave patterns.

Standing waves are very important in discussing room acoustics and the operation of musical instruments. For this reason names have been given to each of the standing wave patterns as shown in Figure 4.16. The standing wave with the lowest frequency, f, is called either the *fundamental*, or *first harmonic*, and it contains only one loop. The second standing wave, whose frequency (2f) is twice the fundamental, is referred to as either the *first overtone* or the *second harmonic*, and it contains two complete loops. The higher frequency waves are similarly named, and some of them are also shown in the figure. We will have many occasions to use this nomenclature throughout the text.

We have seen that standing waves are produced when a rope is vibrated at one of its "right frequencies." In the above discussion these "right frequencies" were 1, 2, 3, 4, etc., Hz. Now a person is certainly free to vibrate the rope at some other frequency, say 3/4 Hz. What happens then? In this case, as illustrated in Figure 4.17, the resulting rope pattern will be a complicated mess, continually changing in appearance, of very low amplitude, and lacking the simple beauty and symmetry of standing waves. When the vibrational frequency is raised to 1 Hz, the fundamental frequency, the whole picture changes, and a standing wave is established with large amplitude vibrations and a stationary appearance

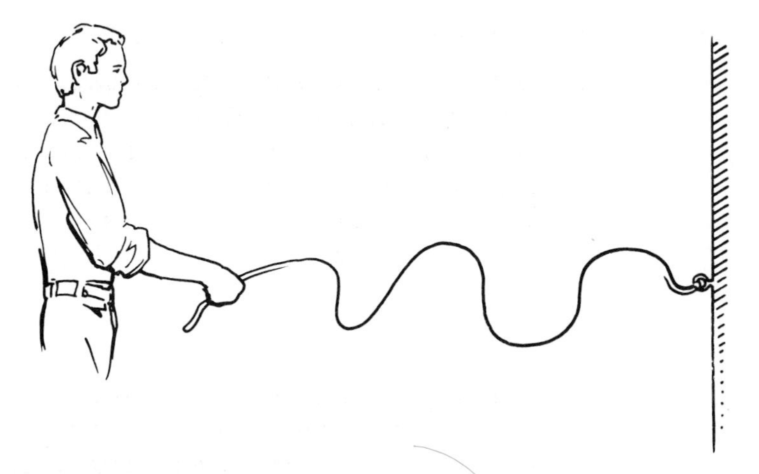

Figure 4.17. Forcing a rope to vibrate at a frequency other than one of its standing wave frequencies produces a pattern which is not "stationary" like those of standing waves. The rope continually distorts and changes its shape from instant to instant.

as shown in Figure 4.15 (A). Raising the frequency to 1.4 Hz again destroys the standing wave and forces the string into low amplitude, complex patterns, similar to that shown in Figure 4.17. At 2 Hz (twice the fundamental standing wave frequency), another standing wave is established as shown in Figure 4.15 (B). Standing wave frequencies are often called the "natural" or "resonant" frequencies of the rope, and they are the frequencies at which the rope "likes" to vibrate with large amplitude oscillations.

Standing Waves Are the Result of Superposition

Standing waves are another result of the principle of linear superposition, and they occur when a wave is reflected back on itself at some sort of a boundary. The simplest example to visualize is that which arises when a forward and a backward wave simultaneously travel along a string. Figure 4.18 shows one cycle of a wave heading toward a wall, to which the right end of the string is attached. The cycle is reflected at the wall and travels back to the left. Only one cycle of the wave is shown for the sake of clarity. In reality, the forward and backward waves are comprised of many cycles which are generated continuously at the left end of the string and then reflected at the wall. Both waves exist on the string at the same time, passing through each other as they move.

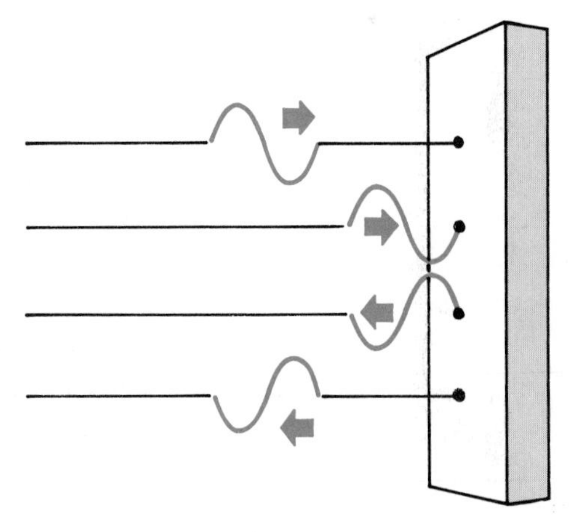

Figure 4.18. A forward traveling wave gives rise to a backward traveling wave because of reflection at the wall.

According to the principle of linear superposition, the total effect of the two waves can be obtained by adding them together. Figure 4.19 depicts the result that is observed as time passes, starting arbitrarily from a moment when the shapes of the two waves happen to coincide. The five sets of illustrations in the picture cover one complete period (T), in time intervals of one-fourth the period. Remember that a

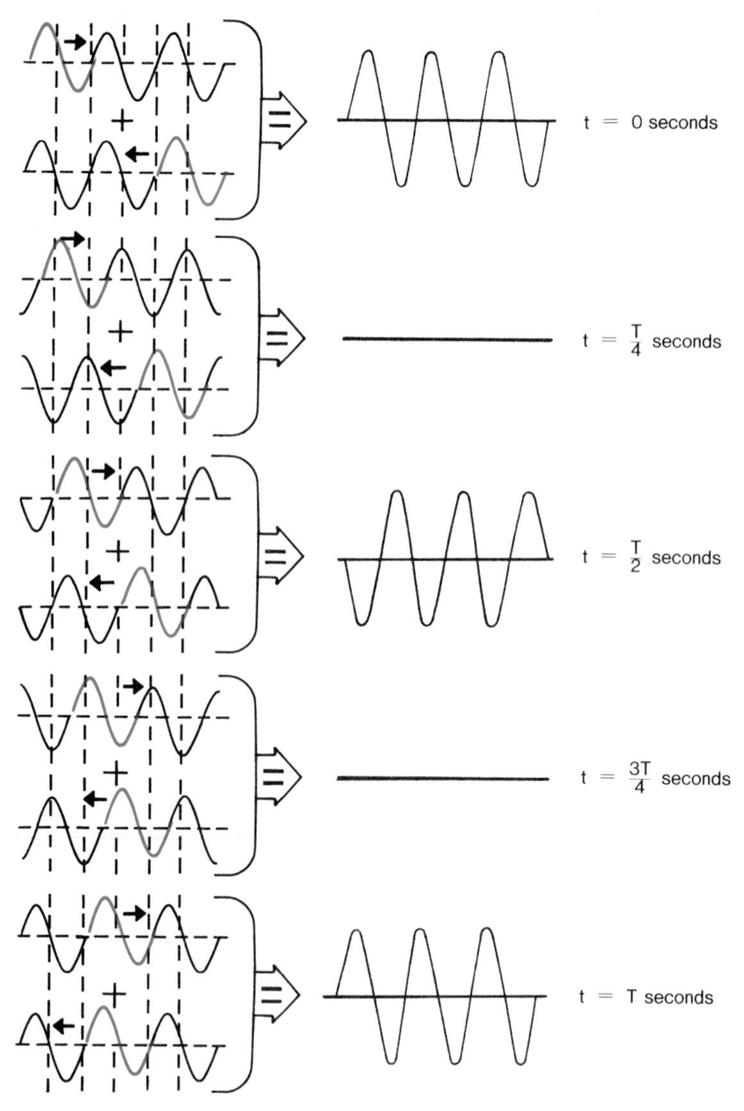

Figure 4.19. A series of illustrations showing how two oppositely traveling waves combine at various times to produce a standing wave. The pictures show the development for each one-fourth of a wave period. T is the period and t represents the time. In a time equal to T each wave moves a distance equal to its wavelength.

wave travels a distance of one wavelength in a time equal to the period. In each case the two oppositely-traveling waves are drawn on the left side of the figure. The movement of each wave can be followed in time by observing the position changes of the colored cycles. To the right of each pair of waves is drawn the actual instantaneous shape of the string, which is obtained by adding together the two waves according to the principle of linear superposition.

By scanning down the right-hand patterns you can see that the shape of the string changes as time passes. For example, at the times of 0, T/2, and T seconds, the two waves constructively reinforce each other, and the string

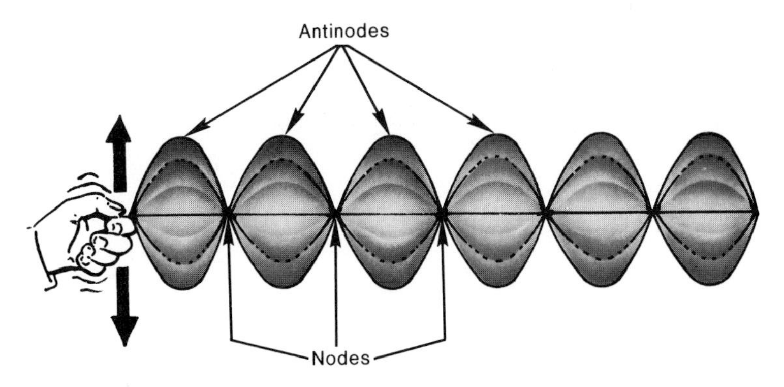

Figure 4.20. A 6-loop standing wave with its nodes and antinodes.

possesses an amplitude which is twice the amplitude of each wave. At the times of T/4 and 3T/4 seconds, however, the two waves exactly cancel each other due to destructive interference, leaving the string momentarily flat! At times other than 0, T/4, T/2, 3T/4, and T seconds the string has a shape which is intermediate between those shown, the shape changing smoothly from one moment to the next. Generally speaking, the shape of the string changes so rapidly that the eye sees only a blurred pattern of "loops," and the resulting motion of the string resembles the standing wave pattern in Figure 4.20, which has six loops with a node point between adjacent loops. The reason that the name "standing waves" is given to this phenomenon is because the overall pattern in Figure 4.20 does not travel to the right or to the left, as do the individual waves. Instead, the string simply vibrates up and down with certain places along the string, the nodes, which never move and are literally "standing still."

Calculating the Conditions Which Produce Standing Waves

There is a simple fact about standing waves that allows one to calculate the fundamental and overtone frequencies. Notice in Figure 4.21 that the distance between two successive nodes is one-half of a wavelength ($\lambda/2$); therefore, each "loop" of the standing wave occupies a horizontal distance of $\lambda/2$. In this picture the length of the string, L, contains two loops. From this information the length of the string is simply the number of loops (2) multiplied by the length of each loop ($\lambda/2$):

$$L = 2\left(\frac{\lambda}{2}\right).$$

From Equation 3.3 the wavelength is related to the frequency and speed of the wave by the relation

$$\lambda = \frac{v}{f}.$$

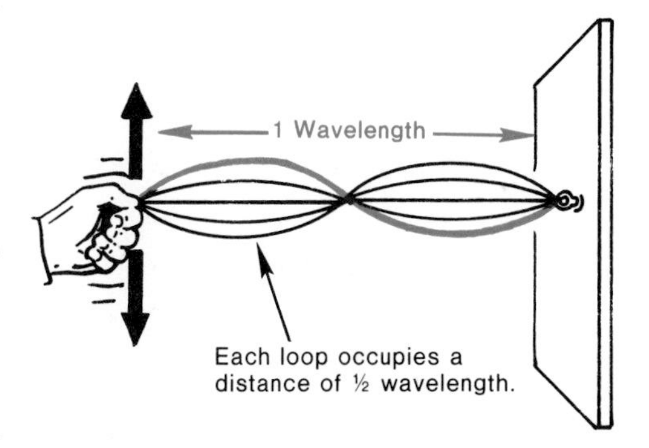

Figure 4.21. A 2-loop standing wave. Notice that the two loops represent one complete cycle, or one wavelength, of the wave.

These two formulas can be combined into a single expression by substituting the value for λ into the first equation, i.e.,

$$L = 2\left(\frac{v}{2f}\right).$$

Solving for f yields the frequency of a two-loop standing wave:

$$f = 2\left(\frac{v}{2L}\right).$$

Notice that the frequency depends on the number of loops (2), the speed of the wave as it moves along the string (v), and the length of the string (L). If the standing wave had a different number of loops, as do the waves illustrated in Figure 4.15, the above formula can be easily modified to give the correct frequency. Let's call the correct frequency f_n, where n denotes the number of loops in the standing wave pattern:

$$f_n = n\left(\frac{v}{2L}\right). \qquad \text{(Equation 4.1)}$$

"n" corresponds to the harmonic frequency at which the string is vibrating: n = 1 is the first harmonic, n = 2 the second harmonic, etc.

For example, when n = 1 the above equation gives the fundamental (or first harmonic) frequency of a vibrating standing wave which contains one loop. Substituting n = 2 into Equation 4.1 yields the frequency of the 2nd harmonic standing wave shown in Figure 4.21. Likewise, higher integer values of n give the frequencies for the higher harmonics. It is interesting to note that the frequencies of the harmonics are simply integer-multiples of the fundamental frequency, a result which was mentioned at the very beginning of this section.

The transverse standing waves described by Equation 4.1 play an important role in hi-fi, for they are involved in the way all string instruments produce musical sound. Musical instruments will be discussed in section 4.9.

Question: What is the fundamental frequency of a string whose length is 0.330 meters and on which transverse waves travel at 291 meters/second? Using n = 1 in Equation 4.1 for the fundamental frequency,

$$f_1 = 1\left(\frac{291 \text{ m/s}}{2 \times 0.33 \text{ m}}\right) = 440 \text{ Hz.}$$

This frequency corresponds to the concert A tone produced by the A string on a violin.

Summary

Transverse standing waves are stationary wave patterns and are formed in a medium, such as a string, when two identical transverse waves pass through the medium in opposite directions. These two waves must have the "right frequency." The allowed frequencies for standing waves on a string are given by the Equation 4.1. Nodes are places on the standing wave at which the medium is not vibrating, whereas antinodes are places at which the medium is vibrating maximally.

4.6 Longitudinal Standing Waves

The most important audio related example of longitudinal standing waves is that involving sound itself. Standing waves of sound arise in a manner like that described above for transverse waves on a string. A forward traveling wave reflects from an obstacle, such as a wall, and produces a backward traveling wave, as illustrated in Figure 4.22. The two oppositely traveling waves add together to produce a standing wave.

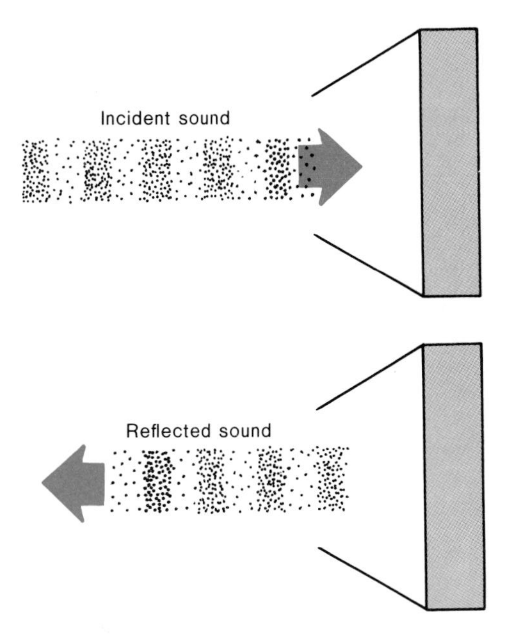

Figure 4.22. A sound wave being reflected by a wall.

Standing waves of sound are difficult to draw and, therefore, Figure 4.23 utilizes a slinky to reveal the salient features of a longitudinal standing wave. The colored dots on the slinky identify the vibrational, or displacement, nodes and antinodes of the "air molecules" which participate in the standing wave. At a displacement node the air molecules do not vibrate, while at a displacement antinode the air molecules vigorously vibrate along the direction of the individual traveling waves. As a result, there are *pressure* nodes and antinodes located along the standing wave. At a pressure node the pressure always remains at a constant, normal level, and your ears would detect no sound at such a place. At a pressure antinode the pressure rises and falls maximally about the normal level; your ears would detect a loud sound at a pressure antinode. Although these pressure nodes and antinodes arise because of the vibration patterns illustrated in Figure 4.23, a pressure node does *not* correspond to a displacement node, and a pressure antinode does *not* correspond to a displacement antinode. In fact, a pressure node corresponds to a displacement antinode and a pressure antinode corresponds to a displacement node. The proof of this not-so-obvious fact is beyond the scope of this book.

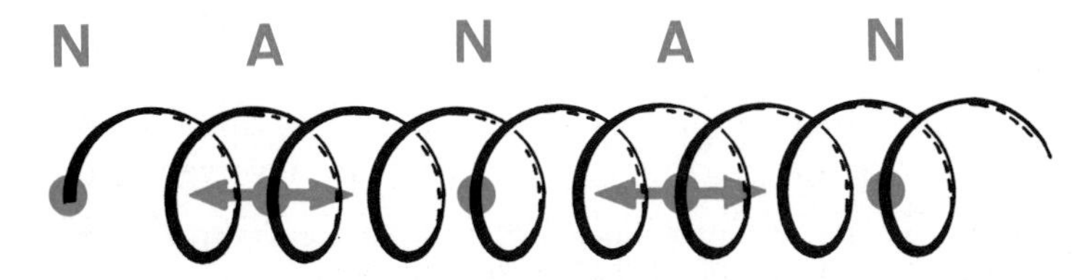

Figure 4.23. A longitudinal standing wave on a slinky, showing the displacement nodes and antinodes.

Standing Sound Waves in a Room

According to the previous discussion, it should be possible to set-up standing waves of sound in a listening room. Suppose, as shown in Figure 4.24, that a loudspeaker is placed at one end of a room and its sound is directed toward the opposite wall. Within the room there will be the forward-traveling sound wave, and the backward-traveling wave

Please note that the colored standing wave patterns which appear in Figures 4.24, 4.26, and 4.27 are drawn as an aid in visualizing the location of the displacement nodes and antinodes. The patterns are not meant to imply that the standing waves are transverse, because they are, in fact, created by longitudinal sound waves.

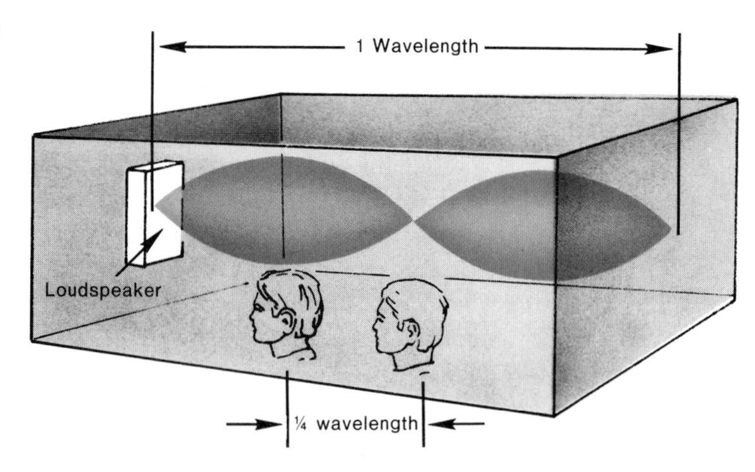

Figure 4.24. In a room containing a standing sound wave a noticeable change in loudness will be heard when the listener moves between a node and an antinode, a distance of one-fourth the sound wavelength.

caused by reflection of the sound from the wall. If the sound frequency is just "right," a standing sound wave will be produced inside the room with its nodes and antinodes. Moving your head back and forth between a node and an antinode would reveal large changes in loudness.

It is possible to calculate the distance between a node and an antinode, which is one-fourth of the wavelength of sound. Since one wavelength of a 1,000 Hz tone is 34 cm

$$\left(\lambda = \frac{v}{f} = \frac{344 \text{ m/s}}{1{,}000 \text{ Hz}} = 0.34 \text{ m}\right),$$

the node-antinode distance is 8.5 cm. Therefore, moving your head a distance of only 8.5 cm would result in a pronounced change in the loudness.

Standing waves of sound can play a large role in producing bad acoustics in a listening room; they make some frequencies in the music appear louder or softer than others. The extent of such loudness changes depends on where the listener sits in the room (node vs. antinode) and on which frequencies are being played. Remember, standing waves cannot be set up within a room at any arbitrary sound frequency. Only the "right" frequencies, as determined by the dimensions of the room and the speed of sound, will give rise to standing waves. To help minimize severe standing wave effects, the listener should angle the speakers toward the center of the room, as shown in Figure 4.25. With the speakers oriented at an angle, the sound waves are not directly reflected back on themselves; rather, they must undergo many reflections before returning back on themselves. Since sound power is lost on each reflection, the returning waves will have lost considerable power, making it more difficult to set up standing waves. For this reason, it is generally a good idea to "angle-in" your speakers whenever possible.

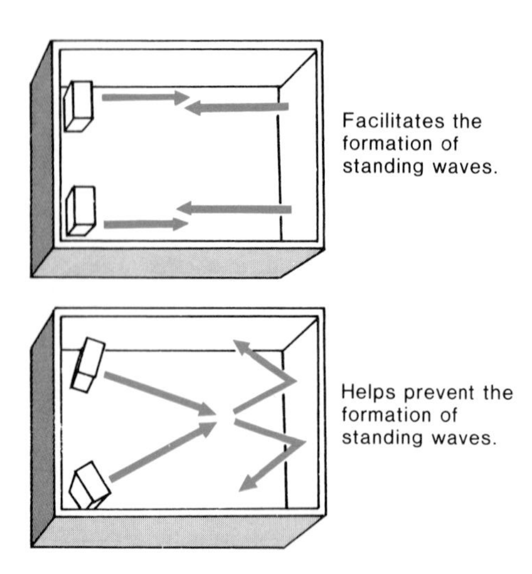

Figure 4.25. Angling the speakers in toward the center of the room helps to minimize the formation of standing waves.

Standing Waves and Musical Instruments

In addition to their role in determining room acoustics, standing sound waves play another important role. They are involved in the way all wind instruments (trumpet, flute, clarinet, etc.) produce sound. Musical instruments will be discussed further in section 4.9, but the basic idea is that all wind instruments are tubes of air molecules. Figure 4.26 shows a cylindrical tube of air which is open at one end and closed at the other. The sound wave, which is generated by the tuning fork in the picture, travels down the tube, reflects from the closed bottom, and travels back up the tube. If the

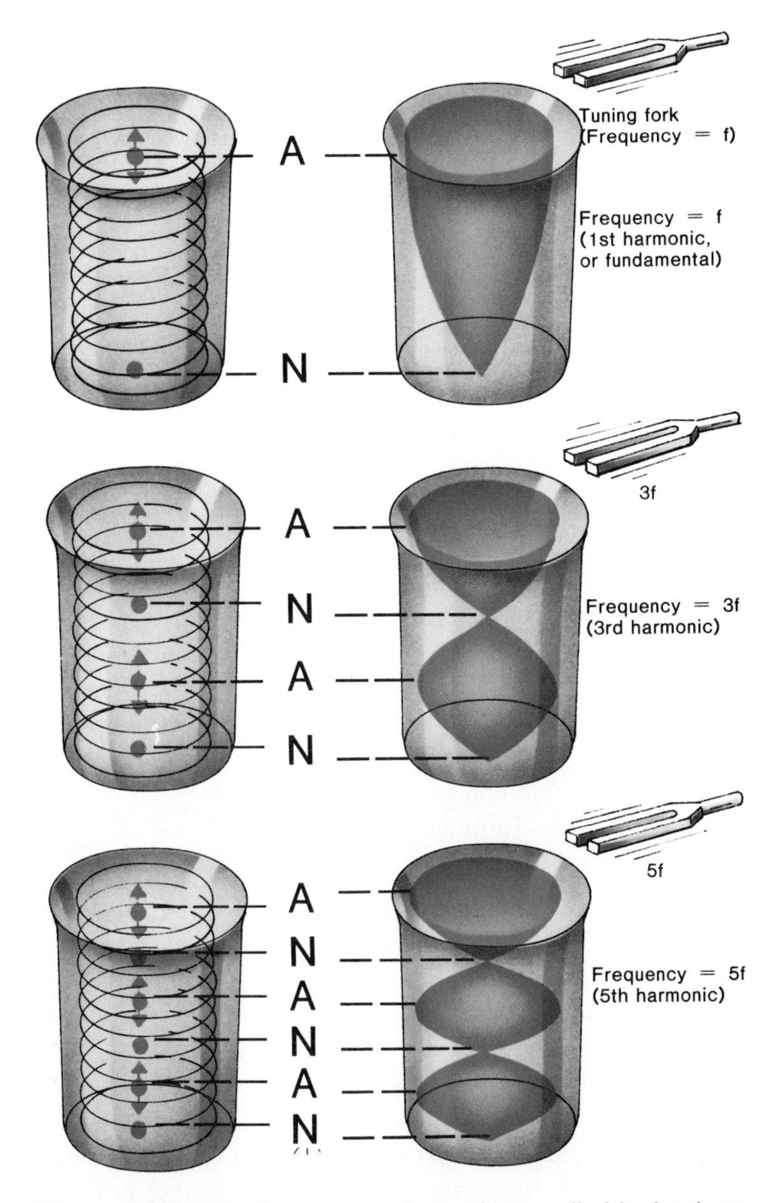

Figure 4.26. Standing waves of sound in a cylindrical column of air, open at one end and closed at the other. Only odd harmonics are present, as indicated by the three examples shown.

frequency of the tuning fork is just "right," the forward and backward traveling waves combine to form a standing wave in the tube. To serve as a visual aid, a slinky has been drawn inside the tube. The colored dots which are attached to the slinky once again represent the motion of the air molecules at the nodes and antinodes of the standing wave. To the right is drawn a "blurred" pattern of the standing wave. This pattern symbolizes the amplitude of the vibrating molecules along the length of the tube; the widest part of the pattern indicates a region of the tube where the molecular vibration is greatest (an antinode), while the narrowest part of the pattern indicates a region of no molecular vibration (a node). The three examples in Figure 4.26 emphasize that more than one standing wave pattern can be set up in the tube of air, corresponding to the fundamental and overtone frequencies. The situation is similar to that discussed earlier for transverse standing waves on a string.

To calculate the fundamental and overtone frequencies, notice that the air molecules at the open end of the tube are free to vibrate in and out of the tube, while those at the closed end are not. Therefore, when any standing wave pattern exists, there must always be a displacement antinode at the open end and a node at the closed end. Since the distance between an antinode and a node is one-fourth of a wavelength ($\lambda/4$), the patterns in Figure 4.26 reveal that only an odd number of quarter-wavelengths can fit into the length of the tube. This condition can be expressed as $L = n(\lambda/4)$, where $n = 1$, or 3, or 5 . . . , etc. An even number of quarter-wavelengths cannot fit into the tube. If they did, there would have to be a node at both ends of the tube or an antinode at both ends. Such a situation is prohibited by the fact that the air molecules are free to vibrate at the open end but cannot vibrate at the closed end of the tube. Since the wavelength is related to frequency and the wave speed ($\lambda = v/f$), the fundamental and overtone frequencies can be calculated by substituting this expression for λ into the above equation for L:

$$L = n\left(\frac{v}{4f}\right).$$

Solving for f yields

$$f_n = n\left(\frac{v}{4L}\right) \text{ where}$$

we have replaced f by the symbol f_n to denote that the allowed standing wave frequencies depend on the number n; n can

Question: What is the fundamental frequency of a cylindrical organ pipe which is 0.1954 meters long and open at one end? Using 344 meters/second as the speed of sound and $n = 1$ for the fundamental, Equation 4.2 becomes

$$f_1 = 1\left(\frac{344 \text{ m/s}}{4 \times 0.1954 \text{ m}}\right)$$

$$f_1 = 440 \text{ Hz}.$$

be an odd-integer number, i.e., n = 1, 3, 5, . . . , etc. Therefore, the possible harmonic frequencies for a one-end-open air column can be calculated by using the relation:

$$f_n = n\left(\frac{v}{4L}\right); \quad n = 1, 3, 5, \ldots, \text{etc.} \quad \text{(Equation 4.2)}$$

Standing waves can also set up in a tube of air even when the bottom, as well as the top, end of the tube is open, as shown in Figure 4.27. This occurs because sound waves traveling down the tube reflect from the mass of stationary air located just outside the bottom opening. However, this situation is different from that in a tube which is closed at the

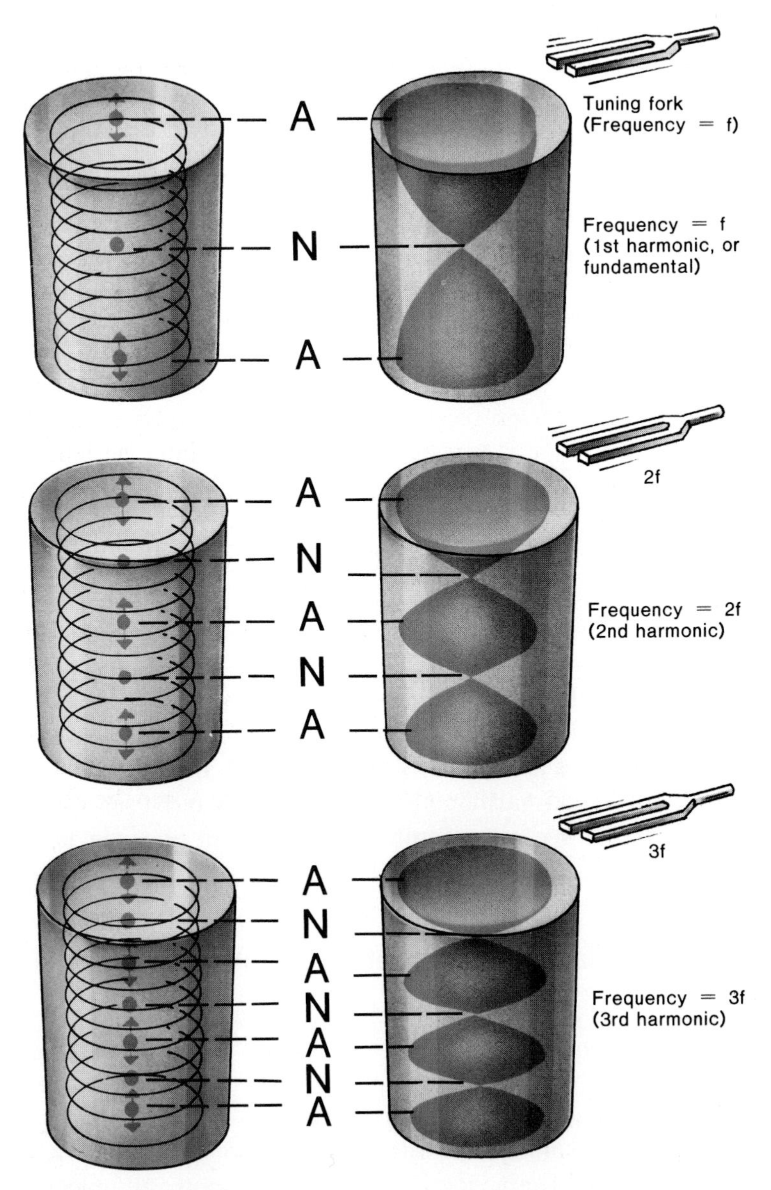

Figure 4.27. Standing waves of sound in a column of air which is open at both ends. Both even and odd harmonics are present, as indicated by the three examples shown.

bottom. When the bottom of the tube is open, the air molecules there are free to move in and out just like those at the top end. Any standing wave pattern which is set up must therefore have an antinode at both ends of the tube. Since the distance between two antinodes is one-half of a wavelength ($\lambda/2$), the patterns in Figure 4.27 reveal that the length of the tube must either be an odd or an even number of half-wavelengths. Therefore, $L = m\left(\frac{\lambda}{2}\right)$, where $m = 1, 2, 3, 4, \ldots$, etc. The fundamental and overtone frequencies are calculated as follows.

$$L = m\left(\frac{\lambda}{2}\right) = m\frac{v}{2f_m}.$$

Solving for f_m yields

$$f_m = m\left(\frac{v}{2L}\right); \quad m = 1, 2, 3, 4, \ldots, \text{etc.}$$

(Equation 4.3)

The above formula gives the standing wave frequencies for an air column with both ends open.

Question: What is the fundamental frequency of an organ pipe which is 0.1954 meters long and is open at both ends? Using 344 meters/second for the speed of sound and m = 1 for the fundamental, Equation 4.3 becomes

$$f_1 = 1\left(\frac{344 \text{ m/s}}{2 \times 0.1954 \text{ m}}\right)$$

$$f_1 = 880 \text{ Hz}.$$

Notice that this is twice the frequency of an identical tube with its bottom end closed.

Summary

A standing wave is the result of two identical, but oppositely traveling waves, which add together in the same region of space. Longitudinal, as well as transverse, waves can produce standing waves. Displacement nodes are places on the standing wave at which the medium is not being vibrated, whereas antinodes are places at which the medium is being maximally vibrated. For cylindrical columns of air, the fundamental and harmonic frequencies which characterize the standing waves are given by Equations 4.2 and 4.3.

Only the odd-harmonic standing waves can be set up on a one-end-closed air column. Both the even and odd harmonics can be established for the both-ends-open air column.

4.7 Natural Frequencies and Resonance

In the previous section we discussed standing waves on a string and in air columns. In each case we saw that the standing waves could be set up only at certain special or natural frequencies, which are related to the length of the string or air column according to Equations 4.1–4.3. This idea of special frequencies is important, because all objects, not just strings and columns of air, have special frequencies—called natural frequencies—at which they "like" to vibrate. There are many familiar examples of such natural frequencies. By jumping up and down at the end of a diving board one can "feel" its natural frequency. A strong downward push on the front fender of a car with bad shock absorbers will start the chassis oscillating at its natural frequency. If an object's natural frequencies lie between 20 Hz and 20 kHz, then its motion will produce audible characteristic sounds whose frequencies will precisely match the natural frequencies. For example, line up two or three different glasses on a table and tap each one on its rim as in Figure 4.28. Each will vibrate at its own natural frequencies and produce its characteristic tones; all the glasses will sound different.

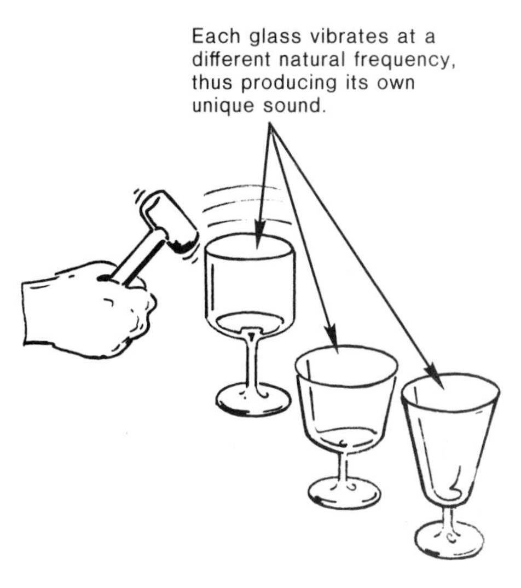

Figure 4.28. Striking an object causes it to vibrate at its own natural, or resonant, frequencies.

Natural frequencies are also called "resonant" frequencies because they are intimately involved in a phenomenon called resonance. *Resonance* is the transfer of large amounts of energy to an object from a source such as a wave, and it occurs when the frequency of the wave coincides with a natural frequency of the object. A familiar example of resonance involves someone on a swing as illustrated in Figure 4.29. It is quite clear that a swing has a characteristic rhythm, because there is a natural frequency at which it prefers to oscillate once it has been set into motion. Pushing another person on a swing should have made you subconsciously aware of resonance. It is very easy to push a swing if your motion is in synchronism with that of the swing. Furthermore, with gentle shoves at the right time, in synchronism with the swing, it is possible to reach a very high amplitude swinging motion. This is the whole idea of resonance. One system (the pusher, or a wave) can readily transfer energy to another system (the swing and the person on it), provided the latter is being pushed at a natural frequency of vibration.

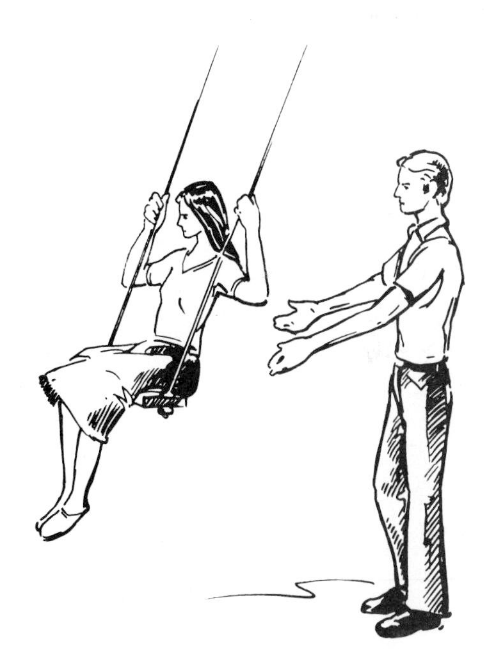

Figure 4.29. It is easy to achieve a high amplitude motion when the swing is pushed in synchronism with its natural frequency of vibration. The transfer of vibratory energy from the pusher to the swing under such conditions is called "resonance."

A more startling example of resonance can be demonstrated by using sound to shatter a glass beaker. Suppose that it has been determined that the glass has a natural frequency of 900 Hz. Remember, this is a frequency at which the rim of the glass will oscillate if tapped, as shown in Figure 4.28. Now a small speaker can be placed close to the rim of the beaker. The speaker is powered by an audio generator/amplifier combination which is set to produce a pure tone of 900 Hz, as shown in Figure 4.30. Because the glass "likes" to vibrate at 900 Hz, it will readily absorb the sound energy. As more and more energy is absorbed, the oscillations of the rim of the beaker eventually become violent

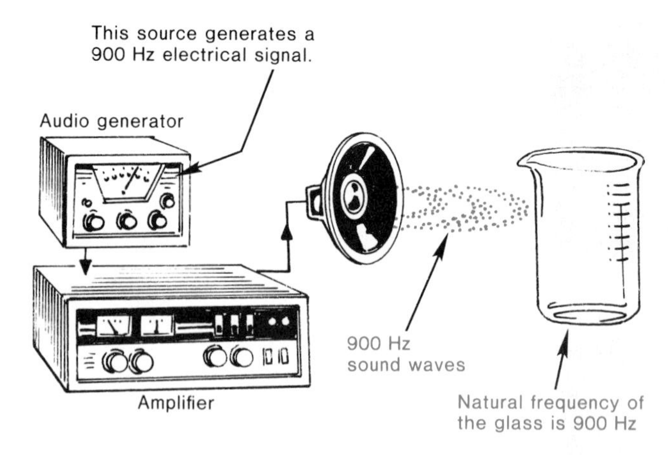

Figure 4.30. An experiment for breaking the glass beaker with sound waves via resonance.

enough to shatter the glass! Figure 4.31 shows actual photographs of a beaker shattering under such conditions. Note that it is critical for the sound to have the *same frequency* as the *natural frequency* of the beaker. If the sound frequency were lowered by only 2 Hz, to 898 Hz, the beaker could not be broken no matter how loud the sound. The two frequencies must be precisely matched.

Perhaps the most spectacular demonstration of resonance occurred at the Tacoma Narrows Bridge, Tacoma, Washington. This enormous steel and concrete bridge (2,800 feet long and 39 feet wide) was opened for traffic on July 1, 1940. Early in the morning on November 7, 1940, the wind was blowing at a velocity of 45 miles/hour—which in itself was no real threat to the bridge. However, the turbulent wind was oscillating back and forth in the Tacoma Narrows at a frequency of about 14 oscillations per minute, or 0.23 Hz, which accidentally happened to match the natural frequency of the bridge! It is very hard to think of a bridge made from concrete and steel, having a mass in the millions of kilograms, as having a natural frequency of vibration. But, as the photos in Figure 4.32 testify, it most certainly did! The resonance condition allowed the bridge to absorb energy from the oscillating wind thereby increasing its own amplitude of vibration. After about an hour of this energy transfer the bridge finally collapsed at a time when it was twisting about the white center stripe approximately 35° from the horizontal! Incredible! Note that the cause of the collapse was *not* that the wind velocity was 45 mph. It was the fact that the turbulent wind was oscillating at the same frequency as the natural frequency of the bridge. The wind oscillations caused a large amount of energy to be transferred to the bridge (just like a person being pushed on a swing), and the subsequent bridge oscillations caused its collapse. A new, but redesigned, bridge was built using the original anchorages and tower foundations, and it still stands today. Evidently the engineers did their homework correctly the second time around.

Figure 4.31. The glass beaker just before (top photo) and after (bottom photo) breaking. Note the small (1") speaker behind the beaker which produces the 900 Hz tone.

Figure 4.32. The dramatic story of the Tacoma Narrows Bridge collapse. The culprit was resonance.

Resonance Applications in Hi-Fi

In hi-fi the examples of resonance are not quite as spectacular as the collapse of the Tacoma Narrows Bridge, but they are no less important. Resonance effects can severely limit the performance of any hi-fi system, and the quality of the music it produces. In the previous section we have already discussed the bad effect which standing sound waves can have on the acoustics of a listening room. The standing waves occur at the sound frequencies which match the resonant frequencies of the room itself. Resonance can also cause ill effects in the process of sound reproduction in speakers and turntables, and we shall discuss these effects in later chapters dealing specifically with these units.

One should not get the idea that all resonance effects are bad. The production of sound by all musical instruments involves resonant frequencies in a very basic way, which will be discussed in section 4.9. In fact, without the resonant frequencies, musical instruments would generate only uninteresting sounds.

Forced Vibrations

Be careful not to confuse resonance vibrations with forced vibrations. Resonance vibrations can occur only when the object is forced to vibrate at one of its natural frequencies. Forced vibrations can occur at any frequency, since any object can be forced to vibrate at frequencies other than its natural frequencies. Returning to the example of the swing, a gentle push will cause the swing to oscillate at its own natural rate of one oscillation per second, for example. It is, however, entirely possible to grab hold of the swing and simply force it to vibrate at any desired frequency, for example, 4 1/2 oscillations per second. These, understandably, are called *forced vibrations,* and it is implied that the pusher must retain a hold on the swing at all times so as to ensure its being forced into the desired vibrational frequency.

Another good example of forced vibrations is the stylus that is being pushed by the record groove at frequencies which can lie anywhere in the audio range. Of course the stylus/tone arm combination has its own natural frequency of vibration, and audio engineers are careful to design this natural frequency well below the audio range, say 10 Hz. It is not difficult to imagine what would happen if its natural frequency were to lie within the audio range—perhaps at 100 Hz. A 100 Hz tone enscribed in the groove would force the stylus/tone arm to oscillate at its natural frequency and extremely large amplitude vibrations would build up (remember the Tacoma Narrows Bridge?). Eventually the stylus could lose contact with the groove and the tone arm might skid across the record.

Summary

Natural or resonant frequencies of an object are the special frequencies at which the object "likes" to vibrate. Resonance is the transfer of large amounts of energy to an object from a source such as a wave. Resonance occurs when the frequency of the wave coincides with a natural frequency of the object.

4.8 Beats

In previous sections we have seen examples of constructive and destructive interference. In all cases, however, the waves possessed the same frequency, and we now wish to consider what happens when two sound waves have *slightly different frequencies*. To demonstrate this effect place two identical tuning forks side by side as shown in Figure 4.33. Attach a small piece of putty on one of the tuning forks so that its frequency will be lowered slightly. When the two forks are sounded simultaneously, the sound loudness will be heard to rise and fall periodically—faint, then loud, then faint, then loud again—and so on. This periodic variation in the sound loudness is called *beats,* and it is caused by the interference between two sound waves which have slightly different frequencies. Figure 4.33 shows the sound pressure patterns which are produced by 440 Hz and 436 Hz tuning forks. As the sound waves leave the tuning forks they spread out and overlap, such that when they reach your ear the sound waves are superimposed and interference results. Notice, in Figure 4.33, that there are points along the waves where constructive and destructive interference result. When a constructive part reaches your ear a loud sound will momentarily be heard and, conversely, when a destructive part arrives, no sound is momentarily heard. Two sound waves which are very close to each other in frequency, within 10 Hz, will be

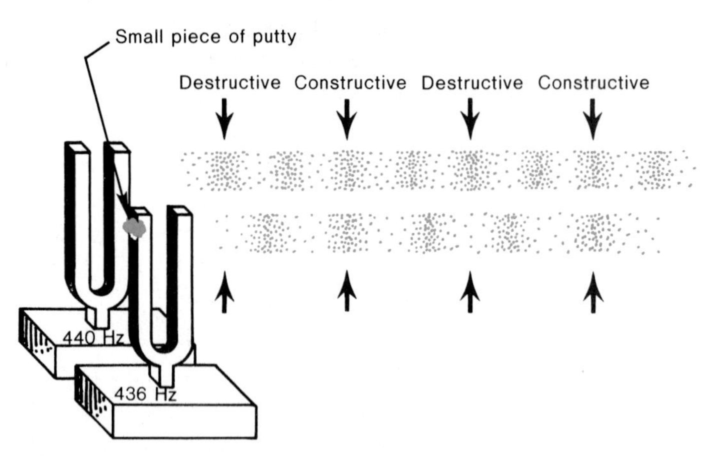

Figure 4.33. The sounding of two tuning forks with slightly different frequencies produces the phenomenon of beats.

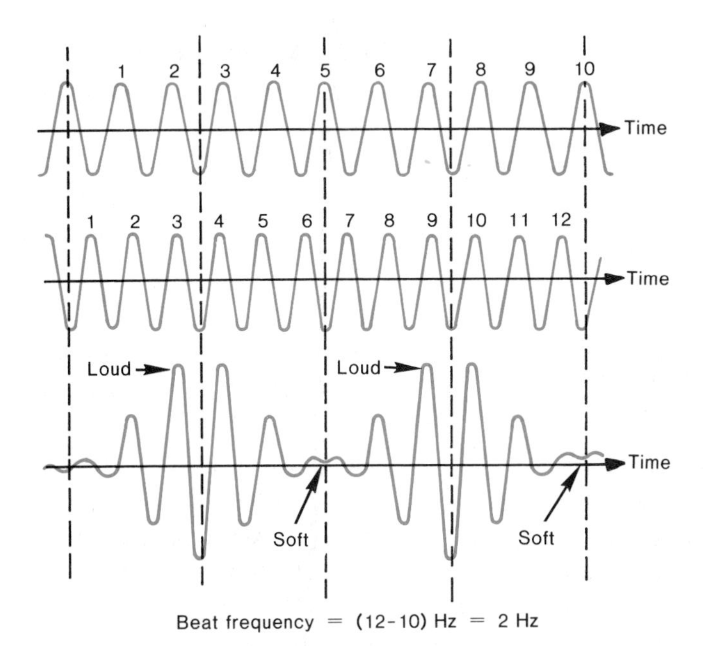

Figure 4.34. The 10 Hz and 12 Hz sound waves produce, by the principle of superposition, a sound wave with a beat frequency of 2 Hz.

heard as a single frequency whose loudness periodically rises and falls. If the frequency difference between the two sounds is greater than about 10 Hz, *both* of the individual frequencies, in addition to the loudness changes, will be heard.

The beat frequency is the number of times each second that the sound loudness rises and falls. A detailed analysis shows that the beat frequency is the *difference* between the two sound frequencies. In the above example the resulting sound level will rise and fall 4 times every second (440 − 436 = 4). Another illustration of beats is shown in Figure 4.34 which depicts a 10 Hz and a 12 Hz wave traveling together. Although these frequencies lie below the audio range they were chosen for convenience of the drawing. The top two drawings depict the pressure variations in a one-second time interval for each of the two waves. The third drawing shows the resulting sound wave which is obtained by adding the first two waves together (principle of linear superposition!). Notice how the amplitude of the resulting wave is no longer a constant as it is in the individual 10 Hz and 12 Hz waves. Instead the amplitude changes from a maximum to a minimum, back to a maximum, etc. At the times when the amplitude is a maximum, the sound will appear to be loud and when the amplitude is a minimum, the sound will be soft. As expected, the sound changes its loudness at a rate of 2 Hz (12−10 = 2) *which is the beat frequency*.

Tuning a String Instrument

Many people often tune string instruments by listening for the beat frequency when playing the same note on two different strings. If the two strings are slightly out of tune with respect to each other, they will vibrate at different frequencies and the listener will hear the characteristic beat note. Adjusting the tension in one of the strings until the beat completely disappears will ensure that the strings are vibrating at the same fundamental frequency and, hence, be in tune.

4.9 Musical Instruments

Musical instruments are one of the most interesting examples of standing waves, natural frequencies and resonance. Standing waves play a fundamental role in the production of sound from musical instruments. For example, if you are playing a guitar, trumpet, or drums, waves are created which travel through a medium and reflect back from some type of a boundary. As a result, standing waves are established in the medium. With string instruments the medium in which the waves move is obviously a stretched string. Wind instruments utilize the air itself. Percussion instruments can use a variety of media such as the stretched membrane of a drum, the metal plates of cymbals, or the bars (wooden, plastic, or metal) of a xylophone.

As an aid in understanding musical instruments, recall the discussion in section 4.5 concerning standing waves on a string. The behavior of the string illustrates two important features which characterize the manner in which all instruments work. First of all, both the string itself and something with which to vibrate the string are required if standing waves are to be established. Similar requirements hold for any musical instrument, as shown in Figure 4.35. There is the instrument itself, the trumpet, guitar, drum, etc. But equally important, there must be something with which to vibrate the air within and around the instrument. A trumpeter buzzes his lips into a mouthpiece, a guitar player plucks the strings, and a drummer strikes the drum head. In each case something is forced into vibration which, in turn, causes the surrounding air to vibrate.

Secondly, standing waves on the string in section 4.5 could be set up only at one of the natural frequencies given by Equation 4.1. Likewise, any musical instrument is characterized by its own set of natural frequencies. It is not possible, however, to write down a simple equation like Equation 4.1, because the shapes of most instruments are too complicated. Nevertheless, the buzzing of a trumpeter's lips cause a trumpet to produce only natural trumpet frequencies, while the

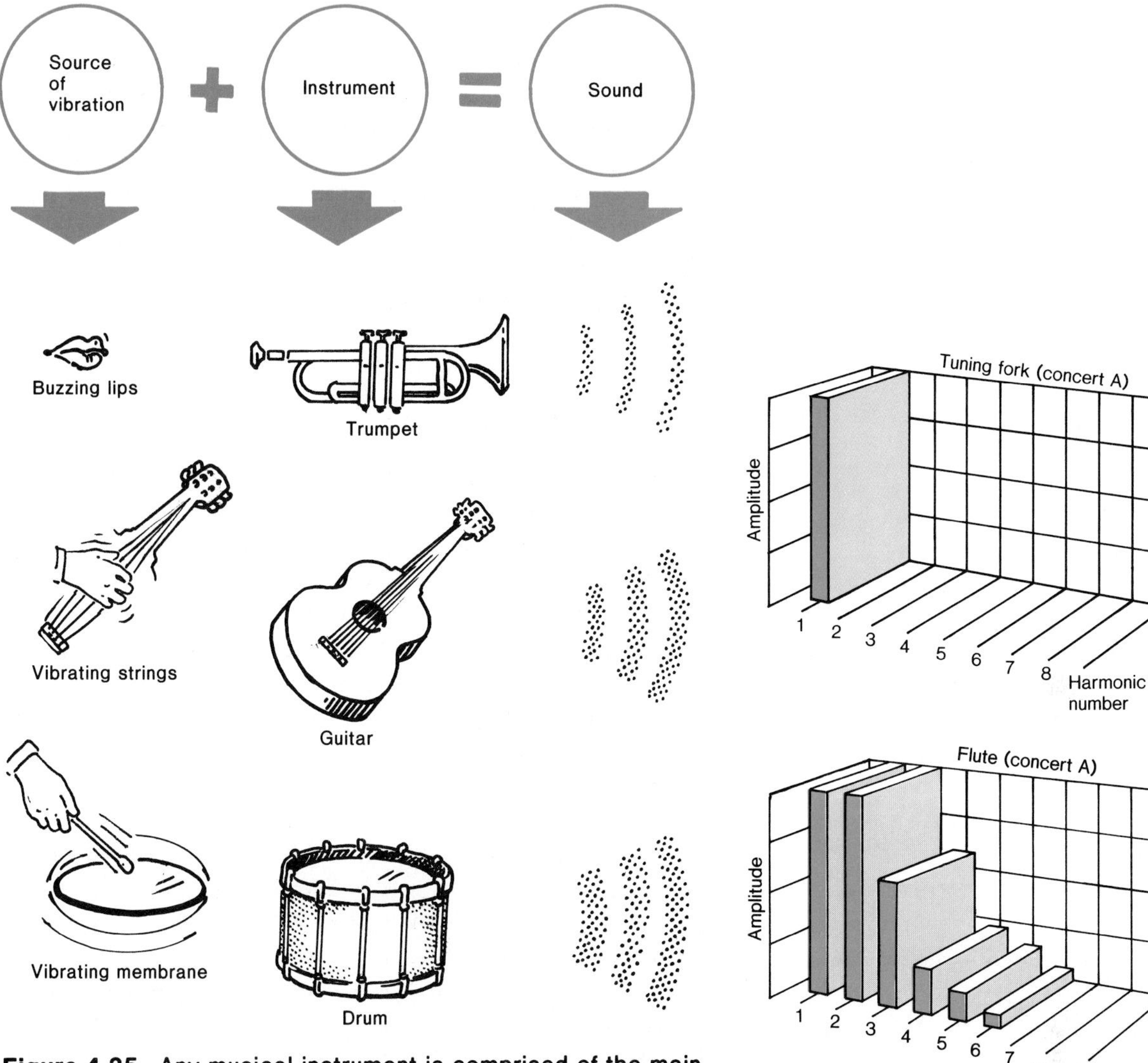

Figure 4.35. Any musical instrument is comprised of the main body of the instrument and a source of vibration. Only sound at the natural frequencies of the instrument is produced.

vibrating guitar strings cause the guitar to produce only natural guitar frequencies, and the vibrating drum head causes the instrument to produce only natural drum frequencies. In general, then, any instrument will only produce sound whose frequencies are determined by the standing wave frequencies which are allowed by the instrument.

It is important to realize that the amplitudes of the natural frequencies of one musical instrument are not the same as the amplitudes of another, even though they are playing the same note. Figure 4.36 uses bar graphs to illustrate the differences between the natural frequencies of an ideal tuning fork, a flute, and a trumpet when each is producing a "concert A" note. The horizontal axis of the graph shows the possible natural frequencies in terms of the fundamental frequency (440 Hz for concert A) and the higher harmonics.

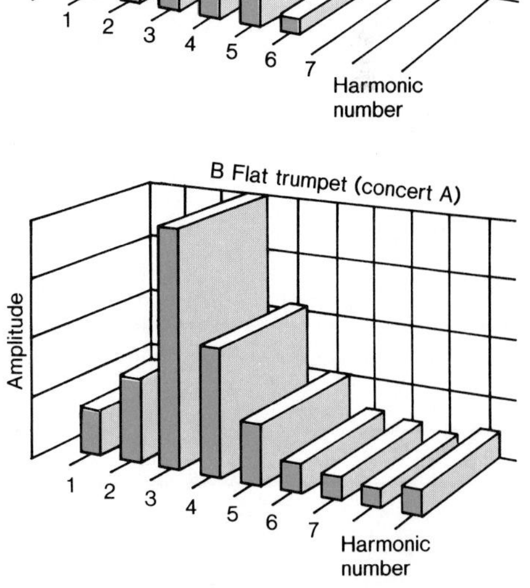

Figure 4.36. Bar graphs showing the natural frequencies present in the sound of a tuning fork, a flute, and a trumpet when each produces a "concert A" tone. The fundamental, or first harmonic, frequency is 440 Hz, the second harmonic frequency is $2 \times 440 = 880$ Hz, etc.

The vertical bars indicate which natural frequencies are produced by the instrument when it is playing "concert A." The heights of the bars give the amplitudes of the corresponding frequencies, thus showing which natural frequencies are louder than others. Notice that the three graphs are not the same, even though each instrument is playing the same note! The upper graph indicates that the sound of the tuning fork contains only a single frequency, the fundamental. By contrast, the two other graphs reveal the remarkable fact that the sound of a flute, as well as the sound of a trumpet, contains a mixture of harmonically-related frequencies! In addition, the heights of the bars in these two graphs shows that the loudness of the flute harmonics is not the same as the loudness of the trumpet harmonics. For example, relative to the fundamental, the third harmonic is noticeably louder for the trumpet than it is for the flute. Also, the seventh, eighth and ninth harmonics are present in the trumpet but not in the flute. Clearly, the natural frequencies of one instrument are not produced with the same amplitudes as those of another.

The differences in amplitudes between the instruments illustrated in Figure 4.36 are significant, because they are the reason why the tuning fork, the flute, and the trumpet sound different, even though each produces the same note. In other words, it is the number and relative amplitudes of the harmonics which give each instrument its characteristic quality or timbre. To understand further this point consider Figures 4.37, 4.38, and 4.39. These pictures show the patterns of pressure fluctuations for each of the sounds whose bar graphs were introduced in Figure 4.36. As we shall now demonstrate, the bar graph provides the "recipe" from which the pattern of pressure fluctuations can be constructed.

The tuning fork bar graph in Figure 4.37 shows that a tuning fork produces only a single frequency, its fundamental at 440 Hz. Such a note is called a *pure tone,* because only the fundamental frequency is present, and all higher harmonics are absent. The familiar pressure pattern associated with this note is drawn below the bar graph. By contrast, the flute and trumpet bar graphs indicate that the notes produced by these instruments are not pure tones.

The flute bar graph in Figure 4.38 reveals that the flute produces a concert A note by adding together the fundamental frequency at 440 Hz with the five next-higher harmonics, each with its proper relative amplitude. (Notice that this is another example of the principle of linear superposition.) The resulting total pressure pattern is shown in the figure, and it is much more complex than that of a pure tone. For comparison, the pressure pattern for a concert A on a trumpet is shown in Figure 4.39. Clearly, the trumpet and flute pressure patterns do not look the same. We now see the

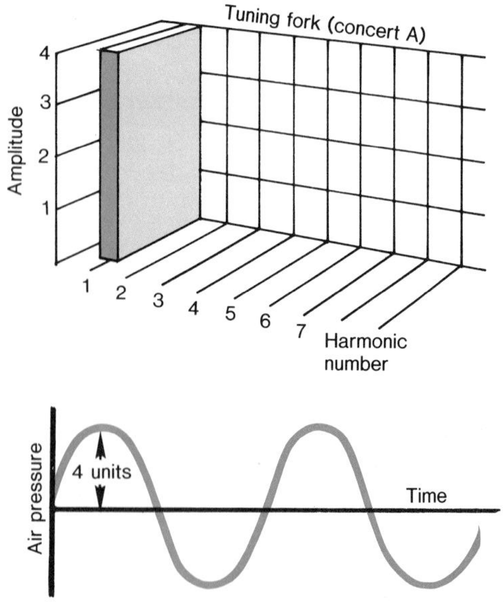

Figure 4.37. A tuning fork playing a concert A note (fundamental frequency = 440 Hz). Upper drawing is the bar graph recipe for the sound. Lower drawing is the corresponding pattern of how the pressure fluctuates in time.

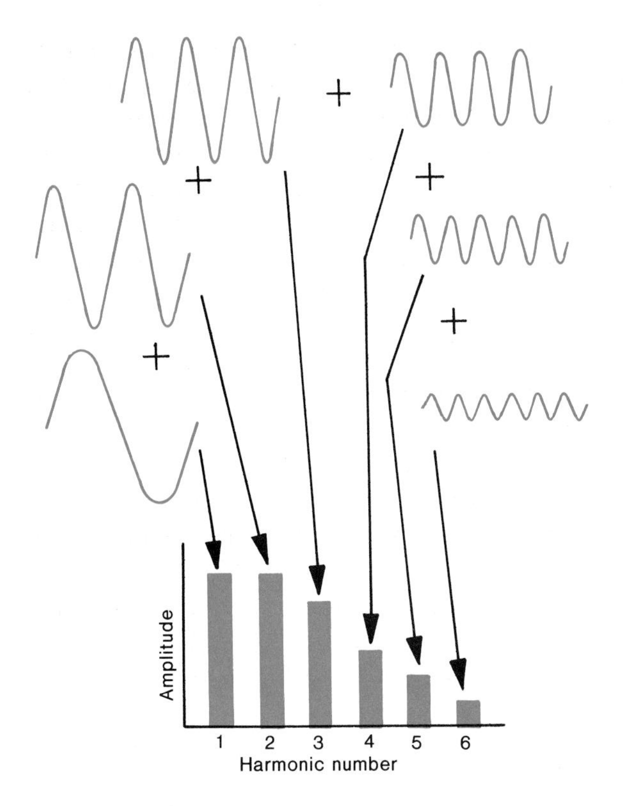

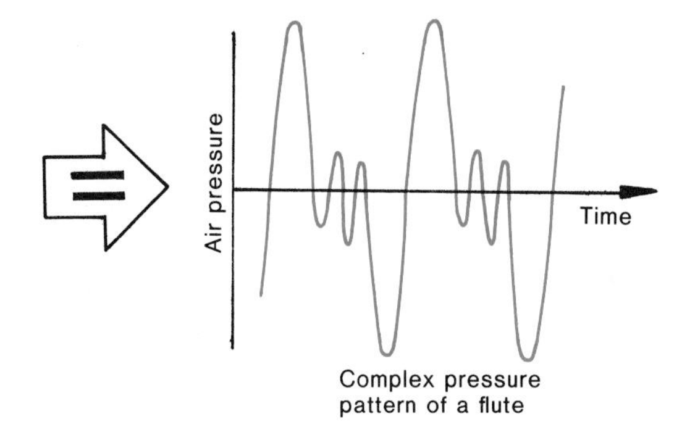

Figure 4.38. The total pattern of how the pressure fluctuates in time when a flute plays a concert A (fundamental = 440 Hz). The total pattern is the sum of the first six harmonics according to the bar graph and the principle of linear superposition. Notice that the relative amplitudes of the harmonics correspond to the heights of the vertical bars in the bar graph.

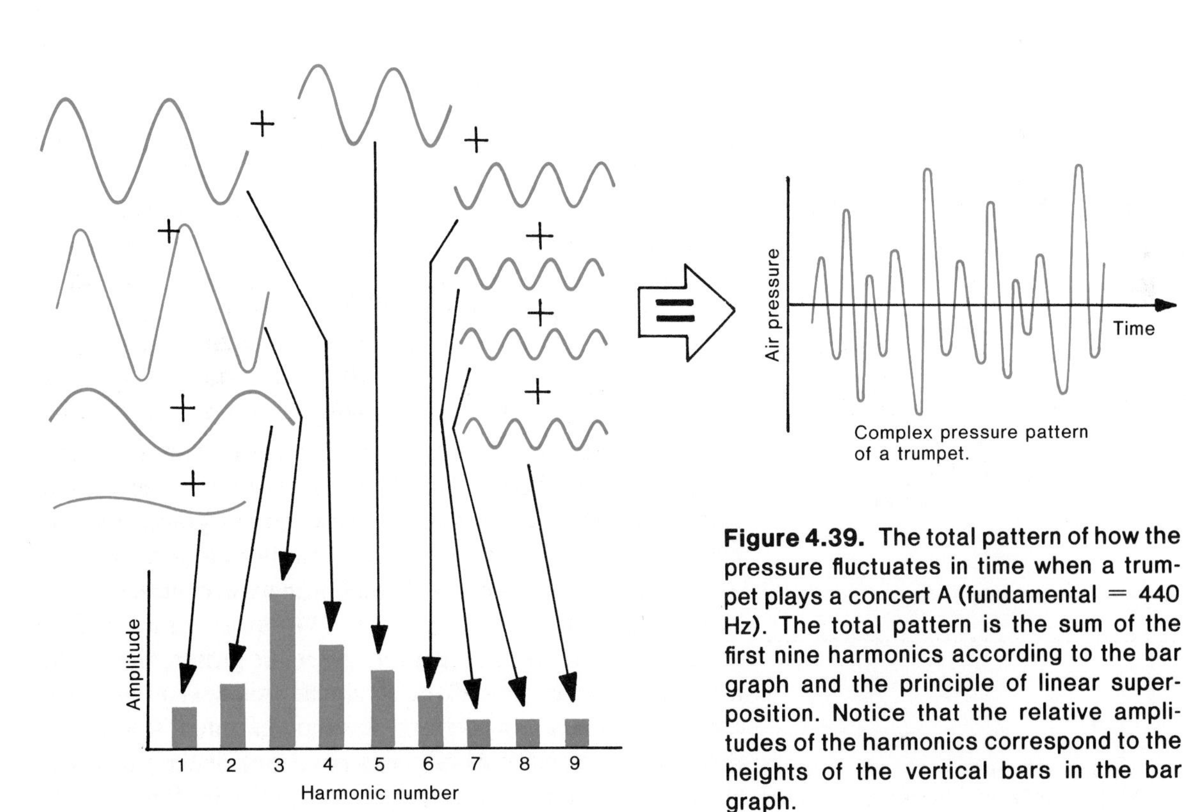

Figure 4.39. The total pattern of how the pressure fluctuates in time when a trumpet plays a concert A (fundamental = 440 Hz). The total pattern is the sum of the first nine harmonics according to the bar graph and the principle of linear superposition. Notice that the relative amplitudes of the harmonics correspond to the heights of the vertical bars in the bar graph.

reason why listeners have no trouble distinguishing the sound of a trumpet from that of a flute; each instrument produces different amplitudes of its harmonic frequencies.

The bar graphs and pressure patterns illustrated above are appropriate only for the concert A note. In general, the corresponding pictures for other notes played on these instruments do not look the same. Each note has its own unique bar graph according to which the instrument generates the various harmonics that make up the sound. Moreover, the experienced and talented performer can alter the amplitudes of the harmonics, and thus change the bar graph recipe, as he attempts to achieve a desired sound quality.

In general, the great variety of musical instruments can be classified under one of the following headings which categorize the type of medium in which the waves propagate:

String Instruments

guitar
violin
cello
piano

Wind Instruments

Wood Winds	**Brass**
flute	trumpet
clarinet	trombone
saxophone	French horn
piccolo	cornet
bassoon	tuba

Percussion Instruments

drum
cymbal
triangle
xylophone

Let us now take a more careful look at each of these groups in order to show explicitly the nature of the standing waves which are produced.

String Instruments

A string instrument is one which utilizes a string stretched between two supports as a means of vibrating the air around the instrument. Familiar examples are the guitar, violin, and piano. A guitarist plucks the string. A violinist draws a bow across the string. In a piano a small hammer strikes the string every time the pianist depresses one of the keys. The plucking, bowing, or hammering action causes the string to vibrate, and standing waves become established according to the string's natural frequencies.

It is difficult for many people to visualize how a string instrument can vibrate at more than one natural frequency at the same time. Figure 4.40 shows a simplified illustration of how the vibrating string might look with only the first and second harmonics present. In this case the string will generate a sound in which both the first and second harmonic frequencies will be heard simultaneously, although each frequency may have a different loudness.

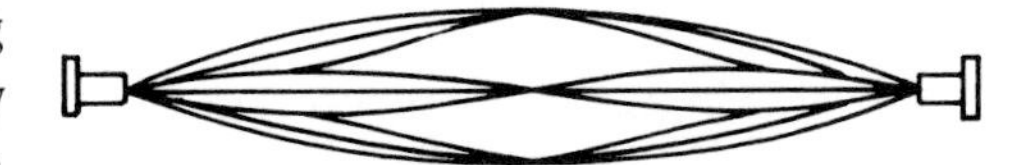

Figure 4.40. A string vibrating simultaneously in two standing wave patterns which represent the first and second harmonics.

To change from one note to another the instrumentalist changes the allowed standing wave frequencies by employing one of the three methods: changing the string's length, changing its tension, or altering its density. All three of these methods are employed when a string instrument is played. The length of the vibrating string can be changed by simply altering the finger placement, as shown in Figure 4.41, where only the fundamental standing wave pattern is shown for the sake of simplicity. Moving the finger down the neck and closer to the sounding board shortens the string and raises the frequency of the tone being produced. Conversely, moving the finger to increase the string's length creates a tone of lower frequency. To understand these changes refer to Equation 4.1 and notice that the length of the string L occurs in the denominator of the expression; it can be seen that shortening L raises the frequency of the standing wave and, conversely, lengthening L lowers the frequency. Changing the tension or density of the string also alters the natural frequencies, because these characteristics change the velocity with which the waves travel back and forth along the length of the string. Turning the appropriate tuning peg either tightens or loosens the string. Tightening the string increases its tension and a higher frequency standing wave is produced. Similarly, loosening the string and lowering its tension results in a lower pitched note being sounded. With regard to the density of the strings, a glance at a guitar will show that its strings all have different thicknesses, i.e., densities. Strumming across the strings will reveal that the relatively massive E string produces the note with the lowest pitch. The progressively lighter strings, for example, A, D, G, and B, yield sounds which are progressively higher in pitch. Therefore, length, tension, and density all play important roles in determining the allowed standing waves and, hence, the sound emitted by a string instrument.

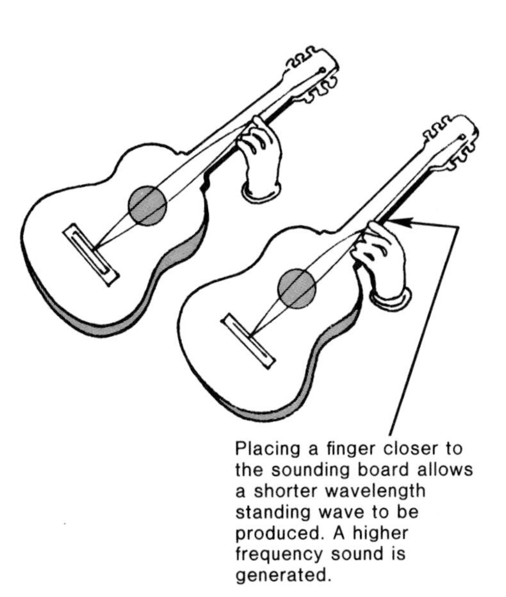

Placing a finger closer to the sounding board allows a shorter wavelength standing wave to be produced. A higher frequency sound is generated.

Figure 4.41. Changing the length of the string changes the frequency and, hence, the pitch of the sound produced by a string instrument. Only a standing wave at the fundamental frequency is shown for the sake of simplicity.

The relationship between the velocity (v) of a wave on a stretched string, the tension (T), and the mass per unit length (ρ) of the string is:

$$v = \sqrt{\frac{T}{\rho}} .$$

Increasing the string's tension will increase the wave velocity which, according to Equation 4.1, results in a higher frequency standing wave. Similarly, the frequency also can be raised by playing a string which has a lower density (i.e., playing on a light string rather than on a relatively massive one).

Wind Instruments

The fundamental difference between sounds produced by "strings" and "winds" is that standing waves in the former are set up in a string under tension, while in the latter they are set up in a column of air. Familiar examples are the

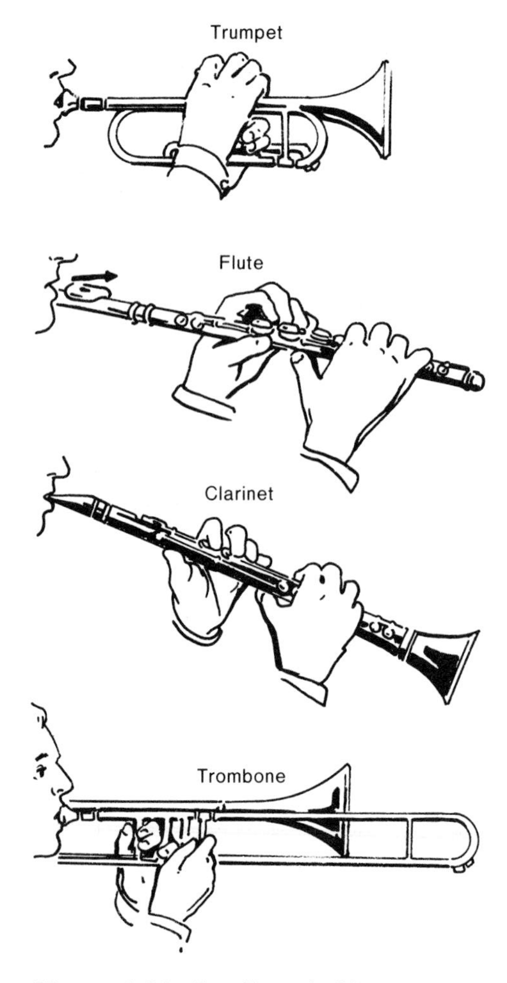

Figure 4.42. Familiar wind instruments.

flute, clarinet, trumpet, and trombone shown in Figure 4.42. A flutist blows across an opening in one end of the flute (much like blowing across the top of a soda bottle to produce a tone) in order to vibrate the air within the instrument. A clarinetist blows into the clarinet across a single blade of cane, called the "reed." The vibrating reed causes the air within the instrument to vibrate. A trombonist, like a trumpeter, blows air through tightly compressed lips into the horn; the vibrating lips cause the air within the instrument to vibrate. In each case standing waves are set up at the natural resonant frequencies of the air column.

The natural frequencies of a wind instrument are altered primarily by changing the length of the air column, and Equations 4.2 and 4.3 imply that the longer the air column, the lower the frequencies, or pitches, produced by the instrument. Although, strictly speaking, these equations apply only to simple cylindrical tubes of air, this conclusion is generally true for all wind instruments—the longer they are, the lower is the pitch of the notes which they play.

On a flute or clarinet the effective length of the instrument is changed by depressing keys which raise or lower covers on holes along the length of the instrument. When all of holes are covered, the effective length is a maximum, and the pitch of the corresponding note is the lowest attainable. Opening a hole shortens the effective length, a fact which is easiest to visualize if we imagine the hole to be so large that the lower end of the instrument is about to fall off! In such a case the natural frequencies are raised to those of a tube whose length extends only as far as the hole. In reality, the holes are not very large, and the actual effective length depends on the hole size and the shape of the instrument.

On wind instruments, such as the trombone and trumpet, the mechanism for changing the length of the vibrating air column differs from that of a flute or a clarinet. The performer changes the effective length of the trombone by sliding a tube back and forth. The lowest pitched note is produced when the sliding tube is extended as far out as possible. The way in which a trumpeter changes the effective length of the instrument and produces different notes is not so obvious, however. A trumpeter depresses valves which insert short lengths of tubing in line with the main tube of the instrument.

Percussion Instruments

The percussion family of instruments is remarkably large and diverse. Two of its best known members are the drums and xylophone. The drum head is a two-dimensional diaphragm which is clamped around its entire perimeter. When struck, two-dimensional vibrational waves are sent out in all directions, and upon reaching the edge they are reflected back toward the center. Thus, standing waves are set up. However, the standing waves and corresponding natural frequencies are more complicated than those on a string or air column. Figure 4.43 shows the fundamental pattern along with 5 higher frequencies, and indicates that the nodes of the standing waves are lines rather than points. The fundamental frequency is labeled f_1, and the higher frequencies are given in terms of f_1. It should be noted that the higher natural frequencies are not integral multiples of f_1 as they were in the string and air column.

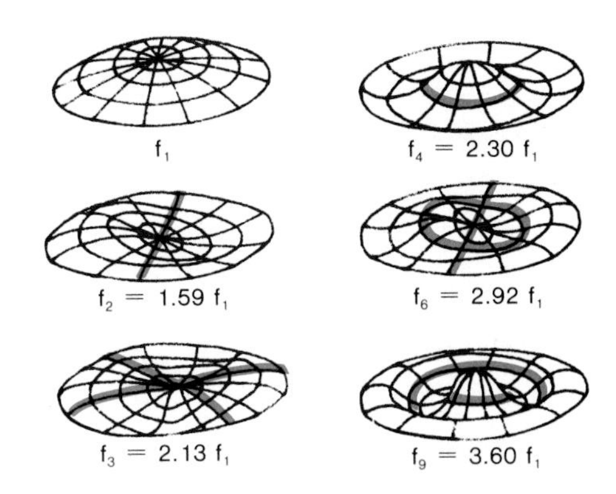

Figure 4.43. The standing wave patterns for a drum membrane. Notice that the higher natural frequencies are not integer multiples of the fundamental, f_1. The node lines are shown in color.

The xylophone utilizes the standing waves and natural frequencies in a bar, much like a guitar uses them in a stretched string. Standing waves in a bar can be established by striking it near its center, and the standing wave pattern of Figure 4.44 (B) is established. The vertical supports are nodal points for the standing wave. Shorter bars create higher frequencies while longer bars generate lower frequencies.

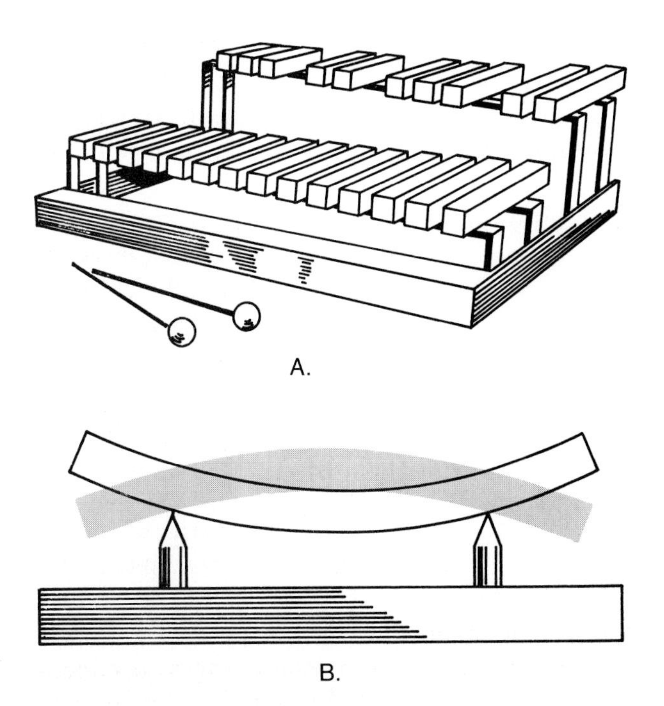

Figure 4.44. The bars of the xylophone vibrate transversely with nodes at the vertical supports.

The Frequency Range of Musical Instruments and the Human Voice

It was mentioned previously that each musical instrument generates its characteristic sound by producing standing waves. And the quality of sound, or timbre, from each instrument depends on the amplitude of the harmonics which it generates. Figure 4.45 shows the frequency ranges of some musical instruments and the human voice. From this graph

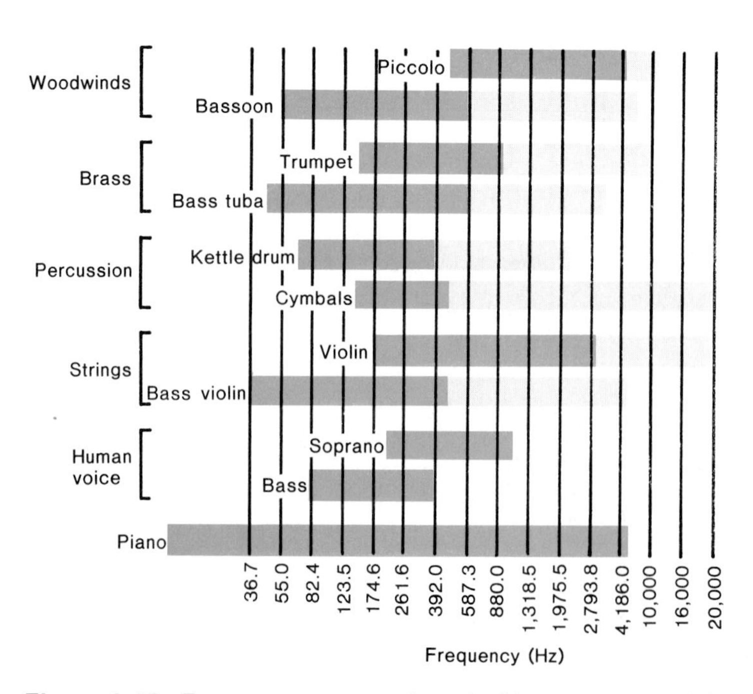

Figure 4.45. Frequency ranges of musical instruments and the human voice. The darker colored bars indicate the range of fundamental frequencies generated by the instruments. The lighter colored bars show the range of the higher harmonics produced by the instruments.

it is interesting to note that the fundamental frequencies generated by these instruments range all the way from 16 Hz to slightly over 4.5 kHz, as represented by the darker colored bars. However, the higher harmonics range, in many cases, all the way out to the limit of hearing at 20 kHz, as indicated by the lighter colored bars. It is important that a hi-fi system be capable of reproducing the high frequency treble notes because a failure to do so might result in a partial loss of the music's harmonic content. Any loss in such content causes the musical instruments to lose their timbre.

Electronic Instruments

The widespread availability of electronic technology has had a tremendous impact on musical instruments. It is now routinely possible to generate any type of sound completely electronically, without involving a traditional musical instrument. Such electronic instruments are called "synthesizers," and one of the best known synthesizers is that developed by Robert Moog.

Synthesizers, like any musical instrument, must involve natural frequencies of vibration. The difference between synthesizers and traditional musical instruments is that synthesizers use purely electronic means to generate these frequencies. The two basic methods by which these frequencies are produced are called "additive" and "subtractive" synthesis.

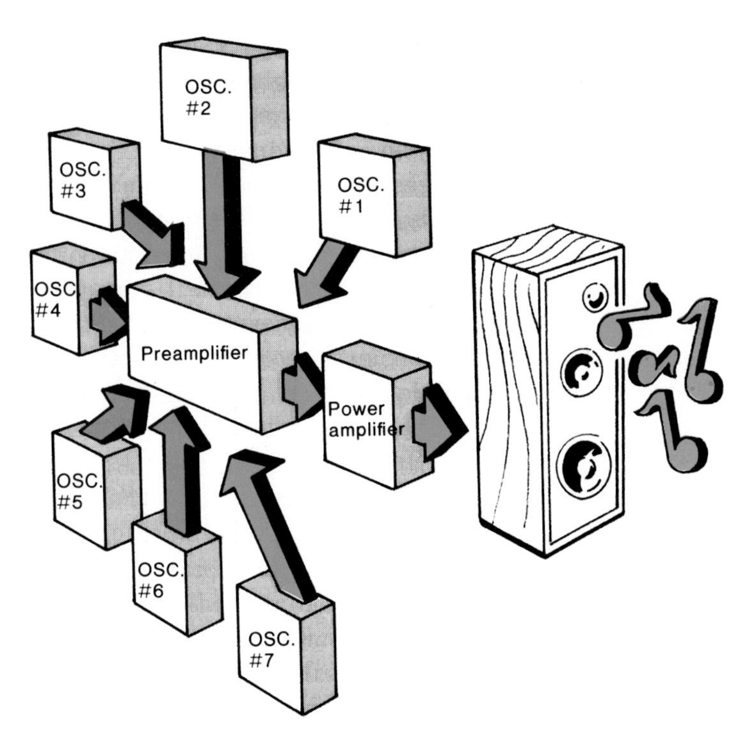

Figure 4.46. An additive synthesizer. "osc." = oscillator.

Additive synthesis is easy to understand. Electronic devices called oscillators, or signal generators, can be built which will produce a single-frequency electrical wave. A front panel control permits the user to select the desired frequency, and another control allows its amplitude to be changed. When this single-frequency signal is applied to a speaker, via the usual preamp/power amp combination, a pure tone corresponding to the selected frequency results. Figure 4.46 shows a number of such oscillators connected in this fashion to form a simplified additive synthesizer. In order to create an electronic note, the performer uses oscillator #1 to generate the fundamental frequency (first harmonic), oscillator #2 to generate the second harmonic, oscillator #3 for the third harmonic, and so on. In this sense, playing a note on this synthesizer consists of selecting the harmonics and their relative amplitudes from those available via the various oscillators. The harmonics are then added together in the preamp to form the musical tone according to the principle of linear superposition. Since the user can select the amplitude for each harmonic, it should be apparent that the notes produced by a synthesizer can sound vastly different than the same notes produced by ordinary musical instruments; the creative possibilities are almost unlimited.

Subtractive synthesis is a useful alternative, because electronic devices can be built which produce electrical waves which inherently contain mixtures of a fundamental and higher harmonics. For example, a so-called "square wave"

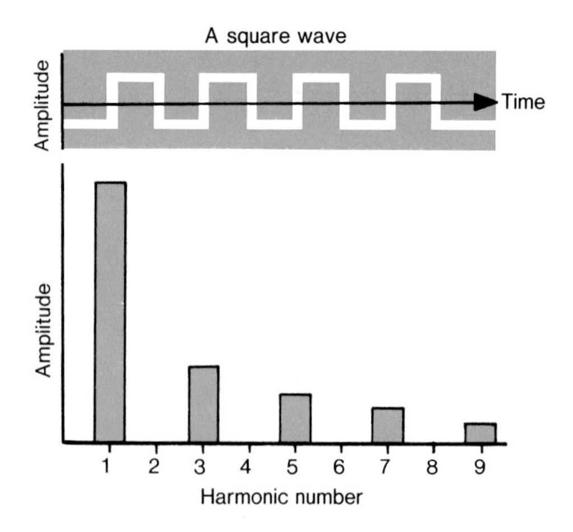

Figure 4.47. A square wave and its bar graph recipe, which indicates the appropriate mixture of harmonics.

oscillator produces an electrical wave pattern like that shown in Figure 4.47. Although it is not at all obvious, the square wave pattern is made primarily from the fundamental frequency and its odd-harmonics, according to the bar graph which is also shown in the drawing. In a similar fashion, there are oscillators which can produce electrical waves of many other geometric shapes, each being a mixture of harmonics according to its own characteristic bar graph. The main idea behind subtractive synthesis is that various components of these mixtures can be removed or subtracted from the signal with electronic filters. The result is that the timbre of the resulting tone depends on which harmonics have been removed, rather than added as in the method of additive synthesis. A subtractive synthesizer is illustrated in Figure 4.48.

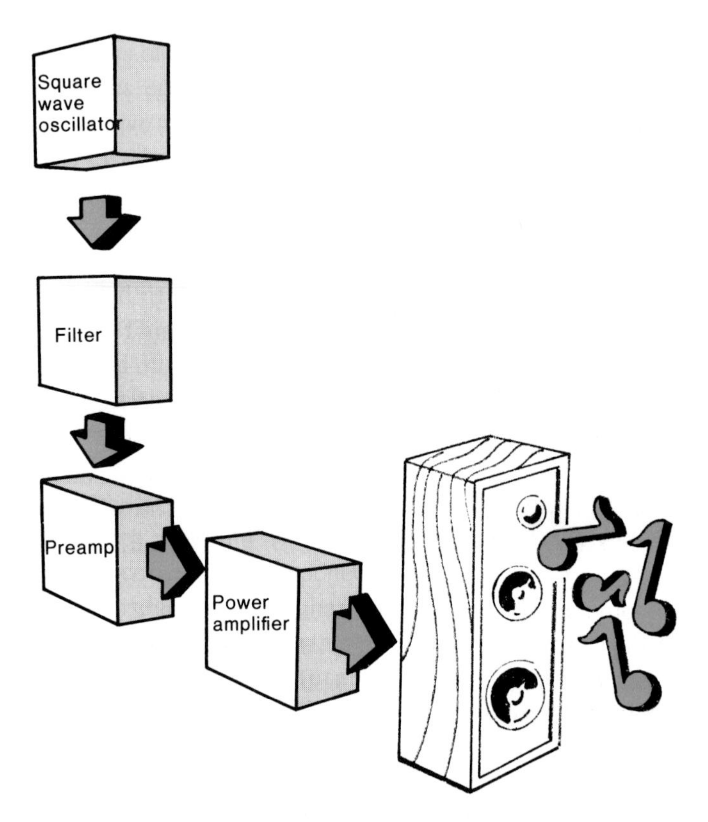

Figure 4.48. A subtractive synthesizer.

Actual synthesizers which are heard on records or in live performances are likely to employ both additive and subtractive synthesis. They may also use these methods together with an electrical wave generated from the sound of any traditional instrument. In fact, it is a rare musical performance today in which the sound has not been electronically adjusted in some fashion.

Summary

All musical instruments generate their sounds by the production of standing waves. Instruments are combined into groups according to the type of media in which the standing waves are established. The three main groups are the string instruments, wind instruments, and the percussion instruments. In general, when an instrument plays a note, the fundamental frequency and various higher harmonics are generated simultaneously. Different instruments, however, which are playing the same note produce different relative amplitudes of the higher harmonics. It is the quantitative differences in these amplitudes that allow the listener to distinguish which type of instrument is playing the note.

The natural frequencies of a string instrument can be changed by altering the length of the string, changing the string's tension, or using a string with a different mass density. The natural frequencies of a wind instrument are changed primarily by altering the length of an air column.

Synthesizers allow the user to create artificial notes by electronically altering the relative amplitudes of the harmonics present in the note. The two main types of electronic synthesis are called additive and subtractive.

4.10 Fourier Analysis

Throughout the previous section we have been discussing complex sound waves and how they may be created from the sum of individual harmonics according to a bar graph recipe. By now you are probably wondering how the bar graphs originate. Certainly no one can tell exactly what bar graph corresponds to the pattern of pressure fluctuations coming from a flute, for example, merely by listening to the sound. Even if someone were to show you the complex total pattern for the flute from Figure 4.38, it is not obvious how to calculate backward and deduce the bar graph according to which the pattern was created. Fortunately, however, in 1822 Joseph Fourier (pronounced: Four-Yea) discovered an incredible method by which a complex wave pattern can be decomposed into a series of harmonically related sine waves. For example, Fourier's method of analysis as applied to the complex wave pattern in Figure 4.49(A) would tell you that it was made up of the six harmonics in Figure 4.49(B). In addition, it would also tell you the amplitude (indicating loudness) of each harmonic, and thus allow the appropriate bar graph to be constructed.

Fourier's method finds widespread applications in the audio industry, and the analysis may be carried out with the aid of a digital computer or a spectrum analyzer. Figure 4.50 shows an example of how a spectrum analyzer works. The sound from a flute is sent into the spectrum analyzer

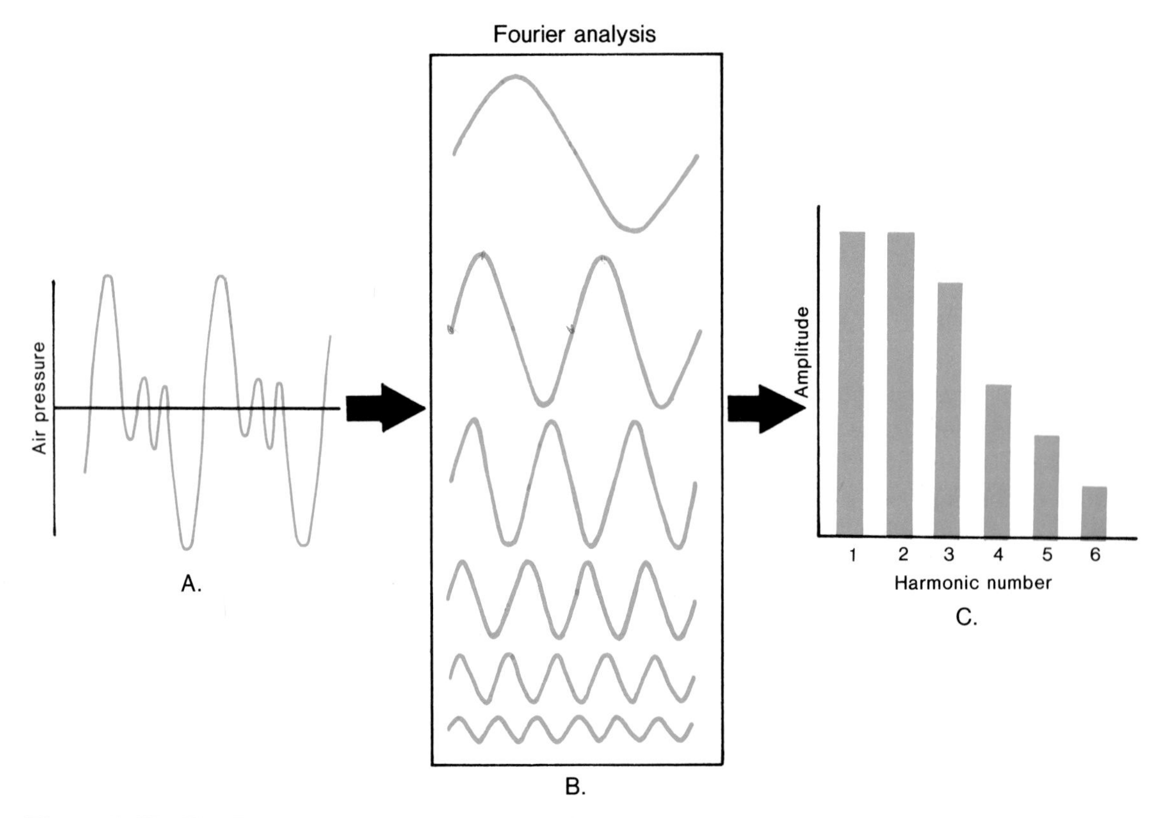

Figure 4.49. The Fourier analysis of the complex wave pattern in (A) shows that it is created from the six harmonics in (B), leading to the bar graph in (C).

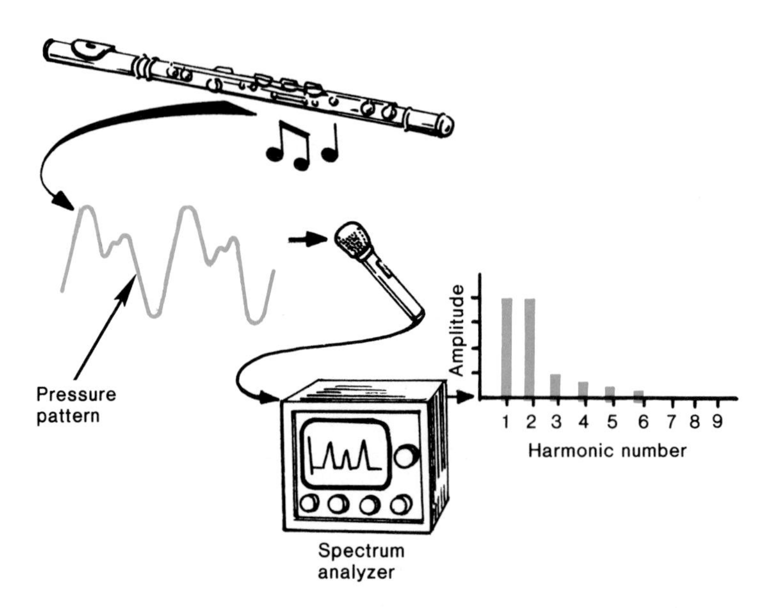

Figure 4.50. A spectrum analyzer receives a complex wave and determines the amplitude and frequency of each harmonic present in the complex wave.

via a microphone. The output display of the analyzer is the familiar bar graph which plots the amplitude versus frequency for each harmonic present in the input waveform. With such a device the harmonic content of the flute can be determined instantly. We shall see in our study of audio components such as amplifiers, tuners, cartridges, etc., that Fourier's method plays a central role in characterizing an important type of audio distortion, called total harmonic distortion, which is produced by each of these components.

Summary

A complex pressure pattern, generated, for example, by a musical instrument playing a note, can be decomposed into a sum of harmonically-related sine waves by Fourier's method. Fourier's method allows one to determine which harmonics are present in the complex wave and the amplitude of each harmonic. Fourier's method may be performed with either an electrical instrument called a spectrum analyzer or a computer.

Summary of Terms

Antinodes—Regions along a standing wave where the medium has its maximum amplitude of vibration.

Beats—A throbbing sound which is heard when the sound alternately becomes loud, soft, loud, soft, etc. Beats are caused by the interference of two overlapping sound waves which have slightly different frequencies. The interference between the two waves results in regions of alternate reinforcements and cancellations which give rise to the variations in the sound loudness.

Complex Wave—A general term which is used to designate any type of wave (sound, water, slinky, etc.) which is composed of more than one sine wave. All voices and musical instruments produce complex sound waves.

Constructive Interference—(See **Interference.**)

Destructive Interference —(See **Interference.**)

Diffraction—The bending of waves around objects or corners.

Dispersion—The ability of a speaker to spread sound uniformly over a wide area.

Forced Vibrations—The setting up of vibrations within an object, either at the object's natural frequency of vibration, or at any other frequency, by a vibrating force.

Fourier Analysis (or Fourier's Theorem)—A method, discovered by Joseph Fourier, that will resolve any complex wave into a sum of simple sine waves which are harmonically related to each other.

Fundamental—The lowest natural frequency of a vibrating object. In music it is the lowest frequency present in a note.

Harmonics—Frequencies which are multiples of the fundamental frequency. The first harmonic is also called the fundamental, the second harmonic has twice the fundamental frequency, etc.

In-phase—(See **Phase.**)

Interference—The adding together of two or more waves, according to the principle of linear superposition, to produce regions of reinforcement and regions of cancellation. Constructive interference results when the crests of one wave meet the crests of another wave, and troughs meet troughs, such that reinforcement of the waves occurs. Destructive interference results when the crests of one wave meet the troughs of another, and vice versa, such that cancellation of the waves occurs.

Natural Frequency—The frequency at which an object or system naturally vibrates. The natural frequency is determined by the object's weight, shape, and type of material. Also called the resonant frequency.

Nodes—Points along a standing wave where the medium does not vibrate.

Note—A musical sound wave which consists of a specific fundamental frequency plus any of its harmonics or overtones. The sound wave corresponding to a note is a complex wave.

Out-of-phase—(See **Phase.**)

Overtones—Frequencies which are multiples of the lowest or fundamental vibrating frequency. The first overtone has twice the frequency of the fundamental, the second overtone has three times the fundamental frequency, etc.

Phase—A term which describes the relative position of one wave with respect to another. Two identical waves are said to be "in-phase" when crests meet crests and troughs meet troughs; constructive interference results when two waves are in-phase. Two identical waves are "out-of-phase" when the crests of one wave meet the troughs of another, and vice versa; destructive interference results when two waves are out-of-phase.

Principle of Linear Superposition—A very important and useful law which states that when two or more waves are present in the same region of space at the same time, the resulting wave is simply the sum of the individual waves.

Pure Tone—A sound wave which consists only of a single frequency, with no higher harmonics. A pure tone is not the same as a musical note (see **Note**).

Resonance—The setting up of vibrations in a body at one of its natural frequencies of vibration by a vibrating force, or wave, which has the same frequency.

Resonant Frequency—(See **Natural Frequency.**)

Sine Wave—A general term which is used to designate any type of wave (sound, water, slinky, etc.) which consists only of a single frequency. Sine waves are represented by very smooth-looking curves.

Sound Quality—(See **Timbre.**)

Spectrum Analyzer—An electronic device which will decompose a complex wave into a sum of harmonically-related sine waves. The spectrum analyzer will yield both the amplitudes and the frequencies of the constituent sine waves.

Standing Waves—Stationary wave patterns, consisting of an alternating sequence of nodes and antinodes, formed in a medium when two identical waves pass through the medium in opposite directions.

Timbre (or Sound Quality)—A term which describes the characteristic tonal quality of the sounds emitted by a musical instrument. The timbre of a musical note is governed by the number and relative amplitudes of the harmonics or overtones which are present.

Review Questions

1. What is meant by the principle of linear superposition?
2. After two waves have passed through one another do they still continue to reinforce or cancel each other, or do they retain their individual shapes?
3. Under what conditions are two waves "out-of-phase" and "in-phase"?
4. What can cause a perfectly functioning microphone to produce very little electrical output at certain frequencies?
5. What causes a standing wave pattern to appear "stationary"?
6. What is the difference, in terms of nodes and antinodes, between an open-ended reflection and a closed-ended reflection?
7. Harmonically-related waves are designated by two different names which are given to each frequency. What are the names which are given to the first four frequencies in the series?
8. Explain how the fundamental frequency of a string instrument or a wind instrument (air column) can be changed.
9. What causes beats and give an example of their use?
10. What is the difference between a loudspeaker which has a wide dispersion and one which has a narrow dispersion?
11. What is meant by the "natural frequencies" of an object?
12. What is resonance, and how is it related to the natural frequencies of an object?
13. Explain, in general terms, what factors govern the frequency of sound which is produced by a string and a wind instrument.
14. What is Fourier analysis, and what kinds of information does it convey about the composition of a musical note?

Exercises

Note: The following questions have up to 5 possible answers. Please select the **one** response which best answers the question.

1. The third overtone of a 1000 Hz fundamental frequency is:
   1. 4000 Hz.
   2. 3000 Hz.
   3. 2500 Hz.
   4. 1333 Hz.

2. Two waves which are always in-phase, adding together, cause:
   1. a standing wave.
   2. constructive interference.
   3. destructive interference.
   4. antinodes.
   5. None of the above answers is correct.
3. If the trough of one wave coincides with the trough of another then the result will be:
   1. beats.
   2. destructive interference.
   3. a condensation.
   4. resonance.
   5. constructive interference.
4. If you were to produce a standing wave in air, and then placed your ear at a **pressure node,** you would:
   1. hear a loud sound of a single frequency.
   2. hear nothing.
   3. hear beats.
   4. hear the separate frequencies of the two waves which produced the standing wave.
   5. not be able to hear any loudness difference between the node and an antinode.
5. Two notes whose frequencies are 500 Hz and 496 Hz are sounded together. The beat frequency is:
   1. 996 Hz.
   2. 4 kHz.
   3. 6 Hz.
   4. 4 Hz.
   5. 96 Hz.
6. The method of breaking up any complicated wave into a series of sine waves is called:
   1. timbre.
   2. resonance.
   3. standing wave.
   4. Fourier analysis.
7. Even though your hi-fi system is working perfectly, you could be sitting in front of your speakers and not hear certain tones which have been cut into the record being played. This could happen because of:
   1. destructive interference.
   2. refraction.
   3. constructive interference.
   4. the presence of pressure antinodes.
   5. traveling waves.
8. For a standing wave, the region of maximum vibration is called the:
   1. displacement antinode.
   2. condensation.
   3. displacement node.
   4. amplitude.
   5. pressure antinode.
9. A standing wave results from:
   1. two identical waves traveling together in-phase.
   2. two identical waves traveling in the opposite direction and passing through each other.
   3. two waves whose beat frequency equals 10 Hz.
   4. two waves which pass through each other and have different frequencies.
10. The timbre of a sound wave is determined by:
    1. its fundamental frequency.
    2. its amplitude.
    3. the number of beats that it produces.
    4. how loud the note sounds.
    5. the number and relative amplitudes of its overtones.
11. When two identical waves are out-of-phase with each other, the result will be:
    1. a loud sound.
    2. nodes.
    3. a reinforcement of the two waves.
    4. a cancellation of the two waves.
    5. antinodes.
12. Interference between two waves occurs:
    1. only with transverse waves.
    2. only with longitudinal waves.
    3. with both transverse and longitudinal waves.
    4. None of the above answers is correct.
13. The distance between two successive antinodes of a standing wave on a string is equal to:
    1. two wavelengths.
    2. one wavelength.
    3. one-half of a wavelength.
    4. one-fourth of a wavelength.
14. Why does a middle C played on a piano and on a guitar sound different?
    1. Because they will produce different beat frequencies.
    2. Because the periods of the overtones may be different.
    3. Because the fundamental frequencies may be different.
    4. Because they both have exactly the same number, and relative amplitudes, of the overtones.
    5. Because the number and relative amplitudes of the overtones may be different.
15. The second overtone of 400 Hz is:
    1. 200 Hz.
    2. 400 Hz.
    3. 800 Hz.
    4. 1200 Hz.
    5. 1600 Hz.

16. Sound of the same frequency is coming from two identical sound sources. If you are sitting at a place where you hear no sound, then the distance between you and each source differs by:
    1. 1/2 of a wavelength.
    2. 1 wavelength.
    3. 2 wavelengths.
    4. There is no difference in the distances.
    5. 1/4 of a wavelength.
17. You play each of three musical instruments separately, sounding the same note on each. The fundamental frequency of the note is 440 Hz. You Fourier analyze the sound from each instrument and get the results in the table below for the percentage of sound at various overtone frequencies. What can you conclude about the instruments? Assume your playing technique is perfect for each one!

Instrument	440 Hz	880 Hz	1,320 Hz
#1	50%	0%	50%
#2	60%	30%	10%
#3	50%	50%	0%

    1. #1 and #2 are identical instruments, but #3 is different.
    2. All three are different instruments.
    3. #1 and #3 are identical instruments, but #2 is different.
    4. All three are identical instruments.
    5. #2 and #3 are identical instruments, but #1 is different.
18. "Beats" are an example of:
    1. resonance.
    2. interference.
    3. natural frequency.
    4. standing waves.
    5. reverberation.
19. Two identical sound sources differ in distance from a listener by one wavelength. The result will be:
    1. no sound at the listener.
    2. sound which is twice as loud as one source.
    3. constructive interference at the listener resulting in a louder sound.
    4. destructive interference at the listener resulting in a softer sound.
20. Do all sounds that have the same fundamental frequency and the same number of overtones sound the same?
    1. No, because the amplitude of the overtones may be different.
    2. No, because the periods of the overtones may be different.
    3. Yes.
    4. No, because the frequencies of the overtones may be different.
    5. No, because the wavelengths of the overtones may be different.
21. When breaking the glass beaker with sound, the frequency of the sound must be:
    1. greater than the natural frequency of the glass.
    2. less than the natural frequency of the glass.
    3. the same as the natural frequency of the glass.
    4. a note that contains all frequencies above the beaker's natural frequency of vibration.
22. The experiment in which a glass beaker is broken with sound waves is a fine example of:
    1. beats.
    2. reverberation.
    3. the Doppler effect.
    4. refraction.
    5. resonance.
23. Striking a tuning fork and then holding the vibrating end of it on a table causes the table to vibrate at the same frequency as the tuning fork. This is an example of:
    1. beats.
    2. the Doppler effect.
    3. forced vibrations.
    4. constructive interference.
    5. diffraction.
24. The natural frequency of an object is determined by:
    1. its size.
    2. its shape.
    3. the material from which it is made.
    4. All of the above answers are correct.
25. The collapse of the Tacoma Narrows Bridge is an example of:
    1. destructive interference.
    2. resonance.
    3. beats.
    4. constructive interference.
    5. None of the above answers is correct.
26. What was the critical factor which caused the Tacoma Narrows Bridge to collapse?
    1. It was a wooden, rather than steel, bridge.
    2. There was too much traffic on it.
    3. The speed of the wind was too high.
    4. The wind was oscillating back and forth in synchronism with the bridge's natural frequency.
    5. Ice had frozen on the suspension wires.

Chapter 5

DECIBELS AND YOU

5.1 Introduction

One of the most important concepts encountered in audio is the notion of a *decibel* (abbreviated as *dB*). Examine the specification sheet for any piece of equipment, be it a turntable, receiver, tuner, cartridge, speaker, or tape deck. You will find that the decibel concept is involved in an extensive way in the technical parameters which describe these devices. Although "decibel" is a widely used term in hi-fi, few people really understand its meaning or the reasons for its utility in the field of communications. In this chapter, then, we address ourselves to the fundamental questions: what is a decibel, and why is it so important in hi-fidelity? In answering these questions we will examine a variety of illustrative examples involving decibels in hi-fidelity.

5.2 How Decibels Are Used in Hi-fi

On an almost daily basis we find ourselves comparing various items by price, weight, size, color, or what-have-you. In the hi-fi world comparisons also play an important role, although it is usually either two power levels or two sound intensity levels which are being compared. A decibel is the unit in which the comparison between the two levels is expressed. In order to illustrate how decibels are used in hi-fi, we shall now consider four important examples.

Example #1—Comparing the Power Ratings of Two Amplifiers

Suppose that there are two amplifiers on the shelf in a store. Amplifier A has a power rating of 40 watts, while amplifier B is rated at 10 watts. The simplest method of comparing the two amplifiers is to say that A is 4 times more powerful than B. Alternatively, it can be said that A is 6 decibels (or 6 dB) more powerful than B. The dB method of comparison is obtained with the aid of Table 5.1 and the following two rules:

a. Divide the larger power by the smaller power (40 watts/10 watts = 4). This ratio is called the *power ratio*, and it is given in the left hand column of Table 5.1.

b. Having found the power ratio, the corresponding decibel comparison may be found in the right hand column of Table 5.1.* In this example a power ratio of 4 corresponds to 6 dB.

Power Ratio or Intensity Ratio	Difference in Decibels
1.00	0 dB
1.26	1 dB
1.58	2 dB
2.00	3 dB
2.50	4 dB
3.16	5 dB
4.00	6 dB
5.00	7 dB
6.30	8 dB
7.90	9 dB
10.0	10 dB
20.0	13 dB
31.6	15 dB
50.0	17 dB
100 = 10^2	20 dB
1,000 = 10^3	30 dB
10^4	40 dB
10^5	50 dB
10^6 = 1 million	60 dB
10^7	70 dB
10^8	80 dB
10^9 = 1 billion	90 dB
10^{10}	100 dB

Table 5.1. The correlation between the ratio of two powers, or two sound intensities, and the corresponding difference in decibels. See the text for the two rules on how to use this table.

*For those readers who wish to use a pocket calculator instead of Table 5.1, the relationship between the power ratio and dB's is given by

$$dB = 10 \log (P_2/P_1),$$

where (P_2/P_1) is the ratio of the two powers, and "log" means the logarithm to the base 10.

Example #2—The dBW

There is a growing trend within the audio industry to express the power rating of an amplifier in decibels by comparing its output power to 1 watt. For example, a 31.6 watt amplifier is 31.6 times more powerful than 1 watt, and a power ratio of 31.6 corresponds to 15 dB, according to Table 5.1. Therefore, the amplifier's power rating can be expressed as 15 dBW, where the "W" is included as a reminder that the amplifier is 15 decibels more powerful than 1 watt. In a similar manner, a 5 watt amplifier can be rated as 7 dBW, because it is 7 decibels stronger than 1 watt. Since all amplifier ratings which use the dBW concept are being compared to 1 watt, the 1 watt level is often called the *reference level*. It is worthwhile to note that an amplifier which has a power rating of 1 watt can also be rated at 0 dBW. This arises because, according to Table 5.1, a power ratio of 1 corresponds to 0 decibels. Notice that a power rating of 0 dBW does not mean that the amplifier delivers zero power!

Example #3—Comparing Two Sound Powers

Suppose that a hi-fi system has been turned on, and that 0.005 watts of sound power are leaving the speakers and entering the room. The VOLUME control on the amplifier is then turned up, so that 0.500 watts of sound power now enter the room. To express the new sound power relative to the old power, in terms of decibels, follow the two rules mentioned in Example #1:

a. The power ratio is 100 (0.500 watts/0.005 watts = 100).
b. A power ratio of 100 corresponds to 20 dB, according to Table 5.1.

Therefore, the 0.500 watt sound power is 20 dB greater than the 0.005 watt level.

Example #4—Comparing Two Sound Intensities

Two different sound intensities can also be compared using the decibel concept in exactly the same manner as the previous examples. The only difference is that the phrase "power ratio" must be replaced by "intensity ratio." Consider the situation where the sound intensity within one room is 15×10^{-12} watts/m^2, while in another room it is 30×10^{-12} watts/m^2.* Since the sound intensity ratio is 2 ($30 \times 10^{-12}/15 \times 10^{-12} = 2$), we see by using Table 5.1 that the sound intensity in the louder room is 3 dB greater than in the quieter room.

*The sound intensities given here, as small as they may seem, are rather typical. Such small intensities can be heard because of the remarkable sensitivity of the ear.

Negative Decibel Numbers

The use of negative decibels (e.g., -3 dB) also occurs in the hi-fi literature, and it signifies that one power level (or intensity) is **less** than that of another. In Example #1, for instance, it is correct to say that the power of amplifier B, relative to amplifier A, is -6 dB. Likewise, in Example #4 the sound intensity in the quiet room, relative to the louder room, may be written as -3 dB.

Summary

It is possible to compare two power levels, or two intensities, using the concept of decibels. To calculate the decibel comparison first compute the ratio of the two powers (or intensities). Then, using Table 5.1, look up the corresponding decibel number. Two power levels which are identically equal have a 0 dB comparison rating. The power of an amplifier is often rated in dBW, e.g., 20 dBW. The "W" in the dBW abbreviation implies that the power rating of this amplifier is 20 dB greater than a 1-watt reference power. Negative decibel numbers are often used to indicate that one power level (or sound intensity) is less than another level.

A.

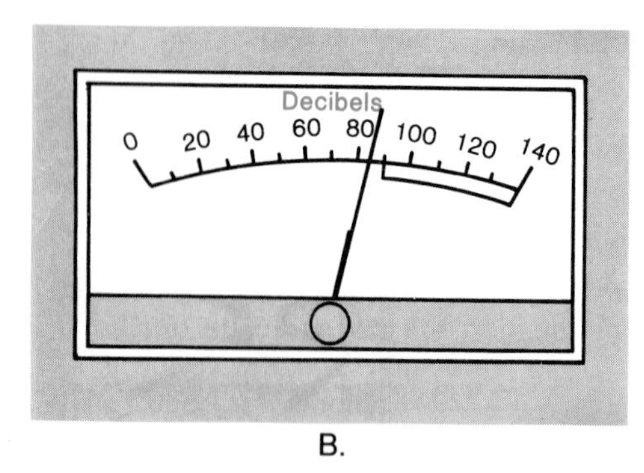

B.

Figure 5.1. (A) A sound level meter and (B) a close-up of its measurement scale. Notice that the scale is calibrated in units called "decibels."

5.3 The Sound Level Meter

One of the most conveniently controllable features of sound is its intensity. It is relatively easy to raise or lower our voices, and the sound intensity emanating from speakers can be changed by a simple turn of the VOLUME control on the preamp. In audio it is important to measure sound intensity in a precise manner, and the *sound level meter*, or *dB meter*, is an electronic unit which has been designed to perform just such a measurement. Figure 5.1 shows one of these meters, along with its measurement scale which is calibrated from 0 to 140 decibels. Recall from the discussion in section 5.2, that decibels provide a way of comparing two sound intensity levels (or two powers). One of the two levels can be regarded as a reference level, and we saw that it was possible to express the power rating of an amplifier with respect to a 1 watt reference power using dBW's. In a similar fashion the sound level meter compares the measured sound intensity with respect to a reference intensity of 1×10^{-12} watts/m^2. In other words, the sound level meter measures the sound intensity which exists at a given location, divides it by the reference intensity of 1×10^{-12} watts/m^2, and electronically converts the ratio into decibel readings according to Table 5.1.

The reference intensity of 1×10^{-12} watts/m^2 is an incredibly small number, and it is chosen because it corresponds to the smallest sound intensity that the average person can hear when the frequency is 1,000 Hz. For this reason, the reference intensity is often called the *threshold of hearing*, and it is indicated as 0 dB on the sound level meter. Therefore, a reading of 90 dB means that the sound intensity is 90 decibels greater than the threshold of hearing level (0 dB). In order to obtain a better feeling for sound levels experienced in everyday life, it is instructive to carry the sound level meter around to various locations and record the sound levels.

Let's start out at a front row seat at a live rock concert. The amplified music in this ear-splitting situation would quickly send the needle to a near-maximum level of 120 dB. The level is so loud, in fact, that it approaches the 130 dB level where sound begins to cause physical pain, as shown in Table 5.2. As you drive home in city traffic, the sound level meter would show a more respectable 90 dB inside your car. And an intimate conversation between you and your date in the quiet of your apartment would register about 65 dB. As the sounds become quieter and quieter, there will come a point when they are inaudible. This point, as mentioned earlier, is called the threshold of hearing and is indicated as 0 dB on the meter. By definition, any increase in loudness above the threshold of hearing would register as audible sound. Remember, that even when the meter reads 0 dB there are still sound waves moving about the room. They are so weak, however, that the average person can not hear them! The decibel scale on the sound level meter is arranged so that typical sounds register somewhere between 0 dB and 140 dB.

	Sound Intensity (dB)
Threshold of hearing	0
Rustling leaves	10
Talking (3 feet)	65
Hair dryer	80
Inside car in city traffic	90
Car without muffler	100
Live rock concert	120
Threshold of pain	130

Table 5.2 The sound intensity levels of various situations, as expressed in decibels.

Summary

The sound level meter, or dB meter, can be used to measure the sound intensity at various locations. The meter always compares the measured intensity to a reference intensity of 1×10^{-12} watts/m^2, and expresses the comparison in decibels. The sound level meter places typical audible sounds on a scale between 0 dB and 140 dB, with

60–80 dB being considered the level for average situations. Anything above 100 dB is quite loud with the threshold of pain starting at 130 dB.

5.4 Decibel Changes Vs. Loudness Changes

It was mentioned in section 3.6, and in subsequent discussions, that the strength of a sound wave can be measured by its intensity. When the wave reaches the listener's ear it is interpreted by the brain as either a loud or soft sound, depending on the wave's intensity; large sound intensities give rise to loud sounds. However, the relationship between intensity changes and the accompanying loudness changes is not a simple proportionality. For example, a doubling of the sound intensity does *not* cause the perceived loudness to double! We shall now discuss the fascinating correlations between intensity changes and loudness changes, and the consequences of these correlations for hi-fi.

A 1 dB Change in the Sound Intensity Barely Changes the Loudness Level

Suppose that you are sitting in front of a hi-fi system which is producing a sound intensity of 90 dB. If the VOLUME control on the amplifier is turned up slightly to produce a 91 dB level, would the accompanying change in loudness be obvious to you? Just barely!!!

Hearing tests with large groups of people have revealed that a one decibel (1 dB) change in sound intensity is approximately the smallest loudness change, or "step," that the average listener can detect.

Each note is 1 dB louder than the preceding one. Therefore, each note is only slightly louder than the preceding one.

Any change smaller than 1 dB is simply not noticed by the average person. This means, for example, that a +4 dB change results in the sound loudness being increased by 4 such audible steps. Similarly, turning down the VOLUME control reduces the loudness, and the dB change is denoted by the use of a minus sign. A reduction from 90 dB to 82 dB is a −8 dB change, and signifies that the loudness has been reduced by 8 audible steps.

A +10 dB Increase in the Sound Intensity Doubles the Sound Loudness

The notion of a decibel as the smallest audible change in loudness is a useful one. The typical user, however, does not count audible steps when changing the sound level of a hi-fi system. The listener simply turns the VOLUME control, and judges when things are loud or soft enough. What do decibels mean, then, to this type of listener? For example, suppose that a record is being played at a sound intensity corresponding to 90 dB, and suddenly the VOLUME control is turned up to produce a level of 100 dB. How does the listener perceive these 10 extra audible steps of loudness? Can we say, perhaps, that the 100 dB level is twice as loud as the 90 dB sound? Indeed we can!

It has been found experimentally that if the sound intensity increases by 10 dB, then the new sound level will appear to be approximately twice as loud.

Notice that it is only the difference in dB which determines relative loudness levels. A 70 dB sound will appear twice as loud as a 60 dB sound; a 40 dB sound will appear twice as loud as a 30 dB sound; and an 80 dB sound will appear twice as soft as a 90 dB sound. Each of these situations involves a difference of 10 decibels and, hence, a factor of two in the perceived loudness.

Note: The 1 dB and 10 dB rules for loudness changes are approximately correct for sound frequencies near 1,000 Hz. The rules deviate slightly for frequencies other than 1,000 Hz.

What about sounds which are 20 dB apart? How many times louder is an 80 dB sound compared to a 60 dB sound? A little inductive reasoning will provide the answer. Going from +60 dB to +70 dB doubles the loudness, while going from +70 dB to +80 dB increases the loudness again by a factor of two. Therefore, a +80 dB level is four times louder (2 × 2) than a +60 dB sound. In this way it is possible to correlate large changes in decibel levels with perceived changes in loudness.

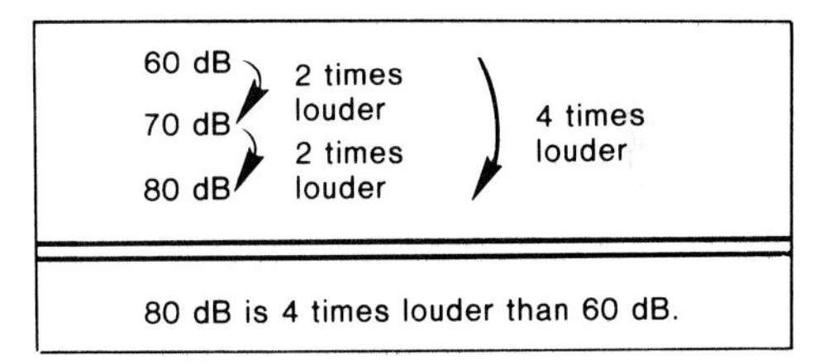

80 dB is 4 times louder than 60 dB.

Summary

1. The smallest change in loudness that can be perceived by the average person is approximately 1 dB.
2. A sound level which is 10 dB greater or smaller than another appears to be twice as loud, or twice as soft, respectively.

5.5 Examples of Decibel Vs. Loudness Changes in Hi-fi

Consider two hi-fi systems, as shown in Figure 5.2, which are identical in every respect except that amplifier "B" delivers twice as much electrical power as amplifier "A." Place a sound level meter at the same distance in front of each of the speakers and then position yourself so you are also equidistant from both speakers. Now ask the following two questions:

1. How much louder will "B" sound compared to "A," and, in particular, will "B" sound twice as loud to you?
2. What will the sound level meters indicate?

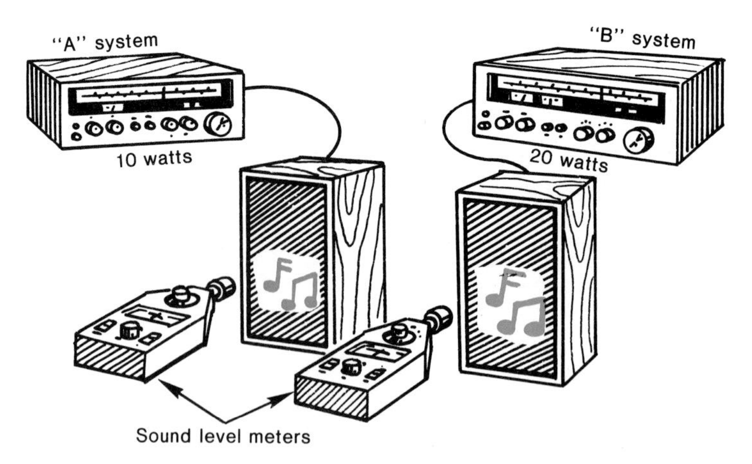

Figure 5.2. Even though the "B" system is delivering twice as much sound power as the "A" system, "B" does not sound twice as loud as "A", but rather sounds only slightly louder (3 dB).

The answer to the first question is quite surprising. "B" will *not* sound twice as loud—not by a long shot! It will, amazingly, sound only about 3 "audible steps" (or +3 dB) louder. This result can be obtained by consulting Table 5.1; a power ratio of 2 (e.g., 20 watts/10 watts = 2) gives rise to a 3 dB difference between the two sound levels.

Since +1 dB is the smallest increment in loudness which the average person can detect, a +3 dB increase is still a very small change in loudness. In other words, the ear has the remarkable property that if the sound intensity reaching it is doubled, the brain preceives only a slight (3 audible steps) increase in loudness. This fact about the human ear is completely unknown to the average person, and even to most audiophiles. It is so startling that most people have a hard time believing it.

Now, what about the sound level meters? You would discover, of course, that the difference in the readings is 3 dB. If the "A" meter registered a level of 80 dB, then the "B" meter would read 83 dB. The actual level indicated on each meter is not the important feature here, because this can be changed by moving the meters either closer to, or farther from, the speakers. The main point is that the difference between the two will always be 3 dB (80 dB vs. 83 dB, 91 dB vs. 94 dB, etc.).

Think about what all of this means when you are purchasing an amplifier. Suppose that you decide to audition two amplifiers which are operating at their rated output powers of 12 watts and 240 watts, as indicated in Figure 5.3. The 240 watt amp is 20 times more powerful, and it certainly will cost considerably more. But will it sound 20 times louder to the listener? No, and you can use a sound level meter to prove it! First measure the sound intensity coming from the speaker driven by the 12 watt amplifier; suppose this reading turns out to be 75 dB. Turn off the 12

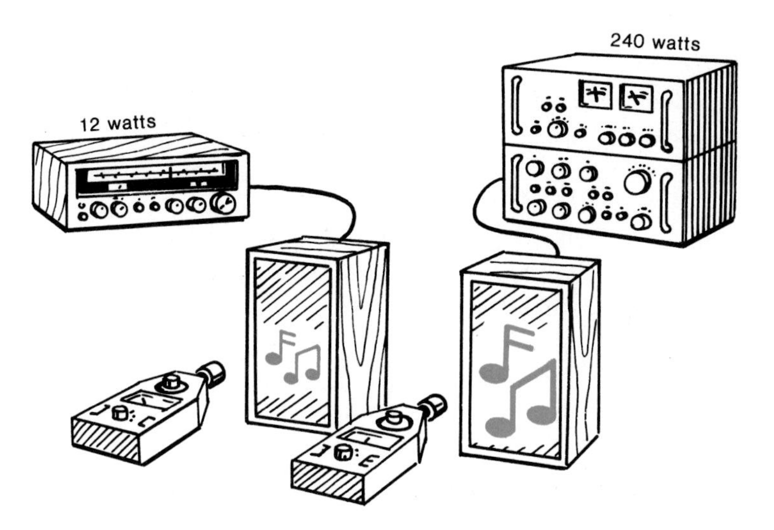

Figure 5.3. For all its power and greater price tag, the hefty 240 watt amplifier sounds only slightly greater than twice as loud (13 dB) as the 12 watt amplifier.

watt amp and switch on the 240 watt amp. The sound level meter will jump to about 88 dB, for a change of roughly +13 dB (240 watt/12 watt = 20. See Table 5.1.). Recall, from section 5.4, that a +10 dB increase is perceived as twice the loudness by a listener. Therefore, the following conclusion is reached; even though a 240 watt amplifier produces 20 times more electrical power than a 12 watt amp, the more powerful amplifier will sound just *slightly more than twice as loud.* Amazing!

From the preceding discussion it should be obvious that one does *not* buy larger amplifiers merely to obtain significantly louder sounds. If you are dissatisfied with your present amplifier because it cannot generate loud enough sounds, then buying an amplifier twice as powerful (at perhaps twice the cost) is not going to help. Your new amplifier will produce sounds which are a meager 3 dB louder. You would need an amplifier at least 10 times more powerful to produce sounds that appear twice as loud, and such a purchase simply may not be practical because of the prohibitive cost. A potentially less expensive solution to the loudness problem is to obtain a more efficient pair of loudspeakers (see Chapter 6). This is not to say that more powerful amplifiers are not desirable. They certainly are, but for reasons other than loudness, as will be discussed in Chapter 8.

Summary

Doubling the power of an amplifier does not double the sound loudness; the loudness increases by the slight amount of only 3 decibel "steps." It requires an amplifier which is at least 10 times more powerful in order to produce a sound which appears twice as loud.

5.6 The Fletcher-Munson Curves

Needless to say, the ear is an incredible device with unique characteristics which play an important role in audio-system design. Correspondingly, the consumer should be aware of the implications that these design features have for the listening experience. Perhaps the most important characteristic is shown by the Fletcher-Munson curve which demonstrates that the threshold of hearing (TOH) is strongly frequency dependent, as shown in Figure 5.4. The horizontal axis represents sound frequencies within the audible range, and the vertical axis portrays the sound intensity as registered in dB by a sound level meter. The TOH level at 1 kHz is arbitrarily taken to be the 0 dB reference level. Remember that the threshold of hearing represents the intensity level at which sound just becomes audible, and Figure 5.4 shows that the TOH is strongly dependent upon the sound frequency. Now consider a tone whose frequency is 200 Hz. According to Figure 5.4 the TOH curve has risen by approximately +30 dB above the 0 dB reference level. On a sound level meter such a tone must register a full 30 dB more than a 1 kHz tone just to be at the threshold of hearing! (Remember, that each 10 dB change means an increase in loudness by a factor of 2.) In other words, the ear becomes more *insensitive* to sounds below 1 kHz, and more power is required to make them audible.

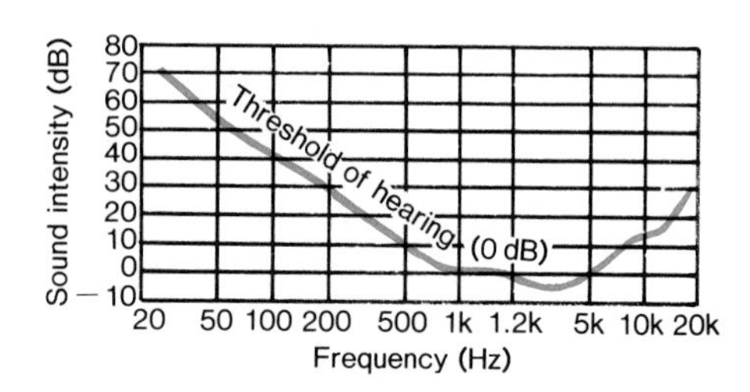

Figure 5.4. The threshold of hearing strongly depends on the frequency. The ear becomes less sensitive at the low and high frequency ends of the audio range, and is most sensitive between 1 kHz and 5 kHz.

Question: How much more sound intensity is a hi-fi system required to generate in order to reproduce a 70 Hz note, compared to 1 kHz, at the TOH?

Solution: Using Figure 5.4, it can be determined that a 70 Hz tone must be 50 dB louder than a 1 kHz tone in order to be at the TOH. Table 5.1 can be used to translate 50 dB into an intensity ratio of 100,000. Therefore, an audio system must generate 100,000 times more intensity at 70 Hz compared to 1 kHz if both tones are to be at the threshold of hearing. That's incredible—to say the least! But, amplifiers and speakers do it all the time!

The loss of the ear's sensitivity for frequencies below 1 kHz has profound consequences for a hi-fi speaker system which is attempting to reproduce the deep bass notes. The speakers are being called upon to generate substantial amounts of sound in order to overcome the ear's severe loss of sensitivity. As shown in the previous example, a bass speaker generating audible low frequencies at the TOH level must produce thousands of times more sound intensity than a comparable speaker producing midrange notes around 1 kHz at the TOH levels. Although this requirement, as we shall presently see, is not nearly so severe for sound levels above the TOH level, the basic idea is the same, and consequently bass speakers must be relatively large and rugged in construction. In general the problem of low frequency sound reproduction is of paramount importance in speaker design (see Chapter 6).

Figure 5.4 also shows that the ear becomes insensitive to high frequency tones above 5 kHz, although the loss of sensitivity is less serious than that occurring at the low frequencies. It is in the middle range from 1 kHz to 5 kHz where the ear is most sensitive to sounds.

Figure 5.5 shows a more complete representation of the Fletcher-Munson data, where a family of curves is drawn, one for each of the various loudness levels from 0 dB all the way up to + 120 dB. As we shall soon discuss, these curves are responsible for the presence of the so-called LOUDNESS switch on most amplifiers. Each curve in Figure 5.5 represents a line of constant loudness. For example, the lowest trace is the TOH shown separately in Figure 5.4, and it tells what intensity is required to keep the sound at a constant loudness corresponding to the threshold of hearing. The second curve from the bottom, the 10 dB curve, was determined by first increasing the sound level at 1 kHz by +10 dB. As the frequency of the sound was changed away from 1 kHz, a group of listeners was asked to adjust the amplifier power until the loudness at the new frequency appeared to match the 10 dB level at 1 kHz. The results indicate that, as in the TOH case, when the frequency is lowered more power must be delivered to the speakers just to keep the perceived loudness the same.

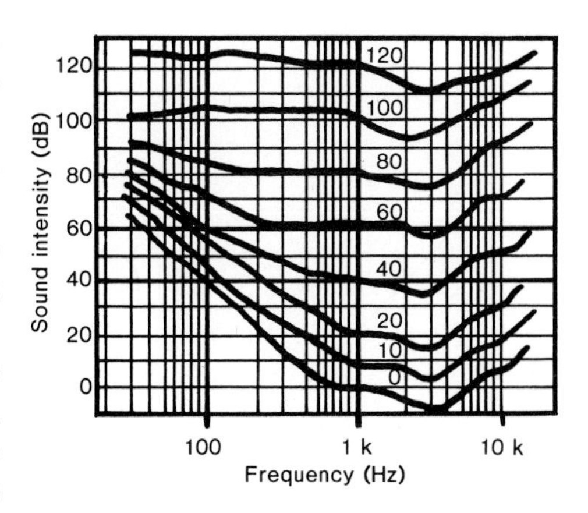

Figure 5.5. The Fletcher-Munson constant loudness contours. Each curve represents the amount of sound intensity required to keep the sound loudness constant as the frequency is changed. The number which labels each curve indicates the loudness level in decibels above the threshold of hearing.

The constant loudness curves of Figure 5.5 are plotted for every 20 dB change and continue all the way up to 120 dB. What is intriguing about these curves from a hi-fi point of view is that their shapes change as the sound level increases from 0 dB to 120 dB. The slopes of the curves below 1 kHz become "flatter" and, for sounds above 100 dB, the curves are nearly horizontal lines. What does this mean to audio reproduction? Plenty! When you are playing the hi-fi at loud levels, say 100 dB or higher, for example, you will hear plenty of bass in relation to the midrange frequencies because at these levels the ear is equally sensitive to the middle and low frequencies. However, as the VOLUME control is turned down to 40 dB, for example, the listener notices that the bass notes are missing. What really happens, according to Figure 5.5, is that the ear at the 40 dB level is less sensitive to the bass frequencies, and it simply doesn't hear them at the same relative loudness as the mid-frequencies. This is why many audio enthusiasts prefer to play their favorite music loud—the ear then hears those apparently "lost bass notes." Figure 5.5 shows that there is also some relative loss of the treble notes at low-sound levels, although this loss is not nearly as severe as that encountered at the bass frequencies.

Notice that the *slope* of the Fletcher-Munson curves above 1 kHz does *not change* very much as the sound loudness is increased from 0 dB up to 120 dB. This means that there will be no relative loss of the highs as the sound level is lowered.

The LOUDNESS Switch on an Amplifier

Many people are accustomed to attending live performances where loudness prevails and, consequently, the full range of low, midrange, and high frequencies is heard in equal proportions (see Fig. 5.5, 100 dB curve). To achieve the same relative tonal balance at home the hi-fi user must set the VOLUME control quite high. Because of external constraints (like neighbors!), however, it may not always be possible to turn up the sound level, and the option of listening to low sound levels without the concomitant "loss" of the bass and treble notes would be a desirable feature. This option is found on most good amplifiers, or receivers, and is provided by the LOUDNESS switch (Fig. 5.6). When the switch is in the ON position, electronic circuitry is activated which automatically increases the amount of bass power produced by the amplifier when the VOLUME control is turned down. Some amplifiers will also boost the treble notes relative to the midrange frequencies but, as pointed out earlier, this boost is usually minor compared with the bass boost. In this manner the LOUDNESS feature compensates for the ear's loss of sensitivity at low listening levels by restoring the full tonal balance between the low, midrange, and high frequencies. Remember, the LOUDNESS control does *not* mean that it makes all frequencies louder. It means that the amplifier will boost only the bass frequencies (and perhaps the treble notes) and only when the volume is turned down. For loud levels the LOUDNESS function has little effect, corresponding to the fact that your ears hear all frequencies more or less uniformly under such conditions. With the switch in the OFF position the LOUDNESS option is disabled completely.

Figure 5.6. Most hi-fidelity amplifiers, or receivers, are equipped with a LOUDNESS switch. When set to the ON position, the LOUDNESS switch allows the amplifier to compensate for the ear's reduced sensitivity at the low and high frequencies when the VOLUME control is set for low sound levels.

Summary

The ear is not equally sensitive to all sound frequencies. At low sound levels the Fletcher-Munson curves indicate that the ear is most sensitive to frequencies in the 1 kHz to 5 kHz range. The ear becomes markedly less sensitive for frequencies below 1 kHz and above 5 kHz. At high sound levels (100 dB and greater) the ear becomes, more or less, equally sensitive for all frequencies which lie below 1 kHz. The LOUDNESS control of an amplifier attempts to compensate for the ear's insensitivity to the bass and treble notes at low sound levels. The LOUDNESS control boosts the intensity of the bass frequencies (and, in some cases, the intensity of the high frequencies) when the sound level is low. The amount of boosting becomes less and less as the user turns "up" the VOLUME control.

5.7 The Loss of Hearing with Age

Another problem which may affect your purchase of a hi-fi system is the loss of hearing with age. Figure 5.7 shows how the ear's ability to hear various frequencies changes with both age and sex. In Figure 5.7 (A) it is assumed that a 25-year-old woman has "normal" hearing which is arbitrarily called "0 dB." This level should not be confused with the threshold of hearing which is also arbitrarily called 0 dB. In Figure 5.7 "0 dB" simply means a tone of normal loudness with respect to which all other tones are referenced. Between the ages of 25 and 40 women experience a hearing loss of about 5 dB at all frequencies. At ages exceeding 40 years, the graph reveals a greater hearing loss at 4 kHz than at 500 Hz. In fact, at age 55 the average woman has lost about 15 dB of hearing at 4 kHz and only 8 dB at 500 Hz.

By contrast, the loss of hearing in men, as indicated in Figure 5.7 (B), is more dramatic. At the young age of 25 they begin losing the ability to hear the high frequencies relative to the low frequencies, and the loss is much more severe than with the women. The average male can expect to lose an astonishing 30 dB of hearing at 4 kHz by the time he reaches 58 years of age. This means that a 4 kHz tone will appear only 1/8 as loud as it did in his younger days. (Forgot where this 1/8 comes from? See section 5.4.) On the other hand, men do not lose the notes below 500 Hz as fast as some women. Some consolation!

There is an old adage which seems to be true. By the time you are old enough to afford a high-quality audio system, you can't even hear it!!!

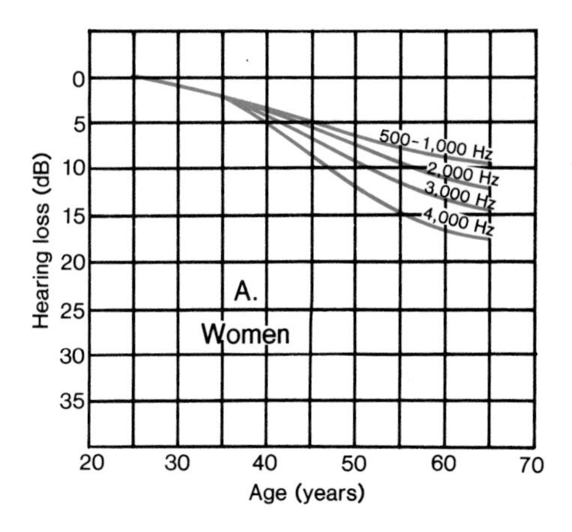

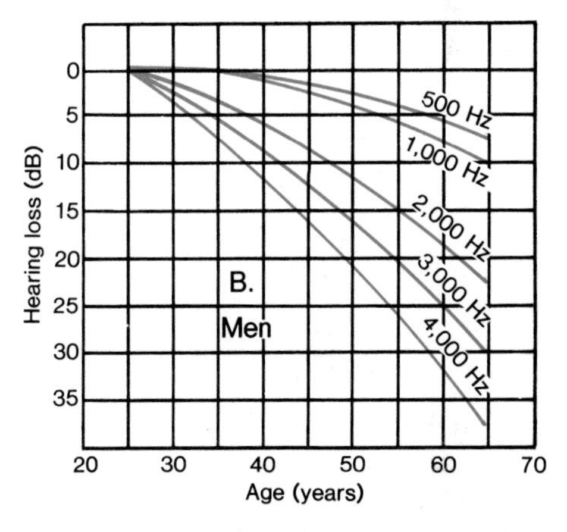

Figure 5.7. Typical hearing losses for women and men as they become older.

5.8 Frequency Response Curves

As a final example, let us consider the application of decibels in characterizing a *frequency response curve*. The concept of frequency response applies to all types of audio equipment, and its pervasiveness mirrors the importance of relative loudness and pitch in music. It is not unusual, for example, that relatively loud, high-pitched sounds from a symphonic string section occur together with softer, rhythmic beats from a kettle drum. Any hi-fi system should be able to reproduce such loudness and pitch relationships accurately. It certainly would be unacceptable if your system generated the sound of a kettle drum with three times its

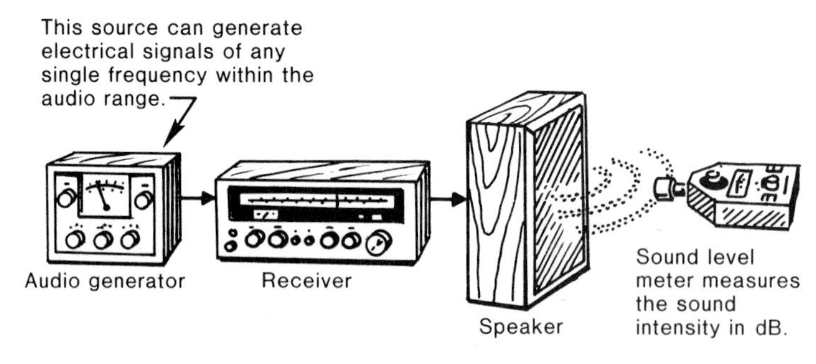

Figure 5.8. An experimental set-up to measure the frequency response curve of a speaker.

original loudness, while reproducing the violins at one-third of their recorded level. Every note should be given its due—no louder or softer relative to other sounds in the music. The ability of each hi-fi component to reproduce impartially all frequencies within the audio range, each according to its recorded loudness, is an absolute necessity for faithful music reproduction, and is described by the frequency response curve.

Speakers are a good example of what a frequency response curve means to the listener. The input to a speaker is, of course, electricity, while the output is sound. As shown in Figure 5.8, the sound intensity emanating from the speaker can be measured in decibels with a sound level meter, while the input is controlled by using an electronic device called an *audio generator*. An audio generator can produce electrical signals of any single frequency in the audio range. The electrical output of the audio generator is sent to the AUX input of the receiver where it acts like any other program source. Obviously, if the audio generator is set to produce electricity with a frequency of 6 kHz, the speaker will produce a 6 kHz sound. By simply turning the dial on the front panel of the generator, the sound frequency can be selected at will. Equally important is the fact that the audio generator can be arranged to supply electricity at *constant amplitude*, regardless of frequency. A *perfect* speaker, driven by such an audio-generator/amplifier combination, would produce the *same sound intensity*, as measured by the sound level meter, at all frequencies. If, for example, the (perfect) speaker produces a sound intensity of 90 dB at 20 Hz, then it also produces 90 dB of intensity at 200 Hz, 2,000 Hz, or any other frequency in the audio range. Figure 5.9 shows a horizontal line that results from connecting the data points and which represents the performance of this perfect speaker. In the hi-fi game this performance graph is called a "flat frequency response" for obvious reasons.

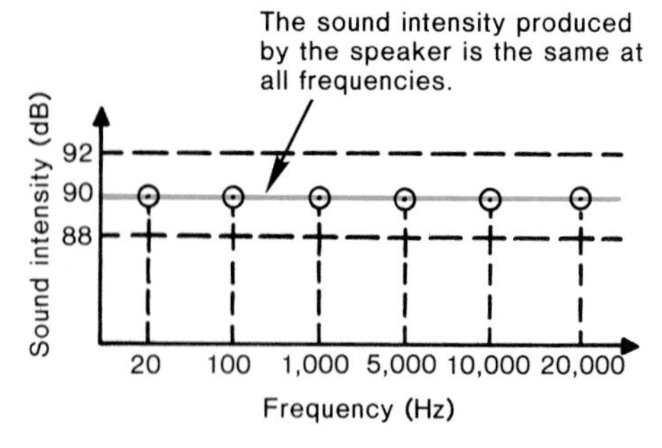

Figure 5.9. An ideal, or perfect, speaker produces a straight-line, or "flat" response, when its output sound intensity is plotted against the sound frequency.

Now for the bad news! No speaker, or for that matter any audio component, can produce a perfect or "flat" frequency response. The output sound intensity of a real speaker would vary above and below the 90 dB level in our example, depending upon the frequency of the electricity being supplied by the audio generator. This is true even though great pains are taken to ensure that the input electricity is kept at a constant amplitude (constant input electrical power) at all frequencies. The inherent imperfections in the speaker design lead to a frequency response curve which typically appears like Figure 5.10. Notice that the sound produced by the speaker at 200 Hz is measured at 92 dB, which is +2 dB above the level at 1 kHz. At 10 kHz the speaker produces slightly less intensity, 89 dB, and so it goes throughout the entire audio range of frequencies.

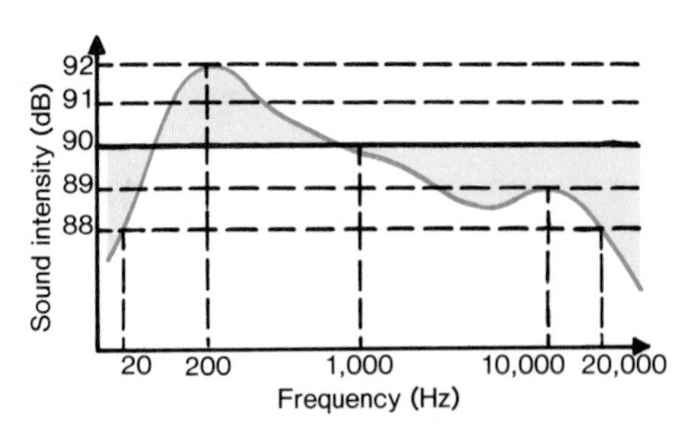

Figure 5.10. The frequency response curve for a real speaker of excellent quality.

The data portrayed in Figure 5.10 are very often written as a technical specification which would read as: Frequency response, 20 Hz–20 kHz, +2 dB, −2 dB. This particular specification tells us that the speaker will produce sound over the entire audio range from 20 Hz to 20 kHz. And, furthermore, the output sound intensity (the "response") will deviate by no more than +2 dB, or less than −2 dB, from the reference level, which, in this case, was arbitrarily taken to be 90 dB at 1 kHz. The frequency response specification, while very helpful, conveys less information than the frequency response curve depicted in Figure 5.10. The specification 20 Hz–20 kHz, +2 dB, −2 dB tells us only that the output sound intensity can deviate by as much as ±2 dB from a reference level, but the specification does *not* tell us the precise frequencies at which the deviations occur. Figure 5.10, being more detailed, shows that the +2 dB deviation occurs at 200 Hz while the −2 dB points are located at both 20 Hz and 20 kHz. Whenever possible, the prospective buyer should always look for the frequency response *curve*, like that shown in Figure 5.10, when considering the purchase of an audio component.

If the frequency response of all audio components deviates from perfect "flatness," how much deviation, then, can be tolerated in hi-quality sound reproduction? What is tolerable depends on the component being considered, the specifics for which we will discuss individually in appropriate later chapters. Figure 5.10 would be considered excellent for speakers, cartridges, and tape decks although somewhat inferior for good amplifiers and tuners.

Summary of Terms

Audio Generator—An electronic component, acting as a sound source, which produces a single-frequency electrical signal. The frequency can be selected by the user and it may lie anywhere in the audio range. Audio generators are used for testing audio components.

dB Meter—(see **Sound Level Meter.**)

dBW—A unit for measuring the power of an amplifier, in terms of decibels, relative to a reference level of 1 watt.

Decibel (dB)—A term which compares two powers or two sound intensities. The decibel is closely correlated with the manner in which the ear perceives loudness changes. One decibel is approximately the smallest change in sound loudness which can be heard.

Flat Frequency Response—A phrase used to describe a theoretically perfect audio component which produces a uniform, or constant, output signal strength at any frequency within a specified frequency range.

Fletcher-Munson Curves—A series of graphs which characterize the ear's sensitivity to various frequencies at different loudness levels.

Frequency Response Curve—A graph which shows how uniformly an audio component reproduces either electrical signals or sound over a specified range of audio frequencies.

LOUDNESS Switch—A preamplifier control which adds predetermined amounts of bass and treble boosts to compensate for the ear's loss of sensitivity at low volume listening levels.

Sound Level Meter—A small portable meter which compares sound levels, measured in decibels, relative to some standard level called 0 dB. The scale on a sound level meter usually ranges from 0 dB to 140 dB.

Threshold of Hearing (TOH)—The smallest sound which can just be heard by a normal ear. Often called the 0 dB level. The threshold of hearing strongly depends upon the sound frequency.

Review Questions

1. In terms of decibels, what is the smallest change in loudness which the ear can perceive? How much change, in decibels, is required in order to double the apparent loudness of a sound?
2. According to the threshold of hearing curve, how much more sound intensity is required to hear a 100 Hz tone than is required for a 1,000 Hz tone which are at the TOH? Express your answer in both dB and in an approximate power ratio (see Table 5.1).
3. What is the purpose of the LOUDNESS control on an amplifier, and why is it useful when you are listening to your hi-fi system?
4. What happens to the sensitivity of the ear as the sound intensity increases?
5. What is a frequency response curve and why is it important when characterizing audio components?

Exercises

NOTE: Select the **one** best response which answers the question.

1. Approximately, the smallest noticeable increment in a change of loudness is:
   1. 0 dB.
   2. 3 dB.
   3. 1 dB.
   4. the threshold of hearing.
2. To compensate for the natural hearing losses which occur at high and low frequencies, at low volume levels, you would use:
   1. the SELECTOR switch of the preamp.
   2. the TAPE MONITOR switch of the preamp.
   3. the ON-OFF switch of the power amp.
   4. the LOUDNESS control of the preamp.
   5. the MUTE control.
3. Two rock bands are playing and one band produces 31.6 times more sound intensity than the other because the amplifiers of one band are not working. This corresponds to a difference of:
   1. 30 dB.
   2. 15 dB.
   3. 20 dB.
   4. 40 dB.
   5. 70 dB.
4. To make the loudness of a sound appear twice as soft, you must decrease the sound intensity by:
   1. 3 dB.
   2. 1 dB.
   3. 10 dB.
   4. a factor of 2.
   5. a factor of 100.
5. If two amplifiers are rated at 12 and 120 watts, respectively, how much louder will the larger amp sound?
   1. 1.0 dB.
   2. 2.0 dB.
   3. 4.0 dB.
   4. 10 dB.
   5. 100 dB.
6. If two speakers each produce 7×10^{-10} watts/m^2 of sound intensity, then they differ in loudness by:
   1. 1 dB.
   2. 0 dB.
   3. 3 dB.
   4. 10 dB.
   5. −1 dB.

7. In order to double the loudness of sound coming from a speaker, one would:
   1. increase the sound intensity by a factor of 2.
   2. increase the sound intensity by a factor of 10.
   3. increase the sound intensity by 3 dB.
   4. increase the sound intensity by 20 dB.
   5. take off the speaker grill cloth.
8. Approximately, what is the sound level of two people talking?
   1. 10 dB.
   2. 30 dB.
   3. 60 dB.
   4. 100 dB.
   5. 130 dB.
9. Suppose that you walked into a hi-fi store and listened to two sound systems. The first one is rated at 30 watts and the second one at 60 watts. The second system sounds:
   1. 60 times louder.
   2. 30 times louder.
   3. twice as loud.
   4. only slightly louder.
   5. None of the above answers is correct.
10. What is the important conclusion concerning hearing as people grow older?
    1. Men lose their ability to hear the treble notes faster than women.
    2. Women lose their ability to hear treble notes faster than men.
    3. Men suffer the smallest loss in the bass notes compared with the treble notes.
    4. Women suffer the greatest loss in the bass notes compared with the treble notes.
    5. None of the above answers is correct.
11. The LOUDNESS switch on a preamplifier:
    1. makes all sounds, regardless of their frequency, louder.
    2. increases the loudness of low-frequency sounds at high-volume levels.
    3. increases the loudness of all frequencies at low-volume levels.
    4. increases the loudness of only the low and high frequencies at low-volume levels.
12. In the audible range of frequencies, 20 Hz to 20 kHz, the frequencies difficult to hear are:
    1. those near 1 kHz.
    2. the middle range frequencies.
    3. only the very high range of frequencies.
    4. only the very low range of frequencies.
    5. both the very high and very low range of frequencies.
13. If an amplifier is producing 10 watts of power and suddenly its VOLUME control is turned up so that it is now producing 50 watts, what is the increase in loudness that accompanies this change?
    1. It is 5 times louder.
    2. It is $(50 - 10) = 40$ dB louder.
    3. It is 7 dB louder.
    4. It is 10 dB louder.
    5. It is 3 dB louder.
14. The threshold of pain at 1,000 Hz is ________ above the threshold of hearing.
    1. 0 dB
    2. 10 dB
    3. 80 dB
    4. 130 dB
    5. 180 dB
15. One speaker sounds 32 times louder than another. This corresponds to an intensity difference of:
    1. 20 dB.
    2. 32 dB.
    3. 40 dB.
    4. 50 dB.
    5. 70 dB.
16. If the output power from an amplifier is **increased** from 72 dB to 92 dB, the sound will appear:
    1. twice as loud.
    2. four times louder.
    3. $(92 - 72) = 20$ times louder.
    4. only slightly louder.
    5. None of the above answers is correct.
17. If you were listening to music which had a loudness of +15 dB above the threshold of hearing, then the music would:
    1. be inaudible.
    2. be rather loud but not unbearable.
    3. cause pain to your hearing.
    4. cause deafness.
    5. be audible, but rather quiet.
18. Suppose that you just traded in your 20 watt amplifier for a 50 watt amplifier. Your new sound system will be:
    1. 2.5 times louder.
    2. 2.5 dB louder.
    3. 4 dB louder.
    4. $50 - 20 = 30$ dB louder.
    5. 10 dB louder.

19. Suppose that you just bought a receiver whose power is rated at 8 dBW. The power of this receiver, in watts, is:
    1. 1 watt.
    2. 2.5 watts.
    3. 6.3 watts.
    4. 8 watts.
    5. 10 watts.
20. A 100 watt power amplifier can also be rated as ___________ dBW.
    1. 0
    2. 1
    3. 10
    4. 20
    5. 30
21. About how much hearing loss can a 40 year old woman expect to incur when compared to a 25 year old woman? (Assume the sound frequency is 4 kHz.)
    1. There will be no hearing loss.
    2. The loss will be about 0 dB.
    3. The loss will be about 5 dB.
    4. The loss will be about 12 dB.
    5. The loss will be about 20 dB.
22. About how much hearing loss can a 40 year old man expect to incur when compared to a 25 year old man? (Assume the sound frequency is 4 kHz.)
    1. There will be no hearing loss.
    2. The loss will be about 0 dB.
    3. The loss will be about 5 dB.
    4. The loss will be about 12 dB.
    5. The loss will be about 20 dB.
23. An amplifier is producing 10 watts of power. Suddenly its VOLUME control is turned up so that it's now producing 100 watts. What is the increase in loudness that accompanies this change?
    1. It is 10 times louder.
    2. It is (100 − 10) = 90 dB louder.
    3. It is 7 dB louder.
    4. It is 10 dB louder.
    5. It is 3 dB louder.
24. Three loudspeakers have the following frequency response specifications:

    A. 40 Hz − 15,000 Hz, + 3 dB, − 2 dB
    B. 30 Hz − 16,000 Hz, + 3 dB, − 1 dB
    C. 20 Hz − 20,000 Hz, + 0.5 dB, − 1 dB.

    Please rank the three loudspeakers, from best to worst.
    1. A, B, C
    2. A, C, B
    3. B, A, C
    4. B, C, A
    5. C, B, A
25. A loudspeaker has a frequency response of 20 Hz − 17,000 Hz, + 2 dB, −1 dB. Assuming that the +2 dB rise occurs at 5 kHz, how much more (or less) sound intensity does the loudspeaker deliver at this frequency than it ideally should have? The loudspeaker delivers ______________ sound intensity.
    1. 2 times more
    2. 2 times less
    3. 1.58 times more
    4. 1.58 times less
    5. 4 times more
26. There are two amplifiers, A and B, sitting on a shelf. Amp A is rated at 140 watts and B at 70 watts. Amplifier A is capable of producing sound that is:
    1. slightly louder than B.
    2. slightly quieter than B.
    3. twice as loud as B.
    4. 10 dB louder than B.
    5. 10 times louder than B.

Chapter 6

LOUDSPEAKERS

6.1 Moving toward Perfection

The goal of a perfect hi-fi system is to reproduce the sound exactly as it was recorded. This seems like an obvious statement, and it implies that each link in the chain—the source(s), amplifiers, and speakers—must flawlessly perform its job. There is much truth in the old adage about a chain and its weakest link. It is generally recognized that the most deficient component is the loudspeaker and, to a large extent, significant departures from perfect reproduction can be traced to this component. There is a very simple experiment which dramatically demonstrates this fact; walk into a hi-fi store and ask the salesperson to let you make a speaker comparison test. Any good store will permit you to use a speaker comparator which is an electronic device that allows a variety of speakers to be auditioned, as illustrated in Figure 6.1.

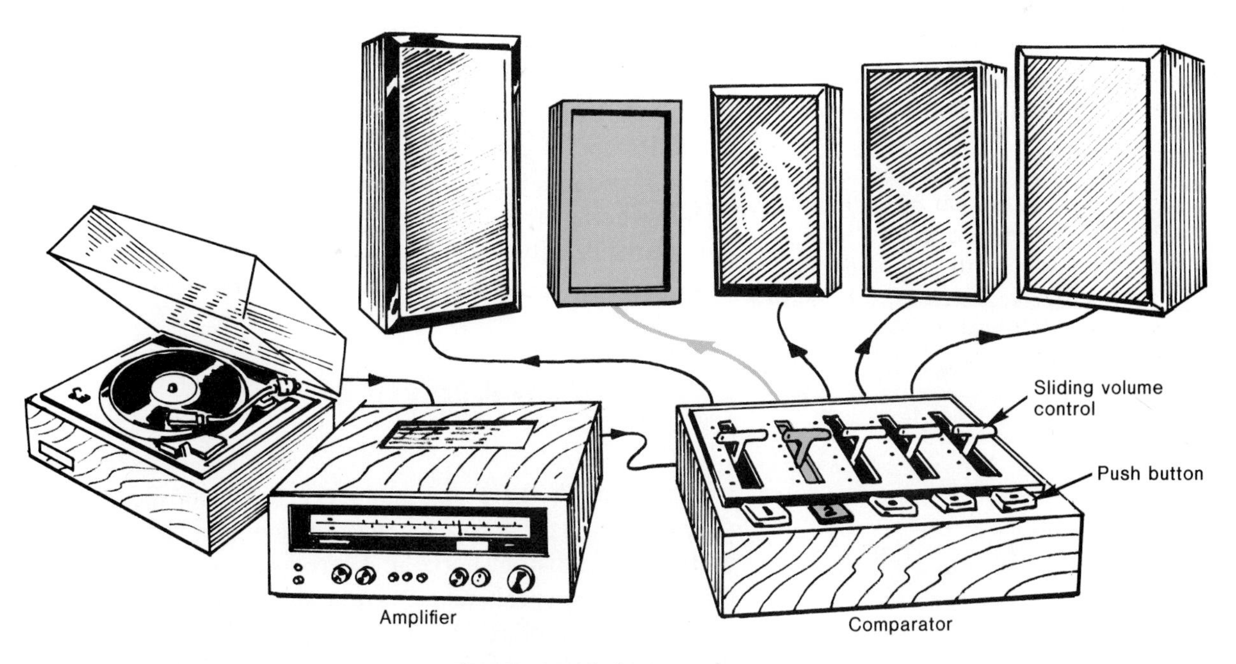

Figure 6.1. A speaker comparator makes it easy to hear sonic differences among loudspeakers. Any speaker can be selected with the appropriate push button, and the loudness of its sound can be adjusted with the corresponding volume control.

By pressing the appropriate switch on the comparator, any speaker can be attached to the same source/amplifier. In the true scientific spirit, everything in this test is kept constant except the speakers. Therefore, differences in the tonal qualities of the sound are attributable to the speakers alone. Upon switching back and forth among the various units, you will be absolutely astounded to hear the sonic variety; for there will exist as many different "sounds" as there are different speakers.

Actually, the combination of speakers plus listening area should be considered collectively as the Achilles' heel of audio. The sound quality of a speaker system will show distinct variations when it is removed from the dealer's showroom and played in your own room. In fact, simply moving the speakers around within the room can have a considerable effect on the sound quality, particularly for the bass notes. We will discuss some of the more important consequences of room acoustics in section 6.12.

What Is a Good Loudspeaker? Everyone Has a Different Answer

Assembling Loudspeaker Drivers
(Photo courtesy of AVID Corporation)

Because speakers are a long way from perfection, there is plenty of room for each designer to give his own interpretation as to what constitutes good reproduction. The result is that a tremendous number of speakers has been spawned on the audio market—far more than any other component. This gives the customer a great variety of choices when purchasing loudspeakers. As a result, some confusion exists because each manufacturer expounds the one or two parameters of loudspeaker design which presumedly make their product "perfect" for every customer. Audio engineers correctly tell us that there are literally dozens of interrelated variables to be considered when designing a good loudspeaker system. Hence it is difficult, if not impossible, to decide whether the optimization of a few of these variables will automatically lead to a good-sounding system. It rarely does.

Moral: Beware of speaker ads.

Even though there are many excellent loudspeakers on the market today, it appears that fashioning a "perfect" one is still on the distant horizon. There are several reasons for this. Although speakers are rather simple looking, they are moderately complicated from an engineering point of view. A score of factors, correlated by the good-ol' laws of physics, influence their performance. In many cases the factors are inversely related, meaning that a maximization of one factor minimizes another. This makes it impossible to optimize simultaneously all the elements of a speaker design. Engineers must then make "trade-offs" which are governed by a knowledge of what consumers want to hear and the price that they are willing to pay for it. Also, loudspeakers are the only components in an audio system which actually produce audible sound, and any evaluation process must take into account exactly how the listener's ears and brain perceive the sound. For example, suppose that you made a comparison test between two different speakers, one of which was "perfect" and the other imperfect. By listening, your auditory senses would probably have no trouble in distinguishing one from the other—they would simply sound different. So far so good. Now ask yourself the $64 question: "Exactly what is it that makes the imperfect speaker sound imperfect?" To answer this question one must be able to correlate audible deficiencies with flaws in the speaker design. No real speaker is free from the various types of sound distortion which result from such flaws. Audio designers need to know more about the relationship between distortion and its effects upon our aural perception. In other words, it will be impossible to design a near perfect speaker until the hearing processes are more fully understood.

It should be remembered that the room acoustics (furniture, walls, drapes, room dimensions, etc.,) and speaker placement have a profound effect on sound reproduction, so that even perfect speakers would yield an inaccurate replica of the original recording. This is not to say that quality speakers are not desirable—they certainly are. However, they are, by themselves, not the total sound picture that a listener experiences.

To answer the question "Why are there so many speaker designs?" we see that designers have great flexibility in selecting a choice of particular parameters which will result in different "sounds." This, coupled with the fact that no one fully knows what perfect reproduction means, is responsible for the large proliferation of speaker designs. In this chapter we are going to take a close look at various loudspeakers—their hows and whys—along with providing some hints on both purchasing and positioning speakers in your home.

6.2 A Close Look at a Work Horse—The Cone Speaker

There are many types of devices designed for the purpose of converting electrical signals into audible sound. Names such as permanent magnet cone speakers, electrostatic speakers, and magnaplanar speakers show the diversity.

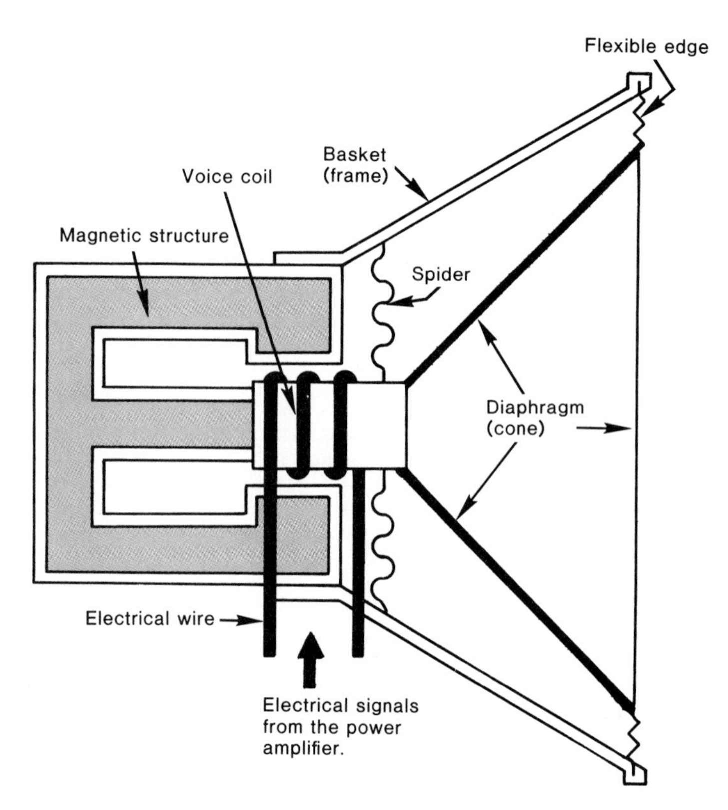

Figure 6.2. The essential parts of a permanent magnet cone speaker.

Only one—the cone speaker—has attained a position of overwhelming dominance in the hi-fi marketplace. Figure 6.2 shows a cross-sectional view of the permanent magnet cone speaker.

The cone speaker is constructed with four major elements which, when integrated, are capable of transcribing the complex electrical signals from the amplifier into audible sound:

1. the diaphragm (or cone).
2. the flexible edge.
3. the voice coil.
4. the permanent magnet.

The entire assembly is held together by an outer metal frame called the *basket*. The *diaphragm* is a cone, usually made from special paper, which creates the sound waves by pushing the air back and forth during its oscillations. The diaphragm is attached to the basket by a *flexible edge* which not only supports the cone within the basket but allows it to move freely. The flexible edge is a key element in loudspeaker design, and the final size and performance characteristics of a speaker are directly linked to it.

The *voice coil*, as shown in Figure 6.3, is a hollow cylinder around which are wrapped coils of wire. The voice coil cylinder is rigidly attached to the apex of the diaphragm and, as the voice coil is forced to vibrate back and forth, the

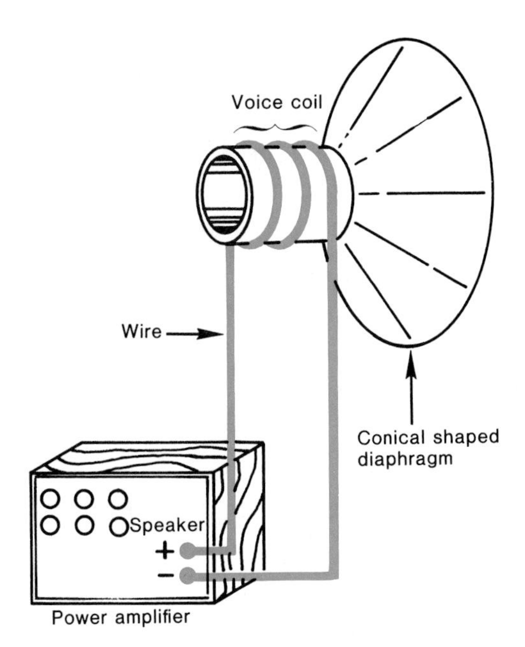

Figure 6.3. The voice coil is the only part of a cone speaker which carries electricity. It is attached to the diaphragm and contains many turns of wire. The two free ends of the wire are connected to the speaker terminals of the power amplifier.

diaphragm must follow in precise synchronism. The two free ends of the wire are connected to the two speaker output terminals of the power amplifier. The electrical signals which actually carry the audio information flow through this wire. It should be mentioned that the voice coil is the only part of the entire speaker assembly which carries the electricity.

The last part is the *magnetic structure*, shown in Figure 6.4, which is usually made of ceramic or alnico materials. Alnico is a combination of aluminum, nickel, and copper. These permanent magnets possess the desired properties of providing strong magnetism which retains its strength over many years.

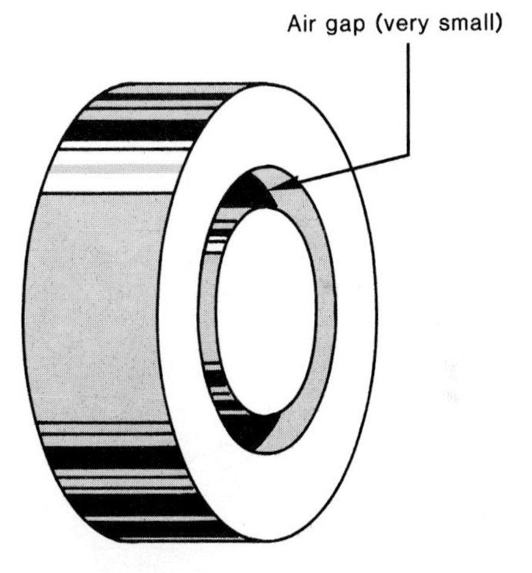

Figure 6.4. A view of the permanent magnetic structure showing the tiny air gap between the inner and outer sections.

The Voice Coil and Permanent Magnet Team Do All the Pushing and Pulling

The hollow voice coil cylinder is constructed so that it can slip over, without touching, the solid inner cylinder of the magnetic structure (Fig. 6.2). The magnetic structure, along with the voice coil, are the parts of the speaker which actually convert the incoming electricity into diaphragm motion. The scientific principles behind this action are quite fascinating and they will be presented in Chapter 11—**Electromagnetism**. Suffice it to say that the electric current causes the voice coil to become an electromagnet (just like the ones you see in the big auto junkyards). Depending upon the direction in which the electricity is moving through the

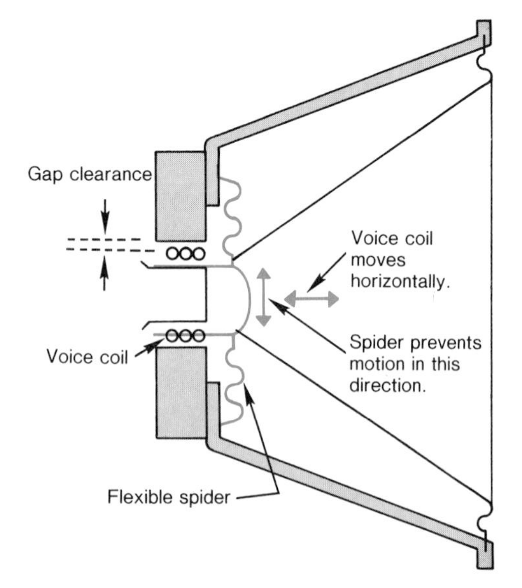

Figure 6.5. The spider keeps the voice coil centered about the inner cylinder of the magnetic structure. The gap clearance is extremely small and the spider eliminates any frictional problems.

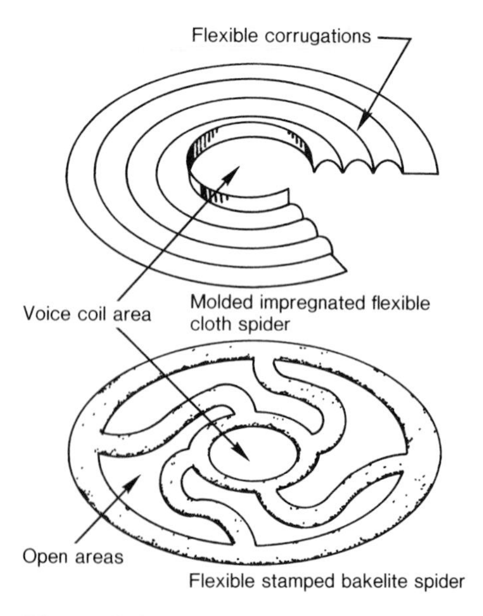

Figure 6.6. Some typical spiders.

helical coil, this electromagnet is either attracted to, or repelled from, the permanent magnet structure. If an attraction is created, the voice coil, and hence diaphragm, is pulled towards the stationary magnetic structure and a sound rarefaction is created in front of the cone. If a repulsion occurs, the voice coil is pushed away from the magnetic structure and a condensation in the air results. For example, incoming electricity which is changing its direction of travel at a rate of 2,000 times each second will produce 1,000 attractions and 1,000 repulsions each second. The result will be the generation of a 1,000 Hz sound wave!

Enter the Spider

The voice coil is exactly centered about the magnetic structure by a support called the "spider" (Fig. 6.5). The spider permits the voice coil to vibrate horizontally without any vertical motion. Any vertical motion, as shown in Figure 6.5, would lead to serious frictional problems which are caused by the voice coil rubbing against the permanent magnet. Figure 6.6 shows some typical spiders. Notice in Figure 6.5 that the gap clearance between the edge of the voice coil and magnet is extremely small. It is approximately 25 thousandths of a cm for large speakers and only 8 to 10 thousandths of a cm for the smaller ones. A small gap increases the efficiency of the magnet/voice coil combination, thus allowing more sound power to be generated with a smaller, less costly, magnet. With this small clearance in mind, you can easily understand the importance of the spider's job.

Summary

The diaphragm is the part of the speaker which moves back and forth and causes the air molecules to vibrate, thus creating the sound wave. The voice coil is attached to the diaphragm. The force between the electrical current in the voice coil and the magnetic structure is what causes the diaphragm to move. The flexible edge is used to attach the outer edge of the diaphragm to the metal basket. The flexible edge allows the diaphragm to vibrate freely. The spider supports and centers the voice coil on the magnetic structure.

6.3 Taking Off the Front Cover

Perhaps the best way to gain an understanding of loudspeakers is to gather up some of the more popular models, take off the front covers, and ask "Why do they look the way they do?" The answer to this question will tell you a lot about loudspeaker design. Figure 6.7 shows five such speakers which are of the popular "bookshelf" variety. From

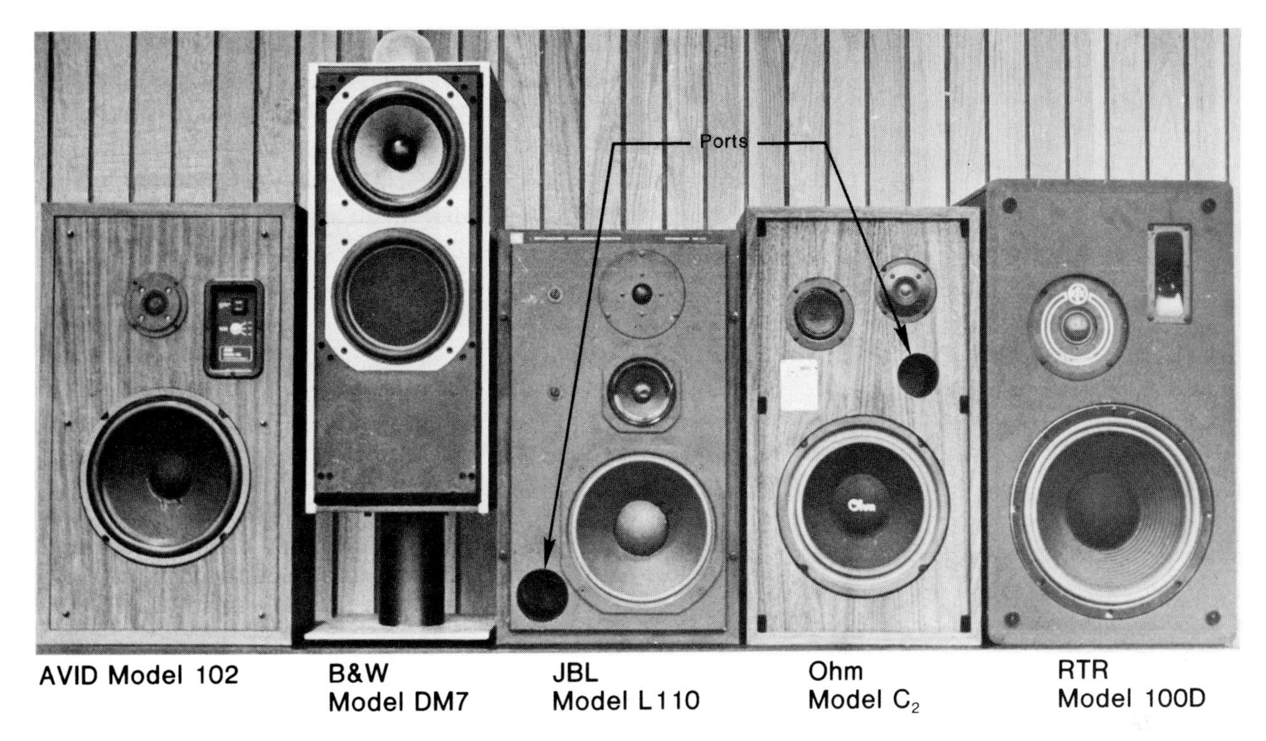

Figure 6.7. Some popular hi-fi loudspeakers with the cloth covers removed. Notice the diversity in the number, size, and shape of the drivers within each enclosure.

this figure there are three obvious deductions that one can immediately make:

1. Each cabinet contains multiple (2 or more) speakers within it, and upon close examination most speakers are of the familiar cone type. Each speaker is often called a "driver." The reasons for multiple speaker arrays will be discussed in section 6.4.
2. The speakers within each enclosure have different sizes. This point will be examined in sections 6.5 and 6.6.
3. The cabinets are completely sealed although the JBL–L110 and Ohm C_2 have cylindrical holes (called "ports") cut into the front faces. The type of enclosure is critical to loudspeaker performance and will be dealt with in sections 6.8 and 6.9.

The diversity among the above loudspeakers should tell you one interesting and important fact; no one manufacturer has the secret to good loudspeaker design. In fact, there are really no secrets because most of the scientific principles of loudspeaker design have been known for some time. There are only compromises and honest differences of opinion on how these compromises should be made. With the implicit understanding that all these loudspeakers can produce excellent sound, let us examine each of the three points mentioned above in order to understand better the design philosophies which lead to these features. From a consumer point of view buying loudspeakers would be a lot simpler if one could categorically say, "A 10-inch woofer enclosed in an infinite baffle always sounds better than two three-inch woofers enclosed in an acoustic suspension cabinet." Alas, such is not the case, because some listeners will obviously prefer one type while others will find another system more to their liking. Nevertheless, an intelligent buyer should be aware of the design differences which exist among different manufacturers.

6.4 Why Do Loudspeaker Enclosures Contain Multiple Drivers?

In Figure 6.7 it was shown that there are many drivers enclosed within a single cabinet. Loudspeakers should make a reasonable attempt to reproduce the audio frequency range with a minimum amount of distortion. Audio engineers have found it difficult to design a single, full-range driver which, by itself, will yield quality sound at loud listening levels. Therefore, the audio frequency range is usually divided into two or three segments with each part being handled by specially designed drivers. It is like partitioning the entire space of a house into smaller segments, e.g., kitchen, bedrooms, living room, etc., such that each room can be optimized for a specific function.

ADS Model 200 Loudspeakers

Three-way Systems and Crossover Frequencies

In a three-way speaker system the deep bass notes are handled by a relatively large cone driver called the *woofer*. The intermediate frequencies are produced by the *midrange* driver, and the highest tones are generated by the *tweeter*. Figure 6.8 schematically shows a three-way system and the reasons for the different driver sizes will be discussed in sections 6.5 and 6.6. Shown in Figure 6.9 are some typical values for the frequency span covered by each of the three drivers. Notice that each type of driver produces sound in a definite frequency region, leaving the remainder of the audio range to be covered by other types which are better suited to do their particular jobs. The three idealized graphs in Figure 6.9, which show the sound intensity versus frequency, are most important in hi-fi, and each one is called the *frequency response curve* for the particular driver. (See section 5.8 for a discussion on frequency response curves.)

The frequency at which one driver drops off and another picks up (500 Hz and 5 kHz in Figure 6.9) is called the *crossover frequency*. It follows that a three-way speaker system has two *crossover frequencies* as shown in Figure 6.9.

Figure 6.8. A three-way speaker system.

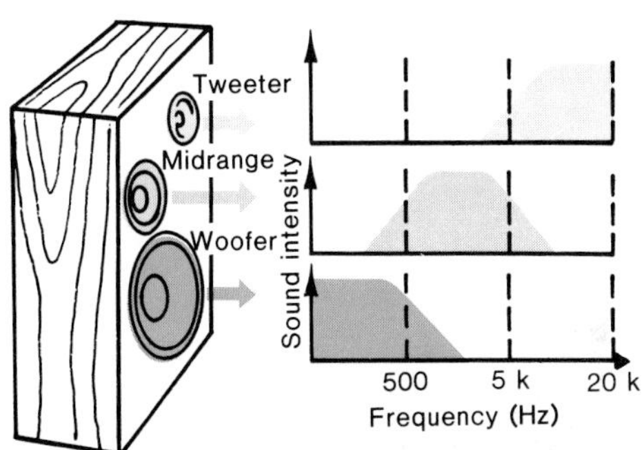

Figure 6.9. A three-way loudspeaker utilizes a woofer, a midrange, and a tweeter to cover the entire range from 20 Hz to 20 kHz. Each driver operates over a limited frequency range, and the frequencies where they overlap are called the "crossover frequencies" (500 Hz and 5 kHz in this example).

Although 500 Hz and 5 kHz have been arbitrarily selected in this example, the actual positions of the crossover frequencies vary with the manufacturer; they can range anywhere from 500 Hz to 3 kHz for the woofer: midrange crossover, and from 3 kHz to 7 kHz for the midrange: tweeter crossover.

Ensuring that each driver handles the audio frequencies for which it is intended is the job of the *frequency crossover network* which is located within the speaker cabinet. The crossover network is an electronic circuit which is necessary because the electrical signals coming from the power amplifier contain all of the audio frequencies (the highs, the middles, and the lows) mixed together. Figure 6.10 indicates that this circuit directs the correct frequencies to the proper drivers.

A few manufacturers also use multiple drivers to cover the same part of the frequency range. For example, instead of one large woofer, two or three smaller woofers may be used. Such designs are sometimes used because of power requirements and other factors, such as spreading out the sound into many directions at once. No frequency crossover network is required for multiple drivers which cover the same frequency range.

Two-way Systems

A re-examination of the left-most speaker in Figure 6.7 shows an example of a two-way system. The large cone is a low-frequency driver which extends out to 1 kHz while the smaller, top cone is, of course, the high-frequency driver.

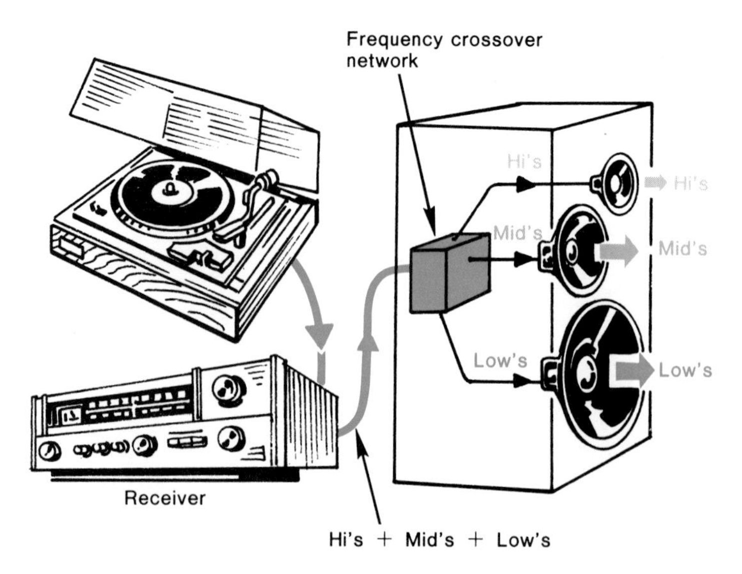

Figure 6.10. The job of the frequency crossover network is to direct the high frequencies to the tweeter, the middle frequencies to the midrange, and the low frequencies to the woofer.

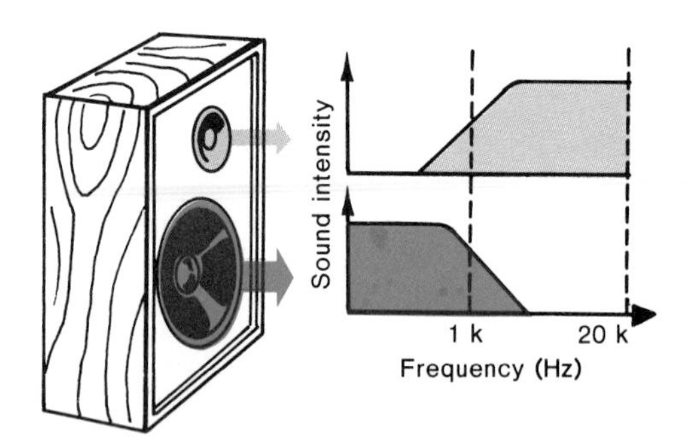

Figure 6.11. A two-way loudspeaker system with a single crossover frequency at 1 kHz.

Figure 6.11 shows the response curves for each driver in a two-way loudspeaker. Once again, the selection of the crossover frequency rests with the manufacturer, and it can vary considerably among different brands.

A Consumer Education Point

A question is often raised: "Do cabinets which contain many drivers sound better than those which have fewer?" Having read the concluding remarks in section 6.1 you should know the answer: **Maybe** and **Maybe not.** There are many, many factors which influence the sound emanating from a loudspeaker and the number of drivers, by itself, is insufficient to guarantee sound quality.

When it comes to loudspeakers remember THE CARDINAL RULE: **When evaluating loudspeakers let your ears—not one or two fancy specifications—be the final arbitrator of sound quality.**

6.5 Woofers Are Large

It was mentioned earlier that the woofers are the largest of the drivers, with diameters ranging from 6″ (15 cm) all the way up to a whopping 30″ (76 cm). These low-frequency drivers are so large because they must push an enormous amount of air in order to create normal listening levels. There are several reasons for this. First, there is a large amount of acoustical energy in the bass region relative to the midrange and highs. It is rather obvious that the boom of a kettle drum or the deep notes of a bassoon generate far more sound energy than the tinkle of a triangle (one could tinkle all day and never get tired). In short, a substantial fraction of the total sound energy in music falls within the bass region, and it has been estimated that up to 40% of it lies below 500 Hz!

Second, the ear becomes much less sensitive at the bass frequencies compared to the midrange, and the speaker is required to create much more sound intensity just in order to be heard. To see this, consider Figure 6.12 which is a reproduction of the famous Fletcher-Munson curves that were discussed in section 5.6. Suppose that a particular loudspeaker is producing a 1 kHz tone which a listener perceives to be 40 dB in loudness. According to Figure 6.12, if the sound frequency is lowered to 100 Hz the speaker must then produce more sound intensity just to keep the perceived loudness on the 40 dB constant loudness curve. How much more intensity is required? Looking at the left-hand scale of the figure we see that it requires about 20 dB (60 − 40 dB) more intensity just to maintain the constant loudness. Now, according to Table 5.1, a 20 dB increase means that the speaker must produce an astounding **100 times** more intensity at 100 Hz than at 1,000 Hz just to keep the loudness the same!!

With these two facts about bass requirements in mind, let us examine the design options which are available for creating large amounts of acoustical energy.

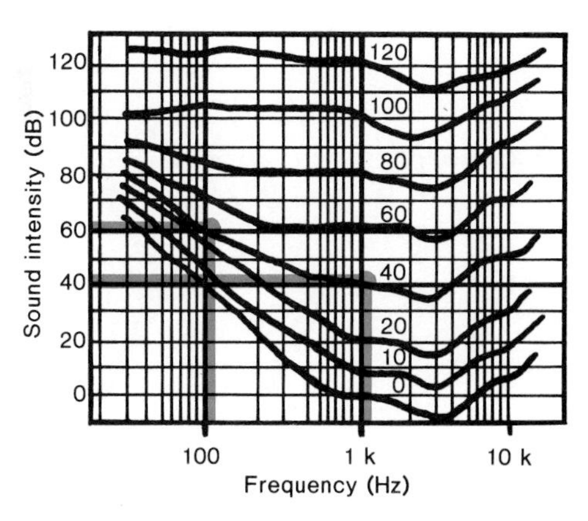

Figure 6.12. The Fletcher-Munson curves show that a speaker must produce much more sound intensity at the low frequencies, compared to the midrange, in order to maintain a constant loudness level. For example, a constant loudness level of 40 dB requires about 60 dB of sound intensity at 100 Hz and only 40 dB at 1 kHz, a difference of 20 dB.

The Amount of Sound Power Produced by a Speaker Depends upon the Total Volume of Air Which Is Moved

In the previous discussion it was pointed out that the bass driver must be capable of generating large amounts of acoustic power in the low-frequency region. A very appropriate question at this time would be:

> "What physical parameter of a driver determines the loudness of sound which it can generate?"
>
> Answer:
>
> As the cone vibrates back and forth, it is the total **volume** of air which is moved during each oscillation that determines the acoustic output of a driver.

In order for woofers to push the large volume of air required for high acoustic output, designers have three options available:

1. They can make the woofer cone larger in diameter.
2. They can keep the diameter relatively small but provide for a longer cone excursion. This implies that the cone must travel faster as it vibrates because it must cover a larger distance in the same amount of time.
3. They can use more than one woofer in a speaker enclosure.

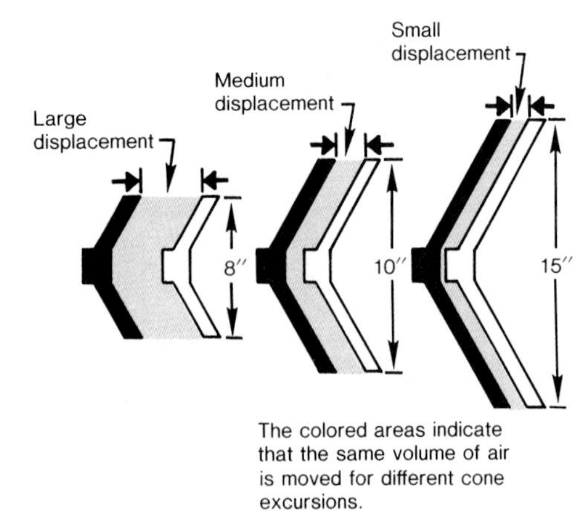

Figure 6.13. The acoustic power produced by a driver is determined by the volume of air that it displaces. All three drivers are pushing the *same volume of air*, and, hence, produce the same loudness. The 8″ cone, having the smallest area, must move further than either the 10″ or 15″ cones.

Figure 6.13 shows three different size drivers which are producing the **same amount of sound power.** The 8″ (20 cm) cone, being the smallest, must travel much further than the larger 15″ (38 cm) cone. Since they both execute the same number of vibrations each second, it should once again be clear that the 8″ cone—with its greater travel—moves much faster. All three cones in Figure 6.13 are displacing the same volume of air, so they will sound equally loud! This situation is very analogous to the window fan. If you want the fan to move a large volume of air, there are two basic choices: (1) buy a fan with large blades (which corresponds to a large diameter woofer) or (2) use a smaller fan but run it at a faster speed (which is exactly what the small, higher speed, long excursion woofers do).

Another Consumer Education Point

The next time you read a speaker ad or walk into a hi-fi store, do not be fooled into thinking that larger woofers automatically provide more bass than smaller ones. Smaller cones can produce just as much sound power by permitting the diaphragm to "throw" over greater distances. In fact, most of the very popular "bookshelf" loudspeakers are examples of the long throw, small diameter, type.

6.6 Tweeters and Midranges

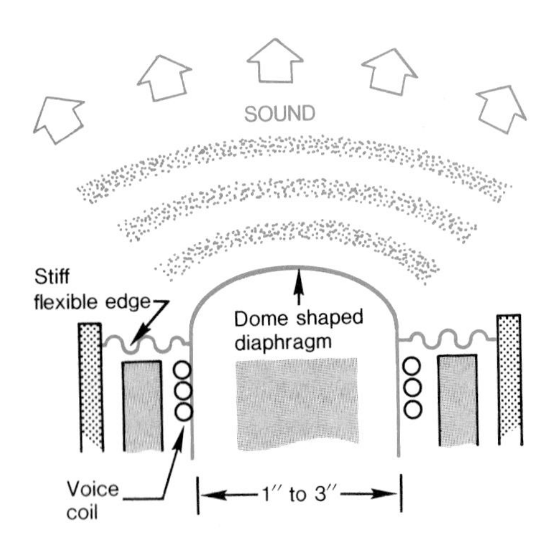

Figure 6.14. Cross-sectional view of a tweeter. Tweeters have small, lightweight diaphragms which are attached to the basket with a stiff flexible edge. The diaphragm is usually dome shaped to help in spreading the treble notes over a wider listening area.

At the high-frequency end of the sound spectrum lies the diminutive tweeter. Its job is to provide good high frequency reproduction and, in order to do so, it must be made extremely light with a rather stiff flexible edge. Figure 6.14 shows a typical tweeter whose diameter may range anywhere from 1″ to 3″ (2 to 8 cm). Notice that tweeters are often dome-shaped. The dome shape increases the sound dispersion because sound waves are radiated at greater angles from a hemispherical surface than from a much "narrower" cone surface. Other tweeter designs either utilize cones which are almost flat (very little taper) or horns. The horn tweeter is nothing more than a small driver which is coupled to the mouth of a horn. The idea is very similar to a person speaking through a megaphone (horn). The use of a horn yields louder sounds and helps to disperse better the high frequencies throughout the listening area.

The weight and suspension stiffness of a driver are analogous to the density and tension of a guitar string. The heavy and relatively loose A string produces the lower frequency notes while the very light and tautly stretched B string easily creates the treble notes. Remember that a tweeter reproducing a 6 kHz signal is vibrating 10 times faster than a woofer producing 600 Hz. It is no wonder that the tweeter must be as light as possible in order to follow successfully these incredibly fast movements.

If it seems rather obvious to you that a small, lightweight tweeter is required to produce the high frequencies, then a good follow-up question would be; "How can a small diameter tweeter push the large amounts of air which are required to produce loud sound levels?" Part of the answer is that nature has been very kind to audio designers because most musical selections contain only very small amounts of high-frequency energy. Hence tweeters are not called upon to push large volumes of air. Also, for reasons which are not at all obvious, engineers know that lightweight, high-frequency radiators are more efficient than heavy low-frequency woofers. This means that, with all other things being equal, a lighter driver will produce more sound than a heavier one for identical input electrical powers. Therefore, any loss in sound power that a tweeter suffers because of its small diameter can be compensated by its increased efficiency due to its reduced mass and higher operating frequency.

Because the midrange driver must bridge the gap between the woofer and tweeter, its acoustic performance must be compatible with both of them. The midrange driver operates in a range where there is a modest amount of sound energy, and its intermediate size is indicative of the fact that it must produce reasonable sound levels. Obviously, the midrange's frequency span must overlap the high end of the woofer and the low end of the tweeter. Most important, the efficiency of the midrange must match that of the other drivers. A midrange driver with an efficiency which is greater than that of either the woofer or tweeter would sound relatively louder, and an imbalance in the sound spectrum would result (Fig. 6.15).

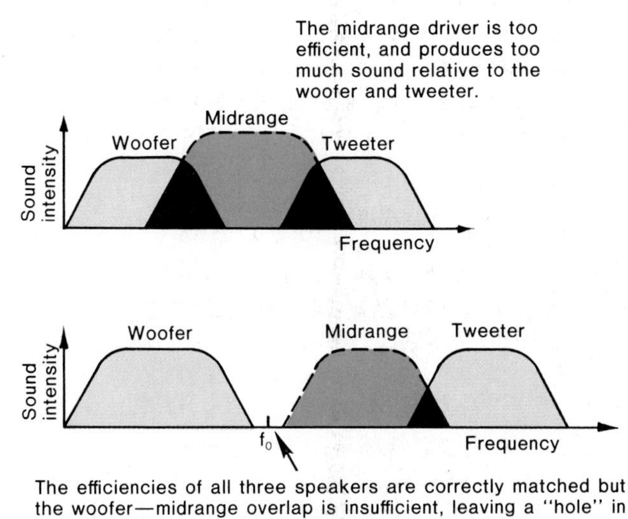

Figure 6.15. Two problems which must be avoided when selecting a midrange driver.

In addition to the usual cone drivers, some manufacturers prefer to use horns or domes for the midrange drivers. These types of drivers can yield better sound dispersion at their high-frequency end than the cone drivers. A more complete discussion of horns, their assets and liabilities, is given in section 6.9.

6.7 The Importance of a Woofer's Natural Frequency of Vibration to Good Bass Reproduction

We have seen that the woofer is a most important element in loudspeaker design because it is delegated the job of producing the difficult bass notes which contain large amounts of musical energy. Since a large fraction of this energy resides below 500 Hz, it is clear that any woofer which cannot produce tones below 200 Hz, for example, seriously degrades the musical quality. What is the lowest frequency that a woofer should be able to reproduce? The answer seems obvious. Since the span of human hearing ranges from 20 Hz to 20 kHz, the lower limit of woofer action should be 20 Hz. However, in practice, there is usually a trade off between this limit and price, and it may be desirable to accept a lower limit of 50–60 Hz. Designers normally use large diaphragms to reproduce the deepest bass notes (below 40 Hz). For acoustic reasons (which will be discussed shortly) these large cones require large enclosures which may stand as high as 1 meter. If you are a student and live in a small apartment, the size and expense of such a loudspeaker may be prohibitive and the smaller "bookshelf" speakers have a greater appeal. However, in many of the "bookshelf" loudspeakers, the bass response extends down to only 50 or 60 Hz. This does not sound as bad as it appears on paper, because most composers (fortunately) use notes whose frequencies lie above this lower limit. The result is that the smaller systems perform quite admirably with almost all types of music. An exception would be organ music which contains the deep, deep bass notes which literally shake the listener.

The woofer's natural frequency of vibration plays a vital role in determining the lower limit of its reproduction capabilities. In fact this parameter is so important that many other design characteristics, including the volume of the loudspeaker enclosure, are dependent upon it. Recall, from section 4.7, that all objects prefer to vibrate at a particular frequency which is called the "natural" or "resonant" frequency of vibration. Drivers are certainly no exception. A gentle tap on the cone, hanging in its suspension, will cause it to oscillate back and forth at its own natural frequency.

This phenomenon can also be visualized by considering what happens to a weight which is attached to the end of a spring (Fig. 6.16). Here the analogy to the driver is quite close. The mass hanging on the spring corresponds to the mass of the diaphragm while the elasticity of the spring correlates with the flexibility of the cone's suspension. The mass hanging on the spring possesses a natural frequency of vibration which can be altered by changing one or two parameters. If the suspended mass is made *heavier,* the system will oscillate more *slowly* with the decreased natural frequency. Using a lighter mass has the opposite effect of raising the system's natural frequency. Also, if the spring is replaced by a *weaker* one, the mass will vibrate *slower,* which implies a *lower* natural frequency. Conversely, a stiffer spring raises this frequency.

With a little thought you will be convinced that the natural frequency of a driver is determined by its mass (or weight), and the elasticity of the suspension which secures it to the metal basket (Fig. 6.17). If the mass of the diaphragm is increased—making it heavier—the diaphragm will vibrate at a slower rate, reflecting its lowered natural frequency. Similarly, making the suspension looser (or less springy) will produce the same effect. This is exactly the same situation as applied to the A and B strings of a guitar as pointed out in section 6.6. What does all this have to do with a woofer's performance? The next few paragraphs will give you the answer.

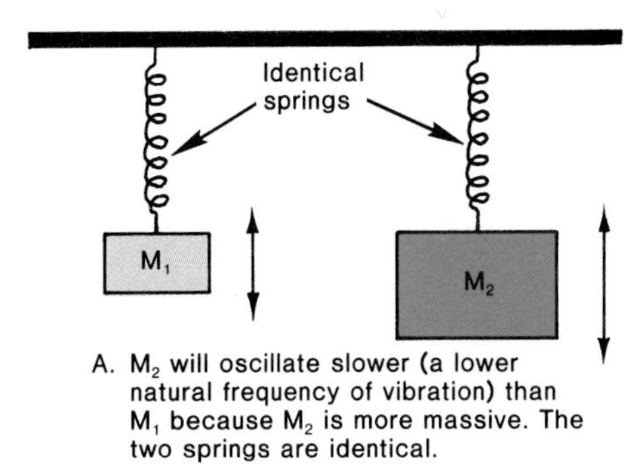

A. M_2 will oscillate slower (a lower natural frequency of vibration) than M_1 because M_2 is more massive. The two springs are identical.

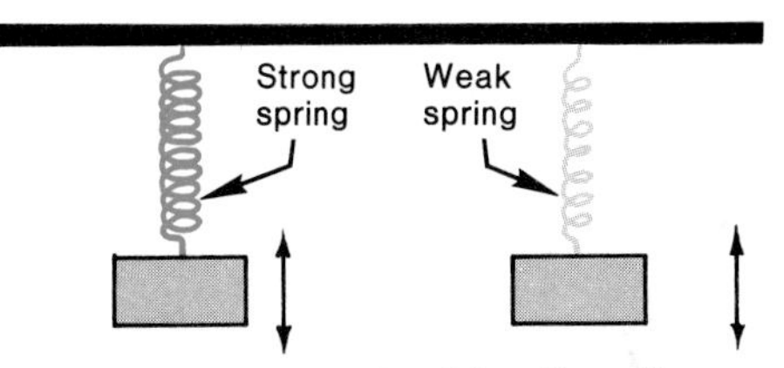

B. The mass on the right will oscillate slower because its spring is weaker (more compliant). Both masses are identical.

Figure 6.16. There are two methods which can lower the natural frequency of a vibrating system. (A) Make it more massive. (B) Use a higher compliance (i.e., weaker) spring.

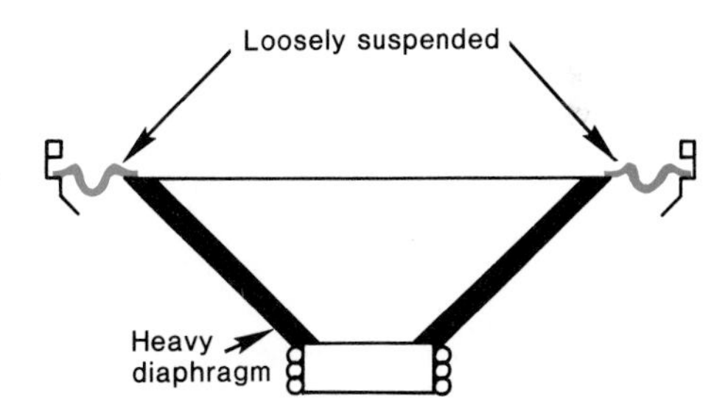

Figure 6.17. A driver cone possesses a natural frequency of vibration which is governed by its mass and the stiffness (or looseness) of the suspension. A loose suspension and/or a large cone mass will lower the natural frequency.

How a Woofer's Performance Is Affected by Its Natural Frequency of Vibration

The importance of a woofer's natural frequency can be demonstrated by performing a very simple experiment. Using an audio generator, apply a signal to a woofer and measure its output sound intensity with a sound level meter. Figure 6.18 shows the set-up. In this experiment the frequency of the audio generator is initially set at 500 Hz. The frequency is then slowly lowered to about 10 Hz while the sound level is continuously measured by the sound level meter. The natural expectation from such an experiment would be that the woofer produces the same sound intensity at all frequencies. If this were the case, life for audio designers would indeed be easy. Figure 6.19, which is a plot of the woofer's sound output, shows that this is not quite the case. Serious problems arise when the sound frequency is at, or below, the woofer's natural frequency of vibration. Close to the woofer's natural frequency the output sound is louder

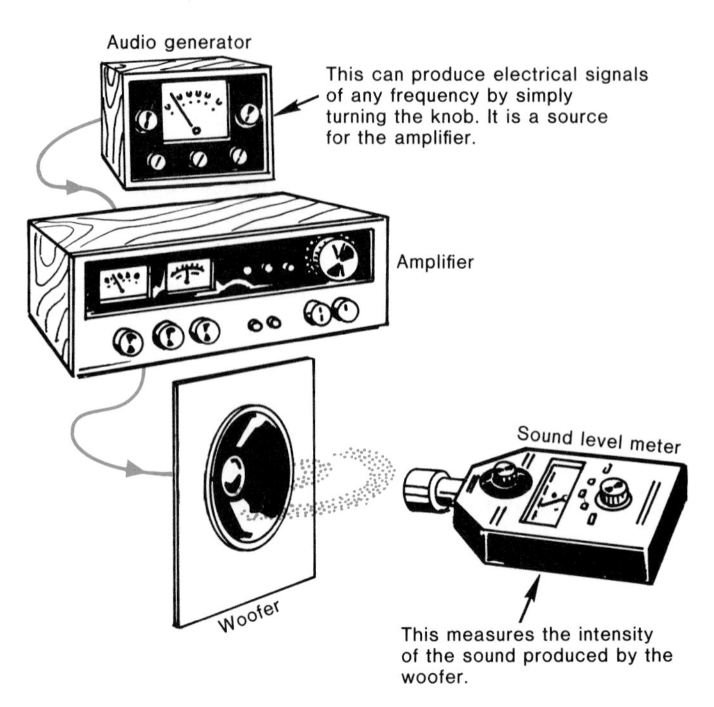

Figure 6.18. An experiment which demonstrates that the sound produced by a woofer diminishes for frequencies which are below its natural frequency.

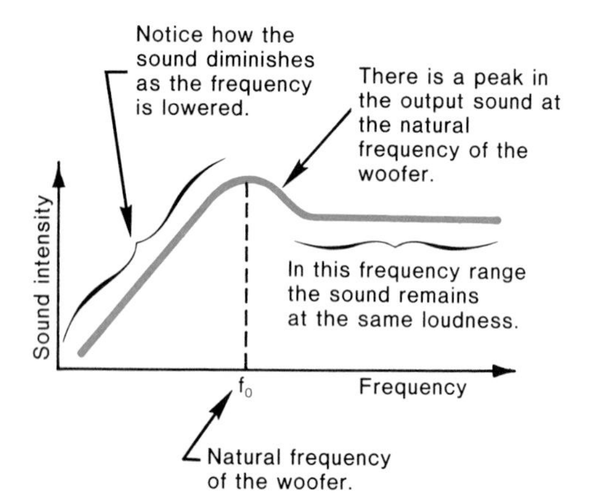

Figure 6.19. The results of the "woofer experiment" from Figure 6.18 are displayed on this graph. The output sound intensity is plotted on the vertical axis and the sound frequency is placed on the horizontal axis.

than normal causing a peak in the curve, as shown in Figure 6.19. The reason for this peak is simple. At its natural frequency—the frequency where the woofer likes to vibrate—the oscillations are quite large as it easily swings back and forth. And, of course, the unusually large displacements produce a louder than normal sound, hence the peak. This peak, often called the *"resonant peak,"* is undesirable because it tends to produce a "boominess" in the bass reproduction. The resonant peak can be subdued with a proper choice of speaker enclosure.

Another real problem, revealed in Figure 6.19, lies at frequencies below the peak where the sound output drastically drops off. This loss dictates that woofers cannot be used much below their natural frequency of vibration. The major conclusion concerning the importance of a woofer's natural frequency is,

The natural frequency determines the lowest note that a driver can effectively reproduce. At frequencies below the natural frequency the sound level falls off too rapidly and it is of little use.

This conclusion applies to all types of drivers, woofers, midranges, or tweeters, although it is of greatest concern for the woofers because it dictates how well they can produce those highly valued bass notes. If you recall from our previous discussion, designers can lower the natural frequency by either building a larger diaphragm (more massive) or providing an extremely loose suspension. We shall see in section 6.9 that each of these two choices will lead to vastly different enclosure designs.

6.8 An Initial Look at Loudspeaker Enclosures: A Baffling Problem

In the previous three sections we have discussed the reasons why there are many drivers within a loudspeaker enclosure and the need for different sizes. We are now in a position to discuss the enclosures themselves but, before we do that, let us dispel a common myth.

Fact: The main function of the cabinet is **not** to combine all the drivers into a good-looking enclosure!

This is a shocking statement and it certainly needs further elaboration. So, we shall begin by discussing one of the enclosure's *real purposes,* that of providing a baffle for the bass notes.

There is a very serious problem associated with the bass sounds generated by a woofer standing alone. This can be easily heard by performing the following simple experiment. Take two identical woofers and connect them to an amplifier such that the music can be switched back and forth between the two. Let one be free-standing and connect the second one to a large rigid board, called a *baffle.* The baffle has a circular hole cut in it which allows the sound to pass through. As shown in Figure 6.20, the baffle can take various shapes: either flat, a partial enclosure with an open back, or a com-

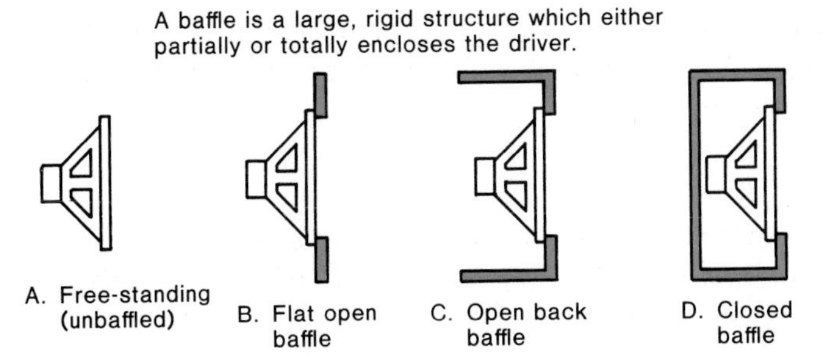

Figure 6.20. The unbaffled driver at the left will produce far less bass than the three baffled drivers shown on the right. Only (D), and variations of it, is used in most hi-fi speakers.

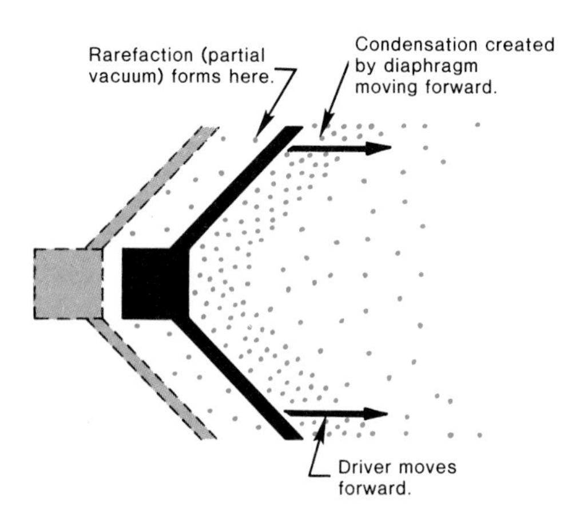

Figure 6.21. As the driver cone moves to the right, it creates a region of higher air pressure immediately ahead of it. The condensation leaves the diaphragm and travels through space at the speed of sound. Behind the cone a rarefaction is simultaneously produced.

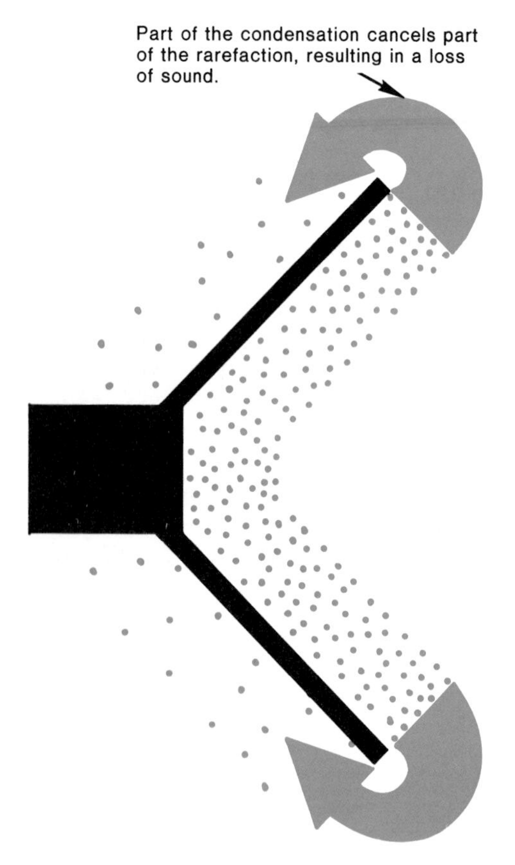

Figure 6.22. With an unbaffled driver part of the pressure condensation in front rushes around the rim and cancels some of the pressure rarefaction behind. Therefore, less sound is available for listening.

pletely enclosed cabinet. In any case the general results will be the same;

> **The baffled driver will produce greater bass sound levels than the unbaffled one.**

As strange as this may seem, this fact of life is really one of the underlying reasons for the use of speaker cabinets in hi-fi systems. Since the problem of speaker baffling is so important, let us now examine why the baffle enhances the bass response.

Why Do Unbaffled Speakers Lack Bass?

The answer to this question goes back to the way in which the diaphragm creates the sound during its back and forth oscillations. As the diaphragm moves forward it compresses the air immediately ahead of it, thus creating a condensation which travels toward the listener at the speed of sound (Fig. 6.21). Simultaneously, at the rear surface, a rarefaction is also being produced which travels to the left—a direction which is opposite to that of the condensation's motion. It often comes as a surprise to most people when they learn that *both* the *front and rear* surfaces of the diaphragm produce sound waves. You will hear sound equally well by standing either in front of or behind an unbaffled driver. Now, here the problem arises. Part of the condensation in the front will bend around the rim due to diffraction and cancel part of the rarefaction behind. This results in less sound being transmitted to the listener and, hence, a loss in loudness (Fig. 6.22). This is a beautiful example of destructive interference between the front and back sound waves.

Sound Cancellation Is a Problem Only with the Bass Notes

This partial cancellation effect is most severe in the lower bass region and it becomes negligible at the higher frequencies (i.e., midrange and treble). At the lower frequencies the sound readily diffracts around the rim of the diaphragm, leading to destructive interference with the rear-surface sound. At the higher frequencies sound diffracts to a much smaller extent (see section 4.4), and, therefore, cancellation with the rear-surface sound is not a serious problem.

The Solution Is to Use a Baffle

The obvious solution to the problem of partial sound cancellation is to block the front-to-back path and prevent the condensations from reaching the rarefactions—hence the use of a baffle as shown in Figure 6.23. The enclosure, therefore, serves a very important acoustic function. The precise manner in which a woofer is baffled will determine, to a large extent, many other characteristics of the loudspeaker system. This subject will be more fully explored in the next section.

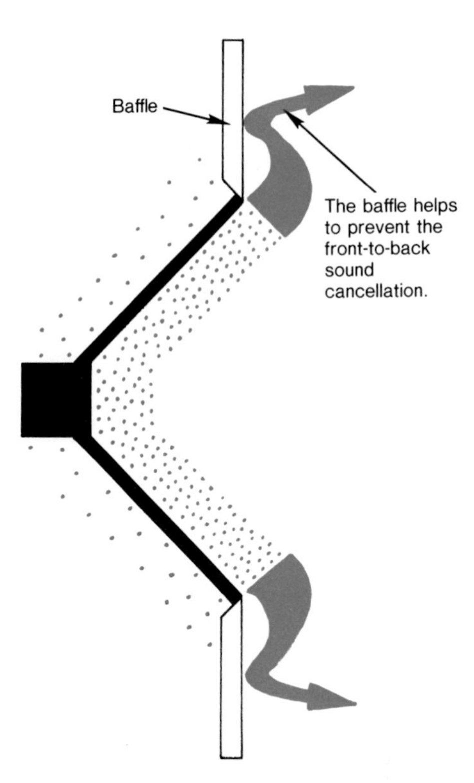

Figure 6.23. A baffle helps to prevent the low frequency sound cancellation.

6.9 A Closer Look at Loudspeaker Enclosures

People who are familiar with loudspeakers are aware that enclosures come in a tremendous variety of sizes. There are the large models, approximately 1 meter in height, and the very popular "bookshelf" speakers which offer big bass in a small enclosure. Figure 6.24 gives an indication of these size variations. In order to understand the acoustic reasons for the different enclosure sizes, recall the following important points which were discussed earlier:

1. The loudness of sound produced by a driver depends upon the volume of air which is pushed. A small cone can produce the same sound levels as a larger cone—the smaller one must simply move its diaphragm further.
2. A driver produces sound from both its front and rear surfaces. At the low frequencies a baffle is required to prevent the partial cancellation of these two waves.
3. The natural, or resonant, frequency of a driver determines approximately the lowest frequency that it can produce. The resonant frequency can be altered by changing either the mass of the diaphragm or the compliance (springiness) of the elastic suspension.

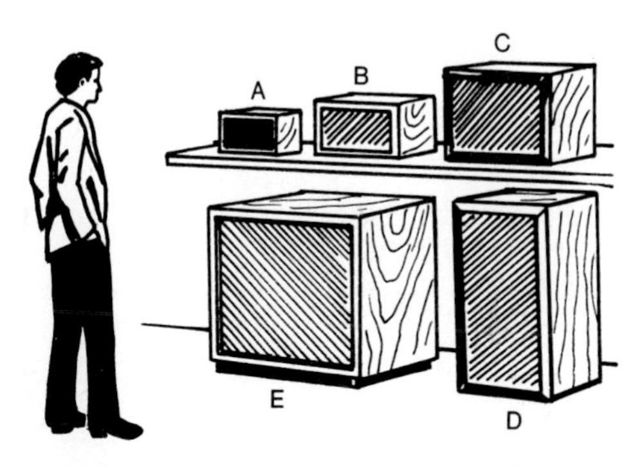

Figure 6.24. The relative sizes of some loudspeaker enclosures, ranging from A(¾ft^3 or 0.02m^3) to E(15ft^3 or 0.43m^3).

With these three all-important facts in mind, let us examine how they are used in explaining some of the more common enclosures. **The theme of the following discussion is to show that the enclosure strongly influences the bass frequencies generated by the woofer, and the enclosure has little influence on the midrange and tweeter speakers.**

There are three basic types of enclosures which we shall discuss in the following order: "sealed," "vented," and "horn" (see Fig. 6.25).

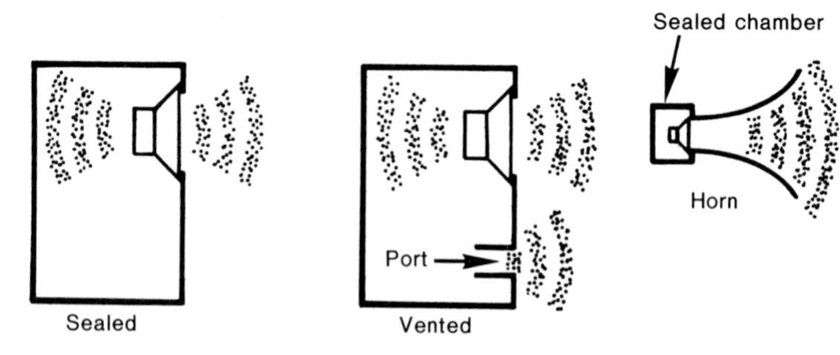

Figure 6.25. The three basic types of loudspeaker enclosures. The sealed and vented are the most commonly used.

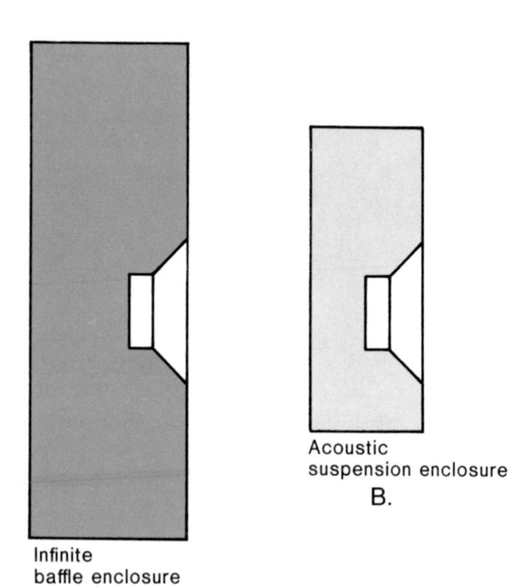

Figure 6.26. The two sealed enclosures represent (A) the infinite baffle enclosure and (B) the smaller acoustic suspension system.

The *sealed enclosures* completely trap the rear wave and prohibit it from escaping. There are two important variations of the sealed enclosure which are found on the market:

the *infinite baffle* enclosure and
the *acoustic suspension* system.

Both of these types are shown in Figure 6.26, A and B.

The *vented enclosure* (often called a bass reflex, tuned port, or phase inverter) permits the rear wave to escape—under controlled conditions—so that it may reinforce the front wave [Fig. 6.25]. In this case the lowest frequencies emanate not only from the woofer's front surface but also from the port.

The *horn* is, perhaps, the oldest type of enclosure and was made famous by Edison's early phonograph. The rear wave is contained within a sealed enclosure while the front sound is first transmitted through a horn before emerging into the room.

In the remainder of this section we shall discuss in detail the design philosophies behind each of these three enclosure types, along with a comparison of their relative advantages and disadvantages. Remember, as we have stressed many times, the quality of sound which emerges from a loudspeaker depends upon many factors and not on the enclosure type alone. Even though Brand X—which has a vented enclosure—sounds great, it certainly does not follow that all vented enclosure loudspeakers will be so highly rated.

The Infinite Baffle, One Kind of Sealed Enclosure

As its name implies, the idea behind an infinite baffle enclosure is to provide a very large baffle which prevents the front-rear sound cancellation. One obvious way to accomplish this, although not very practical, is either to mount the woofer in a wall which divides two large rooms or insert it in the door of a large closet. In either case the rear sound is radiated into a large room which isolates it from the front surface sound waves. Becoming more practical, one can place the woofer in a completely enclosed box, which will certainly reduce the undesirable bass cancellation. While this technique solves one problem, it immediately raises another. The air entrapped within the enclosure acts just like an acoustic "spring" upon the cone. As the diaphragm moves inward, the air within the enclosure is compressed, and it tries to push the cone outward. Similarly, as the cone is pushed outward it lowers the pressure within the enclosure, and the outside air, being higher in pressure, attempts to push the cone inward. So the entrapped air acts as if an imaginary spring were attached between the rear surface of the cone and the back of the box (Fig. 6.27). Therefore, the motion of the diaphragm is under the control of two springlike influences: (a) the elastic suspension holding it to the basket and (b) the "acoustic spring" of the entrapped air. These two, acting together, result in a "stiffer" suspension which will *raise* the resonant frequency of the woofer (Fig. 6.28). The change in the woofer's resonant frequency depends upon the size of the enclosure relative to the size of the woofer. As the woofer vibrates it displaces a certain volume of air, as shown in Figure 6.13. If the volume of the displaced air is an appreciable fraction of the total enclosure volume, then the accompanying air pressure changes within the enclosure will also be large. These large pressure changes tend to restrict the cone's movement and thus act like a very stiff spring. Therefore, a large woofer mounted in a tiny cabinet will undergo the largest changes in its natural frequency. On the other hand, if the volume of the displaced air is only a tiny fraction of the enclosure's volume, the air pressure changes will also be small. In this case the entrapped air does not exert much influence on the cone's motion. A small woofer, for example, which is mounted in a large enclosure does not experience any significant change in its resonant frequency.

The infinite baffle enclosures generally require larger cabinets. These enclosures are based on the philosophy of placing the woofer in a relatively large cabinet such that the

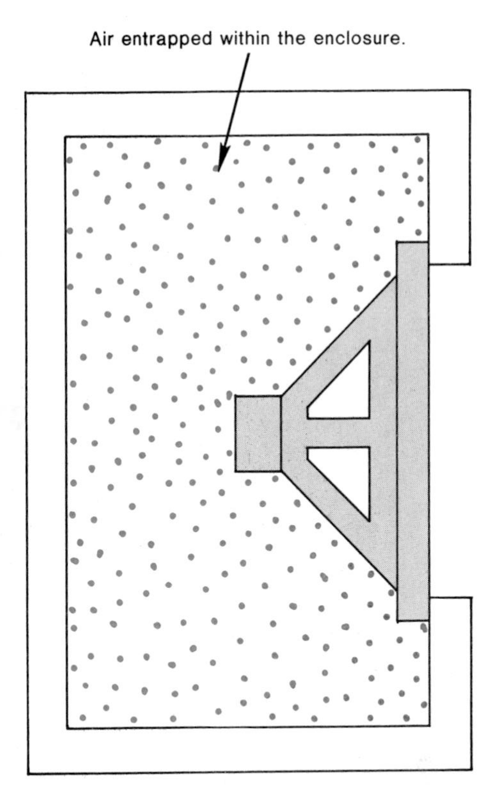

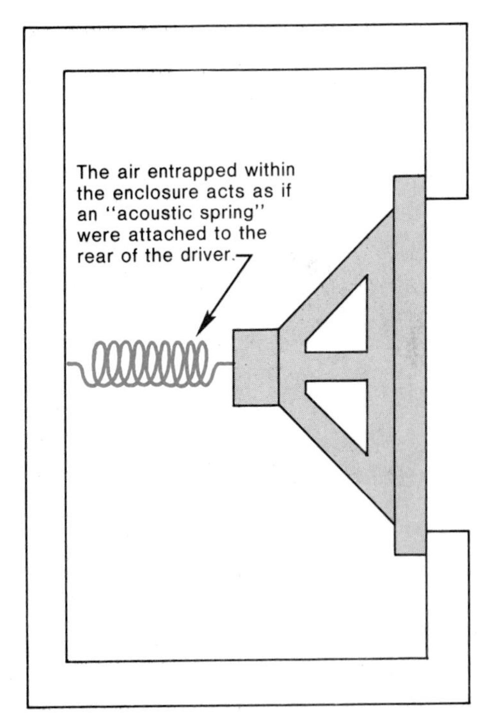

Figure 6.27. The air trapped within a speaker enclosure acts like a spring on the cone. Making the enclosure smaller has the effect of attaching a stronger spring to the cone.

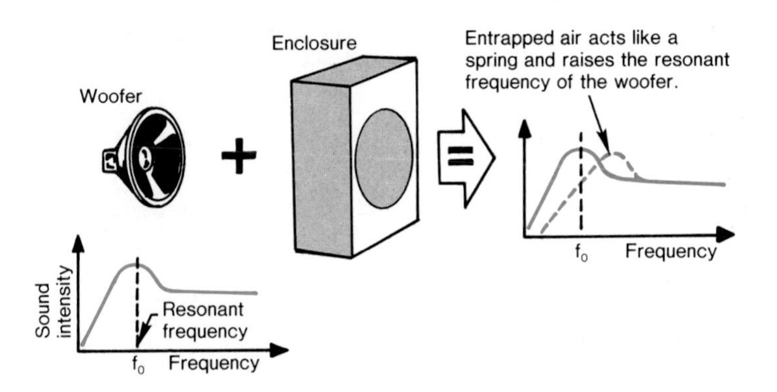

Figure 6.28. Placing a woofer in an enclosure raises its natural frequency of vibration, because the trapped air behaves like an additional spring acting on the woofer.

entrapped air raises the resonant frequency by only a small amount, say 10%. In this case the bass response is determined primarily by the mass of the large woofer and the stiffness, or compliance, of its suspension. The resonant frequency is not changed appreciably by the air within the box. How large are these enclosures? It depends upon the woofer size, and Table 6.1 gives some representative values.

Woofer Diameter	Size of Enclosure	Typical Resonant Frequency
20 cm (8″)	0.1 m^3 (3 1/2 ft^3)	100 Hz
30 cm (12″)	0.24 m^3 (8 1/2 ft^3)	65 Hz
38 cm (15″)	0.43 m^3 (15 ft^3)	50 Hz

Table 6.1 Some typical loudspeaker volumes for infinite baffle enclosures.

Note, using Table 6.1, that these enclosures can be quite large, especially for the largest woofers. The large diaphragm size means that its movement can be relatively small in order to produce loud listening levels. Mainly because of their large size, these infinite baffle loudspeakers have been replaced in popularity by the smaller acoustic suspension systems which will be discussed later on in this section.

The sound wave generated at the back of the diaphragm is absorbed in the infinite baffle design. The infinite baffle enclosures contain sound absorbing material to prevent the back-surface sound from experiencing internal reflections and thus interfering with the cone's motion. These reflections, if not absorbed, affect the cone's motion and distortion can be produced. This brings up a disadvantage of the infinite baffle enclosure, namely, that half of the sound power (the rear wave) is absorbed and is not available for listening. It would be nice if the rear wave could be put to good use and thus raise the efficiency of the loudspeaker. We will see, shortly, that the vented enclosure is designed to accomplish this.

Conclusion for the Infinite Baffle Enclosures

The infinite baffle enclosure utilizes a large diameter, short throw, woofer which is contained in a large volume enclosure. The trapped air does not appreciably affect the diaphragm's motion. The rear wave is lost to sound absorbing material which is contained within the enclosure.

The Acoustic Suspension (or Air Suspension) Loudspeakers, Another Kind of Sealed Enclosure

Acoustic designers have always dreamed of producing big bass sounds from small enclosures. With the popularity of stereo rapidly growing in the early 60s, it became clear that small, portable loudspeakers were needed which, in sound quality, could compete favorably with the larger systems. It just did not seem possible to achieve good bass from a small enclosure because, up to that time, everyone was thinking of the large woofer, large enclosure scheme of producing good bass. As usual, there is always more than one way to "skin a cat," and manufacturers began playing with the idea of using smaller woofers with large "throws" (section 6.5).

Enter the "bookshelf" speaker with its air suspension idea. Now, as we all know, there are two ways to produce loud sounds:

1. use a large diaphragm woofer with a small excursion (throw), or
2. employ a smaller diaphragm with a large excursion.

The first method is the principle behind the infinite baffle enclosure and the second method is the central theme of the acoustic suspension speakers. If a long throw is to be achieved, then the suspension must be extremely loose (highly compliant) in order to permit the long excursions. Acoustic suspension speakers are easy to spot. The suspension is usually a half-round roll which "gives" very easily when touched (Fig. 6.29). The speaker is like the mass vibrating on the spring (Fig. 6.16). If the weight is to move

ADS L710/SERIES II Acoustic suspension loudspeaker

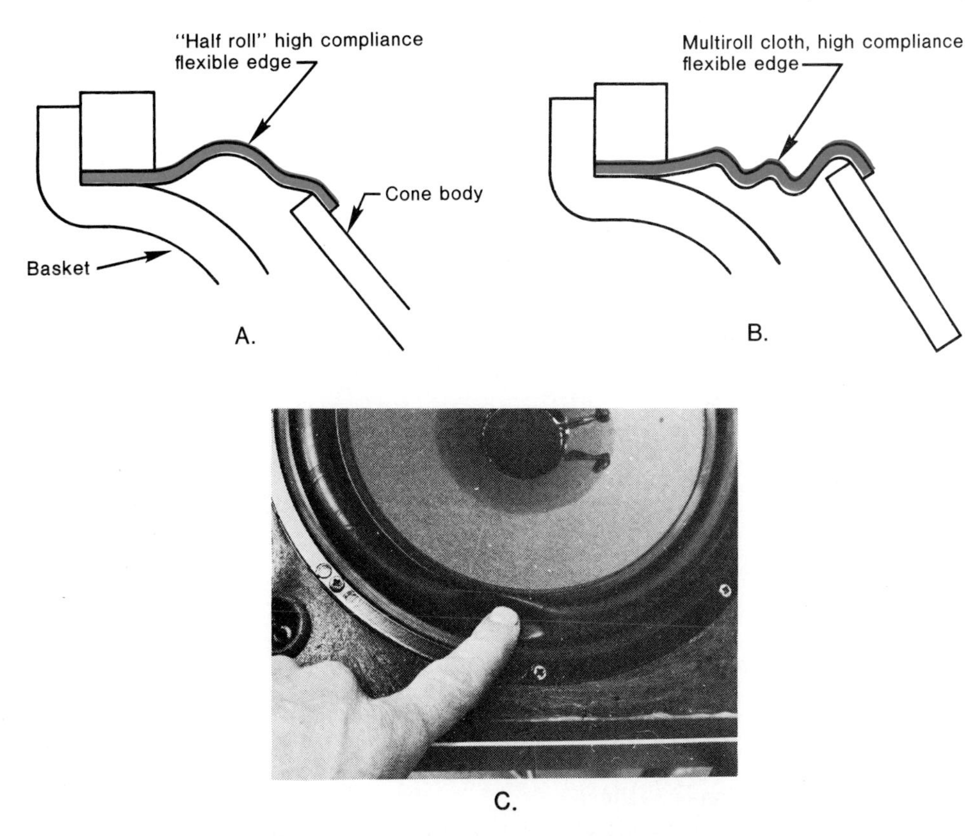

Figure 6.29. (A) and (B); Two types of high compliance flexible edges which are used in acoustic suspension loudspeakers. (C) The "soft" touch.

up and down with large displacements, then the spring must be extremely weak or highly compliant. Even though the cone is relatively small, it must be sturdily built, and hence rather massive, because it must withstand high speeds and great accelerating forces associated with long throws. A cone undergoing large displacements must be more rigidly constructed than one moving through relatively small excursions.

The heart of the acoustic suspension idea is in how the air within the sealed cabinet is used. The result of an extremely compliant suspension plus a rather massive cone (needed for mechanical strength) means that the resonant frequency of these cones is very *low*, normally in the neighborhood of 15 to 20 Hz. In principle, the bass reproduction should extend this far down. Now comes the enclosure. If the enclosure were to remain large, the low frequency resonance would remain around 20 Hz. However, the idea is to get good bass out of a *small enclosure*, so let us see what happens as it is made smaller. According to Figure 6.28, placing the highly compliant cone into a small enclosure will raise its resonant frequency to some value greater than 20 Hz, perhaps 35 to 50 Hz (Fig. 6.30). Now you can see that the design objectives of the famous "bookshelf" speaker have been met—good bass in a small enclosure. It is true that the cabinet has raised the woofer's natural frequency from 20 Hz to 50 Hz with some loss in bass reproduction. However, even 50 Hz is quite excellent because, as mentioned before, very few composers write music with notes below 50 Hz. The name

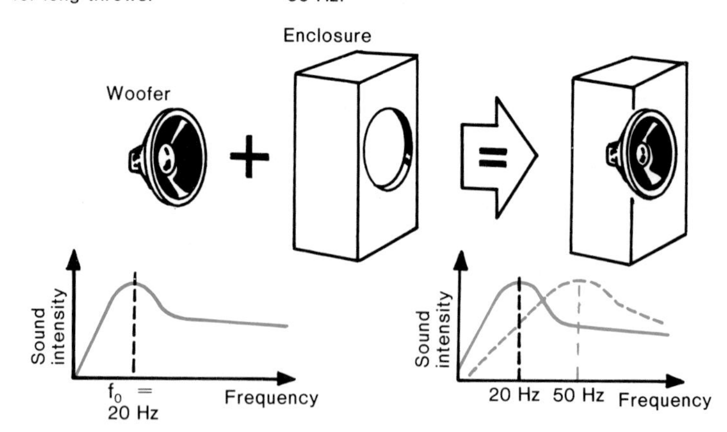

Figure 6.30. The acoustic suspension loudspeaker places a small, highly compliant woofer in a small cabinet to yield good bass response.

"Acoustic Suspension" or "Air Suspension" should now be apparent. The entrapped air literally provides a great deal of the suspension springiness.

Conclusion for the Acoustic Suspension Speakers

Using a highly compliant, long throw woofer which is placed in a compact cabinet, the popular acoustic suspension speaker has emerged which gives excellent bass (around 50 Hz) from a small enclosure. Of course, being a sealed enclosure, the acoustic suspension forfeits the rear sound wave by having the enclosure filled with sound absorbing material.

The Vented Enclosures ("Bass Reflex," "Tuned Port," or "Phase Inverter")

The vented enclosure is a very popular type of loudspeaker which attempts to increase the bass response by taking advantage of the sound produced from the back surface of the diaphragm. In its basic form the vented enclosure consists of an opening, or port, in the front of the enclosure through which the rear wave can exit (Fig. 6.31). In principle the supplemental rear wave could double the sound power produced by the speaker. At this point things should seem a little strange. If you will recall, the main purpose of the enclosure, in the first place, was to act as a baffle which prevented the cancellation between the front and rear waves. All of a sudden, how is it possible to let the rear wave out and not run headlong into this problem? It is possible to utilize the rear wave by being very clever and using the concept of interference to our advantage; only this time we shall use constructive, rather than destructive, interference.

The key idea behind the bass reflex design is to force the rear wave to travel an effective* distance of one-half of a wavelength before exiting through the port and combining with the front sound. At the time of creation the two waves are one-half of a wavelength out-of-phase because when the front surface is producing a condensation the rear surface creates a rarefaction. By delaying the rear wave by one-half of a wavelength, it will be in-phase with the front wave when they are combined.

ELECTRO-VOICE Interface: 1 Series II vented loudspeaker

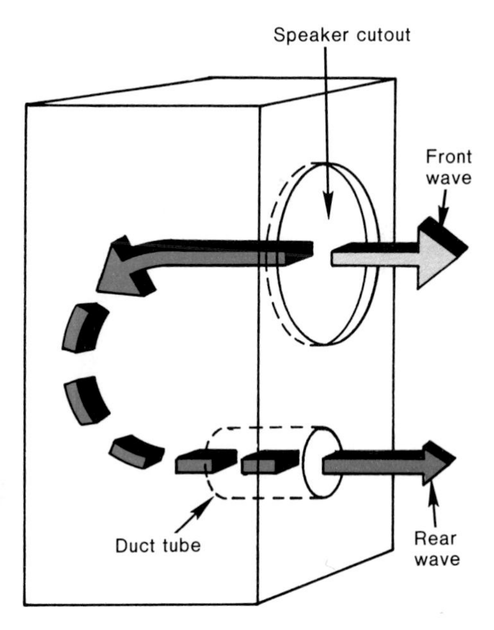

Figure 6.31. The vented enclosure utilizes the rear sound wave to enhance the bass response of the loudspeaker.

*In reality the bass reflex process is much more complicated than indicated in this discussion. Certain characteristics, like port size, enclosure size, and type of port, are all involved in changing the phase of the rear wave.

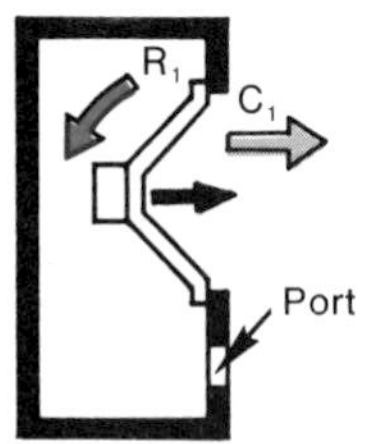

As the cone moves forward it creates a condensation C_1 and a rarefaction R_1 which travel in opposite directions at the speed of sound.

A.

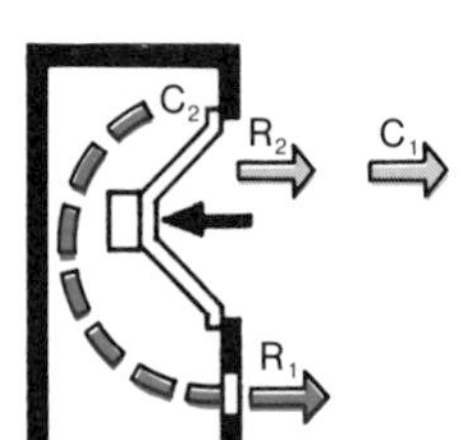

A short time later the rarefaction R_1 will arrive at the port or opening. If R_1 emerges at exactly the same time as the rarefaction R_2 is being created, then constructive interference will result, and the bass notes will be enhanced.

B.

Scale ᶜ

Figure 6.32. Details of the bass reflex action. The black arrow denotes the direction in which the diaphragm is moving.

Good feature: peaks are not nearly as pronounced.

Resonance peak of the driver without bass reflex action.

Sound intensity

Frequency

Note the extended bass of the bass reflex.

Figure 6.33. The effect of the bass reflex action, shown by the dotted curve, produces an extended and smoother bass response when compared to the response of the driver by itself.

Consider Figure 6.32 which shows the interesting details of bass reflex action:

1. Part (A) shows a condensation C_1 and a rarefaction R_1 being simultaneously created as the cone moves forward.
2. The rarefaction R_1 travels inside the enclosure at the speed of sound and, at some time later, emerges through the port, as shown in part (B).
3. Of course the speaker diaphragm continues to vibrate during the time required for R_1 to reach the port. Suppose—and this is the heart of bass reflex action—that the cone just happens to be moving *backward*, thus creating a new rarefaction R_2 at its front surface at the precise moment when R_1 leaves the port. Now R_1 and R_2, both being rarefactions, are *"marching together" as they leave the loudspeaker.* **A constructive interference results with the front sound being reinforced by the in-phase rear wave.** A slick trick!

Under these ideal conditions one can visually follow the bass reflex action. As the front surface of the diaphragm creates a condensation (rarefaction), there will be another condensation (rarefaction) simultaneously emerging from the port—*a condensation (rarefaction) which was created one-half a cycle earlier from the rear surface.* Reinforcing condensations will be produced, followed by reinforcing rarefactions, such that the rear wave which exits from the port is always in-phase with the front wave. Theoretically, the utilization of the rear wave should double the total sound power. Now you can see why many people become excited over bass reflex speakers.

The sound reinforcement acts only on the bass frequencies. It should be pointed out that the bass reflex action of constructive interference is greatest at the particular sound frequency where the path traveled by the rear wave is effectively one-half of a wavelength. For normal sized enclosures this frequency is usually less than 100 Hz. In practice the bass reflex action has the very nice effect of also smoothing out the resonance peak of the driver by producing two, much smaller peaks while extending the bass down to lower frequencies (Fig. 6.33). The bass reflex action is, admittedly, not an easy concept for most people to understand fully. Here are a few nagging points that might be bothering you:

1. In Figure 6.32 (B) it is shown that the constructive interference occurs between two condensations, or two rarefactions, each of which was created by the diaphragm at different times, e.g., R_2 and R_1, etc.

Isn't the sound quality, in some way, altered because the vented enclosure combines condensations (and rarefactions) which belong to different cycles of the sound wave? Not really. Stop and consider that the two interfering parts, R_1 and R_2 for example, are only one-half of a cycle apart. On a time scale this means that there is a delay between R_1 and R_2 of only 1/200 seconds for a 100 Hz bass note. This time differential is so short that the music has not changed between the time it takes to create these two rarefactions. Therefore, the two rarefactions can be added together with little or no change in the resulting sound quality.

2. Why don't the front and rear waves begin to cancel at frequencies other than the one for which constructive interference occurs? Normally they would, except speaker designers are very clever. The enclosure, with its port, is so designed that the higher frequencies are prohibited from leaving the port, thereby eliminating any possibility of destructive interference between the front and rear waves and the concomitant loss of sound. Only a small band of frequencies near the desired frequency are allowed to pass through, thus ensuring only constructive interference.

The Passive Radiators

There are bass reflex loudspeakers in which the open port is replaced by another driver, called the "slave" or "drone" driver. These loudspeakers are referred to as *passive radiators.* In such a design the slave driver is passive in the sense that it is not electrically connected to the power amplifier. Instead, it is driven by the sound wave coming from the rear surface of the electrically driven woofer. Using a slave driver permits audio designers to achieve the bass reflex action in enclosures that are smaller than those associated with the open-duct type of enclosures.

The Horn Enclosure

Everyone is familiar with the basic ideas behind the horn enclosure because we have all seen the use of a megaphone at one time or another. There are, in fact, two properties of a megaphone that are immediately apparent to a listener; one is good and the other is not so good. First, using a megaphone results in a much louder sound being produced. The horn has the very nice property of increasing the sound level with the same vocal effort than if it were not used. The megaphone itself does not actively amplify the sound like an electronic amplifier because it is just a plain old piece of cardboard all rolled up. However, its presence makes it easier for the sound waves to be radiated into the surrounding air, allowing the vocal cords to produce louder notes with relative ease. In loudspeaker terms, this means that the driver, when radiating into a horn, does not have to be as large in order to produce reasonable loudness levels (Fig. 6.34). This scheme is so efficient, in fact, that typical horn drivers used for midrange reproduction are only about one-third the diameter as compared with what would otherwise be required. When compared to other types of enclosures—such as the sealed and vented enclosures—the horns are, by far, the most efficient sound producers.

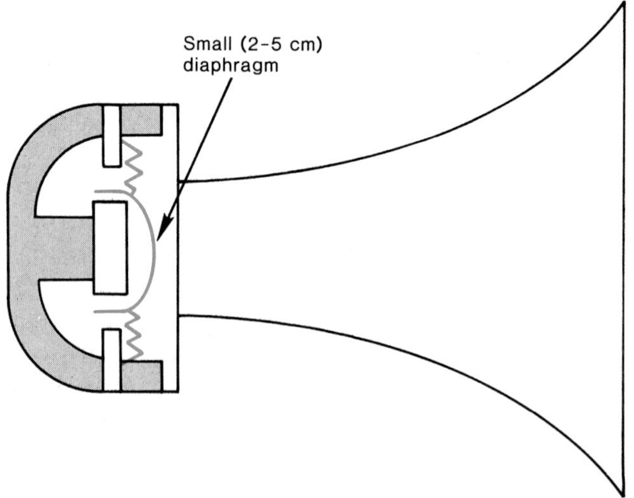

Figure 6.34. The driver in a horn loudspeaker contains a small moving diaphragm (shown in color) which feeds the sound waves directly into the throat of a horn. More sound is produced with less electricity.

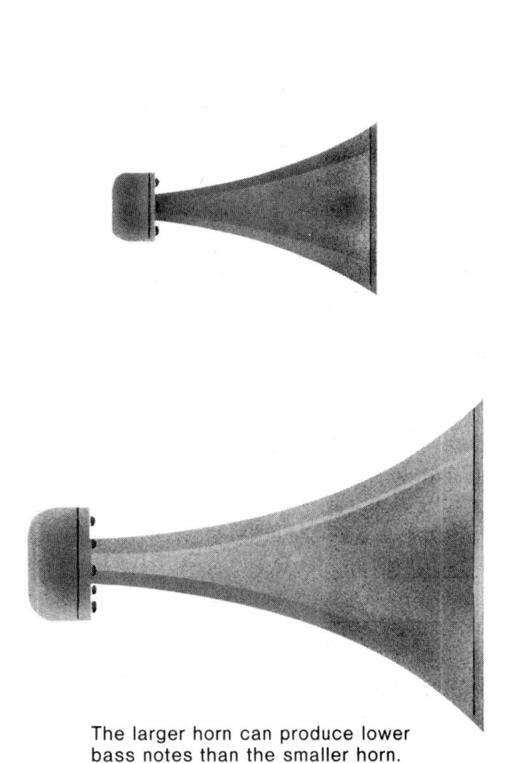

The larger horn can produce lower bass notes than the smaller horn.

Figure 6.35. The low frequency cutoff of a horn speaker depends on its size. Larger horns can reproduce deeper bass notes.

Now for the bad news. Besides a louder sound, careful listening will also show that the sounds emerging from the megaphone are somewhat shrill and hollow, caused by the lack of bass notes. All horn speakers have what is called a *cutoff frequency*, below which the horn will not reproduce any sound. The cutoff frequency, according to loudspeaker theory, depends upon the dimensions of the horn, with larger horns being able to reproduce lower bass notes than smaller ones (Fig. 6.35). The main reason that horns are seldom used for woofer applications is that they must be extremely large in order to reproduce the low notes. For example, the reproduction of a 40 Hz tone requires a horn which is 3 to 4 meters in diameter (at the mouth) and 4 to 5 meters long. Therefore, horns are used mainly as midranges and tweeters where only a reasonable size is needed.

6.10 Loudspeaker Specifications

Loudspeakers, like all audio components can be characterized in terms of technical specifications. Although almost all speaker manufacturers promote one or more of these specifications in their ads, it is generally recognized that no one parameter, or set of parameters, can be uniquely correlated with the sonic performance of loudspeakers. As we have mentioned many times, loudspeakers are rather complex entities with many interdependent variables affecting their performance. When shopping around for loudspeakers it is generally a good idea to have some understanding of these specifications. However, your final judgment on speaker *quality* should be based mostly on careful listening

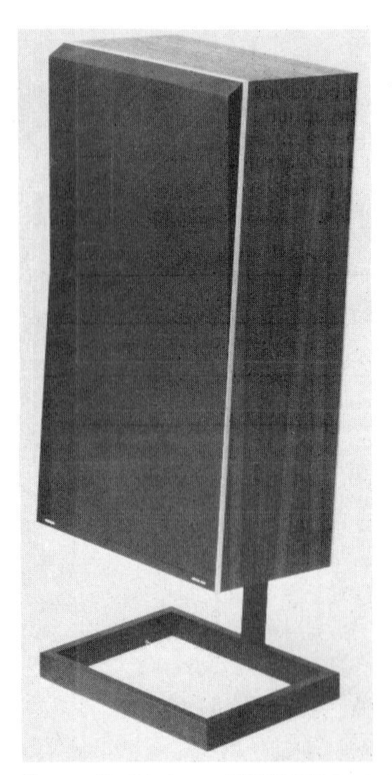

Bang & Olufsen BEOVOX M100-2 Loudspeaker

comparisons rather than on specifications alone. In this section we shall outline some of the more important loudspeaker specifications and the possible consequences which they might have on the listening experience. Such "specs" will include loudspeaker power ratings and sensitivity, frequency response, and dispersion. There are a host of other specifications which you will also see from time to time; specifications which have absolutely no relation to the quality of sound produced by a speaker. We shall have some comments on these items as well.

Power Ratings of Loudspeakers

The minimum recommended power rating of a loudspeaker is the smallest amount of electrical power from which the loudspeaker can produce quality music (i.e., low distortion) at reasonably loud listening levels in an average room. In a hi-fi system the electrical power sent to the loudspeakers comes from the power amplifier, and it must be able to supply at least the minimum recommended power to the speakers. Figure 6.36(A) depicts a speaker with a large minimum recommended power rating that is connected to an under-powered amplifier. In an attempt to achieve normal listening levels, the listener is forced to turn up the VOLUME control close to its maximum setting. When a loud transient pulse comes along, e.g., a cymbal crash, the amplifier has no more reserve power to accommodate the pulse and a serious distortion called "clipping" is produced. As surprising as it may seem, the distorted musical peaks produced by an under-powered amplifier can actually burn out a driver, particularly the tweeter. The distortion caused by "clipping" generates lots of high harmonics which are absent when there is no distortion. The high harmonics are routed to the tweeter via the crossover network and, since these harmonics do carry electrical power, they can damage the tweeter. Figure 6.36 (B) shows an amplifier which has an adequate power rating, and can drive the speaker without a significant amount of clipping distortion.

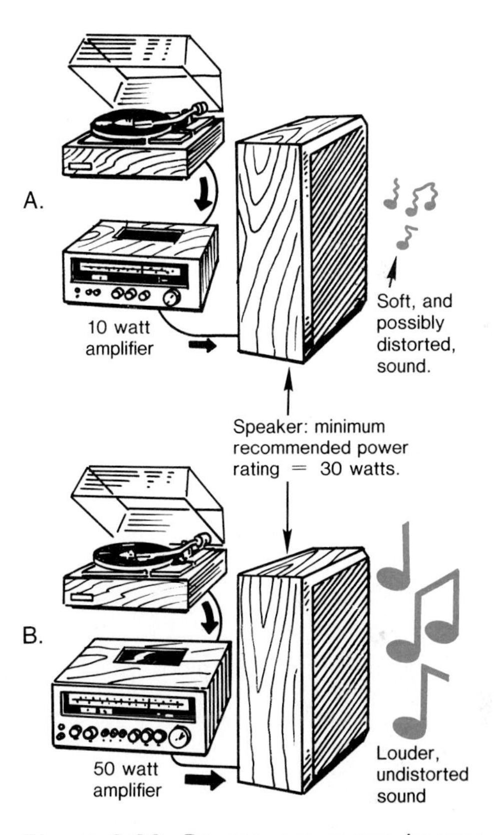

Figure 6.36. Do not use power hungry speakers with weak amplifiers.

The minimum recommended power rating is closely related to how efficiently the loudspeaker converts the electrical power into sound power. The more efficient a speaker is, the less its minimum recommended power rating will be. The reason is that an efficient speaker needs less electrical power input than an inefficient speaker to produce the same sound power output. The expression given for percent efficiency in Equation 6.1 nicely illustrates this point:

$$\%\ \text{Efficiency} = \left(\frac{\text{Output sound power (in watts)}}{\text{Input electrical power (in watts)}}\right) \times 100. \qquad \text{(Equation 6.1)}$$

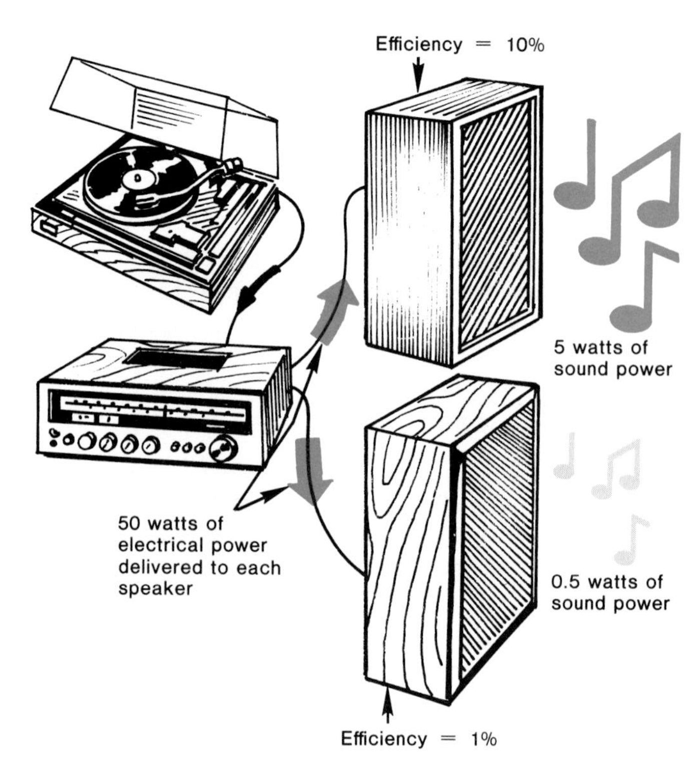

Figure 6.37. More efficient loudspeakers produce more sound power from the same amount of electrical power.

Suppose, as in Figure 6.37, that 50 watts of electrical power are sent to each of the two speakers. The speakers are identical except that one has an efficiency of 10% while the other has an efficiency of 1%. From Equation 6.1 we can calculate the sound power to be expected from each speaker:

$$10\% = \frac{\text{Output sound power}}{50 \text{ watts}} \times 100,$$

$$\text{Output sound power} = 5 \text{ watts},$$

and

$$1\% = \frac{\text{Output sound power}}{50 \text{ watts}} \times 100,$$

$$\text{Output sound power} = 0.5 \text{ watts}.$$

The more efficient speaker clearly gives more sound power for the same electrical input.

You may be wondering what happens to the electrical power which is not converted into sound power by the speaker. The answer is that it is degraded to heat inside of the speaker. In the above example 10% and 1% were selected because they represent the approximate maximum and minimum speaker efficiencies obtainable on today's market. It is amazing, but true, that most speakers waste at least 90%

of the electrical power they receive. The 10% speaker in our previous example wastes 45 watts (50 − 5 = 45), while the 1% speaker wastes 49.5 watts (50 − 0.5 = 49.5), as shown in Figure. 6.38. It has been said, tongue-in-cheek, that a hi-fi loudspeaker is really just an efficient heater in disguise, which "wastes" a little of its input power in the form of sound!

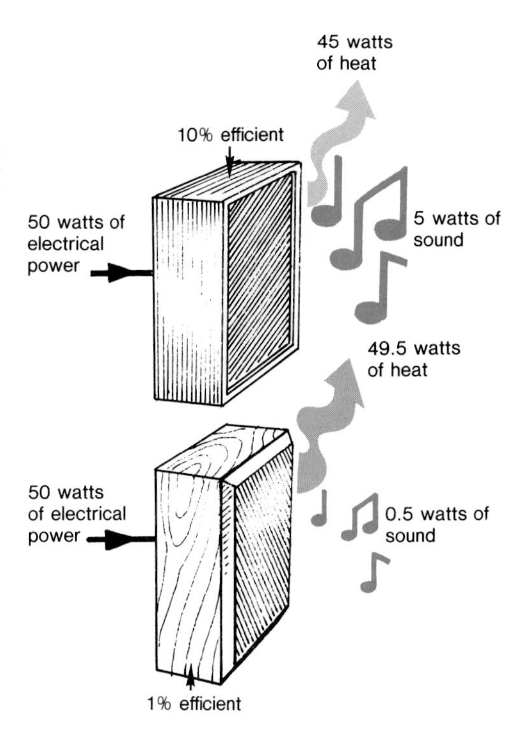

Figure 6.38. All loudspeakers waste most of the electrical input power as heat; very little gets converted into sound.

The maximum power rating of a loudspeaker is the largest amount of electrical power the speaker can receive without being damaged. If the amplifying system can deliver more power than the speaker can safely handle, then the drivers can be easily blown out by simply turning up the VOLUME control on the amplifier. If the power output capability of the amplifier should happen to exceed the maximum power rating of the speakers, it may be wise to install fuses (see Chapter 7) in order to protect the speakers from the disaster shown in Figure 6.39.

The above discussion about the minimum recommended power and the maximum power rating reveals that the amplifying system and the speakers should be matched as far as power is concerned. In general, it is a good idea for the power rating of your amplifier to fall somewhere between these two specifications. For example, suppose that you have selected a pair of speakers with the following ratings:

Minimum Recommended Power Rating: 10 watts
Maximum Power Rating: 40 watts.

A 25 to 30 watt amplifier would be very appropriate, because it is unlikely (being below the 40 watt maximum power rating) to blow out the speakers, and it affords some reserve power above the 10 watts to handle the sudden peaks in the music. Notice that your selection of speakers dictates, to a certain extent, the power, and hence cost, of the amplifier. Speakers with larger minimum power ratings require more powerful amplifiers to drive them, and this invariably translates into a higher amplifier cost.

If you already own an amp which is rated, for example, at 20 watts, then you would do well to consider speakers whose power ratings might fall in the range

Minimum Recommended Power Rating: 1 to 5 watts
Maximum Power Rating: greater than 25 watts.

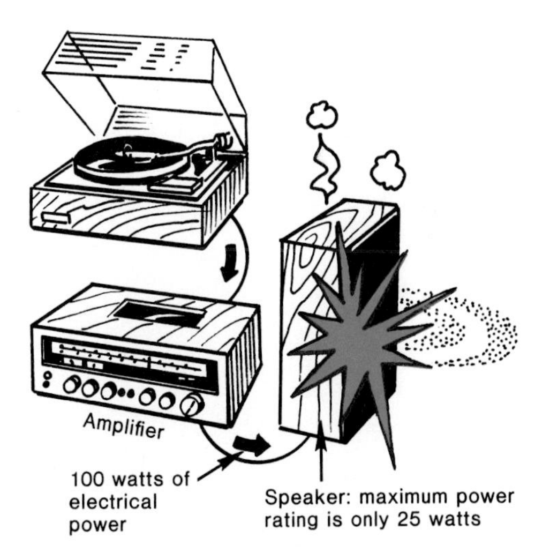

Figure 6.39. Do not exceed the maximum power rating of a loudspeaker by putting more electrical power into it than it can handle.

Notice that nothing has been said regarding the power ratings vs. sound quality of speakers. The reason is because there is no correlation. The power ratings simply give the buyer some indication as to the amount of amplifier power which is required to drive the speaker. They are in no way related to how good the speaker sounds.

The sensitivity rating of a loudspeaker is a measure of the sound intensity, expressed in dB, which a speaker will produce when it is driven with a specified input power. For example, a sensitivity rating of

85 dB at 1 meter, 1 watt input,

Bose Loudspeakers

means that 1 watt of input electrical power will produce a sound intensity of 85 dB when the listener is located 1 meter (approximately 3 feet) in front of the speaker. Sensitivity ratings, therefore, are an indication of the sound loudness that can be produced by a loudspeaker. Larger values (90 dB, 95 dB, etc.) signify that the loudspeaker will sound louder, but not necessarily "better," than a speaker which has a smaller sensitivity. Unfortunately, sensitivity ratings may vary from manufacturer to manufacturer, who quote either different input powers, e.g., 10 watts, or different speaker-listener distances, e.g., 3 meters. It is difficult to compare two sensitivity ratings if they are not stated in precisely the same manner.

As discussed in Chapter 5, ten times more input power is required to produce a +10 dB increase in the sound intensity (see Table 5.1). Therefore, a speaker with a sensitivity rating of 85 dB for a 1 watt input would require 10 watts of input power to produce a 95 dB intensity level, and 100 watts for 105 dB. If you enjoy listening to music at the 105 dB level, which is very loud, then the sensitivity rating of the speaker tells you (after a little calculation) that you would need a 100 watt amplifier to do the job properly.

Frequency Response

The frequency response, as discussed in section 5.8, indicates the frequency range over which the speaker can be expected to deliver its sound power. Of course, as with any frequency response measurement, the deviation from perfect "flatness" (expressed in decibels) must also be included. For example, a properly stated frequency response specification would be written as: 40 Hz to 16,000 Hz ± 4 dB. Sometimes a specification will read "flat, 40 Hz to 16,000 Hz." This implies that the output sound power will vary by no more than +1 dB, or less than −1 dB, over the specified frequency range. Very few speakers on the market today are "flat" to within ±1 dB over this frequency range, so beware of ads, especially for lower priced units, which make this type of claim. Usually, however, the phrases "flat" or "±4 dB" are missing altogether and manufacturers only specify the *frequency range*, such as 30 Hz to 20,000 Hz. In such cases it is safe to assume that the sound level at the two extremes of 30 Hz and 20,000 Hz has dropped at least −10 dB from its nominal value, usually taken to be the sound level at 1 kHz.

> **NOTE:** Sometimes "frequency response" is incorrectly used to mean "frequency range." Whenever a specification for the frequency response does not include the ± dB factor, it is not a true frequency response specification. It represents the frequency range.

Do not confuse "frequency response" with "frequency range." Frequency response is more descriptive of the speaker's performance because it states the maximum variation which can be expected in the output sound over the specified frequency limits. The frequency range only shows the frequency limits at which the output sound level has dropped by −10 dB; it says nothing about the sound level variations within this range. The "frequency range" is almost a meaningless specification.

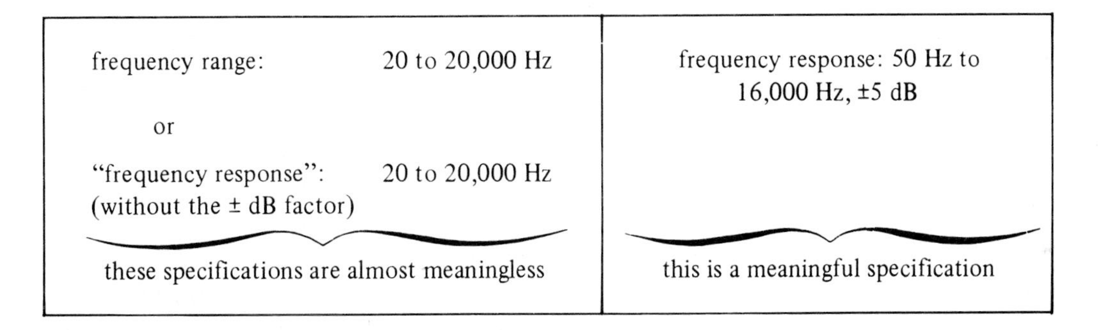

frequency range: 20 to 20,000 Hz

or

"frequency response": 20 to 20,000 Hz
(without the ± dB factor)

these specifications are almost meaningless

frequency response: 50 Hz to 16,000 Hz, ±5 dB

this is a meaningful specification

It might appear that the buyer should attempt to purchase speakers whose frequency response matches that of the ear: for example, 20 Hz to 20,000 Hz ±5 dB. Such wide range speakers are available although they tend to be relatively expensive. Besides price there is another good reason why speakers with a smaller frequency response might be acceptable. Virtually all records cut today do not contain frequencies above 16 kHz nor frequencies which lie below 50 Hz. Therefore, it might be quite reasonable to purchase a loudspeaker which has a response from 50 Hz to 16,000 Hz ±5 dB.

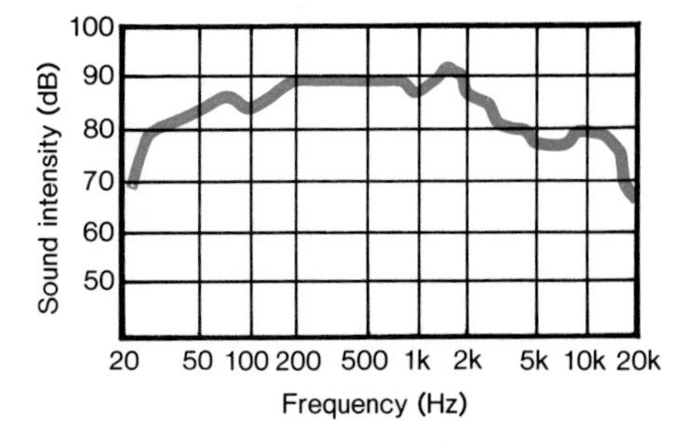

Figure 6.40. A typical loudspeaker frequency response curve.

A much better method of displaying the frequency response characteristics of a speaker is to plot a graph as shown in Figure 6.40. Such a curve not only displays the maximum deviations in the speaker's response, but it also

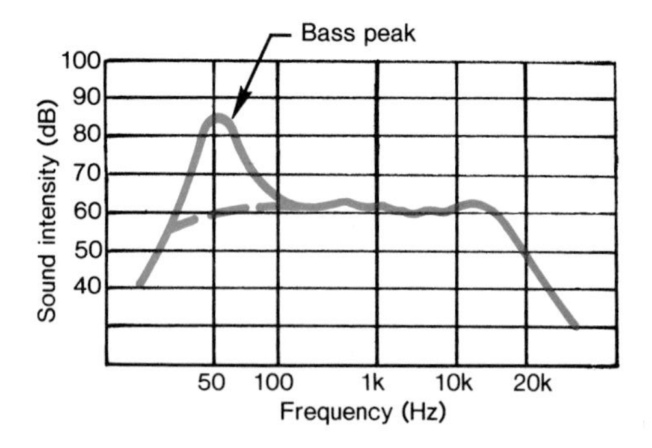

Figure 6.41. The severe peak in the response curve near 50 Hz causes the speaker to sound "boomy."

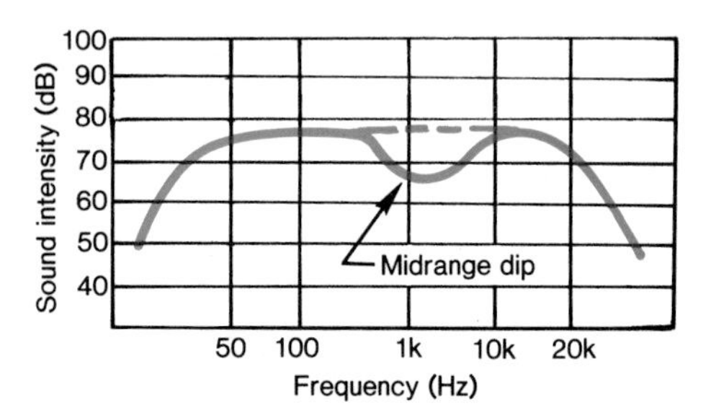

Figure 6.42. A severe midrange dip in the speaker's response will cause a loss of "presence."

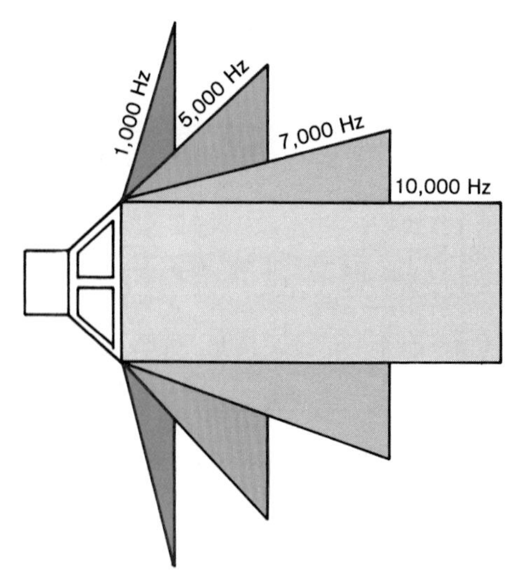

Figure 6.43. High frequency sound spreads out little after leaving the loudspeaker. The higher frequencies are said to have less dispersion than the lower frequencies.

indicates the frequencies where they occur. Sometimes these curves appear rather jagged-looking, as in Figure 6.40, although one should not be alarmed at the presence of small dips and bumps. These slight variations are sometimes present in even the finest loudspeakers and tend to be inaudible. It is the wide dips or large bumps that have the greatest influence on the quality of sound that you hear. Good speakers should produce all portions of the sound spectrum with no significant dips and peaks. A pronounced peak in the bass region, as shown in Figure 6.41, can lead to a "boominess" in the music. If you have ever listened to a juke box, the bass is immediately noticeable and it appears to jump out at you. This is caused by an unnatural bump in the bass region and most people readily tire of this "sound" after a few hours.

In a very real sense a considerable amount of acoustic energy lies in the midrange between approximately 500 Hz and 5,000 Hz. Most recordings contain the vocals "up front" with respect to the accompanying instruments. If speakers which you are auditioning place the vocals behind the drums, for example, a midrange "dip" is usually present as shown in Figure 6.42. Such speakers are said to lack "presence." In general, the pronounced features of a frequency response curve become very apparent to the ear when A/B comparisons are being made with a speaker which has a relatively flat frequency response.

Speaker Dispersion

It was mentioned in section 4.4 that all sounds do not spread out uniformly into the listening area once they leave the speaker. Because of diffraction the low frequencies spread out much more than the higher frequencies (Fig. 6.43). Therefore, the lower frequencies are audible throughout the entire room, but the higher frequencies are heard best directly in front of the loudspeaker. The higher the frequency, the narrower the beam of sound which the speaker emits. The differential spreading of the sound waves is called speaker dispersion, and it is important because the listener will begin to lose the high frequency notes when walking in an arc away from the center axis of the speaker. A well-designed speaker should distribute the high frequency sounds over as wide an area as possible (good dispersion). Speaker manufacturers will often include a specification to denote this property. Such a specification might read:

120 degree dispersion, ±6 dB between
50 Hz and 16,000 Hz.

This specification suggests that if you move along an arc anywhere within 60 degrees on either side of the speaker's center axis, the sound loudness will essentially remain the same (to within ±6 dB) over the frequency range from 50 Hz to 16,000 Hz. Like the specification for frequency response, the dispersion specification needs the ± dB notation to have full meaning. Sometimes a speaker manufacturer will only state the dispersion as: 120 degrees. This abbreviated form is almost meaningless because it neither specifies the variation in loudness (± dB) nor does it state the frequency range over which the 120 degree dispersion is applicable.

A much better method of revealing speaker dispersion is to include a so-called "polar diagram." A polar diagram, as shown in Figure 6.44, is a loudness curve drawn on a 360-degree circular grid, where the distance from the center of the circles indicates the relative loudness level. The loudspeaker is assumed to be located at the center of the circles and facing along the 0° line. Three different frequencies, 100 Hz, 1.5 kHz, and 5 kHz are shown in Figure 6.44. Note that the 100 Hz curve is almost a perfect circle, which means that this note will appear equally loud when the listener walks completely around the speaker. The 1.5 kHz tone shows poorer dispersion because its loudness is greatest along the speaker axis, 0°, while the loudness progressively decreases as one moves around and behind the loudspeaker. For example, when the listener walks through an angle of 135° from the speaker axis the loudness of the 1.5 kHz tone has dropped almost to zero. The 5 kHz curve obviously shows the least amount of sound dispersion, and the sound pattern is definitely "beamed" along the speaker axis. For good dispersion the highest frequency shown on such a polar diagram should fall off by no more than 6 to 8 dB for points which are 60 degrees on either side of the speaker's axis.

A simple at-home test for dispersion is to use a single speaker while listening to an FM station. Tune between two stations with the muting switch turned OFF, such that the interstation "noise" or "static" is heard through the loudspeaker. The noise that you will hear is a fairly even blend of all audio frequencies, and it is a good source for evaluating speaker dispersion. Now, using this interstation noise, walk around the speaker in an arc, starting from its center axis, and keep the distance between you and the speaker always the same. As you walk around, listen for when the high frequencies begin to decrease appreciably in loudness, and note the approximate angle at which this occurs. This test will give you a rough estimate of the speaker's dispersive properties. If moving 50° to 60° off axis, as shown in Figure 6.45, does not drastically change the balance between the high, midrange, and low frequencies, then the speaker's dispersion is excellent.

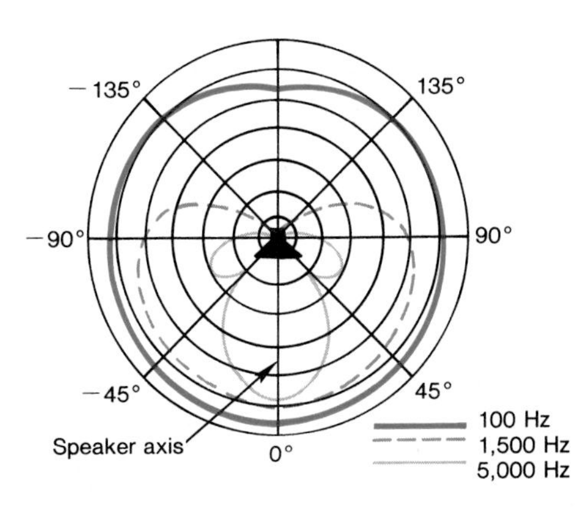

Figure 6.44. A polar diagram shows how the different sound frequencies are spread out, or dispersed, in various directions from a loudspeaker. The speaker is assumed to be located at the center of the circles and facing along the 0° line. The distance from the center of the circular grid indicates the loudness level.

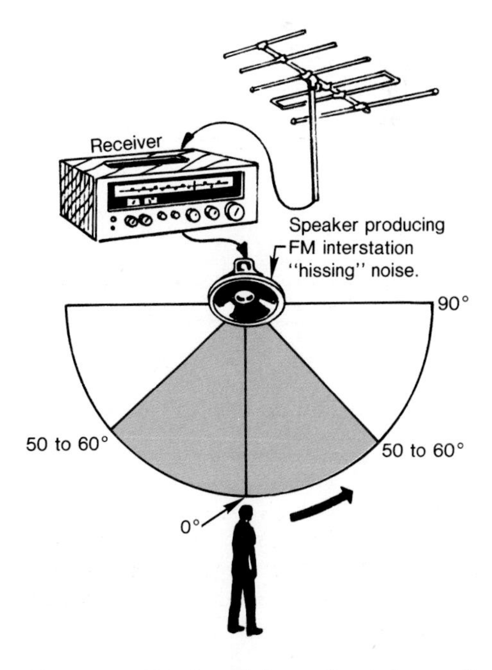

Figure 6.45. A quick method for estimating loudspeaker dispersion.

Additional Loudspeaker Specifications

There are many parameters of loudspeaker design which, by themselves, have *very little bearing* on the sonic performance of the loudspeakers. Contrary to this fact, speaker manufacturers often promote these parameters in their advertisements, hoping that this "specsmanship" will entice prospective buyers into choosing their products. We shall present a list of these entities with the hope that you will not equate their presence in a speaker "ad" with *sound quality*.

1. **Speaker Efficiency.** This specification is important when you are interested in matching the speaker to the power rating of your amplifier. Efficient speakers produce louder sounds from a given amount of input electrical power. As discussed above, efficiency is related to the minimum recommended power rating. As long as the amplifier's power rating exceeds the minimum recommended power rating of the speaker, it does not necessarily follow that more efficient speakers will sound *better*, although they certainly will sound louder.
2. **Number of Drivers Housed Within the Enclosure.** Depending upon the manufacturer, loudspeaker enclosures may contain anywhere from two to ten drivers (see section 6.4). It is always tempting to postulate that an enclosure with eight drivers, for example, will sound better than one containing only four. It may or it may not. After careful listening, do not be alarmed if you have chosen a loudspeaker with three drivers and the "rejects" have more.
3. **Diameter of the Drivers.** Woofers come in a variety of diameters ranging all the way from the 76 cm (30″) elements to the diminutive 15 cm (6″) ones (see section 6.5). Although the larger drivers certainly look more impressive, size alone does not guarantee good sound. The diameters of the midranges and tweeters are even less significant in predicting sound quality.
4. **Crossover Frequencies.** There is considerable latitude among manufacturers as to where the crossover frequencies should occur in two-, three-, and even four-way systems. It is by no means clear that the selection of 500 Hz and 3.5 kHz crossover frequencies in a three-way system is any better or worse than 700 Hz and 4 kHz. There are simply too many other variables which must also be considered when designing a good system.

5. **Magnet Structure and Voice Coil Diameter.** Sometimes you will see manufacturers quoting either the weight of the permanent magnet or the diameter of the voice coil. While these two parameters are important to audio engineers who are designing an overall loudspeaker, heavier magnets and larger voice coils, by themselves, do not ensure a "better bass" sound.
6. **Speaker Impedance.** Speaker specifications always include a parameter called "impedance." Most speakers on the market are rated at 8 ohms of impedance although there are some which are rated as low as 4 ohms, or as high as 16 ohms. As we shall discuss in Chapter 7 the impedance of a speaker is a measure of how much electrical power it can "draw" from the power amplifier. Smaller impedance ratings imply that the amplifier can deliver more power to the speakers. Again, do not confuse quantity with quality. Speaker impedance plays no role in a loudspeaker's sonic performance.

6.11 Hints on Buying Loudspeakers

Because of the large number of loudspeakers on the market today—each with its own unique sound—the selection of a speaker can be a time-consuming, although enjoyable, task. Unlike all other audio components there are no hard and fast rules which will automatically ensure both quality reproduction and a "sound" suited to the buyer's taste. However, the selection process can be aided by some general guidelines sprinkled with a few warnings about common loudspeaker myths.

Forming an Initial Opinion

Before you ever walk into a hi-fi store, it is always a good idea to have some general ideas regarding both models and prices; otherwise you are at the complete mercy of the salesperson. Remember, salespeople neither have to pay your bills nor do they have your ears for listening, which are the two things that this speaker selection game is all about.

Let's face it, the bottom line for most of us is the price. Since there is such a diversity of speakers, ranging all the way from $30 to $1,500 **each**, it is always a good idea to keep in mind an affordable price range. This can not only keep you out of debtor's prison, but it can help limit the number of speakers which must be auditioned. Some previous knowledge of loudspeakers helps both you and the salesperson. Just how much should be spent is obviously a

EPICURE model 3.0 Loudspeaker

personal matter, but remember that any price difference between two speakers will be magnified by a factor of two when purchasing a stereo pair. The difference between $200 and $250 may not seem like a large amount but the pairs would cost $400 and $500 with the difference now being a substantial $100. Also, within certain limits, a higher priced loudspeaker may not sound better to you than a lower priced one, so do not automatically assume that price means quality. It is certainly true that an $800 speaker will be an all-around better performer than a $75 one. However, speakers being what they are, it does not follow that a $250 speaker will sound better—even after careful and prolonged listening—than a $200 one. As is always the case with loudspeakers, let your ears, with careful listening, tell you which one is the best suited to your taste.

Perhaps the best way to form an initial opinion about speakers is to listen to the systems owned by your friends. The word "listen" is very important, because it implies that you have spent considerable time—several hours or more—hearing the loudspeakers produce your favorite music. Listening to speakers for long periods of time is important because a new speaker may initially impress you, but after a short while you may quickly tire of the "sound." This is commonly known as "listener's fatigue," and it results from the subtle effect of the speaker producing excessive distortion which begins to irritate the listener after a period of time. "Listener fatigue" will subtly suggest that you turn off the sound.

Good Buying Hint

If you can listen to and enjoy the speakers for hours on end without any apparent fatigue or sonic discomfort, then the speakers pass an important test in the selection process.

Rarely does one get the opportunity at a store to spend hours listening to one or two different speakers, so listening to friends' speakers can be a major factor in aiding the selection process. Simply ask yourself: "Do I get tired of listening to these speakers, or could I listen to them all day?"

Auditioning at the Hi-fi Store

Auditioning speakers at the store means that side-by-side comparisons can be easily made using a comparator. Of course, as discussed earlier, you generally do not have enough time to determine if listener fatigue will be a problem. However, the store experience will let you narrow the field down to one or two speakers which you can, hopefully, take home on a trial basis for more extensive listening. Here are some guidelines that can help you make a reasonable choice among a large variety of candidates.

Using a comparator, like the one illustrated in Figure 6.46, is the best way to compare loudspeakers. The comparator is an electronic switching device which allows the listener to switch rapidly between two or more loudspeakers while playing the same music. Rapid switching is essential, because it is important to hear how loudspeaker "A" reproduces the music *at approximately the same time* as loudspeaker "B." Our audio memories tend to be rather short and can fool us even after several minutes have elapsed. Ask that all speaker models which you might prefer be linked to the comparator. Since most comparators can handle 5 to 8 stereo pairs, you might also ask that additional speakers which lie outside of your price range be included, just for the sake of interest. The store should have a quiet, isolated, listening room so that you can audition speakers without distracting "store noise," such as conversations and other music.

Figure 6.46. The best way to compare speakers is to use a comparator.

Use an amplifying system which has a power output similar to the one you will be using. If you are planning to use an amplifier whose power output is 10 watts, then you may be misled using a 100 watt amplifier to listen to speakers in the store. Some really great speakers simply cannot perform well unless they get a lot of power from the amplifier. At any rate, check with the salesman so that you know at least if the performance characteristics of his amplifier are very different than yours.

Bring in one or more of your favorite albums—no tapes or FM broadcasts—with which you are totally familiar. Using your own records makes it much easier to compare the performances of different speakers. Generally speaking, tape recordings and FM stations are not as good as records for auditioning speakers. Almost all tapes and FM programs are derived originally from records. Why not play your own records and get "first generation" quality without the extra noise and distortion which has been introduced in the broadcast or in the tape recording. It makes good sense to audition speakers using music with which you are totally familiar. In order to judge the merits of a loudspeaker one must first have a knowledge of how the music should sound. This means, acoustically speaking, that you know what to listen for in terms of instrumental and vocal sound quality.

When using the comparator only compare two speakers at any one time—the so-called A/B listening test. Do not attempt to compare a third speaker until you have determined which one of the originals, A or B, you prefer. Comparing three speakers at one time will only confuse you and make the selection more difficult. When you are convinced that A is better than B, then repeat the entire comparison test using speakers A and C. Always compare two at a time using the previous "best" with the current speaker. In this manner you will arrive at a final candidate for possible purchase.

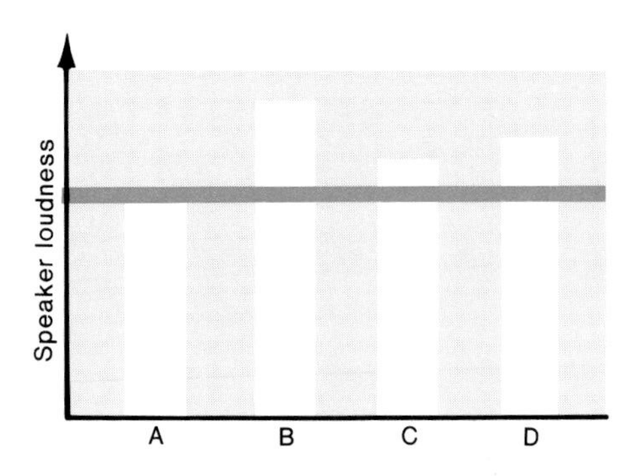

A. The ear gives speaker B an unfair advantage over the others because B is the loudest.

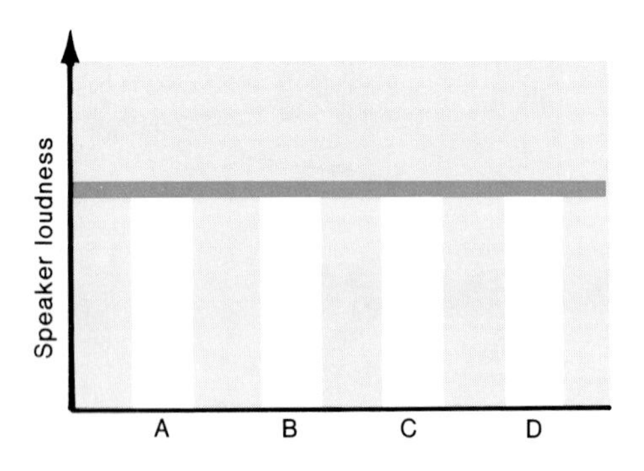

B. No speaker has an unfair advantage when they are all equally loud.

Figure 6.47. Use the comparator to adjust all speakers to the same loudness. A more accurate comparison of speaker quality can then be made.

Be sure that the loudness of each loudspeaker is the same. Remember the Fletcher-Munson curves (see Figure 6.12); at low sound levels the ear loses its ability to hear the high and low notes. Therefore, speaker "A," being played at low volume, can sound worse than speaker "B" which is being played at a higher sound level. However, with equal loudness, "A" may sound better than "B." In short, speaker comparisons are valid only when the speakers are playing equally loud, because the ear is naturally biased in favor of louder sounds, as indicated in Figure 6.47. To eliminate such a bias, adjust the volume controls on the comparator so that all speakers sound equally loud. Set the loudness at about the sound level you plan to use at home.

Keep the listening times for the speakers approximately equal. Listen to a speaker 10 to 30 seconds before switching to the second one. Keep the listening times roughly the same because the brain sometimes "likes" one sound better than another simply because it has heard one for a longer time.

Listen for the things you like best in the music, and judge which speaker reproduces these features better. Telling someone what to listen for in a speaker makes about as much sense as telling someone what to look for in a husband or wife.

If you ask a professional reviewer, such as those who write articles in hi-fi magazines, they will describe speakers with words such as "thin," "gutsy," "presence," "glassy," "bright," "heavy," etc. We are sure that these sonic terms have very definite meanings to each individual reviewer, but it is nearly impossible to translate them into something upon which we would all agree. For the average person who is buying speakers for the first or second time these terms have little meaning.

You will not go far astray if you simply ask yourself "*Is there something missing from the music*?" Or, more positively, "*Does this speaker offer audible advantages over the other one*?" This is where your knowledge of the type of music becomes important, because only you can define what is meant by "something missing" or "audible advantages." It could be a lack of bass, a dull thud instead of a crisp boom, too much midrange emphasis relative to the bass and treble, or sharp transients which do not die away as fast as they should. In any case it is a very personal and subjective judgement. While listening, make an attempt to compare the speaker's high frequency, midrange, and bass responses. With a little practice a person can selectively choose a region of the audio spectrum in which to listen. Remember, also, that a speaker is somewhat like its "distant cousin" the musical instrument, because it adds certain tonal qualities of its own to the music. Like musical instruments no two speakers—even mates of a stereo pair—have the exact same tonal quality.

Ask if you can take your first choice of speakers home for a trial listening period. In that way you can determine if listener fatigue will be a problem, and if the speakers still sound good in your room. Remember that room acoustics can markedly affect sound, and those speakers which sounded superb in the dealer's showroom may be a disappointment when you take them home. In all candor there are not many dealers who cheerfully lend out speakers, but it is worth asking about.

One final note. Any time and effort spent in auditioning speakers is well worth it, because a hi-fi system sounds only as good as its speakers!

6.12 Loudspeaker Placement

It is well-known that concerts can sound vastly different in one auditorium as compared to another. Likewise, the sound from loudspeakers is greatly influenced by the room in which they are located. In fact, simply moving the speakers about the same room can have a marked effect on the sound. Therefore, the question of speaker placement in the listening room is an important one. However, there are no

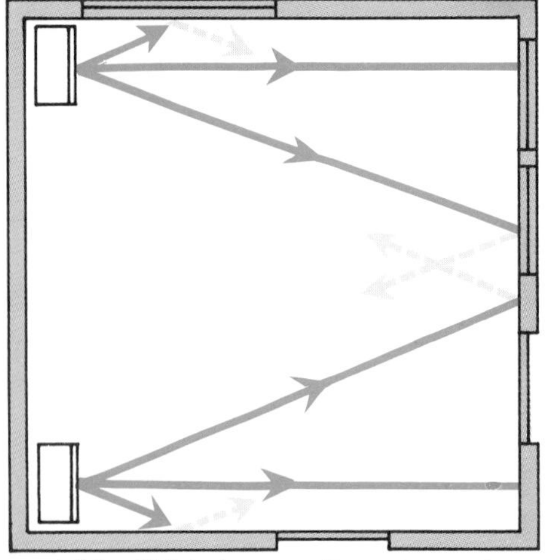
A "dead" listening room (few reflections)

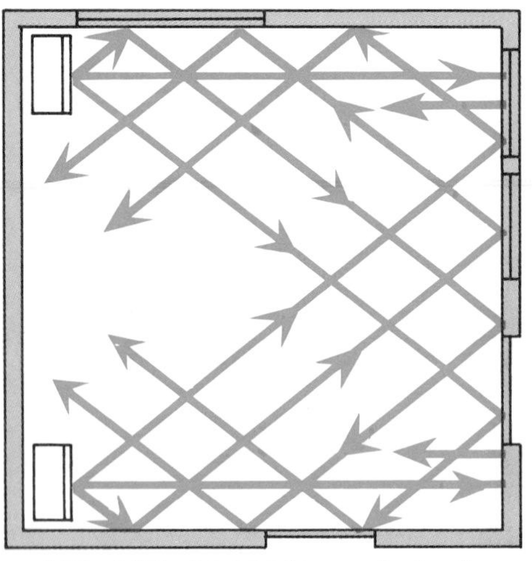
A "live" listening room (many reflections)

Figure 6.48. The two extremes in a listening room which can lead to poor acoustics; a "dead" room and a "live" room.

hard and fast rules by which to decide where to place speakers. About the closest that one can come to a rigid rule is to say "try it, you might like it," because a pragmatic attitude is very useful when experimenting with the various possibilities. While there are no rigid rules, here are some guide lines which might be useful for obtaining the best possible sound from a hi-fi system.

Try to avoid extremes when choosing a listening room. Many of the effects of speaker placement are related to the acoustic properties of the listening room, and words such as "live" and "dead" are sometimes used to describe the room. An acoustically "live" room is one in which there is little or no sound absorption. The best example of a live room is a tile-lined bathroom. The hard, smooth surfaces reflect the sound with very little absorption. Accordingly, the sound bounces around for quite a long time before it finally dies away, as shown in Figure 6.48. A live room has a large reverberation time, and it always sounds louder because of the sustained build-up of the sound waves. (See section 3.8) The opposite of a live room is a "dead" room. A completely "dead" room is also shown in Figure 6.48 for comparison, and in such a room there are very few reflections. Most of the sound is absorbed by heavy rugs, drapery, furniture, etc. It is wise to furnish the listening room so that it falls between the two extremes, as neither one makes for good sounding music. In the "live" room the music will take on an echo-like quality as the sound reverberates, with the notes lasting long after the speakers have stopped producing them. All of the individual features of the music will be blurred into one another, much like a photograph in which the motion was too fast for the camera to freeze. As we will see shortly, some reflection of sound is beneficial, however. In a completely "dead" room the music will have a dull and lifeless quality. In fact, since the sound reaches the ears only along a direct path when there are no reflections, the listener would have to sit directly in front of the speakers to hear anything at all in a completely "dead" room!

The corners of the room may be used to create "acoustic images" of the speakers. To understand what acoustic images are and how proper speaker placement can give you their benefits, imagine that the floor of the listening room is a mirror. What would you see? Of course, you would see the speaker itself sitting on the floor. However, you would also see an image of the speaker reflected from the mirrored floor, as shown in Figure 6.49. Listening to speakers is like "seeing" with your ears. As far as sound is concerned, the floor really is a mirror, particularly for the bass tones. Remember that the low frequencies spread out or disperse readily because of diffraction, and they leave the speaker in many directions. This diffractive property causes an abundance of

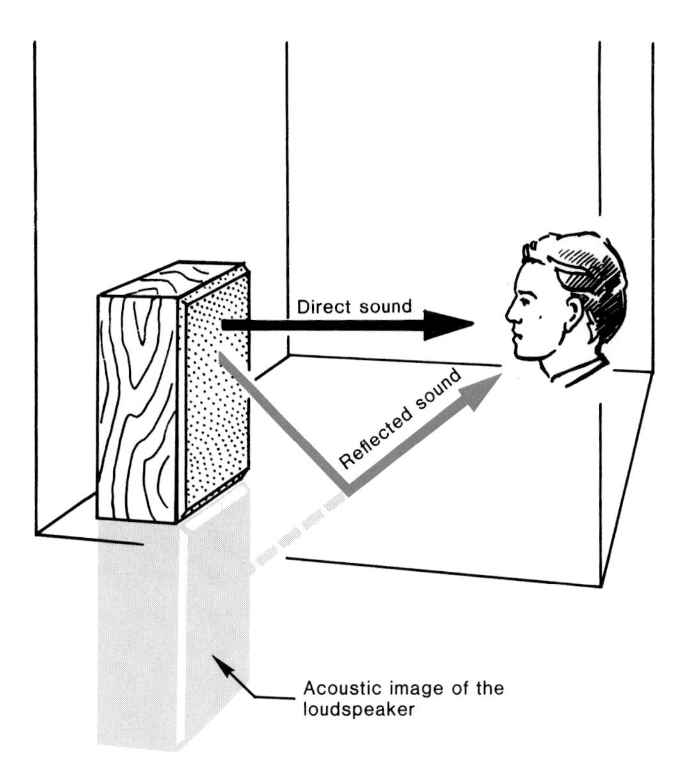

Figure 6.49. The sound reflected from the floor appears to the ear as if it came from another loudspeaker, the acoustic "image," located beneath the floor.

bass tones to arrive at the floor and other surfaces such as the walls. The bass tones are easily reflected from these surfaces, and some of the reflected sound reaches the listener. Thus, the total sound that the listener hears comes from two places. As indicated in Figure 6.49, there is the direct sound which travels directly to the listener after leaving the speaker. There is also the reflected sound which reaches the listener after being reflected from the floor. The dotted line in the picture shows that the listener thinks that the reflected sound is coming from the "acoustic image" beneath the real speaker. In this fashion the listener gains the impression that the floor of the room has been lowered to include a second speaker. As a result, the room sounds larger than it really is, and this feeling of spaciousness is psychologically beneficial.

If one acoustic image boosts the sound intensity, then two images will increase it even further. Figure 6.50 shows how the benefits of two acoustic images can be obtained by simply placing the speaker in a corner of the room. As far as the total sound is concerned, the side wall of the room will seem to recede because of the acoustic image located behind it. The additional acoustic image gives the room an even more spacious sound. So, using the corners will help to increase the intensity of the bass notes while making the listening room sound larger than it actually is.

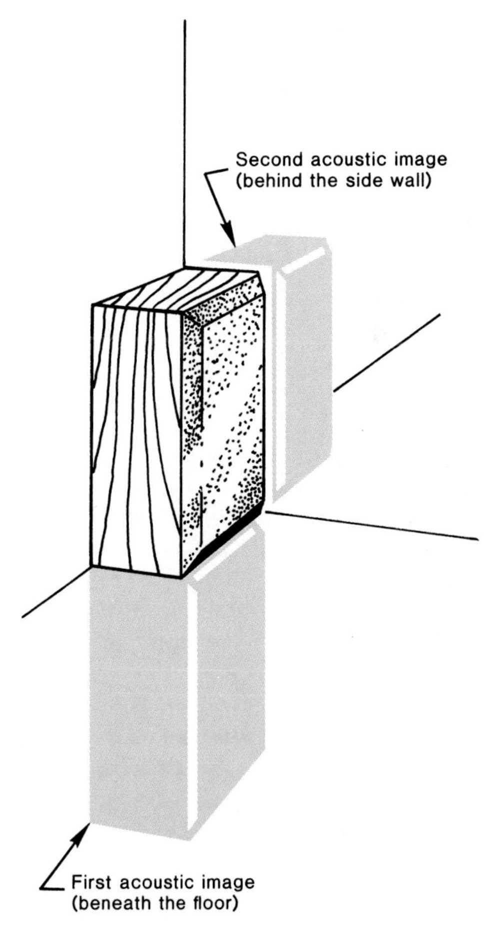

Figure 6.50. Corner placement creates two acoustic images.

Keeping the speakers on the floor has certain advantages. If you desire a lot of bass it is generally not a good idea to suspend hi-fi speakers far above the floor, whether against the center of a wall or in corners. It may make for a stylish image to have speakers above the floor on handsome bookshelves. But it makes for weak acoustic images, and leads to a loss of the nice, solid sounding bass which the floor and acoustic images can promote.

In addition to wasting acoustic images, locating the speakers far above the floor can cause another problem. Figure 6.51 shows the reason, which has to do with the way the speaker spreads out its sound. High treble tones are notoriously bad in this respect, for they tend to emerge in narrow beams that do not spread out very much. The treble notes are important, however, because they are responsible for how realistically the speakers can reproduce bright and sharp sounds, like those of a cymbal, a violin, wood block, or snare drum. A listener certainly does not want to miss these important frequencies because of sitting one or two meters below the speakers. As we shall soon see, missing the high treble tones leads to yet another problem, one having to do with stereo separation. If the speakers must be located over your head, at least angle them downward to help the high frequencies reach the listening area.

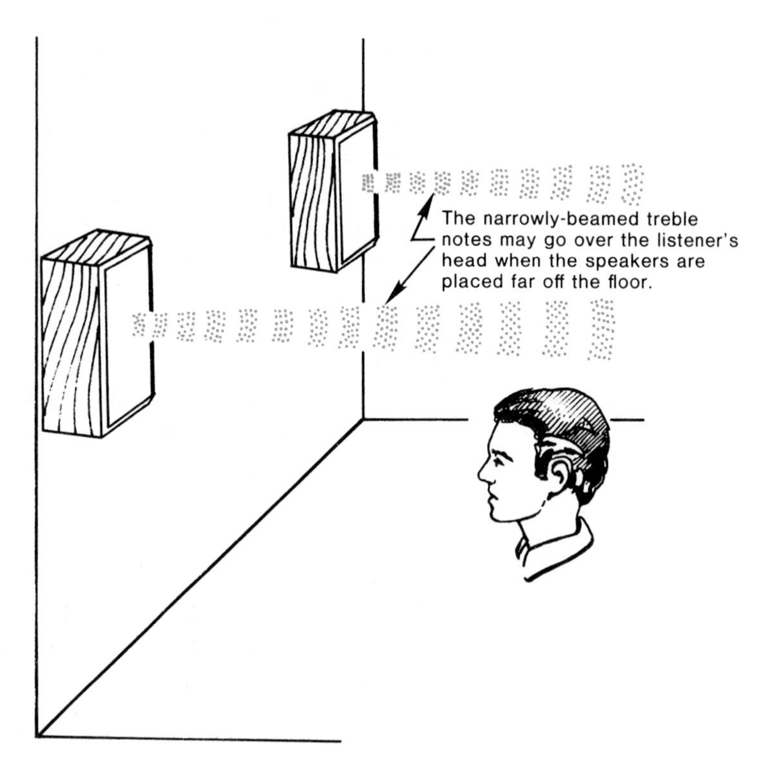

Figure 6.51. It may not be wise not to place speakers up on the wall. Such placement can lead to a loss of bass and cause the high frequency sound to be beamed over the listener's head.

The speakers should be placed to give the best possible "stereo effect." The "stereo effect" of a system is what allows a listener to distinguish the left from the right side of the band. It is this effect which is mainly responsible for the current widespread popularity of hi-fi. But yet, many who buy stereo systems are unaware that badly arranged speakers at home can eliminate much of the stereo effect for which they paid money at the store. What allows the speakers to tell you that a guitar is playing on the left and a set of drums is playing on the right? Figure 6.52 points out one sure thing; it is not the bass tones. They emerge from the speaker and spread out readily to encompass such a wide angle that they bounce off of everything. By the time the bass tones reach the ears, they have been reflected so often that the listener can not distinguish whether the sound originated from the left or the right speaker. Figure 6.52 also reveals the contrasting directionality of the high pitched treble tones. We have mentioned before that the treble notes emerge from the speaker in a relatively tight beam, and do not spread out much. Their tendency not to spread out means that they reach the listener without reflecting from every available surface. Thus, the origin of the sound can be discerned as being either from the right or from the left. If you want to achieve the best possible stereo effect, then do nothing that would prevent the very directional high pitched treble tones from reaching your ears. Avoid placing the speakers far above the floor where the "highs" might go over your head. Do not put obstructions in front of the speakers, because, although they do not affect bass tones very much, they will

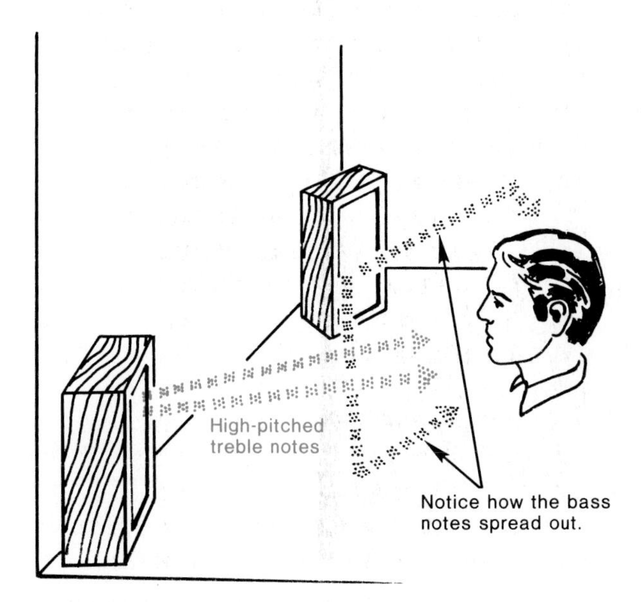

Figure 6.52. It is primarily the directionality of the high-pitched treble notes which allows them to give the stereo effect.

block the "highs." Do not place your speakers closer than about 2 meters, or you will lose the stereo effect completely. Finally, do not separate the speakers too much. About 5 meters is maximum, or else the directional "highs" will pass by you on either side, and the sound will appear to have a "hole in the middle."

Experiment with different arrangements to find what you like best. Precisely how to place your speakers depends on the kind of speakers you have and the room in which they will be. Figure 6.53 shows a good arrangement with which to start: corner placement, with about 3 meters separation between the speakers, and speakers angled 10–15 degrees in toward the center. The slight angle inward provides a more direct path to the listener for the directional "highs" which are responsible for the stereo effect, and it also helps to minimize standing waves, as discussed in section 4.6. Try other arrangements until the best results are obtained for your taste. You may even want to buy a special test record with exaggerated stereo effects with which to experiment. As you try various arrangements, remember that the more separated the speakers are, the more they should be angled inward. With a little effort you will be able to find an arrangement which makes the speakers sound their best.

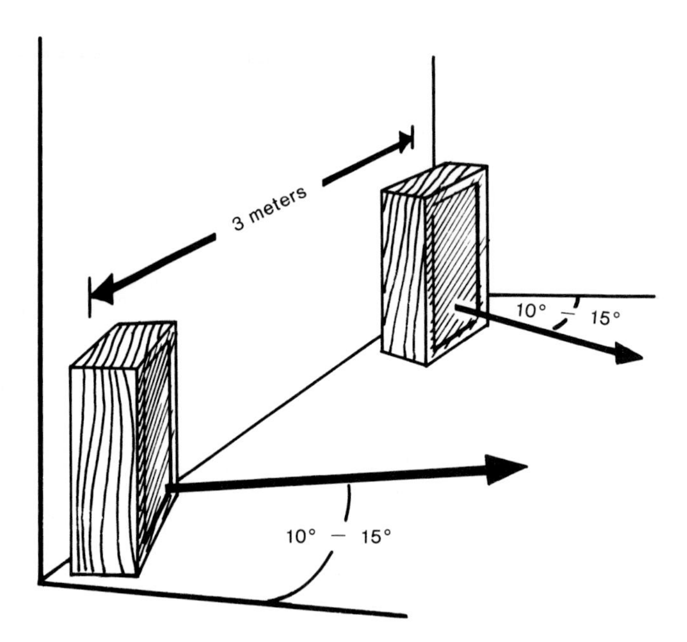

Figure 6.53. Optimize the sound of your speakers by trial and error, starting with the arrangement shown in the picture.

Summary of Terms

A/B Comparison Test—The direct comparison between the sounds of two similar components, such as loudspeakers, which is made by switching back and forth between them.

Acoustic Image—When sound is reflected from a wall, it appears to originate from a location which is behind the wall. This apparent location is called an acoustic image and it is similar to a visual image seen in a mirror.

Acoustic Suspension (Air Suspension)—A loudspeaker design which permits good bass reproduction from a relatively small-sized enclosure. The acoustic suspension systems utilize a floppy, highly compliant cone suspension and the trapped air to provide the proper suspension characteristics. They are generally of low-efficiency design, requiring relatively high amplifier powers compared to the vented and horn enclosures.

Air Suspension—(See **Acoustic Suspension.**)

Baffle—Any structure which prevents the front and rear surface sound waves of a speaker diaphragm from interfering with each other and causing possible sound cancellation. It is the panel on which the driver is mounted, although the term "baffle" sometimes applies to the entire enclosure that houses the speaker.

Basket—The metal frame of a speaker which holds together the cone, spider, voice coil, and magnetic structure.

Bass Reflex (Vented Enclosure, Phase Inverter, or Tuned Port)—A loudspeaker design in which the cone's "back-surface wave" is sent out through a port in the face of the enclosure to reinforce the bass output.

Bookshelf Loudspeaker—A loudspeaker which possesses a relatively small enclosure.

Compliance—A measure of a mechanical system's ability to move. A highly compliant phono stylus, or speaker diaphragm, is very flexible and it can move over relatively large distances with only a small applied force.

Cone (See **Diaphragm.**)

Crossover Frequency—The frequency at which the crossover network begins to route the signal to a different driver. In a two-way speaker system the crossover frequency is the frequency at which the woofer and high-frequency responses are divided.

Crossover Network—An electronic device used in loudspeakers which is designed to route the appropriate frequencies to the proper drivers; e.g., woofers, midranges, and tweeters.

Cutoff Frequency—The frequency below which a horn-type enclosure will no longer produce sound.

Diaphragm (or Cone)—The part of a driver which pushes on the air to produce the sound. A cone is a specially shaped diaphragm commonly used in drivers.

Dispersion, Speaker—The ability of a loudspeaker to spread sound into a wide listening area. Dispersion decreases as the frequency increases.

Dome Speaker—A speaker whose diaphragm is dome-shaped. Dome speakers increase the sound dispersion.

Driver—A term that is applied to any sound-producing device which is installed in an enclosure. Sometimes called a speaker.

Efficiency—The ratio, expressed as a percentage, of output signal to input signal; it is often used to estimate the electrical power needed to drive a loudspeaker.

Enclosure, Loudspeaker—An acoustically designed cabinet for the driver(s). The enclosure interacts with the driver(s) and it can have a large effect on the sound production, especially at the low frequencies.

Flexible Edge—(See **Suspension.**)

Frequency Range—The range of frequencies which a loudspeaker can reproduce. Usually the sound level at the extreme ends of the frequency range is −10 dB below some midrange (usually 1 kHz) level. Sometimes confused with frequency response, the frequency range does not give the deviations from perfect flatness, as denoted by a "±dB" term.

Horn Speaker—A type of speaker which contains a "megaphone" (horn) attached to a driver. The megaphone is used to increase the sound level from the driver.

Impedance—Resistance to the flow of alternating current, expressed in ohms.

Infinite Baffle—A type of loudspeaker design in which the woofer is mounted in either a large enclosure or in a large wall. The enclosure isolates the "back-surface" wave from the "front-surface" wave. The enclosure is so large that it does not appreciably change the resonant frequency of the driver.

Listener Fatigue—An auditory exhaustion resulting from loud sound levels together with high distortion. Listener fatigue can be caused by the speakers.

Loudspeaker—The combination of an enclosure plus drivers.

Magnetic Structure—The part of a driver which contains a magnet and the necessary iron required to produce a permanent magnetic field around the voice coil.

Maximum Power Rating—A loudspeaker specification which states the maximum electrical power that the loudspeaker can accept before serious damage is incurred.

Midrange Speaker—A driver which is used to reproduce the midrange frequencies, usually between 500 and 5,000 Hz.

Minimum Recommended Power Rating—A loudspeaker specification which states the input electrical power which is required to produce reasonable sound levels in an average room.

Passive Radiator—A type of loudspeaker design which is similar to the bass reflex design except that the cabinet is not openly vented. Instead, a second woofer, which is not electrically driven, covers the vent and it is called the passive resonator.

Phase Inverter—(See **Bass Reflex**.)

Resonance Peak—The large output sound power which is generated when a driver is vibrating at its natural, or resonant, frequency.

Sensitivity Rating—The level of the sound intensity, measured in dB, that a loudspeaker produces at a distance of (usually) 1 meter when it receives 1 watt of input electrical power.

Spider—A part of the speaker which ensures that the voice coil is properly centered about the magnetic structure, thus preventing rubbing between the two.

Suspension (Flexible Edge)—The elastic material which fastens the edge of the diaphragm to the rim of the basket. The suspension is an important part of a loudspeaker design.

Three-way System—A loudspeaker which uses three types of drivers to cover the audio range: woofer, midrange, and tweeter.

Throw—A term which is used to describe the maximum displacement of a speaker diaphragm as it vibrates; e.g., a long throw speaker.

Tuned Port—(See **Bass Reflex**.)

Tweeter—A driver which is designed to reproduce the high-frequency sound, usually between 5 kHz and 20 kHz.

Two-way System—A loudspeaker which uses two types of drivers to cover the audio range: a woofer and a high-frequency driver.

Vent—An opening or port in a loudspeaker enclosure.

Vented Enclosures—(See **Bass Reflex**.)

Voice Coil—The part of the speaker which carries the electric current.

Woofer—A low-frequency driver, usually reproducing sound frequencies between 20 Hz and 500 Hz.

Review Questions

1. Why are there so many different loudspeaker designs on the market today?
2. Name the four major parts of a cone speaker and explain the importance of their jobs.
3. When the cloth cover on a loudspeaker is removed, does the number and types of drivers mounted in the enclosure tell you anything about how good the loudspeaker "sounds"?
4. Why are woofers normally larger than either midrange or tweeter drivers? How can a small diameter woofer generate the same loudness levels as a larger woofer?
5. How does a baffle improve the sound produced by a woofer? Which frequencies are affected the most by the presence of a baffle?
6. Why should the resonant frequency of the woofer-plus-enclosure combination be kept as low as possible?
7. Explain the importance of the enclosures in the following three types of loudspeakers: infinite baffle, acoustic suspension, and bass reflex.
8. Describe an infinite baffle loudspeaker and explain its advantages and disadvantages.
9. Describe the principles of operation of an acoustic suspension loudspeaker. How can it be easily recognized?
10. Explain how a vented enclosure loudspeaker enhances the level of the bass frequencies.
11. What are the properties, both good and bad, of a horn enclosure which affect its use as a loudspeaker?
12. Briefly, describe a recommended procedure for buying loudspeakers.
13. Explain the difference between frequency response and frequency range.
14. What is speaker dispersion, and how does it affect the sound quality as one moves around the room?
15. List the speaker specifications which are not necessarily related to sound quality.
16. Describe how speaker placement in different locations within your room can produce various amounts of bass boost.

Exercises

NOTE: The following questions have up to 5 possible answers. Please select the **one** response which best answers the question.

1. The role of the permanent magnet in a speaker is:
   1. to keep any external magnetism from interfering with the speaker operation.
   2. to lower the impedance of the speaker.
   3. to provide a stationary magnetic pole which the moving coil can be either attracted to or repelled from.
   4. to improve the frequency response of the speaker.
2. Of the following types of speaker systems, the one considered to be the most efficient is the:
   1. bass reflex.
   2. acoustic suspension.
   3. large, infinite baffle.
   4. horn.
   5. sealed enclosure.
3. What is the most serious deficiency of a horn loudspeaker?
   1. It cannot produce enough loudness.
   2. It is too inefficient.
   3. It cannot reproduce the deep bass notes without being excessively large.
   4. It cannot be used as a tweeter.
   5. It cannot be used with crossover networks.
4. In a bass reflex loudspeaker what is the "ideal" relationship between the bass sound being produced by the front surface of the speaker and that coming through the port?
   1. They are in-phase and produce less bass.
   2. They are out-of-phase and produce less bass.
   3. They are in-phase and produce more bass.
   4. They are out-of-phase and produce more bass.
   5. There is no port in a bass reflex speaker.
5. Which one of the answers below does not represent an actual type of loudspeaker?
   1. A bass reflex.
   2. A horn.
   3. A midrange reflex.
   4. An infinite baffle enclosure.
   5. An acoustic suspension.
6. Which one of the following loudspeakers will sound "best"? The "best" one will contain:
   1. 1 woofer and 1 midrange-tweeter speaker.
   2. 1 woofer, 1 midrange, and 1 tweeter speaker.
   3. 2 woofers, 1 midrange, and 1 tweeter speaker.
   4. 2 woofers, 2 midrange, and 1 tweeter speaker.
   5. (None of the above systems will necessarily give a better sound.)
7. The main purpose of all loudspeaker enclosures is to:
   1. protect the diaphragm from people touching it.
   2. add beauty to the loudspeakers.
   3. separate the woofer from the tweeter.
   4. separate the front-sound from the back-sound which has been produced by the diaphragm.
   5. physically hold the drivers in place.
8. Which one of the following is not a part of a cone driver?
   1. A port.
   2. A diaphragm.
   3. Magnet.
   4. Voice coil.
   5. Flexible edge.
9. A three-way speaker system means:
   1. there are three speaker cabinets.
   2. the speakers are connected to three separate amplifiers.
   3. the attenuation switch on the back of the speaker has three positions.
   4. the speaker has three times the normal speaker power.
   5. there are three types of speakers in each cabinet, the woofer, midrange, and tweeter.
10. The tweeter driver in a three-way system will generally reproduce the sound frequencies in the range:
    1. 20–20,000 Hz.
    2. 200–500 Hz.
    3. 500–1,000 Hz.
    4. 1 kHz–5 kHz.
    5. 5 kHz–20 kHz.
11. A loudspeaker which contains a woofer, midrange, tweeter, and a super tweeter has _______ crossover points.
    1. 1
    2. 2
    3. 3
    4. 4
    5. 5
12. A driver designed for the reproduction of high frequencies is called a:
    1. woofer.
    2. baffle.
    3. tweeter.
    4. midrange speaker.
    5. crossover.

13. The __________ is a driver used to reproduce low frequency sounds.
    1. woofer
    2. midrange
    3. horn
    4. tweeter
    5. baffle
14. Which one of the following makes a better sounding loudspeaker?
    1. Higher crossover frequencies.
    2. Larger magnet.
    3. Higher maximum power rating.
    4. Lower impedance.
    5. None of the above answers necessarily results in better sounding loudspeakers.
15. Which type of loudspeaker generally has the largest diaphragms which move through the smallest distances?
    1. The infinite baffle.
    2. Acoustic suspension.
    3. Horns.
16. What is the part(s) of a driver which actually carries the electricity?
    1. The basket.
    2. The diaphragm and magnet.
    3. The spider and diaphragm.
    4. The magnet only.
    5. The voice coil.
17. The part of a driver which pushes the air is called the:
    1. baffle.
    2. diaphragm.
    3. voice coil.
    4. rim.
    5. permanent magnet.
18. The type of loudspeaker enclosure which necessarily uses the entrapped air to "stiffen" the motion of speaker diaphragm is called:
    1. acoustic suspension.
    2. bass reflex.
    3. phase inverter.
    4. infinite baffle enclosure.
    5. vented enclosure.
19. The speaker system which uses the back-radiated sound to reinforce the front-radiated sound is called:
    1. a sealed enclosure.
    2. a folded horn.
    3. a bass reflex enclosure.
    4. an acoustic suspension enclosure.
    5. an infinite baffle enclosure.
20. An unbaffled speaker is typically plagued by which one of the following problems?
    1. Poor crossover characteristics.
    2. Too much bass response.
    3. Poor bass response.
    4. Excessive high frequency dispersion.
    5. The speaker enclosure must be large.
21. The purpose of a speaker baffle is:
    1. to confuse, or baffle, the listeners.
    2. to keep the diaphragm from vibrating too much.
    3. to promote the cancellation of the separate sound waves generated by the front and back surfaces of the diaphragm.
    4. to give strength to the speaker enclosure.
    5. to prevent the cancellation of the separate sound waves generated by the front and back surfaces of the diaphragm.
22. Which of the following is NOT a design feature which must be considered in all multidriver hi-fi loudspeakers?
    1. Proper baffling.
    2. Good dispersion of high frequencies.
    3. Good power handling ability.
    4. Bass response.
    5. Dispersion of low frequencies.
23. Which one of the following items will always result in a better sounding loudspeaker?
    1. More speakers in an enclosure (3 or more).
    2. Larger diameters for the woofers.
    3. Greater speaker efficiency.
    4. Less speaker impedance.
    5. None of the above answers necessarily results in better sounding speakers.
24. Which one of the following speakers is most efficient?

	Amp. power to speaker	**Sound power from speaker**
1.	200 watts	2 watts
2.	100 watts	10 watts
3.	20 watts	1.6 watts
4.	5 watts	0.2 watts
5.	2 watts	0.02 watts

25. 200 watts of electrical power are fed into a speaker whose efficiency is 10%. How much power is lost in heating the speaker?
    1. 200 watts.
    2. 20 watts.
    3. 10 watts.
    4. 100 watts.
    5. 180 watts.

26. Which one of the following systems would sound the loudest?

	Amp. power to speaker	Speaker efficiency
1.	200 watts	0.4%
2.	75 watts	1%
3.	30 watts	5%
4.	10 watts	10%
5.	3 watts	25%

27. As you increase the efficiency of a speaker, the amplifier power needed to drive that speaker at a constant sound loudness must __________ .
    1. remain the same
    2. decrease
    3. increase

28. The speaker specification which tells the power needed to damage a speaker is called:
    1. minimum recommended power.
    2. impedance.
    3. maximum power rating.
    4. dispersion.
    5. crossover frequency.

29. If the minimum recommended power rating for a speaker is high, the speaker is:
    1. easy to damage.
    2. very inefficient.
    3. a low quality speaker.
    4. hard to damage.
    5. very efficient.

30. You own an integrated amplifier with a 40 watts per channel power rating. Which one of the following power-rated speakers should you buy?

	Minimum Recommended Power	Maximum Power Rating
1.	20 watts	80 watts
2.	50 watts	120 watts
3.	10 watts	35 watts
4.	35 watts	90 watts
5.	5 watts	30 watts

31. Sound radiated from a speaker is least directional (spreads out more) at:
    1. the low frequencies.
    2. the mid frequencies.
    3. the high frequencies.
    4. (Sound spreads out equally well at all frequencies.)
    5. None of the above answers is correct.

32. As the frequency of the sound radiated by a loudspeaker increases, the sound pattern:
    1. becomes more directional.
    2. becomes less directional.
    3. stays the same for all frequencies.

33. The general property which tells how loudspeakers send sound waves in various directions at different frequencies is called:
    1. dynamic range.
    2. speaker reflex.
    3. speaker efficiency.
    4. speaker selectivity.
    5. speaker dispersion.

34. The frequency response of a loudspeaker:
    1. tells how fast the diaphragm is moving at each frequency.
    2. is equal to the efficiency of the loudspeaker.
    3. tells how many drivers are located within the cabinet.
    4. is directly proportional to the position of the crossover frequencies.
    5. tells how well the loudspeaker reproduces each frequency in relation to all others.

35. The cancellation effect between the front and rear surfaces of an unbaffled speaker:
    1. is most severe at the low frequencies.
    2. is most severe in the midrange frequencies.
    3. is most severe at the high frequencies.
    4. is equally severe at all frequencies.
    5. is not a problem in sound reproduction.

36. When a speaker cone is moving forward a ____ is created at the back surface.
    1. high pressure region.
    2. rarefaction.
    3. condensation.
    4. node.
    5. standing wave.

37. Approximately how much bass sound will reach your ears when one acoustic image is present, compared to that produced without the image?
    1. More bass will be produced.
    2. Less bass will be produced.
    3. The same amount of bass will be produced.

38. Crossover networks are used in loudspeakers to:
    1. provide baffling.
    2. separate the sound into different frequency bands, and then send each band to the appropriate driver.
    3. increase the dispersion of the woofer.
    4. help create acoustic images.
    5. increase the minimum power rating.

39. Two loudspeakers, A and B, are identical except that A is baffled and B is unbaffled. What is the difference in the sound produced by the two loudspeakers?
    1. The frequency response curve of A shows a large high frequency loss when compared to that of B, due to the cancellation of the high frequency sound waves generated at the front and back surfaces of A's diaphragm.
    2. The frequency response curve of B shows a large amount of low frequency loss when compared to that of A, due to the cancellation of the low frequency sound waves generated at the front and back surfaces of B's diaphragm.
    3. There is no difference between the sound produced by the two loudspeakers.
    4. The frequency response curve of A shows a large amount of low frequency loss when compared to that of B, due to the cancellation of the low frequency sound waves generated at the front and back surfaces of A's diaphragm.
    5. The frequency response curve of B shows a large amount of high frequency loss when compared to that of A, due to the cancellation of the high frequency sound waves generated at the front and back surfaces of B's diaphragm.

40. If you wanted to build a speaker driver that moved the largest possible amount of air (thus being capable of producing a loud sound), which of the following would you choose?
    1. A large diameter diaphragm which is mounted with a low compliance suspension.
    2. A large diameter diaphragm which is mounted with a high compliance suspension.
    3. A small diameter diaphragm which is mounted with a low compliance suspension.
    4. A small diameter diaphragm which is mounted with a high compliance suspension.

41. Which type of loudspeaker needs to be unreasonably large in order to reproduce the lowest bass frequencies?
    1. The infinite baffle.
    2. The acoustic suspension.
    3. The horn.
    4. The bass reflex.
    5. The sealed enclosure.

Chapter 7

ELECTRICITY—THE BASIC INGREDIENT IN HI-FI SYSTEMS

7.1 Introduction

The first record player, invented by Thomas Edison, was completely mechanical in its reproduction of sound and used no electricity. The vibrating needle moved a natural mica diaphragm which was attached to the now familiar horn. Even the turntable was driven by a hand-wound spring. The electronic age has changed all that, and the use of electricity now lies at the heart of all hi-fi systems. Each of the four building blocks mentioned in Chapter 2 makes extensive use of electricity, and a good background in electrical principles is essential for an understanding of modern systems.

Electricity is really not new for its presence was known to the Greeks as far back as 600 B.C. However, a true understanding of its origin and the development of practical applications is only a scant 150 years old. Today we take the importance of electricity for granted, although few people understand its origin or how it is manipulated to do all the nice things expected of it. It is the purpose of this chapter to provide the necessary insights into these facets in order to explain why electricity is the cornerstone of all audio systems.

7.2 An Atomic View of Matter and Electricity

All substances, be it water, copper, or chocolate pudding, are created from elementary units called *atoms*. Each atom is comprised of a relatively massive nucleus about which electrons orbit in much the same manner as the planets revolve around the sun, as illustrated in Figure 7.1.

It is known that electricity is found in two basic forms, called *positive* and *negative* charges, which appear throughout nature in equal abundance. The electron carries the negative charge and it represents the smallest unit of negative electricity found in nature. The *nucleus* contains two types of particles, *protons* and *neutrons,* which have about the

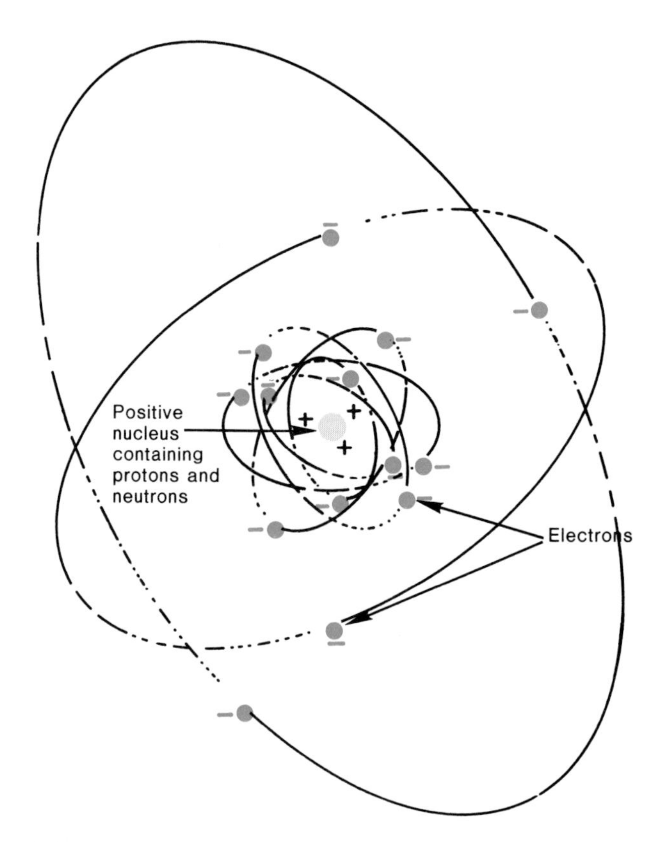

Figure 7.1. The atom contains protons and neutrons within its nucleus. The electrons orbit at various distances about the nucleus.

same mass (weight) but differ in their charge. The neutrons, as the name implies, are electrically neutral and carry no net positive or negative charge. The proton, on the other hand, is the basic carrier of *positive electricity* and it contains an amount of positive charge which is precisely equal, but opposite, to the charge on the electron. Even though the electron and proton are "partners" in an electrical sense, they do not share any other characteristics such as mass or "living quarters." The protons are 1,840 times more massive than electrons and spend almost all of their time within the nucleus. The electrons, on the other hand, reside outside the nucleus.

Electrons are of particular importance to hi-fi enthusiasts because they are the entities which actually constitute the flow of electricity in components such as amplifiers, tuners, cartridges, and speakers. When the atom is part of an electrical conductor, such as copper, an electron or two may actually become detached from each parent atom and wander freely throughout the metal. It should be noted that only the electrons move through a conductor. The remainder of the atom is rigidly constrained to immobility; it then follows that the nuclei and the protons therein are also immobile.

In general, atoms are electrically neutral which implies that the number of protons and electrons is identically equal.

Element	Atomic Number (number of protons in the nucleus)	Number of Electrons in Orbit About the Nucleus	Number of Neutrons in the Nucleus (Most Abundant Isotope)
Hydrogen	1	1	0
Helium	2	2	2
Lithium	3	3	4
Beryllium	4	4	5
Boron	5	5	6
Carbon	6	6	6
Nitrogen	7	7	7
Oxygen	8	8	8
Fluorine	9	9	10
Neon	10	10	10
Sodium	11	11	12
Magnesium	12	12	12
Aluminum	13	13	14
Silicon	14	14	14

Table 7.1. A partial listing of the lighter elements which shows the number of electrons, protons, and neutrons contained within each element.

In the periodic table of the elements, atoms are classified by their *atomic number*, which represents the number of protons found within the nucleus. All atoms with the *same atomic number* constitute an *element* and some of the simpler elements are shown in Table 7.1. Therefore, the main characteristic that distinguishes one element from another is the number of protons in the atom. As the Table 7.1 shows, hydrogen is the simplest of the elements with a single electron orbiting a single proton. Helium has two electrons, two protons, and two neutrons. Notice that each element in the list contains an equal number of electrons and protons, although there is no strict rule which governs the number of neutrons found within the nucleus.

As the atomic number of an atom increases, more and more electrons must be placed into orbits about the nucleus. The modern theory of how electrons and protons behave under such circumstances is called quantum mechanics. While quantum mechanics is a highly mathematical theory, it does make some general statements regarding the distribution of electrons about the nucleus. Electrons move about

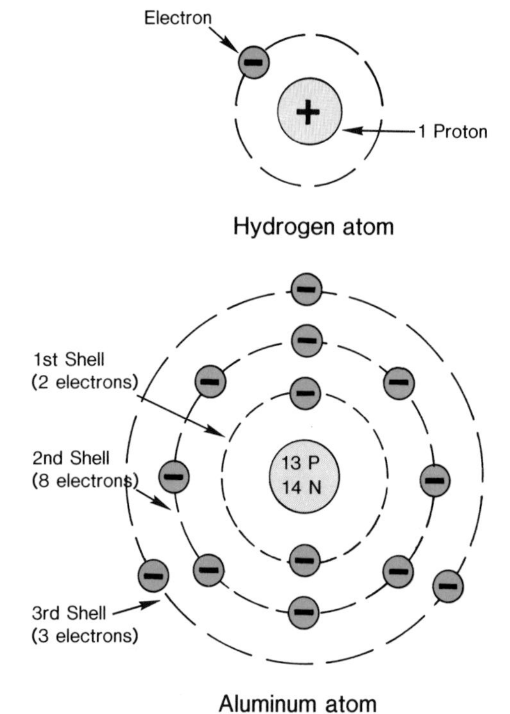

Figure 7.2. A neutral atom has an equal number of protons and electrons. According to modern atomic theory, each orbit can only contain a certain number of electrons.

the nucleus in orbits or shells, and only a certain number are permitted in any one orbit. Figure 7.2 shows a very simplified picture for the two elements hydrogen and aluminum. Aluminum is an element with thirteen electrons and the distribution of these electrons about the nucleus is as follows: two electrons fit into the innermost orbit, eight in the second shell and three in the third shell. A solid piece of aluminum is a large number (a *very* large number) of aluminum atoms packed closely together. As we shall see later on it is possible, by a variety of methods, to completely remove some of the outer electrons, thereby leaving behind positively charged aluminum atoms. Having been displaced, the electrons will attempt to find their way back to the positively charged atoms because of the mutual attraction between unlike charges. This process of separation and return is the basis for all electrical action.

Summary

Electrons and protons represent, respectively, the smallest known units of negative and positive charge. Both of these charges are found within the atom; the protons reside in the nucleus and the electrons orbit about it. Each element in the periodic table is uniquely classified by the number of protons which exist in the nucleus.

7.3 The Forces between Electrically Charged Particles

We are now in a position to state a very basic and important law concerning the nature of the electrical forces which exist between charged particles:

Like charges repel, and unlike charges attract.

In other words, the negatively charged electrons repel other electrons while protons repel protons. However, an electron and a proton are attracted toward each other. Figure 7.3 illustrates these attractive and repulsive forces.

The strength of the electric force between charged particles is larger when the particles are closer together and weaker when they are farther apart. In this sense the electrical force is somewhat like the gravitational force between an astronaut and the earth; the farther that the astronaut travels away from the earth, the weaker the gravitational force becomes, and vice versa. When a force between two objects decreases as the distance between them increases, the force is said to be *inversely related* to the distance. In fact, the electrical force of attraction or repulsion is inversely proportional to the *square* of the distance between the ob-

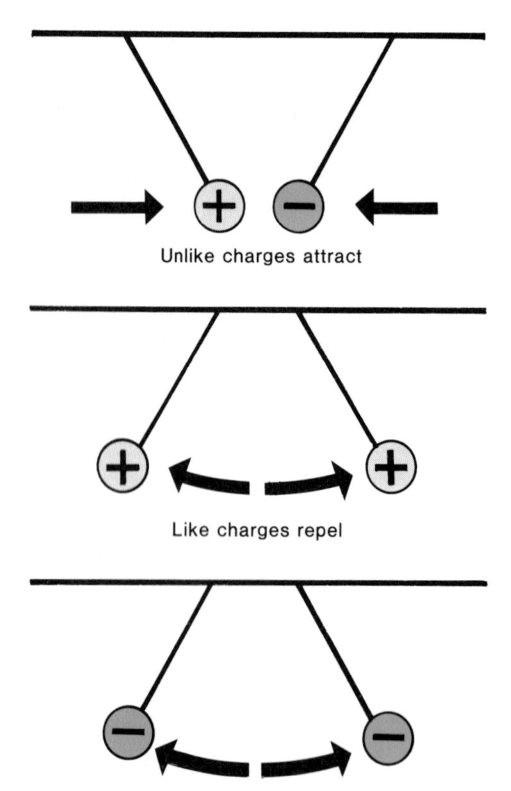

Figure 7.3. An illustration which describes the forces between electrical charges.

jects. This means, for example, if the distance is doubled the force will be reduced to only one-fourth of its original strength, $[(1/2)^2 = 1/4]$. Likewise, bringing the objects three times closer produces a nine-fold ($3^2 = 9$) increase in the force.

The strength of the attractive or repulsive force also depends on how much electrical charge each object contains. The greater the amount of charge, the larger is the force; the strength of the force is directly proportional to the amount of charge on each object.

As mentioned earlier, an atom can be thought of as a collection of electrons whirling around the nucleus. All of the electrons, being negatively charged, mutually repel each other, although they are attracted to the protons within the nucleus. The protons, of course, are repelled from each other although there is a very powerful nuclear force which prevents them from flying apart. This nuclear force holds the protons together within the nucleus in spite of the strong repulsive electrical forces. The intricate interplay of attractive and repulsive electrical forces, along with the nuclear force, is responsible for holding together all the constituents of the atom.

The force F between two electrically charged particles can be described mathematically by an equation which is called *Coulomb's law*. Using Q_1 and Q_2 to denote the amount of electrical charge on each object (see section 7.4), r to denote the separation distance, and k to indicate a constant of proportionality

$$\left(k = 9\times10^9 \frac{\text{Newton}\cdot\text{m}^2}{\text{Coulomb}^2}\right),$$

Coulomb's law is

$$F = kQ_1Q_2/r^2.$$

Notice that r^2 appears in the denominator; this causes F to decrease when r increases, and vice versa.

Summary

The fundamental fact, upon which all electrical systems are based, is that like charges repel and unlike charges attract. The force of repulsion or attraction is inversely proportional to the square of the distance between two particles, and it is directly proportional to the amount of electrical charge on each object.

7.4 Electrical Current

We have just seen that electrons which travel around the nucleus are held in their orbits by an electrical attraction to the protons within the nucleus. If, somehow, one or more of these electrons could be forced out of orbit and placed somewhere else—say in another material—then a charge imbalance would result. The "stripped" atom, being deficient in electrons, will possess a net positive charge, and the new location of the electrons will naturally have an excess of negative charges. In general, then, matter can exist in three electrical states: neutral (or uncharged), negatively charged, and positively charged, as shown in Figure 7.4. An uncharged material contains equal numbers of electrons and protons. A body which contains a *negative charge* possesses an *excess* of electrons, while a *positively charged* object has a *deficiency* of electrons. Because of the way in which charged objects have been defined, it stands to reason that for every negatively charged body there must be, somewhere, other objects which are missing electrons.

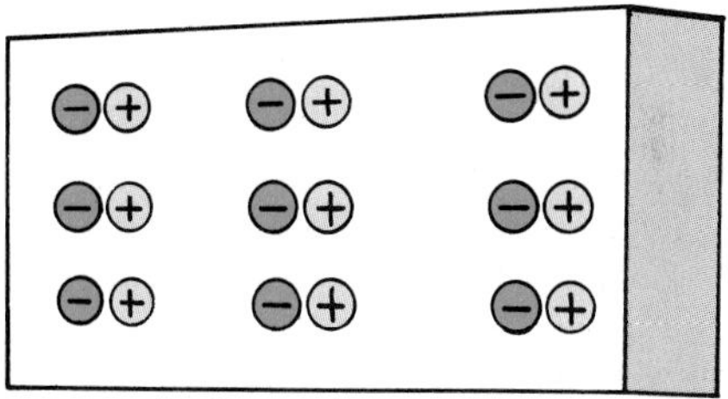

Uncharged (or neutral) bar

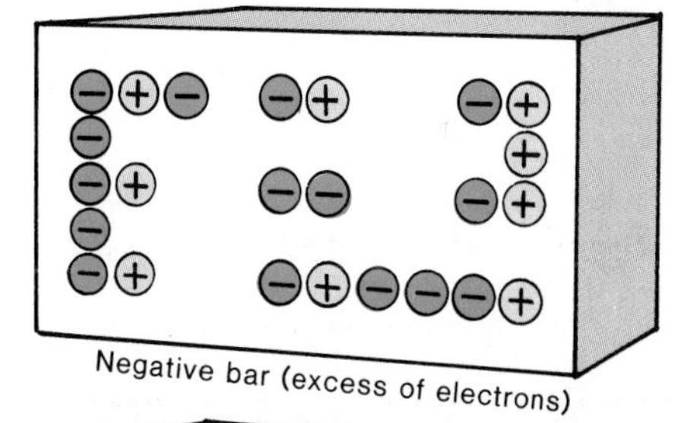

Negative bar (excess of electrons)

Positive bar (lack of electrons)

Figure 7.4. The atoms which constitute a solid piece of material are not free to move. Electrons, on the other hand, can be either added to, or removed from, a material. Removing electrons leaves the substance with a positive charge; adding electrons creates a negatively charged substance.

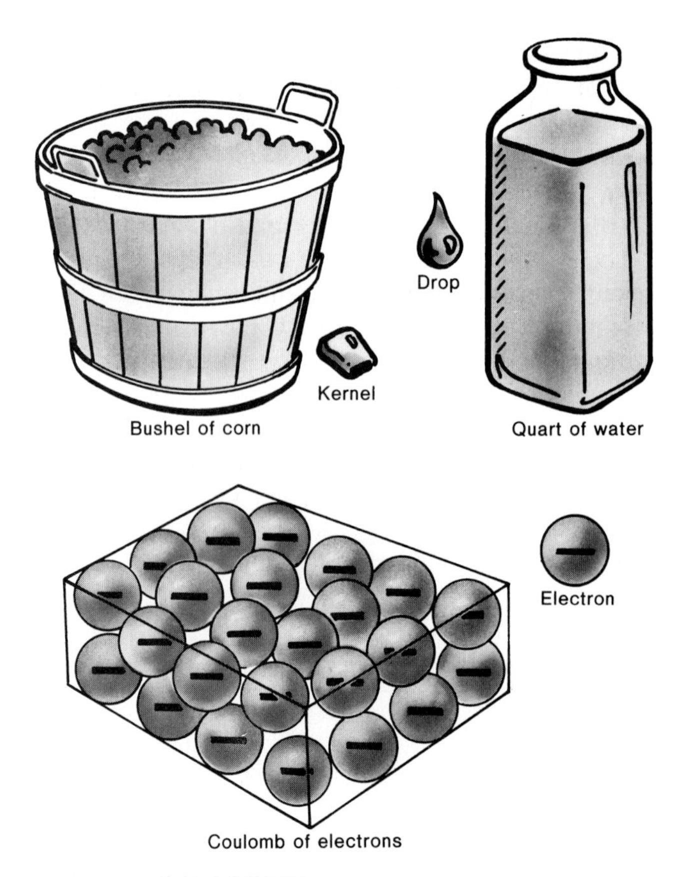

Figure 7.5. The coulomb is a convenient measuring unit for representing a large number of electrons.

In working with electrical charges a convenient unit is needed for expressing large numbers of electrons because one electron, by itself, is just too small. As shown in Figure 7.5, it is a lot easier to speak of a "bushel of corn" rather than saying "50 thousand kernels." Likewise, we define a *coulomb of electricity* to be the total charge contained by 6.24 million, million, million, electrons. This is a staggering number by any ordinary standard although it is relatively easy, in practice, to utilize this number of electrons.

> One coulomb = the total charge contained by 6.24×10^{18} electrons.

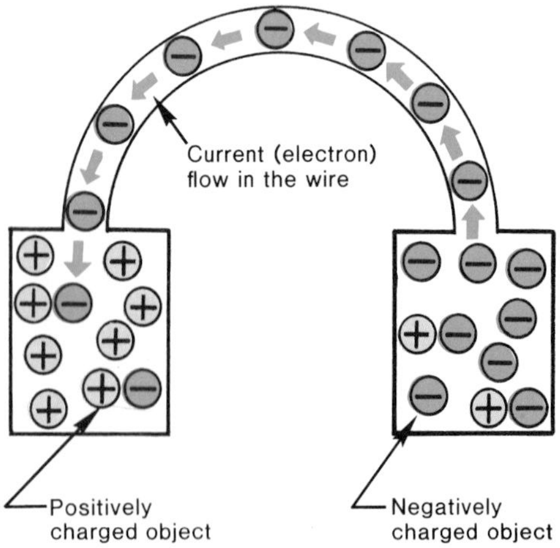

Figure 7.6. The excess electrons on the right side flow through the wire because they are attracted to the excess positive charge on the left side. The flow of electrons is called a "current."

Whenever two oppositely charged objects are connected by a wire, the excess electrons are pulled toward the positive charges due to the attractive force between unlike charges (Fig. 7.6). There would be a momentary flow of the excess electrons until the two bodies became electrically neutral, at which time the electron transfer would stop. The flow of electrons is called a *current* and, as we shall show, it is the flow of electrons between oppositely charged objects which gives rise to all the marvelous tasks that electricity can perform. There is a unit, called an *ampere*, which expresses the amount of electron flow. For example, a phrase like "5,000 gallons per second" expresses the size of the current flow in a river. Electrical current is measured in an identical fashion. Since the coulomb is a quantity of electricity, just like a gallon is a quantity of water, one coulomb per second is a measure of electrical flow. One coulomb per second is called "one ampere." One ampere represents the total charge on 6.24×10^{18} electrons flowing past a particular point in the wire each second (Fig. 7.7). Many common devices, such as light bulbs, toasters, and speakers, use about one ampere of current in their operation. The sound intensity which a speaker produces is directly related to the amount of current being sent to it from the power amplifier—more current generates a louder sound. In all these devices the electrical current is the central theme upon which their principles of operation are based, and the ability to control the current is of great importance.

Note: Please do not confuse the coulomb with the ampere. The coulomb is simply a unit which represents the charge on a collection of 6.24×10^{18} electrons. When the electrons begin to move within a wire a current is formed. The strength of the current is measured in a unit called an ampere; one ampere represents one coulomb of charge flowing through the wire each second.

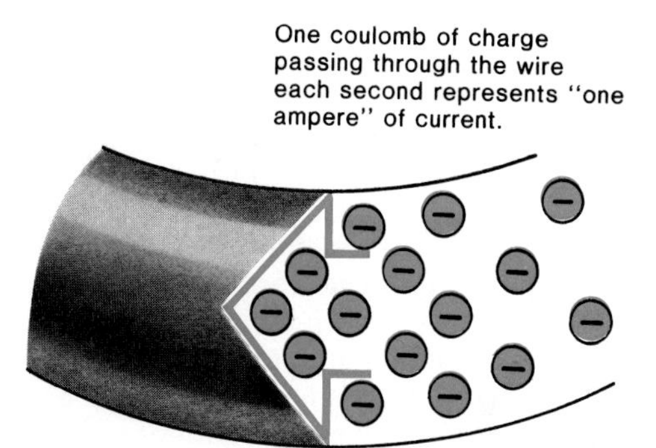

Figure 7.7. One ampere of current is defined to be one coulomb of charge (6.24×10^{18} electrons) flowing through a wire each second.

Summary

Electrons flowing between oppositely charged objects constitute an electrical current. The unit of current is called an ampere which represents one coulomb of charge (6.24×10^{18} electrons) passing through the wire each second.

7.5 Charge Separation and Electrical Voltage

A very common method of producing charged objects is by friction. Everyone has experienced the shock when walking over a nylon carpet. The friction, generated by walking, is sufficient to transfer electrons between the carpet and yourself, thus creating two oppositely charged bodies. Connecting a metal object between you and the room (a door knob does very nicely) causes an electrical "shock" as the electrons quickly flow back to the positive charge. Although this is usually a one-shot affair there is really no reason why a continuous flow of current couldn't be maintained; just keep shuffling while holding on to the door knob! The friction separates the charges, and the door knob provides an easy path for them to reunite again. Not very exciting or useful. But it would work.

A battery is a much more practical solution to this problem. Batteries, of course, come in all sizes and shapes from the small 1.5 volt flashlight cells to the heavy 12-volt lead-acid car batteries. In all cases a battery has two terminals,

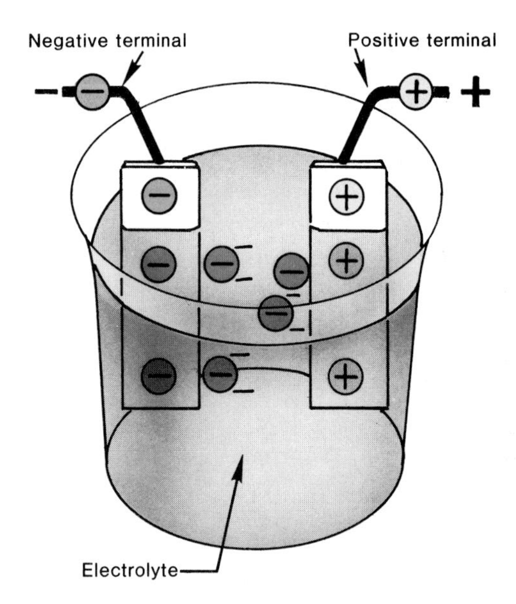

Figure 7.8. The electrolyte, in cooperation with the two terminals, removes electrons from the positive terminal and places them on the negative terminal. The electrolyte continues this action until the negative terminal will accept no more electrons.

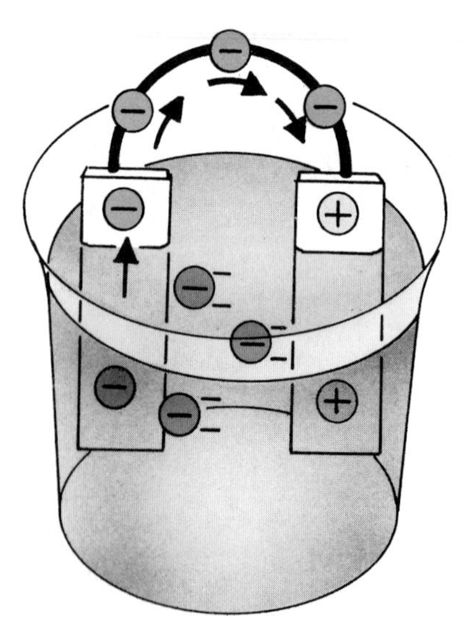

Figure 7.9. When a wire is connected between the two terminals of a battery, the excess electrons on the negative terminal rush through the wire to the positive side. Inside the battery the electrolyte maintains a continous flow by returning the electrons to the negative terminal.

called the positive and negative, and its interior is filled with a chemical substance known as an electrolyte. The electrolyte is of crucial importance because it removes electrons from one terminal, leaving it with a net positive charge, and deposits them on the other (Fig. 7.8). The electrolyte can be a liquid, such as found in car batteries, or in the form of a thick paste which is used in the manufacturing of flashlight batteries. A chemical reaction occurs between the electrolyte and the terminals in which the stored up chemical energy of the electrolyte is consumed in the process of separating the electrical charges. The excess electrons, having been placed on the negative terminal, would very much like to recombine with the opposite charge on the positive terminal. However, as long as the electrolyte has a reserve of chemical energy it will completely counterbalance the force of attraction and hold the excess negative charge on the negative terminal. The electrolyte pushes electrons onto the negative terminal until it will accept no more. At this time both plates are fully charged and no more electrons are brought over.

If a wire is connected between the two terminals, as shown in Figure 7.9, the electrons will leave the negative terminal and flow through the wire to the positive terminal. As the electrons leave the negative terminal, the electrolyte separates more electrons from the positive side and carries them to the negative side. As long as the electrolyte is "active" (not dead) a continuous flow of electrons will be maintained in the wire. This type of electron flow, exemplified by a battery, is called *direct current* (DC) for the obvious reason that the electron motion is always in one direction and it never reverses itself. We shall see later in this chapter that a second type exists, called alternating current, which also finds great usage in electrical applications.

We have seen that a continuous flow of current implies that the electrolyte must continually expend its reserve chemical energy in separating the electrical charge. There is a famous law of which you may have heard. It states that energy cannot be created nor destroyed; it can only be changed from one form to another. This law means that if the electrolyte *loses* its chemical energy, then the electrons which it moved onto the negative terminal must have *gained* it. You can think of energy in the scientific world as being analogous to money in the financial world. It is no secret that the process of paying your bills will leave you poorer and, correspondingly, someone else richer. Of course, it is possible to increase your own wealth without leaving someone else poorer; simply make counterfeit money. Nature does not recognize this as a viable solution, however, because she will not permit the "counterfeiting" of energy.

The *voltage* of a battery is the amount of energy that each coulomb of electrons receives from the electrolyte. In other words,

$$\left\{\begin{matrix}\text{energy given to} \\ \text{the separated} \\ \text{electrons (in} \\ \text{joules*) by the} \\ \text{electrolyte}\end{matrix}\right\} = \left\{\begin{matrix}\text{number of} \\ \text{coulombs} \\ \text{separated}\end{matrix}\right\} \times \left\{\begin{matrix}\text{voltage of the} \\ \text{battery}\end{matrix}\right\}$$

For example, suppose that a 12-volt car battery has just finished placing 3 coulombs of electrons onto the negative terminal. It costs the electrolyte

$$(3 \text{ coulombs}) \times (12 \text{ volts}) = 36 \text{ joules}$$

of energy to perform this process. Although the electrolyte loses 36 joules of energy, the transferred electrons will be richer in energy by the same amount. Batteries with larger voltage ratings, say, 100 volts as compared to 12 volts, are capable of giving more energy to the negative charges.

The process of charge separation by the electrolyte, followed by the return of the electrons through the wire, may seem like a "dog chasing its tail." It is not at all obvious what good may come of this. The following section, which introduces the concept of resistance, will illustrate the usefulness of electrical current.

Summary

The battery is a common type of "electrical pump," and its "strength" is rated in volts. The battery achieves its "pumping action" by separating and holding the positive and negative charges onto two separate terminals. When an external wire is connected between the two terminals, the electrical force of attraction draws the electrons to the positive terminal, thus creating a current in the wire. The process of charge separation results in an energy transfer between the electrolyte and the electrons. Inside a 1.5-volt battery, for example, the electrolyte loses 1.5 joules of energy for each coulomb of electrons that it separates. However, the law of energy conservation requires that the electrons gain this energy in the form of electrical energy. When the electrons leave the negative terminal they may, in turn, expend this energy into heat or some other form of energy such as that carried by sound waves.

*A joule is a unit of energy. Approximately 2,000 joules of energy are consumed when an adult walks up a 3 meter flight of stairs.

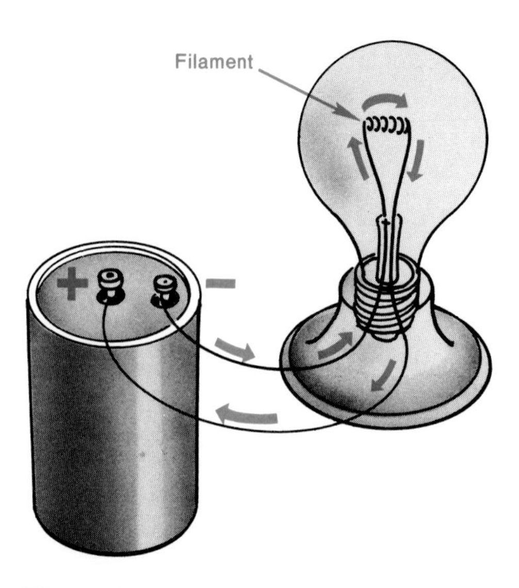

Figure 7.10. When a light bulb is inserted into the path of the moving electrons, the resistance of the filament causes it to become "white" hot.

7.6 Electrical Resistance

Consider a light bulb which has been inserted in the wire path of the moving electrons as shown in Figure 7.10. The electrons enter the light bulb and spiral through a thin tungsten wire called the filament. As they move through the filament, they encounter "friction" between themselves and the tungsten atoms. Like rubbing your hands together on a cold day, the friction causes the tungsten to heat up. The tungsten heats up so well, in fact, that it becomes "white hot" thus emitting the characteristic white light. The tungsten filament offers *resistance* to the current because it causes the electrons to slow down and loose their energy via friction (Fig. 7.11). Therefore, the energy that the electrons initially acquired from the electrolyte is dissipated into heat and light. When the electrons finally reach the positive terminal their energy is gone. They can begin another round trip through the light bulb when they are given another "shot" of energy from the electrolyte. This cycle is illustrated in Figure 7.12. Many of the appliances found around the home operate on the resistance-heating idea although they use alternating current rather than direct current. Electric ovens, toasters, coffee makers, dryers, and furnaces all function on the principle that current which flows through a relatively high resistance material, called the heating element, causes it to heat up.

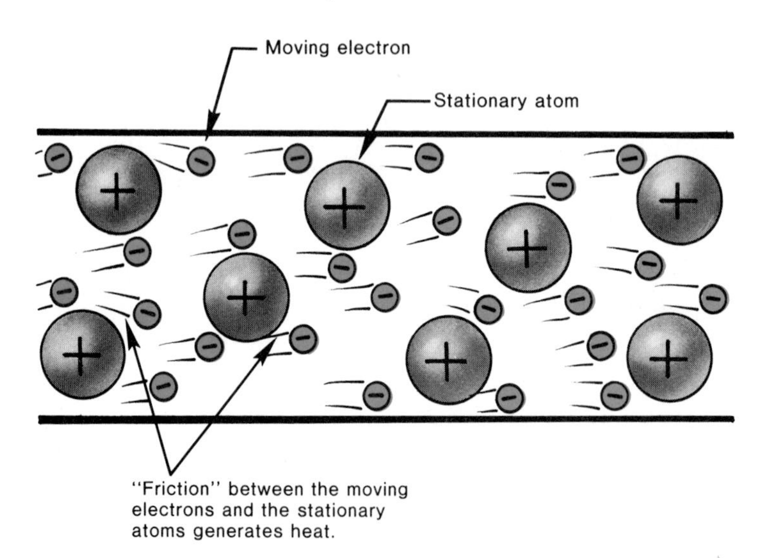

Figure 7.11. "Friction" between the moving electrons and the stationary atoms causes the material to offer "resistance" to the current flow. This friction, or resistance, causes the production of heat.

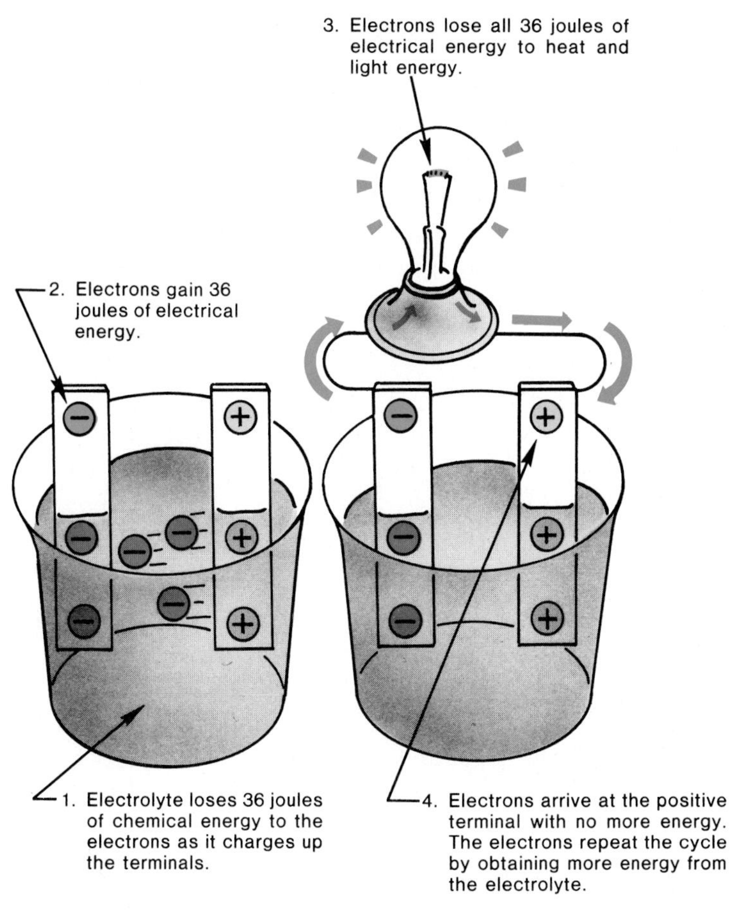

Figure 7.12. One complete cycle of electron flow through a battery/light bulb system. For the sake of illustration it is assumed that 36 joules of chemical energy are given to 3 coulombs of electrons by the 12 volt battery (3 coulombs × 12 volts = 36 joules).

More on Resistance

The electrical resistance of a wire is measured in a quantity called *ohms* (abbreviated as Ω). There are three major factors which determine how much resistance a material offers to the current.

1. **The resistance of a wire is directly proportional to its length**. Longer wires offer more resistance than shorter ones, as shown in Figure 7.13, because there is more opportunity in the longer wire for the moving electrons to lose energy via "friction" with the stationary atoms. Thus, if 100 meters of copper wire has a resistance of 2 ohms, then 200 meters of the same wire will have a resistance of 4 ohms.
2. **The resistance of a wire is inversely proportional to its cross-sectional area**. Wires which have a larger

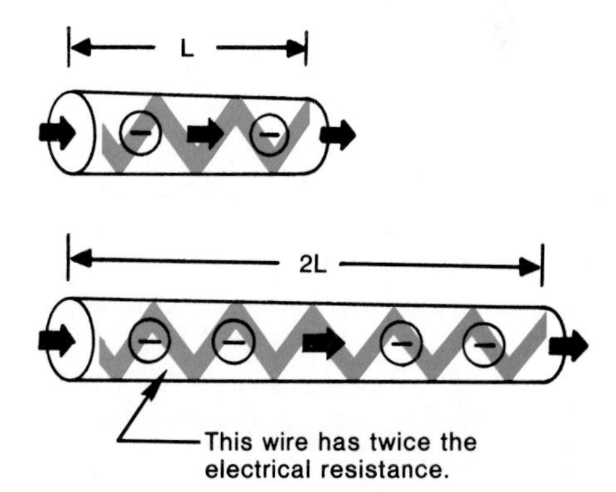

Figure 7.13. The resistance of a wire is directly proportional to its length. Longer wires offer more resistance than do shorter wires. Audio engineers use a zigzag line (\/\/\/) as a symbol for resistance.

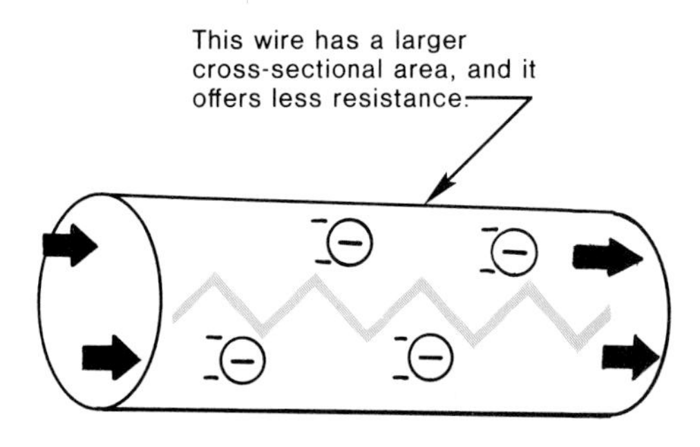

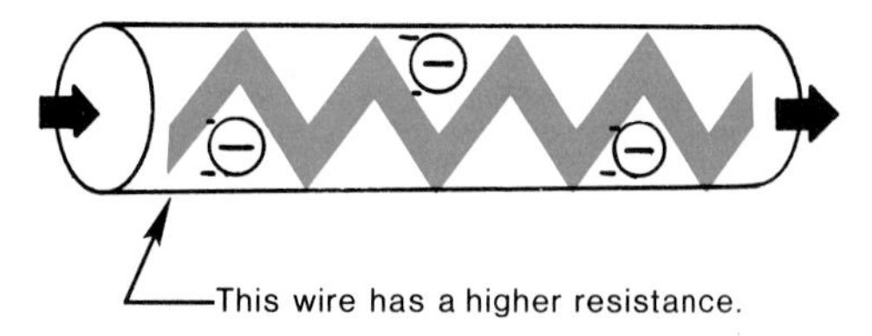

Figure 7.14. The electrical resistance of a wire also depends on its cross-sectional area. Thicker wires have less resistance than do thinner wires. The resistance of a wire is inversely proportional to its cross-sectional area.

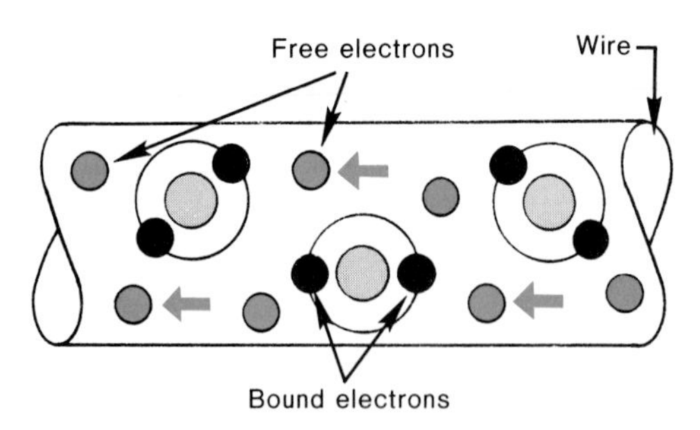

Figure 7.15. The electrons which become detached from the outer orbits of the atoms provide the electron flow in a conductor. In contrast to these "free" electrons, the electrons which remain in their orbits are called "bound" electrons.

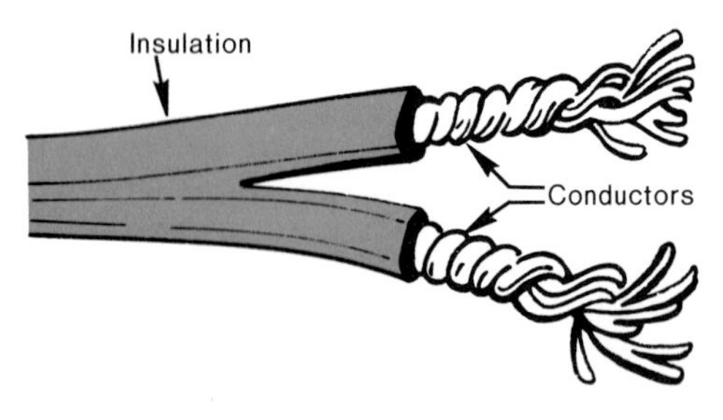

Figure 7.16. The ordinary lamp cord is comprised of two wire conductors which are separated by either rubber or plastic insulation.

cross-sectional area, i.e., thicker wires, offer less resistance as shown in Figure 7.14. The reason is that more electrons can flow through a thicker wire than a thinner one because of the additional "room" in the thicker wire. Saying that a thicker wire allows more current to flow is tantamount to saying that it has less resistance. This relationship between cross-sectional area and the amount of flow is also true for water pipes. Larger pipes permit more water to flow than do smaller pipes; the larger pipes, therefore, have less resistance. If a copper wire with a cross-sectional area of 0.01 cm^2 has a resistance of 6 ohms, then a thicker wire of the same length and material, but with twice the cross-sectional area (0.02 cm^2) will have one-half the resistance, or 3 ohms.

3. **Different types of materials have different resistances.** Materials are divided into three general classes according to their ability to conduct electricity: conductors, insulators, and semiconductors. *Conductors,* as their name implies, readily allow electrical currents to flow through them. Such common elements as copper, silver, gold, and aluminum are examples of good electrical conductors. Within a conductor the electrons which are located in the outermost orbits of the atoms become detached and wander through the material. These "freed" electrons encounter relatively little resistance as they flow over large distances (Fig. 7.15). On the other hand, materials such as rubber, glass, and plastics possess extremely large resistances, and these substances are collectively called *insulators.* Within the insulators the outer electrons do not become detached from the parent atom and they remain bound to it. Since there is no pool of freely wandering electrons, the insulators are unable to conduct electricity. At first glance, it would seem like insulators play only a minor role in the transport of electrical current. This is not so. The job which they perform is as important as that of the conductors, because the two of them act in unison to route the electricity to the precise locations where it is needed. The conductors provide an easy pathway for the electrons while the insulators, on the other hand, act like stop signs by forbidding the electrons from entering their "space." The ordinary lamp cord consists of two conducting copper wires separated by rubber insulation (Fig. 7.16). *Semiconductors,* as their name implies, have resistance characteristics which are intermediate between conductors and insulators.

The Symbol for Resistance

As mentioned earlier, the electrical resistance of a substance is measured in ohms. When drawing electrical circuits audio engineers use a zig-zag line (⩘⩗) as the symbol for electrical resistance, with the number of ohms written above it. For example, a typical 60 watt light bulb has a resistance of approximately 240 ohms. The light bulb circuit illustrated in Figure 7.10 would be schematically drawn as shown in Figure 7.17.

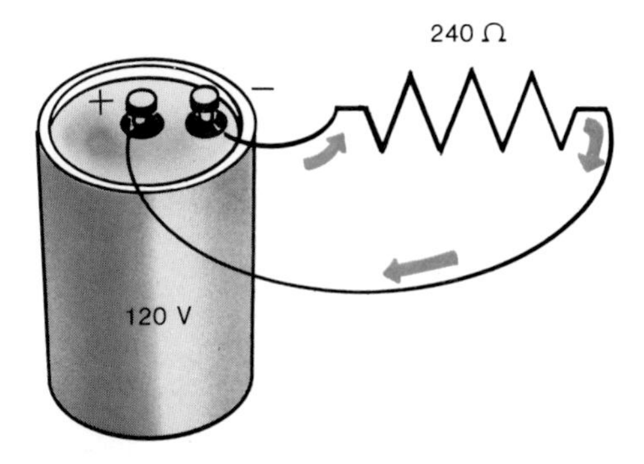

Figure 7.17. The resistance of a normal 60 watt light bulb is approximately 240 ohms, which is represented by the zig-zag symbol.

Summary

Resistance is the opposition to electron movement, and it is measured in ohms. Resistance is caused by the electrical "friction" between the moving electrons and the stationary atoms. Resistance produces heat, and many household devices such as lights, heaters, and ovens work on this principle. The resistance of a wire is directly proportional to the length of the wire, inversely proportional to the cross-sectional area of the wire, and depends on the type of material from which the wire is made.

7.7 Ohm's Law

There are three major concepts that are involved in the use of electricity: current, voltage, and resistance. These concepts are related by a very important law known as "Ohm's law." In order to understand Ohm's law, it is helpful to point out the similarity between electrical flow and water flow. To begin, Figure 7.18 illustrates water being pumped around a closed pipe. The water current, measured in gallons/sec, is analogous to electrical current, measured in coulombs/sec. The pump, which supplies the energy to move the water, corresponds to the battery in an electrical circuit. Figure 7.18 (A) shows that a strong pump causes a large amount of water current to flow through the pipe. Likewise, a battery with a large voltage rating produces a larger current than does a weaker battery. In addition, the actual amount of water flow also depends on how obstructed the pipe has become due to mineral deposits. Any such build-up causes the pipe to become narrower, and less water flows than would have with an unobstructed pipe. As indicated in Figure 7.18 (B), the more resistance which is offered by such deposits, the smaller is the current flow. In an electric circuit a similar situation exists. The greater the number of ohms of resistance, the smaller is the electrical current. Therefore, by

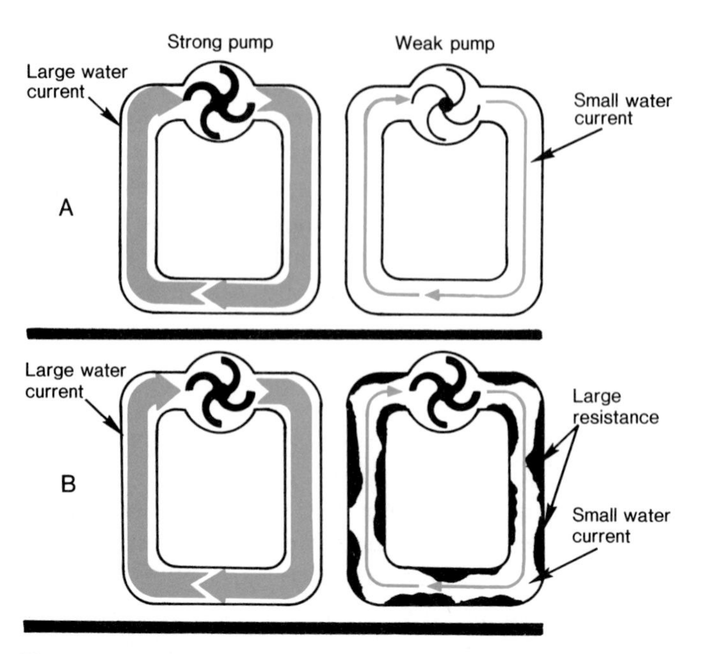

Figure 7.18. (A) A stronger pump pushes more gallons per second of water through a pipe than does a weaker pump. (B) More resistance produces less water current.

analogy with water flow in a closed circuit, we make the following statements concerning electricity flowing in a closed circuit:

1. The electrical (water) current is directly proportional to the size of the voltage (pump). Larger voltages (pumps) produce larger currents.
2. The electrical (water) current is inversely proportional to the resistance of the wire (pipe). Larger wire (pipe) resistances lead to less current.

These two statements are very important in electricity, and they constitute what is known as Ohm's law. Ohm's law can be written as a simple formula,

$$\text{Current} = \frac{\text{Voltage}}{\text{Resistance}}. \qquad \text{(Equation 7.1)}$$

With this equation the amount of current flowing through the 240 ohm "light bulb" in Figure 7.17 can be calculated if the strength of the battery is 120 volts. The resulting current is 0.5 amperes, as shown below:

$$\text{Current} = \frac{120 \text{ volts}}{240 \text{ ohms}} = 0.5 \text{ amps}.$$

Summary

The flow of electricity in an electrical circuit is analogous to the flow of water through a pipe. In both cases the current which flows is determined by the size of the pump and the amount of resistance encountered. Increasing the pump size or decreasing the resistance will produce larger currents—a fact which is known as Ohm's law. Ohm's law states that current is directly proportional to voltage and is inversely proportional to resistance. In other words, Current = Voltage/Resistance.

7.8 Generators and Alternating Current

Alternating current generators are similar to batteries in the fact that they also separate charge onto positive and negative terminals, although the type of charge on each terminal alternates with time. At one instant a terminal will have a positive charge and a short time later it will acquire a negative charge: positive, negative, positive, negative, etc. The alternating charge on the terminals produces a current which also alternates back and forth. Generators, like batteries, must expend energy, and they use nuclear, fossil (oil, coal), or hydro (waterfalls) energy to achieve the charge separation. Of course, the electric company ultimately charges us, the consumers, for this energy. The electricity for most industrial and home use is almost exclusively produced by high speed turbine generators. Wires from each of the two generator terminals are run into your home where they appear as the well-known 120-volt electrical outlets (Fig. 7.19).

Many "miniature" generators are used in hi-fi components routinely. A phono cartridge is nothing more than an alternating voltage electrical generator which is capable of providing small amounts of separated charges. It derives its energy from the rotating record which forces the stylus to move up and down, back and forth. A microphone is also a generator which utilizes the energy of the incoming sound waves to produce charge separation. And finally, a playback head on a tape deck is an electrical generator which is activated by the energy carried by the moving tape. In each of these three cases the generator develops an alternating voltage whose frequency is determined by the frequency of the source providing the energy. If, for example, the record groove forces the stylus up and down at a rate of 5,000 Hz, then the cartridge will develop an alternating voltage at its terminals whose frequency is also 5,000 Hz. The alternating voltage produces an alternating current which is delivered to the preamp for amplification.

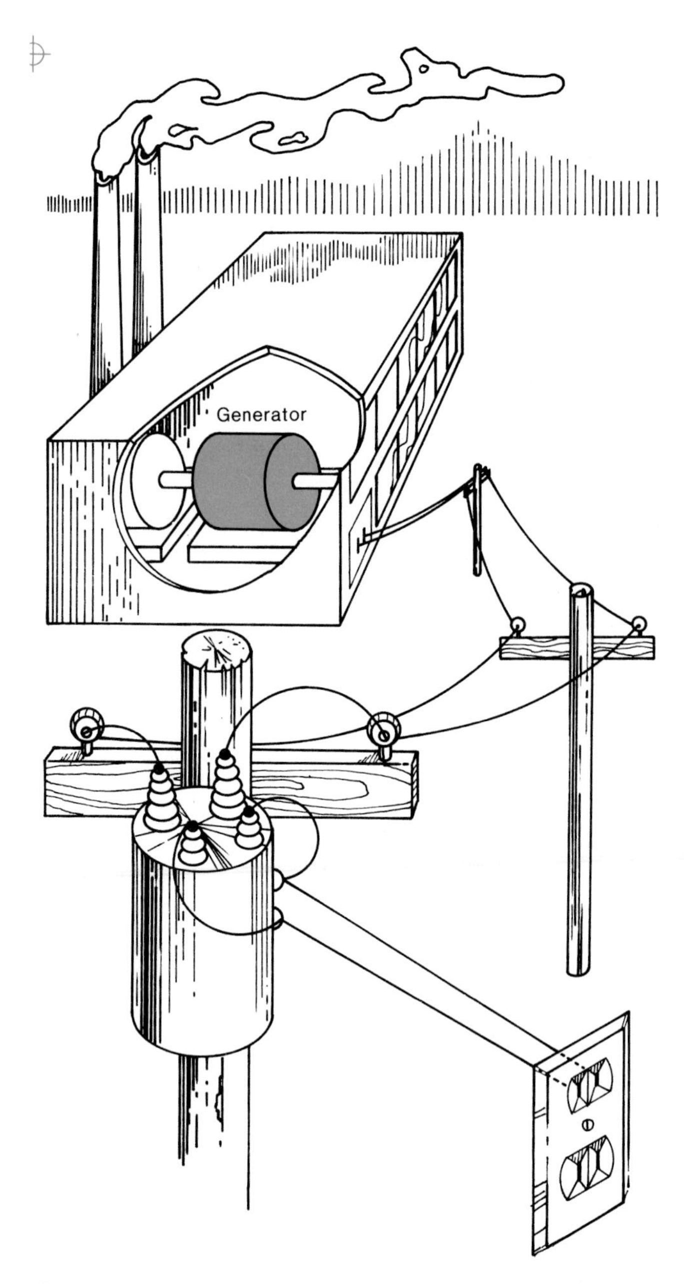

Figure 7.19. The typical 120 volt electrical outlet consists of two oppositely charged terminals. The charges have been separated by a generator, and the amount and type of charge on each terminal varies with time.

As you can see, the idea of creating an alternating charge separation by generators is of fundamental importance in audio systems. For the moment we shall not be concerned with how a generator works, but we will explore some of its interesting consequences in electrical applications. Chapter 11 will present the details of how cartridges and tape heads work.

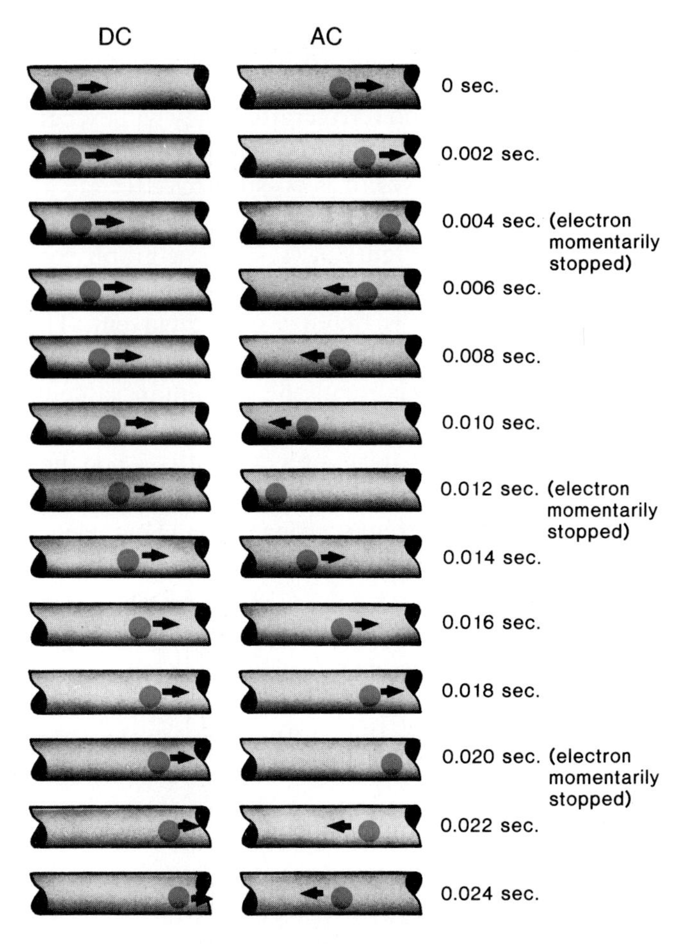

Figure 7.20. A comparison between DC and AC current in a wire. The colored dot represents an electron, and the arrows indicate the direction of the current flow. The AC frequency is 60 Hz. The absence of an arrow in the AC pictures at 0.004, 0.012, and 0.020 seconds indicates that the electron has momentarily stopped, prior to reversing its direction of travel.

Alternating Current

Alternating current, as its name implies, flows back and forth inside the wire. The electrons first move in one direction, say to the right, and after a short time they stop, reverse their motion, and flow to the left. Again, after a short period, they halt their motion and begin a movement to the right, and the entire process is repeated over and over. Figure 7.20 shows a comparison between DC and AC currents. The salient feature of this diagram is that alternating current reverses its motion every half-cycle just like the air particles which make up a longitudinal sound wave (see Fig. 3.19). By analogy with sound waves one can speak of the *period* of an alternating current as the *time* required for the electrons to complete one back and forth cycle. And the *frequency* is the *number of cycles* which are *completed each second.* For example, the alternating current found in homes

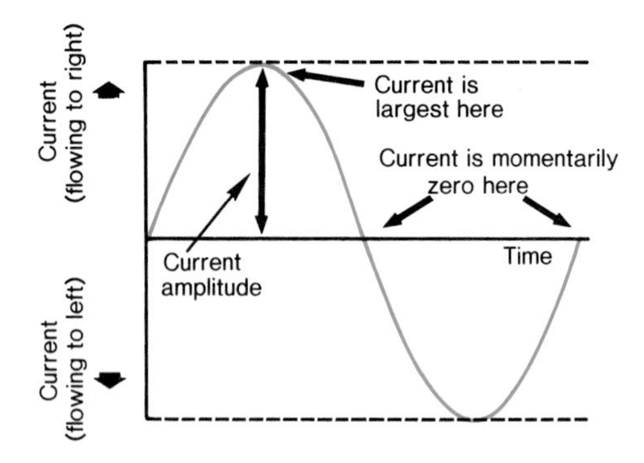

Figure 7.21. An AC waveform shows that alternating current is continually changing in both amount and direction.

has a frequency of 60 Hz which implies a period of 1/60 (= 0.017) seconds. As the electrons travel through various appliances, such as lights and refrigerators, they will execute one complete cycle of oscillatory motion every 0.017 seconds.

As with longitudinal waves, it is possible to associate all of the other wave parameters, such as amplitude and speed, with alternating current. Figure 7.21 shows a graph containing one cycle of AC where the horizontal axis represents time and the vertical axis is the size of the current. The drawing also shows that an AC waveform has an *amplitude* associated with it which represents the maximum value of the current. The curve does *not* imply that alternating current is a transverse wave. Alternating current is a bonafide longitudinal wave, because the electron's motion is parallel to the direction of the current's propagation. However, a graph of the current versus time artificially makes it look like a transverse wave.

In Figure 7.21 the current starts at 0 amps and begins moving to the right (+ current) until it reaches its maximum value one quarter of a cycle later. Once past this point the current begins to decrease (but still traveling to the right) until it again becomes momentarily zero at the half-cycle point. This zero current point is reached when the electrons must momentarily stop in order to reverse their direction. During the second half-cycle the same type of motion occurs, except that the electrons are now traveling to the left.

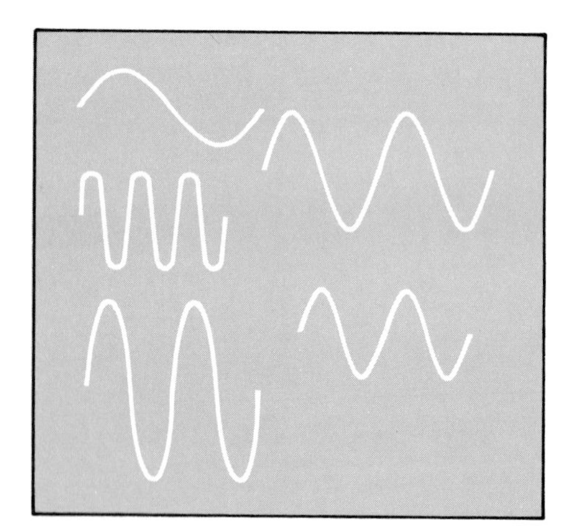

Figure 7.22. Alternating current waveforms, just like those of sound waves, can be generated with various amplitudes and frequencies.

As shown in Figure 7.22, AC waveforms can be produced in all amplitudes and frequencies, and they are identical in appearance to the simple waveforms which represent the pressure variations in sound waves. This similarity is of crucial importance because it is then possible, via the use of a microphone, to convert all the parameters of an incoming sound wave—like amplitude and frequency—into an alternating current which is an exact replica. The only difference between the two waves is that air pressure has been replaced by electric current, as shown in Figure 7.23.

Once the sound wave has been duplicated as an alternating current, its amplitude can be magnified by amplifiers before being sent to the speakers. The speaker converts the incoming AC waveform into a sound wave which is a replica of the alternating current waveform. The speaker can be thought of as a microphone working in reverse.

What happens when the sound pressure is a complex pattern as most are? The microphone once again converts the sound into a complex electrical waveform which is an exact replica as shown in Figure 7.24. Therefore, an AC current can be produced with any complicated shape in order to reflect accurately the incident sound pattern. Figure 7.24 also shows how a record is cut. The electrical signals from the amplifier are applied to a cutting stylus which cuts the

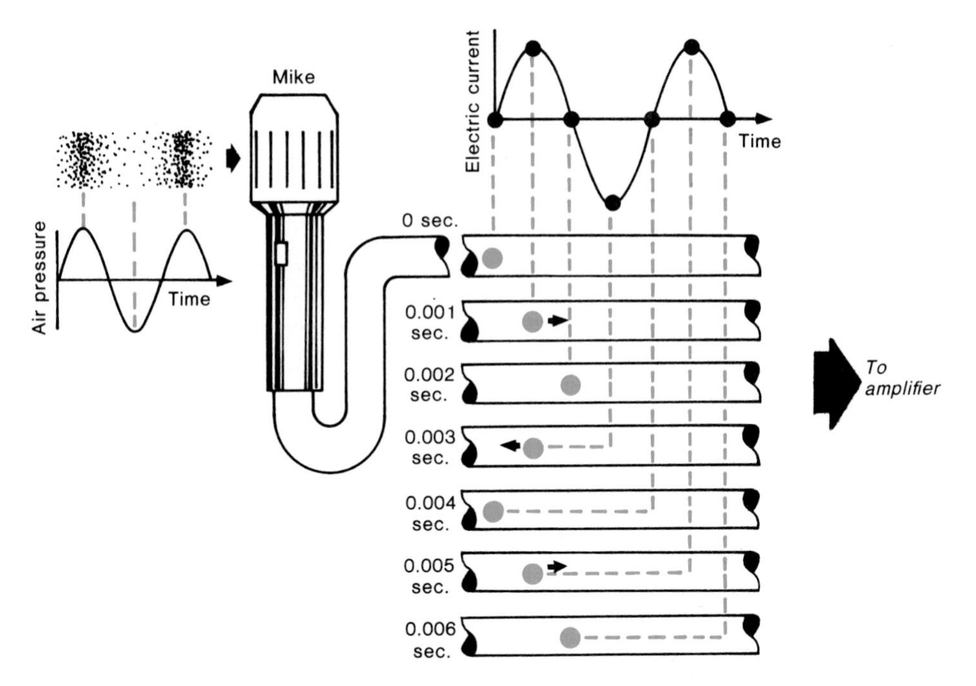

Figure 7.23. The microphone receives the incident sound (pressure) wave and generates an AC electrical signal. The AC waveform is an exact replica of the sound pressure waveform. The frequency of the tone is assumed to be 250 Hz. The motion of the electron (colored dot) is shown at various times. At 0, 0.002, 0.004, and 0.006 sec the electron has momentarily stopped, prior to reversing its direction of travel.

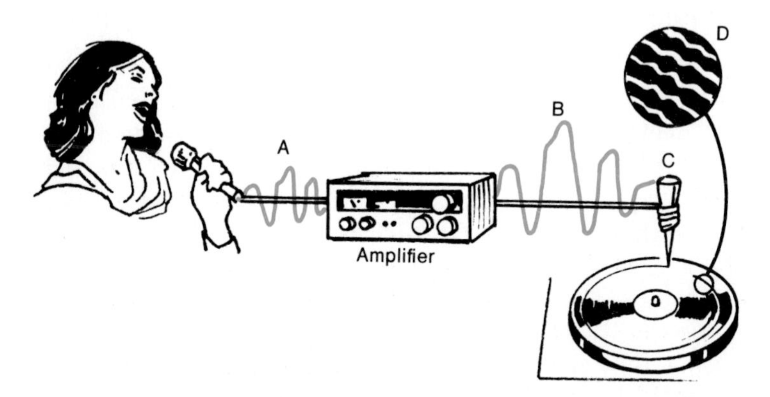

Figure 7.24 (A) The singer creates a complex sound wave which is converted by the microphone into a small alternating current waveform. (B) The amplifier produces an enlarged replica of the electrical signal which is sent to a (C) cutting stylus. (D) The stylus cuts a mechanical replica of the electrical signal into the disc.

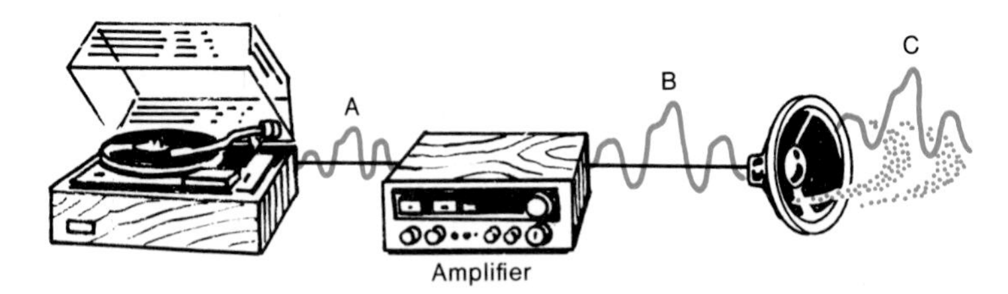

Figure 7.25. (A) The stylus tracks the record groove and a small alternating current is produced by the cartridge. (B) The current is amplified in order to drive the speakers which (C) recreate the original sound.

wave pattern into the disc. Figure 7.25 shows the playback process whereby the longitudinal electricity waves are converted into sound by the speaker.

Now we see the importance of characterizing sound via the wave concept. Hi-fi systems convert the sound waves into electrical waves. Although these waves exist in a physically different medium, they possess all the wave characteristics (frequency, amplitude, period, etc.) of the original sound wave. Thus electrical waves "carry" all the necessary information about the sound. Just how accurately a system can make these transformations determines whether it is low-fi or hi-fi. As we shall see in later chapters, most of the technical specifications of audio components deal with the problem of accurate conversions.

Summary

Alternating current occurs when the electrons within a wire oscillate back and forth about their equilibrium positions. Their motion is exactly analogous to the motion of the particles which comprise a longitudinal slinky or a sound wave. Therefore, an alternating current is also a longitudinal wave with all the properties such as frequency, period, wavelength, and amplitude. This fact is of critical importance in audio systems because it allows devices such as microphones, speakers, cartridges, and tape decks to transcribe one type of wave into another. Although the two transcribed waves may be physically different—sound/electrical, electrical/sound, mechanical movement/electrical, magnetic/electrical—they preserve the audio information by retaining the same frequency (pitch) and amplitude (loudness) relationships. An audio system can be viewed as a series of transformations which convert one type of wave to another. How accurately each device can perform the transformations will dictate its fidelity.

7.9 Some Interesting Examples of Ohm's Law and AC Current

AC electricity is used extensively in both household appliances and hi-fi systems. AC, like DC, also obeys Ohm's law, although a special average value must be used for current and voltage, because AC is constantly changing in time. This average, called the RMS* average, has been found to give the correct value for the average power consumed in AC circuits. The "120 volts" which labels an AC household electrical outlet refers to its RMS voltage. Likewise, we shall always use RMS values for currents and voltages without always explicitly mentioning it. Now for some interesting examples.

The Electric Toaster

Figure 7.26 shows a typical electric toaster plugged into a 120-volt AC electrical outlet. As mentioned in section 7.8, the current into the two terminals is reversing 120 times every second. The drawing shows a "snapshot" when the right terminal happens to be negative and the left terminal is positive; 1/120 seconds later (one-half of the period) the charge on the two will be reversed. If the resistance of the

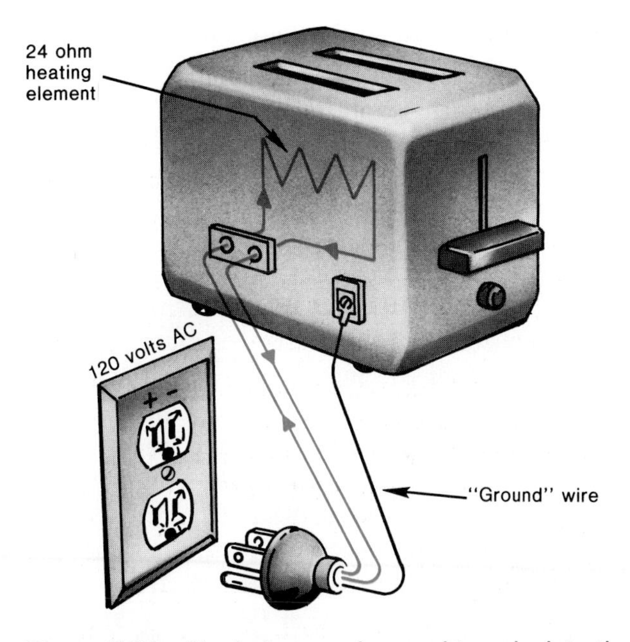

Figure 7.26. Ohm's law can be used to calculate the average current drawn from the 120 volt outlet by the toaster.

*RMS means *R*oot *M*ean *S*quare, which is a mathematical method for calculating average values.

toaster's heating element is 24 ohms, how much alternating current will flow? Using Ohm's law we can calculate the current:

$$\text{Current} = \frac{120 \text{ volts}}{24 \text{ ohms}} = 5 \text{ amps.}$$

The copper wires connecting the toaster to the outlet have negligible resistance, so that they do not heat up to any great extent. The heating element of the toaster, on the other hand, has a relatively high resistance and therefore produces practically all of the heat. This arrangement of resistances is accomplished by using a good conductor for the power cord, such as copper, which has a relatively large cross-sectional area. The heating element, on the other hand, is made of a much thinner and poorer conducting material such as tungsten. These two factors produce a high resistance filament which, in turn, generates most of the heat in the circuit. Alternating current is just as effective as direct current when it comes to generating heat because equal amounts of electrical "friction" are produced when the current is moving to the left or to the right. It is like rubbing your hands together back and forth on a cold day to get a little heat. In other words, the longitudinal wave of AC current carries energy to the toaster just as well or better than DC current.

In Figure 7.26 notice that the plug has three prongs, whereas only two are needed to provide a path for the current. The third prong is a safety feature and is called an electrical "ground," which is connected directly to the metal case of the toaster via a third copper wire. Figure 7.27 (A) shows a toaster operating under normal conditions, and the electrical ground, while attached to the toaster, is not being utilized. It is only when the toaster functions *abnormally* that the electrical ground becomes important. Part (B) illustrates what happens when the heater element comes loose and accidentally touches the metal case while the user is also touching it. The electricity will flow from the electrified case, through the ground wire, and into the wall where a safe path is provided for it to flow into the ground. If the ground wire were not present, as shown in (C), the electrical current would flow through the user's hand (and body) as it returns to the ground, thus causing an electrical shock. The grounding wire works because of Ohm's law which, in effect, states that more current will flow along a path of low resistance, like the copper "ground" wire, rather than along a path of much higher resistance like the user's body! When the toaster becomes defective, the current will flow through the "ground" wire rather than through the user, thus preventing a severe shock.

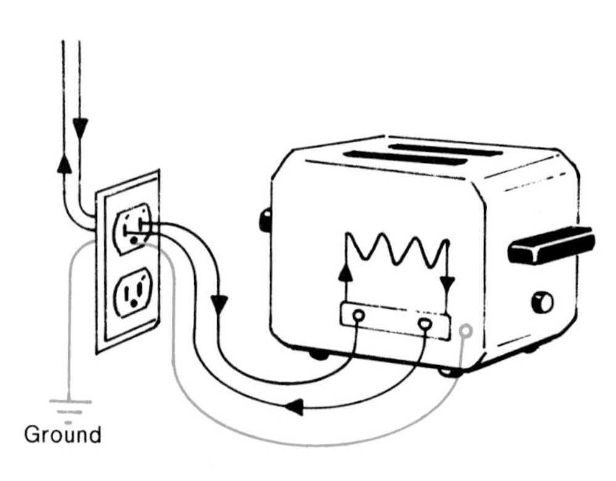

A. Normal operation.

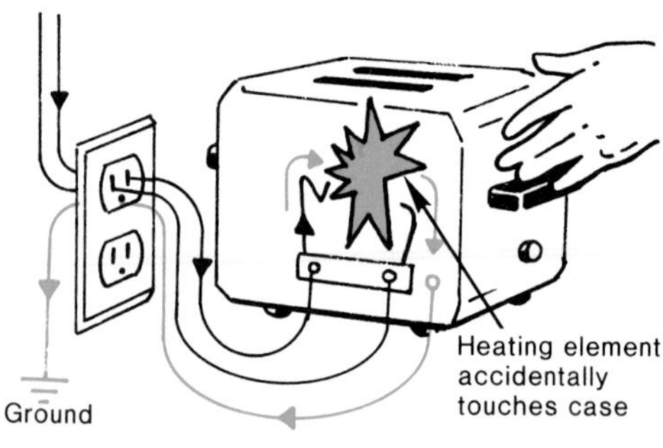

B. Abnormal operation, but toaster safely grounded.

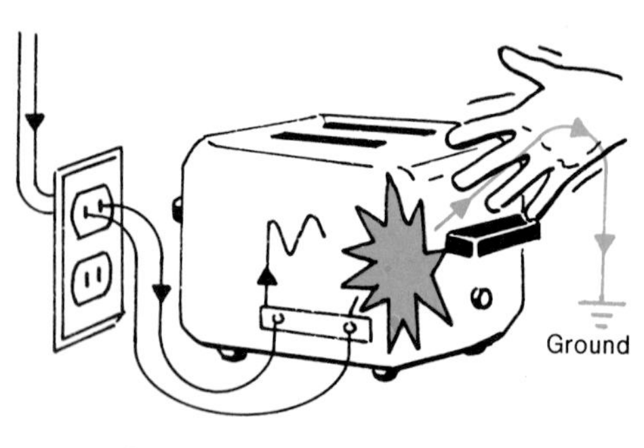

C. Abnormal operation, causing shock to the user.

Figure 7.27. The 3-prong plug means that the toaster is safely "grounded." The use of a 2-prong plug can lead to an electrical shock if the toaster malfunctions.

The Hi-Fi Speaker and Amplifier

Another example of Ohm's law is the amount of current being delivered to the speakers by the power amplifier (Fig. 7.28). The amplifier produces, at the speaker output jacks, an alternating current which is an exact replica of the original musical pressure variations. If, at a given instant in time, the amplifier is producing 24 volts, then the 8 ohm resistance (sometimes called impedance) of the speaker voice coil will permit 24 volts/8 ohms = 3 amps of current to flow through the circuit. The 3 amps traveling through the voice coil cause it to heat up, just like a light bulb, in addition to forcing the cone into the proper vibratory motion. In this case the current produces two effects, rather than pure heating, and the term "impedance" is substituted for "resistance" to denote this difference.

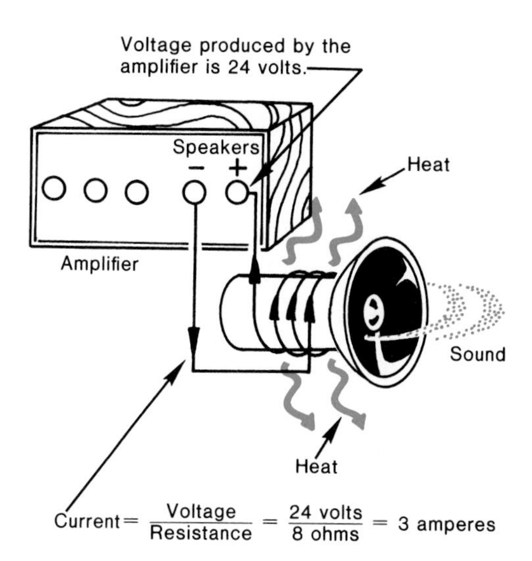

Figure 7.28. The current delivered by the power amplifier causes the speaker diaphragm to vibrate. In addition, heat is produced within the voice coil by the interaction of the current with the resistance of the voice coil.

The VOLUME Control In the Preamplifier

The *VOLUME control* located on the front panel of all amplifiers is nothing more than a variable resistor which determines the amount of current flowing through the system. The sound intensity produced by a loudspeaker is proportional to the current in the voice coil, and controlling the current effectively controls the sound intensity. As shown in Figure 7.29, the VOLUME control is a variable resistor whose value is changed whenever the VOLUME control

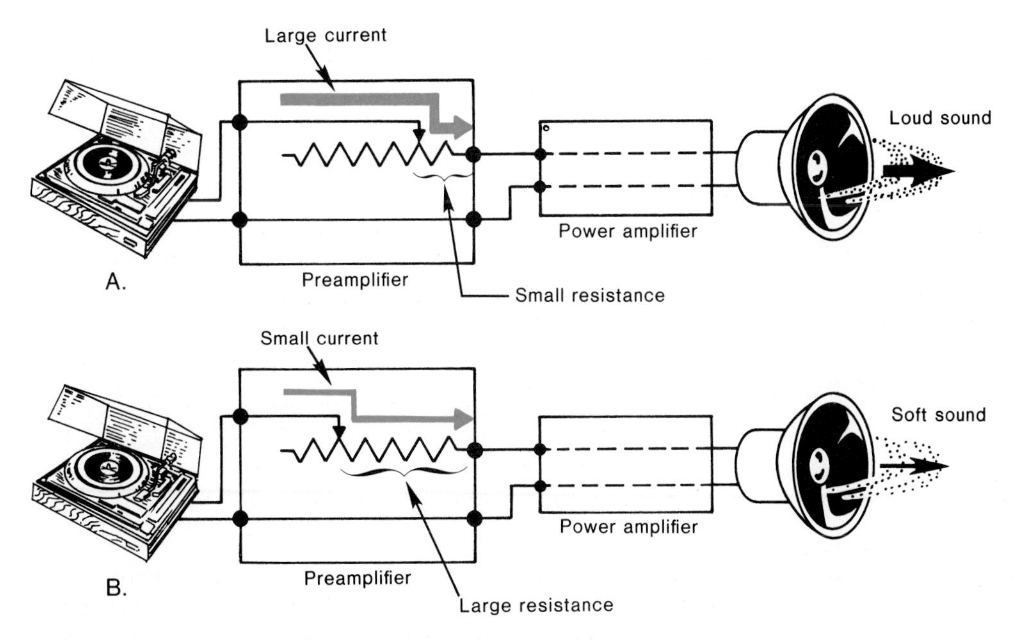

Figure 7.29. (A) The VOLUME control (a variable resistor) is set for minimum resistance, thereby allowing a large current to pass through the system. (B) The resistance of the VOLUME control has been increased, which results in a decrease in the current; less current causes the speaker to produce quieter sounds.

knob is rotated.* Diagram (A) shows the case where the control is turned all the way "up" so that fairly loud sounds would be heard. Here the variable resistor introduces very little additional resistance into the circuit, thus permitting larger currents to flow. Ultimately these larger currents are translated into louder sounds by the speaker. In Figure 7.29 (B) the VOLUME control has placed substantially more resistance into the circuit which causes a corresponding reduction in the current. This produces the quieter sounds when the volume is turned "down."

The BALANCE Control In the Preamplifier

The *BALANCE control*** is functionally very similar to the VOLUME control. It provides a method for making one speaker louder while reducing the loudness of the other. The main purpose of the BALANCE control is to regulate the currents in the left and right channels. The most common type of BALANCE control is constructed such that when the current in one channel is increased (decreased), the current in the other channel is decreased (increased). The BALANCE control in Figure 7.30 (A) is set to the "right." The

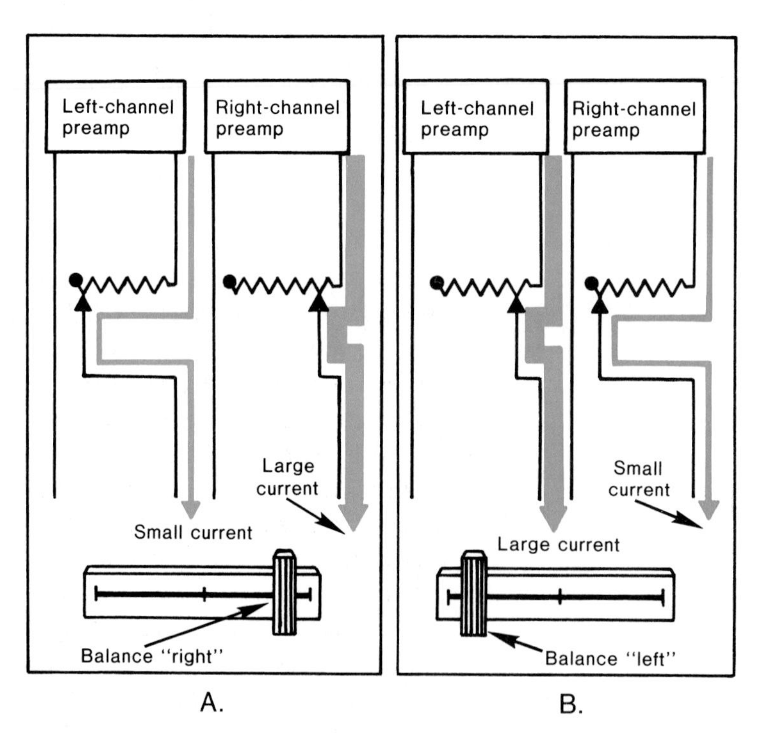

Figure 7.30. A BALANCE control which consists of two variable resistors—one for each stereo channel. (A) The preamp is balanced to the "right." (B) The preamp is balanced to the "left."

*The VOLUME control shown in Figure 7.29 is for illustrative purposes only. An actual VOLUME control is somewhat more involved, but the differences need not concern us here.

**The BALANCE control found on an actual amplifier may operate in a more complicated manner. Our discussion is only for illustrating how resistors can be used to control the current.

variable resistor in the right channel introduces only a small amount of resistance thus permitting a relatively large current to flow. In the left channel, however, the large amount of resistance allows only a small amount of current to pass. The final result is that the sound level is smallest in the left channel. The Balance control method, shown in the drawing, assumes that a single lever simultaneously controls both the left and right variable resistors. Some preamplifiers may have two separate Balance controls, one for each channel. Figure 7.30 (B) shows the opposite case where the system is balanced to the "left."

7.10 Electric Power

Power is defined as the amount of energy (in joules) that a device either delivers or consumes divided by the time (in seconds) that the device is operating;

$$\text{Power} = \frac{\text{Energy delivered or consumed}}{\text{Time}}. \quad \text{(Equation 7.2)}$$

The unit of power is called the "watt." For example, if a power amplifier delivers a total of 1,200 joules of energy to the speakers in 6 seconds, then the power would be

$$\text{Power} = \frac{1200 \text{ joules}}{6 \text{ sec.}} = 200 \frac{\text{joules}}{\text{sec.}}$$
$$\text{Power} = 200 \text{ watts.}$$

Therefore, this amplifier is delivering 200 joules of electrical energy each second to the speaker. Practically all electrical devices come with a power rating. A most familiar example is the 60 watt light bulb which consumes 60 joules of energy per second of use. Similarly, a 900 watt hair dryer consumes 900 joules of electrical energy in each second.

There is another, and very useful, expression for the power which involves the current and voltage. This expression is

$$\text{Power} = (\text{Current}) \times (\text{Voltage}) \quad \text{(Equation 7.3)}$$

The above formula can be explained as follows. Recall, that the voltage rating of a generator, such as an amplifier, tells how much electrical energy has been given to each coulomb of charge which is sent out of the amplifier. The current is the number of coulombs being sent out each second. The product of the voltage and the current then gives the total amount of electrical energy being supplied by the amplifier each second which, according to Equation 7.2, is the power being delivered. Therefore, Equation 7.3 correctly predicts the power if the current and voltage are known.

As another example of using Equation 7.2, consider a speaker which is receiving 25 watts of electrical power for 10 minutes. How much energy is consumed during this time? Equation 7.2 answers this question very nicely:

$$\text{Power} = 25 \text{ watts}$$
$$= \frac{\text{Energy}}{10 \text{ min} \times 60 \text{ sec/min}}$$

or

$$\text{Energy} = 15{,}000 \text{ joules.}$$

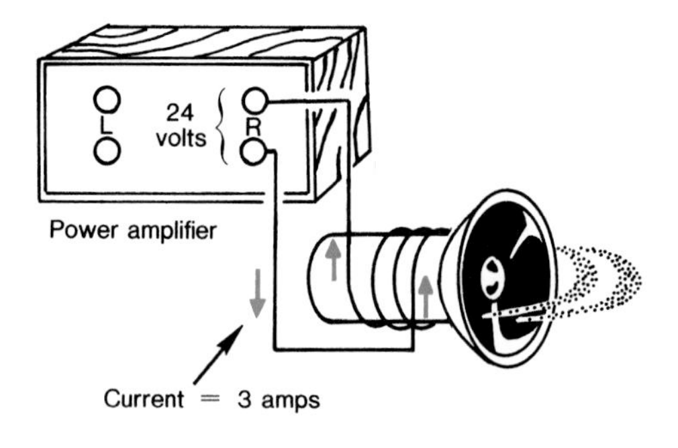

Figure 7.31. The electrical power being delivered from the amplifier to the speaker is the product of the current times the voltage.

Suppose, for example, that a power amplifier is producing 24 volts at the chosen volume setting, and suppose that 3 amperes of current are flowing to a speaker as shown in Figure 7.31. The power in watts delivered by the amplifier to the speaker can be calculated from Equation 7.3:

$$\text{Power} = (3 \text{ amperes}) \times (24 \text{ volts}),$$
$$\text{Power} = 72 \text{ watts}.$$

Very often it is important to calculate the power when the current and resistance are known, rather than the current and voltage. It is possible to modify Equation 7.3, with the help of Ohm's law, in order to obtain a convenient expression for the power. Using Ohm's law,

$$\text{Voltage} = (\text{Current}) \times (\text{Resistance}), \quad \text{(Equation 7.1)}$$

we can substitute this expression for the voltage into the power formula of Equation 7.3:

$$\text{Power} = (\text{Current}) \times [(\text{Current}) \times (\text{Resistance})]$$

or

$$\text{Power} = (\text{Current})^2 \times (\text{Resistance}). \quad \text{(Equation 7.4)}$$

This expression for the power involves the product of the squared-current and the resistance. We shall now consider a practical example of Equation 7.4.

Example of Fusing Speakers

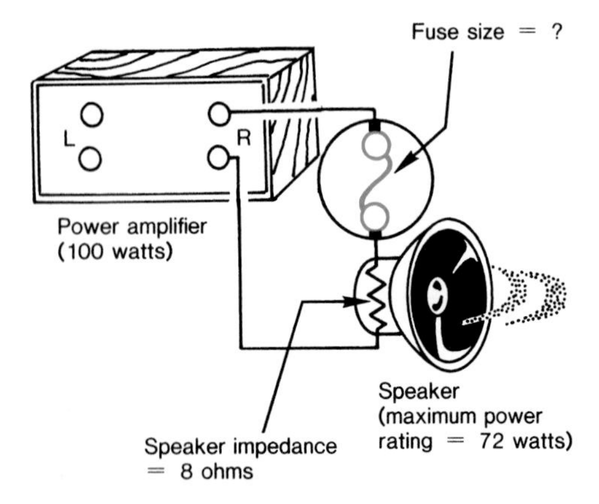

Figure 7.32. The formula, Power = (Current)^2 × (Resistance), can be used to calculate the correct fuse size for protecting expensive speakers.

Situations may arise when the amplifier's output power rating exceeds the speaker's maximum power rating as shown in Figure 7.32. As long as the amplifier's volume is kept low there is no danger of damaging the speakers. However, if someone should accidentally turn up the volume such that the amplifier output power exceeds the maximum power rating of the speakers . . . well, you will see grown people cry as the voice coil quickly heats up and melts. This costly mistake can be avoided by inserting fuses into both the left and right channel speaker wires. If the current exceeds the fuse "size," then the fuse will burn up and stop the current flow long before the speakers become damaged. Since a fuse costs approximately 25¢, and most drivers cost anywhere between $10 and $200, it is obviously far cheaper to replace a fuse than a driver. Assume that the speaker impedance is 8 ohms and its maximum power rating is 72 watts (both of these numbers are furnished by the manufacturer). The maximum allowed current can be calculated via the following method:

$$\text{Power} = (\text{Current})^2 \times (\text{Resistance}) \quad \text{(Equation 7.4)}$$
$$72 \text{ watts} = (\text{Current})^2 \times (8\ \Omega)$$
$$\text{Current} = 3 \text{ amps}.$$

Therefore, choose a 3 amp, or less, fuse to protect the speakers. Notice that one only needs to know two pieces of information about the speakers themselves: their impedance and their maximum power rating.

Large and Small Units of Power

Power, like most other quantities, comes in large and small amounts. Very often abbreviations are used to signify these units and some of the more common ones are listed below.

(a) 1 kilowatt = 1,000 watts
1 kw = 1,000 watts

(b) 1 milliwatt = 1/1,000 watt
1 mw = 1/1,000 watt

(c) 1 microwatt = 1/1,000,000 watt
1 μw = 1/1,000,000 watt

Another Application of the Power Concept: How to Compute Your Electric Bill

It comes as no surprise that the usage of your hi-fi system is always followed by a bill from the power company. The bill charges you for the total amount of electrical energy consumed over some period of time, perhaps a month. When you plug your hi-fi into the wall socket, the electric company is providing a certain amount of power to operate the system. Since power is defined to be the amount of electrical energy consumed each second, the total energy used by your hi-fi system is the product of the power it consumes and the duration of time that the set is turned on;

$$(\text{Total electrical energy used}) = (\text{Power}) \times (\text{Time}).$$

(Equation 7.2)

To compute your cost, first express the power in **kilowatts** and the time in **hours** so that the total energy, as calculated by Equation 7.2, will be expressed in kilowatt-hours. Next, mutiply the energy consumed by the cost per kilowatt-hour:

$$\text{Total cost} = \begin{pmatrix}\text{Energy consumed}\\ \text{(in kilowatt-hours)}\end{pmatrix} \times \begin{pmatrix}\text{Cost per}\\ \text{kilowatt-hour}\end{pmatrix}$$

(Equation 7.5)

For example, suppose that your hi-fi draws 125 watts of power from the electrical outlet and that you use it on the average of 80 hours per month. The total energy consumed is then

$$\text{Energy} = (0.125 \text{ kilowatts}) \times (80 \text{ hours}) \\ = 10 \text{ kilowatt-hours.}$$

If the power company charges 6¢ per kilowatt-hour of energy consumption then the monthly cost for using your stereo will be

$$\text{Cost} = (10 \text{ kilowatt-hours}) \times (6¢ \text{ per kilowatt-hour}) \\ = 60¢.$$

Table 7.2 gives the average energy consumption and cost for many common electrical appliances.

	Estimated Electrical Energy Consumption in 1 Month (in kilowatt-hours)	**Cost for 1 Month (rate = $0.06 per kilowatt-hour)**
Hi-fi	10	$0.60
Air conditioner (room)	220	13.20
Fan (window)	14	0.84
Hair dryer	1.2	0.07
Shaver	0.15	0.01
TV (color)	42	2.52
Dishwasher	30	1.80
Range	100	6.00
Toaster	2.8	0.17
14 cu. ft. refrigerator/ freezer	95	5.70
Washing machine	9	0.54
Clothes dryer	85	5.10

Table 7.2. The average costs per month for running some home appliances.

Summary

Power expresses the amount of energy which is either consumed or delivered by an electrical device for each second of its operation. Electrical power is commonly measured in watts, and one watt is equal to one joule of energy per second. The important equations for power are:

$$\text{Power} = \frac{\text{Energy}}{\text{Time}},$$

$$\text{Power} = (\text{Voltage}) \times (\text{Current}),$$

$$\text{Power} = (\text{Current})^2 \times (\text{Resistance}).$$

The last equation is useful when calculating the size of a fuse needed for speaker protection. The cost of using electricity can be determined by the following relation:

$$\text{Cost} = (\text{Power consumption (in kilowatts)}) \times \\ (\text{Time of consumption (in hours)}) \times \\ (\text{Cost per kilowatt-hour}).$$

7.11 Series and Parallel Circuits

Typically, in most households more than one electrical device is used at the same time. Several light bulbs and a stereo may be simultaneously turned on. It is even possible to operate several pairs of speakers, called the "main" and "remote" speakers, from the same amplifier. There are two methods for wiring multiple electrical devices together, series wiring and parallel wiring.

Series Wiring

Connecting electrical devices in series means that they are connected one after the other, so that there is only one pathway along which the electrons can flow. The very same current flows through each device. Using the analogy of flowing water, this concept is illustrated in Figure 7.33, where two sections of pipe are connected in series with the pump. Notice that each pipe has accumulated mineral deposits to a different extent. Therefore, each offers a different resistance, R_1 or R_2, to the flow. As the water moves through each pipe

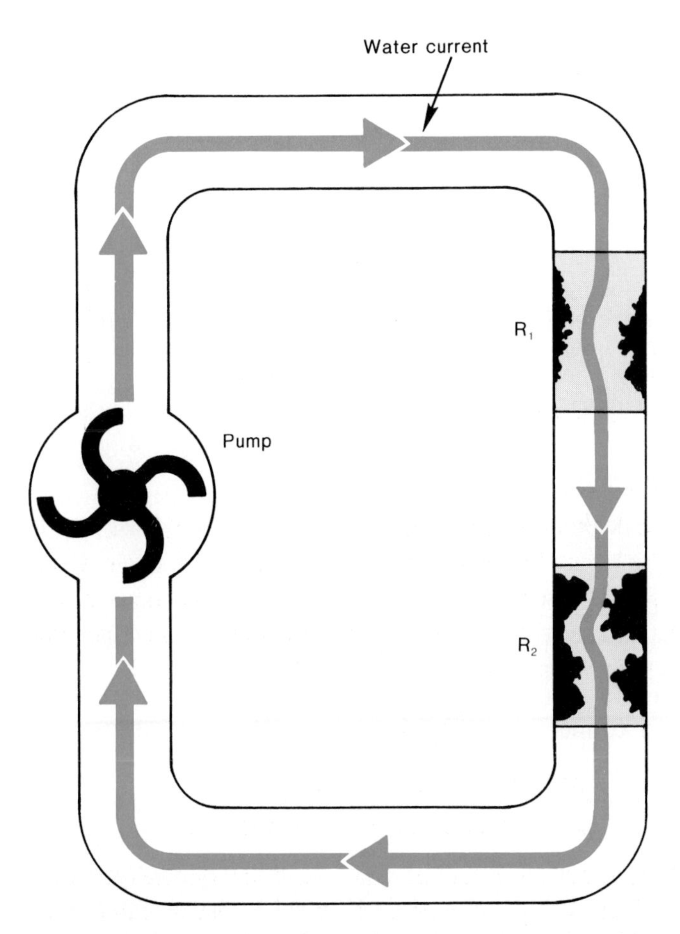

Figure 7.33. When two water pipes are connected in series, the same water flows through each. R_1 and R_2 denote the resistance of the two pipe sections.

in sequence, the pump must enable the water to overcome the first resistance R_1 and then the resistance R_2. In other words, the total resistance of the water circuit is $R_1 + R_2$. The same situation exists when electric current flows through more than two devices connected in series. The source of voltage must enable the current to pass through a total effective resistance R_T given by the sum of each resistance:

$$R_T = R_1 + R_2 + R_3 + \cdots . \qquad \text{(Equation 7.6)}$$

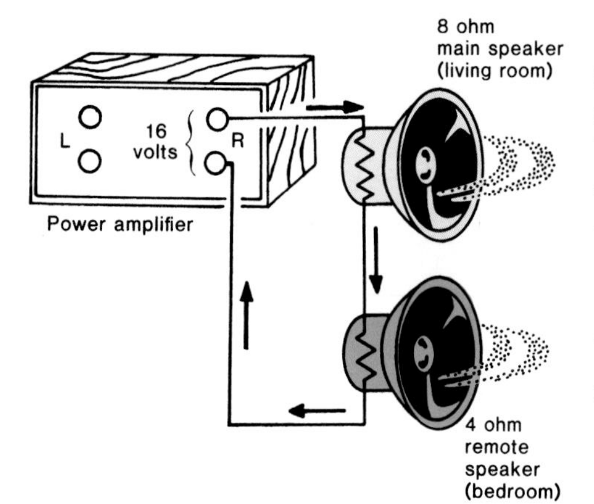

Figure 7.34. Two speakers which are wired in series.

Figure 7.34 shows an 8Ω main speaker and a 4Ω remote speaker connected in series to the right channel of a power amp. The amplifier is supplying 16 volts at the chosen VOLUME control setting. Ohm's law can be used to calculate the current flowing through each speaker if the total resistance is first determined from Equation 7.6:

$$R_T = 8\Omega + 4\Omega = 12\Omega.$$

Then the current can be obtained from the voltage and total resistance by using Ohm's law:

$$\text{Current} = \frac{\text{Voltage}}{\text{Resistance}} = \frac{16 \text{ volts}}{12 \text{ ohms}},$$

$$\text{Current} = 1.33 \text{ amperes}.$$

Because the two speakers are connected in series the same current, 1.33 amperes, flows through each one. Once the current flowing through each speaker is known, the total power delivered by the amplifier can be determined with the aid of Equation 7.4:

Main Speaker

$$\text{Power} = (\text{Current})^2 \times \text{Resistance},$$
$$\text{Power} = (1.33 \text{ amperes})^2 \times 8\Omega,$$
$$\text{Power} = 14.2 \text{ watts}$$

Remote Speaker

$$\text{Power} = (1.33 \text{ amperes})^2 \times 4\Omega,$$
$$\text{Power} = 7.1 \text{ watts},$$

$$\left(\begin{array}{l}\text{Total Power Delivered}\\ \text{by the Amplifier}\end{array}\right) = 14.2 + 7.1 = 21.3 \text{ watts}.$$

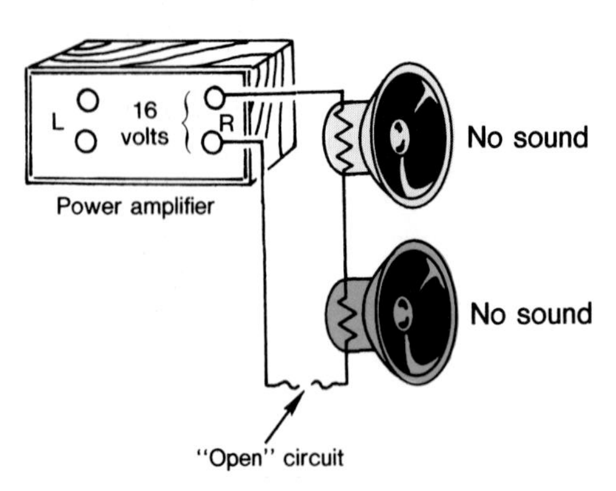

Figure 7.35. In a series connection neither speaker produces sound when the circuit is "open."

There are two important features which are common to all series connections. First, there is only one path along which the electrons can travel. A break in the path, such as occurs when one of the speakers "burns out," produces an "open" circuit, and all current ceases. Therefore, the other speaker will stop producing sound, as shown in Figure 7.35. Secondly, the sound intensity produced by each speaker is reduced when the speaker is wired in series with another speaker. The total resistance of the two speakers is larger than that of either speaker alone, and this increased resistance causes less current to flow in the circuit. Thus, each speaker in a series connection receives less current, and, hence, produces less sound.

It is interesting to note that the 16 volts provided by the power amplifier is "shared" by the two speakers, with neither receiving the full amount. The "voltage sharing" happens when devices are connected in series, and is predicted by Ohm's law, as shown below:

Main Speaker Voltage = Current × Resistance
Voltage = 1.33 amperes × 8Ω
Voltage = 10.67 volts
Remote Speaker Voltage = 1.33 amperes × 4Ω
Voltage = 5.33 volts

Total Voltage = 10.67 + 5.33
= 16 volts.

Parallel Wiring

There is another method of connecting electrical devices, called parallel wiring, that avoids the difficulties of series wiring. Connecting electrical appliances in parallel means that each device is attached directly to the voltage source, so that each receives the full voltage being supplied, rather than just a portion of it as in a series connection. Figure 7.36 shows two speakers connected in parallel to the right channel of a power amp, which is supplying 16 volts at the chosen volume setting. Notice the difference between this hook-up and that in Figure 7.34. In Figure 7.36 the amplifier provides each speaker with the full 16 volts. However, since each speaker has a different resistance, the amount of current flowing through the main speaker will not be equal to the current flowing through the remote speaker. This fact follows directly from Ohm's law:

Main Speaker $\text{Current} = \dfrac{\text{Voltage}}{\text{Resistance}}$

$\text{Current} = \dfrac{16 \text{ volts}}{8 \text{ ohms}} = 2 \text{ amperes}$

Remote Speaker $\text{Current} = \dfrac{16 \text{ volts}}{4 \text{ ohms}} = 4 \text{ amperes}$

$\left(\begin{array}{l}\text{Total Current Supplied} \\ \text{by the Amplifier}\end{array}\right) = 2 + 4 = 6 \text{ amperes}$

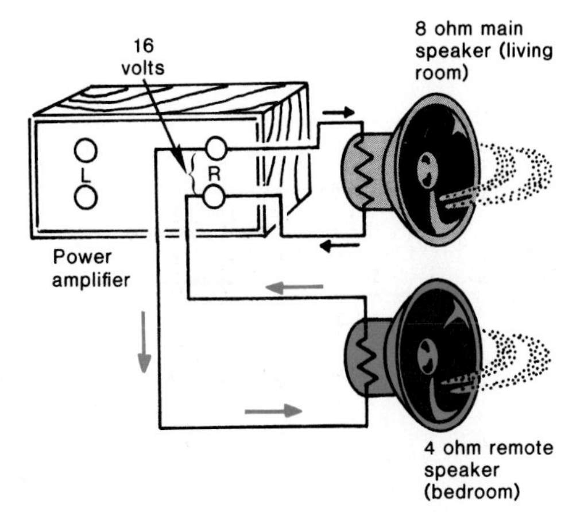

Figure 7.36. Two speakers wired in parallel.

Remember that the electrons which make up the 2 amperes of flow through the main speaker are not the same electrons which comprise the 4 amperes of flow through the remote speaker in Figure 7.36. This basic difference between

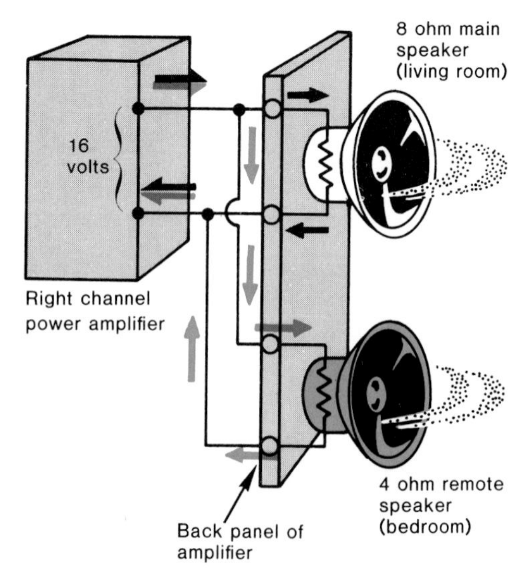

Figure 7.37. Many amplifiers have "remote" speaker jacks which allow an additional speaker to be connected to each stereo channel. The remote speaker is wired in parallel with the "main" speaker.

series and parallel wiring is emphasized by indicating the current arrows for one speaker in color and for the other speaker in black.

The back panels of many modern stereo amplifiers are equipped with an extra set of speaker terminals for connecting a second pair of speakers (sometimes called the remote speakers). Although the remote terminals are physically displaced from the main speaker terminals on the back panel, they are, in fact, connected in parallel with the main terminals, as shown in Figure 7.37. In this diagram we shall assume that an 8Ω speaker is placed in the living room and that a 4Ω speaker has been connected to the remote speaker terminals for use in the bedroom. Of course, there will be an identical set-up for the speakers attached to the left stereo channel. Notice that the only difference between this and Figure 7.36 is that the "black" current and the "blue" current travel a short way together before splitting up and going their separate ways.

Once the current flowing through each speaker is known, the power delivered by the amplifier can be calculated using the convenience of Equation 7.4:

Main Speaker Power = $(\text{Current})^2 \times \text{Resistance}$
Power = $(2 \text{ amperes})^2 \times 8\Omega$
Power = 32 watts

Remote Speaker Power = $(4 \text{ amperes})^2 \times 4\Omega$
Power = 64 watts

$$\left(\begin{array}{l}\text{Total Power Delivered}\\ \text{by the Amplifier}\end{array}\right) = 32 + 64 = 96 \text{ watts}$$

The total power drawn from the power amplifier is 96 watts (32 + 64 = 96). Notice that the main speaker will draw 32 watts whether or not the remote speaker is connected. When the remote speaker is attached in parallel, the amplifier must produce the extra 64 watts, providing, of course, that the total power does not exceed the amp's maximum power rating. The maximum power output of the amplifier in Figure 7.37 had better be at least 96 watts, or it could be damaged trying to supply more power than its design permits.

The above calculation can be done in another way, by first determining the total effective resistance R_T which the amplifier "thinks" is connected to it. The proper formula for any number of devices connected in parallel is given below.

$$\frac{1}{R_T} = \frac{1}{R_1} + \frac{1}{R_2} + \frac{1}{R_3} + \cdots \qquad \text{(Equation 7.7)}$$

In our case the amplifier really doesn't know that an 8Ω and a 4Ω speaker exist separately, but only that there is a total effective resistance R_T:

$$\frac{1}{R_T} = \frac{1}{8 \text{ ohms}} + \frac{1}{4 \text{ ohms}}$$

$$\frac{1}{R_T} = \frac{1}{8} + \frac{2}{8} = \frac{3}{8 \text{ ohms}}$$

$$R_T = \frac{8}{3} \text{ ohms.}$$

Using Ohm's law with this total resistance of 8/3 ohms and a voltage of 16 volts shows that the total current provided by the amplifier is 6 amperes, as we already know:

$$\text{Current} = \frac{\text{Voltage}}{\text{Resistance}},$$

$$\text{Current} = \frac{16 \text{ volts}}{8/3 \text{ ohms}} = 6 \text{ amperes.}$$

It may seem strange that the total resistance of a parallel arrangement of two speakers is less than the resistance of any one speaker alone. This paradox can be resolved as follows. As the number of parallel speakers is increased, the amplifier must provide more current to supply the additional devices. Even though the amp is now providing more current, it has no way of knowing whether more speakers are connected or that, somehow, the total resistance has been lowered. Remember, Ohm's law predicts that a decrease in the total resistance will produce an increase in the current. Therefore, as far as the amp is concerned, more speakers drawing more current has the same effect as a decrease in the total resistance! Parallel connections always result in decreased resistance.

Figure 7.38 returns to the flowing water analogy as an aid in understanding why parallel connections result in decreased resistance. In part (A), the two sections of pipe, #1 and #2, are connected in parallel with the pump. In (B) the two pipe sections are replaced with a single, equivalent section of pipe which has a cross-sectional area that is equal to the sum of the cross-sectional areas of pipes #1 and #2. Recall from section 7.6 that a pipe with a larger cross-sectional area has a lower resistance to water flow. Therefore, the effective resistance of the pipe in (B) is lower than the resistance of either pipe #1 or pipe #2 alone. This is tantamount to saying that the total resistance R_T of a parallel circuit is less than any one of the individual resistances.

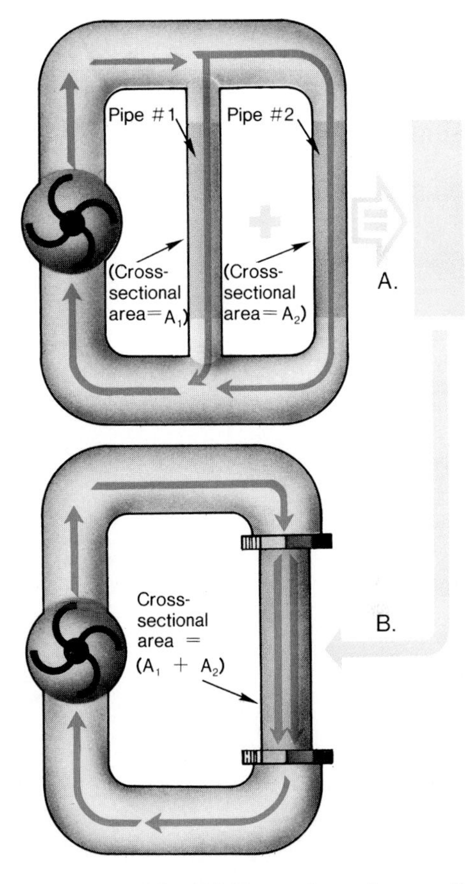

Figure 7.38. (A) Two pipes in parallel can be combined to form a single pipe (B) which has the combined cross-sectional area of pipes #1 and #2. The resistance of the pipe in (B) is less than the resistance of either #1 or #2.

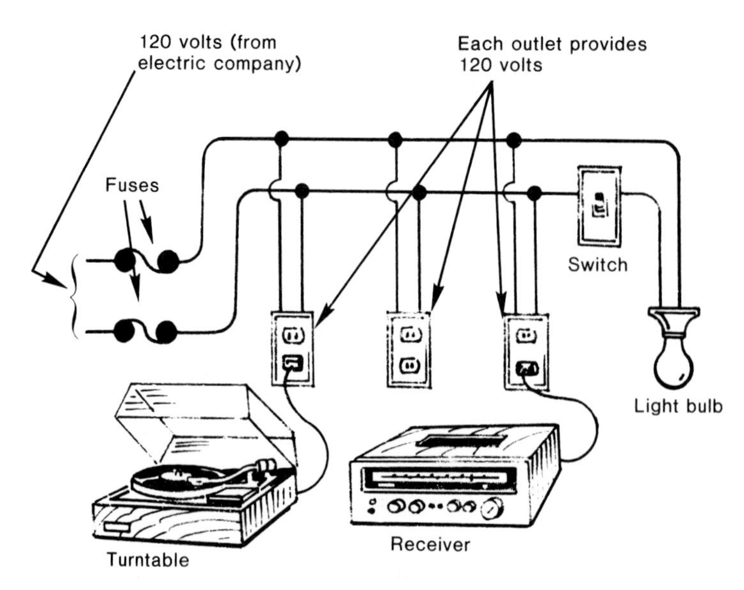

Figure 7.39. All home appliances are wired in parallel. The current going to any appliance is not affected by the presence or absence of the others.

Actually, parallel wiring is so common that everytime you plug additional appliances into wall sockets parallel connections are made. Figure 7.39 shows a typical application of parallel wiring. In this case three devices, the turntable, amplifier, and light bulb, are connected to the two lines from the power company. The principal feature of parallel wiring is that each appliance is connected "between" the wires, and each device takes whatever current it needs from the total current flow provided by the power company. Part of the total current is sent to the light bulb while the remainder is divided between the turntable and amplifier. An important advantage of parallel wiring is that if one appliance should burn out, or simply be turned off, the other devices continue to receive their normal current. Notice, in Figure 7.39, the fact that there is no current flowing into the second outlet does not hamper the normal functioning of the other outlets.

Summary

The same current flows through devices wired together in series, and each device receives only a portion of the total voltage. The total effective series resistance is given by $R_T = R_1 + R_2 + R_3 + \cdots$. An "open" circuit halts the current flow in a series connection, and all electrical devices turn off. When devices are wired in parallel, each device receives the total voltage, but a different current flows through each according to its resistance. The total effective parallel resistance is given by $\frac{1}{R_T} = \frac{1}{R_1} + \frac{1}{R_2} + \frac{1}{R_3} + \cdots$.

When resistors are added in parallel the total resistance decreases. If one electrical device is turned off for any reason in a parallel circuit, the other devices can continue to operate normally.

7.12 Capacitors and Inductors in Hi-fi

A hi-fi system has many features which depend on its ability to distinguish between the various frequencies of the audio signal. For example, the frequency crossover network in a loudspeaker sends the high frequencies to the tweeter, the middle frequencies to the midrange, and the low frequencies to the woofer, as discussed in section 6.4. Clearly, there must be electrical components within the crossover network which can separate out these frequency "bands" from the incoming signal and route each one to the appropriate driver. In addition, the typical preamp contains BASS and TREBLE tone controls which allow the user to enhance or diminish certain frequency ranges relative to others. Both of these examples utilize two important electrical components, *capacitors* and *inductors,* whose behavior depends on the frequency of the AC current.

When capacitors and inductors are employed in designing AC electric circuits they are referred to with special symbols, just as a resistor is indicated by a jagged line (-/\/\/\/-). Figure 7.40 (A) shows that the symbol for a capacitor is two vertical parallel lines (—||—) because it is essentially two

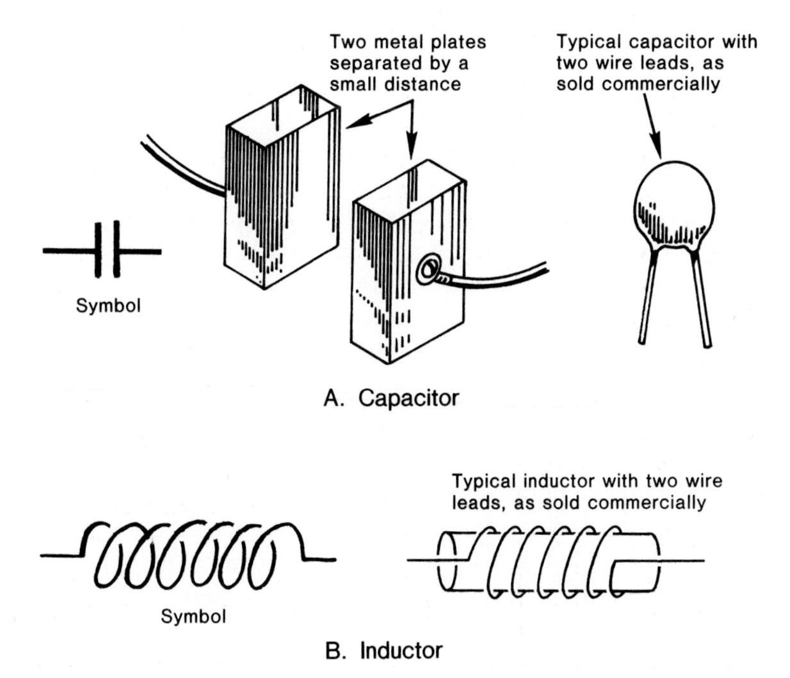

Figure 7.40. Typical capacitor and inductor, along with their symbols.

metal plates separated by a small distance. An inductor is made from a coiled wire and its symbol (-ꝏꝏ-) represents this fact, as illustrated in part (B).

Capacitors and inductors, like resistors, impede the flow of AC current in the electric circuits where they are used. In other words, capacitors and inductors are another source of "ohms." However, as we shall soon see, they behave somewhat differently than resistors, and the word "impedance," rather than "resistance," is used to refer to the ohms which they contribute. Do not be confused by fancy words, however. The main idea behind Ohm's law still applies; the more ohms there are in a circuit to impede the flow of electrons, the less current there will be. Remember this fact, and you will be able to understand how capacitors and inductors work.

Before discussing the interesting properties of capacitors and inductors, we shall first consider the related properties of resistors in order to understand the relationships among the three elements.

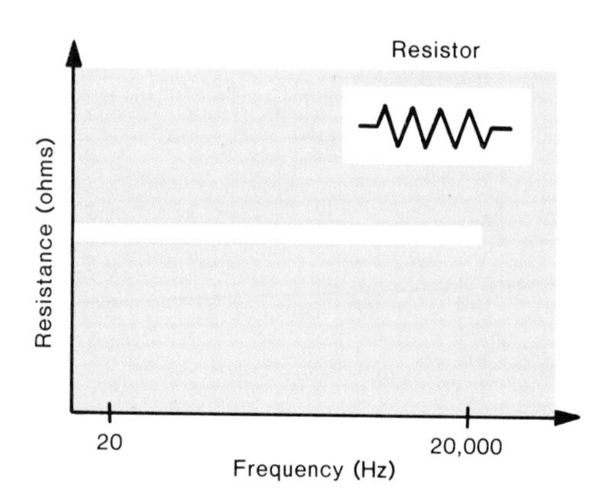

Figure 7.41. A plot of resistance versus frequency is a horizontal straight line because the resistance of a straight wire, or a resistor, does not depend on the frequency of the AC current.

Resistors

As mentioned in section 7.6, the electrical resistance of a wire is a measure of the wire's opposition to the current flowing through it. Resistance is measured in ohms, and it depends on the length and cross-sectional area of the wire, as well as the material from which it is made. The resistance, however, does *not* depend on the frequency of the AC current which passes through it. Figure 7.41 illustrates this fact by plotting the resistance of the wire along the vertical axis and the frequency of the AC current along the horizontal axis. The graph indicates that the number of ohms of resistance is the *same* for all frequencies.

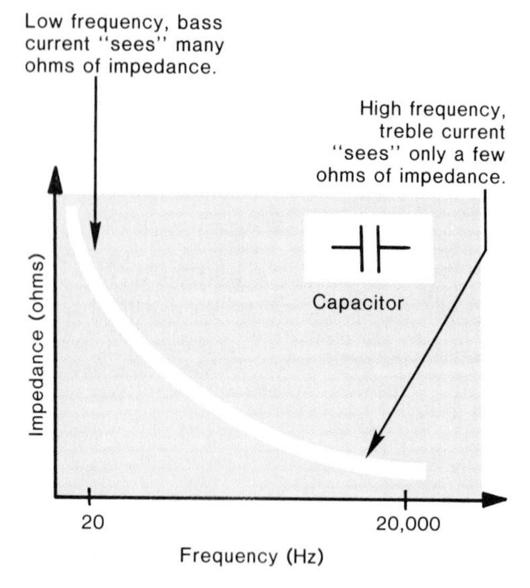

Figure 7.42. A capacitor contributes impedance to an electric circuit, but the amount that it contributes decreases as the frequency of the AC current increases.

Capacitors

The main feature of a capacitor is that its impedance changes as the frequency of the AC current changes. Figure 7.42 illustrates a graph of a capacitor's impedance versus the frequency. This curve shows that the impedance decreases as the frequency increases. Since the impedance is small when the frequency is high, AC current for the high pitched treble notes will be allowed to flow easily; it will not be impeded or reduced very much by the capacitor. On the other hand, the figure shows that the impedance of a capacitor increases as the frequency decreases. Therefore, the flow of AC current for the low pitched bass notes will be reduced by a capacitor. In a sense, a capacitor "passes" the high frequencies and "blocks" the low frequencies.

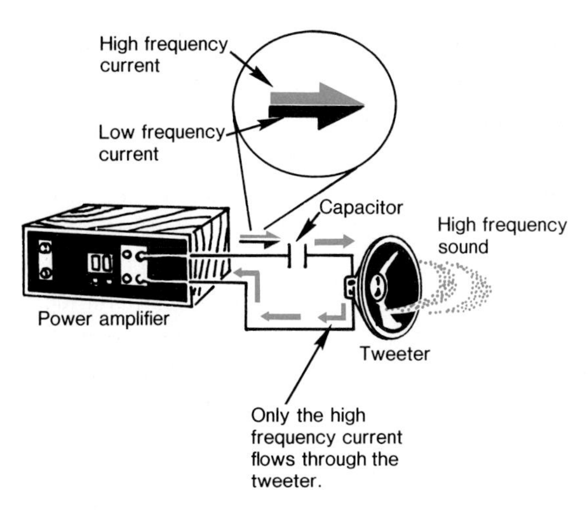

Figure 7.43. A capacitor in series with the tweeter offers a small impedance to the high frequency current and a large impedance to the low frequency current. Therefore, the high frequency current reaches the tweeter, while the low frequency current does not.

Figure 7.43 shows a capacitor which is connected in series with a tweeter. The capacitor permits the high frequency notes to be sent to the tweeter, because the capacitor only offers a small impedance to this current. On the other hand, the capacitor prevents the bass notes from reaching the tweeter, since the capacitor offers a large impedance to the flow of low frequency current.

Inductors

The main feature of an inductor is that its impedance increases as the frequency of the AC current increases. This behavior, as shown by the curve* in Figure 7.44, is the opposite of that shown in Figure 7.42 for a capacitor. Since the impedance of an inductor is small when the frequency is low, the AC current for the bass tones will be allowed to flow easily; it will not be impeded or reduced very much by the inductor. However, the curve shows that the impedance increases as the frequency increases. Therefore, the flow of high frequency AC current will be reduced by an inductor. An inductor "passes" the low frequencies and "blocks" the high frequencies.

Figure 7.45 shows an inductor allowing the low frequencies to reach a woofer, while keeping out the high frequencies.

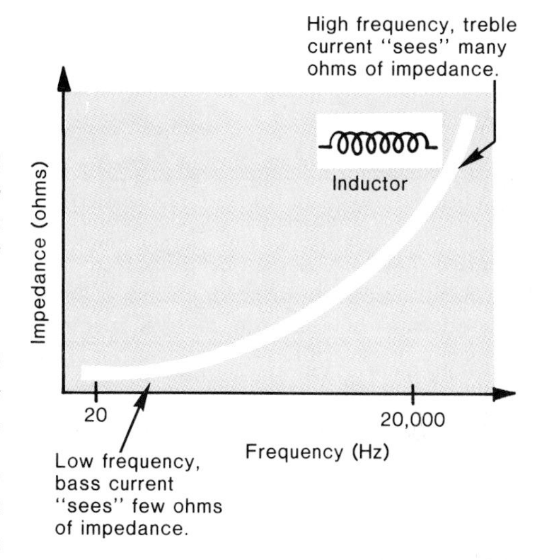

Figure 7.44. An inductor contributes impedance to an electric circuit, and the amount that it contributes increases as the frequency of the AC current increases.

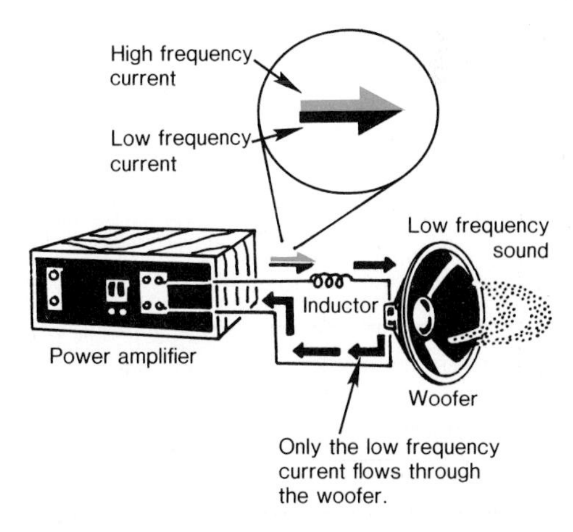

Figure 7.45. An inductor in series with the woofer offers a small impedance to the low frequency current and a large impedance to the high frequency current. Therefore, the low frequency current reaches the woofer, while the high frequency current does not.

Capacitor/Inductor Combination

The useful features shown by the curves in Figure 7.42 and 7.44 can be combined by using a capacitor and an inductor connected together in series. Notice the interesting curve that results when two elements are combined, as il-

*This curve shows the impedance of an actual inductor and includes the effect of the resistance of the wire.

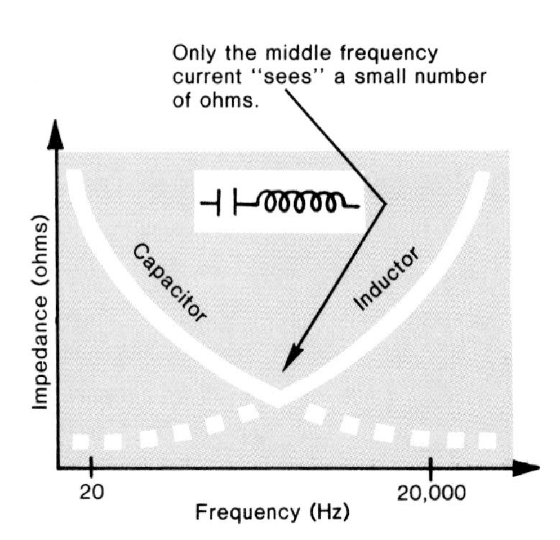

Figure 7.46. A series combination of a capacitor and an inductor offers a small impedance only at the middle frequencies.

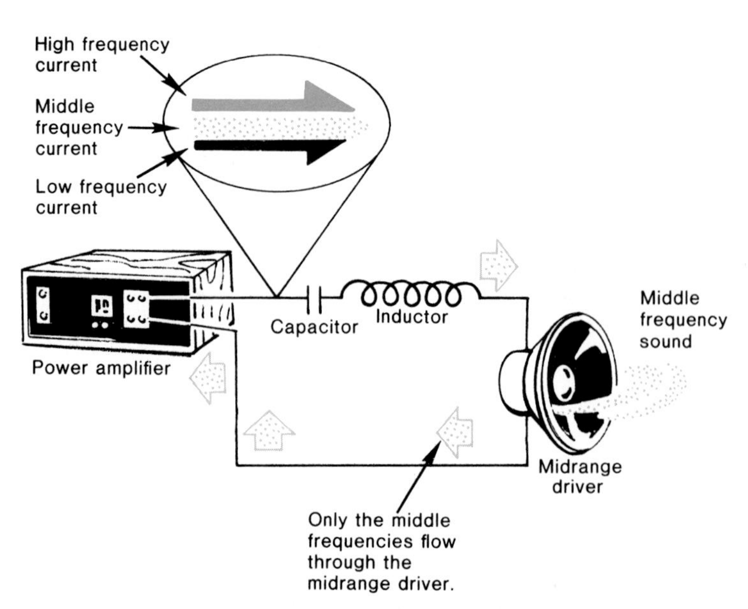

Figure 7.47. A capacitor/inductor combination in series with a midrange driver offers a small impedance only to the middle frequencies. To high or low frequencies the combination offers a large impedance. Therefore, the middle frequency current reaches the midrange driver, while the high and low frequencies do not.

lustrated in Figure 7.46. The impedance of the combination is large when the frequency of the AC current is either low or high. The impedance is small only for the middle frequencies. Therefore, only AC current for the midrange notes will flow easily with little reduction by the capacitor/inductor combination. Figure 7.47 shows that this combination allows only the middle frequencies to reach the midrange driver, while keeping out both the high and low frequencies.

A Frequency Crossover Network

One of the main features of the modern hi-fi loudspeaker is its frequency crossover network. The crossover network in Figure 7.48 is made from capacitors and inductors, and simultaneously uses the three examples which have been discussed above.* The figure shows three drivers (woofer, midrange, and tweeter) attached to an amplifier. Notice that the three driver circuits are wired in parallel with each other. Each driver circuit contains either a capacitor or an inductor, or both, that allows only the proper range of frequencies to reach the driver.

*Figure 7.48 is intended for illustration purposes only. Actually, crossover networks come in many designs and involve additional considerations which need not concern us here.

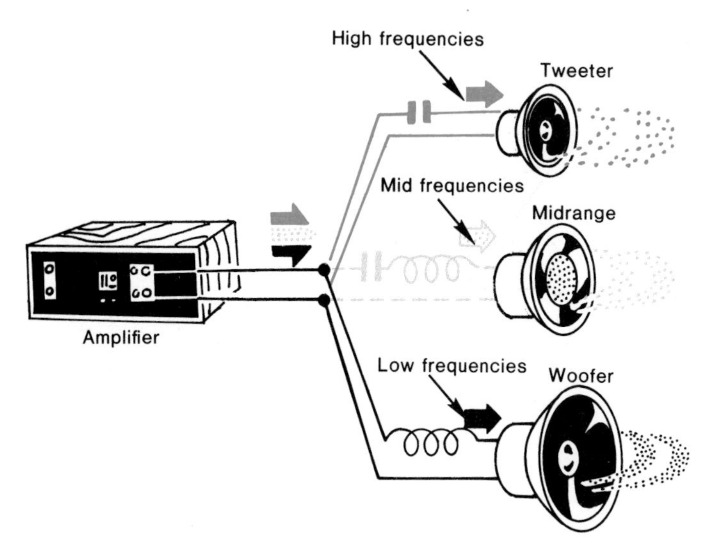

Figure 7.48. Using capacitors and inductors, the crossover network directs the highs to the tweeter, the middle notes to the midrange driver, and the lows to the woofer.

Summary

Both capacitors and inductors add impedance to an AC electric circuit. As the AC frequency increases, the impedance of a capacitor decreases, while that of an inductor increases. Capacitors and inductors may be used separately or in combination for specific purposes. A frequency crossover network is a good example of their application.

7.13 How an Amplifier Works

One of the preamplifier's functions is to receive the relatively weak electrical signals from a program source and enlarge (amplify) them to a much greater level. This property of enlargement is also true for the power amplifier which must produce signals that are strong enough to actuate the speakers. At one time or another many of us have looked inside a real amplifier, or perhaps a T.V. set, and noted the apparent complexity of these devices. Although electronic circuitry does appear rather complicated to the untrained eye, we shall show that the basic principles of amplifier operation are relatively easy to understand. An integrated amplifier can be thought of as an "electronic box" which interacts with the outside world via three connections (see Fig. 7.49):

1. the power cord which electrically attaches the amplifier to the 120 volt, 60 Hz, wall receptacle,
2. the input jack which receives the relatively weak signals from the program source,
3. the output jack which sends the amplified signals to the speaker(s).

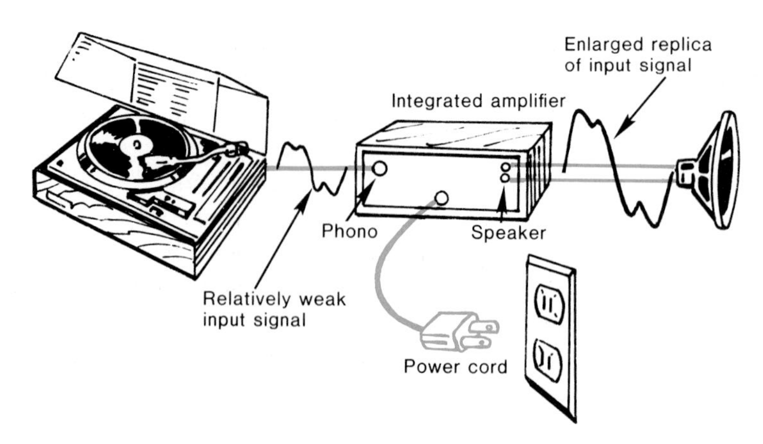

Figure 7.49. An amplifier utilizes the electrical energy supplied by the 120 volt wall receptacle to create an output signal which is an enlarged replica of the input signal.

These three connections are most important from a conceptual point of view. Once the amplifier circuits receive the weak input signal, they produce an enlarged *replica* which is then sent to the output jacks. Of course, energy is required for the amplifier to create the output signal. It receives the necessary electrical energy from the power plug which is attached to the 120-volt wall outlet.

It is possible to show how an amplifier circuit actually creates the output signal by first considering Figure 7.50. The circuit in Figure 7.50 does not represent a bonafide amplifier, because there is no provision for the source input signal. We shall rectify this situation shortly. First of all, the 120-volt, 60 Hz electricity from the wall socket is sent to the section of the amplifier called the *power supply.* As schematically indicated in the figure the power supply converts the 120-volt AC into what is essentially a *direct current*

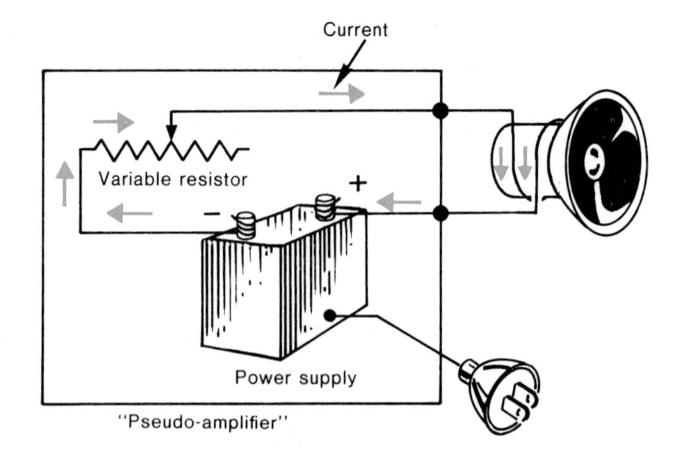

Figure 7.50. This pseudo-amplifier has all the features of a real amplifier, except that there is no provision for a source input signal. The power supply converts the 120 volt, 60 Hz AC into (about) 30 volts of DC. The output voltage from a power supply is just like that from an ordinary battery.

"battery." The output voltage of this "battery" is usually less than the 120-volt input, and it normally ranges from 10 to 40 volts depending upon the requirements of the amplifier. If the power supply is designed, for example, for a 30-volt operation, then it is imperative that the 30 volts remain constant at all times during the operation of the amplifier. It is interesting to observe that electricity is transmitted to your home in the form of AC although the amplifier essentially converts it into a DC voltage. (It is cheaper to send AC rather than DC.) Perhaps you are wondering; "Could one unplug the amplifier's power cord and replace the power supply with an ordinary battery?" Sure, we do it all the time with portable radios!

The amount of current which flows in the circuit of Figure 7.50 depends upon the resistance of the variable resistor. If the movable slide is positioned such that it is at the left of the resistor, most of the resistance is by-passed and a large current will flow. This is exactly analogous to the variable resistor used for the VOLUME control, as discussed in section 7.9. The distance through which the speaker cone moves depends upon the amount of current which passes through its voice coil; larger currents produce greater displacements. Imagine yourself inside the "amplifier" of Figure 7.50. By simply moving the variable resistor slide left and right you can readily change the amount of current being drawn from the power supply and sent to the speaker. Suppose, for example, that you moved the slide left and right at the ambitious rate of 100 times each second. Then the current in the circuit would also change its value at the rate of 100 Hz which, in turn, would cause the speaker to move back and forth in synchronism. Therefore, you can produce sound of any frequency by simply moving the variable resistance slide back and forth at the same frequency. You are effectively acting as a program source for this system because the "input" is the sliding motion of the pointer. It is important to realize that the current which is driving the speaker is coming from the "battery-like" power supply. You, the "program source," are not supplying the current which goes to the speaker. You are merely controlling it by changing the resistance.

In order to convert this pseudoamplifier into a real amplifier the variable resistor must be operated by someone, or something, other than people. The ideal situation would allow the small current from the program source, such as a turntable, to somehow regulate the variable resistor. Such a device does exist and it is called a transistor! The transistor acts just like a variable resistor whose resistance is controlled by the input current. Return to the analogy of water flowing

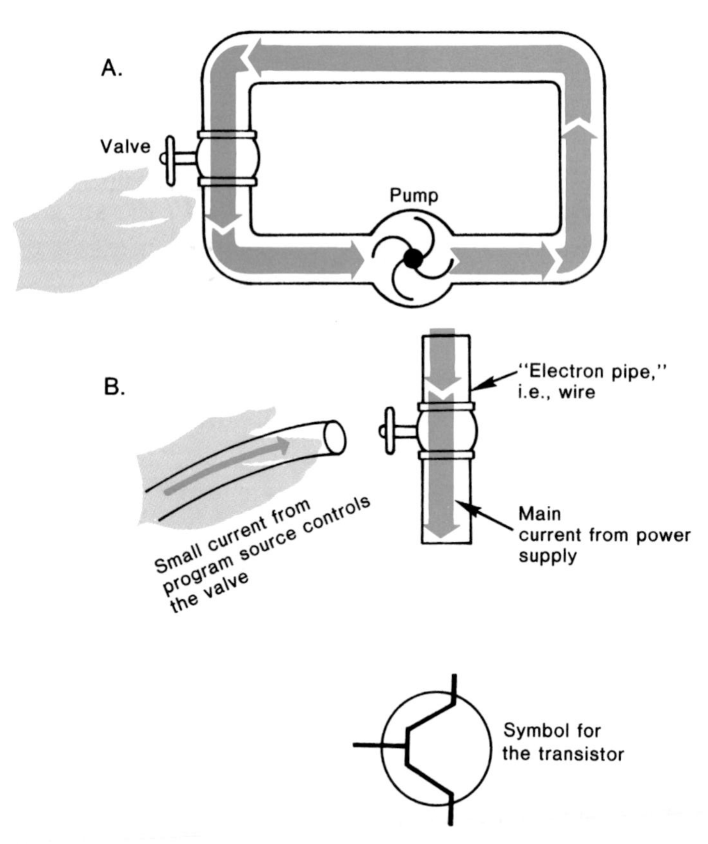

Figure 7.51. (A) A water line which contains a pump and a valve. (B) A transistor uses the small source current to achieve its valving action.

through a pipe. Figure 7.51 (A) shows a complete circuit which includes the pump (the "power supply") and a valve. The valve can be opened and closed with very little energy, but it can control the large flow of water being pumped through the pipe. This situation is analogous to a transistor, as shown in part (B). The small current from the program source causes the valving action which controls the main current being pumped through the speaker by the power supply. With a transistor as the valve, the small source current changes the resistance in the speaker circuit, thus producing the valving action.

The diagram in Figure 7.52 shows that the variable resistor has been replaced with a transistor. Notice that there is now a provision for an input signal from the program source. Remember, the cartridge is a small electrical generator and a complete path must exist between its two terminals in order for a current to flow. The small current leaves the negative terminal of the cartridge, flows through the transistor, and then finds its way back to the positive cartridge terminal (Fig. 7.52). A transistor amplifier is comprised of two basic circuits: the "source circuit" whose current is produced by the cartridge (source), and the

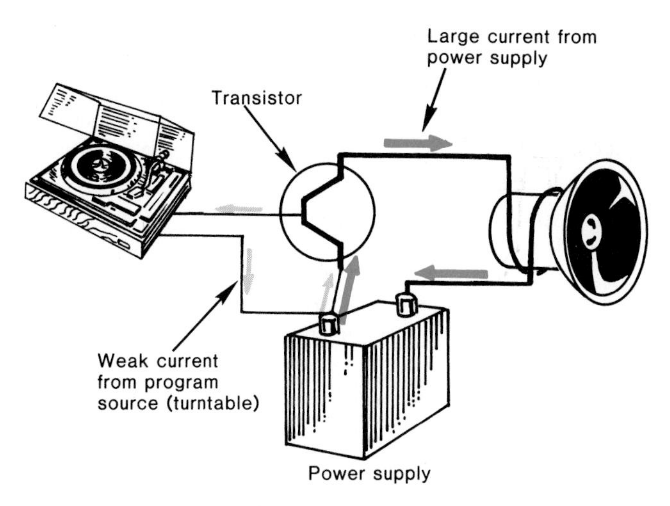

Figure 7.52. The small current which flows from the phono cartridge controls the resistance of the transistor. The transistor's resistance, in turn, governs the amount of current which is sent to the speakers from the power supply.

"speaker circuit" whose current is produced by the power supply. Even though the source current is very small, it is large enough to produce significant changes in the resistance of the speaker circuit via the valving action of the transistor. Figure 7.53 shows a more explicit picture of the transistor's action. When the source input signal is increasing, as shown in part (A), the transistor becomes a very low resistance, and a large current flows in the speaker circuit. The interesting property of a transistor is that it only requires a very small input current in order to reduce greatly the resistance of the transistor. On the other hand, part (B) shows the input current when it is decreasing. The smaller input current causes the transistor to increase greatly its resistance, thus reducing the current to the speaker. If the incoming signal has a frequency of 1 kHz, then the transistor will change its resistance at the same rate. The change in resistance causes the power supply to generate and deliver a 1 kHz current to the speaker which, in turn, produces the sound by the vibrating cone. The whole thing, from a technological point of view, is really quite beautiful. Of course, there are many modifications which are found in modern circuits that we have not shown in Figure 7.53 for the sake of simplicity.

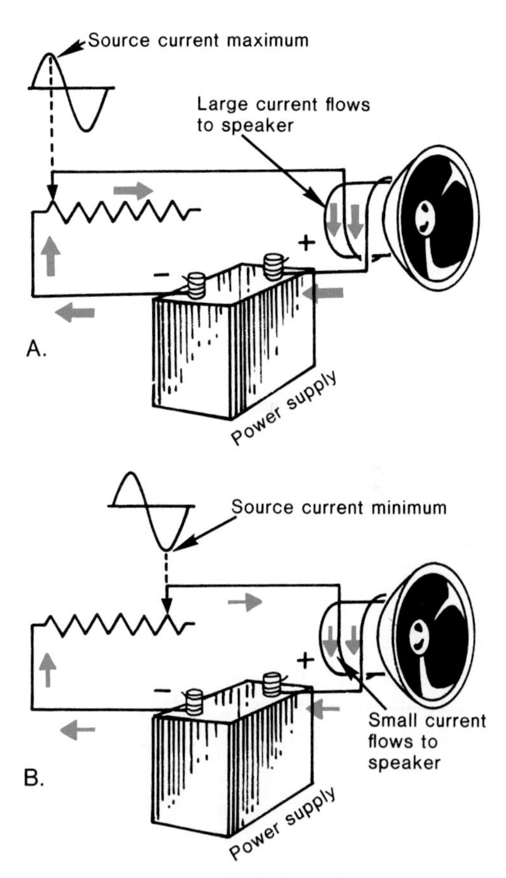

Figure 7.53. (A) As the source input current becomes a maximum positive valve, the resistance of the transistor becomes smaller, and more current flows in the speaker circuit. (B) A large negative input current results in a smaller speaker current, because the resistance of the transistor becomes larger.

Summary

The transistor acts like a variable resistor which is controlled by the input current. Small changes in the input current produce, through the transistor, large changes in the resistance of the speaker circuit. This, in turn, results in large changes in the current which is being sent to the speaker from the power supply; hence, the amplifier achieves its amplifying action.

Although the size of the output current is controlled by the size of the input current, the two currents are completely separate. The output is, therefore, only a replica of the input. The ouput current comes from the power supply whereas the input current is derived from the source generator, e.g., a phono cartridge.

Summary of Terms

Ampere—A measure of electric current. One ampere of current results when one coulomb of charge flows every second.

Amplifier—An electronic device which is used for increasing the size (amplitude) of electrical signals.

Atom—The smallest "particle" of an element. An atom contains atomic units called protons, neutrons, and electrons. The protons are positively charged and they are located inside the nucleus, along with the electrically-neutral neutrons. The negatively charged electrons orbit the nucleus in specific shells or orbits. The protons and neutrons have equal masses and they are about 1,840 times more massive than an electron.

Atomic Number—The number of protons in the nucleus of an atom. Different elements are distinguished by different atomic numbers. It is also the number of electrons in an electrically neutral atom.

BALANCE Control—A control located in the preamp/control center which regulates the relative amount of electrical power in each stereo channel. The BALANCE control is a variable resistor which has been inserted into the amplifier circuits.

Bound Electrons—Electrons which are bound tightly to the nucleus, and are not free to wander about the material of which they are a part.

Capacitor—An electrical component whose impedance decreases as the frequency of the AC current increases. The impedance of a capacitor is measured in ohms.

Charges (Electrical)—A positive or negative quantity of electricity. The smallest amount of negative electrical charge is carried by an electron. A proton carries the smallest unit of positive charge and it is equal, but opposite, to that carried by the electron.

Conductor—A material which has a low resistance to the flow of electrical current.

Coulomb—An amount of electric charge that contains 6.24×10^{18} electrons (or protons).

Coulomb's Law—A law which gives the force F between two electric charges Q_1 and Q_2, separated by a distance r; $F = kQ_1Q_2/r^2$, where k is a proportionality constant.

Current (AC and DC)—The electric current is the flow of electrons through a wire. In direct current (DC) the electrons move in one direction only. In an alternating current (AC) the electrons simply oscillate back and forth about an equilibrium position. AC and DC are both used extensively in hi-fi systems. Current is measured in amperes.

Electrolyte—The chemical in a battery which reacts with the plates and separates electric charges onto the positive and negative plates.

Electron—(See **Atom and Charges**.)

Element—A material whose atoms all have the same atomic number.

Free (Unbound) Electrons—Electrons which are only loosely bound to the nucleus, and may be easily detached to wander about the material of which they are a part.

Fuse—A special thin conductor designed to melt when a predetermined current is passed through it. This stops the current and protects an electrical device from receiving an excess amount of current.

Generator—Any device which is capable of separating electric charge. If the two terminals of a generator are connected by a wire, a current will flow. Common generators in hi-fi are: phonograph cartridges, tape playback heads, and microphones.

Ground (Electrical)—A term which refers to the third prong on a three prong plug. It is a safety feature which protects the user against being shocked.

Impedance—A general term which characterizes the ability of resistors, capacitors, and inductors to "impede" or to "resist" the flow of current in electrical circuits. Impedance is measured in ohms.

Inductor—An electrical component whose impedance increases as the frequency of the AC current increases. The impedance of an inductor is measured in ohms.

Insulator—A material which has a high resistance to the flow of electricity. On an atomic scale, the atoms of an insulator do not contribute any "free" electrons which are capable of wandering through the material. The electrons of an insulator are all "bound" to their respective atoms.

Joule—A unit of energy.

Kilowatt-Hour—A measure of the total amount of electrical energy which is used. It is the product of the power consumed and the total time of consumption. The electric bill is computed using kilowatt-hours.

Neutron—(See **Atom.**)

Nucleus—(See **Atom.**)

Ohm—A unit of electrical resistance or impedance.

Ohm's Law—An important law in electricity which states that voltage equals the resistance times the current:

Voltage = (Resistance) × (Current).

Parallel Connection—A method of connecting two or more electrical devices to the same source of voltage such that each device receives the full voltage being supplied.

Power—The amount of energy generated or used per second. Power is measured in watts.

Power Law—A law which states that power equals voltage times current:

Power = (Voltage) × (Current).

Power Supply—An electrical unit found inside amplifiers which provides a source of current for the operation of an amplifier. They are always DC sources.

Proton—(See **Atom and Charges.**)

Resistance (Electrical)—The opposition to the flow of electrical current. Resistance is a result of "electrical friction" between the atoms of a material and the flowing electrons. Resistance is measured in ohms.

Resistor—An electronic component which offers resistance to the flow of electrical current.

Series Connection—A method of connecting two or more electrical devices such that all devices have the same current flowing through them.

Transistor—A solid state device in which the input current controls the amount of current in the output circuit. A transistor is designed such that small changes in the input current lead to large changes in the output current.

Unbound Electrons—(See **Free Electrons.**)

Variable Resistor—A device whose resistance can be changed by moving a sliding contact on a resistor. Variable resistors are used for VOLUME and BALANCE controls.

Voltage—A measure of the amount of electrical energy that a charge possesses. A 1 volt source, such as a battery or generator, gives one joule of electrical energy to each coulomb of charge which passes through it.

VOLUME Control—A control on the preamp/control center which regulates the overall sound level from both stereo channels. The VOLUME control is a variable resistor which has been inserted into the electronic circuits.

Watt—A unit of power. One watt equals one joule of energy per second.

Review Questions

1. Explain, in terms of electrons and protons, how positive and negative objects are formed.
2. Inside a piece of conducting material, what is meant by a "free" electron? What is a "bound" electron?
3. Explain the nature of the forces between like and unlike charges.
4. What is a coulomb of electrons?
5. In terms of free and bound electrons, differentiate between a conductor and an insulator.
6. What is an ampere? How is it different from a coulomb?
7. Define, in terms of energy, a volt. How is it possible to have a large voltage and very little current?
8. What law governs the relationship between current, voltage, and resistance?
9. For a conducting wire, like copper, what two geometrical factors determine its resistance?
10. Why does an electric current flow through a conductor but not through an insulator?
11. Explain the difference between an alternating current and a direct current. Name some AC generators in hi-fi.
12. Why is alternating current so important in hi-fi for transmitting information? (Hint: what parameters of an AC waveform are necessary to convey loudness and pitch?)
13. Explain several uses for variable resistors in the preamplifier/control center.

14. How is electrical power related to current and voltage? To current and resistance? What are the units of power?
15. What factors must you know in order to fuse correctly your speakers?
16. If two loudspeakers are connected in parallel, what happens to the sound from one of them if the other "burns out"? Does the amplifier deliver more, or less, current when two operating loudspeakers are connected in parallel as compared to series?
17. If two loudspeakers are connected in series, what happens to the sound from one of them if the other "burns out"? Does a loudspeaker become louder or softer when it is connected in series with other loudspeakers?
18. How does a transistor amplifier produce an enlarged replica of the input signal?
19. Explain why a capacitor allows only the high frequency current to reach a tweeter. Explain why an inductor allows only the low frequency current to reach a woofer.
20. Explain how to construct a three-way crossover network using capacitors and inductors.

Exercises

NOTE: The following questions have up to 5 possible answers. Please select the **one** response which best answers the question.

1. Three speakers are connected in parallel. Their resistances are 8 ohm, 8 ohm, and 4 ohm respectively. The total resistance of the three is:
   1. 16 ohms.
   2. 20 ohms.
   3. 4 ohms.
   4. 12 ohms.
   5. 2 ohms.
2. If three 4 ohm speakers are wired in series the total resistance becomes:
   1. 2 ohms.
   2. 4 ohms.
   3. 8 ohms.
   4. 12 ohms.
   5. 16 ohms.
3. What happens to the sound coming from the remaining loudspeakers arranged in a parallel circuit when one of them burns out?
   1. All the remaining speakers stop producing sound.
   2. Each remaining speaker becomes louder.
   3. Each remaining speaker becomes less loud.
   4. Nothing happens—the remaining speakers produce the same loudness as before.
4. If two speakers are connected in series and one of them burns out, then the sound from the other speaker will:
   1. become twice as loud.
   2. become slightly louder.
   3. not change in loudness.
   4. become one-half as loud.
   5. be turned off.
5. If two resistors are added together, in parallel, then the total resistance:
   1. decreases.
   2. increases.
   3. remain the same.
   4. One cannot say what will happen unless the values of the resistors are given.
6. Two objects, one of which has a deficiency of electrons, and the other has an excess of electrons, will:
   1. repel each other.
   2. attract each other.
   3. do nothing.
   4. repel each other at first and then become attracted.
   5. None of the above answers is correct.
7. Two 16 ohm speakers, hooked in parallel across the left speaker terminals of your integrated amplifier (the right speaker terminals being unused), appear to the amplifier as a single __________ speaker.
   1. 32 ohm
   2. 16 ohm
   3. 1/8 ohm
   4. 1/16 ohm
   5. 8 ohm
8. The rate at which energy is dissipated is called power. Electrical power is equal to the:
   1. current times the voltage.
   2. current times the resistance.
   3. current divided by the voltage.
   4. current divided by the resistance.

9. Two 8 ohm speakers are connected in series to one channel of a power amplifier. The amplifier delivers 64 watts. The current being supplied by the amplifier is:
   1. 2 amps.
   2. 4 amps.
   3. 8 amps.
   4. $\sqrt{8}$ amps.
   5. 16 amps.
10. What is the maximum size fuse which you can use to protect a 4 ohm speaker whose maximum power rating is specified at 100 watts?
    1. 2 amp.
    2. 5 amp.
    3. 6 amp.
    4. 8 amp.
    5. 10 amp.
11. If 24 volts provides 6 amperes of current to a 4 ohm speaker, how much electrical power is being consumed?
    1. 6/4 watts.
    2. 4/6 watts.
    3. 96 watts.
    4. 144 watts.
    5. 6 watts.
12. Power is the energy being consumed each second, and it is:
    1. identical to current.
    2. identical to voltage.
    3. related to voltage and current.
    4. independent of voltage and current.
13. How much voltage is necessary to push 2 amps through a 50 ohm resistor?
    1. 10 volts.
    2. 16.6 volts.
    3. 100 volts.
    4. 50 volts.
    5. 150 volts.
14. What size battery is needed to push 4 amps through a resistance of 5 ohms?
    1. 20 volts.
    2. 5 volts.
    3. 1.25 volts.
    4. .8 volts.
    5. None of the above answers is correct.
15. Which is **not** a factor in determining the resistance of a wire?
    1. Length.
    2. Diameter.
    3. Thickness.
    4. The type of insulation around the wire.
    5. The type of metal in the wire.
16. A wire has a resistance of 4 ohms. If its length is increased 9 times, what is the new value of the resistance?
    1. $4 \times 9 = 36$ ohms.
    2. $4 + 9 = 13$ ohms.
    3. 4/9 ohms.
    4. 9/4 ohms.
    5. None of the above answers is correct.
17. A current of 2 amps is passing through a resistance of 200 ohms. The voltage is:
    1. 400 volts.
    2. 100 volts.
    3. 800 volts.
    4. $200 - 2 = 198$ volts.
    5. None of the above answers is correct.
18. More current can be produced by:
    1. raising the voltage and/or lowering the resistance.
    2. raising the resistance.
    3. lowering the voltage.
    4. pushing fewer electrons through the wire.
19. The ohm is a unit of:
    1. current.
    2. voltage.
    3. speaker efficiency.
    4. resistance.
    5. power.
20. If you make the cross-sectional area of a wire four (4) times as great, you will make its resistance ____________as large.
    1. 4 times
    2. 2 times
    3. one-half
    4. one-fourth
    5. one-eighth
21. If one triples the length of a piece of wire its resistance will be:
    1. 2/3 as great.
    2. 1/3 as great.
    3. 3 times as great.
    4. 1.5 times as great.
    5. There will be no change in its resistance.
22. The property of a material to impede the flow of electric current is called:
    1. current.
    2. coulombs.
    3. power.
    4. voltage.
    5. resistance.

23. Why are thick wires rather than thin wires normally used to carry current?
    1. Thick wires offer less resistance to the current.
    2. Thick wires offer more resistance to the current.
    3. Alternating current, such as that used in homes, will only pass through thick wires.
    4. More volts can be carried by thick wires.
    5. None of the above answers is correct.
24. Electric current that continually reverses in direction is called:
    1. alternating current.
    2. battery current.
    3. direct current.
    4. to-and-fro current.
    5. None of the above answers is correct.
25. If an electric current makes 100 complete cycles each second then it is called a ______________.
    1. 100 Hz transverse current.
    2. 100 Hz ohms.
    3. 100 Hz DC current.
    4. 100 Hz AC current.
    5. 100 Hz parallel current.
26. Which wave properties does an alternating current possess?
    1. Frequency.
    2. Amplitude.
    3. Period.
    4. Wavelength.
    5. It possesses all of the above properties.
27. If the electrons in a wire vibrate back and forth about a fixed position one has __________ current flowing in the wire.
    1. AC
    2. BC
    3. DC
    4. CD
    5. CA
28. The reason an insulator hinders the flow of current is that:
    1. it is too thick.
    2. the protons are bound too tightly to the atom.
    3. it has too few atoms to allow current flow.
    4. there are no free electrons.
    5. its resistance is too low.
29. The current scientific explanation of how a conductor conducts electricity is that:
    1. some atoms wander freely through the conductor and they carry the current.
    2. a few electrons are detached from each atom and are then free to wander throughout the entire conductor.
    3. all the electrons are detached from each atom and wander throughout the conductor carrying the current.
    4. all electrons remain attached to their parent atom and these electrons carry the current.
30. If you measured the electric current through a speaker, you would be measuring:
    1. the number of ohms passing through each second.
    2. the number of coulombs passing through each second.
    3. the number of volts passing through each second.
    4. the number of watts passing through each second.
    5. None of the above answers is correct.
31. The amount of energy that one coulomb of charge carries is called the:
    1. ohm.
    2. power.
    3. coulomb.
    4. voltage.
    5. resistance.
32. If 3 amperes are flowing through a conductor for 4 seconds how much charge has passed through the conductor?
    1. 3 coulombs.
    2. 4 coulombs.
    3. 8 coulombs.
    4. 9 coulombs.
    5. 12 coulombs.
33. What is a mV?
    1. 1,000 volts.
    2. 10,000 volts.
    3. 1/100 volt.
    4. 1/1,000 volt.
    5. 1/1,000,000 volt.
34. The term microvolt (μ V) means:
    1. 1,000,000 volts.
    2. 1/100 of a volt.
    3. 1/1,000 of a volt.
    4. 1/10,000 of a volt.
    5. 1/1,000,000 of a volt.

35. The fundamental law which describes the basic force between charged bodies is:
    1. like charges repel, unlike charges also repel.
    2. like charges repel, unlike charges attract.
    3. like charges attract, unlike charges repel.
    4. There is no force between charged objects.
    5. Ohm's law.
36. If an object has fewer electrons than protons then the object is said to be:
    1. positively charged.
    2. a conductor.
    3. an insulator.
    4. negatively charged.
    5. electrically neutral.
37. Atom #1 has an equal number of protons and electrons, and so does atom #2. In order to make these atoms exert an electrical force of attraction on each other:
    1. leave them as they are.
    2. take away an electron from #1 and a proton from #2.
    3. take away an electron from #1 and from #2.
    4. take away an electron from #1 but do nothing to #2.
    5. take away a proton from #2 but do nothing to #1.
38. If 25 coulombs pass through a conductor in 10 seconds what is the current?
    1. $25 + 10 = 35$ amps.
    2. $25 - 10 = 15$ amps.
    3. $25/10 = 2.5$ amps.
    4. $25 \times 10 = 250$ amps.
    5. None of the above is correct.
39. A positive electrical charge is produced when:
    1. neutrons flow through a conductor.
    2. neutral atoms flow through a conductor.
    3. an atom gives up a negatively-charged electron.
    4. an atom gives up a positively-charged electron.
40. Atoms normally have __________ which make them electrically neutral.
    1. neutrons
    2. the same number of neutrons and protons
    3. the same number of electrons and neutrons
    4. the same number of electrons and protons
41. Electric current is produced when:
    1. the atoms in a conductor move toward a positive terminal of a battery.
    2. the "bound" electrons in the conductor move toward the positive terminal.
    3. the "free" electrons in the conductor move toward the negative terminal.
    4. the "free" electrons move toward the positive terminal.
42. Electrical resistance is a property of materials which:
    1. causes a conductor to get hot when an electric current flows through the material.
    2. prevents or reduces the flow of current through the material.
    3. is caused by friction.
    4. is larger in a thin wire than in a thicker wire.
    5. All of the above answers are correct.
43. An amplifier works by:
    1. a process where the current in one electric circuit controls the current in another.
    2. a process where the source current flows through the speaker circuit.
    3. a process where the speaker current flows through the source.
    4. the source current providing energy to the power supply.
44. "Amperes" and "coulombs" are two terms which are used in electrical terminology. The difference between these two terms is that:
    1. amperes measure the amount of electrical charge, while coulombs measure the rate of charge flow.
    2. amperes measure the rate of charge flow, while coulombs measure the amount of charge.
    3. amperes measure the rate of charge flow, while coulombs measure the amount of energy carried by the charge.
    4. amperes measure the amount of energy carried by the charge, while coulombs measure the amount of charge.
    5. There is no difference between amperes and coulombs.
45. Two speakers, whose impedances are 4 ohms and 8 ohms, are wired in series. Which one of the following statements applies to this situation?
    1. The current flowing through each speaker is different, and, therefore, they each receive a different amount of electrical power.
    2. The current flowing through each speaker is the same, and, therefore, they each receive the same amount of electrical power.
    3. The current flowing through each speaker is the same, but each receives a different amount of electrical power.
    4. The current flowing through each speaker is different, and each receives the same amount of electrical power.

46. Two speakers, which have the same resistance, are connected in series across the output terminals of an amplifier. If the amplifier is delivering a total of 68 watts of power, how much power does each speaker receive?
   1. 68 watts.
   2. 34 watts.
   3. 20 watts.
   4. 17 watts.
   5. $(68 \times 2) = 136$ watts.
47. A two-ampere speaker fuse blows when the voltage applied to the speaker terminals is 16 volts. The power being sent to the speaker is:
   1. 8 watts.
   2. 18 watts.
   3. 32 watts.
   4. 64 watts.
   5. 14 watts.
48. Hi-fi system A has two 8 ohm speakers connected in parallel across the output terminals of a monaural power amplifier. Hi-fi system B has a single 4 ohm speaker connected across the output terminals of an identical monaural power amplifier. The VOLUME controls of both amps are set to produce identical voltages. How does the total current supplied by amplifier A compare to that supplied by amplifier B?
   1. Current A is twice current B.
   2. Current A equals current B.
   3. Current A is four times current B.
   4. Current A is one-fourth current B.
   5. Current A is one-half current B.
49. Two resistors are connected in parallel, and this combination has a total resistance of 6 ohms. If the same two resistors were connected in series the total resistance would:
   1. be less than 6 ohms.
   2. be greater than 6 ohms.
   3. be equal to 6 ohms.
   4. be zero.
50. An 8 ohm speaker can accept a maximum of 50 watts before it burns out. What size fuse is needed in order to protect the speaker from damage?
   1. 2.5 amps.
   2. 25 amps.
   3. 50 amps.
   4. 100 amps.
   5. The correct answer cannot be determined from the above data.
51. A speaker is connected to a power amplifier. Doubling the voltage output by the power amplifier ______________ the power consumed by the speaker.
   1. doubles
   2. cuts in half
   3. quadruples
   4. does not change
   5. increases by a factor of 1.414
52. An amplifier is producing 64 watts of electrical power. What is the current through a four ohm resistor which is connected to the amplifier?
   1. 4 amps.
   2. 2 amps.
   3. 1/4 amp.
   4. 256 amps.
   5. 16 amps.
53. Suppose that an amplifier is delivering current to a 8 ohm speaker. If the output voltage of the amplifier is 12 volts, what is the current being sent to the speaker?
   1. 8 amps.
   2. 1.5 amps.
   3. 96 amps.
   4. 20 amps.
   5. The current cannot be calculated from the above data.
54. What is the condition necessary for current to flow from one end of a wire to another?
   1. The wire must be a good insulator.
   2. Both ends must have the same positive charge.
   3. Both ends must have the same negative charge.
   4. A voltage must exist between the two ends of the wire.
   5. The resistance of the wire must be less than 16 ohms.
55. Amplifiers use a combination of AC and DC in their operation. Why does the AC part carry the audio information instead of the DC part?
   1. DC will not flow through the wires to the speakers.
   2. AC is convenient because it is readily available from the 120 volt wall receptacles.
   3. All the wave properties (like amplitude and frequency) of the music can be conveyed by the AC part.
   4. DC is not steady enough.

56. Each of two objects is electrically charged, and they exert a force on one another. As the two objects are moved closer together, the force between them:
    1. does not change.
    2. becomes weaker.
    3. changes from attraction to repulsion.
    4. becomes stronger.
    5. changes from repulsion to attraction.
57. An inductor and a capacitor are connected in series. The impedance of this combination is greatest at the:
    1. midrange frequencies.
    2. midrange and treble frequencies.
    3. midrange and bass frequencies.
    4. bass and treble frequencies.
    5. The impedance is the same for all frequencies.
58. A capacitor:
    1. is used in electric circuits to modify the circuit's frequency response.
    2. is constructed from a coil of wire, and it is used to "pass" low frequency current in an electrical circuit.
    3. is a pair of parallel plates which offers a high impedance to the flow of high frequency current.
    4. is used in woofer circuits.
59. The impedance offered by a capacitor is large when:
    1. the frequency of the current is high.
    2. the frequency of the current is low.
    3. the amplitude of the AC current is large.
    4. the amplitude of the AC current is small.
    5. the period of the AC current is small.
60. Assume that an inductor is made by wrapping many turns of wire around a cylindrical spool. When this "homemade" inductor is connected in series with a speaker:
    1. all AC current frequencies will reach the speaker without being reduced in amplitude.
    2. no AC current will reach the speaker, and, hence, no sound will be produced.
    3. the speaker will produce mostly bass tones.
    4. the speaker will produce mostly treble tones.
    5. the speaker will produce mostly midrange tones.
61. The electrical signal from a sound source (e.g., a turntable) is small, and it is the job of an amplifier to convert it into a large signal. The amplifier accomplishes this by:
    1. using the small current from the sound source for the amplifier's power supply current.
    2. using the small current from the sound source to control the amplifier's power supply current with the aid of a transistor.
    3. using the amplifier's power supply current to control the small current from the sound source with the aid of a transistor.
    4. using an inductor in the path of the power supply current.
62. A transistor can be thought of as ____________ whose value can be changed by the __________.
    1. an inductor/frequency of the source current.
    2. a capacitor/frequency of the source current.
    3. variable resistor/frequency of the source current.
    4. a variable resistor/amplitude of the source current.
    5. variable resistor/frequency of the power supply current.
63. Suppose that a turntable is connected to an amplifier. Where does the amplifier obtain the electrical power that is sent to the speakers? The power is obtained from:
    1. the small AC signal produced by the cartridge.
    2. the turntable's power supply.
    3. the RIAA equalization preamplifier.
    4. the amplifier's power supply.
    5. the tone control section.

Chapter 8

AMPLIFIERS

8.1 Introduction

Amplifiers play such an important role in sound reproduction that they are found extensively in almost all phases of audio equipment. In Chapter 2 we became familiar with the preamp/control center and the power amplifier which constitute two of the four basic building blocks of a sound system. These two blocks can be purchased separately, or as a single unit with a tuner (a receiver) or without a tuner (integrated amplifier). Amplifier circuits are also associated with other blocks, particularly the sources. The "source preamplifiers," as these circuits are called, may or may not be an integral part of the source—a point which we shall discuss in the next section. Generally speaking, all of the above units, be they source preamplifiers, integrated amplifiers, power amplifiers, etc., fall under the general classification of "amplifiers." We shall devote this chapter to a study of both the use and performance of "amplifiers" in hi-fidelity.

8.2 The Source Preamplifiers

All sources of music, be it a tuner, phono cartridge, tape deck, or a microphone have associated with them a **source preamplifier.*** For example, a tuner contains a built-in amplifier, often called a "tuner preamp," that receives the relatively weak signals from the electronic circuits which extract the audio information from the radio wave. As shown in Figure 8.1 (A), the tuner preamp enlarges the voltage to a value of about 0.5 volts before passing it on to the preamp/control center. Tuner preamps also perform another impor-

*Do not confuse the source preamplifier with the preamp/control center. As we shall now see, the two have different functions within a sound system. In order to avoid confusion we shall explicitly use phrases like "tuner preamp," "tape preamp," or "phono preamp" to denote the source preamplifiers, whereas the second building block will be called the preamp/control center.

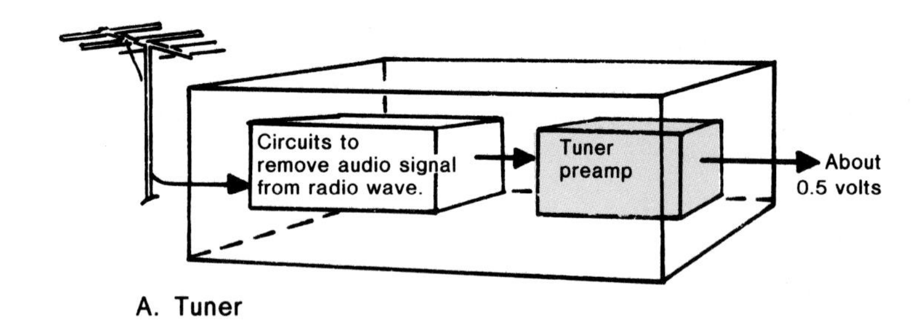

A. Tuner

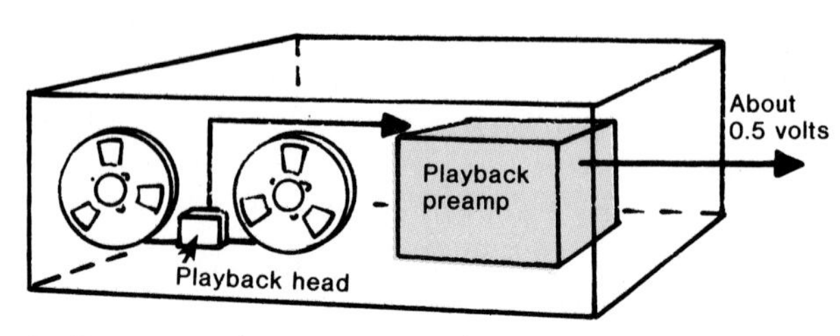

B. Tape deck (playback mode)

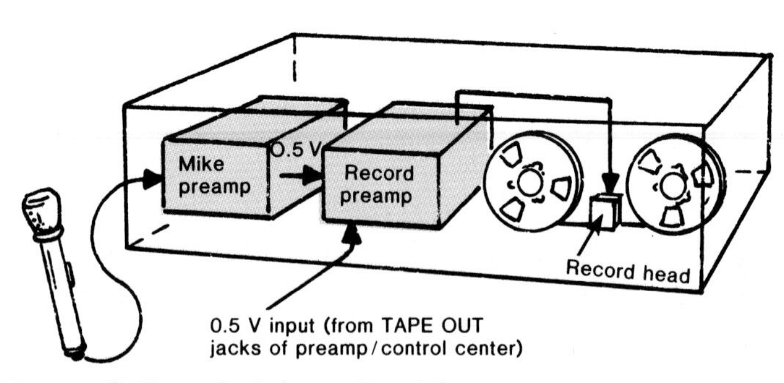

C. Tape deck (record mode)

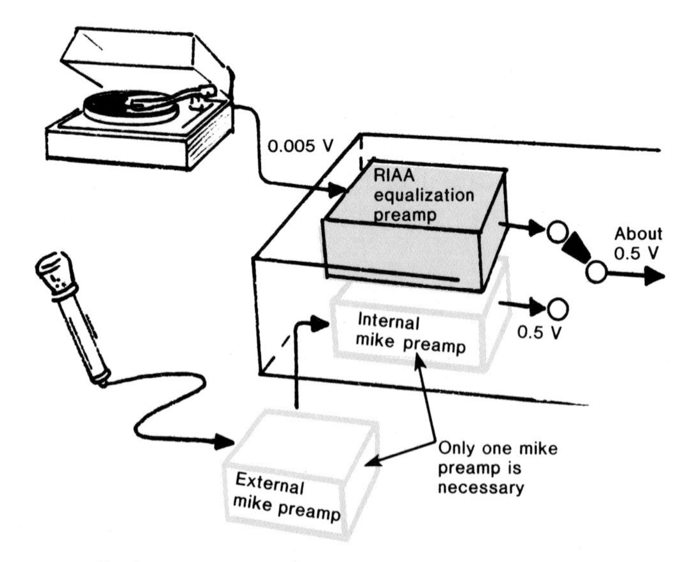

D. Preamp/control center

Figure 8.1. All sources of sound utilize a source preamplifier to enlarge and equalize the audio signals. The source preamplifier may be found within the source itself, in the preamp/control center, or it may occur as a separate unit.

tant function called "equalization."* Equalization helps to reduce the amount of noise which can accompany an FM broadcast, and the details of this feature are presented in Chapter 10.

Tape decks contain several source preamps whose functions depend on whether the deck is being used for recording or playback. Upon playback [Fig. 8.1 (B)] the playback preamp enlarges the small voltage produced by the playback head to a level of about 0.5 volts. This voltage is large enough to actuate a pair of headphones in case the listener does not wish to use the main speakers. Like the tuner preamp, the tape preamp also equalizes the signal, although for different reasons. Chapter 13 has a further discussion on this point. When a tape recording is being made, there are two possible inputs; one from a microphone and the other from the TAPE OUT jacks of the preamp/control center as shown in Figure 8.1 (C). Almost all quality tape decks contain a built-in microphone preamp which boosts the approximately 0.5 mV signals from the mike to about 0.5 V (a thousand fold amplification). Figure 8.1 (C) also shows that the signal which leaves the microphone preamp is sent to the record preamp which equalizes the signal and controls the amount of voltage being sent to the record head. This voltage level is regulated by the user through the LEVEL controls located on the front panel of the tape deck. Adjusting the voltage level allows the listener to compensate for the different sound levels associated with various musical selections.

Figure 8.1 (D) shows how a microphone is used as a source for the preamp/control center. Once again a mike preamp must be used to raise the voltage level from approximately 0.5 mV to 500 mV, and to equalize the signal. In most cases the mike preamp must be purchased separately, although a few audio components already have it built in.

Finally, the cartridge output voltage (approximately 5 mV) from a turntable is sent to the cartridge preamp, which is often called the "RIAA Phono Equalization Preamplifier." This preamplifier, located within the preamp/control center, also performs the dual task of amplification and equalization. This important preamp is discussed in Chapter 12.

We see from the previous discussion that each source has its own preamp which usually serves in the dual roles of amplification and equalization. In the case of tape decks and tuners the preamps are located within the sources themselves, while the phono preamp is always found inside the

*In general "equalization" means that the amplitude of the audio signal has been either increased or decreased in certain frequency ranges. For example, "tuner equalization" gradually boosts all the frequencies above 1 kHz, such that a 15 kHz signal has been enlarged 17 dB above the level at 1 kHz.

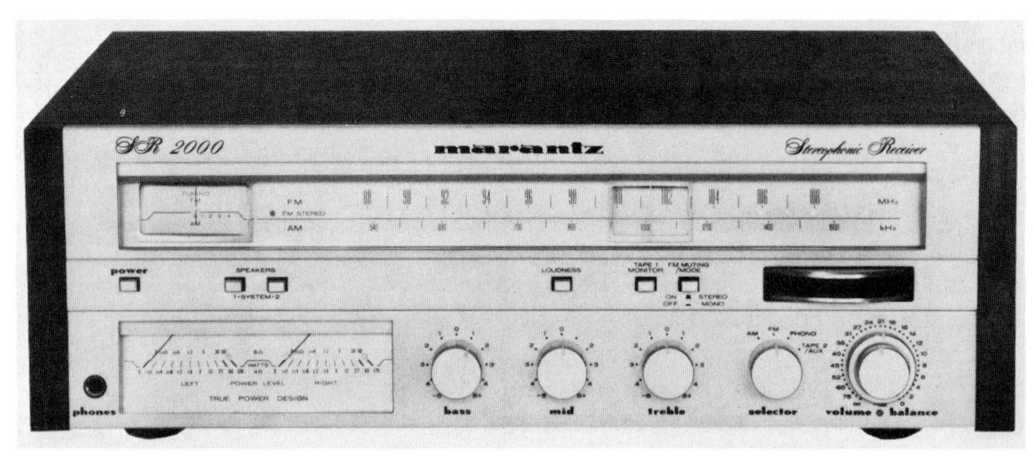

Marantz Model SR 2000 Receiver

preamp/control center. In the case of microphones their preamps may be purchased as a separate unit or, in a few instances, they are contained within the preamp/control center.

8.3 Front Panel Controls on the Preamp/Control Center

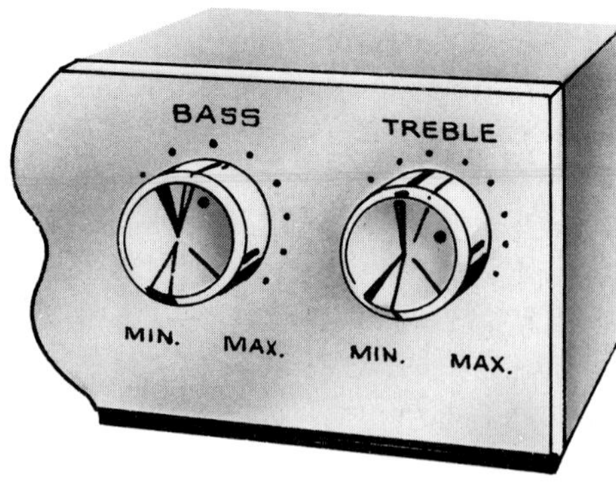

In Chapter 2 it was mentioned that the preamp/control center served the dual roles of managing the tonal qualities of the sound as well as routing the signal through the system (the electronic traffic cop). In this section we wish to explore the tone control facilities and show exactly what effect they have on the sound quality which emerges from the speakers. In particular we shall be concerned with the VOLUME, BALANCE, BASS and TREBLE tone controls, the LOUDNESS control, and the LOW and HIGH filters. Figure 8.2 shows some of these controls on a typical amplifier.

All of these tonal controls, with the exception of the VOLUME, are used primarily to boost or cut the sound power preferentially in selected frequency ranges. In other words, tone controls modify the frequency response of the system. As discussed in section 5.8, audio engineers take pains to ensure that a quality audio system correctly reproduces all the tones with the proper relative output—i.e., a "flat" frequency response. Why, then, does the preamp offer tone controls which change the response of the system? While audio designers strive to make hi-fi systems better and better they have no control over factors which are external to the components themselves. For example, your records may be old and scratchy, or the tape recording may contain too much background noise called "hiss." Furthermore, the room acoustics, such as the presence of heavy drapes and lots of stuffed furniture, may lead to a severe loss of the high

Figure 8.2. Some tone controls which are often found on amplifiers.

frequencies before they reach your ear. And speaking of ears, your own hearing may be deficient and some sort of acoustic emphasis from your hi-fi system could help to overcome this. Or many people prefer different "sounds" as a matter of personal taste. In any case, the tone controls are placed on the preamplifier/control center to help the listener overcome these external problems.

The VOLUME Control

The VOLUME control is used to regulate the amount of sound power emanating from the speakers. As shown in section 7.9, this control is actually a variable resistor which is located within the preamp. The resistor controls the amount of current transmitted from the preamp to the power amplifier which, in effect, controls the amount of current which is sent to the speakers.

BASS and TREBLE Controls

The BASS and TREBLE controls are somewhat like the VOLUME control except that they are used to either boost or cut the sound intensity of the low and high frequencies. Figure 8.3 shows a set of typical response curves when these two controls are set at various positions. In this example the TREBLE control determines the response curve for frequencies greater than 800 Hz, while the BASS control determines the curve for frequencies less than 800 Hz. The controls may be used together or separately. Notice that when the BASS and TREBLE controls are positioned at "12 o'clock" (middle curve) the overall response of the amplifier is "flat," i.e., there is no boosting or cutting of the sound at any frequency. The top curve shows what happens

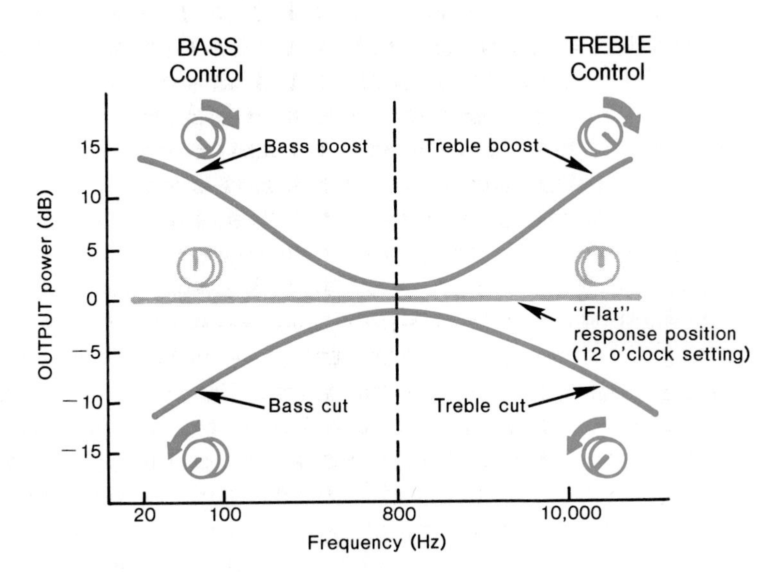

Figure 8.3. The action of the BASS and TREBLE controls is either to boost or cut the sound level at the low and high frequencies.

when both the BASS and TREBLE controls are turned up to their maximum levels. Notice that there is no boosting of the output for frequencies near 800 Hz. However, for frequencies both above and below 800 Hz the amplifier gradually boosts the output to a maximum value of approximately +13 dB at both ends of the audio range. It is typical of tone controls that only the bass and treble frequencies are changed, and the midrange frequencies remain unaffected.*

The lower curve in Figure 8.3 shows the effect of maximum BASS and TREBLE cuts. Once again, the midrange frequencies are unaffected, and the cutting action gradually increases to −10 dB near the two ends of the frequency scale. In this case the two controls halve the original loudness; remember that a −10 dB change produces sounds which are one-half as loud (see section 5.4).

The use of the BASS and TREBLE tone controls to change the tonal quality of music is obviously a matter of personal taste, and many people prefer to leave them in their "flat" positions. Often listeners will use the TREBLE control to cut the amount of record noise or static from a weak FM station. Of course this procedure will also attenuate the high frequency tones in the music itself, because the cutting action of the TREBLE control has no way of distinguishing between music and noise. The listener must ultimately be the judge of which is the lesser of the two evils.

HIGH and LOW Filters

Amplifiers are sometimes equipped with FILTER switches which are usually found in pairs and are called "high" ("scratch") and "low" ("rumble"). The high, or scratch filter is useful for removing the scratchy sounds associated with old and worn records. The scratchy sounds are more correctly called surface noise. The low, or rumble, filter helps to eliminate the low frequency noise generated by some turntables. The platter on which the record lies is mechanically suspended such that it is relatively free from the influence of external vibrations. If the isolation is not sufficiently complete, the platter will vibrate. This type of vibration, called rumble, is inherently low frequency and it will be picked up by the cartridge. In addition to rumble sometimes a small amount of 60 Hz line voltage can leak into the phono circuits, thus producing a 60 Hz low frequency sound in the output. A rumble filter normally works below 100 Hz and helps to eliminate both of these problems.

Subsonic Filters
When the amplifier has a frequency response which extends well below 20 Hz, a SUBSONIC filter can be extremely important. This filter is similar to a RUMBLE filter, except that it is usually designed to remove only those frequencies below the audible range. Tone arm resonance can cause excessive vibrations at frequencies less than 20 Hz, and these can be removed wih either a SUBSONIC or a RUMBLE filter. However, because a SUBSONIC filter removes only inaudible frequencies, it does not lose any of the audible bass tones that a RUMBLE filter might remove.

*More expensive preamps are available which divide the audio range into as many as twelve sections, each with its own tone control similar to the BASS and TREBLE controls discussed here.

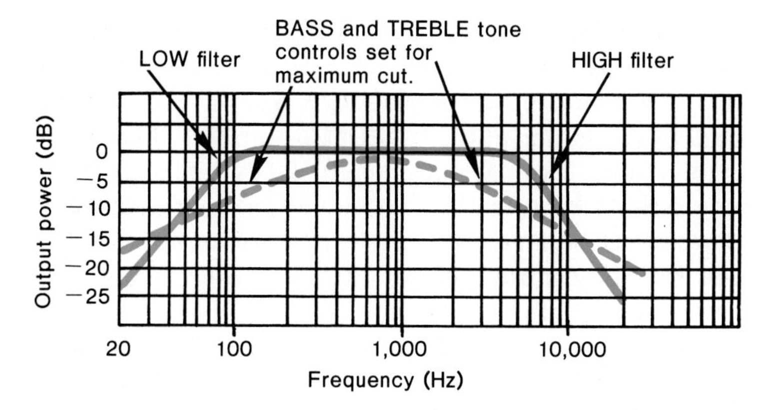

Figure 8.4. The action of the HIGH and LOW filters is much sharper than that of the BASS and TREBLE controls.

Perhaps you are wondering what the difference is between the HIGH and LOW filters versus the TREBLE and BASS controls. Figure 8.4 shows that the tone controls gradually attenuate the sound level on either side of 800 Hz. The filter action does not begin until a definite "cut-on" frequency is reached; then it attenuates the signal rather rapidly. In Figure 8.4 the filters start to work at approximately 100 Hz and 4 kHz, but these frequencies will vary from manufacturer to manufacturer. The idea behind the filter type of action is to keep the response as flat as possible until the frequency range which requires correction is reached. Presumedly a 4 kHz cut-on frequency will not drastically affect the musical quality (remember most of the acoustic energy lies below 1 kHz), yet it will produce a substantial reduction in the high frequency noise. Another difference between filters and the BASS and TREBLE controls is that the filters are either on or off. They do not have intermediate positions as do the BASS and TREBLE controls.

If the FILTERS might be of importance to you, look carefully when buying an amplifier because they are not a universal feature.

LOUDNESS Control

When reproducing music, one does not always wish to listen at the same level as the original sound. This creates problems because the ear loses its ability to hear the bass and treble notes as the sound level decreases. The LOUDNESS control, discussed in section 5.6, is designed to compensate automatically for the ear's insensitivity to the bass and treble notes at low volume levels. At high levels the ear possesses a relatively flat frequency response, and it hears the low and high frequencies in the correct relationship to the midrange frequencies. However, at low listening levels the ear's response becomes considerably "nonflat," and the listener has more difficulty in hearing those bass and treble

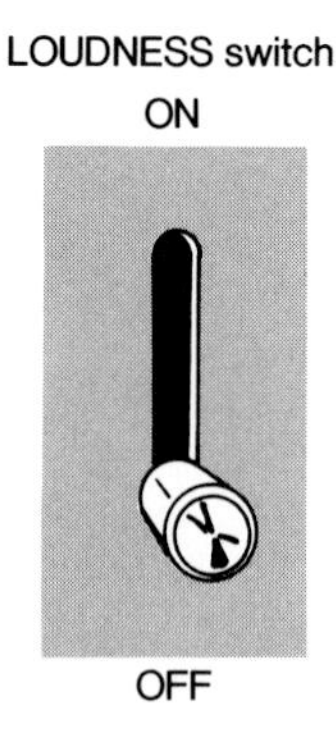

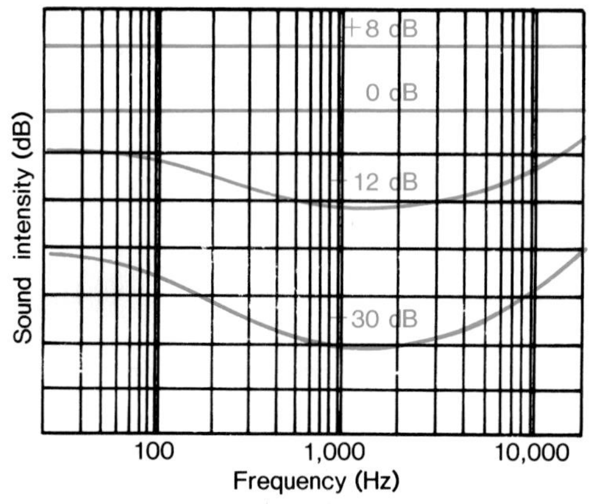

A. The LOUDNESS response curves of an amplifier

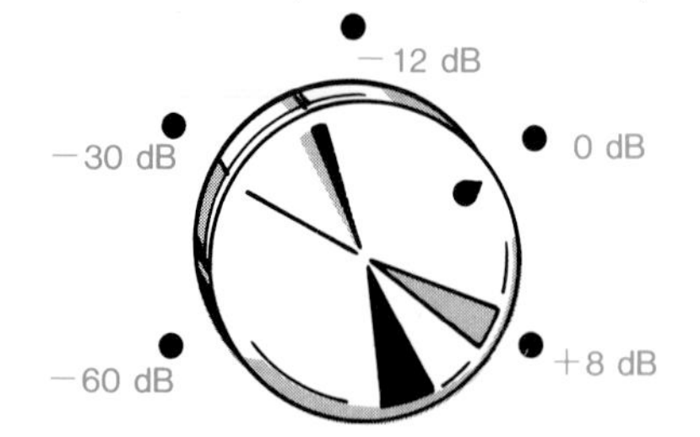

B. The LOUDNESS markings on the VOLUME control

Figure 8.5. The LOUDNESS control attempts to restore the full tonal balance between the high, midrange, and low frequencies, as heard by the listener, regardless of the VOLUME setting.

notes. The LOUDNESS switch, when activated, attempts to overcome this problem by automatically varying the frequency response of the amplifier in a manner which is dictated by the setting of the VOLUME control. For example, as the volume is turned down, the LOUDNESS control will progressively boost the bass and treble frequencies in order to overcome the ear's insensitivity. In other words, when the LOUDNESS control is "ON," the amplifier has a different frequency response for each setting of the VOLUME control.

The LOUDNESS control attempts to maintain an aurally flat response for the listener. Figure 8.5 (A) shows typical LOUDNESS contours of an amplifier and how they are determined by the setting of the VOLUME control. Notice that there are dB markings placed around the VOLUME control in part (B) (these markings may or may not actually be written on your amplifier, and their presence depends upon the manufacturer). Each dB marking tells the LOUDNESS contour to which the amplifier is responding. For example, if the VOLUME control is positioned at the −30 dB mark, then the corresponding curve in Figure 8.5 (A) shows the response of the amplifier. Notice, once again, as the VOLUME is turned up the response of the amplifier becomes flatter.

While many listeners prefer the LOUDNESS feature, there are others who prefer to use the BASS and TREBLE controls as a means of compensating for the ear's loss of sensitivity. Amplifier manufacturers do not know what type of speakers you are going to use, so they design the LOUDNESS contours for speakers of average efficiency. If you should own very inefficient speakers, the sound power will be relatively low and the LOUDNESS control may not compensate enough. Adjusting the BASS and TREBLE controls may be a better solution. On the other hand, very efficient speakers give relatively loud sounds and the LOUDNESS feature may boost the lows and highs too much. That is why this feature can always be switched on or off.

BALANCE Control

The BALANCE control is provided on the preamp such that the sound emanating from each stereo channel can be adjusted to the same level. Slight differences in speaker efficiencies, different acoustic environments, and asymmetric speaker placement with respect to the listener, are reasons why one speaker might sound louder than another. In any case the proper method for adjusting the BALANCE control is to set the amplifier to the MONO mode and adjust the BALANCE until the sound is equally strong from both speakers, or until it appears to be coming from a point midway between the speakers. Return the amplifier to the STEREO mode and you are all set to listen! Section 7.9 presented a discussion on how this control works.

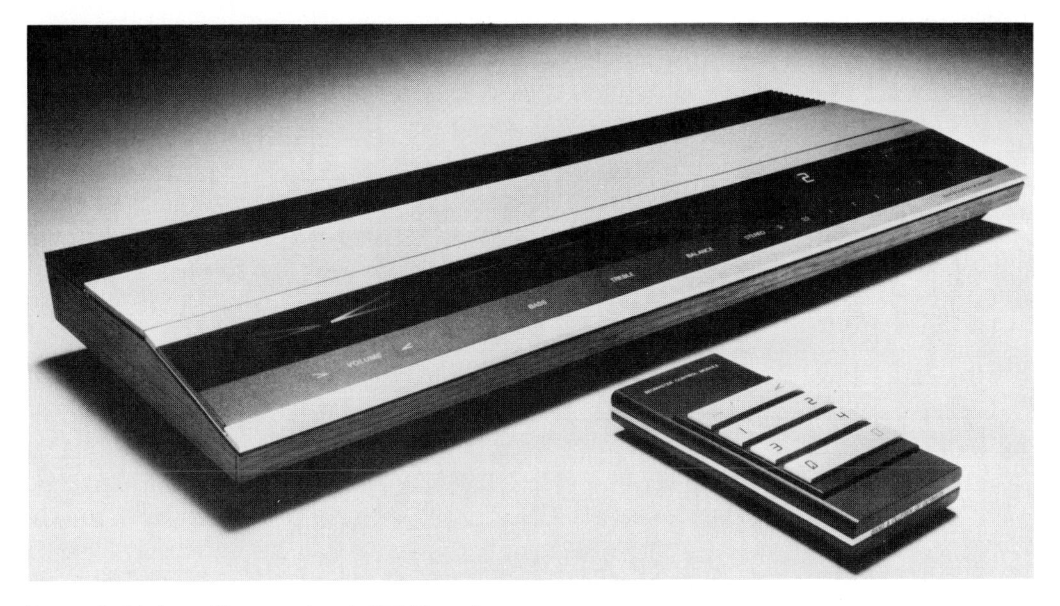

Bang & Olufsen Beomaster 2400 Receiver

8.4 The Back Panel of an Integrated Amplifier

The preamp/control center performs several important tasks as it processes the electrical signals. As mentioned in the previous section, the preamp can change the tonal properties of the sound by use of the tone controls. The preamp can also be thought of as an electronic traffic cop which can accept signals from many different sources and route them to a variety of auxiliary devices such as noise reduction systems, reverberation units, tape decks for recording, and the like.

Figure 8.6 shows the back panel of a typical integrated amplifier which, of course, has a preamp/control center incorporated into it; a receiver's back panel would also look very similar. The drawing exhibits the different source units which the integrated amplifier can accommodate. Of course there are the usual jacks for the PHONO and TUNER inputs, and the sound from your TV set could be sent through the system via the AUX inputs. This particular amplifier has one TAPE MONITOR switch which can accommodate a tape deck. Since integrated amplifiers and receivers can be purchased with either one or two TAPE MONITOR facilities, you should decide in advance which number best suits your overall needs. Chapter 2 presented a detailed account of the usage of the sources and the TAPE MONITOR switch; we shall not delve into this point any further here.

Other nice features, which may or may not be on your amplifier, are the PREAMP OUT and MAIN AMP IN

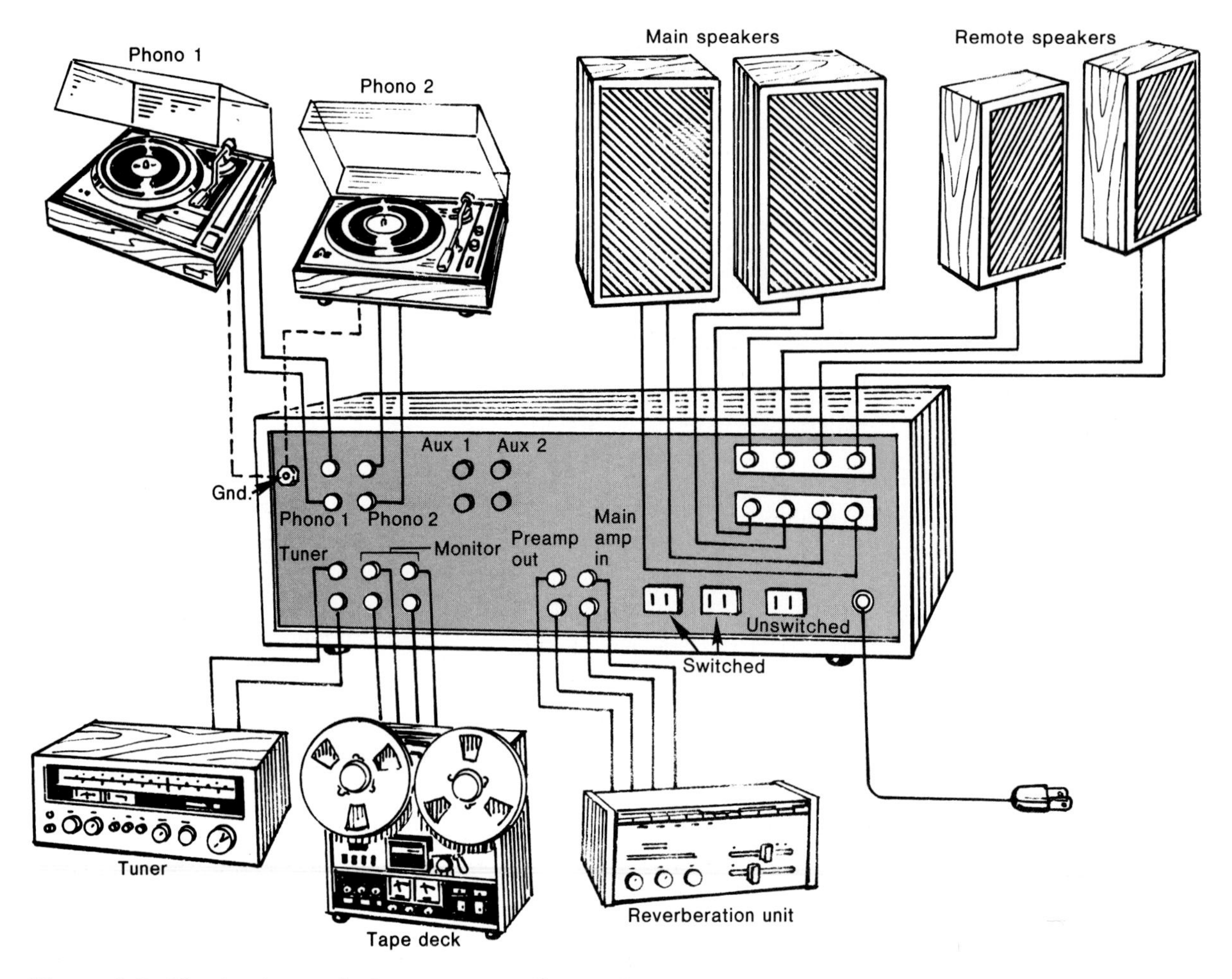

Figure 8.6. The back panel of a very versatile amplifier.

jacks shown in Figure 8.6. The PREAMP OUT jacks allow the signal to be extracted after it has passed through the entire preamplifier and just before it enters the power amplifier through the MAIN AMP IN jacks. If you should own, for example, a receiver with these additional jacks, it would be possible to insert devices into the circuit such as equalizers, reverberation units, noise reduction systems, etc. Placing these components in the PREAMP OUT/MAIN AMP IN jacks frees the TAPE MONITOR jacks for other devices such as tape decks. It must be realized though, that when the signal arrives at the PREAMP OUT jacks it has already passed through the TONE CONTROL section and has been subjected to its influence.

Headphones can be plugged into the amplifier, usually through the front panel, so that the music may be heard without the use of the main or remote speakers. This feature is found on virtually all amplifiers with the exception of separate power amplifiers.

The last feature on the back panel is the presence of the *switched* and *unswitched* 120-volt outlets. Devices, such as turntables and tape decks, can be plugged into these outlets. The switched outlet only has power in it when the amplifier itself is turned on, while the unswitched outlet has power available regardless of whether the amplifier is turned on or off.

8.5 Negative Feedback—A Beneficial Application of Destructive Interference

The main purpose of an amplifier is to amplify. This statement seems trivial, but there is more to it than meets the eye. Not only must an amplifier produce an output signal which is an enlarged copy of its input signal, but it must also produce a completely accurate and undistorted copy. Figure 8.7 illustrates what is meant by the words "accurate and undistorted." Part (A) shows a perfect amplifier which enlarges each portion of the input signal by the same factor of two, for example. In other words, the size of the signal at point #1 on the output is two times greater than it is at point # 1 on the input; the exact same thing can be said for point #2. For contrast, part (B) of the picture shows an imperfect amplifier, which enlarges the signal at point #1 by a factor of two, and at point #2 by a factor of four. The output signal is certainly an enlarged version of the input, but the amplifier has not enlarged each part of the output by the same amount. The amplifier in part (B) produces a lot of "distortion."

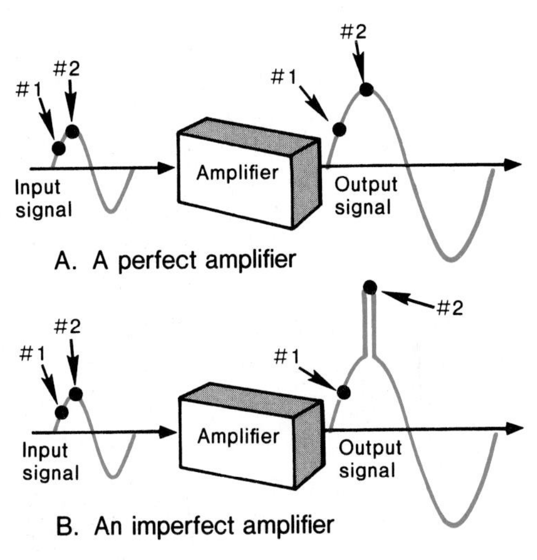

Figure 8.7. (A) The perfect amplifier enlarges the signal at points #1 and #2 by the same factor. (B) The imperfect amplifier enlarges the signal at points #1 and #2 by different relative amounts.

Negative feedback is important, because it is a technique for reducing distortion. Actually you are already familiar with the benefits of negative feedback in every-day life. Everyone who drives, for example, has probably misjudged a turn in the road and found himself heading toward the side of the road. The moment that the visual image of the rapidly approaching roadside "feeds back" to the brain, the steering is corrected, as shown in Figure 8.8. The steering correction minimizes the driving error (or "distortion") and keeps the vehicle on the road. Negative feedback works the same way in amplifiers, and it keeps the audio signal more on the undistorted "road". Virtually all modern amplifiers utilize feedback. To understand the use of feedback in amplifiers and the reason for calling it "negative feedback," recall that the audio signals within an amplifier are waves. It is helpful to remember this fact, because negative feedback is a beautiful example of how *destructive interference* between two waves can be used beneficially. Figure 8.9 offers an explanation of the main idea behind negative feedback.

Figure 8.8. Negative feedback in action. The visual image of the approaching roadside "feeds back" into the driver's brain, enabling him to correct the steering and avoid going off the road.

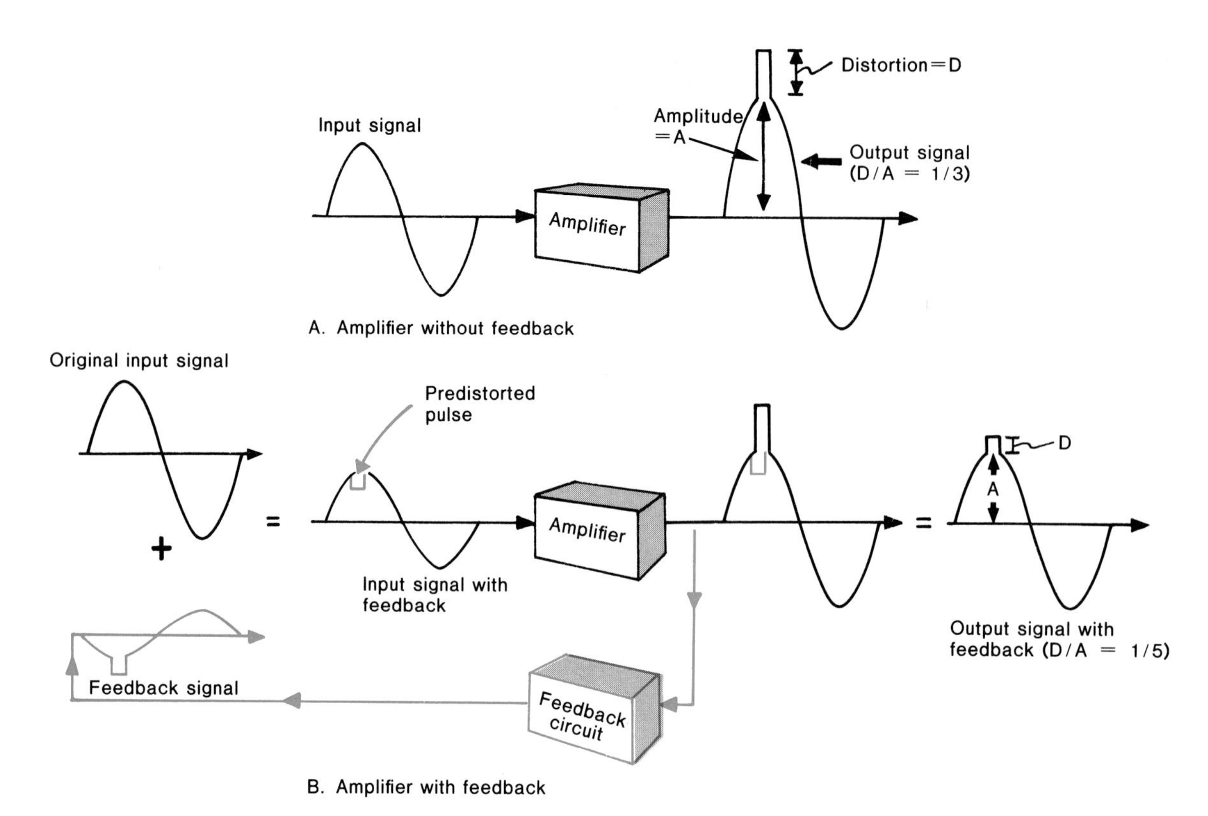

Figure 8.9. (A) Amplifier without negative feedback produces a large amount of distortion. (B) Amplifier with negative feedback (shown in color) produces an output which has relatively less distortion.

Part (A) of Figure 8.9 shows an amplifier which distorts the input signal by adding an unwanted rectangular pulse to the crest of the output signal. This distortion can be reduced substantially by employing a negative feedback circuit. In part (B) negative feedback is introduced, whereby a small fraction of the output signal is *inverted* so that a crest becomes a trough and a trough becomes a crest. This inverted fraction of the output signal is routed back, or "fed-back," to the input, where it is combined with the original input signal. The "combining" is an example of destructive interference, since the two signals are out-of-phase. Since the fed-back signal is effectively subtracted from the original input signal, the process is called "negative feedback."

It should be pointed out that the time required for a signal to pass through an amplifier is almost instantaneous. Therefore, the time delay between the original input signal and the slightly delayed feedback signal is too small to be audibly detected.

One obvious consequence of negative feedback is that the combined input signal to the amplifier is somewhat reduced, and the figure shows that the input signal in part (B) is not quite as large as it is in part (A). Therefore, the *output* signal in part (B) is not quite as large as it is in part (A).

True enough, some output signal is lost when negative feedback is employed; but look at what is gained. The combined input signal with feedback carries along the rectangular pulse (colored) which is pointing down. Just when the amplifier tries to distort the signal by adding an upward pointing rectangle, the downward pointing rectangle comes along and partially cancels the distortion. The beneficial effect of the negative feedback is to "tell" the input signal how and when the amplifier is going to distort it. The feedback signal predistorts the input signal in a manner which is opposite to the distortion introduced by the amplifier. The result is that a relatively smaller amount of distortion occurs in the output signal. Figure 8.9 (A) illustrates an example where the height of the rectangular distortion pulse, D, is arbitrarily drawn at 3 units. The amplitude, A, of the output sine wave signal is arbitrarily drawn at 9 units. Without feedback the ratio of the distortion to the output signal amplitude is $D/A = 3/9 = 1/3$. With feedback, part (B) of the figure shows that the D/A ratio is 1/5. Therefore, feedback reduces the amount of distortion relative to the output sine wave signal.

Negative feedback is also used to reduce the sound distortion caused by any erratic motion of a speaker's diaphragm. The negative feedback signal from the speaker to the amplifier allows for the partial correction of this type of distortion. This type of feedback is related to the "damping factor" specification of the amplifier. An amplifier with a large damping factor has a large amount of negative feedback, and such an amplifier can control much of the diaphragm's erratic motion. However, the amplifier's ability to produce loud sounds from the speaker is decreased. The amount of negative feedback used by an amplifier is a compromise between less distortion and less output power.

Summary

Negative feedback involves taking a small portion of the output signal of an amplifier, turning it around so that it is exactly out-of-phase with the input signal, and feeding it back into the input. Negative feedback is beneficial because it reduces distortion. It is an example of destructive interference between waves. Negative feedback is also used to reduce the sound distortion produced by speakers.

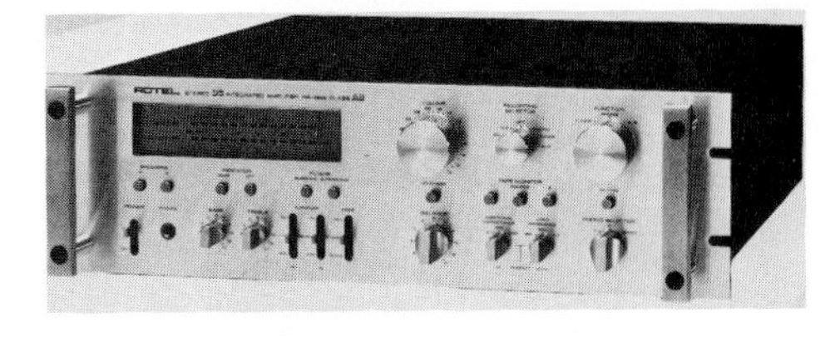

Rotel Model RA-2040 Integrated Amplifier

8.6 Photography—A Helpful Analogy for Understanding Amplifier Specifications

The technical specifications of an amplifier are numbers which tell how well the amplifier will perform. Specifications are essential for comparing amplifiers, or any other audio

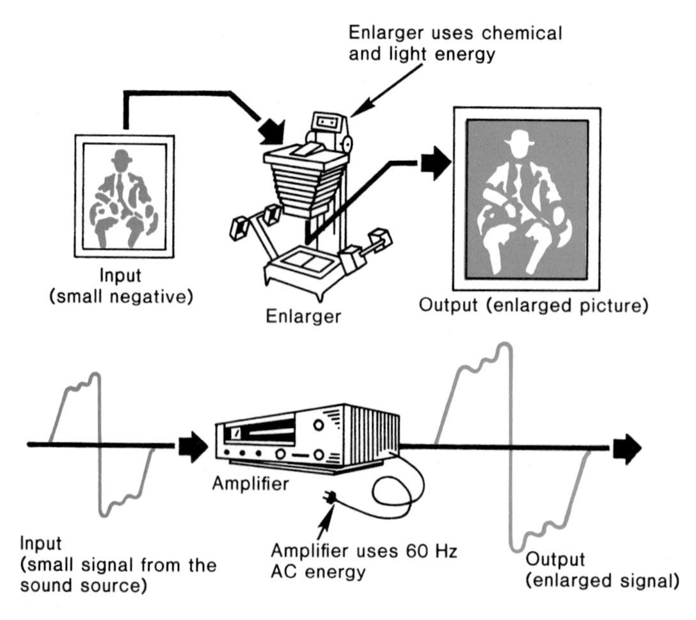

Figure 8.10. Photography is a helpful analogy with which to understand the technical specifications of an amplifier.

components, and they allow the well-informed consumer to purchase the best quality hi-fi equipment for the least money. One way to appreciate the meaning of amplifier specifications is to consider photography as a helpful analogy. Figure 8.10 illustrates the similarity between the photographic process and audio amplification.

Photographic equipment uses chemical and light energy to capture a small image on film and convert it into an enlarged picture. Likewise, as discussed in section 7.13, an amplifier uses electrical energy to convert a small input signal into an enlarged output signal. The output signal, except for its enlarged size, is supposed to be an exact *replica* of the input signal. No reproduction process is perfect, however, and we can only expect that amplifier's output signal bears a close resemblance to the input signal; obviously, the closer the better. There are five features of the photographic process which have a direct counterpart in the technical specifications of an amplifier, and these will now be discussed.

Distortion

One of the most beautiful aspects of photography is color. A basic requirement of color photography is that the photographic process should not introduce new colors; a red rose should look red, and not reddish-yellow. The different colors correspond to the different frequencies of light waves, so that the addition of unwanted colors means that the photographic process is producing unwanted light frequencies. Likewise, an amplifier should not generate additional frequencies of its own which were not contained in the original music. The *distortion* specification of an amplifier indicates the extent to which the amplifier contributes unwanted frequencies to the music. *Total harmonic distortion* (*THD*) and *intermodulation distortion* (*IM*) are two specifications of audio components which deal with the accuracy of frequency reproduction.

Frequency Response

A photographic enlargement must obviously preserve the relative sizes of the objects. If, for example, the negative shows that one person is 10% taller than another, then an enlargement should also show the same 10% differential. A familiar example of exaggerated size distortion occurs when looking into those wavy fun mirrors at the amusement park.

In audio, the relative loudness and softness of musical sounds is an important characteristic of music. It is not unusual that the soft sound of a symphonic string section is accompanied by a loud burst of a kettle drum, and it is important that any sound system be able to reproduce accurately such loudness relationships. It certainly would be unsatisfactory if the percussion sounds came through with only one-half of their original loudness while the system reproduced the violins accurately.

Every note should be given its due, no louder or softer than intended in the music. Therefore each tone, regardless of its frequency, should be enlarged by precisely the same factor as it passes through the amplifier. The *frequency response* specification indicates how well the amplifier can uniformly amplify all the frequencies in the sound. In other words, an amplifier with a good frequency response preserves the loudness relationships among the musical instruments.

Signal-to-Noise Ratio

Photographs sometimes contain an excessive number of blemishes, such as water marks, scratches, black spots, etc. An excessive amount of such blemishes causes so much visual "noise" that it detracts from the subject of the photo. Amplified music is also accompanied by an unwanted component called "noise." This type of audio noise is familiar to anyone who has dialed between two FM stations and heard the "static" or "hiss" which is always present. When you play *any* source through your system, the noise generated by the source preamplifier is always present, although a well-designed component keeps it at an inaudible level. Noise is generated by the electronic components, such as transistors and resistors which make up the amplifier circuits, and proper design can significantly reduce this noise. The *signal-to-noise ratio* of an amplifier is a specification which tells the user how much noise the amplifier generates relative to the musical signal.

Input Sensitivity

Every photographer knows that it is difficult to take pictures in dim light. In such a situation photographers may use highly sensitive film which is capable of responding to small amounts of light. Likewise, amplifiers deal with small input signals, such as those generated by magnetic phono cartridges. Amplifiers must also be sensitive enough to respond to the small input signals. The *input sensitivity* specification of an amplifier indicates how small the input voltage can be and still cause the amplifier to respond well.

Maximum Input Signal

One of the mistakes that a photographer must avoid is over exposing the film in the presence of bright light. Over exposed pictures have that characteristic washed-out look, lacking proper detail, because the amount of light was simply too much for the film to tolerate. This situation is similar to an "overloaded" amplifier. An arbitrarily large number of volts cannot be sent into an amplifier without severe distortion occuring in the output signal. The *maximum input signal* specification is the maximum number of input volts that the amplifier can safely handle without being "overloaded."

Summary

Photographic Analogy	Amplifier Counterpart	Amplifier Specification
Colors are added to those already present.	Audio frequencies are added to those already present.	Distortion
Picture reproduces the correct relative sizes of all objects.	The sizes of electrical waves of all frequencies must be enlarged by the same amount.	Frequency response
Blemishes should not obscure the subject of the picture.	Noise should not obscure the music.	Signal-to-noise ratio
Film can respond to small amounts of light.	Amp can respond to a small number of input volts.	Input sensitivity
Too much light over exposes film.	Amp will distort the sound because of excessive input voltage.	Maximum input signal

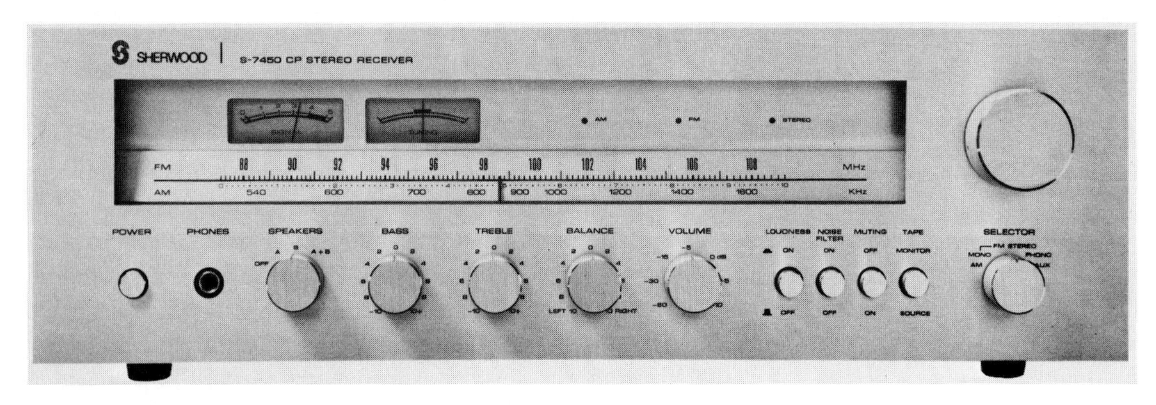

Sherwood Model S-7450 CP Receiver

8.7 Amplifier Specifications

In many respects, making comparisons among amplifiers can be made much more precise than making comparisons among speakers. By "amplifier" we mean the combination of the preamp/control center and power amplifier sections of any audio system. It was pointed out in Chapter 6 that there are no hard and fast rules which govern speaker selection, and that it was only possible to outline some very general guidelines to aid the prospective buyer. Fortunately, the performance of an amplifier—or any other audio component except the speakers—can be characterized by a half-dozen or so technical specifications which can immensely aid the buyer in obtaining a good product. If the buyer knows what specifications are important and how to interpret them, it is possible to purchase an amplifier which will give a quality performance.

In this section we will give a rather complete description of the important amplifier specifications, and then state an opinion as to what values constitute high quality. In arriving at such opinions there is always a trade-off between quality and price, and the values we recommend for amplifier specifications should ensure quality sound reproduction at a reasonable price. Of course, it is always possible to find amplifiers whose specs are far superior than the ones which we are suggesting; however, the price tag will usually be superior too. A summary of all the recommended amplifier specifications is presented at the end of this chapter.

Our discussion will follow the relatively new (1978) standards proposed by the Institute of High Fidelity (IHF), an organization of high-fidelity component manufacturers. These new standards are designed to update and expand upon the amplifier specifications which existed before 1978. Today, many manufacturers continue to use the pre-1978 amplifier specifications, and we will point out, wherever appropriate, the differences between the two sets of standards.

Unless otherwise noted, however, all of our discussions will be based on the new 1978 standards.

The total number of specifications recommended by the IHF is twenty-eight, and a discussion of all of them is beyond the scope of this book.* The IHF distinguishes between specifications which are of primary importance and those which are of secondary importance. We will discuss all the primary specifications and a few of the more important secondary ones. With this information the buyer will be well armed when comparing the relative merits of different amplifiers. Listed below are the specifications that we will now consider.

Primary Specifications

1. Total Harmonic Distortion
2. Continuous Average Power Output
3. Dynamic Headroom
4. Frequency Response
5. Signal-to-Noise Ratio
6. Input Sensitivity
7. Maximum Input Signal

Secondary Specifications

1. Intermodulation Distortion
2. Stereo Separation
3. Damping Factor

Total Harmonic Distortion—THD (Primary Specification)

The harmonics produced by musical instruments are essential tonal features because their presence gives the sound its characteristic timbre. Like musical instruments all audio components, such as tape decks, cartridges, tuners, amplifiers, speakers, etc., also produce harmonics of a fundamental note. However, in audio components the presence of these harmonics is an undesirable feature, and they constitute one form of audio distortion called (appropriately) "total harmonic distortion." In order to illustrate this problem consider Figure 8.11 in which a single frequency of 1,500 Hz is being sent to an amplifier from an audio generator. In principle the output waveform of the amplifier is supposed to be an *exact replica*, only enlarged, of the input electrical wave. This implies that the output waveform must have a frequency of precisely 1,500 Hz with no other frequencies present. Surprisingly, the output waveform is not an exact duplicate of the input but may have the distorted shape which is shown, somewhat exaggerated, in Figure 8.11. As

*The interested reader may obtain detailed information about all the specifications in the IHF publication A–202, 1978, "Standard Methods of Measurement for Audio Amplifiers." This publication is available from the Institute of High Fidelity, Inc., 489 Fifth Ave., New York, NY 10017.

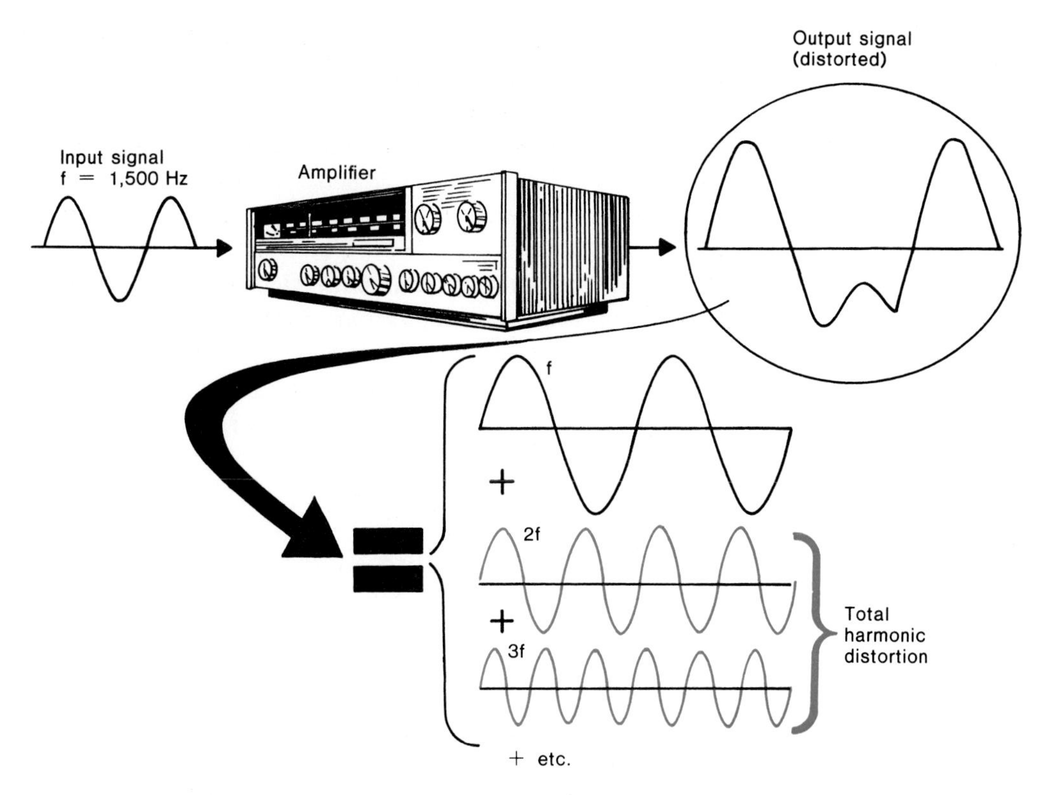

Figure 8.11. Total harmonic distortion occurs when an amplifier produces unwanted harmonics of a fundamental frequency. A quality amplifier keeps THD to an inaudible level.

mentioned above, amplifiers are not the only hi-fi components which are guilty of producing a distorted output. All audio components produce distortion although components which are good enough to be called hi-fidelity are designed to keep the distortion to a minimum level.

According to Fourier's method of analysis presented in section 4.10, the complex output wave shape is comprised of a fundamental sine wave, whose frequency is 1,500 Hz, *plus all the higher harmonics*, as shown in Figure 8.11. Thus we see that it is the presence of the higher harmonics at 3,000 Hz, 4,500 Hz, etc., which are responsible for the distorted appearance of the output waveform. Since these higher harmonics were not present in the input wave, they were created by the amplifier and added to the output signal.

Total Harmonic Distortion (THD): Total harmonic distortion is produced when higher harmonics are added to the fundamental frequency as it passes through an amplifier. The THD specification expresses, as a percentage, the voltage of the unwanted harmonics relative to the voltage of the fundamental frequency.

It is beyond the scope of this book to explain why these unwanted harmonics are generated. Suffice it to say that, in the case of tuners and amplifiers, the transistor circuits are responsible for the THD. Since all amplifiers produce harmonic distortion, it is only the better ones which can keep the level of the unwanted harmonics to an acceptable minimum.

The THD produced by an amplifier depends on the amount of power that it is producing. Figure 8.12 illustrates a typical distortion vs. power graph for an amplifier. For this amplifier the THD remains less than 0.5% until a power level of 50 watts is reached. Beyond 50 watts the THD begins to increase rapidly.

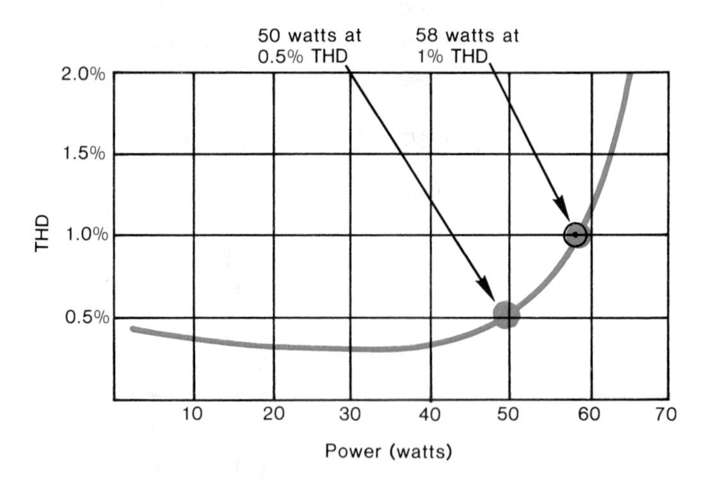

Figure 8.12. THD becomes very large as the power output is increased beyond 50 watts for this particular amplifier.

What is interesting about Figure 8.12 is that it shows how an amplifier manufacturer selects the power rating of the component. The manufacturer must first decide on what is the maximum value of the distortion to be allowed. Once this has been determined, the graph in the figure then gives the amplifier's power. For example, suppose that the manufacturer decided that the maximum acceptable value for the THD is 1.0%. The graph then shows that the amplifier must be rated at 58 watts. If the manufacturer wanted to advertise distortion values which were lower than 1%, then the power rating must also be lowered. For example, a maximum THD rating of 0.5% implies that the amplifier can only be rated at about 50 watts. Therefore, manufacturers must make trade-offs when writing the specifications for the power amplifier sections of a system. They can increase the power rating of an amplifier only if they also increase its distortion rating.*

*This statement does *not* mean that amplifiers with high power ratings produce more distortion than do amplifiers with lower power ratings. The statement means that, for a specific amplifier, any attempt to draw more power—beyond a certain level—will be accompanied by substantial increases in distortion.

What is considered a good value for THD? Generally speaking, anything less than 0.5% is considered good for amplifiers and tuners, and many medium priced amplifiers generate less than 0.1% THD when they produce 1 watt or less. Most other hi-fi components produce more than 0.5% THD. For example, the THD of phono cartridges and tape decks may range from 1% to 5%, while in speakers it may be even higher. In principle, it is desirable to keep the THD level for any audio component as low as possible, consistent with affordable prices.

In terms of listening, a 1% THD value means that 99% of the amplified output voltage will contain all the frequencies which were originally present in the music. And 1% of the output voltage will be the unwanted higher harmonics of the input frequencies, which were generated by the amplifier. In general it takes very, very careful listening in order to "hear" these harmonics, so that devices producing THD values of 1% or less are, in themselves, producing very little distortion.

Of course, what is really important is the *overall* THD which is produced by the *entire* audio system. To a good approximation the overall THD is the sum of the THD's produced by each component. For example, if you own the following system then you could expect the combined THD to be near 6%: cartridge (THD = 2%), integrated amp (THD = 1%), speakers (THD = 3%). Now 6% THD is getting near the point where most people begin to notice it. Thus it is important to keep the THD level of each component as low as possible. Most certainly one should be aware of amplifiers for which the maximum distortion is not stated in the specifications.

Amplifier Buying Guide

Total Harmonic Distortion: Look for a THD of 0.5% or less, at the full rated power of the amplifier.

Continuous Average Power Output (Primary Specification)

One of the most important specifications of an amplifier is its power rating because, in large measure, the cost is directly related to it. It goes without saying that a 50-watt integrated amplifier (or receiver) will be more expensive than a 10-watt unit. Perhaps the 50-watt amp will not be five times more expensive, but the cost differential will certainly be noticeable to your pocketbook. The power rating of an amplifier, by itself, is *not* an indication of the amplifier's

Nikko Model ALPHA VI Power Amplifier

overall quality because the 10-watt unit may be a better performer than the 50-watt amp. However, since the power rating does dictate price, it is always a good idea to know your approximate power requirements. The power requirements are governed by the efficiency of your loudspeakers, as discussed in section 6.10. More efficient speakers require smaller amounts of amplifier power than the less efficient ones. It may seem that efficient loudspeakers are the solution to keeping down the cost of amplifiers because of the smaller power levels which are required. This is not necessarily the case for several reasons. First, remember that there is no relation between the efficiency of a loudspeaker and how good it sounds to you. It may well be that you are unable to find a loudspeaker which is both efficient and suits your personal listening taste. Second, the size limitations of your listening room may force you to consider the smaller "bookshelf" loudspeakers and these usually tend to be very inefficient. In any case, the loudspeakers, in large measure, determine your amplifier power needs. Many loudspeaker manufacturers will give the MINIMUM RECOMMENDED POWER RATING (see section 6.10) which is required to produce reasonable sound levels from the speakers. How much more power you wish to purchase, above this minimum level, is obviously up to you, and the ability of your speakers to handle the power.

Continuous Average Power Output: The average amount of power that an amplifier, operating under certain specified conditions discussed below, can deliver on an uninterrupted, steady basis. Often called the "power rating" of an amplifier.

Below is a complete and accurate power rating for a 50-watt per channel amplifier. Due to a 1975 FTC ruling, manufacturers are required to include all of these conditions if they wish to state the power rating of their product.*

> 50 watts (continuous) per channel,
> into 8 ohm speakers,
> both channels simultaneously driven,
> from 20 Hz to 20,000 Hz at less than
> 0.2% THD at the rated power.

The various aspects of this example will now be discussed in greater detail.

1. **"50-watts (continuous) per channel"** means that the amplifier will deliver 50 watts into each stereo channel for a total of 100 watts of power. A 4-channel system would require 50 watts to be sent to each of the four channels. In the days before the FTC ruling some manufacturers would omit the "per channel" and the 50 watts looked bigger than it really was because, almost invariably, this was the sum of the powers in each channel. In this case each stereo channel would be delivering only 25 watts. Any time that the "per channel," or its equivalent, phrase is missing from an amplifier advertisement or spec sheet, always assume that the stated power is the *total* of both channels.

   The continuous power is the power that an amplifier will continuously deliver, day and night, without exceeding its rated THD distortion. Sometimes "continuous average power" is inaccurately called "rms power."**
2. **"into 8 ohm speakers"** refers to the impedance of the speakers being used. It is important that the power rating include the impedance of the speakers that you are planning on using. In sections 7.6 and 7.7 it was pointed out that smaller speaker impedances (resistances) allow more current to flow which, in turn, means more power will be delivered to the speakers.

The output power may also be expressed in dBW relative to a reference level of 1 watt. The procedure is exactly the same as that discussed in section 5.2, where the power ratio is computed and converted to decibels via Table 5.1. In our example 50 watts of power gives a power ratio of 50 watts/1 watt = 50. Table 5.1 shows that this corresponds to a level of 17 dBW above the 1 watt level.

*The catch here is the phrase ". . . *if* they wish to state the power rating. . . ." A manufacturer may *not* wish to state the power rating at all and, therefore, the above list may be omitted altogether. This omission often occurs when buying compacts or consoles, and the buyer has no idea what the power rating is; usually it is very low.

**rms means "root mean square," and it is a mathematical expression for computing a particular type of average power which an amplifier can deliver.

Therefore, the maximum power that an amplifier can deliver depends on the impedance of the speakers. Often the manufacturer will specify the power rating for several speaker impedances, e.g.,

42 watts (continuous) per channel into 16 Ω speakers.
50 watts (continuous) per channel into 8 Ω speakers.
64 watts (continuous) per channel into 4 Ω speakers.

Therefore, the correct power rating of an amplifier can be determined only when the impedance of your speakers is known.

3. **"both channels simultaneously driven"** indicates that stereo programs require both speakers to produce sound at the same time. Suppose, as in the above example, the amplifier could deliver a maximum of 50 watts to each speaker under these conditions. Now if one speaker was disconnected, you would find that the amplifier could deliver slightly more power, say 54 watts, into the remaining speaker. (We shall not go into the reason why this happens.) The point here is that a manufacturer could make the amp appear slightly more powerful than it actually is when operating under normal conditions, i.e., with both channels working. The phrase "both channels simultaneously driven" ensures that the power rating yields a realistic value.
4. **"from 20 Hz to 20,000 Hz at less than 0.2% THD at the rated power,"** means that the amplifier can deliver its rated power (50 watts per channel) over the entire audio range from 20 Hz to 20 kHz. And, the output signal contains less than 0.2% total harmonic distortion. It is true that very few musical sources actually utilize this complete range; FM transmission and most records have an upper limit of

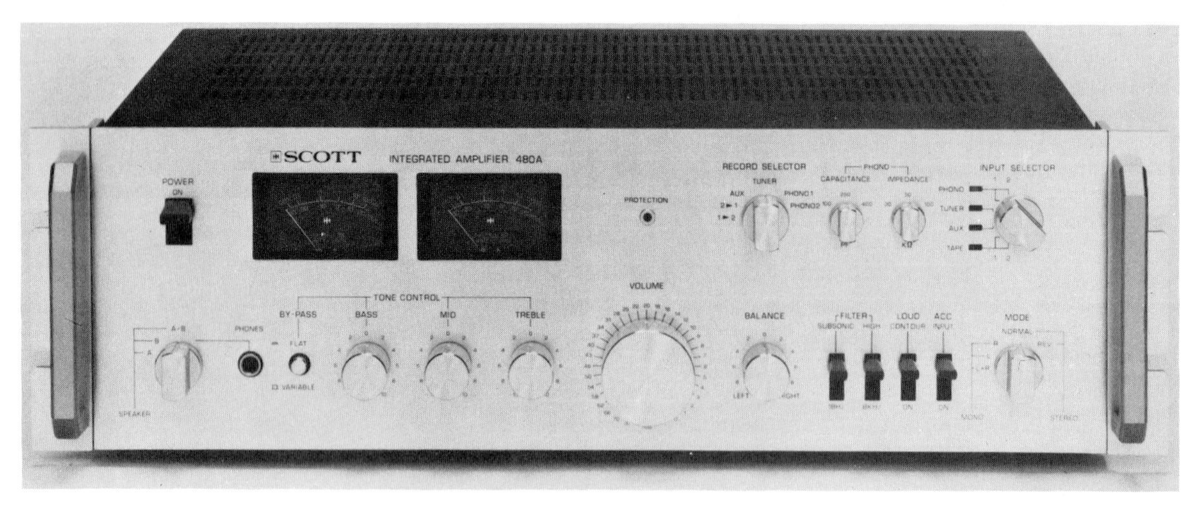

Scott Model 480A Integrated Amplifier

only 15 kHz. Nonetheless, you should still insist that an amplifier be able to deliver its rated power into all frequencies which at least span the complete, 20 Hz to 20 kHz, range. This is not a difficult task for most amplifiers.

Amplifier Buying Guide

Continuous Average Power Output:

1. Be sure that this specification is stated accurately as required by the 1975 FTC ruling:
   "X" watts (continuous) per channel, into "Y" ohm speakers, both channels simultaneously driven, from 20 Hz to 20 kHz at less than 0.5% THD at the rated power. (Note: The buyer must substitute the desired values for "X" and "Y.")
2. Match the power rating of the amplifier to the corresponding needs of the loudspeakers, as discussed in section 6.10.

Dynamic Headroom (Primary Specification)

Consider driving your car on a long trip. Most cars have a maximum cruising speed at which they can run for long periods of time. Even with a maximum cruising speed of 120 km/hr, it may be possible to accelerate the car to 150 km/hr for a short period of time if, for example, you wish to pass another car. Your car could not sustain the 150 km/hr speed all day, however, because it would probably become severely damaged. The same is true with amplifiers. For brief periods of time an amplifier may well be able to generate more than its rated power without exceeding its distortion rating. Dynamic headroom is a specification which tells something about such an ability, and it is stated as follows:

Dynamic Headroom: The ratio, expressed in decibels, of the short term power output which an amplifier can deliver relative to its continuous average power output. A "short period of time" is approximately 0.02 seconds.

If an amplifier has a 50 watt rating for its continuous average power output, but it can generate 100 watts (still within its distortion limits) for brief periods, then the power ratio is 100 watts/ 50 watts = 2. According to Table 5.1 this power ratio corresponds to a dynamic headroom of 3 dB. Dynamic headroom is of importance to many listeners because most high-fidelity program material contains low average levels accompanied by occasional high level peaks, which are usually less than 0.02 seconds in duration. When an amplifier is being operated at or near its rated power a dynamic headroom of 3 dB would allow a sudden, quick musical peak which contains *twice* the average power (3 dB) to be reproduced without serious distortion.

Amplifier Buying Guide

Dynamic Headroom: There is no general agreement as to what constitutes "good" versus "bad" dynamic headroom. Most amplifiers have dynamic headrooms which fall between 0 dB and 3 dB.

Frequency Response (Primary Specification)

The frequency response of an amplifier shows how much the different frequencies are amplified relative to each other (see section 5.8 for a detailed discussion on frequency response). The usual method of measuring the frequency response begins by attaching an audio generator to either a high-level (TUNER, TAPE, or AUX) or a low-level (PHONO) input of the amplifier. The generator is initially set to produce a 1 kHz signal whose (RMS) voltage level is 0.5 V for a high-level input or 5.0 mV for a low-level input. The VOLUME control on the amplifier is adjusted so as to produce an output power of 1 watt. The output power of the amplifier is then measured as the frequency of the input signal is changed, while keeping the voltage level of the input signal constant at all times. (Of course, all filters and the LOUDNESS control must be turned off, and the BASS and TREBLE tone controls must be set to their "flat" positions during the measurement.) As the frequency is changed, the amplifier's output power will deviate from the 1 watt reference power. The amount of deviation will depend on the quality of the amplifier, with smaller deviations occurring with the better amplifiers. Usually, the output power vs. frequency results are plotted on a graph, known as the *frequency response graph,* whose definition is stated as follows:

Frequency Response Graph: A graph of an amplifier's output power, plotted in dB relative to 1 watt, versus the input frequency.

Figure 8.13 illustrates the frequency response graphs for two different amplifiers.

Very often manufacturers will simply give a frequency response specification, rather than a graph. Take, for example, the specification 20 Hz → 20,000 Hz, +2 dB, −1 dB. This frequency response specification means that the amplifier will amplify all frequencies in the range from 20 Hz to 20,000 Hz, and that no frequency will be amplified by more than +2 dB, or less than −1 dB, from the reference power of 1 watt. Of course the problem with such a specification, versus a graph, is that the "spec" does not tell *where* in the frequency range the power deviations occur. For example, the two curves in Figure 8.13 have the same frequency response specification, 20 Hz → 20,000 Hz, +2 dB, −1 dB, yet they look vastly different. The curve depicted in Figure 8.13 (A) has a low frequency peak, and it is somewhat lacking at the high end. Figure 8.13 (B) shows a frequency response curve for another amplifier which lacks bass response and overemphasizes the highs. Because of this, two amplifiers which have identical frequency response specifications may actually sound different, since they may not have identical graphs.

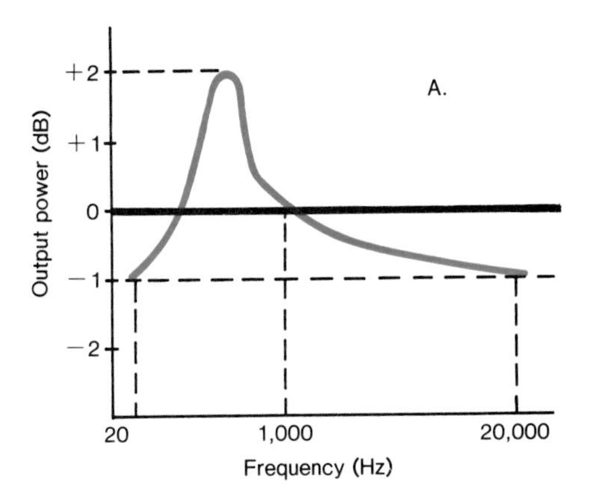

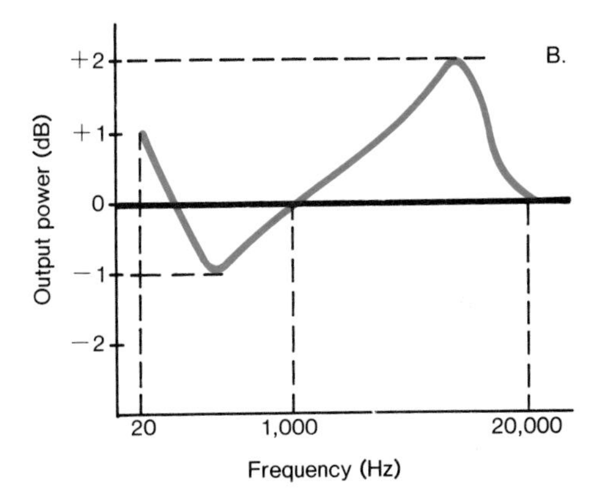

Figure 8.13. Two frequency response graphs which are each consistent with the frequency response specification of 20 Hz → 20,000 Hz, + 2 dB, −1 dB.

When judging frequency response specifications, remember that the wider the frequency range, the better. Also, the smaller the dB numbers, the better:

good: 20 Hz → 20,000 Hz, +0.3 dB, −0.5 dB.
better: 20 Hz → 20,000 Hz, +0.2 dB, −0.2 dB.*
perfect: 20 Hz → 20,000 Hz, +0 dB, −0 dB.

Distrust all frequency response specifications which do not carry the dB figures!

The illustration given above for the frequency response specification actually refers to a high-level input (TUNER, TAPE, AUX). For the PHONO input the specification is stated slightly differently because of the RIAA equalization process which occurs within the amplifier. This specification is given, for example, as

RIAA(+2 dB, −1 dB).

The frequency range is always assumed to be 20 Hz → 20,000 Hz, and, in this case, the output power will vary by no more than +2 dB, or less than −1 dB, from that of a perfect amplifier. The reason for specifying the PHONO frequency response in this manner is directly linked to the nature of the RIAA equalization amplifier found in all preamplifier/control centers. This point will be discussed further in section 12.9.

*Frequency response specifications like 20 Hz → 20,000 Hz, +0.2 dB, −0.2 dB, are often abbreviated as 20 Hz → 20,000 Hz, ±0.2 dB.

Amplifier Buying Guide

Frequency Response:

1. **TUNER, TAPE, AUX Inputs:** Look for a frequency response of at least 20 Hz → 20,000 Hz, ±0.5 dB. The wider the frequency range, the better; the smaller the dB numbers, the better.
2. **PHONO Inputs:** Look for a frequency response of RIAA (±1.0 dB). The smaller the dB numbers, the better.

Signal-To-Noise Ratio—S/N (Primary Specification)

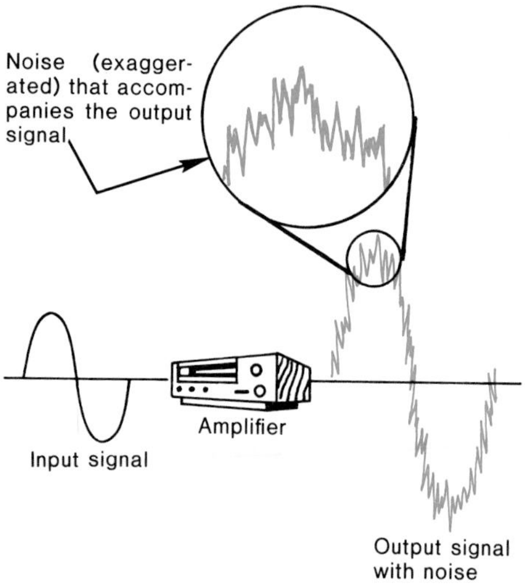

Unfortunately, all amplifiers add a certain amount of "noise" to the signal, which all of us have heard at one time or another. If you tune in a weak station on the radio, there is the ever present background hiss or static which originates in the tuner's preamplifier. In general, audio engineers call this annoying disturbance "noise," and it is created by the electronic components, such as transistors and resistors, from which the amplifier is built. The amount of noise present in the output of an amplifier should be kept as low as possible, as indicated in Figure 8.14. To characterize the relative strength of the noise, designers have invented a *Signal-to-Noise ratio* (abbreviated S/N). By "signal" we mean anything which is "good," i.e., the music. Of course, "noise" represents contributions to the sound which are not so good.

Figure 8.14. All amplifiers add a certain amount of noise to the ouput signal.

The word "ratio" provides the key to understanding this specification, because essentially it is just the output signal power divided by the noise power. The ratio is expressed in terms of decibels, as discussed in section 5.2. In order to allow comparisons among amplifiers, a standard output power of 1 watt and a standard frequency of 1 kHz have been agreed upon. These references are the same ones used above in other specifications. The definition of the Signal-to-Noise ratio is thus stated as follows:

Signal-to-Noise (S/N) Ratio: The ratio of 1 watt of output signal power (at 1 kHz) to the A-weighted noise power produced by the amplifier. The ratio is expressed in decibels.

Notice that the decibels used in the S/N specification are "A-weighted," rather than the "just-plain" decibels discussed in Chapter 5. Figure 8.15 shows the meaning of "A-weighted noise" by illustrating how the S/N ratio is measured. In part (A) an audio generator sends a 1 kHz signal into the amplifier. (The voltage level which is chosen for the input signal depends on which input is being used, and it is

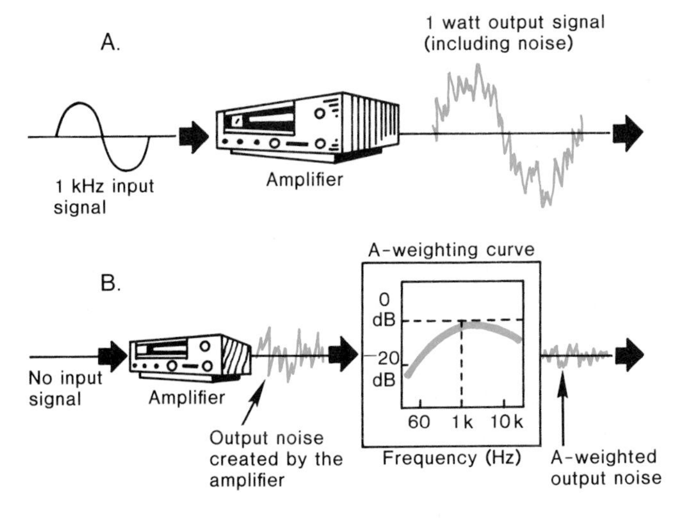

Figure 8.15. **(A) The 1 kHz output contains the signal plus noise. (B) An A-weighted noise measurement.**

specified by the IHF publication mentioned earlier.) The VOLUME control on the amplifier is then adjusted until the power of the output signal is exactly 1 watt. To measure the noise generated by the amplifier the input signal is set to zero, as illustrated in part (B) of the figure. With a zero input signal the output contains only the noise. However, before being measured, the noise is sent to an electronic device which performs the "A-weighting." This device *reduces* the amount of noise present at each frequency according to the curve shown in the figure. For example, according to the A-weighting curve, the measured noise at 60 Hz is reduced by about 20 dB relative to the noise at 1 kHz. Similarly, the noise level at 10 kHz is reduced by about 5 dB relative to that at 1 kHz. Therefore, the total noise power which leaves the A-weighting circuit is *less* than the noise that enters it. To obtain the S/N ratio for the amplifier, the 1 watt reference power is divided by the A-weighted noise power, and the ratio is converted into decibels as indicated in section 5.2.

Manufacturers prefer to measure "A-weighted" noise, because our ears do not respond equally well to all sound frequencies. Our hearing is most sensitive to the midrange frequencies, and becomes less sensitive at both the low and high frequencies. (Remember the Fletcher-Munson curves in section 5.6 which describe the ear's sensitivity to various frequencies.) Therefore, the typical listener is not bothered as much by noise at either 60 Hz or 10 kHz when compared to the noise at 1 kHz. A sensible S/N specification should reflect this fact by giving the annoying midrange noise a greater "weight" than the low and high frequency noise. Hence, an A-weighted noise measurement is useful, because it relates to how the ear actually perceives the noise; A-weighting gives "real life" audibility to the S/N ratio.

It should be emphasized that the A-weighting electronic circuit is only used by the manufacturer when measuring the S/N ratio of the amplifier. This circuit is *not* found in any amplifier that you would buy in the store.

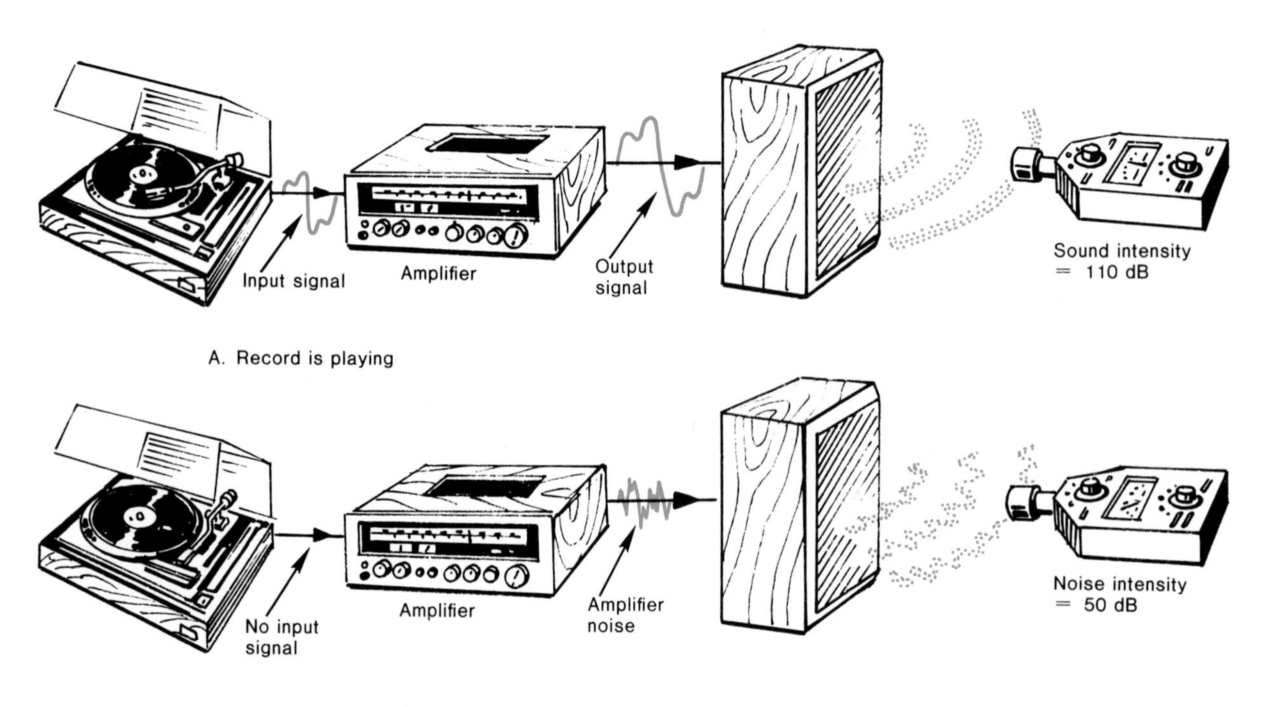

Figure 8.16. If the signal (music) is 110 dB and the noise generated by the amplifier is 50 dB, then the S/N ratio is (110-50) = 60 dB.

Figure 8.16 illustrates the importance of the S/N specification to one's listening pleasure. Suppose a record is being played, and the VOLUME control has been adjusted such that the output registers 110 dB on a sound level meter (A-weighted). How much noise would be heard if the S/N ratio of the amplifier is rated at 60 dB (A-weighted)? Since 60 dB is, by definition, the *difference* in decibels between the signal and the noise, there must be (110 − 60) = 50 dB of noise coming through the speakers. Therefore, 50 dB of noise would be heard simultaneously along with 110 dB of music. Under normal conditions one would not be aware of this noise in the presence of loud music, but during quiet passages in the program there would be a certain amount of audible noise. Therefore, it should be obvious why larger numbers for the S/N ratio are better. Larger S/N numbers mean a greater difference between the level of the music and the noise.

The S/N ratio is specified separately for the various inputs. In general, the PHONO input rating is poorer (smaller) than the rating for a high-level input. This is because the signal from the cartridge is extremely weak (1 to 5 mV) and must be first amplified by the RIAA equalization amp, whereas the signals from the TUNER, TAPE, or AUX inputs are already at a high level (about 500 mV) and do not require as much additional amplification. The additional amplification required for the phono signal introduces additional noise, thus lowering the S/N ratio.

The pre-1978 S/N ratio was measured by choosing the signal level as the full rated power output of the amplifier, rather than 1 watt; in addition, A-weighting was not used, and a standard input voltage was not specified. Because of these differences, it is not possible to correlate directly the pre-1978 S/N ratio with the new specification. In general, however, the buyer can usually tell which specification is being used by the manufacturer, because the new S/N ratio is often written as, for example, S/N: 70 dB (Re. 1 watt output, A-weighting). The phrase "Re. 1 watt output, A-weighting" means that the S/N ratio was measured using a **re**ference output power of 1 watt, and the noise power was measured using an A-weighting circuit.

Amplifier Buying Guide

Signal-to-Noise Ratio (S/N):

1. **Post-1978 Specification**
   a. **PHONO Input:** Look for a S/N ratio (A-weighted) of 70 dB or higher (75 dB, 80 dB, etc.).
   b. **TUNER, TAPE or AUX Inputs:** Look for a S/N ratio (A-weighted) of 80 dB or higher.
2. **Pre-1978 Specification**
   a. **PHONO Input:** Look for a S/N ratio of 60 dB or higher.
   b. **TUNER, TAPE, or AUX Inputs:** Look for a S/N ratio of 70 dB or higher.

Sansui Model AU-719 DD/CC Integrated Amplifier

Input Sensitivity (Primary Specification)

All amplifiers can accept one or more sources for their input; these sources may be phono cartridges, tuners, tape decks, or microphones. In a hi-fi system amplifiers respond to a small input signal by ultimately delivering an enlarged copy of it to the output speaker terminals, thereby furnishing the speakers with sufficient power to produce the sound. Some amplifiers are *more sensitive* than others, because they require *less input signal* from the source to produce the same output power.

In order to measure the input sensitivity of an amplifier, manufacturers use the following procedure. An audio generator, which acts as the source, is connected to one of the input jacks of the amplifier; the audio generator is set to produce a 1 kHz tone. The VOLUME control on the amplifier is turned to its maximum level, and a power meter is attached to the output speaker terminals. Starting from zero, the input voltage from the audio generator is slowly increased until the amplifier produces 1 watt of output power. The value of the input voltage which produces the 1 watt of output power is, by definition, the input sensitivity rating of the amplifier.

Input Sensitivity: The input voltage required (using a 1 kHz signal) to produce an output power of 1 watt when the VOLUME control is turned to its maximum position.

For example, suppose that two amplifiers have the following input sensitivities for the PHONO input:

Amplifier A: 0.2 mV,
Amplifier B: 0.05 mV.

Amplifier A requires 0.2 mV of input signal in order to produce 1 watt of output power. Amplifier B requires *less* input signal to produce the same 1 watt of output; therefore, B is the more sensitive amplifier. Notice that nothing is said about the maximum power that each amplifier is capable of producing. It may well be that "A" has a 100 watt power rating and "B" has a 90 watt rating. The input sensitivity rating only is concerned with the value of the input voltage that is required to produce 1 watt of output power. Input signals which are substantially larger than the input sensitivity rating will cause an amplifier to produce its full rated power.

Different amplifier inputs (PHONO, TUNER, TAPE, AUX) generally have different sensitivity ratings. The PHONO input is the most sensitive, with typical values for magnetic cartridge inputs lying between 0.1 mV and 2.0 mV. The reason for these low values is simply that a magnetic phono cartridge generates a noticeably smaller voltage than any other source; for this reason, PHONO inputs are sometimes called *low-level inputs*. By contrast, the TUNER, TAPE, and AUX inputs are referred to as *high-level inputs*, since they require substantially larger input voltages to produce the 1 watt reference output power. Typical input sensitivities for high-level inputs lie between 10 mV and 100 mV. The larger voltages required by the high-level inputs correspond to the fact that tuners and tape decks generate more source voltage than do magnetic phono cartridges.

It should be emphasized that the input sensitivity rating of an amplifier is not by itself an indication of its quality. Using the previous example, it does not follow that amplifier B, which is more sensitive than amplifier A, is a better quality amplifier. The input sensitivity rating is important, however, as an aid in properly "matching" a source to the amplifier. As an illustration, let us consider how to match a magnetic phono cartridge to an amplifier.

When a record is playing, the output voltage from the cartridge is never constant. The voltage rises when the music becomes loud, and it falls when the music becomes soft. Most phono cartridge manufacturers, however, specify the *average output voltage* that their cartridges will produce under specified playing conditions (see section 12.8). When the cartridge is connected to the amplifier, it is the average output voltage of the cartridge which dictates the average power level being delivered by the amplifier to the speakers. If you wish to obtain the full rated power from the amplifier, it is not sufficient just to turn the VOLUME control full-up; in addition, the cartridge's average output voltage must be large enough. What happens if the phono cartridge can never generate enough voltage to cause the amplifier to produce its full rated power, even though the VOLUME control is set to its maximum? Clearly, in such a situation the amplifier cannot utilize all of its power producing capacity, and you will not be getting all the power that you paid for. The phono input sensitivity of the amplifier indicates how to avoid this problem. How? It tells the buyer what the smallest average output voltage of the cartridge should be in order to cause the amplifier to produce its full rated power. Here's how it is done.

Remember, the PHONO input sensitivity rating of an amplifier tells how many volts are required to produce 1 watt of power. A proportionality can be used to calculate how many input volts are required to produce the maximum rated power of the amplifier. Suppose, for example, that the PHONO input sensitivity rating of the amplifier is 0.3 mV, and the maximum rated power of the amplifier is 100 watts. From Equations 7.1 and 7.3, it can be shown* that the voltage is proportional to the *square root* of the power, and the required proportionality is

$$\frac{\text{Cartridge average output voltage}}{0.3\text{ mV}} = \frac{\sqrt{100\text{ watts}}}{\sqrt{1\text{ watt}}} = 10,$$

or

$$\text{Cartridge average output voltage} = 3\text{ mV}.$$

Therefore, you would want to buy a cartridge whose average output voltage is 3 mV, or greater, although it should not be significantly larger for reasons which will be discussed shortly. The average output voltage should not be less than 3 mV because it would be insufficient to cause the amplifier to deliver its rated power—even though the VOLUME control is turned to its maximum level.

Table 8.1 offers a convenient chart that relates the PHONO input sensitivity to the power output rating of an amplifier. To use the Table locate the power level which matches, as closely as possible, the amplifier's power output rating. Then, using the left-hand column, find the PHONO

Phono Input Sensitivity (mV)	Power Output Rating (watts/channel)							
	9	16	25	36	49	64	81	100
0.1	0.3	0.4	0.5	0.6	0.7	0.8	0.9	1.0
0.2	0.6	0.8	1.0	1.2	1.4	1.6	1.8	2.0
0.3	0.9	1.2	1.5	1.8	2.1	2.4	2.7	3.0
0.4	1.2	1.6	2.0	2.4	2.8	3.2	3.6	4.0
0.5	1.5	2.0	2.5	3.0	3.5	4.0	4.5	5.0
0.6	1.8	2.4	3.0	3.6	4.2	4.8	5.4	6.0
0.7	2.1	2.8	3.5	4.2	4.9	5.6	6.3	7.0
0.8	2.4	3.2	4.0	4.8	5.6	6.4	7.2	8.0
0.9	2.7	3.6	4.5	5.4	6.3	7.2	8.1	9.0
1.0	3.0	4.0	5.0	6.0	7.0	8.0	9.0	10.0

Table 8.1. The numbers in the table are cartridge average output voltages (mV) needed to cause an amplifier with a given PHONO input sensitivity (mV) to produce its maximum rated power.

*Solving Equation 7.1 for the current and substituting it into Equation 7.3 yields $P = V^2/R$, or $V = \sqrt{PR}$.

input sensitivity that approximately matches the rating of the amplifier. The intersection of the corresponding column and row within the chart yields the minimum value of the cartridge average output voltage which should be used with the amplifier. For example, an amplifier with a power output rating of 85 watts and a PHONO input sensitivity rating of 0.72 mV would require a phono cartridge whose average output voltage is at least 6.3 mV.

Matching high-level sources to the amplifier is performed in exactly the same manner as outlined for the phono cartridge. Fortunately, there is a great deal of compatibility between the voltage outputs of high-level sources and the high-level input sensitivity ratings of the amplifier. Consequently, the prospective buyer almost never has to worry about matching the two.

The pre-1978 input sensitivity was defined to be the amount of input voltage required to produce the full rated power (not just 1 watt) of the amplifier when the VOLUME control was set at its maximum setting. Selecting the proper match between the cartridge and the amplifier is much easier with the pre-1978 specification because there are no formulas or tables to consult. You simply select a cartridge whose average output voltage *exceeds* the PHONO input sensitivity rating of the amplifier. For example, suppose that your receiver has a pre-1978 PHONO input sensitivity of 4 mV. You would not want to buy a magnetic cartridge whose average output voltage was 2 mV, because this voltage is insufficient to cause the receiver to deliver its full rated power—even though the VOLUME control is full "up." You should choose a cartridge whose average output voltage is 4 mV or larger. The pre-1978 PHONO input sensitivities for most amplifiers range between 1 and 5 mV. In addition, the high-level input sensitivities fall between 50 mV and 300 mV, depending on the manufacturer.

In general, the prospective buyer can usually tell whether the PHONO input sensitivity rating is based on the new or the old standard. The new standard will usually be written as, for example, "0.3 mV (Re: 1 watt output)," where the phrase "Re: 1 watt output" means that the PHONO input sensitivity rating is made using a 1 watt output **re**ference power. If this phrase is not present the consumer can infer which standard is being used by examining the size of the voltage. If the PHONO input sensitivity is less than 1 mV the new standard is probably being used; values greater than 1 mV imply the old standard. When in doubt, ask the salesperson.

Amplifier Buying Guide

Input Sensitivity

1. **Post-1978 Specification**
   a. **PHONO Input**: Look for an input sensitivity between 0.1 mV and 0.5 mV (not critical). Match the average output voltage of a magnetic cartridge to the amplifier's input sensitivity according to Table 8.1 or the equivalent calculation.
   b. **TUNER, TAPE, or AUX Inputs**: Look for input sensitivities between 10 mV and 40 mV (not critical).
2. **Pre-1978 Specification**
   a. **PHONO Input**: Look for an input sensitivity between 1 mV and 6 mV (not critical). Be sure the average output voltage of a magnetic cartridge is greater than the amplifier's input sensitivity.
   b. **TUNER, TAPE, or AUX Inputs**: Look for input sensitivities between 50 mV and 300 mV (not critical).

Maximum Input Signal (Primary Specification)

A. No output clipping

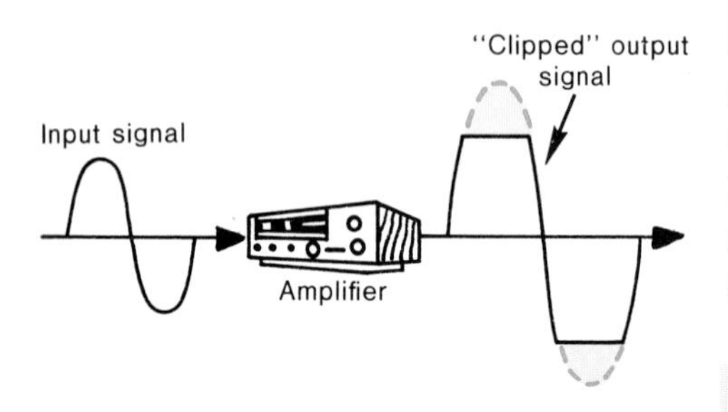

B. Amplifier "clips" the output signal

Figure 8.17. Amplifier clipping.

When an amplifier receives more input voltage than it can handle, it stops doing its job of amplifying. Figure 8.17(A) shows an amplifier properly amplifying each part of the input electrical wave. No difficulty arises because even the crests of the wave are not large enough to exceed the amplifying capacity of the amplifier. However, part (B) of the picture shows what happens when the input electrical wave is too large. The most extreme parts, the crests and troughs, are "clipped" off, as if by a pair of scissors; these parts of the wave are simply too large for the amplifier to handle. Excessive clipping severely distorts the music and results in a very unpleasant sound. The definition of maximum input signal is stated as follows:

Maximum Input Signal: The largest input voltage that an amplifier can accept without clipping the output signal (with the VOLUME control turned to a low level).

Not all amplifier inputs will tolerate large signals to the same degree. The high-level inputs (TUNER, TAPE, AUX) will normally accept input signals greater than 10 volts before clipping occurs, whereas the PHONO input is usually overloaded with signals which range from 130 mV to 300 mV.

To understand the importance of the maximum input signal specification, think about what happens when you are listening to a hi-fi system. Normally, you do not listen with the VOLUME control full-up. For one thing, most musical selections consist of passages of average sound levels, as well as peaks which may rise many times above the average level. An amplifier which is delivering its full power under average conditions has no reserve power in order to reproduce the large musical peaks. As a result, these transient peaks are "clipped" and serious distortion occurs. Of course, one solution is to turn down the VOLUME control so that less average power is being delivered. When a peak comes along the amplifier now has enough reserve power in order to reproduce it faithfully. The maximum input signal specification gives you some idea about the range of input voltages that an amplifier can handle without clipping the output signal when the VOLUME control is set low.

For example, suppose that a 3 mV phono input signal causes the amplifier to produce its full rated power, as illustrated in Figure 8.18(A). Furthermore, suppose that the amplifier has a maximum PHONO input signal specification of 160 mV, as indicated in part (B) of the illustration. Remember, 160 mV is the largest phono input signal that the amplifier can accept, with its VOLUME control turned to a low level, and still produce its full rated power without serious clipping. With this particular amplifier it would be wise to use a phono cartridge whose average output voltage

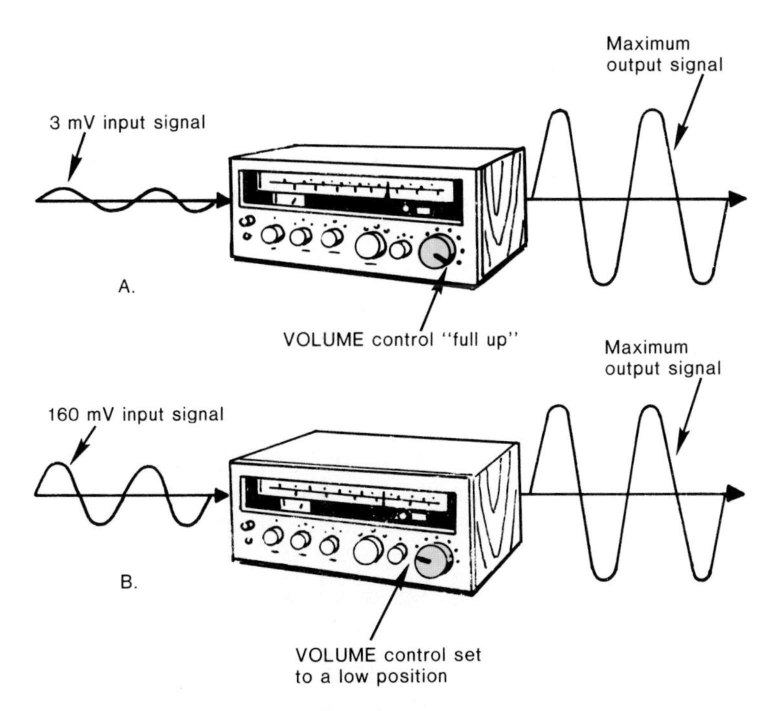

Figure 8.18. (A) A 3 mV input signal will cause the amplifier to produce its full rated power when the VOLUME control is turned full-up. (B) 160 mV produces the full rated power when the VOLUME control is turned to a low level, since 160 mV is the maximum input signal specification for the amplifier.

lies between 3 mV and 8 mV. Any cartridge whose average output voltage falls within this range is certainly capable of causing the amplifier to produce its full rated power. Furthermore, under listening conditions where the VOLUME control is set at a low level, a musical peak of 160 mV could safely be handled by the amplifier with no clipping problems. Therefore, a large maximum input signal specification ensures that clipping distortion is kept at a minimum under average listening conditions. For most medium priced amplifiers the maximum PHONO input signal specification lies between 130 mV and 300 mV, which is more than adequate to protect against undue clipping.

The maximum input signal for the high-level input is usually greater than 10 volts. In most cases, 10 volts is adequate protection against clipping at the high-level inputs.

Amplifier Buying Guide

Maximum Input Signal

1. **PHONO Input:** Look for a maximum input signal at least 20 times greater (the larger, the better) than the average output voltage of the magnetic cartridge. Most medium priced amplifiers have values between 130 mV and 300 mV.
2. **TUNER, TAPE, or AUX Inputs:** Look for a maximum input signal greater than 10 volts (not critical).

Intermodulation Distortion (Secondary Specification)

Intermodulation distortion, or IM as it is called, is another type of frequency distortion produced by all audio components. However, unlike THD the unwanted frequencies are not harmonically related to the input frequencies. As we have discussed in section 4.9, complex musical waves are comprised of many simple waves of different frequencies and amplitudes, all of which must simultaneously pass through the amplifier. When the amplifier reproduces the complex pattern, it generates harmonic distortion for each of the individual simple waves as discussed above. In addition, the amplifier produces IM distortion because of the simultaneous presence of two, or more, waves.

Figure 8.19 shows an example of two such waves passing through an amplifier. These frequencies have been arbitrarily chosen to be 800 Hz and 1,300 Hz. While the harmonics of 800 Hz and 1,300 Hz will be generated by the amplifier (total harmonic distortion), other unwanted frequencies will also appear in the output. Two of these waves will have

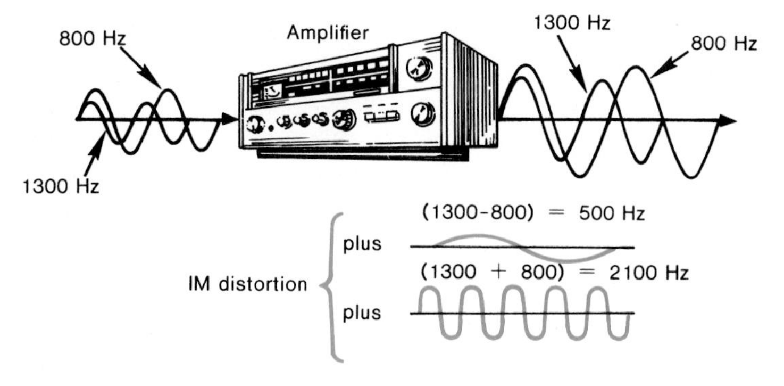

Figure 8.19. Intermodulation distortion occurs when two or more waves of different frequencies are simultaneously passing through an amplifier. The output contains the original frequencies plus the "sum" and "difference" frequencies created by the intermodulation distortion.

frequencies which are the sum and difference, respectively, of the two input frequencies. In Figure 8.19 these two waves have frequencies of (800 + 1,300) = 2,100 Hz and (1,300 − 800) = 500 Hz. These two unwanted waves are the result of intermodulation distortion, and it should be clear that their frequencies are not harmonics of the input waves. The additional frequencies would sound relatively dissonant in the presence of the undistorted music. If the extra unwanted part of the output in this example is 1% of the desired 800 Hz and 1,300 Hz output, the IM distortion would be specified as 1%. IM distortion is defined as follows:

Intermodulation Distortion: The undesired sum and difference frequencies which are produced when two frequencies pass through an amplifier at the same time. It is expressed as a percentage, relative to the strength of the original two frequencies.

Like THD, IM distortion increases substantially when the amplifier's rated power is exceeded. As is also the case with THD, *all* audio components produce IM. It is important that manufacturers keep the level of the IM waves as small as possible, since the combined intermodulation distortion in a complete audio system is the sum of the IM figures for each component.

Amplifier Buying Guide

Intermodulation Distortion: Look for IM values less than 0.5% at the full rated power of the amplifier.

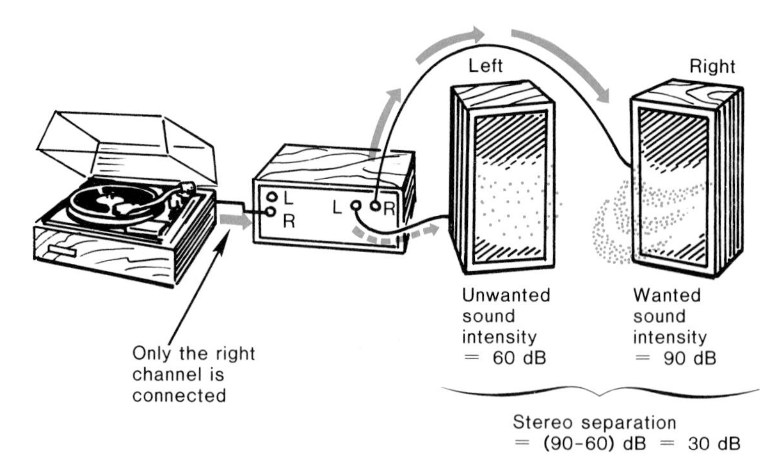

Figure 8.20. The stereo separation of an amplifier is the difference between the sound level emanating from the right speaker and that which has leaked into the left channel.

Stereo Separation (Secondary Specification)

Stereo separation can best be illustrated by considering the following demonstration which you can perform at home (Fig. 8.20). Disconnect the *left phono* lead of your stereo system. You would then expect the signal to travel through the right amplifier and finally emerge from the right speaker. *No* sound should emanate from the left speaker. However, your expectations would only be partially fulfilled, because some of the right signal will "leak over" into the left channel of the amplifier and produce sound from the left speaker. This leakage into the left speaker is a form of noise in the sense that it is unwanted. Suppose, as in Figure 8.20, that you turned up the VOLUME control such that the sound level from the right channel is 90 dB. And, furthermore, suppose that the amount of sound which has leaked over into the left channel is 60 dB. The difference between these two numbers, $(90 - 60) = 30$ dB, is called the *stereo separation* of the amplifier. Notice that the stereo separation is defined exactly like the S/N ratio. Thus, the stereo separation represents the difference in decibels between the wanted signal (right speaker) and the unwanted signal (the right channel sound which has leaked over into the left channel). Like the S/N ratio, the stereo separation should be as large as possible. The definition of stereo separation reflects this similarity to the S/N ratio.

Stereo Separation: The ratio of the power of the desired signal in one channel to the power of this signal which has leaked over to the other channel. This ratio is expressed in decibels.

A large stereo separation guarantees that music which is meant to come through only one speaker will not be heard from the other one.

You may be wondering how the signal destined for one channel can ever find its way into the other channel, since a stereo amplifier really consists of two separate amplifiers within it. Well, the two channels are not quite separate because the two amplifiers share a common power supply (see section 7.13) which provides the electrical energy necessary to produce the amplified output signals. It is through the common power supply that signals can make their way into the alternate channel. A well-designed power supply does a reasonably good job of preventing the signal in one channel from partially transferring into the other, although the isolation is far from perfect and this leads to the situation shown in Figure 8.21. Power supplies are an expensive part of any amplifier, and cost prevents the obvious solution of providing separate power supplies for each stereo channel. Another cause of leakage occurs when the electrical components, such as switches and wires, are placed too close to each other. Close placement promotes cross-channel signal leakage and a well-designed amplifier allows for spacious component placement.

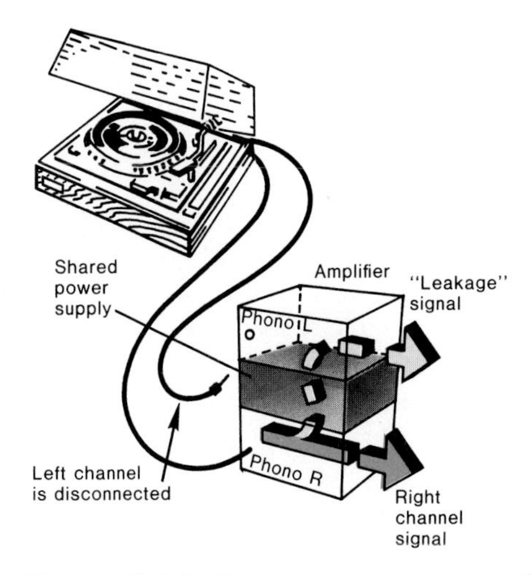

Figure 8.21. A common power supply for both stereo channels provides the main path for leakage between channels.

Amplifier Buying Guide

Stereo Separation: Look for a stereo separation of 30 dB or higher (35 dB, 40 dB, etc.).

Damping Factor (Secondary Specification)

The damping factor describes a very interesting feature of amplifiers which allows them to control precisely the movement of the speaker cone. For example, suppose that you are listening to a musical passage and a trumpet is playing a series of short, sharp staccato notes. As each staccato note is played through your hi-fi system a short burst of electrical power is delivered to the speakers, causing the cones to vibrate. Immediately after the note is finished the electrical power ceases, but the cones tend to keep vibrating. The fact that the speakers tend to produce sound after the note has stopped is called "overhang," and it leads to a dull, rather than a sharp, or crisp, reproduction of the original recording. This is analogous to a car with bad shock absorbers. Push down on the front fender and you will notice that the car continues to oscillate up and down long after you have stopped pushing. Shock absorbers, of course, are designed to damp out these oscillations and quickly return the

Bose Model 550 Receiver

car to a rest position. Likewise, all power amplifiers have built into them "electronic shock absorbers" which quickly bring the speaker cone to a halt when the signal has stopped. The electronic shock absorber substantially reduces the problem of sound overhang, and it results in a sharper reproduction. The technical specification which describes the damping ability of an amplifier is called the "damping factor."

Damping Factor: A number which describes an amplifier's ability to control the movement of the speaker's diaphragm.

Unfortunately, there is no easy way for us to interpret this number as was the case for the S/N, frequency response, stereo separation, etc., parameters. Values for the damping factor between 20 and 30 are usually considered satisfactory.

Amplifier Buying Guide

Damping Factor: Look for a damping factor of 20 (no units) or higher (25, 30, etc.).

Summary of Amplifier Specifications and Buying Guides

Total Harmonic Distortion (THD) (Primary Specification)

Look for a THD of 0.5% or less, at the full rated power of the amplifier.

Continuous Average Power Output (Primary Specification)

1. Be sure that this specification is stated accurately as required by the 1975 FTC ruling:

   "X" watts (continuous) per channel, into "Y" ohm speakers, both channels simultaneously driven from 20 Hz to 20 kHz at less than 0.5% THD at the rated power. (Note: The buyer must substitute the desired values for "X" and "Y.")

2. Match the power rating of the amplifier to the corresponding needs of the loudspeakers, as discussed in section 6.10.

Dynamic Headroom (Primary Specification)

There is no general agreement as to what constitutes "good" versus "bad" dynamic headroom. Most amplifiers have dynamic headrooms which fall between 0 dB and 3 dB.

Frequency Response (Primary Specification)

1. **TUNER, TAPE, AUX Inputs:** Look for a frequency response of at least 20 Hz→20,000 Hz, ±0.5 dB. The wider the frequency range, the better; the smaller the dB numbers, the better.
2. **PHONO Inputs:** Look for a frequency response of RIAA (±1.0 dB). The smaller the dB numbers, the better.

Signal-to-Noise Ratio (S/N) (Primary Specification)

1. **Post-1978 Specification**
   a. **PHONO Input:** Look for a S/N ratio (A-weighted) of 70 dB or higher (75 dB, 80 dB, etc.).
   b. **TUNER, TAPE or AUX Inputs:** Look for a S/N ratio (A-weighted) of 80 dB or higher.
2. **Pre-1978 Specification**
   a. **PHONO Input:** Look for a S/N ratio of 60 dB or higher.
   b. **TUNER, TAPE, or AUX Inputs:** Look for a S/N ratio of 70 dB or higher.

Input Sensitivity (Primary Specification)

1. **Post-1978 Specification**
   a. **PHONO input:** Look for an input sensitivity between 0.1 mV and 0.5 mV (not critical). Match the average

output voltage of a magnetic cartridge to the amplifier's input sensitivity according to Table 8.1 or the equivalent calculation.

b. **TUNER, TAPE, or AUX Inputs:** Look for input sensitivities between 10 mV and 40 mV (not critical).

2. **Pre-1978 Specification**

   a. **PHONO Input:** Look for an input sensitivity between 1 mV and 6 mV (not critical). Be sure the average output voltage of a magnetic cartridge is greater than the amplifier's input sensitivity.

   b. **TUNER, TAPE, or AUX Inputs:** Look for input sensitivities between 50 mV and 300 mV (not critical).

Maximum Input Signal (Primary Specification)

1. **PHONO Input:** Look for a maximum input signal at least 20 times greater (the larger, the better) than the average output voltage of the magnetic cartridge. Most medium priced amplifiers have values between 130 mV and 300 mV.
2. **TUNER, TAPE, or AUX Inputs:** Look for a maximum input signal greater than 10 volts (not critical).

Intermodulation Distortion (IM) (Secondary Specification)

Look for IM values less than 0.5% at the full rated power of the amplifier.

Stereo Separation (Secondary Specification)

Look for a stereo separation of 30 dB or higher (35 dB, 40 dB, etc.).

Damping Factor (Secondary Specification)

Look for a damping factor of 20 (no units) or higher (25, 30, etc.).

Summary of Terms

A-Weighting—A method commonly used when measuring noise levels which deemphasizes the low and high frequency noise relative to the midrange noise near 1 kHz. An A-weighted noise measurement attempts to account for the annoyance factor of the noise.

BALANCE Control—A front panel control which adjusts the output sound level from each channel.

BASS/TREBLE Controls—Continuously adjustable front panel controls which change the output sound level in selected frequency ranges. The BASS control can either boost or cut the low frequencies. The TREBLE control performs a similar function for the high frequencies.

Clipping—A phenomenon which occurs when the input signal to an amplifier is so large that the amplifier is not able to amplify properly the crests and troughs of the signal. As a result, the crests and troughs of the output electrical wave appear flattened, as if "clipped" off by a pair of scissors. Severe clipping introduces a large amount of distortion into the output wave.

Continuous Average Power Output—A specification which accurately describes how much power an amplifier can deliver on an uninterrupted basis. In order to have full meaning this specification must give the continuous watts of power per channel, the speaker impedance into which the rated power is delivered, the number of channels which are working simultaneously, the frequency range over which the rated power can be delivered, and the maximum % THD at the rated power.

Damping Factor—A measure of an amplifier's ability to damp unwanted vibrations of the speaker cone.

dBW—A unit for measuring the power of an amplifier, in terms of decibels, relative to a reference level of one watt.

Dynamic Headroom—The ratio of the power output which an amplifier can deliver for short periods of time to the continuous average power output, expressed in decibels.

Equalization—The process of electronically boosting or cutting the level of audio signals in certain frequency ranges relative to those in other frequency ranges.

Frequency Response—A specification which indicates how uniformly an audio component reproduces either signals or sound over a specified range of audio frequencies. In order to have full meaning, this specification must include both the frequency range and the ± decibel deviation from perfect flatness, e.g., 20 Hz → 20,000 Hz, +0.5 dB, −0.8 dB.

HIGH (SCRATCH) Filter—A high frequency filter which is used to reduce the noise from records.

High-Level Inputs—A phrase which is used to refer to the TUNER, TAPE, and AUX inputs of an amplifier, because they can accept higher level input voltages than does the PHONO input.

Input Sensitivity—The input voltage to an amplifier that is required to produce a power output of one watt when the VOLUME control is set to a maximum.

Intermodulation Distortion (IM)—A distortion of the audio signal that results when two different frequencies are simultaneously passed through an amplifier, or any hi-fi component. IM distortion causes two new frequencies to be added to the two original frequencies. The two new frequencies which are created are the sum and difference of the original frequencies. IM distortion is measured by dividing the combined voltage of the unwanted signals by the combined voltage of the original signals, and expressing the ratio as a percentage.

LOUDNESS Control—A control which is either on or off according to the setting of a front panel switch, and works in conjunction with the VOLUME control. The LOUDNESS control boosts the low frequencies (and sometimes the highs, too) at low listening levels in order to compensate for the ear's loss of hearing.

LOW (RUMBLE) Filter—A low frequency filter which usually starts to work below 100 Hz, and is used to reduce turntable rumble.

MAIN AMP IN Jack—An electrical connector on either a receiver, an integrated amplifier or a power amplifier, which can be used to connect an external preamplifier/control center or other accessories to the power amplifier. This jack may, or may not be present—depends on the manufacturer.

Maximum Input Signal—The largest input voltage that an amplifier can accept, when its VOLUME control is set low, without producing a clipped output signal.

Negative Feedback—A technique for reducing amplifier distortion, which involves taking a small portion of the output signal, turning it around so that it is exactly out of phase with the input signal, and feeding it back into the input.

Power Rating—(See **Continuous Average Power Output**)

PREAMP OUT Jack—An electrical connector from which the preamp output signal may be extracted. This signal can then be used with another power amplifier or sent to other accessories like an equalizer. This jack may, or may not be present—depends on the manufacturer.

RIAA Equalization—The specific equalization process used for phonograph records. When a record is cut, the bass frequencies are deemphasized and the high frequencies are boosted relative to the midrange frequencies. On playback, the RIAA phono equalization preamplifier reverses these effects to produce a flat frequency response. The advantages of the RIAA equalization process are that it reduces record noise and allows about 20 minutes or more of audio to be recorded on each side of a record.

RIAA Phono Equalization Preamplifier—(See **RIAA Equalization**.)

RMS—An abbreviation which stands for "root mean square," a kind of mathematical process for calculating average AC voltage or average AC current. Sometimes the continuous average power output is referred to (incorrectly) as the RMS power.

RUMBLE (LOW) Filter—(See **LOW Filter**.)

SCRATCH Filter—(See **HIGH Filter**.)

Signal-to-Noise Ratio (S/N) of An Amplifier—The ratio of 1 watt of output signal power to the A-weighted noise power, expressed in terms of decibels.

Source Preamplifier—A name used in this text to denote a class of amplifiers which equalize and amplify the small source signals to a value of about 0.5 volt. Source preamplifiers are used in tuners, in tape decks, with microphones, and with phono cartridges.

Stereo Separation—The ratio of the power of the desired signal in one channel to the power of this signal which has leaked over to the other channel, expressed in decibels.

SUBSONIC Filter—A low frequency filter which usually starts to work below 20 Hz, and is used to reduce the effects of tone arm resonance and rumble.

Total Harmonic Distortion (THD)—A distortion of the audio signal which results when a single frequency is passing through an amplifier, or any hi-fi component. THD causes new frequencies to be added to the original frequency. The new frequencies are the higher harmonics or overtones of the original frequency. THD is measured as a percentage relative to the strength of the original frequency.

TREBLE Control—(See **BASS/TREBLE Controls**.)

VOLUME Control—A control which is used to regulate the output sound level of a hi-fi system.

Review Questions

1. What are source preamplifiers, and why are they necessary? Name four source preamplifiers.
2. Explain the functions of the HIGH filter, LOW filter, BASS control, and TREBLE control. What is the difference between filters and bass/treble controls?
3. What is the difference between the VOLUME control and the LOUDNESS control?
4. Explain how negative feedback works in an amplifier. Why can negative feedback reduce the amount of distortion produced by the amplifier?
5. Describe, in terms of input and output frequencies, the difference between THD and IM distortion.
6. What conditions must be stated in order to specify correctly the continuous average power output rating of an amplifier?
7. What does the dynamic headroom of an amplifier represent? How is the dynamic headroom related to the continuous average power output rating of the amplifier?
8. Describe how the frequency response specification of an amplifier should be stated. Explain the meaning of each term in the specification.
9. Explain why the S/N ratio is an important amplifier specification. What is A-weighted noise?
10. What does the input sensitivity rating of an amplifier represent? How is it related to the VOLUME control setting and the power produced by the amplifier?
11. How does the maximum input signal specification differ from the input sensitivity specification?
12. What is the meaning of the stereo separation specification of an amplifier?

Exercises

NOTE: The following questions have up to 5 possible answers. Please select the **one** response which best answers the question.

1. What control allows an amplifier to compensate for the ear's loss of low frequency sensitivity when the VOLUME control is turned down?
   1. The LOW filter.
   2. The HIGH filter.
   3. The BALANCE control.
   4. The LOUDNESS control.
   5. The TREBLE control.
2. Which one of the following amplifier specifications is generally related to the quality of an amplifier?
   1. Its frequency response.
   2. Its THD value.
   3. Its S/N value.
   4. Its IM value.
   5. All of the above specifications are related to an amplifier's quality.

3. Which one of the following controls is *not* found in the preamplifier/control center?
   1. BASS.
   2. TREBLE.
   3. BALANCE.
   4. LOUDNESS.
   5. All of the above controls are found in the preamplifier/control center.
4. Which one of the following is *not* a valid amplifier specification?
   1. Frequency response.
   2. (S/N) ratio.
   3. THD.
   4. Damping factor.
   5. Dispersion.
5. In order to reduce high frequency noise it is often better to use the SCRATCH filter rather than turning down the TREBLE control. The reason is that:
   1. The TREBLE control adds additional high frequency noise to the music, while the SCRATCH filter does not.
   2. The SCRATCH filter causes less of the high frequency music to be lost while still removing much of the noise.
   3. The TREBLE control cannot reduce the high frequency noise.
   4. The SCRATCH filter also boosts the low frequency sound in order to compensate for any loss of the high frequency sound.
   5. The TREBLE control is not located in the preamplifier/control center while the SCRATCH filter is.
6. The STEREO/MONO switch on an amplifier:
   1. Converts a stereo program into a mono program.
   2. Converts a mono program into a stereo program.
   3. Turns off the sound to one of the speakers when set to the MONO position.
   4. Can only be used with FM broadcasts and not with records.
   5. Will not play mono programs when set to the STEREO position.
7. The following specification for an amplifier's continuous average power output rating is incomplete:

   75 (continuous) watts, with both channels simultaneously driven, from 20 Hz → 20,000 Hz.

   What is missing?
   1. The speaker's impedance is not stated.
   2. The specification should be stated for only one channel driven.
   3. The THD level is not stated.
   4. Both the speaker's impedance and the THD rating are not stated.
   5. The specification is complete as it stands.
8. All amplifiers possess the property of creating unwanted output signals whose frequencies are the sum and difference frequencies of the input signals. This undesirable trait is measured by the amplifier's:
   1. S/N ratio.
   2. frequency response.
   3. THD rating.
   4. IM rating.
   5. stereo separation rating.
9. Suppose that an amplifier has a continuous average power output rating of 100 watts, and a dynamic headroom rating of 3 dB. If needed, how much power can the amplifier deliver on a short term basis? (You may wish to consult Table 5.1.)
   1. 100 watts.
   2. 3 watts.
   3. (100+3) = 103 watts.
   4. 100×3 = 300 watts.
   5. 200 watts.
10. An amplifier has a continuous average power output rating of 80 watts. For a short period of time, it can deliver 127 watts without exceeding its rated distortion. Using Table 5.1, what is the dynamic headroom rating of this amplifier?
    1. 1 dB.
    2. 2 dB.
    3. 3 dB.
    4. 1.6 dB.
    5. 5 dB.
11. Suppose that three frequencies, 1 kHz, 3 kHz, and 8 kHz, are simultaneously sent through an amplifier. IM distortion produces the following *unwanted* frequencies:
    1. 1 kHz, 3 kHz, 8 kHz.
    2. all the higher harmonics of 1 kHz, plus all the higher harmonics of 3 kHz, plus all the higher harmonics of 8 kHz.
    3. (3−1) kHz = 2 kHz,
(3+1) kHz = 4 kHz,
(8−1) kHz = 7 kHz,
(8+1) kHz = 9 kHz,
(8−3) kHz = 5 kHz,
(8+3) kHz = 11 kHz.
    4. (1+3+8) kHz = 12 kHz.
    5. (8−3−1) kHz = 4 kHz.

12. Three amplifiers have the following frequency response specifications:
    Amplifier A: 50 Hz → 15,000 Hz, ± 1.2 dB.
    Amplifier B: 20 Hz → 15,000 Hz, ± 1.0 dB.
    Amplifier C: 15 Hz → 22,000 Hz, ± 0.2 dB.
    In terms of their frequency responses, which one of the following statements is true?
    1. B's frequency response is better than A's, and B meets the standard for hi-fidelity performance.
    2. B's frequency response is better than C's frequency response.
    3. A's frequency response is better than C's frequency response.
    4. C's frequency response is better than either A's or B's frequency response, and C meets the standard for hi-fidelity performance.
    5. None of the amplifiers meets the standard for hi-fidelity performance.
13. Suppose that amplifier "A" has a S/N ratio of 80 dB and "B" has a S/N ratio of 50 dB. Then the noise generated by "A" appears to be about ______________________________ than the noise produced by "B".
    1. 30 times softer
    2. 30 times louder
    3. 8 times softer
    4. 8 times louder
    5. 4 times softer
14. Which one of the following answers represents both an accurate and a high quality frequency response specification for an amplifier?
    1. 10 Hz → 25,000 Hz.
    2. ±0.3 dB.
    3. 20,000 Hz.
    4. 40 → 14,000 Hz, ± 2 dB.
    5. 15 Hz → 22,000 Hz, ± 0.4 dB.
15. When an amplifier amplifies all audio frequencies by exactly the same amount it is said to have:
    1. a flat response.
    2. a ± 3 dB response.
    3. a ± 1 dB response.
    4. a small damping factor.
    5. a large S/N ratio.
16. Which one of the following is the *best* frequency response for an amplifier?
    1. 20 → 20,000 Hz ± 3 dB.
    2. 20 → 20,000 Hz ± 1 dB.
    3. 15 → 25,000 Hz ± 0.3 dB.
    4. 15 → 25,000 Hz ± 3 dB.
    5. 15 → 25,000 Hz ± 5 dB.
17. The amount of A-weighted noise power produced by an amplifier is measured to be 1×10^{-6} watts. What is the S/N ratio of this amplifier? (Hint. You may wish to consult Table 5.1.)
    1. 30 dB.
    2. 40 dB.
    3. 50 dB.
    4. 60 dB.
    5. 70 dB.
18. Amplifier A has a PHONO input S/N ratio of 65 dB. Amplifier B has a PHONO input S/N ratio of 80 dB. Which one of the following statements is true? (Use the post-1978 S/N specifications.)
    1. Neither A nor B meets the standard for hi-fidelity performance.
    2. Both meet the standard for hi-fidelity performance, but A is better than B.
    3. A meets the standard for hi-fidelity performance, but B does not.
    4. B meets the standard for hi-fidelity performance, but A does not.
    5. Both meet the standard for hi-fidelity performance, but B is better than A.
19. Amplifier A has a TAPE input S/N ratio of 85 dB. Amplifier B has a TAPE input S/N ratio of 80 dB. Which one of the following statements is true? (Use the post-1978 S/N specifications.)
    1. Neither A nor B meets the standard for hi-fidelity performance.
    2. Both meet the standard for hi-fidelity performance, but A is better than B.
    3. A meets the standard for hi-fidelity performance, but B does not.
    4. B meets the standard for hi-fidelity performance, but A does not.
    5. Both meet the standard for hi-fidelity performance, but B is better than A.
20. Amplifier A has a THD rating of 0.3%, and amplifier B has a THD rating of 1.4%. Which one of the following statements is true?
    1. Neither A nor B meets the standard for hi-fidelity performance.
    2. Both meet the standard for hi-fidelity performance, but A is better than B.
    3. A meets the standard for hi-fidelity performance, but B does not.
    4. B meets the standard for hi-fidelity performance, but A does not.
    5. Both meet the standard for hi-fidelity performance, but B is better than A.

21. Which one of the following specifications tells how an amplifier can control the natural tendency of a speaker to vibrate after the music has stopped?
    1. (S/N) ratio.
    2. Frequency response.
    3. Intermodulation distortion.
    4. Damping factor.
    5. Dynamic headroom.
22. A power rating of an amplifier (e.g., 50 watts) is meaningless by itself. The rating must also include:
    1. the frequency range over which the power is delivered.
    2. the total harmonic distortion produced at the power rating.
    3. the operating conditions, i.e., both channels driven or one channel driven.
    4. the speaker impedance into which the power is delivered.
    5. All of the above answers must be included.
23. A 900 Hz signal is being sent to an amplifier. The output signal contains the following frequencies: 900 Hz, 1,800 Hz, 2,700 Hz, etc. The amplifier is producing:
    1. intermodulation distortion.
    2. S/N distortion.
    3. total harmonic distortion.
    4. damping distortion.
    5. stereo separation distortion.
24. A 100 watt amplifier has a phono input sensitivity rating of 0.6 mV. What should be the minimum value of the cartridge's average output voltage so that it can cause the amplifier to produce its full 100 watts?
    1. 0.03 mV.
    2. 0.3 mV.
    3. 3.0 mV.
    4. 6.0 mV.
    5. 30.0 mV.
25. A 64 watt amplifier has a phono input sensitivity rating of 0.2 mV and a phono maximum input signal rating of 40 mV. Which one of the following cartridges should be used with this amplifier?
    1. One having an average output voltage of 0.2 mV.
    2. One having an average output voltage of 0.3 mV.
    3. One having an average output voltage of 2.0 mV.
    4. One having an average output voltage of 4.0 mV.
    5. One having an average output voltage of 10.0 mV.
26. What is a typical value for the average output voltage produced by a magnetic phono cartridge?
    1. About 0.01 mV.
    2. About 0.001 mV.
    3. About 2 mV.
    4. About 20 mV.
    5. About 200 mV.
27. An electronic device which produces large currents that can drive speakers is called:
    1. a tuner.
    2. a tape deck.
    3. a multiplexer.
    4. a preamplifier.
    5. None of the above answers is correct.
28. Which amplifier specification tells the voltage that is required at the PHONO input jacks in order to cause the amplifier to produce one watt of output power?
    1. Phono input impedance.
    2. Phono power rating.
    3. Phono maximum input signal.
    4. Phono S/N ratio.
    5. Phono input sensitivity.
29. Which specification gives the largest phono input voltage that the amplifier can accept without causing the output signal to be clipped?
    1. Phono maximum input signal.
    2. Phono frequency response.
    3. Phono S/N ratio.
    4. Phono damping factor.
    5. Phono input sensitivity.
30. An amplifier has a stereo separation of 38 dB. What is the level of the (unwanted) music coming through the alternate channel if the main channel has a sound level of 65 dB?
    1. 38/65 dB.
    2. 38 dB.
    3. 103 dB.
    4. 27 dB.
    5. 65/38 dB.
31. Which input on an amplifier contains the RIAA equalization amplifier?
    1. No input contains this amplifier.
    2. The PHONO input.
    3. The TUNER input.
    4. The AUX input.
    5. The TAPE IN input.

32. Suppose that you were listening to your hi-fi which was producing very loud music. However, you notice that the high frequencies are missing because the room decor absorbs them excessively. What could you do with the amplifier to remedy this situation?
    1. Turn on the MUTE switch.
    2. Adjust the BALANCE control.
    3. Turn on the LOUDNESS control.
    4. Turn on the HIGH filter.
    5. None of the above answers would rectify the situation.
33. 0.5% THD and IM distortion are inaudible to most people. Nonetheless, it is important to buy an amplifier with this distortion level, or less, because?
    1. It is cheaper.
    2. Low distortion in your amplifier means less distortion from your cartridge.
    3. Low distortion means the tone controls work better.
    4. Distortion in the sound you hear also comes from other parts of your hi-fi system and the effect is cumulative.
    5. Low distortion means you need less power to drive your speakers.
34. If the power output from an amplifier is increased beyond its rated value, the THD and IM will:
    1. increase.
    2. decrease.
    3. remain the same.
35. A 70 watt amplifier has a phono maximum input signal rating of 100 mV. Listed below are five cartridges with different average output voltages. Which cartridge should be used with the amplifier?
    1. Cartridge A: average output voltage = 5 mV.
    2. Cartridge B: average output voltage = 10 mV.
    3. Cartridge C: average output voltage = 20 mV.
    4. Cartridge D: average output voltage = 50 mV.
    5. Cartridge E: average output voltage = 100 mV.
36. An amplifier is delivering 40 watts of electrical power to the right speaker. Due to "leakage" inside the amplifier, 0.4 watts of electrical power, containing right channel music, is sent to the left speaker. What is the stereo separation of this amplifier?
    1. 40 dB.
    2. 30 dB.
    3. 20 dB.
    4. 10 dB.
    5. 0.4 dB.

Please answer the next 7 questions using the data given below.

	Amplifier A	Amplifier B
Continuous average power output	50 watts (continuous) per channel, into 8 ohm speakers, both channels simultaneously driven, from 20 Hz to 20 kHz at less than 0.3% THD at the rated power.	50 watts (continuous) per channel, both channels simultaneously driven, from 10 Hz to 30 kHz at less than 0.4% THD at the rated power.
Frequency response	10 Hz to 22 kHz ±0.1 dB	10 Hz to 35 kHz ±0.05 dB
S/N ratio (Post-1978 spec.)	PHONO: 70 dB AUX: 82 dB	PHONO: 65 dB AUX: 70 dB
Stereo separation	20 dB	35 dB
Damping factor	25	20
IM at rated power	0.3%	0.2%
THD at rated power	0.3%	0.4%

These are the answers for the next 7 questions:

1. A meets hi-fidelity standards, but B does not.
2. A does not meet hi-fidelity standards, but B does.
3. Both A and B meet hi-fidelity standards, although A is better.
4. Both A and B meet hi-fidelity standards, although A is worse.
5. Neither A nor B meets hi-fidelity standards.

37. In terms of the continuous average power output, (use one of the 5 preceding answers) __1__.
38. In terms of frequency response, __4__.
39. In terms of S/N ratio, __1__.
40. In terms of stereo separation, __2__.
41. In terms of damping factor, __3__.
42. In terms of IM, __4__.
43. In terms of THD, __3__.

Chapter 9

HEAT IN HI-FIDELITY

9.1 Introduction

Most people are aware of the important roles that sound, electricity, and magnetism play in the operation of hi-fi systems. Few people, however, realize that heat is also an important concept. Excessive heat affects the performance of many audio components, particularly that of power amplifiers, speakers, styli, and magnetic tapes. It was not too long ago when most amplifiers were of the tube type, and one could look inside the case, see the tubes glow red, and feel the heat being radiated from them. Today tubes are an almost extinct species, for they have been replaced by the familiar solid state transistors. While transistors produce far less heat than their tube predecessors, modern power amplifiers will generate reasonable amounts of heat which can often be felt, especially on the more powerful units, by simply placing your hand near the power transistors. In general, too much heat can seriously degrade the performance of even the best amplifier systems, and it is a major cause of transistor failure. At $10 to $20 each, plus labor, the failure of large power transistors can be very expensive. Obviously, electronic component manufacturers take considerable pains to ensure that their products are adequately cooled. Any electronic component can be damaged by excessive heating if the device is not properly ventilated. Placing audio components within totally sealed cabinets can be a bad practice. Stacking books, paper, records, or other components on top of power amplifiers such that the ventilation holes are restricted can easily cause thermal failure.

A second major area for potential heat problems lies inside the speaker enclosures. As pointed out in section 7.9, (see Fig. 7.28) heat is developed in the speaker voice coil due to the interaction between the electric current and the resistance of the voice coil. In addition to this "electrical friction," heat is also generated by "mechanical friction," because the moving diaphragm and voice coil are always rubbing against other parts, be they parts of the speaker or simply the air alone. In any case, it is not unusual for the combined action

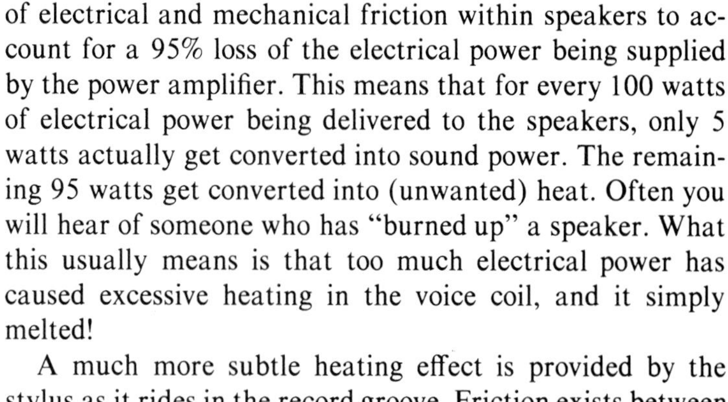

of electrical and mechanical friction within speakers to account for a 95% loss of the electrical power being supplied by the power amplifier. This means that for every 100 watts of electrical power being delivered to the speakers, only 5 watts actually get converted into sound power. The remaining 95 watts get converted into (unwanted) heat. Often you will hear of someone who has "burned up" a speaker. What this usually means is that too much electrical power has caused excessive heating in the voice coil, and it simply melted!

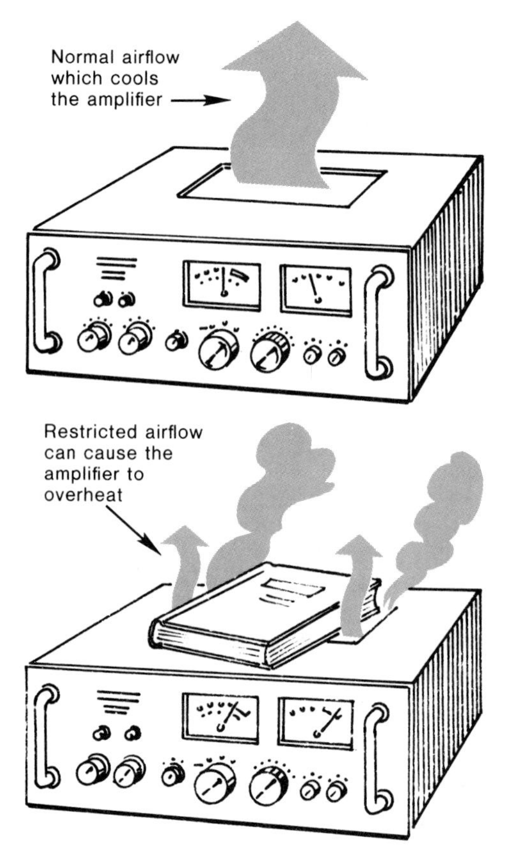

A much more subtle heating effect is provided by the stylus as it rides in the record groove. Friction exists between the tip of the stylus and the moving vinyl, and this leads to the generation of heat. It may not seem likely that the very light tracking forces used with most cartridges, 1 to 3 grams, can lead to significant amounts of friction. However, the stylus tip is exceedingly small, and very small tracking forces can cause the stylus to exert enormous pressures on the groove. It is like sticking yourself with a sharp-pointed needle; it does not require much force in order to generate enough pressure to puncture your skin. It has been estimated that the local temperature of the stylus may be as high as 2,000° F! The friction-generated heat is largely responsible for both the record and the stylus wear. The remaining sections of this chapter will show some of the physical concepts involved with heat and heat transfer.

9.2 Temperature

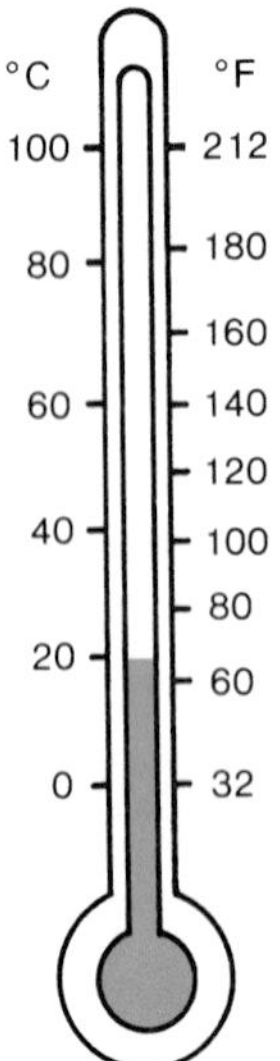

Figure 9.1. The glass thermometer is the most common kind of thermometer. The liquid in the bulb is usually methanol or mercury.

Temperature is a concept which is familiar to everyone; it is a measure of how hot or how cold an object is. Temperature is measured with a thermometer, and the most common type is the glass thermometer illustrated in Figure 9.1. In such a device the liquid (mercury or methanol, for example) is contained in a small storage bulb. When the thermometer is heated, the liquid expands into a narrow capillary tube. Since the amount of expansion is directly proportional to the change in temperature, the temperature can be read from a scale marked on the tube. Typically, either the Fahrenheit (°F) or the Celsius (°C) scale is used to measure the temperature, and the conversion from °F to °C, and vice versa, can be made by using Equations 9.1 and 9.2;

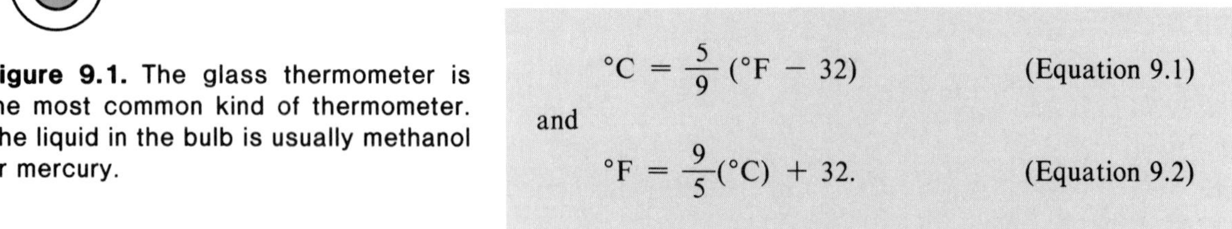

$$^{\circ}\mathrm{C} = \frac{5}{9}(^{\circ}\mathrm{F} - 32) \qquad \text{(Equation 9.1)}$$

and

$$^{\circ}\mathrm{F} = \frac{9}{5}(^{\circ}\mathrm{C}) + 32. \qquad \text{(Equation 9.2)}$$

While the glass thermometer is the most common type, thermometers can be built by using any substance which has some property which changes when the temperature changes. For example, a thermometer often found in power amplifiers uses a heat sensitive resistor, sometimes called a *thermistor*, as illustrated in Figure 9.2. The thermistor is actually a solid state resistor whose resistance changes as the temperature changes. This resistor is placed close to the power transistors, and it is part of a separate electric circuit which contains a small amount of current. If the transistor overheats, the resistance of the thermistor changes, and, in accordance with Ohm's law, the current in this "thermometer circuit" changes. When the current changes beyond a certain predetermined amount, either the amplifier is shut down automatically or its output current is reduced to safe levels. In this manner, expensive power transistors are protected from irreparable damage.

Equation 9.1 can be used to convert a room temperature of 68°F into the corresponding Celsius temperature of 20°C:

$$°C = \frac{5}{9}(68 - 32) = 20.$$

Likewise, by using Equation 9.2 it can be shown that 15°C corresponds to 59°F:

$$°F = \frac{9}{5}(15) + 32 = 59.$$

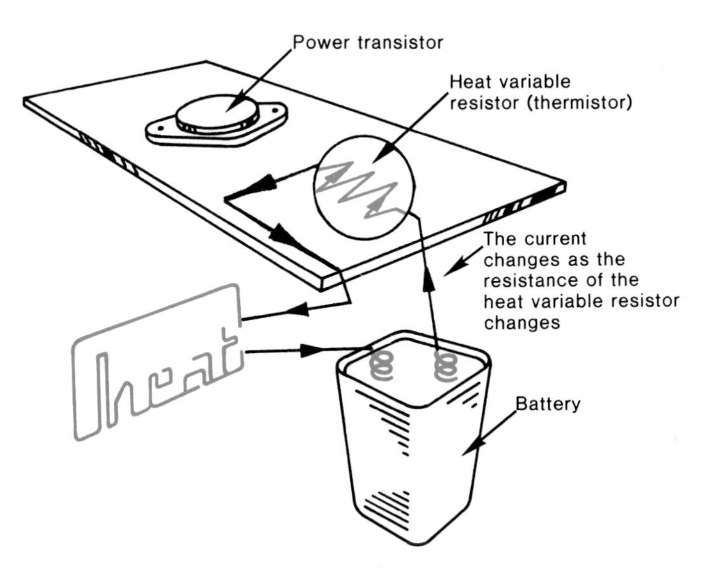

Figure 9.2. The heat sensitive resistor (thermistor) "thermometer" is used to sense overheating of the power transistors.

Thermistors are also used during the initial warm-up period when an amplifier is first turned on. As the amplifier heats up, its amplifying characteristics can change quite markedly. The thermistor is often used in "correction" circuits, which ensure that the amplifier maintains hi-fidelity performance even during the warm-up period.

Summary

A thermometer can be built by using any substance that has some property which changes with temperature. A glass thermometer utilizes the thermal expansion and contraction of a liquid column like mercury. The heat sensitive resistor discussed above depends on the change in electrical resistance caused by a change in its temperature.

Thermometers can be calibrated so that they measure the temperature in either °F or °C; Equations 9.1 and 9.2 give the proper conversion between the two.

9.3 Heat and the Conservation of Energy

Many different forms of energy exist which are rather familiar to most of us. There is heat energy, chemical energy (such as that stored in a battery or a gallon of gasoline) nuclear energy, electrical energy, acoustic energy (sound waves), electromagnetic energy (light and radio waves), and many other types. When an object is moving, such as a magnetic tape, or a rotating turntable, it is said to have mechanical energy .

The basic law which governs all of these energy forms is called the *law of conservation of energy*.

Law of Conservation of Energy: Energy can neither be created nor destroyed. It can, however, be converted from one form to another.

In hi-fi systems energy is continuously being converted from one form to another, and heat energy is always involved. Here are some examples:

1. Electricity is sent to a motor which is used to rotate a turntable platter. The motor also gets hot because of the electrical resistance inside the motor and friction.

ELECTRICAL ENERGY (from wall receptacle)	=	MECHANICAL ENERGY (rotating platter)	+	HEAT ENERGY (motor gets hot)

2. The record groove forces the stylus to vibrate which, in turn, generates the output electrical signal. The friction between the stylus and the groove also generates heat.

MECHANICAL ENERGY (rotating record)	=	MECHANICAL ENERGY (vibrating stylus)	+	HEAT ENERGY (friction)	+	ELECTRICAL ENERGY (output signal)

3. The power amplifier receives electrical energy from the 120 volt power cord and converts it into the output audio signal plus heat.

ELECTRICAL ENERGY (from wall receptacle)	=	ELECTRICAL ENERGY (output signal)	+	HEAT ENERGY (amplifier becomes hot)

4. The electrical output signal from the power amplifier is sent to the loudspeakers. The signals cause the speaker cone to vibrate which produces the sound.

ELECTRICAL ENERGY (from power amplifier)	=	MECHANICAL ENERGY (Vibrating diaphragm)	+	ACOUSTIC ENERGY (sound waves)	+	HEAT ENERGY (speaker gets hot)

Notice that all of these transformations result in heat energy being generated in addition to other types of energy. In many cases the efficiency of a system is rated according to the lost heat energy. The efficiency can be expressed via the following relation:

$$\text{Efficiency} = \frac{\text{Useful output energy}}{\text{Input energy}} \times 100\% \qquad \text{(Equation 9.3)}$$

The "useful output energy" is considered to be the energy form which is wanted by the user;* for a speaker the useful output energy would be the sound energy. The "input energy" is the amount of energy which is delivered to the device. The input energy is always greater than the useful output energy because:

Input Energy $\longrightarrow$ Heat (unwanted) + Useful Output Energy (wanted).

As an example let us consider a "bookshelf" loudspeaker which has a typical efficiency of 1%. Then, for every 50 watts of input electrical power the useful output power (the sound) can be calculated using Equation 9.3:

$$1\% = \frac{\text{Useful output power}}{50 \text{ watts}} \times 100\%$$

or, useful output power = 0.5 watts! Of the 50 watts of input power only 0.5 watts is converted into sound with the remaining 49.5 watts being dissipated as heat. Electric motors, on the other hand, are very efficient; their efficiency rating is somewhere near 95%. This implies that 95 out of every 100 watts of input power gets converted into rotational motion of the motor with only a scant 5 watts being lost to heat.

Summary

There are many forms of energy used in hi-fi systems: electrical, mechanical, heat, and acoustical. The law of conservation of energy states that energy can neither be created nor destroyed; energy can only be transformed from one form

*Very often "power" is substituted for "energy" in Equation 9.3. Although power and energy are slightly different concepts (see section 7.10), they can be used interchangeably here.

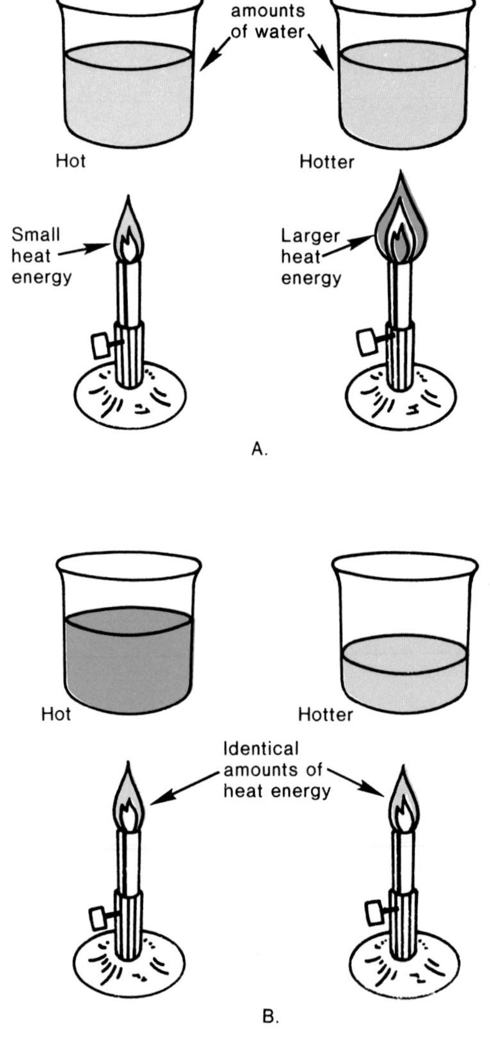

Figure 9.3. (A) Of the two identical objects, the one receiving more heat becomes hotter. (B) The less massive an object is, the hotter it becomes for a given amount of heat input. Thus, the water on the right, having less mass, becomes hotter than the water on the left.

to another. Energy is routinely transformed into other forms during the operation of hi-fi equipment. Efficiency is the useful output energy (or power) produced by a device, expressed as a percentage relative to the energy (or power) put into the device.

9.4 The Relationship between Heat Energy and Temperature Changes

Heat energy and temperature are related in a very common-sense way, because when an object is heated, its temperature rises. The exact amount of the temperature rise will depend on three conditions: how much heat energy is put in, the mass of the object, and the type of material from which the object is made.

To illustrate these conditions, consider Figure 9.3. Part (A) shows a small flame and a large flame heating identical beakers of water. Since the larger flame delivers more heat energy per second, the temperature of its beaker of water will rise faster. This occurs because the change in an object's temperature is directly proportional to the amount of heat energy added to it. In part (B) of the figure the situation is somewhat different; both flames produce identical amounts of heat energy, although the left beaker contains more mass (i.e., more water). Being more massive, the left beaker does not experience as rapid a rise in temperature as does the right beaker. In fact, the heat energy required to raise the temperature of an object by a fixed amount is directly proportional to the mass of the object. Finally, if the water in Figure 9.3 were replaced by another substance, e.g., alcohol, the temperature rise would be different than that experienced by the water.

The information discussed above can be summarized by the following equation:

$$H = sm\Delta t. \qquad \text{(Equation 9.4)}$$

In this equation H is the amount of heat energy, which is measured in a unit called the kilocalorie* (kcal), m is the mass of the object, and the symbol "Δt" is used to indicate *a change* in the value of the temperature. "s" is called the specific heat of the object, and its value depends on the type

*Although heat energy is measured in units of kilocalories, electrical energy—as discussed in section 7.5—is measured in units of joules. The conversion factor which allows one to convert between kcal and joules is 1 kcal = 4,184 joules.

of material being heated. Setting m = 1 kg and Δt = 1 °C in Equation 9.4, we see that the specific heat is just the amount of heat energy (in kilocalories) needed to raise the temperature of a 1 kg piece of material by 1 °C. Table 9.1 gives the specific heats of some common materials.

The power transistors in a power amplifier are typically mounted on massive aluminum plates, so that the heat generated by the transistor is absorbed by the large mass of metal and not by the small transistor itself. However, Table 9.1 indicates that metals generally have low specific heats. To compensate for a small value of "s" in Equation 9.4 audio engineers simply use a large mass "m." This increased mass is one reason why power amplifiers are heavy.

Material	Specific Heat (kcal/kg°C)
Hydrogen Gas	3.4
Water	1.0
Ice	0.49
Aluminum	0.22
Glass	0.20
Copper	0.093
Mercury	0.033

Table 9.1. Specific heats of some common materials.

How much heat, for example, can a 0.3 kg aluminum plate absorb from power transistors without increasing its temperature by more than 1 °C (e.g., 20°C → 21 °C)?

$H = sm\Delta t$

$H = \left(0.22 \frac{\text{kcal}}{\text{kg °C}}\right) \times (0.3\ \text{kg}) \times (1\ °\text{C})$

$H = 0.066\ \text{kcal} = 66\ \text{calories}$

Summary

The amount of heat, H, which is necessary to raise the temperature of m kilograms of mass by an amount Δt is given by the equation

$$H = sm\Delta t.$$

"s" is called the specific heat of the material.

9.5 Heat Transfer

The excessive amount of heat energy which is generated by power amplifiers and loudspeakers must be removed because high temperatures can easily damage the components. In general, there are three methods by which heat energy can be transferred from one place to another, and all three are used in audio components: conduction, convection, and radiation.

Conduction is the method by which heat energy flows through a material. Figure 9.4 shows heat energy being conducted through a large aluminum plate to which a power transistor is attached. In general, power transistors control the rather large amounts of current being delivered to the loudspeakers. Because transistors inherently possess a certain amount of electrical resistance, they become quite hot when the large currents are passed through them. The excessive heat must be continually removed from the power transistors. Transistors "dump" their excess heat energy into the large metallic plate to which they are attached, whereby the heat is removed from the transistor by the process of conduction. For this reason, the metal plate is often called a "heat sink."

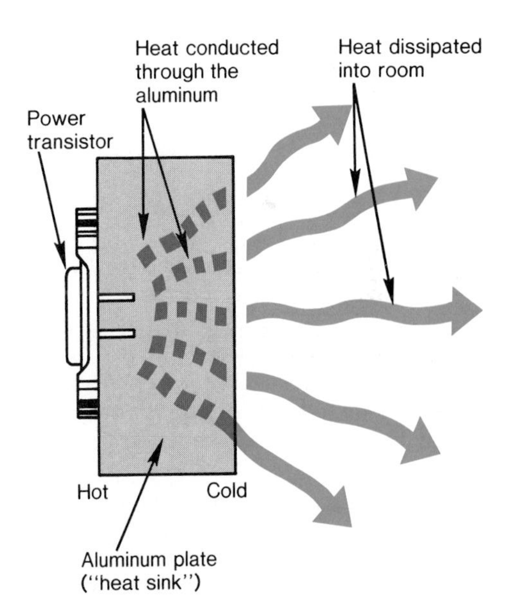

Figure 9.4. The output power transistors are mounted on an aluminum heat sink. The transistor dumps its excessive heat energy into the "sink." The heat then travels through the aluminum by conduction and is dissipated into the listening room.

In Figure 9.4 the side of the aluminum plate which is touching the transistor becomes hot, while the other side stays relatively cool. Because this temperature difference is

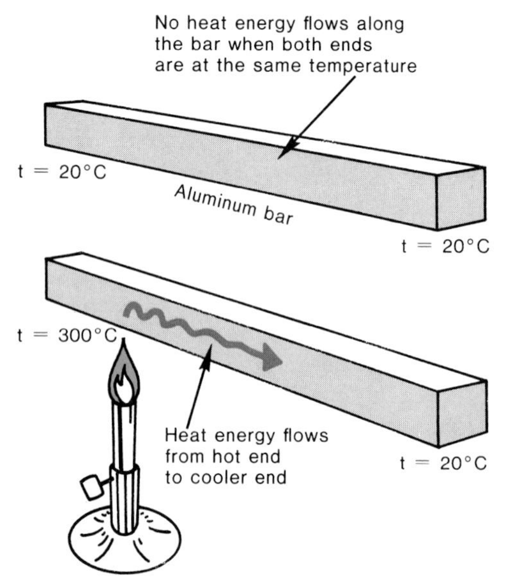

Figure 9.5. Heat energy flows only when there is a temperature differential between the ends of the bar.

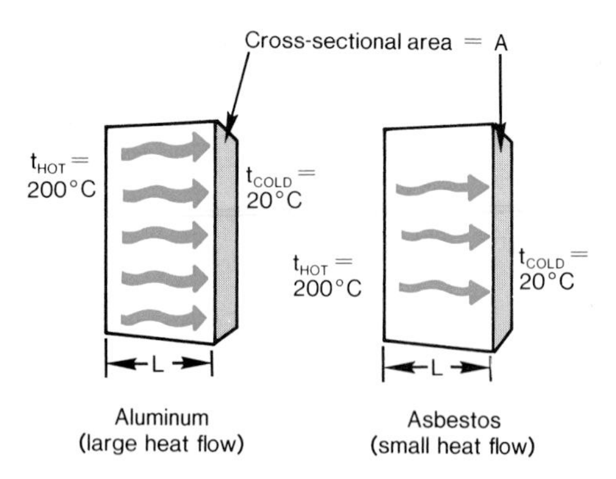

Figure 9.6. The amount of heat which flows through a material depends on the type of material, as well as the temperature difference between its two ends. Materials like aluminum conduct heat well and are said to have a large thermal conductivity, while those like asbestos conduct heat poorly and are said to have a small thermal conductivity.

Material	Thermal Conductivity (kcal/meter·sec·°C)
Copper	9.2×10^{-2}
Aluminum	4.9×10^{-2}
Ice	4×10^{-4}
Glass	2×10^{-4}
Asbestos	2×10^{-5}
Air	5.7×10^{-6}

Table 9.2. The thermal conductivities of some common materials.

important for conduction, it is reemphasized in Figure 9.5. This picture shows that whenever a difference in temperature is maintained between the two ends of a substance, the heat energy always flows from the higher temperature to the lower temperature. And, of course, all heat flow ceases when the two ends have the same temperature. Therefore, after the heat is removed from the power transistors and conducted to the other side of the aluminum plate, it must be dissipated into the room. If this dissipation did not take place, the transistor and both sides of the aluminum heat sink would eventually reach the same temperature, and heat conduction would stop. The result would be a burned out transistor!

The concepts of temperature and heat flow are very similar to that of voltage and current. Electrical current can flow only where there exists a difference in electrical voltage between the ends of a wire. Likewise, heat can flow only when there exists a difference in temperature between the ends of an object. In the case of electricity the amount of current which flows depends on how conductive the circuit is (i.e., the value of its resistance), as well as the size of the applied voltage. Similarly, the flow of heat energy depends on the "thermal conductivity" of the material as well as the temperature difference, as indicated in Figure 9.6. For example, most metals, such as aluminum or copper, have large values for thermal conductivity, because heat energy readily flows through them. Materials such as asbestos have small values for thermal conductivity, and they make good heat insulators.

The amount of heat energy per second which flows from the hot end (t_{hot}) of a material to the cold end (t_{cold}) is given by

$$\text{Heat flow per second} = \frac{KA(t_{hot} - t_{cold})}{L} \qquad \text{(Equation 9.5)}$$

where A is the cross-sectional area of the material, L is the length of the material between the two ends, and t refers to the temperature (See Figure 9.6.). "K" is a proportionality constant, and is called the *thermal conductivity* of the material. Some typical values of K are given in Table 9.2. It can be seen that metals have the highest conductivity while insulators, such as glass and asbestos, have relatively low values.

Convection is the method of heat transfer in which the heat energy is carried from place to place by the mass movement of molecules. For instance, when the heat from the power transistors arrives at the cooler end of the aluminum heat sink in Figure 9.4, the air in contact with the aluminum is warmed. As we all know, warm air rises, and it carries with it the heat energy from the sink. The process of heat removal by warm air flow is one example of *convection*, and the rising warm air can be felt by placing your hand over the ventilation slots of an amplifier. As the power of the amplifier increases, more warm air will be produced. Heat sinks are provided with "fins," as shown in Figure 9.7, in order to increase the surface area in contact with the air and allow a better transfer of heat to the surrounding air molecules. Convection is a common type of heating which is found in homes that have hot water radiators or baseboard heating. The hot water heats the radiator from the interior, and the outside air transfers the energy into the room by convection.

Radiation is the method of energy transfer in which the energy is carried by electromagnetic waves. Examples of electromagnetic waves are: light waves, radio waves, microwaves, infrared waves, and ultraviolet waves. Infrared waves are considered to be the "heat waves." Since electromagnetic energy travels at the speed of light, radiation is the fastest method of heat transfer. When an object becomes hot it emits electromagnetic waves of various frequencies, and if the material becomes hot enough, visible light is emitted. We see such an object as glowing "red hot." Even though the power transistors and their heat sink are not red hot, they are hot, and thus emit radiation. However, the emitted frequencies lie in the infrared region and are too low to be seen. The amount of radiation emitted by an object, surprisingly, depends on the color of its surface. A black surface tends to be one of the best radiators while a shiny silver surface is one of the poorest. That is why heat sinks are painted flat black, as illustrated in Figure 9.7.

Suppose that a transistor is mounted on an aluminum plate whose cross-sectional area is 0.03 m^2 and whose thickness, L, is 0.02 m. If the temperature of the hot side is 80 °C, and the temperature of the cooler side is 20 °C, calculate the rate of heat flow through the plate using Equation 9.5.

$$\text{Heat flow per second} = \frac{(4.9 \times 10^{-2})(.03)(80 - 20)}{(.02)},$$

Heat flow per second = 4.4 kcal/sec.

Figure 9.7. Heat energy which has been conducted to the edge of the heat sink is removed by air convection. Note the "fins," which improve convection, and the black color, which improves radiation.

Summary

Heat energy can be transferred from place to place by one of three methods: conduction, convection, and radiation. In conduction the heat energy flows through a material in a manner which is determined by the difference in temperature between its two surfaces, the cross-sectional area of the surface, the distance between the surfaces, and the thermal conductivity of the material. Convection is the process where the heat energy is carried by the mass movement of molecules, e.g., rising hot air. Radiation removes the heat energy by electromagnetic waves.

> *Water*
> Water is the most important exception to the rule that heating causes expansion, while cooling causes contraction. When cooled below 4°C water actually expands. This fact has profound implications for life on our planet.

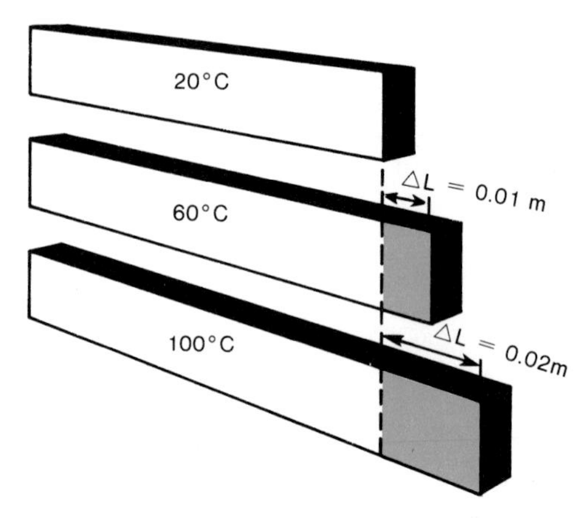

Figure 9.8. The change in length ΔL of a material is proportional to the change in temperature.

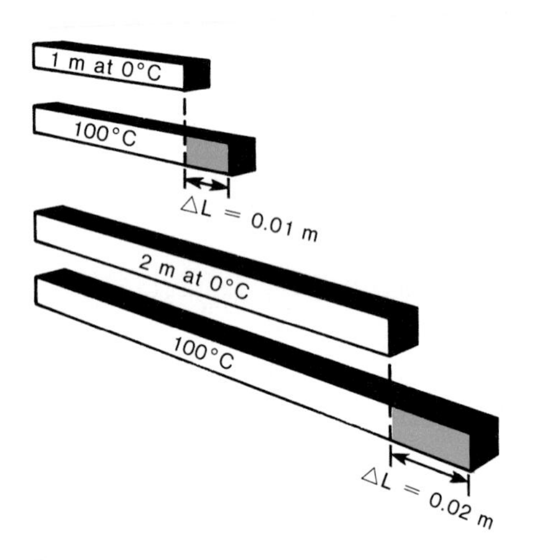

Figure 9.9. The change in length ΔL of a material is proportional to the original length of the material.

Material	α (per °C)
Aluminum	22×10^{-6}
Brass	18×10^{-6}
Steel	11×10^{-6}
Tungsten	4.4×10^{-6}
Pyrex Glass	3.2×10^{-6}

Table 9.3. Coefficients of thermal expansion of different materials.

9.6 Thermal Expansion

With a few notable exceptions, most materials expand when they are heated. The amount that an object expands depends on three factors: the change in its temperature, its original length, and the type of material from which the object is made. Figure 9.8 depicts the first factor involved in thermal expansion—namely, the more the temperature of a material increases, the more it expands. For example, increasing the temperature by 80°C (20°C → 100°C) will cause twice as much expansion as does a 40°C increment (20°C → 60°C). In other words, the change in length, ΔL, is proportional to the change in temperature, Δt. (Once again, the Greek letter "Δ" (delta) means "change.") Figure 9.9 shows the second feature associated with thermal expansion. The amount of expansion depends on the original length of the object. A 2 meter length of material will expand twice as much as a 1 meter length for the same amount of temperature change. Thus, ΔL is also proportional to the original length of material, L_0. Finally, the amount of thermal expansion depends on the type of material from which the object is made. A steel bar will expand by a different amount than, for example, an aluminum bar. Therefore, each material is assigned a number, called the *coefficient of thermal expansion,* which is a proportionality constant representing its ability to expand when heated.

The three factors just discussed can be incorporated into a single equation, as written below:

$$\Delta L = \alpha L_0 \Delta t, \qquad \text{(Equation 9.6)}$$

where α (Greek letter "alpha") is the coefficient of thermal expansion. When $L_0 = 1$ meter, and $\Delta t = 1$°C in Equation 9.6, we see that the coefficient of thermal expansion is just the number of meters by which a 1 meter length of material expands when its temperature is raised by 1°C. Table 9.3 gives some typical values of α, and it can be seen that different materials expand by different amounts.

To convince yourself that thermal expansion is not a trivial matter in the real world, consider a steel bridge which is 2 kilometers long. Suppose the temperature drops to 0°C (32°F) on the coldest winter day, while it reaches 40°C (104°F) on the hottest summer day. What is the maximum amount by which the bridge will expand as the seasons change from winter to summer? The answer is surprising, and it can be obtained easily from Equation 9.6 by using the value of α for steel from Table 9.3:

$$\Delta L = \alpha L_0 \Delta t$$
$$\Delta L = (11 \times 10^{-6} \text{ per } ^\circ C) \times (2{,}000 m) \times (40^\circ C - 0^\circ C)$$
$$\Delta L = 0.88 \text{ meter}$$

The bridge expands almost a whole meter! This is a lot of expansion, and the design of the bridge better take it into account. When driving across a long bridge, have you ever noticed that there are breaks in the roadbed every so often, as shown in Figure 9.10? These are placed in the roadbed to give the steel structural beams the necessary room in which to expand.

Figure 9.10. The expansion joints in the roadbed of a long bridge come in various shapes.

The Bimetallic Strip

In hi-fi there is a fascinating application of thermal expansion which utilizes a simple device called a *bimetallic strip*. A bimetallic strip, as the name implies, uses two different metals. These two metals are permanently fused together to form a sandwich-like strip, as shown in Figure 9.11. The metals are chosen to have different coefficients of thermal expansion. Therefore, when the bimetallic strip is heated, one half of the bimetallic strip will expand more than the other. Since the two halves are fused together, the strip will bend into an arc. The material on the outside of the arc will be that one which expands the most. A look at Table 9.3 will reveal that brass and steel would be possible choices for the two metals.

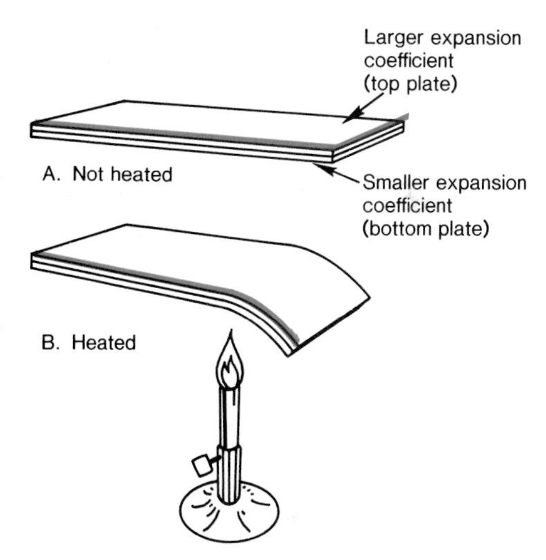

Figure 9.11. A bimetallic strip bends into an arc when heated, with the material having the greater expansion on the outside of the arc.

Many circuit breakers which are used in either power amplifiers or speakers employ bimetallic strips. A circuit breaker is a safety device which stops the current from flowing when it reaches levels that could damage the equipment. In this respect it is like a fuse, but there is a big difference. When a circuit breaker shuts off the current, you simply wait a bit, check to see that everything is ok, and then reset the circuit breaker by pushing a button.

Figure 9.12 shows how a circuit breaker works by using a coil of wire wrapped around a bimetallic strip. The breaker is series wired into the circuit that it is supposed to protect. Therefore, the same current which flows in the circuit enters one terminal of the breaker, passes through the coil of wire, as shown in part (A), and exits from the other terminal. In part (B) of the picture the result of an excessively large current is shown. Since the coil of wire possesses a predetermined amount of resistance, it will heat up in response to the large current and raise the temperature of the bimetallic

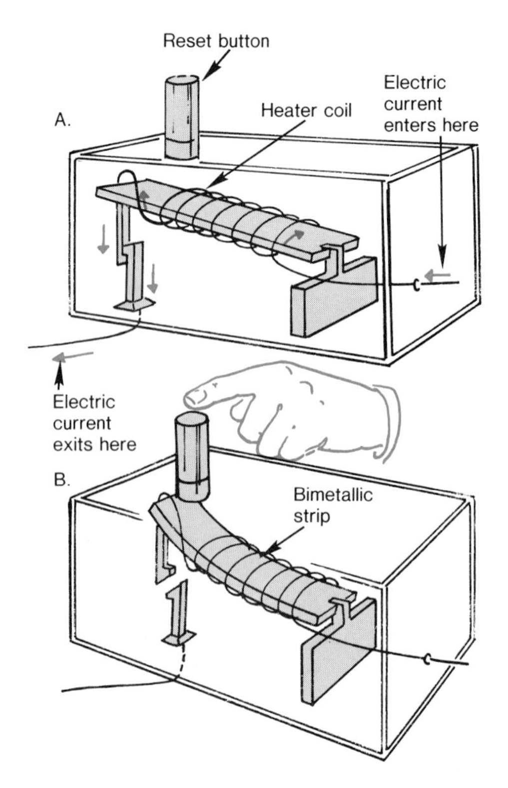

Figure 9.12. (A) A circuit breaker passing current which is below the danger level. (B) A circuit breaker which has opened or broken the circuit to which it is attached, thus stopping the current.

strip. When the bimetallic strip is hot enough, its tendency to bend will cause it to "pop" loose from the latch holding it in place. Thus, the circuit is broken, and the current stops flowing. After the bimetallic strip has cooled, the breaker can be reset by pushing the reset button. Pushing this button simply reconnects the bimetallic strip into the circuit. The actual current level at which the breaker "breaks" will depend on the metals in the bimetallic strip, the amount of heat generated by the coil of wire, and the force holding the latch in place. A neat gadget, isn't it!

Summary

The amount, ΔL, by which a material changes length when its temperature is raised is proportional to the change in temperature, Δt, the original length L_0, and the coefficient of linear expansion α. The amount of expansion is given by

$$\Delta L = \alpha L_0 \Delta t.$$

The bimetallic strip used in circuit breakers is an interesting example of how expansion effects may be used.

9.7 Safety Tips

All electrical components transform energy from one form to another, and heat energy is always one of the products of this transformation. The largest heat problems occur with the stylus, the power amplifier, and the speakers. The amplifier's normal operating temperature varies from model to model, and in some units which are rated over 100 watts per channel, the unit may normally run warm or hot. Here are a few safety tips which can help to prevent overheating in your hi-fi system:

1. Make sure that all audio components are placed such that air can easily flow around them.
2. Do not lay items like paper or album covers on top of components, especially a receiver or power amplifier.
3. Set the cartridge tracking force to the smallest amount that will allow it to track properly. Larger tracking forces produce excess wear on both the record and the stylus.
4. Avoid attaching too many parallel-connected speakers to the amplifier. This would cause the amplifier to produce excessive amounts of current and overheating could occur.
5. Properly protect both amplifiers and speakers from excessive current levels by using fuses or circuit breakers.

Summary of Terms

Acoustic Energy—The energy carried by a sound wave.

Bimetallic Strip—Two pieces of metal permanently bonded together to form a single strip, which bends in an arc when heated because the two metals are chosen to have different coefficients of thermal expansion.

Circuit Breaker—A device for protecting circuits against excessively large electric currents. Unlike a fuse, it may be reset and used repeatedly.

Coefficient of Thermal Expansion—The number of meters by which a one meter length of material will expand when its temperature is increased by one Celsius degree.

Conduction—A process by which heat energy flows through a material via the activity of its molecules, although the material as a whole does not move.

Conservation of Energy—A very important law which states that energy can neither be created nor destroyed; energy can only be transformed from one form to another.

Convection—A process where heat energy is carried by the mass motion of a medium like the rising of warm air. The molecules of the medium actually move over appreciable distances.

Electrical Energy—The energy which is given to electrical charges from either a battery or a generator.

Friction—The force which is exerted when two objects are either rubbing or moving against each other. The frictional force always opposes the motion and causes the objects to slow down. Heat energy is produced by friction.

Heat—(See **Heat Energy.**)

Heat Energy—A form of energy which causes the molecular motion of a substance to increase. A change in an object's heat energy content is detected by a change in its temperature. The amount of heat energy necessary to raise an object's temperature by 1°C depends on its mass and type of material.

Heat Sensitive Resistance Thermometer (Thermistor)—A thermometer constructed from a solid state "semiconductor" whose resistance changes as the temperature changes.

Heat Sink—A metal heat conductor upon which the power transistors are mounted. The heat sink conducts the heat energy away from the transistors.

Heat Transfer—The movement of heat energy from place to place. It can occur via conduction, convection and radiation.

Kilocalorie—A unit for measuring an amount of heat energy. It is the amount of heat needed to raise the temperature of one kilogram of water by one Celsius degree. One kilocalorie equals 4,184 joules.

Mechanical Energy—Energy which is associated with the motion of a body. For example, a rotating turntable platter and a vibrating stylus have mechanical energy.

Radiation—A process in which energy is transferred by an electromagnetic wave such as infrared or light waves.

Specific Heat—The number of kilocalories of heat needed to raise the temperature of one kilogram of a material by one Celsius degree.

Temperature—A measure of a substance's molecular activity caused by the presence of heat energy. The most common scales on which temperature is measured are the Centigrade, or Celsius, scale and the Fahrenheit scale.

Thermal Conductivity—The ability of a material to conduct heat. In other words, the kilocalories of heat flowing per second through a distance of one meter of material, whose cross-sectional area is one square meter, and between whose ends is maintained a temperature difference of one degree Celsius.

Thermistor—(See *Heat Sensitive Thermometer.*)

Review Questions

1. Name some hi-fi components in which excessive heat may be a problem.
2. Explain three methods by which excess heat can be removed from a transistor.
3. What relationship describes how much of the input electrical power to the speaker is converted into sound power?
4. What type of energy is always formed in any energy conversion process?
5. Explain what is meant by the conservation of energy.
6. What type of energy transformation does friction cause?
7. What does the temperature of an object indicate? What factors determine how much an object's temperature changes when it absorbs heat energy?
8. What type of temperature measuring device is used in power amplifiers?

9. List some safety tips which help to prevent serious overheating problems in a hi-fi system.
10. What factors determine how much heat energy per second flows from one end of a bar to the other end via conduction?
11. What factors determine how much a heated bar will expand?

Exercises

NOTE: The following questions have up to 5 possible answers. Please select the **one** response which best answers the questions.

1. What is the efficiency of a loudspeaker that receives 50 watts of electrical power and wastes 49.8 watts as heat?
   1. 50%.
   2. 49.8%.
   3. 1.0%.
   4. 0.4%.
   5. 0.2%.
2. The transistors of a power amplifier produce heat because:
   1. the transistors have resistance.
   2. the power supply does not produce enough current.
   3. the transistors have no resistance.
   4. the heat sink is connected to them.
   5. the heat flows from the heat sink into the transistors.
3. Heat transfer by conduction:
   1. is a process where heat flows through a solid.
   2. is the flow of heat towards the hotter end of a bar.
   3. does not occur in hi-fi.
   4. is the reason why heat sinks are painted black.
   5. occurs when a bar has the same temperature at both ends.
4. Radiation transfers heat energy by:
   1. electromagnetic waves.
   2. convection.
   3. conduction.
   4. air flow.
   5. heat flow through a heat sink.
5. Wear on a stylus is primarily due to the heat generated by:
   1. the record.
   2. the pressure on the record.
   3. the friction between the stylus and the record which can be increased by a greater tracking force.
   4. the stylus.
   5. the cartridge.
6. A change in an object's heat energy is detected by:
   1. its temperature.
   2. a change in its temperature.
   3. friction.
   4. a heat sink.
   5. the process of convection.
7. Devices which can be used for measuring the temperature are:
   1. glass thermometers.
   2. thermistors.
   3. heat variable resistors.
   4. All of the above answers are correct.
8. The law of conservation of energy:
   1. means that the total amount of energy in the universe is a constant; it is only transformed into other forms.
   2. means that you should save energy.
   3. is for the purpose of saving natural resources.
   4. means that heat energy cannot exist.
   5. means that one should avoid destroying energy if at all possible.
9. Convection is a process:
   1. where the air is uniformly warm.
   2. where the natural flow of air carries heat.
   3. where heat flows through a metal heat sink.
   4. in which the transistor becomes hot.
   5. where heat can travel in a vacuum.
10. Heat is the major cause of damage in:
    1. styli.
    2. power amplifiers.
    3. loudspeakers.
    4. fuses.
    5. All of the above answers are correct.
11. The temperature of a room is 77°F. What would this temperature measure on a Celsius scale?
    1. 25°C.
    2. 65°C.
    3. 45°C.
    4. 81°C.
    5. 196°C.
12. One surface of a heat sink on an amplifier has a temperature of 37°C. What would this temperature measure on a Fahrenheit scale?
    1. 67°F.
    2. 99°F.
    3. 88°F.
    4. 112°F.
    5. 123°F.

13. An inventor claims to have developed a loudspeaker which produces three times more sound energy per second than it receives in electrical energy. Is this possible?
    1. Yes, because three units of sound energy are equivalent to one unit of electrical energy.
    2. Yes, because two units of heat energy are added to each unit of electrical energy to produce three units of sound energy.
    3. Yes, because there is no law of nature which governs how many units of sound energy can be produced from one unit of electrical energy.
    4. No, because the loudspeaker would violate the law of conservation of energy by creating more sound energy than it received in electrical energy.
    5. No, because the loudspeaker would violate the law of conservation of energy by destroying more sound energy than it received in electrical energy.
14. Object A, which has a mass of 50 kg, is heated from 40°C to 70°C. Object B, which has a mass of 60 kg, is heated from 50°C to 90°C. Which object had the most amount of heat energy added to it?
    1. Object B, because its temperature rose higher than that of A.
    2. Object B, because it had a greater mass.
    3. Object A, because it was colder than B when the heat energy was first applied.
    4. Object B, because its change in temperature (90°C − 50°C = 40°C) was greater than A's change in temperature (70°C − 40°C = 30°C).
    5. There is not enough information to answer this question because the specific heat of each object is not given.
15. How much heat must be added to a 2 kg block of copper in order to raise its temperature by 25°C?
    1. 4.65 kcal.
    2. 50.0 kcal.
    3. 2.0 kcal.
    4. 4,650 kcal.
    5. 0.4 kcal.
16. A steel bridge is 2,000 meters long. When the bridge warms from 30°C to 40°C, how much does it expand?
    1. 0.022 meters.
    2. 0.880 meters.
    3. 22 meters.
    4. 11×10^{-6} meters.
    5. 0.22 meters.
17. A bimetallic strip bends when heated because:
    1. both metal strips expand by an identical amount.
    2. the two metal strips have different temperatures.
    3. the two metal strips have different initial lengths.
    4. the two metal strips have different masses.
    5. the two metal strips expand by different amounts.
18. If a phono cartridge has an excessive amount of tracking force applied to it, the large amount of friction between the stylus and the record will cause:
    1. clipping.
    2. the stylus to become hot.
    3. the amplifier to become hot.
19. How long will it take for a solid copper bar to transfer 10 kcal of heat from 100°C to 10°C, if it is 0.5 meters long and has a cross-sectional area of 10^{-4} m^2?
    1. 1 minute.
    2. 10 minutes.
    3. 101 minutes.
    4. 1,010 minutes.
    5. 10,100 minutes.

Chapter 10

TUNERS

10.1 Tuners and (Almost) Free Entertainment

The tuner is a neat little device which performs, with the help of the antenna, a seemingly impossible task of producing enjoyable sound from radio waves which can neither be seen nor heard. This aspect of a tuner vividly demonstrates a most remarkable ability of science—namely, that it can recognize and manipulate a phenomenon called radio waves which lie far beyond our natural senses. Tuners provide an inexhaustible supply of free entertainment ("free" that is, if you ignore an ingenious invention called advertising), and they come in two well-known types: the FM tuner (which stands for *F*requency *M*odulation) and the AM (*A*mplitude *M*odulation) tuner. The AM/FM tuners are often combined into one package which, when added to an integrated amplifier and speakers, leads to the familiar AM/FM radio.

To anyone who has listened to AM and FM broadcasts, there are some rather obvious (and some not-so-obvious) differences which uniquely favor FM for the transmission of hi-fidelity programming; FM enjoys freedom from electrical noise, a greater range of audio frequencies, and stereo transmission capabilities. Conversely, AM is much more susceptible to outside electrical interference, be it thunderstorms, auto ignitions, or hair dryers. As we shall see later, this difference is critically tied to the method by which AM and FM waves are transmitted and received—that is, to the technical merits of frequency versus amplitude modulation. Score one for FM! FM stations are allowed to transmit, by the FCC, a much wider range of musical frequencies than AM. FM transmits frequencies from 30 Hz to 15 kHz (the highest frequency cut into most records today), while the upper limit for AM is set at 5 kHz. Since the limits of human hearing extend from 20 Hz → 20 kHz, it is obvious that FM, while not perfect, has a vastly superior frequency range compared to AM. However, FM can be regarded as a source of "hi-fidelity" sound because its frequency range sufficiently approximates the limits of human hearing. Since only

FM is considered to be hi-fidelity, you will find that only FM tuners are reviewed in the new products sections of audio and hi-fi magazines; AM tuners are rarely mentioned. The very limited AM frequency range arises out of historical reasons. When amplitude modulation was first introduced, its primary function was to transmit voice communications and, consequently, it was limited to audio frequencies between 200 Hz and 5 kHz. This was done also to limit the high- and low-frequency noise which often accompanies AM transmission. AM is still better than FM for communications because it has a much longer range. AM radio waves are reflected from a layer in the upper atmosphere and bent back toward the earth to distant receivers. FM does not benefit from this range increasing phenomenon.

FM has also enjoyed great popularity because it can carry stereo programming, but stereo broadcasts were never developed for AM transmission. However, there is now considerable interest in securing a stereo capability for AM stations. When AM was first established, the concept of stereo was yet unborn so that the FCC allocated just enough "frequency space" for only a mono program. FM, on the other hand, gained its popularity during the early 60s when it was used primarily for music. FM was allowed adequate space for a station to broadcast stereo and/or quad. This added attraction, along with FM's low noise and wide audio frequency band, ensured quality music reproduction which could compete favorably with the experience of a live performance.

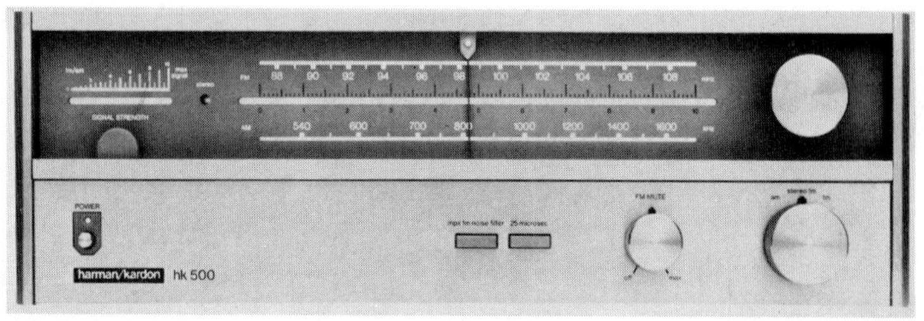

Model hk 500 Harman/Kardon tuner

There is yet another service which is provided by some FM stations; it is called the "SCA (Subsidiary Communications Authorization) subcarrier." Have you ever been in an airline terminal, dentist's office, or restaurant and listened to the very soothing background music which is virtually uninterrupted (i.e., no advertising!)? This is probably an SCA transmission. The music, while very pleasing, is not hi-fidelity quality because its audio bandwidth is about the same as that of an AM radio. This type of easy-listening music is transmitted right along with the station's normal programming, although your receiver cannot reproduce the SCA program without a special adaptor which can be rented from the FM station. However, if you want to own such a decoder for your own personal use, it is perfectly legal to build one and attach it to the tuner section of your receiver.*

Table 10.1 shows a comparison between AM and FM broadcasting.

Feature	AM	FM
Modulation technique	The amplitude of the radio wave changes	The frequency of the radio wave changes
Range	Up to several hundred miles	Usually less than 100 miles
Audio frequencies transmitted	200–5,000 Hz	30–15,000 Hz
Best use	Voice communications	Voice and music
Noise	Can be very noisy	Very little with good tuners
Type of programs transmitted	Mono (stereo in the future)	Mono, stereo, quad, SCA

Table 10.1 The main features of AM and FM broadcasting.

10.2 What Is Modulation? And, Why Is It So Essential for Conveying Information?

Since both AM and FM have "modulation" as part of their names, modulation appears to be a rather important concept, and perhaps it would be wise to look a little deeper into its meaning. It turns out, in fact, that modulation is THE method by which ALL information, be it sight, sound, touch, or smell, is conveyed to us. Before going on to radio waves, let us see how our senses respond to modulation in order to get a better understanding of its importance.

"Modulation," as we shall use it, means "change," and modulation conveys information to all of our senses by some type of change. For example, a straight line by itself conveys very little information other than the fact that it is a straight line. However, if one begins to *change* (or modulate) the line by adding wiggles and curves, then it can convey any information one wishes by simply changing the ordering of the wiggles. We call it handwriting, and it is illustrated in Figure

*See *Popular Electronics*, October 1974.

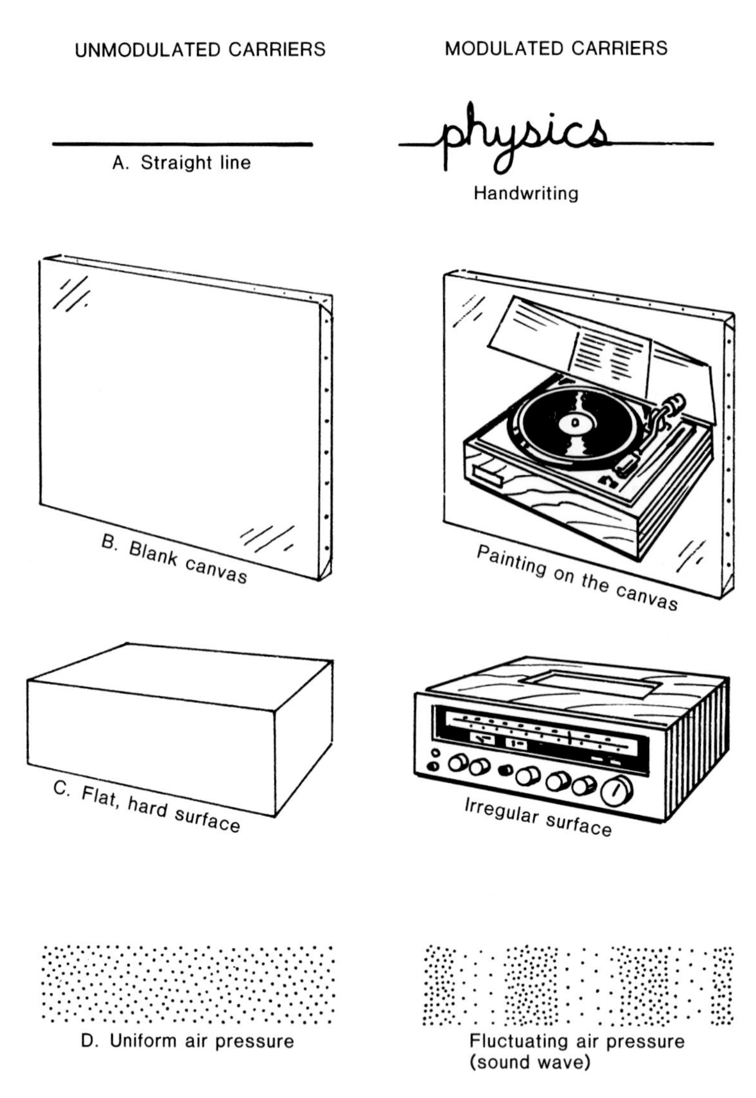

Figure 10.1. Unmodulated carriers convey no information. Modulated carriers carry information.

10.1 (A). In scientific jargon the straight line is called an "unmodulated carrier," and you will notice that it conveys no information. It is only when you wiggle (modulate) the carrier that exciting things begin to happen.

An *un*modulated carrier carries *no* information.
A modulated carrier carries information.

Visual perception can also be thought of as a type of modulation, and the painting in Figure 10.1 (B) illustrates this concept nicely. A white canvas is analogous to the straight line mentioned above because, by itself, it conveys no visual information; the canvas is an unmodulated carrier! However, a painter can now change the appearance of the canvas simply by adding dabs of different colors in the right places and, as if by magic, a picture results. The picture is conveyed because the colors modulate the white background.

The sense of touch also relies on modulation, as illustrated in part (C) of Figure 10.1. Try closing your eyes and running your hand over a hard, flat surface. Not very interesting. Now let it run over a "modulated" surface with its protruding controls and switches, and one immediately senses that the object might be a stereo receiver.

Sound is also a fine example of modulation. If you will recall from Chapter 3, sound is nothing more than a series of condensations and rarefactions that travel at the speed of sound. In this case it is the unperturbed air pressure of 14.7 psi that acts as the unmodulated carrier, as shown in Figure 10.1 (D). In a room where there is complete silence, the air pressure is everywhere the same. However, if someone is whistling a 600 Hz note, then the air pressure rises and falls at the rate of 600 times each second about the constant pressure of 14.7 psi. It is precisely this rising and falling that rhythmically forces your ear drum in and out, and leads to the sensation of sound. A *steady* (unmodulated) *pressure* produces no sound.

10.3 Electromagnetic Waves

It comes as no surprise that all of the program material transmitted by a radio station is carried to your home on a radio wave. This is true for both AM and FM transmissions. It does come as a surprise to most people that radio waves are practically identical to light waves. The only difference is that we cannot see radio waves because they possess a much lower frequency than light waves and therefore cannot be detected by the eye. In fact, both radio and light waves belong to a class of waves called "electromagnetic" waves. This class also includes among its members such well known types as X-rays, microwaves, and infrared waves. The only real difference among all of them is that their frequencies are different as shown in Figure 10.2. At the low end of this electromagnetic range, 10^4 to 10^8 Hz, are found the familiar AM and FM radio waves, while frequencies near 10^9 Hz represent the waves used for radar and microwave ovens. At the highest end, where frequencies exceed an astounding 10^{17} Hz, are found the X-rays and gamma rays. It is very interesting to note that the visible light waves occupy only a very tiny region, near 10^{15} Hz, of the entire electromagnetic range. The credit goes to modern science for developing instruments which can both generate and detect the waves which lie outside the visible region.

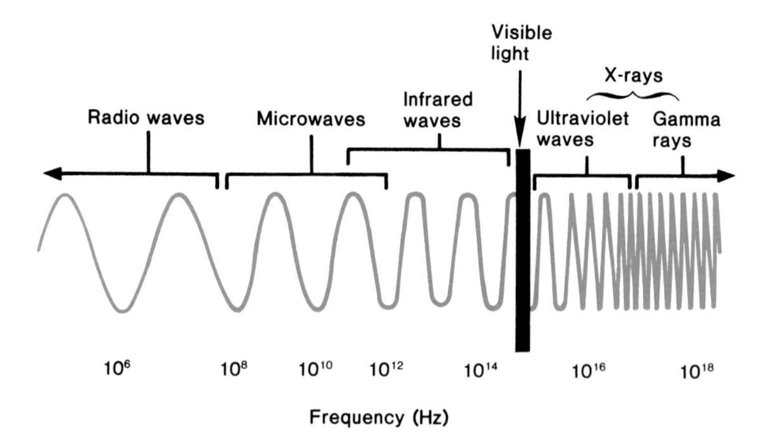

Figure 10.2. Electromagnetic waves range in frequency from below 10^4 Hz (radio waves) to 10^{18} Hz and above (gamma rays). Notice that visible light occupies only a small slice in the entire electromagnetic frequency range.

Even though vision is restricted to only a narrow portion of the electromagnetic frequency range, the eye can detect small frequency changes and it interprets them as changes in color. For example, the colors of red and blue are both caused by electromagnetic waves except that their respective frequencies are different. Red light has a frequency of approximately 4.7×10^{14} Hz while the frequency of blue light is somewhat higher at 6.0×10^{14} Hz.

The frequency range, designated as "radio waves" in Figure 10.2, has been divided into bands, each of which serves a defined communications need. Table 10.2 show the various groupings which begin at 30 kHz and extend to frequencies as high as 300,000 MHz. Of interest is the fact that the entire FM band, 88–108 MHz, falls between TV channels 6 and 7. Therefore, your TV antenna can also be used to pick up FM stations.

Electromagnetic waves possess many of the properties of other types of waves, such as water and sound waves, which we encountered in Chapter 3: amplitude, frequency, period, speed, and wavelength. Electromagnetic waves, unlike sound, do not require a medium as they travel well in material such as glass, or in the void of a complete vacuum. Light from the sun and radio communications with lunar astronauts are illustrative of how easily electromagnetic waves propagate through the vacuum of outer space. In a vacuum all electromagnetic waves, regardless of their frequency, travel at "the speed of light" which is an enormous 186,000 miles per second (or 3×10^8 meters per second)! At this rate radio waves could circumscribe the equator about 8 times every second, which explains why radio communications appear to be almost instantaneous. Using Figure 10.2 it is possible to calculate the wavelength of

Band	Allocation
30–535 kHz	Maritime communications and navigation, aeronautical radio navigation.
535–1,605 kHz	AM radio broadcast band.
1,605 kHz–30 MHz	Amateur radio, government radio, international shortwave broadcast, fixed and mobile communications.
30–50 MHz	Government and nongovernment, fixed and mobile. Includes police, fire, forestry, highway, and railroad services.
50–54 MHz	Amateur.
54–72 MHz	Television broadcast channels 2 to 4.
72–76 MHz	Government and nongovernment services. Aeronautical marker beacon on 75 MHz.
76–88 MHz	Television broadcast channels 5 and 6.
88–108 MHz	FM broadcast.
108–122 MHz	Aeronautical navigation. Localizers, radio range, and airport control.
122–174 MHz	Government and nongovernment, fixed and mobile, amateur broadcast.
174–216 MHz	Television broadcast channels 7 to 13.
216–470 MHz	Amateur, government and nongovernment, fixed and mobile, aeronautical navigation, citizens' radio.
470–890 MHz	Television broadcasting. UHF television broadcast channels 14 to 83.
890–3,000 MHz	Aeronautical radio navigation, amateur broadcast, studio-transmitter relay, government and nongovernment, fixed and mobile. Radar bands 1,300–1,600 MHz.
3,000–30,000 MHz	Government and nongovernment, fixed and mobile, amateur broadcast, radio navigation.
30,000–300,000 MHz	Experimental, government, amateur

Table 10.2. FCC frequency allocations from 30 kHz to 300,000 MHz.

electromagnetic waves from Equation 3.3. For example, the frequency of an AM radio wave is approximately 10^6 Hz, and its wavelength can be determined as follows:

$$(\text{Wave speed}) = (\text{Frequency}) \times (\text{Wavelength}) \qquad (\text{Equation 3.3})$$

$$(3 \times 10^8 \text{ meters/second}) = (10^6 \text{ Hz}) \times (\text{Wavelength})$$

or Wavelength = 300 meters.

Thus the wavelength of a typical AM radio wave is quite large and, in fact, its length is about 3 football fields! Table 10.3 shows the range of both frequencies and wavelengths associated with various electromagnetic waves. It is quite amazing to see the great differences in the wavelengths of the low frequencies when compared to those of the very high frequencies.

Type of Electromagnetic Wave	Approximate Frequency	Wavelength (calculated using Equation 3.3)	Comments
AM radio waves	10^6 Hz	300 meters	3 football fields long
FM radio waves	10^8 Hz	3 meters	About 10 feet long
Microwaves	10^{10} Hz	$\frac{3}{100}$ meters	About 1 inch long
Visible light	10^{15} Hz	$\frac{3}{10{,}000{,}000}$ meters	10 millionths of an inch!
X-rays	10^{17} Hz	$\frac{3}{1{,}000{,}000{,}000}$ meters	30 times the diameter of the hydrogen atom

Table 10.3 The frequencies and wavelengths of some common electromagnetic waves. The speed, in vacuum, of all these waves is 3×10^8 meters/sec.

Electromagnetic waves are transverse waves (see section 3.3), and they only interact with charged objects such as electrons and protons. When a radio wave impinges on a metal antenna, for example, the electrons are forced to move up and down in a direction which is perpendicular (transverse) to the direction in which the wave is traveling. This phenomenon was shown in Figure 3.10, and it is exactly analogous to the cork bobbing up and down under the influence of a water wave. Therefore, radio waves have the very nice property of being able to force electrons into motion as the waves pass. Remember that the protons within a solid are held immobile, and only the free electrons are capable of moving through a conductor. Of course any oscillatory motion by the electrons constitutes an AC current which is sent into the tuner via the antenna leads. It is the job of the tuner to extract the musical information carried by the incoming high frequency current.

10.4 Amplitude Modulation (AM)

AM radio stations have been around a lot longer than FM stations, but AM has some serious drawbacks in terms of both hi-fidelity and stereo broadcasting. A discussion of amplitude modulation will be worthwhile because it is elegantly simple and has widespread use.

Each station broadcasting AM is assigned a broadcast frequency by the FCC. This frequency is the frequency of the electromagnetic carrier wave upon which the programming material is carried over the airways. Therefore, it is sometimes called the "carrier frequency." AM band stations are assigned frequencies between 535 and 1,605 kHz on the dial, and they are spaced every 10 kHz apart. Starting at

540 kHz, AM stations are assigned carrier frequencies of 540, 550, 560, etc.—up to 1,600 kHz. A little algebra shows that 107 different frequency assignments are available, so that there can be at most 107 AM stations operating in any geographic locality without mutual interference. Of course stations located sufficiently far apart can be, and are, assigned the same frequency. The AM transmission range is much greater at night, and in order to avoid conjested airways the number of stations must be decreased. For this reason many AM station must go off the air from dusk to dawn.

How Does Amplitude Modulation Work?

In order to explain how a radio station can transmit its broadcasts via AM radio waves, let us consider an example of a station which is operating on an assigned frequency of 1,200 kHz. Figure 10.3 shows the case where a DJ at the station is speaking into a microphone, and it also shows the appearance of the AM wave which is being transmitted. The part of the wave labeled "unmodulated radio wave" is produced when there is a momentary pause in the DJ's voice, and no audio information is present. In this case the station emits a 1,200 kHz radio wave *whose amplitude is constant.* When the DJ begins to speak, Figure 10.3 shows that the radio wave becomes a complex pattern, which is called the "AM modulated radio wave" because it is the amplitude of the radio wave which is changing.

When the AM wave is intercepted by the antenna, a radio frequency current is produced which enters the tuner. The electronic circuits of the AM tuner are constructed to detect only the *amplitude variations* of the incoming electric signal.

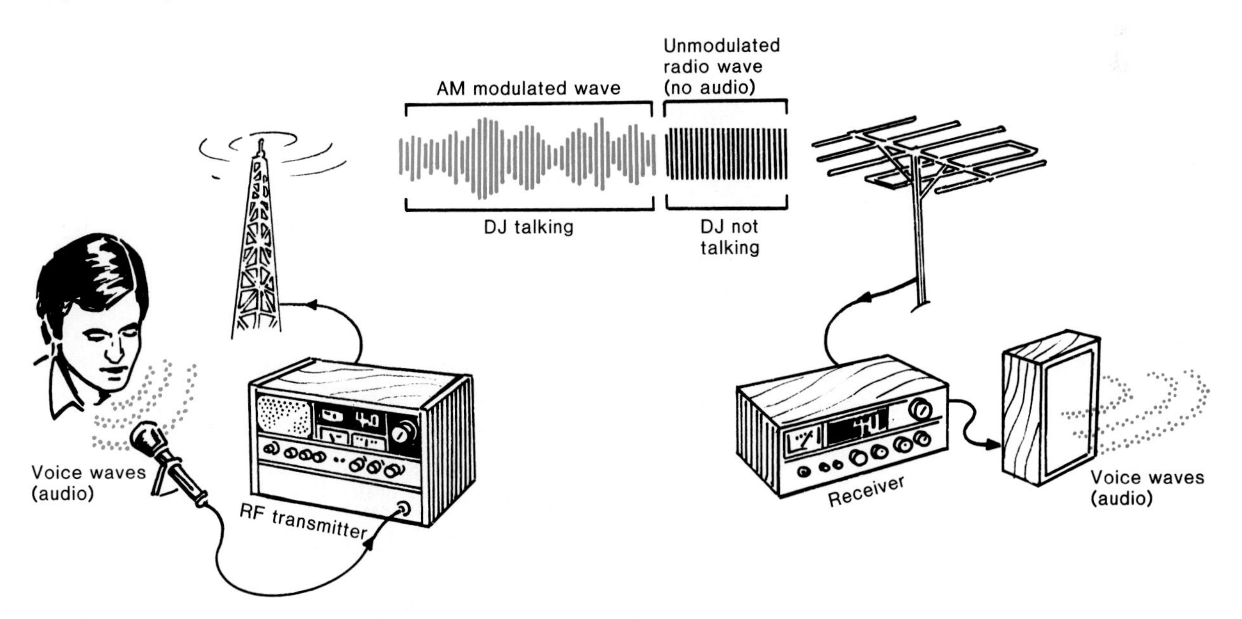

Figure 10.3. A picture of AM radio transmission when the audio information is both present and absent.

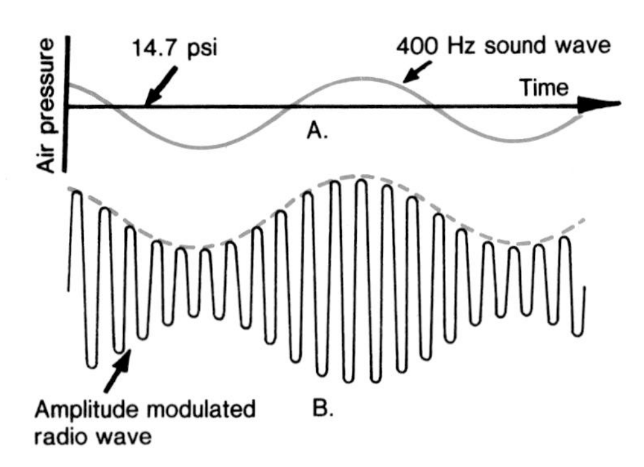

Figure 10.4. The amplitude variations of an AM radio wave in part (B) are an exact replica of the pressure variations in the original sound wave in part (A).

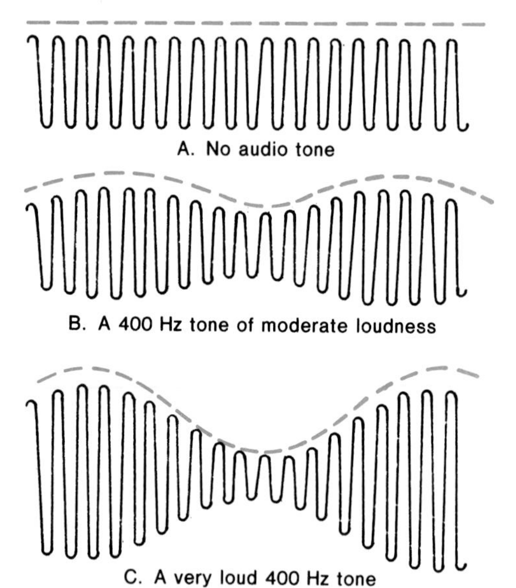

Figure 10.5. A comparison of three AM radio waves carrying a 400 Hz audio tone (in color) which has (A) zero, (B) medium, and (C) maximum loudness.

If the amplitude does not change, the tuner will produce no audio output signal and the speakers are silent. However, if the amplitude of the wave is changing, the tuner is alerted that audio information is present, and it will convert this information into an electrical wave whose frequency lies in the audio range. We will then hear the DJ's voice through the speakers.

We can demonstrate exactly how an amplitude modulated wave carries the audio information by first considering what happens when the DJ whistles a pure tone of 400 Hz into the mike. Figure 10.4 shows the appearance of both the original tone and the AM wave which is transmitted. Figure 10.4 (A) shows the pressure variations of the 400 Hz sound wave as originally whistled by the DJ. Part (B) shows the resulting amplitude modulated wave whose amplitude varies in a manner which *replicates* (see the dotted line) the 400 Hz audio tone. If the 400 Hz tone becomes louder, then its amplitude increases and the corresponding amplitude variations of the carrier wave would also become greater; hence loudness information can also be conveyed by the AM wave. Figure 10.5 illustrates the appearance of the AM wave for three loudness conditions of the original 400 Hz sound wave.

In a sense, the 400 Hz audio tone rides "piggy-back" on the carrier wave. It is important to remember that the sound itself is *not* sent over the air. The only thing which is sent is the carrier wave. This radio wave carries the *information* contained in the audio signal, because the amplitude variations of the radio wave exactly replicate the audio signal.

Of course the pressure patterns for voice and music are complex as discussed in section 4.9. Figure 10.6 illustrates a segment of a voice pattern and how the amplitude modulated carrier appears. As with the simpler case of the 400 Hz tone discussed above, we see that the amplitude variations of the carrier wave correspond exactly to the pressure variations in the sound. The AM tuner must receive the

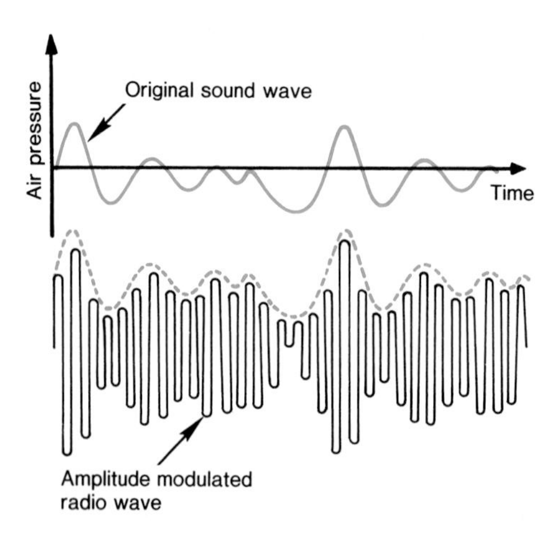

Figure 10.6. The complex sound patterns generated by music and speech can also be carried by an AM radio wave.

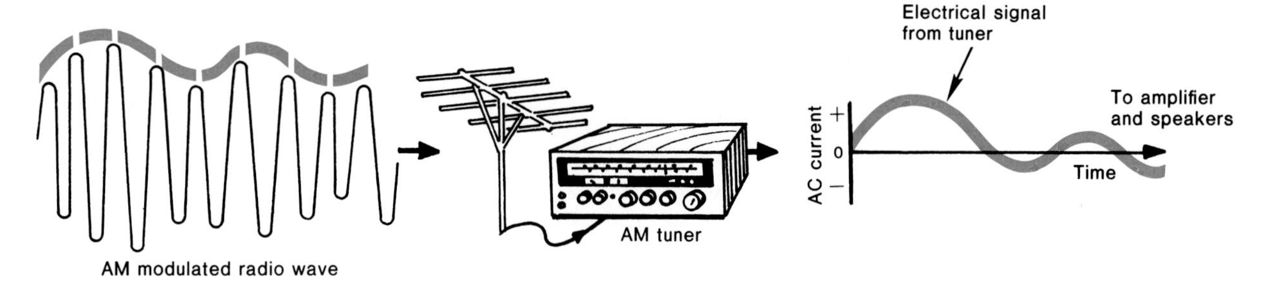

Figure 10.7. The output AC current from the AM tuner possesses a waveform which is an exact replica of the amplitude changes of the AM radio wave.

radio waves and create an audio frequency electrical signal which is an exact replica of the information carried by the radio wave, as shown in Figure 10.7. Only after the audio signal is passed on to the amplifiers and speakers is it finally converted back to sound!

> *Please note:* The listener does not hear the AM radio wave itself. Only the audio information which is carried by the radio wave is converted into sound by a radio tuned to an AM station.

What Would Happen if You Could Actually "See" AM Radio Waves?

Here is an interesting "thought" experiment which will, perhaps, shed some more light on amplitude modulation. We mentioned earlier that radio waves are essentially identical to light waves except that radio waves have a much smaller frequency, one to which the eye is insensitive. Suppose that your eyes *could see* radio waves (which is analogous to what the tuner does)—and then inquire as to what you would "see" when the radio wave is amplitude modulated. Recall that the amplitude of a sound wave is associated with its loudness. In a similar manner, the amplitude of a radio wave (and a light wave, too) corresponds to its brightness. Therefore if you "look" at an amplitude modulated radio wave, you would see its brightness changing in accordance with the musical content that it was carrying; when the music becomes loud, the radio wave becomes "brighter" and when the music becomes softer, the carrier wave becomes "dimmer." All the while the brightness is changing, the "color" of the carrier wave remains the same, since the frequency is what determines the color of visible light, and the frequency of the AM wave does not change. Simple enough!

AM Radio Reception Can Be Noisy

AM transmission can be very noisy especially during a thunderstorm, because lightning generates short bursts of radio waves, some of which have the right frequency to be picked up by the AM tuner. As shown in Figure 10.8 the tuner processes the electrically-generated noise spikes right along with the broadcast information. The tuner has no way

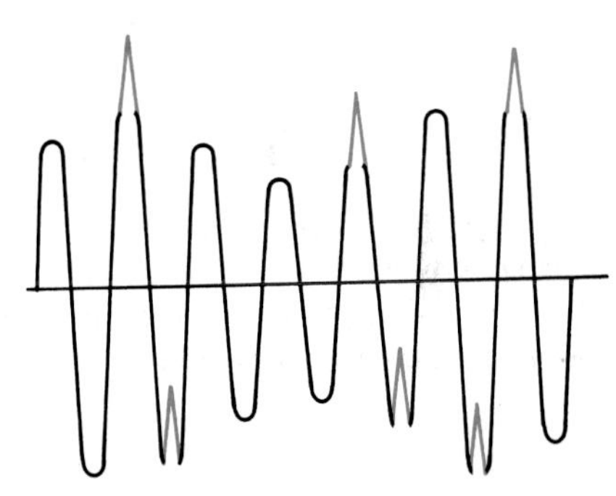

A. An AM radio wave containing noise spikes (in color) produced by lightning or an electrical appliance.

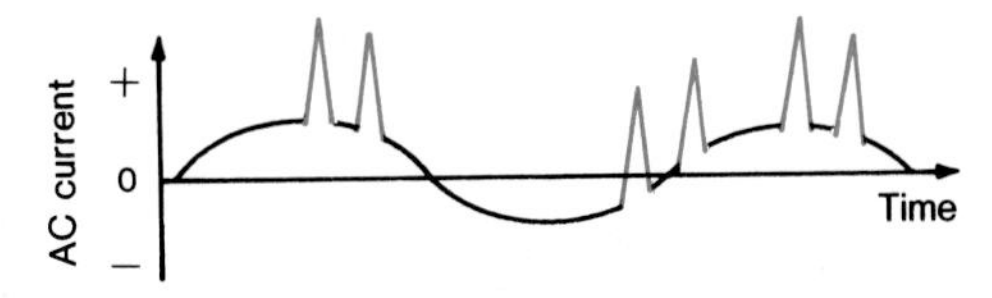

B. The electrical output audio signal from the AM tuner showing the presence of the noise spikes.

Figure 10.8. Any electrical noise which is present in the incoming radio wave is also contained in the output audio signal. This makes AM transmission rather noisy.

of knowing what part of the incoming electrical signal (from the antenna) is due to the broadcast signal and what part is caused by the thunderstorm. It processes both parts equally with the result that the output audio signal contains a lot of annoying electrical "static" and "noise" along with the normal program content. As we shall see in the next section, FM can almost totally eliminate this problem and, hence, it is an excellent noise-free method for transmitting information.

10.5 Frequency Modulation (FM)

Frequency modulation, unlike amplitude modulation, is a method for transmitting information over the airways by changing (modulating) the frequency of the radio wave. AM accomplishes the modulation task by varying the amplitude of the carrier wave while keeping its frequency constant. FM does just the opposite; FM keeps the amplitude of the carrier wave constant at all times but allows the frequency to vary in accordance with the audio information that it is carrying. The FCC has set the frequency band from 88 MHz to 108 MHz for FM broadcasting and assigns a definite frequency to each station. FM stations are spaced 200 kHz apart on the dial. Beginning at 88.1 MHz, FM stations are assigned carrier frequencies of 88.3, 88.5, 88.7, etc., MHz—up to 107.9 MHz. Within this range there can be 100 different frequency assignments. However, when an FM station is assigned a frequency of 100 MHz, for example, the station does *not* always transmit at this exact broadcast frequency. At any instant in time the transmitted FM radio wave may have a frequency which is slightly above or below this assigned value. Strange as it may seem, the FM frequency is continually changing and this process is an excellent method for transmitting music!

How Does Frequency Modulation Work?

Perhaps the easiest way to understand how frequency modulation works is to "see" what such a wave looks like when it is carrying audio information. Figure 10.9 shows an example of an FM station broadcasting at 100 MHz. The part of the radio wave in Figure 10.9 labeled "unmodulated radio wave" is produced when the DJ momentarily stops talking and no audio information is present. Each vertical line represents many cycles of the radio wave. In this case the unmodulated FM wave looks exactly like the unmodulated AM wave shown in Figure 10.3. The FM transmission, of course, uses a much higher assigned frequency, 100 MHz, compared to the 1,200 kHz AM wave. The unmodulated radio wave has both a constant amplitude and a constant frequency.

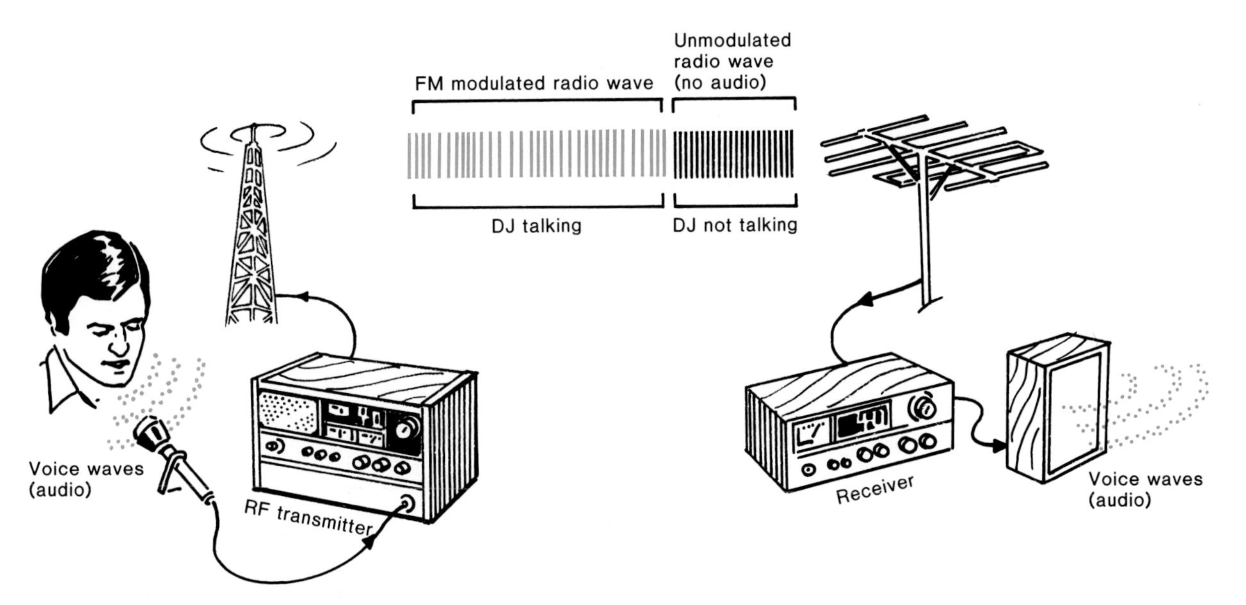

Figure 10.9. A picture of FM radio transmission when the audio information is both present and absent.

When the DJ begins to speak or music is played, the "FM modulated radio wave" has the appearance shown in Figure 10.9. Notice that the frequency of the radio wave is now changing because there are regions where the wave cycles are "compressed" (higher frequency) and other areas in which the cycles are "farther apart" (lower frequency). When an FM radio wave is intercepted by an antenna, an *FM electrical* wave is created within the antenna which travels to the input of the tuner. The tuner, through its electronic circuits, detects only frequency changes in the FM electrical wave. Since an unmodulated wave has a constant frequency, the tuner senses no change in the carrier's frequency, and, correspondingly, produces no audio output signal. On the other hand, an FM modulated wave is, by definition, a wave whose frequency is continually changing. The tuner, sensing the changes in frequency, produces an audio output signal. We will now show exactly how the audio information is encoded onto an FM carrier.

Figure 10.10 shows the appearance of a wave which is being frequency modulated by a 400 Hz audio signal. As expected, when no audio information is present the carrier wave has a constant frequency, as illustrated in part (A). In part (B) an audio wave of moderate amplitude is presented. Notice that, unlike AM, the amplitude of the FM wave remains constant at all times although its frequency is continually changing. The frequency of the modulated wave is changing in accordance with the fluctuations in the 400 Hz tone. When the audio signal rises above the horizontal axis the FM wave "compresses" with a concomitant increase in its frequency. Similarly, when the 400 Hz signal falls below

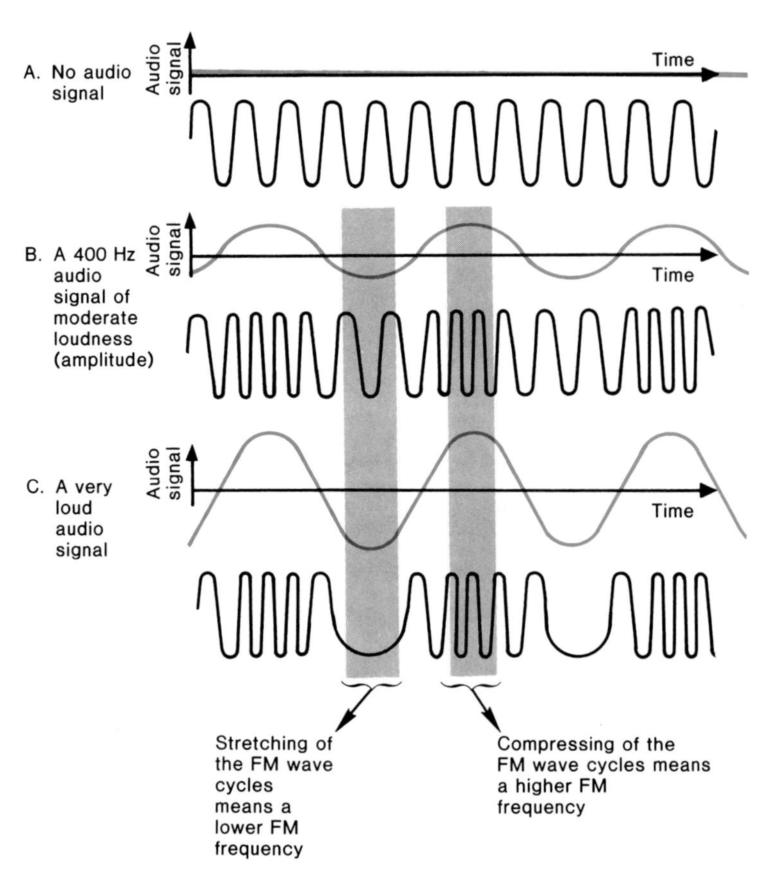

Figure 10.10. A comparison of three FM modulated waves carrying a 400 Hz audio tone (in color) whose amplitude varies from zero [part (A)] to its maximum value [part (C)].

the horizontal axis, the frequency of the FM wave decreases. In this manner the frequency of the FM wave increases and decreases 400 times each second. Part (C) illustrates a very loud (large amplitude) 400 Hz audio signal. Once again, the frequency of the FM wave increases and decreases 400 times each second. However, the compressing and stretching of the FM wave are more severe than in part (B). The severity of the compressing and stretching conveys the loudness of the audio signal.

If the audio signal which is to be transmitted has a complex pattern, then the FM wave will have a more complex appearance to reflect this fact. Such a situation is illustrated in Figure 10.11. Let us take a specific example in order to show exactly how the frequency of an FM radio wave is changed by the audio signal. Assume that the station is broadcasting on an assigned frequency of 100 MHz. The 100 MHz represents the frequency of an unmodulated radio wave, and its frequency will rise and fall about this value depending on the audio content which the wave is carrying. How much can the frequency change? The FCC has ruled that the carrier frequency of an FM wave can change by no more than 75 kHz (=0.075 MHz) either above or below the

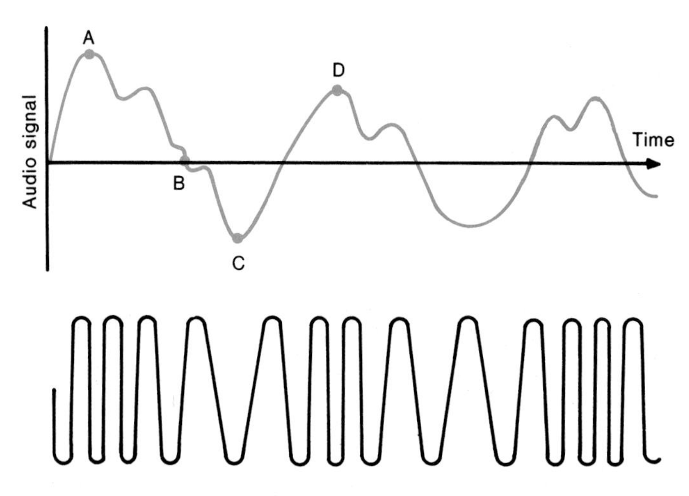

Figure 10.11. A complex audio signal, such as would occur in music, produces a complex frequency modulated wave.

assigned frequency of the station. Considering that the frequency of the unmodulated carrier is 100 million Hertz in our example, and the maximum frequency change allowed by the FCC is only 0.075 MHz, we see that the frequency of the carrier changes comparatively little as it carries the audio information. Consider Figure 10.12 which shows part of the complex audio signal from Figure 10.11 superimposed on a grid of horizontal lines. This graph indicates the frequency of the FM radio wave which is carrying the audio information. For example, "A" is the point of largest amplitude in the sound wave, and the frequency of the FM radio wave will be at its maximum allowed value, 100 MHz + 75 kHz = 100.075 MHz. At "B" the amplitude momentarily returns to a value of zero, and the station now broadcasts at its assigned frequency of 100 MHz. Point "C" shows the amplitude of the audio wave falling below zero and, correspondingly, the frequency of the FM wave falls a proportionate amount to a value of 99.955 MHz (100 MHz − 45 kHz = 99.955 MHz). "D" represents another point on the audio signal which is larger than zero. However,

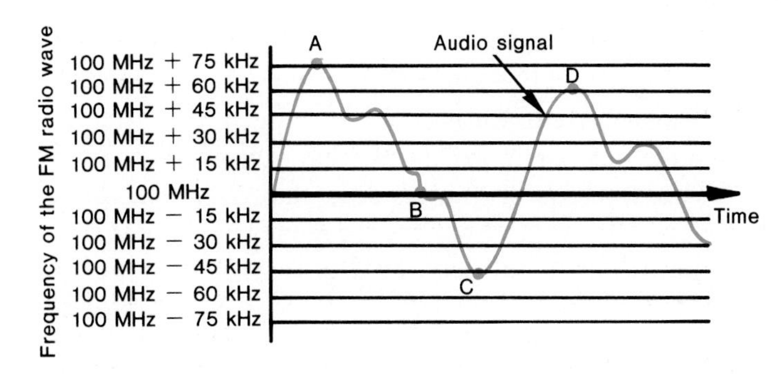

Figure 10.12. The graph shows the frequency of an FM radio wave which is associated with each point along the audio signal.

> *Please note:* As with AM, the listener does not hear the FM radio wave itself. Only the audio information, which is carried by the radio wave, is converted into sound by a radio tuned to an FM station.

its value at "D" is not as large as that at "A" and the frequency of the FM wave will not be as large. In this case the frequency of the radio wave would be 100.060 MHz (100 MHz + 60 kHz = 100.060 MHz) at "D" compared to a higher value of 100.075 MHz at "A." Therefore, Figure 10.12 allows one to calculate precisely how the shape of the audio signal causes the frequency of the radio wave to either increase or decrease about its nominal 100 MHz value. Although frequency modulation is somewhat more complicated than AM, it is a bonafide method of carrying audio information because its frequency parameter is changing in accordance with the instantaneous pressure variations of the sound.

Figure 10.13 illustrates an FM tuner intercepting an FM wave. The electronic circuits within the tuner sense the frequency changes occurring within the incoming FM wave. Using this frequency-change information, the tuner creates an audio signal. The audio signal is then routed out of the tuner, through the amplifier, and to the speakers.

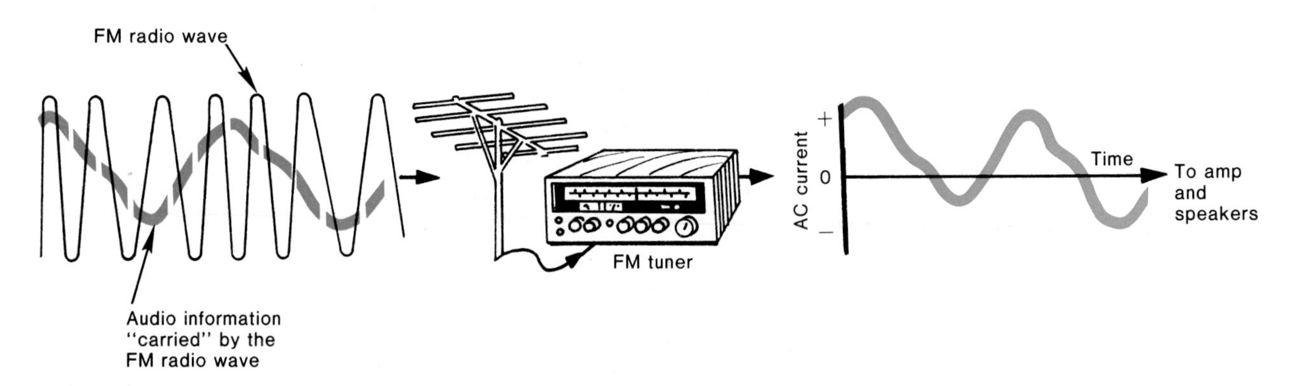

Figure 10.13. The output audio current (signal) from the FM tuner has a waveform which is determined by the frequency changes of the FM radio wave.

What Would Happen if You Could "See" an FM Radio Wave?

Let us again describe, as we did with AM waves, what an FM wave would look like if we could actually see it. Since the amplitude of an FM wave is always constant, you would "see" a wave whose brightness never changed. This situation is, of course, just the opposite of what you would see with an AM wave whose brightness does change in accordance with its audio content. Since the frequency of the FM wave is continually changing, this would cause the "color" of the wave to change, as mentioned in section 10.3. Therefore, if you could see an FM radio wave you would see its "color" continuously changing although the brightness of the various colors would always remain the same. It is this change in "color" by which the audio information is "recognized" by the FM tuner.

Why FM Is Comparatively Noise Free

When an FM wave is being transmitted during a thunderstorm, or in the presence of noise from electrical systems, additional "spikes" will be added to the radio wave just like those shown for AM waves in Figure 10.8. The radio wave which is received at the antenna will contain both the FM program information plus the electrical noise and, at this point, there is no difference in noise between FM and AM. However, once inside the tuner the FM signal encounters a series of electronic circuits called "limiters" which clip off the electrical noise pulses and leave the FM signal reasonably free of noise, as shown in Figure 10.14. It is permissible to clip the amplitude spikes off an FM wave, because the audio information is contained in the frequency oscillations and not in the amplitude. It is not possible to use limiters in AM tuners, because limiting the amplitude of an AM wave would partially remove the amplitude variations with a partial loss of the audio information! Therefore, FM reception is characteristically noise free, and it produces a better sound quality than its AM counterpart.

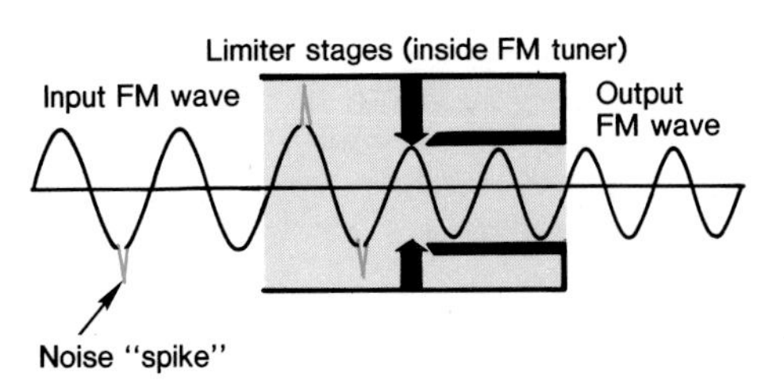

Figure 10.14. The limiter stages inside an FM tuner remove the noise spikes from the RF electrical signal.

10.6 Broadcasting and Receiving FM

It was mentioned in section 10.5 that the frequency of the FM broadcast varies from its assigned frequency. The actual frequency of the radio wave which is transmitted may, at any instant in time, be higher or lower than the assigned frequency by as much as 75 kHz. Of course it is important that each station broadcast within these limits in order to prevent overlap, or interference, with the adjacent stations. With this in mind the FCC has set up the following rules which govern the FM frequency assignments:

1. An FM station must adjust the amplitude, or level, of the audio program so that it does not cause the frequency of the FM radio wave to deviate by more than 75 kHz from its assigned frequency. When an audio signal is strong enough to cause the radio wave to change its frequency by the maximum allowed value of 75 kHz, the FM radio wave is said to be "100% modulated."
2. The closest that two adjacent FM stations can be separated in frequency is 0.2 MHz (which is 200 kHz). This allows a station to change its frequency by either +75 kHz or −75 kHz plus allowing a 25 kHz "guard band" on either side. The guard band, as shown in Figure 10.15, is a safety feature which helps to prevent possible interference between adjacent stations.

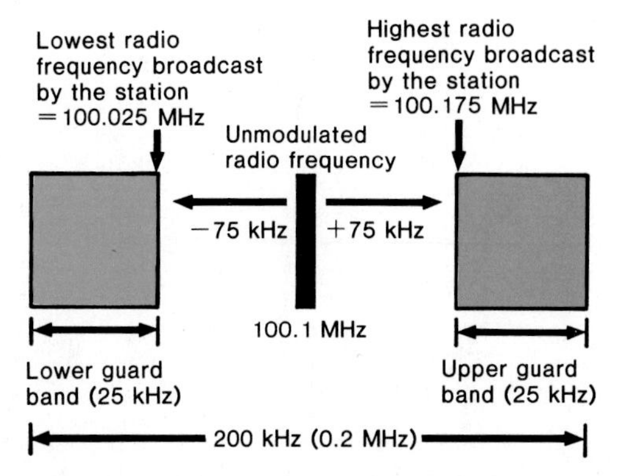

Figure 10.15. Each FM station is allowed 200 kHz of bandwidth for transmitting its program material. The 25 kHz guard bands are a safety feature which help to prevent overlap with adjacent stations.

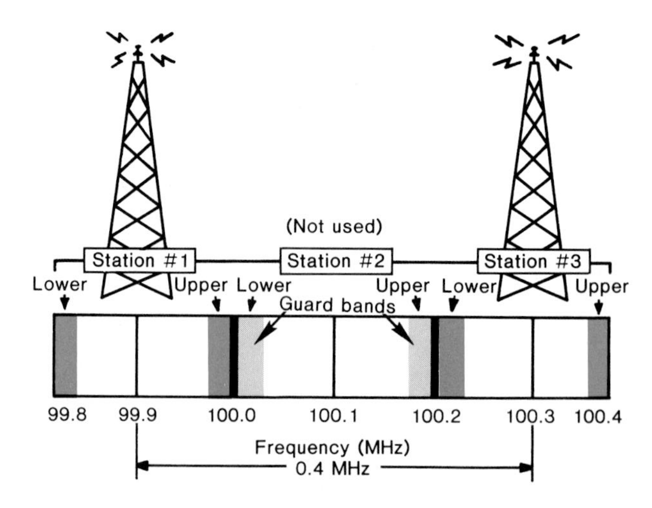

Figure 10.16. In metropolitan areas, where there is a high density of FM stations, FM stations are assigned frequencies which are no closer than 400 kHz, thereby creating an unused channel between two operating stations. The guard bands are shown in color.

3. In metropolitan areas where there is a high density of FM stations, the FCC usually assigns stations to alternate channels, thus separating the closest stations by twice the minimum spacing, or 400 kHz. Figure 10.16 shows a schematic drawing of this alternate-channel arrangement.

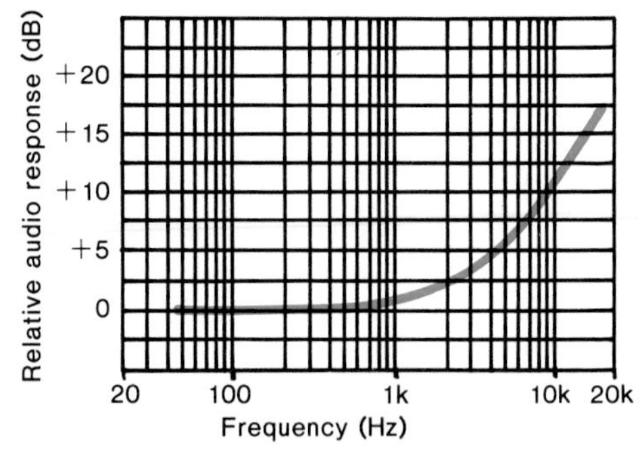

A. Pre-emphasis of the audio signal is performed at the FM station before it is transmitted over the airways.

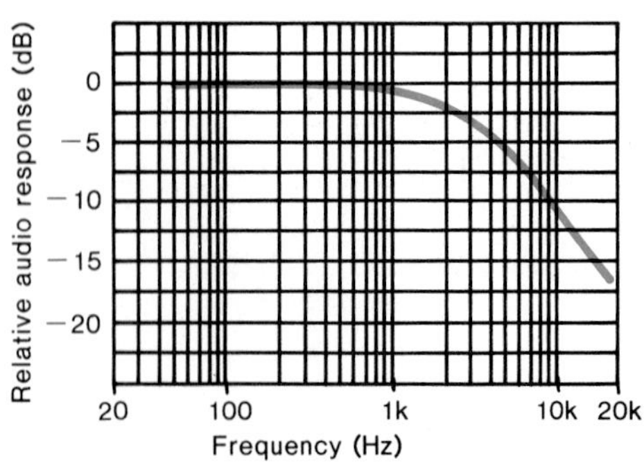

B. De-emphasis of the audio signal is performed in the tuner after the audio information has been extracted from the radio wave.

Figure 10.17. (a) Pre-emphasis and (B) de-emphasis are used in FM to reduce the amount of background noise which occurs during transmission and reception.

Pre-emphasis and De-emphasis in FM Broadcasting

If you have ever listened carefully to an FM broadcast, you will notice that sometimes there is a certain amount of background noise, or "static," which is particularly audible when a soft passage is being played. This noise, which is produced by either the transmission process or the tuner circuits, cannot be eliminated by the limiters alone. Usually this static is comprised of frequencies higher than 1 kHz, and FM transmission keeps the noise to an absolute minimum by a technique called pre-emphasis and de-emphasis. At the FM station the frequencies in the audio signal are boosted according to the curve in Figure 10.17 (A). This boosting is called *pre-emphasis*, and it occurs before the audio signal modulates the radio wave. Notice that the pre-emphasis gradually increases the signal strength, starting at a frequency around 1 kHz, until a maximum boost of +17.5 dB is achieved at 15 kHz. Thus the audio information which is conveyed by the FM radio wave is *not* an exact replica of either the DJ's voice or the record being played. The highs have been deliberately boosted according to the pre-emphasis curve; FM transmission does not alter the bass frequencies.

After the tuner extracts the audio information from the FM wave it creates an electrical audio signal which is still pre-emphasized. Up to this point the process of modulating, transmission, and demodulating has introduced a certain amount of noise into the audio signal. This noise is primarily in the high frequency end of the audio range. The audio signal is then *de-emphasized*, according to Figure 10.17 (B), which exactly reverses the pre-emphasis process by gradually cutting the amplitudes of all frequencies above 1 kHz. Both the audio signal and the accompanying noise are reduced because the de-emphasis circuits have no way of distinguishing between the two. However, since the high frequency audio signals were boosted to begin with, they are returned to their original level, while the high frequency noise is reduced!

Notice that the idea of pre-emphasis and de-emphasis only reduces the noise which has crept into the transmission process *after* the signal has been pre-emphasized and *before* it is de-emphasized. Only the noise which has *not* been pre-emphasized will be reduced by this process. The audio signal from the DJ, on the other hand, has been both pre-emphasized and de-emphasized so that, in the end, it has neither been increased nor decreased relative to its original level. This same type of reasoning applies to the high frequency equalization of records, as we shall see in Chapter 12, and it is also used to reduce the tape hiss associated with the tape recording process (Chapter 13).

FM Stereo

One of the most popular attractions of an FM tuner lies in the possibility of stereophonic reception. The broadcasting of FM stereo began in the early sixties, and most hi-fi tuners are designed to receive both mono and stereo programs. When tuning across the dial a light usually turns on when the tuner detects a stereo program. Many tuners have a

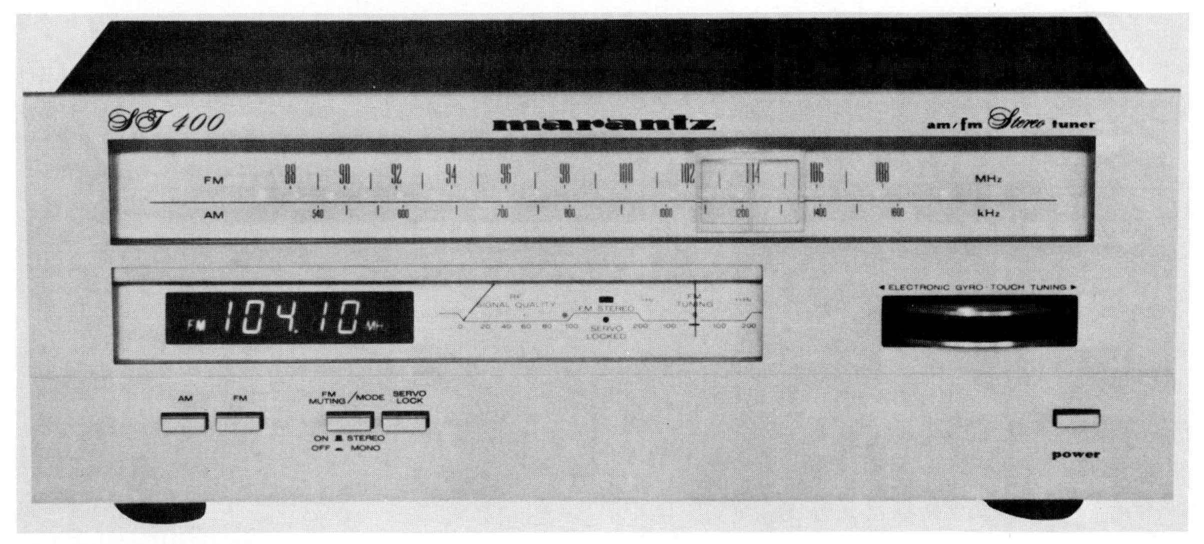

Model ST-400 Marantz tuner

STEREO/MONO switch which, when in the MONO position, converts all stereo programs into mono. Some tuners have an additional switch which allows the listener to receive only stereo stations and no sound is produced when the dial is tuned to a monophonic station; this ensures that all stations received are produced in stereo.

Generally speaking, stereo reproduction is noisier than mono, and this is particularly true when the RF signal strength becomes weak, as would be the case if you tried to tune in a distant station. It also should be mentioned that an FM stereo broadcast has only about half the reach of a mono program transmitted with the same power. There is also another feature of which you may or may not be aware. Most tuners will automatically switch from stereo reception to a less noisy mono when the RF (Radio Frequency) signal becomes too weak. The idea is that most people would rather hear a quiet mono broadcast than a noisy stereo program. In addition, the stereo separation becomes poorer as the RF signal strength becomes weaker. Even though FM stereo is somewhat noisier than mono, the stereo effect more than compensates for the increased noise.

It is not at all obvious how FM stations manage to broadcast two separate stereo channels on one carrier wave, and Figure 10.18 illustrates the problem by using a cargo-carrying truck as an example. In part (A) the truck is carrying hundreds of packages to a central distribution point, somewhat like the carrier wave which carries the audio information corresponding to an FM monaural broadcast. No difficulty arises if the packages are mixed together during the trip, so long as all of them are headed ultimately for the same destination. The situation in a monaural FM broadcast is analogous. No difficulty arises when the "packages" of sound from each instrument in the band are mixed together during their trip on the carrier wave, since they are all headed for the same single speaker anyway. However, suppose the packages were initially separated into two groups, with roughly half headed for one destination (the left speaker) and half for another (the right speaker). Without some method to prevent them from being mixed together during the trip, the packages would arrive at the central distribution point (the tuner) in one hopelessly jumbled pile. The post office would no doubt solve this problem by simply resorting the packages. However, the "packages" of sound can not be resorted inside of the tuner unless the tuner has a way of distinguishing "left" from "right." In both examples the sorting job is simple if the packages are kept completely separate during their trip on the carrier wave.

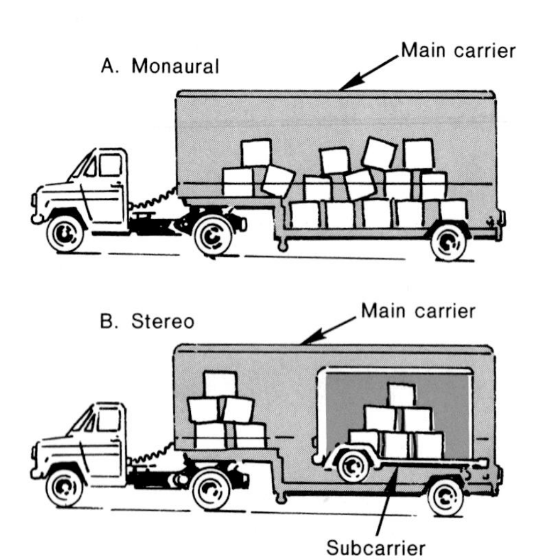

Figure 10.18. (A) In a monaural FM broadcast all of the audio information is mixed together when it is carried by the carrier. (B) In a stereo FM broadcast the main carrier carries part of the audio information, and a subcarrier carries the remainder of the audio information. The subcarrier keeps the remainder of the information separate from the first part.

Figure 10.18(B) offers a solution to the problem described above. The packages destined for one location are carried in the usual, scrambled way. The packages destined for the second location can be loaded into a small trailer; then the small trailer is loaded onto the large truck. The truck is the main carrier, since it carries both the first group of packages as well as the small trailer. However, the small trailer is carrying its own separate load, and, in this sense, it is also a carrier. To distinguish the small trailer from the main truck, the small trailer is called a *subcarrier*.

The situation in an FM stereo broadcast is very similar to Figure 10.18(B). One part of the audio information is carried in the normal manner via FM modulation of the main carrier wave. However, the main carrier wave also carries a subcarrier wave, by means of the process of FM modulation. This subcarrier carries the second part of the audio information and keeps it separate from the first part.

When the main truck arrives at the central distribution point, the small trailer is unloaded, hitched to another cab, and driven to its final destination. Likewise, the group of packages remaining inside the main truck is sent on to its final destination. In a like manner, the main carrier wave arrives at the stereo FM tuner, where the packages of sound are separated and sent to their final destinations—the left and right speakers.

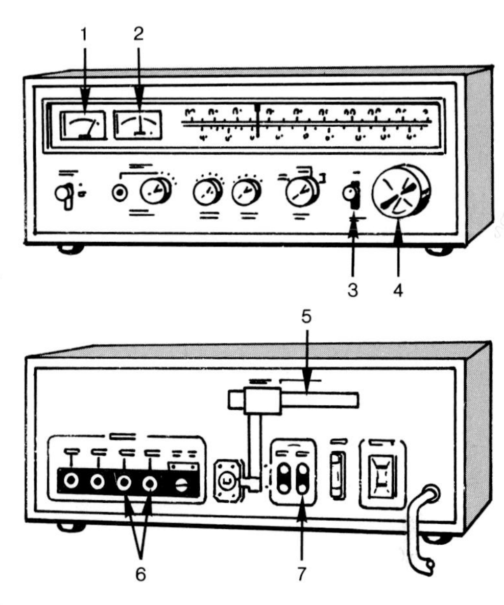

1. Signal strength meter
2. Tuning meter
3. FM muting switch
4. Tuning knob
5. AM Ferrite antenna
6. FM antenna terminals
7. Audio output jacks

Figure 10.19. The front and back panels of a typical tuner.

10.7 Tuning Aids and Other Features

The usefulness of an FM tuner is determined not only by the quality of its electronic circuits, but by its operating convenience as well. Most tuners are equipped with several features which considerably aid the user in selecting a favorite station. Figure 10.19 shows the front and back panels of a tuner in which some of its more important traits are indicated. Although a separate tuner is shown, many of these features are also available when the tuner is incorporated into a receiver. In this section we shall discuss each of these features in turn.

The FM Antenna

The first rule of FM reception is to use a good antenna. A high quality antenna is the best aid for receiving high quality FM stereo, and the performance of even the finest tuners can be seriously degraded with a low quality antenna. An investment in an antenna need only be a small fraction of your total component budget, but it will return listening dividends far in excess of its original price. A tuner's performance depends upon a strong RF electrical signal being delivered to its antenna terminals. A good FM antenna has the ability to intercept the radio waves and efficiently produce a relatively large voltage. By a "good" antenna we do not mean those "T"-shaped plastic ribbon antennas which are often packed with most tuners or receivers, as shown in Figure 10.20. These antennas are usually installed inside

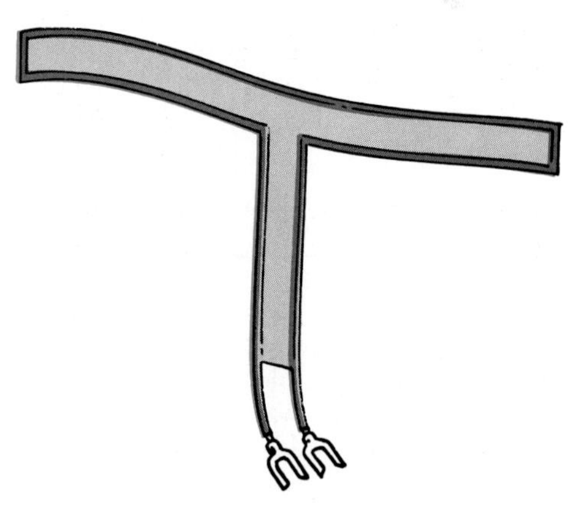

Figure 10.20. An inexpensive FM "T"-shaped antenna, often packed with new tuners or receivers and good only for picking up nearby stations.

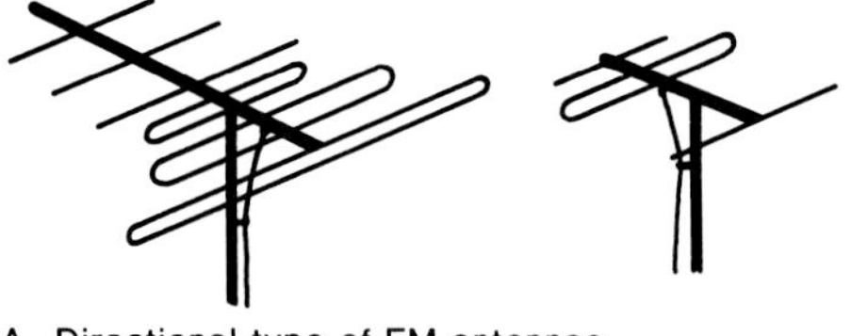

A. Directional type of FM antennas

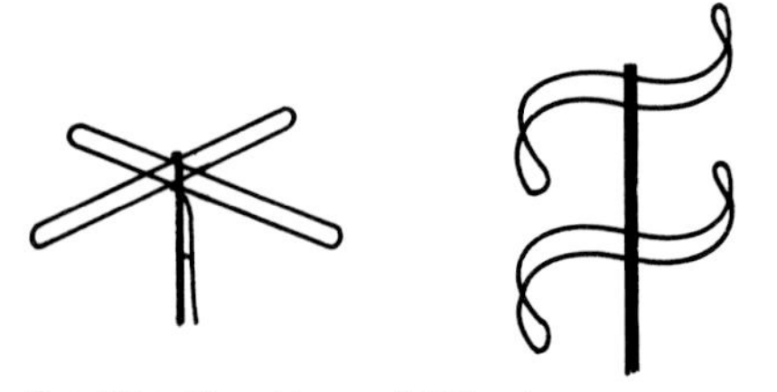

B. Omnidirectional type of FM antennas

Figure 10.21. Two types of FM receiving antennas: (A) Directional, (B) Omnidirectional.

your room, and they are adequate if you live within ten or twenty miles of the station. An antenna which can pull in those distant stations (fifty to eighty miles) should be mounted outdoors and on, or above, the roof. The four types of outdoor antennas shown in Figure 10.21 are classified as either directional or as omnidirectional. The omnidirectional antennas pick up FM stations equally well from all directions while the "directionals" exhibit pickup characteristics which depend on the antenna's orientation. The directional antennas are very helpful in reducing unwanted reflections from buildings and other tall objects which can lead to a form of audio distortion called "multipath distortion."

It was previously mentioned that the entire FM band, 88 MHz to 108 MHz, is located between TV channels 6 and 7. Therefore, it is permissible to connect a tuner to an existing TV antenna, providing that it does not contain an "FM trap" circuit. This circuit removes the "FM" signals so that they will not interfere with the TV reception. These traps are often found in a master antenna system such as that used in an apartment building.

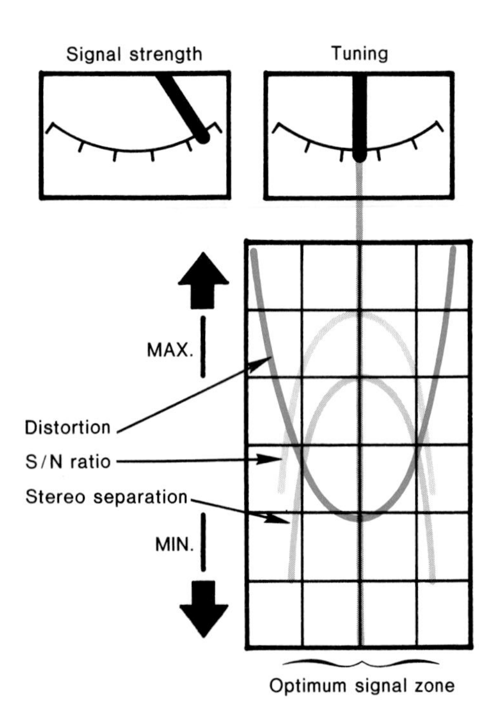

Figure 10.22. The signal strength and tuning meters allow precise tuning of a station. Accurate tuning results in minimum distortion, a maximum S/N ratio, and the greatest stereo separation.

Signal Strength and Tuning Meters

Most hi-fi tuners are equipped with a signal strength meter which indicates the voltage level of the received RF signal. The signal strength meter deflects to the right when the station is tuned; the greater the deflection, the stronger is the received signal (see Fig. 10.22). Often, but not always, a "tuning" meter will be included with the signal strength meter. The tuning meter moves to a center position in order to indicate accurate station tuning. If the tuner is not accurately tuned to the station, there occurs an undesirable increase in the distortion (both THD and IM), a poorer S/N ratio, and a loss of stereo separation. The dependence of these parameters on proper tuning is graphically depicted in Figure 10.22.

The FM Mute Control

The FM muting feature is a noise suppression circuit which eliminates the annoying noise, or hiss, encountered when dialing between FM stations. The muting circuits also suppress weak stations along with the interstation noise, so that the front-panel muting switch must be turned off if such a station is to be received. The muting feature is sometimes called "FM Rush" or "Squelch."

Automatic Frequency Control (AFC)

The function of the automatic frequency control is to lock the tuning circuits onto the desired FM station and prevent the tuner from detuning. Virtually all hi-fidelity tuners are equipped with the AFC feature, although some tuners have a switch on the front panel which allows the AFC to be switched off. Such a switch might be desirable if you were listening to a weak station which was closely situated to a stronger one. In this case the AFC might detune the weaker station in an attempt to "lock on" to the stronger adjacent station. Turning off the AFC would allow you to listen to the weaker station.

High-Blend Switch

Some tuners have a "high-blend" feature that helps to reduce the background hiss which sometimes accompanies a distant FM stereo broadcast. This circuit operates by blending a certain amount of the high-frequency sounds from both the right and left channels. The blending reduces the hiss although it also reduces the stereo separation.

10.8 Tuner Specifications

The technical standards for measuring tuner performance were established by the Institute of High Fidelity (IHF) in 1958. The principal weakness in these standards, as applied to today's tuners, was the lack of measurements for the stereo mode. Since FM stereo was not fully developed at that time, the stereo standards were not set. In 1975 a new and more comprehensive set of tuner standards was established.* Today tuner manufacturers often quote both the new and old specifications and we shall point out the differences between the two standards where it is appropriate.

The specification sheet for a tuner contains considerable information on how well it will perform under various circumstances. In general, the performance of a tuner can be divided into two parts. The first part expresses how well the tuner can select the desired station from all the others. The important specifications which deal with this aspect are:

Sensitivity
Selectivity
Capture Ratio.

The second part of a tuner's performance indicates how closely the output audio signal resembles the original broadcast material, i.e., the reproduction accuracy. Remember that the tuner uses the information which is carried by the radio wave to create a *replica* of the original audio signal. During the replication process the electronic circuits of the tuner introduce a certain amount of unwanted noise and distortion. In addition, the tuner may change the relative

*The new tuner standards can be found in the publication IHF-T-200, 1975, "Standard Methods of Testing Frequency Modulation Broadcast Receivers." This publication is available from the IHF, Inc., 489 Fifth Ave., New York, N.Y. 10017.

balance among the audio frequencies, which will be reflected in its frequency response specification. The tuner may also alter the relative position of the musical instruments in the band. Better tuners obviously do a better job of recreating the original program material, and the important specifications which relate to this aspect are:

Signal-to-Noise Ratio (S/N)
Total Harmonic Distortion
Frequency Response
Stereo Separation.

In this section we shall examine all phases of a tuner's performance and offer some guidelines to aid you in purchasing a tuner. As always, the particular specifications that we suggest are only meant to be a helpful guide in selecting a fine performing tuner at a reasonable price.

Usable Sensitivity or IHF Sensitivity

A good FM tuner should bring in the desired station—whether strong or weak—and produce an output audio signal with an absolute minimum of noise and distortion. The specification which tells how well a tuner can convert radio signals into a satisfactory audio signal is called its *sensitivity*. When applied to the human senses, "sensitivity" is a general term which is familiar to all of us. A person with sensitive hearing can hear distant and faint sounds. A sensitive "eye" implies an ability to see well after normal light levels begin to fade. Similarly, a sensitive tuner can bring in weak and distant stations.

The definition of monaural usable sensitivity (sometimes called IHF sensitivity) is stated as follows:

Monaural Usable Sensitivity or IHF Sensitivity: The smallest radio frequency (RF) input signal that will cause the tuner to produce a total audio output which is 30 dB greater than the background noise and distortion.

It should be noted that the definition of monaural usable sensitivity, like other tuner specifications which we will encounter, involves two entirely different types of signals; an *input* radio signal, and an *output* audio signal. The first signal is the radio wave which travels from the broadcast tower to the tuner's antenna; its frequency lies within the FM broadcast band which extends from 88 MHz to 108 MHz. The second signal is the audio signal which the tuner creates from the information carried by the radio wave. The audio signal is sent out of the tuner to the preamplifier/control center, and its frequency lies in the range from 30 Hz to 15 kHz. The strength of the input radio wave is generally very weak, and it is measured in millionths of a volt.* The strength of the output audio signal is considerably stronger, with a typical value being approximately 0.5 volts.

As an example of the monaural usable sensitivity rating, consider the following specification: 3.0 μV for 30 dB "quieting." To interpret this sensitivity remember that the antenna picks up the radio wave and converts it into a voltage which is sent to the input terminals of the tuner. If the input voltage is only 3 millionths of a volt (a very weak signal indeed), the tuner's output signal will contain a combination of noise and harmonic distortion whose level will be 30 dB below the audio level. The noise and distortion level is said to be 30 dB "quieter" than the audio. An audio frequency of 1,000 Hz is commonly used to measure this specification.

The usable sensitivities of different tuners can easily be compared as shown in the following illustration. Consider the three tuners:

	Usable Sensitivity
Tuner #1	1.8 μV for 30 dB quieting
Tuner #2	3 μV for 30 dB quieting
Tuner #3	10 μV for 30 dB quieting

Tuner #1 has the best sensitivity rating while tuner #3 has the poorest. Tuner #1 can receive a much weaker input RF signal compared to tuner #3 (1.8 μ V vs. 10 μ V), and still produce the *same* quality output audio signal. In practice this implies that tuner #1 can be farther from the transmitting station and still pick it up with the same ease. Therefore, tuner #1 is more "sensitive" and, in this respect, it is better than the other two tuners. However, slight differences in usable sensitivity (e.g., 1.8 μ V vs. 2.0 μ V) do not significantly affect the tuner's quality.

Among the more noticeable changes which have been promulgated by the newer 1975 tuner standards is that the input RF signal strength is now measured in a unit called "dBf" rather than in microvolts. Without going into all the technical details which lie behind this change, we shall simply present a chart (Fig. 10.23) which shows the numerical correlation between the two terms.** A sensitivity rating of 3 μ V for 30 dB quieting would now read, after consulting

*One millionth of a volt is called a "microvolt" (abbreviated 1 μV). Therefore, 1 μV = 1/1,000,000 volt.

**The mathematical expressions which convert microvolts into dBf units are given in the Appendix.

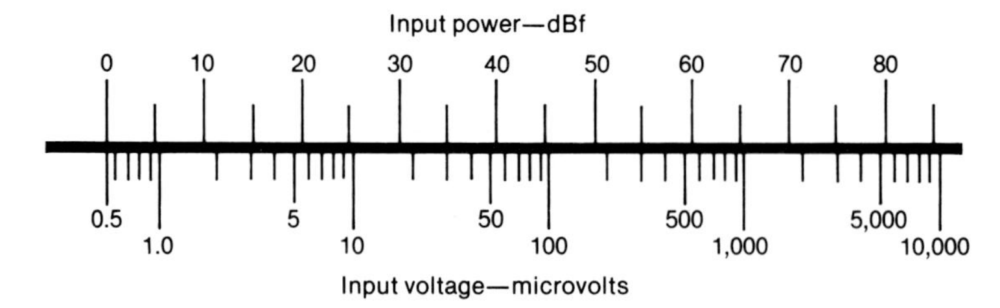

Figure 10.23. A chart which shows the relationship between microvolts and dBf for a 300 ohm antenna.

Figure 10.23, approximately 15 dBf for 30 dB quieting. Manufacturers will continue to list both of these equivalent ratings until the newer standard becomes more familiar to the public.

Tuner Buying Guide

Usable (IHF) Sensitivity: Look for a usable sensitivity of 3 μ V or less for 30 dB quieting, or, alternatively, 15 dBf or less for 30 dB quieting.

NOTE: The usable sensitivity ratings are only strictly applicable for monaural reception. The new standards, as discussed below, also include a stereo sensitivity rating.

50 dB Quieting Sensitivity

The new tuner standards call for the monaural sensitivity rating to be specified for a 50 dB quieting rather than 30 dB. It was felt that the old usable sensitivity figure was never representative of truly good listening conditions, since a noise-plus-distortion level which is 30 dB below the audio is unacceptably noisy. Remember, from Chapter 5, that a 30 dB difference means that the music, during its loudest passages, is only about 8 times louder ($2^3 = 8$) than the combined effects of noise and distortion. Anyone who listened to such an FM broadcast would be aware of the background noise, particularly during the quieter program passages when the music no longer masked the ever-present noise. The 50 dB Quieting Sensitivity is defined as follows:

50 dB Quieting Sensitivity: The smallest RF input signal which will cause the tuner to produce a total audio output which is 50 dB greater than the background noise.

The 50 dB difference implies that the music is 32 (2^5) times louder than the noise. Such an audio signal is quite listenable, although the background noise might be slightly noticeable during the quietest of passages.

As mentioned earlier, the new tuner standards also specify the sensitivity rating for stereo as well as mono reception. The background noise associated with FM stereo is significantly higher than mono, so that the stereo sensitivity rating will be poorer than the mono sensitivity. In other words, more input signal is required in order to achieve a 50 dB signal-to-noise ratio (S/N) in stereo than in mono. Most tuners automatically switch from stereo to mono when the input signal drops below a preset threshold level. This switching is done in order to take advantage of the better noise conditions which prevail with mono reception. The mono 50 dB quieting sensitivity will usually lie between 15 and 25 dBf (3.1 to 9.7 μ V) while the corresponding stereo figures will fall between 33 and 38 dBf (25 to 44 μ V).

Tuner Buying Guide

50 dB Quieting Sensitivity

MONO: Look for a MONO 50 dB quieting sensitivity of 25 dBf (9.7 μ V) or less.

STEREO: Look for a STEREO 50 dB quieting sensitivity of 38 dBf (44 μ V) or less.

The Sensitivity Curves

Any tuner performs better when the radio frequency (RF) waves impinging on the antenna are strong. Most of us are familiar with the inferior reception that often occurs when listening to distant stations. Weak radio waves simply do not allow the tuner to perform optimally, and, therefore, many tuner specifications are related to the strength of the RF waves. The effect of the input RF strength on tuner performance is commonly given by a set of graphs known as *sensitivity curves,* which are illustrated for a typical monaural FM tuner in Figure 10.24. Many tuner specifications, such as the usable sensitivity, the 50 dB quieting sensitivity, and the S/N ratio, are obtained from these curves.

The first thing to notice about the sensitivity curves is that the strength of the RF wave is marked along the horizontal axis in dBf or microvolts, the two ways which are used by tuner manufacturers for specifying the strength of the RF wave. We include a chart just below the horizontal axis which gives the relationship between the two sets of units. The second thing to notice about the sensitivity curves is that the vertical axis records the decibel level of the audio output signal produced by the tuner.

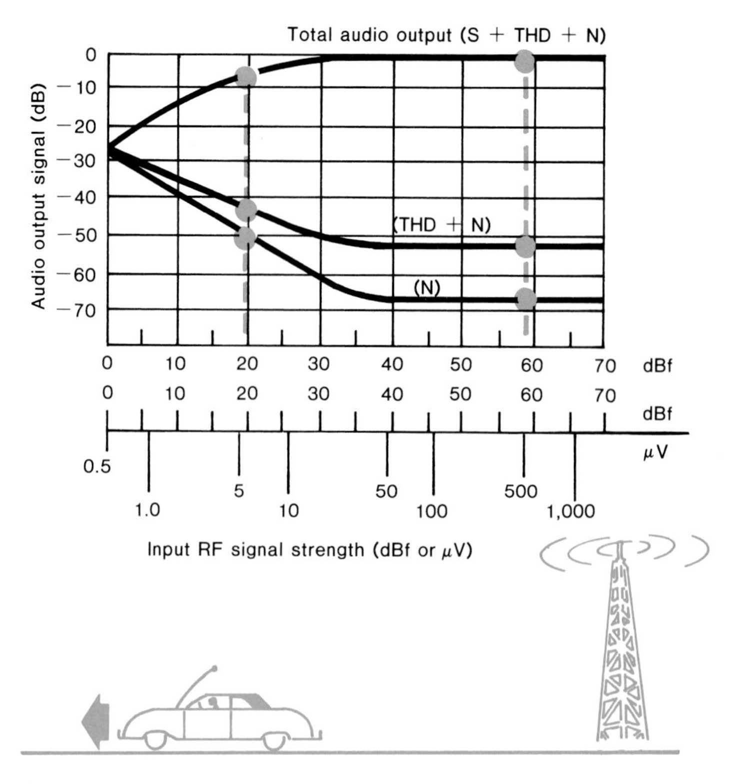

Figure 10.24. FM monaural sensitivity curves for a typical tuner. N = noise output, THD = total harmonic distortion, and S = 1 kHz audio signal. The conversion chart between dBf and μV is for a 300 Ω antenna.

The three curves in Figure 10.24 represent (from bottom to top): (a) the noise (N) generated by the tuner, (b) the total harmonic distortion plus the noise (THD + N) created by the tuner when it is processing a 1 kHz audio tone, and (c) the total audio output of the tuner (S + THD + N), which includes the 1 kHz audio signal (S), the THD, and the noise.

The sensitivity curves can be understood by visualizing what happens to the FM reception of a car radio as the car is driven away from the broadcasting station. When the car is very near the station, the RF wave is strong and the broadcast comes in clear. For example, assume that the car in Figure 10.24 is close to the station, and the RF signal strength is approximately 70 dBf. At this point the total audio output (top curve) from the car's tuner is about 52 dB greater than the combined THD and noise generated by the tuner (middle curve). Under this condition, the listener hears a strong, clean audio sound which contains very little THD and noise. As the car drives away from the station, the RF signal reaching the antenna becomes weaker and weaker. It is a surprising fact that most tuners are able, up to a point, to produce a high quality audio signal, even when the RF signal strength is decreasing. Notice, from Figure 10.24, that both the (THD+N) and (N) curves still remain at very low levels even though the RF signal strength decreases from 70 dBf to 40 dBf. This corresponds to the fact that the audio quality of an FM station does not change appreciably as you drive around in your car—provided that you do not drive too far (20 miles, or so) from the station.

However, as the RF signal strength becomes less than 40 dBf the reception begins to deteriorate because the total audio output begins to decrease, while the "hissing" noise and distortion begin to increase significantly. For example, when the strength of the RF signal decreases from 70 dBf to 20 dBf in Figure 10.24, the total audio output (S+THD+N) from the tuner *decreases* from 0 dB to −5 dB, while the (THD+N) *increases* from −53 dB to −42 dB, and the noise (N) also *increases* from −68 dB to −50 dB. Clearly the reception has deteriorated, because there is 5 dB less audio output (i.e., less music) and 11 dB more distortion and noise. Eventually, if the RF signal strength becomes too weak, you may not be able to hear the music over the noise and distortion.

As pointed out in section 10.7, a good FM antenna is a great asset for improving any tuner's performance, because it can increase the RF signal strength to the tuner when compared to an inferior antenna.

Figure 10.25 shows how to determine the usable sensitivity and the 50 dB quieting sensitivity from the sensitivity

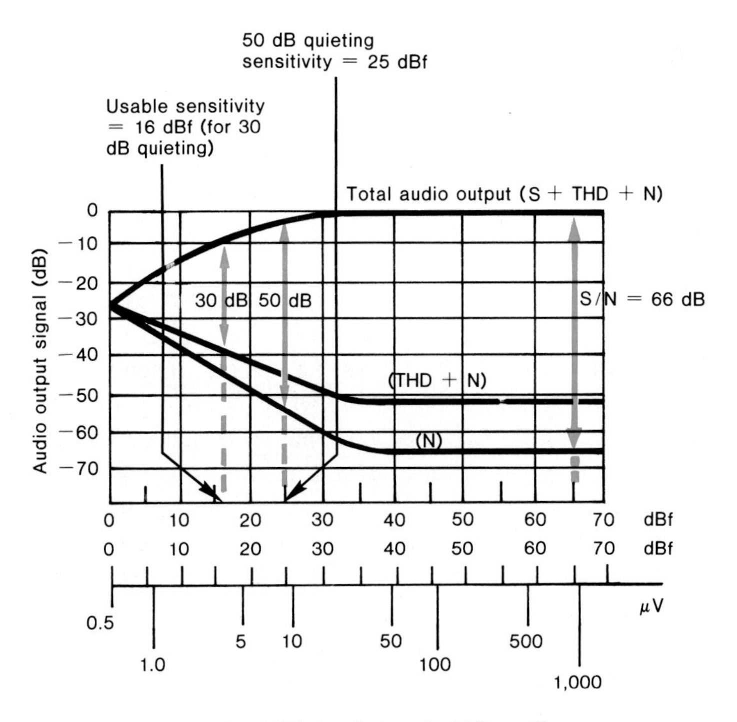

Figure 10.25. Monaural FM sensitivity curves showing the usable sensitivity, the 50 dB quieting sensitivity, and the S/N ratio.

curves. The usable sensitivity occurs when the audio program level (top curve) and the THD-plus-noise level (second curve) are 30 dB apart. When this occurs, the horizontal scale of the graph reads 16 dBf, and that is the usable sensitivity rating of this tuner. The 50 dB quieting sensitivity occurs when the audio program curve and the noise level (the bottom curve) are 50 dB apart. The input signal to the tuner is 25 dBf when this occurs.

Figure 10.26 illustrates a typical set of stereophonic sensitivity curves, and it is apparent that there are some features which are different from the monophonic curves. First, the distortion and noise levels are generally higher in stereo than they are for mono. For example, the mono distortion level for 65 dBf input is approximately −53 dB in Figure 10.25, while it is somewhat higher, −48 dB, for stereo in Figure 10.26. Second, there is an abrupt change in the stereo curves near 25 dBf. Notice, as the input RF signal decreases from 40 dBf to 25 dBf, that both the distortion-plus-noise (THD + N) curve and the noise (N) curve begin to rise. In this region stereo reception begins to deteriorate until the "stereo threshold" is reached (20 dBf in Fig. 10.26). At this point the tuner will follow one of two courses of action, depending upon the manufacturer. Some tuners will automatically switch back to mono reception. Being less noisy, the (THD + N) and (N) curves experience a significant drop below the 25 dBf threshold when the mono circuitry is activated. The mono audio output follows the curve marked "A" in Figure 10.26. Therefore, all reception below the

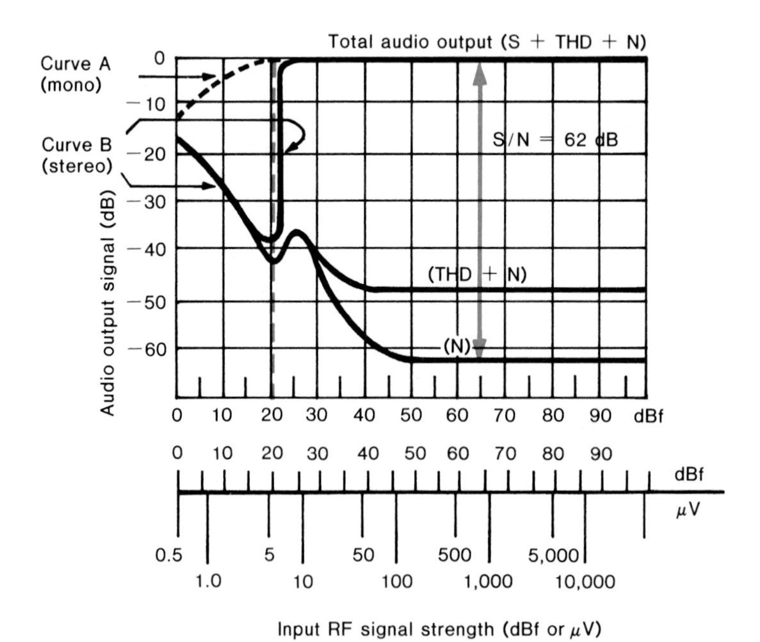

Figure 10.26. Stereo FM sensitivity curves for a typical tuner. N = noise, THD = total harmonic distortion, and S = 1 kHz audio signal. The conversion chart between dBf and μV is for a 300 Ω antenna.

Model 570T Scott tuner

stereo threshold is switched to mono in order to take full advantage of its greater noise-free reception. Some tuners, on the other hand, completely shut off the audio below the stereo threshold, and their output follows curve "B." For input signals whose strength lies below the threshold only distortion and noise would be heard.

The Signal-to-Noise Ratio (S/N)

The total audio output of a tuner must be sufficiently greater than the accompanying noise if listening is to be enjoyable. The signal-to-noise ratio is a specification that expresses how well a tuner can produce audio signals with little noise. In fact, the signal-to-noise ratio expresses the best that a tuner can do in producing loud music and quiet noise, because it is measured with a strong RF input signal to simulate closeness to the broadcasting tower. The S/N ratio is defined as follows:

Signal-to-Noise Ratio (S/N): The ratio of the total audio output level to the noise level, expressed in decibels, when the strength of the RF input signal is 65 dBf.

Figures 10.25 and 10.26 illustrate how the S/N ratios for both mono and stereo modes are determined from the sensitivity curves. Notice that the S/N ratio is the difference, in decibels, between the top (S+THD+N) and bottom (N) curves at the position along the horizontal axis where the RF signal strength is 65 dBf. In these drawings the mono S/N is 66 dB, while the stereo S/N is 62 dB. Notice that the stereo S/N is poorer than the mono S/N.

If a tuner has a S/N ratio of 66 dB, for example, then under the best reception conditions, the noise will be 66 dB below the total audio level. Suppose that you are listening to an FM broadcast. The VOLUME control on your receiver is turned up such that the loudest musical passages register 100 dB on a sound level meter. The level of FM noise that you hear is 66 dB *below* the 100 dB, or 34 dB. A noise level of 34 dB in your room is quite unobtrusive and makes for fine FM listening.

Tuner Buying Guide

S/N Ratio

MONO: Look for a MONO S/N ratio of 65 dB or higher (70 dB, 75 dB, etc.).

STEREO: Look for a STEREO S/N ratio of 60 dB or higher.

Total Harmonic Distortion

A tuner, like an amplifier, generates a certain amount of unwanted higher harmonics of each fundamental note. Total harmonic distortion, which has been previously defined and discussed in section 8.7, is produced by the electronic circuitry within the tuner, and it must be kept to an absolute minimum. Since a tuner's THD is somewhat frequency dependent, it is common to see THD figures quoted for various audio frequencies such as 100 Hz, 1 kHz, and 6 kHz. Frequencies around 1 kHz usually give the smallest amounts of distortion. As we have indicated before, stereo THD is typically worse (higher percentage) than monaural THD. Also, the actual THD level depends on the strength of the RF input signal (see the middle curves in Figures 10.25 and 10.26). Unless otherwise noted, the THD figures refer to a strong RF input signal (65 dBf) which is carrying a 1 kHz audio signal.

Tuner Buying Guide

Total Harmonic Distortion (THD)

MONO: Look for a MONO THD of 0.5% or less.

STEREO: Look for a STEREO THD of 1.0% or less.

Frequency Response

The frequency response of a tuner, like that of any other audio component, tells how well the amplitudes (or loudness) of the audio frequencies are reproduced in relation to those in the original program, as discussed and defined in sections 5.8 and 8.7. In the case of FM, the range of transmitted audio frequencies extends from 30 Hz to 15 kHz so that you can expect the frequency response to cover this range: 30 Hz to 15 kHz, ±1 dB, for example. This means that no frequency in the specified range will be boosted or attenuated, relative to any other frequency, by more than 1 dB. The greater the ± dB number, the less the uniformity and the poorer the response.

Tuner Buying Guide

Frequency Response: Look for a frequency response of 30 Hz to 15 kHz, ± 1 dB or better. The smaller the "± dB," the better. The wider the frequency range, the better.

Stereo Separation

Stereo separation means the same thing for tuners as it did for amplifiers. It indicates the amount of the left channel signal that leaks into the right channel and vice versa, as defined in section 8.7. In theory there should be no leakage but, as discussed in section 8.7, there is always some. If the unwanted signal level in a given channel is 30 dB lower than the wanted signal, then we say that the stereo separation is 30 dB. Suppose that a stereo broadcast is being received such that the sound level from the left channel is 90 dB. The amount of the left channel sound which had leaked onto the right channel would, by definition, be 30 dB below this level. Therefore, a listener would also hear 60 dB (90–30) of material which was originally destined for the left channel appearing through the right speaker. The stereo separation of a tuner should be greater than 20 dB over much of the audible frequency range (100 Hz to 10,000 Hz), and frequently values higher than 30 dB are found at frequencies near 1,000 Hz. Stereo separation decreases as the audio frequency increases. If only one stereo separation figure is quoted, it was probably measured at a midrange frequency near 1 kHz.

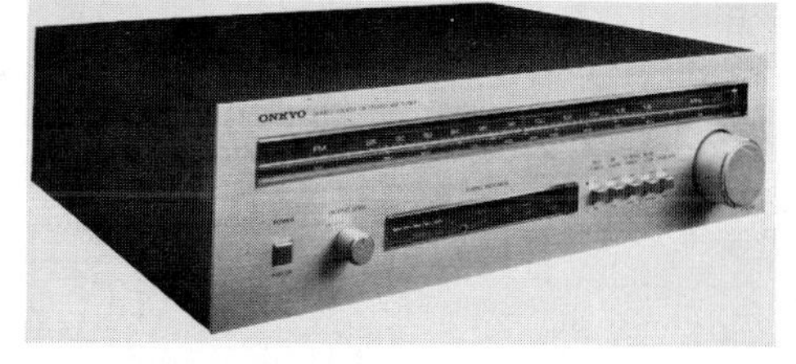

Model T-4090 Onkyo tuner

Some tuners have midrange frequency stereo separations of 40 dB and higher, although it is doubtful that such high figures would enhance your listening experience. First of all, the DJ plays records with a cartridge which, itself, has a stereo separation of only 20 to 25 dB. Second, the FCC requires that a broadcaster maintain a minimum separation of 30 dB at all frequencies during the transmission process. Most stations transmit with this minimum value, so that a tuner with more than 30 dB of stereo separation will have very little effect on the sound quality.

Tuner Buying Guide

Stereo Separation:

1. Look for a stereo separation of 30 dB, or higher, at midfrequencies (near 1 kHz).
2. Look for a stereo separation of 20 dB, or higher, over much of the audible range (100 Hz to 10 kHz).

Alternate-Channel Selectivity

Selectivity represents the tuner's ability to slice out the desired station from the hodgepodge of radio waves reaching it from the antenna. In metropolitan areas there is usually a large number of FM stations, and the dial becomes very "crowded." There may be a good chance of interference between the station that you have tuned in and others which are nearby on the dial. Theoretically, two stations which are adjacent to each other on the FM dial should produce no mutual interference problems. In practice, a certain portion of the "neighboring" broadcast does manage to find its way into the tuner's circuitry and finally emerge through your system. The FCC, as mentioned in section 10.6, attempts to minimize this problem by (usually) assigning stations in any given area to channels which are 0.4 MHz apart; the "alternate channel" spacing (see Fig. 10.16). If you are located close to a powerful FM station and you wish to receive a weaker station which is only 0.4 MHz removed, you will need a tuner with a high alternate-channel selectivity.

Alternate-Channel Selectivity: The number of decibels by which the RF level of an undesired station must exceed the RF level of a tuned in station in order that the undesired audio program is 30 dB *below* the desired audio program. The word "alternate" means that the undesired station is located 0.4 MHz away from the desired station on the FM dial.

Suppose that you are listening to a station at 94.5 MHz, and there is possible interference from a strong station which is 0.4 MHz away (either at 94.1 MHz or 94.9 MHz). An alternate-channel selectivity of 50 dB means that the interfering RF signal must be 50 dB stronger (when picked up by the antenna) than the desired signal in order to produce an audio interference which is 30 dB below the desired program (see Fig. 10.27). According to Table 5.1, a signal which is 50 dB stronger is 100,000 times stronger. Therefore, the interfering RF signal (at either 94.1 MHz or 94.9 MHz) must be 100,000 times stronger than the 94.5 MHz signal in order to produce any significant interference!

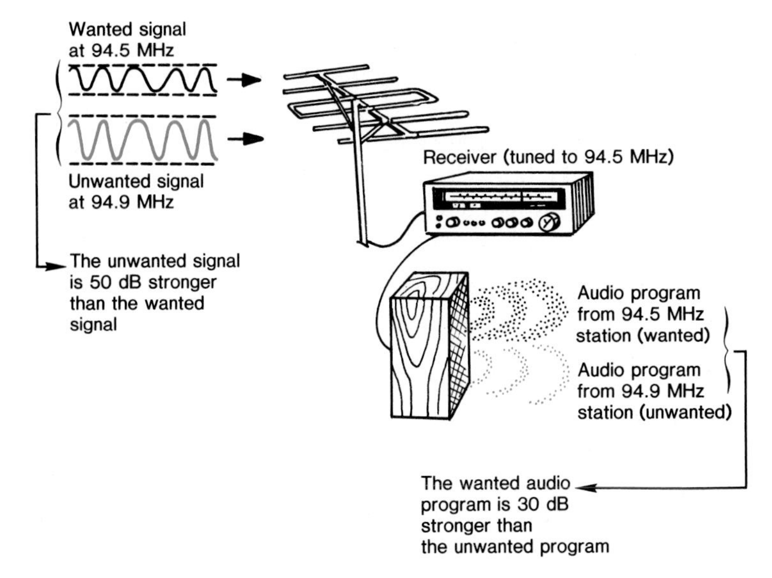

Figure 10.27 An alternate-channel selectivity of 50 dB is illustrated in this drawing.

The 30 dB audio difference is the standard reference level which is always used when comparing the selectivities of different tuners. Different tuners, however, require different RF input levels from the alternate channel in order to achieve this 30 dB difference. These various input levels, expressed in dB, represent the selectivities of the tuners. Most tuners have selectivity ratings of 60 dB or greater. Also, a highly directional antenna can be of great help in selecting out the weaker station without receiving interference from the stronger one; this is particularly true if the two stations lie in different directions.

Tuner Buying Guide

Alternate-Channel Selectivity: Look for an alternate-channel selectivity of 60 dB or higher (70 dB, 75 dB, etc.).

Capture Ratio

A very common television problem often arises when "ghosts" appear on the screen. Ghosts are caused when the television signal arrives at the antenna by more than one path from the station. This phenomenon is called "multipath interference" (see Fig. 10.28). Since the reflected wave travels a longer path than does the direct wave, it will arrive at the antenna a short time after the direct wave and produce the familiar ghost. Ghosts constitute a form of visual distortion, and they often can be greatly reduced by using a

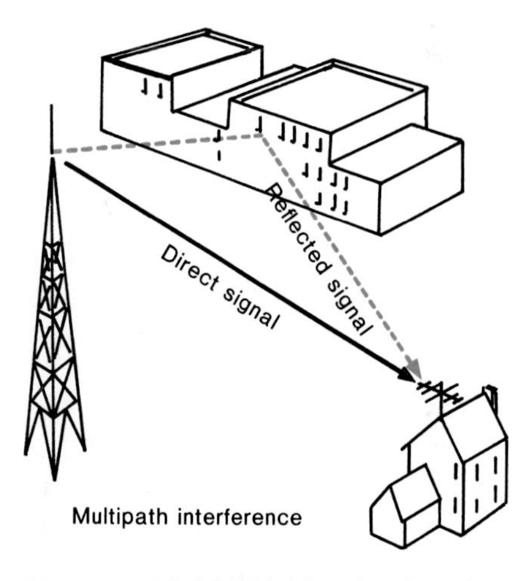

Figure 10.28. Multipath interference can arise when a reflected signal, in addition to the direct signal, reaches the antenna.

directional antenna. Such an antenna, when properly aimed, possesses maximum sensitivity in the direction of the direct wave, and a reduced sensitivity in the direction of the reflected wave. Tuners, like TVs, are also subject to multipath interference which produces audio, rather than visual, distortion. In addition to employing a good directional FM antenna, multipath distortion can be reduced by using a tuner which has a good *capture ratio*. Capture ratio is the property of an FM tuner which allows it to receive only the stronger of two signals which have the *same RF frequency*.

Capture Ratio: The number of decibels by which a stronger RF signal must exceed a weaker RF signal in order for the audio output from the stronger signal to be 30 dB greater than that from the weaker signal. The two RF signals have the same radio frequency.

Suppose that a tuner has a capture ratio of 3 dB. Furthermore, let the strength of the direct signal (at 94.5 MHz) be only 3 dB stronger than that of the reflected signal (also at 94.5 MHz), as shown in Figure 10.29. The tuner's circuitry will suppress the audio output of the reflected signal to the point where it is 30 dB below that of the direct signal's output. A slight additional increase in the strong, direct signal's strength will completely eliminate the interfering signal.

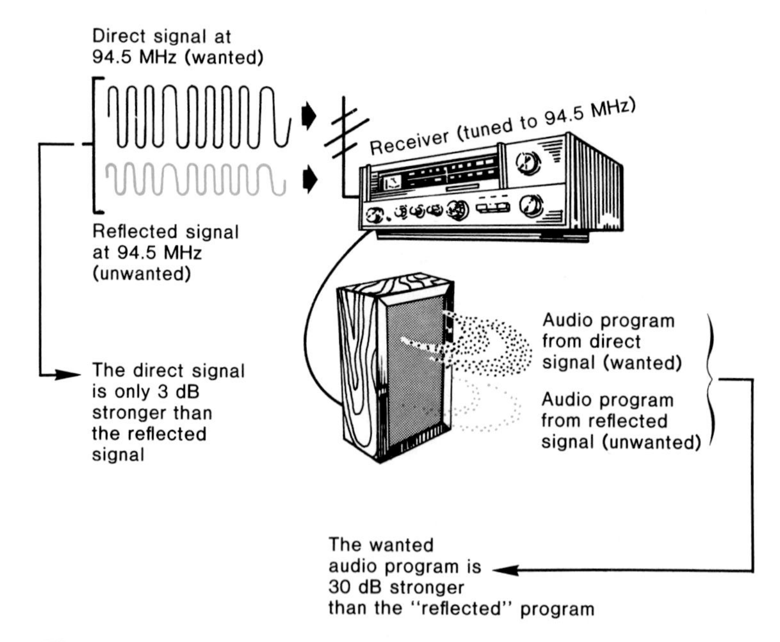

Figure 10.29. A capture ratio of 3 dB is illustrated in this drawing.

Like the selectivity specification, the capture ratios of different tuners are measured by always maintaining a 30 dB difference between the wanted audio signal and the unwanted one. Different tuners require different RF input levels from the reflected wave in order to produce this 30 dB difference. The capture ratios of these tuners are just these input levels, expressed in dB.

Tuner Buying Guide

Capture Ratio: Look for a capture ratio of 3 dB or smaller (2.5 dB, 2.0 dB, etc.).

Additional Tuner Specifications

There are a few additional tuner specifications which you might run into, especially if you carefully read specification sheets or tuner reviews, such as those found in hi-fi magazines. **Image rejection, IF rejection,** and **spurious response rejection,** are lesser known specifications ; all should be 60 dB or higher. They measure the tuner's ability to reject unwanted signals which occasionally appear on the dial. Unwanted commercial aircraft or other mobile transmissions can be reduced, if not eliminated, by a tuner with good image rejection qualities. **AM suppression** is a measure of the tuner's ability to reject the extraneous AM modulation in an FM radio wave, along with such potential noise-producing sources as fluorescent lamps, automobile ignitions, etc. AM suppression is expressed in dB with values greater than 50 dB being acceptable.

Tuner Buying Guide

Look for the following values for these specifications:
- **Image Rejection:** 60 dB or higher.
- **IF Rejection:** 60 dB or higher.
- **Spurious Response Rejection:** 60 dB or higher.
- **AM Suppression:** 50 dB or higher.

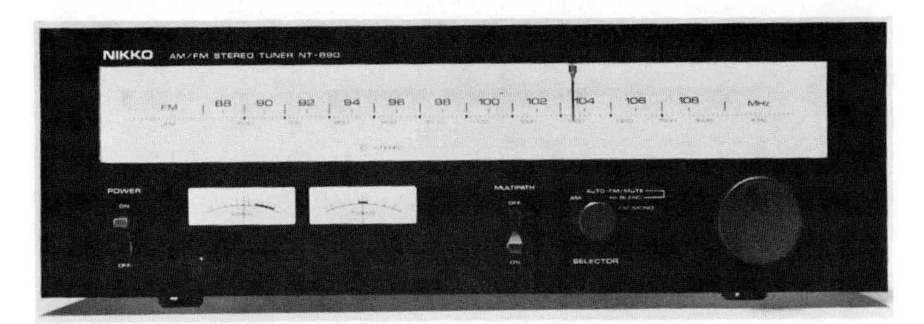

Model NT-890 Nikko tuner

Summary of FM Tuner Specifications and Buying Guides

Usable (IHF) Sensitivity

Look for a usable sensitivity of 3 μV or less for 30 dB quieting, or, alternatively, 15 dBf or less for 30 dB quieting.

50 dB Quieting Sensitivity

MONO: Look for a MONO 50 dB quieting sensitivity of 25 dBf (9.7 μV) or less.
STEREO: Look for a STEREO 50 dB quieting sensitivity of 38 dBf (44 μV) or less.

Signal-to-Noise Ratio (S/N)

MONO: Look for a MONO S/N ratio of 65 dB or higher (70 dB, 75 dB, etc.).
STEREO: Look for a STEREO S/N ratio of 60 dB or higher.

Total Harmonic Distortion (THD)

MONO: Look for a MONO THD of 0.5% or less.
STEREO: Look for a STEREO THD of 1.0% or less.

Frequency Response

Look for a frequency response of 30 Hz to 15 kHz, ± 1 dB or better. The smaller the "±dB," the better. The wider the frequency range, the better.

Stereo Separation

1. Look for a stereo separation of 30 dB, or higher, at midfrequencies (near 1 kHz).
2. Look for a stereo separation of 20 dB, or higher, over much of the audible range (100 Hz to 10 kHz).

Alternate-Channel Selectivity

Look for an alternate-channel selectivity of 60 dB or higher (70 dB, 75 dB, etc.).

Capture Ratio

Look for a capture ratio of 3 dB or smaller (2.5 dB, 2.0 dB, etc.).

Image Rejection

Look for an image rejection of 60 dB or higher.

IF Rejection

Look for an IF rejection of 60 dB or higher.

Spurious Response Rejection

Look for a spurious response rejection of 60 dB or higher.

AM Suppression

Look for an AM suppression of 50 dB or higher.

Summary of Terms

AFC—(See **Automatic Frequency Control**.)

Alternate-Channel Selectivity—The ability of a tuner to receive only the desired station while rejecting the alternate-channel stations which are not wanted. Measured in dB, with larger numbers being better. Technically it is the number of decibels by which the RF signal strength of an undesired station must exceed the RF strength of a tuned in station in order to force through an undesired audio program which is 30 dB below the desired audio program. The word "alternate" means that the undesired station is located on the dial 0.4 MHz away from the desired station.

AM—(See **Amplitude Modulation**.)

Amplitude Modulation (AM)—A type of modulation in which the audio information is conveyed by changing the amplitude of the carrier wave. Large audio levels produce large amplitude changes of the carrier wave. AM radio is the most common form of broadcasting, although it does have some defects which limit its use for hi-fidelity listening.

AM Suppression—The ability of an FM tuner to suppress changes in the amplitude of the received signals, thereby improving the S/N ratio by rejecting unwanted noise and interference.

Antenna—A structure of metallic rods or wires, with which the radio waves interact to produce radio frequency AC electricity. This electricity is sent as input to the tuner.

Audio Signal—The wave of AC electricity which is processed by hi-fi components. It is the electrical replica of the sound wave, and, as such, its frequencies lie in the range from 20 Hz to 20 kHz.

Automatic Frequency Control (AFC)—A circuit which automatically keeps the tuner fine tuned to the selected station. Most hi-fi tuners have an AFC feature, but some provide a front panel switch which allows it to be turned off.

Bandwidth—A range of frequencies which may be usable for a particular purpose. For example, the audio bandwidth for FM transmission is from 30 Hz to 15 kHz; the FM broadcast bandwidth is from 88 MHz to 108 MHz; the bandwidth of human hearing is from 20 Hz to 20 kHz, etc.

Capture Ratio—The ability of a tuner to reject the weaker of two RF signals which have the same frequency. Measured in dB, with smaller numbers being better. Technically it is the number of decibels by which a stronger RF signal must exceed a weaker RF signal in order for the audio output from the stronger signal to be 30 dB greater than that from the weaker signal.

Carrier—In hi-fi it is the radio wave which "carries" the audio information from a transmitter. The frequency of the carrier wave is assigned to the station by the FCC, and corresponds to the number of the station "on the dial."

dBf—A unit for measuring RF signal levels which expresses tuner input levels in terms of decibels rather than microvolts. The "f" in dBf stands for "femtowatt" (10^{-15} watts), and it represents the reference power level used for the 0 dB point. See the Appendix.

De-emphasis—(See **Pre-emphasis.**).

Demodulation—(See **Modulation.**)

Electromagnetic Waves—A class of waves which travel at the speed of light (3×10^8 meters/sec. or 186,000 miles/sec.) in a vacuum. Members of this class include such well-known types as: AM and FM radio waves, visible light, X-rays, and infrared waves.

FCC—Federal Communications Commission.

Femtowatt—10^{-15} watts.

Fifty (50) dB Quieting Sensitivity—A specification which indicates how well a tuner can convert weak radio waves into a satisfactory audio signal. Technically, it is the smallest RF signal which causes the tuner to produce a total audio output which is 50 dB greater than the background noise. It is measured in microvolts (μV) or in dBf.

FM—(See **Frequency Modulation.**)

FM Band—A range of radio frequencies, from 88 MHz to 108 MHz, set aside for FM broadcasts.

FM MUTE (RUSH, SQUELCH)—A control on a tuner which can be used to suppress the "hissing" noise heard when dialing between stations.

FM Trap—A filtering device which is sometimes installed on TV antenna systems to prevent FM radio waves from interfering with TV reception.

Frequency Modulation (FM)—A type of modulation of radio waves in which the audio information is conveyed by changing the frequency, and not the amplitude, of the radio wave. Large level audio signals cause large changes in the carrier's frequency. FM broadcasting is a hi-fidelity medium used for transmitting both stereo and monophonic programs.

Frequency Response—A specification which indicates how uniformly a tuner reproduces the amplitude of one audio frequency compared to another. In order to have full meaning this specification must include both the frequency range and the ± decibel deviation from perfect flatness, e.g., 30 Hz ——> 15,000 Hz, +1 dB, −1 dB.

Guard Band—A range of frequencies which helps prevent stations (which are adjacent on the dial) from interfering with each other. Each FM station is allowed to change the assigned carrier frequency by as much as ±75 kHz in order to convey the audio information. An additional 25 kHz, called the guard band, is allocated at each extreme.

HIGH-BLEND—A feature provided on some tuners which mixes some of the high-frequency sounds from both channels, thereby reducing the noise level. The high-blend feature results in some loss of stereo separation.

IF Rejection—(See **Image Rejection.**)

IHF Sensitivity—(See **Usable Sensitivity.**)

Image Rejection (IF Rejection, Spurious Response Rejection)—One of several specifications which, in general, measure a tuner's ability to reject extraneous signals which occasionally pop up on the FM dial. Two other specifications of this type are IF rejection and spurious response rejection. These specifications are measured in dB, with higher values being better.

Limiter—A circuit in an FM tuner that removes any amplitude variations, such as spikes, caused by atmospheric or ignition noise. FM limiters are largely responsible for producing noise free FM reception.

Microvolt (μV)—One millionth (10^{-6}) of a volt.

Modulation (Demodulation)—A technique by which information is conveyed by a carrier wave in which some property of the wave is allowed to change: e.g., its amplitude or its frequency. The opposite technique is called demodulation, whereby the information is recovered from the carrier wave. Tuners demodulate the modulated carrier waves to recover the audio information.

Multipath Interference—The interference resulting between two radio waves which have identical RF frequencies, and which arrive at the antenna via two or more different paths. One path is a direct path, and the others exist due to reflections of the wave from buildings and other objects.

MUTE—(See **FM MUTE.**)

Pre-emphasis (De-emphasis)—The boosting of a specified range of input frequencies to a component such as a tuner, turntable, or a tape deck. A reverse process, called de-emphasis, occurs in the output which restores the audio signal to its original level. Any noise introduced between the input and output, which was not pre-emphasized, is reduced by the de-emphasis, thus producing a larger S/N ratio. The process is slightly different for each type of source, and it can be thought of as a type of equalization.

Radio—A unit which contains a tuner, a preamplifier/control center, a power amplifier, and speakers. A radio and a tuner are not the same thing.

Radio Frequency (RF) Signal—An electromagnetic wave which can carry information, either by AM or FM, from a transmitter to the tuner. RF signals in the AM broadcast band range from 535 kHz to 1605 kHz, while those in the FM band possess frequencies from 88 MHz to 108 MHz.

RF—Radio frequency.

RUSH—(See **FM MUTE.**)

SCA—(See **Subsidiary Communications Authorization.**)

Sensitivity—(See **Fifty (50) dB Quieting Sensitivity and Usable Sensitivity.**)

Signal Strength Meter—A meter which measures the RF level entering the tuner. It is used as an aid when tuning in the desired station.

Signal-to-Noise Ratio (S/N)—A specification which tells how much audio signal there is relative to the noise under the most favorable conditions. Technically, it is the ratio of the total audio output level to the noise level, expressed in decibels, when the RF input level is 1,000 μV (65 dBf).

Spurious Response Rejection—(See **Image Rejection.**)

SQUELCH—(See **FM MUTE.**)

Stereo Separation—A specification which tells how much sound has leaked from the channel for which it was intended into the other channel for which it was not intended. Technically, it is the ratio of the power of the audio signal, in the channel for which it is intended, to the power of this signal which has leaked over to the other channel. This ratio is expressed in decibels.

Stereo Threshold—The minimum RF signal level needed at the tuner input for the tuner to be able to produce a stereo program. It is measured in microvolts or in dBf.

Subcarrier—A 38 kHz electrical signal which is "carried" by the main FM radio wave when a stereo program is being broadcast. This extra signal is a subcarrier in the sense that it carries the extra audio information from which a stereo tuner extracts the second channel of sound. The prefix "sub" is used to distinguish it from the main carrier wave.

Subsidiary Communications Authorization (SCA)—A monaural broadcast (200 – 5,000 Hz) which is used as background music in restaurants, airline terminals, offices, etc. The SCA program is transmitted by many FM stations right along with their normal stereo programming; however, a special decoder must be added to the tuner in order to receive the SCA program.

Total Harmonic Distortion (THD)—The undesired harmonic frequencies which are added to a fundamental frequency as it passes through a tuner. It is expressed as a percentage relative to the total audio output.

Tuner—A part of a receiver, or a separate unit, which receives radio broadcasts and converts them into audio frequency signals.

Tuning Meter—A meter, sometimes found in a tuner, which can be used in the tuning process. It will indicate the exact center frequency of the FM signal being received, i.e., the exact carrier wave frequency.

Usable Sensitivity (IHF Sensitivity)—A specification which indicates how well a tuner can convert weak radio signals into a satisfactory audio signal. Usable sensitivity (or IHF sensitivity) is the smallest RF signal that will produce an audio output signal which is 30 dB above the distortion and noise. It is measured in microvolts or in dBf.

Review Questions

1. List the differences between AM and FM broadcasting.
2. What is the meaning of modulation and how can it be used to convey information?
3. Compare the various types of electromagnetic waves and describe some of their special wave properties, such as speed, wavelength, frequency, etc.
4. Define amplitude modulation and explain how it is used to transmit information in AM broadcasts.
5. Explain how an FM modulated radio wave can convey an audio signal.
6. Describe the FM frequency assignments, guard bands, and allowed frequency changes of the carrier.
7. Describe the pre-emphasis and de-emphasis in FM transmission and reception processes. What is their purpose?

8. List some common tuning aids and describe how each helps the user in obtaining good quality FM reception.
9. What is the meaning of the usable sensitivity specification of a tuner?
10. How is the 50 dB quieting sensitivity determined? How is the 50 dB quieting sensitivity different from the usable sensitivity?
11. How is the stereo separation of a tuner measured, and what is considered a good value?
12. Explain the meaning of the alternate-channel selectivity specification.
13. When is a good capture ratio necessary, and what is an acceptable value for it?
14. What is an SCA transmission?
15. How far apart in frequency are AM stations assigned?
16. Why is AM more noisy than FM?
17. Why is a good antenna necessary for FM reception?
18. What is the purpose of the MUTE control?

Exercises

NOTE: The following questions have up to 5 possible answers. Please select the **one** response which best answers the question.

1. Which one of the following specifications is stated correctly for a tuner's capture ratio?
   1. 30 Hz to 18 kHz ± 1 dB.
   2. 1%.
   3. 3 dB.
   4. 2 μ V.
   5. 60 dBf.
2. Which one of the following specifications is correctly stated for a tuner's usable sensitivity?
   1. 30 Hz to 15 kHz ± 1 dB.
   2. 2 μ V for 1% THD.
   3. 2 μ V (or 11 dBf) for 30 dB quieting.
   4. 3 dB.
   5. 3%, or less.
3. If a tuner has a usable sensitivity of 3 μ V, what will be the **ultimate** (or best) S/N ratio that the tuner can produce?
   1. 3 μ V (or 15 dBf).
   2. 30 dB.
   3. 20 dB.
   4. 40 dB.
   5. The S/N cannot be determined from the above data alone.
4. If a 400 Hz signal and a 1,000 Hz signal are introduced simultaneously into a tuner with 4% IM (and **no** THD) which of the following frequencies will be produced? (See section 8.7 for a discussion on IM distortion.)
   1. 400 and 1,000 Hz only.
   2. 400, 1,000, 600, 1,400 Hz.
   3. 400, 800, 1,000, 2,000 etc., Hz.
   4. 1,400 Hz only.
5. The usable sensitivity of a tuner is determined where:
   1. the noise and distortion are 30 dB below the total audio output signal.
   2. the THD is less than 0.1%.
   3. the total audio output signal is 30 dB below the noise.
   4. the S/N ratio is greater than 60 dB.
   5. the THD is greater than 1%.
6. If you live very far from broadcast transmitters, you should buy a tuner with a good ________ .
   1. muting ratio
   2. selectivity
   3. stereo separation
   4. sensitivity
   5. frequency response
7. The ability of a tuner to pick up very weak or distant signals and turn them into listenable audio programs is called the tuner's:
   1. frequency response.
   2. S/N ratio.
   3. sensitivity.
   4. selectivity.
   5. THD distortion.
8. What are acceptable values of THD and IM for an amplifier (not a tuner)?
   1. Both less than 10%.
   2. Both less than 5%.
   3. THD less than 1%, IM less than 5%.
   4. Both less than 0.5%.
   5. Both less than 1 dB.
9. What is the technical specification which describes the tuner's undesirable property of adding overtones to a pure tone which has been transmitted over the FM airways?
   1. IM distortion.
   2. Selectivity.
   3. Capture ratio.
   4. THD.
   5. Frequency response.

10. The tuner specification which expresses the total audio output signal strength compared to the background noise is called the:
    1. N/S ratio.
    2. S/N ratio.
    3. capture ratio.
    4. THD.
    5. selectivity.
11. You are listening to an FM station and the sound level is 90 dB. What is the **noise** level if the tuner's signal-to-noise specification is 65 dB?
    1. (90 + 65) = 155 dB.
    2. (90 − 65) = 25 dB.
    3. 90/65 dB.
    4. 65/90 dB.
    5. None of the above answers is correct.
12. A tuner has a signal-to-noise ratio of 80 dB. This means that, considering only the tuner in your hi-fi system, the total audio output signal you hear will sound ____________ times louder than the noise.
    1. 320
    2. 256
    3. 160
    4. 128
    5. 80
13. As the FM signal becomes progressively **weaker,** what happens to the noise generated by the tuner?
    1. It becomes larger.
    2. It becomes smaller.
    3. It does not change.
14. If you lived near a lot of tall buildings and had a lot of reflected "ghosts" coming into your FM tuner, you would want to buy a tuner with a good ___________ rating.
    1. selectivity
    2. sensitivity
    3. S/N
    4. capture ratio
    5. distortion
15. For tuners, the specification which determines how well the tuner will reject the weaker of two signals broadcast at the same frequency is called the:
    1. capture ratio.
    2. selectivity.
    3. signal-to-noise ratio.
    4. sensitivity.
    5. frequency response.
16. In any given region, say the Chicago area, what is the closest that the FCC will normally assign two FM radio stations?
    1. 10 kHz.
    2. 0.4 MHz.
    3. 0.6 MHz.
    4. 30 kHz.
    5. 5 kHz.
17. If there is a large number of stations broadcasting in your area, the tuner specification you should look closely at is:
    1. capture ratio.
    2. selectivity.
    3. signal-to-noise ratio.
    4. sensitivity.
    5. frequency response.
18. The term which describes the ability of a tuner to pick out a desired FM signal, while at the same time rejecting signals from adjacent stations operating at different frequencies, is the:
    1. capture ratio.
    2. sensitivity.
    3. selectivity.
    4. frequency response.
    5. mute.
19. What tuner specification does the following refer to: 30 to 15,000 Hz ± 1 dB?
    1. Frequency response.
    2. Sensitivity.
    3. Capture ratio.
    4. Selectivity.
    5. None of the above answers is correct.
20. The FCC requires a stereo separation in FM transmission of:
    1. at least 10 dB.
    2. at least 60 dB.
    3. at least 30 dB.
    4. at least 70 dB.
    5. at least 80 dB.
21. The voltage at the antenna of a tuner is usually between:
    1. 1 mV and 1,000 mV.
    2. 1 volt and 1,000 volts.
    3. 1 μ V and 1,000 μ V.
    4. ± 500 dB.
    5. None of the above answers is correct.
22. Which of the following methods of broadcasting is assigned the longest wavelength carrier?
    1. FM.
    2. AM.
    3. They both have the same wavelength.
    4. None of the above answers is correct.

23. The switch on the front panel of a tuner which allows one to tune between stations without encountering the annoying harsh noise (or static) that you would otherwise hear is the:
    1. STEREO/MONO switch.
    2. the S/N switch.
    3. the AFC switch.
    4. the MUTE switch.
    5. None of the above answers is correct.
24. If you wished to convert a mono FM tuner to a stereo tuner, you would add a:
    1. second speaker.
    2. multiplexer.
    3. FM MUTE control.
    4. stereo light.
    5. second antenna.
25. Adjusting the direction of a directional FM antenna:
    1. has no effect on FM reception.
    2. should be done only when the tuner is turned off.
    3. helps in obtaining maximum RF signal strength.
    4. should be done only when the speakers are turned off.
    5. should be done only in the MONO mode.
26. For FM broadcasts, the higher the pitch of the audio tone the __________.
    1. more times per second the frequency of the carrier must change
    2. fewer times per second the frequency of the carrier must change
    3. more times per second the amplitude of the carrier must change
    4. fewer times per second the amplitude of the carrier must change
    5. None of the above answers is correct.
27. To broadcast a 500 Hz tone, one would vary the frequency of the carrier 500 times a second. This statement refers to:
    1. AM transmission.
    2. flat frequency response.
    3. selectivity.
    4. FM transmission.
    5. None of the above answers is correct.
28. When transmitting information by FM, if the audio increases in loudness, the ___________ .
    1. carrier will have a larger amplitude change
    2. carrier will have a smaller amplitude change
    3. carrier will have a larger shift in frequency
    4. carrier will have a smaller shift in frequency
    5. No change will occur in the carrier.
29. The louder the audio signal that is to be broadcast via AM the __________ .
    1. greater the frequency change of the carrier
    2. smaller the frequency change of the carrier
    3. greater the amplitude change of the carrier
    4. smaller the amplitude change of the carrier
    5. None of the above answers is correct.
30. To broadcast a 500 Hz tone, the amplitude of the radio wave is varied 500 times per second. This statement refers to:
    1. AM transmission.
    2. flat frequency response.
    3. selectivity.
    4. FM transmission.
    5. None of the above answers is correct.
31. A disadvantage of FM broadcasting, compared to AM, is that:
    1. FM tuners use more power than AM tuners.
    2. FM tuners are usually heavier than AM tuners.
    3. FM tuners have less hi-fidelity than AM tuners.
    4. FM broadcasts do not transmit as far as AM.
    5. FM broadcasts have more commercials.
32. AM is considered a low-fidelity broadcasting medium because:
    1. the typical transmitting towers are located in deep valleys.
    2. the range of broadcasted audio frequencies extends to 15 kHz.
    3. the range of broadcasted audio frequencies extends to 10 kHz.
    4. the range of broadcasted audio frequencies extends to 5 kHz.
    5. The range of broadcasted audio frequencies extends to 1605 kHz.
33. The range of transmission frequencies over which FM is broadcast is:
    1. 55–160 MHz.
    2. 535–1605 kHz.
    3. 100–200 MHz.
    4. 88–108 MHz.
    5. 88–108 kHz.
34. What is the highest audio frequency transmitted by AM?
    1. 2 kHz.
    2. 5 kHz.
    3. 10 kHz.
    4. 15 kHz.
    5. 20 kHz.

35. FM stations do not broadcast audio frequencies above:
    1. 1 kHz.
    2. 5 kHz.
    3. 12 kHz.
    4. 15 kHz.
    5. 18 kHz.

36. Which one of the following is an advantage of FM over AM?
    1. FM is relatively noise free.
    2. FM has a greater audio frequency range.
    3. There are many more FM stereo stations.
    4. All of the above are advantages of FM.

37. As the FM signal strength becomes weaker, the S/N ratio becomes smaller. The tuner's sensitivity, which the IHF adopted in 1975, is:
    1. the strength of the FM radio wave which will produce a S/N ratio of 50 dB.
    2. the strength of the FM radio wave which will produce a S/N ratio of 30 dB.
    3. 20 dBf for all mono tuners.
    4. 50 dB for all tuners.
    5. a measure of the tuning dial's selectivity.

38. The AFC is:
    1. not important in FM tuners.
    2. very important in FM tuners for locking onto the station's frequency.
    3. important in AM tuners to prevent interference between stations.
    4. a method for broadcasting FM.
    5. a specification which has nothing to do with a tuner.

39. The purpose of the guard band in FM transmission is to:
    1. prevent interference between adjacent stations.
    2. prevent interference between alternate stations.
    3. prevent interference due to noise spikes.
    4. stop AM stations from being received on the FM band.
    5. separate any two FM or AM stations in order to prevent their interference.

40. If an FM station is assigned a frequency of 98.3 MHz, the frequency of the carrier wave may be changed, by the audio signal, to any frequency within the range of __________ .
    1. 96.300 to 100.300 MHz
    2. 98.100 to 98.500 MHz
    3. 98.200 to 98.400 MHz
    4. 98.225 to 98.375 MHz
    5. None of the above answers is correct.

41. Two meters, called signal strength and tuning:
    1. accomplish the same function in the tuning process.
    2. aid in the tuning process because one meter measures how much radio signal is being received, and the other tells if the tuner is set exactly to the proper frequency.
    3. aid in the tuning process because the tuning meter measures the amount of radio signal while the signal strength meter measures the audio signal strength.
    4. aid in the tuning process because one meter measures mono broadcasts while the other measures stereo broadcasts.
    5. aid in the tuning process because both meters measure the audio output level.

42. When a radio station sends out an FM broadcast, it __________ .
    1. pre-emphasizes the low frequencies
    2. pre-emphasizes the high frequencies
    3. de-emphasizes the high frequencies
    4. transmits a flat frequency response
    5. neither pre-emphasizes nor de-emphasizes the audio signal

43. The wavelength of a typical AM band radio wave is:
    1. about 10 times longer than that of an FM radio wave.
    2. about 100 times longer than that of an FM radio wave.
    3. about 1,000 times longer than that of an FM radio wave.
    4. about 1,000 times shorter than that of visible light waves.
    5. about one million times shorter than that of visible light waves.

98.300
.075
98.225

98.3
.075

44. A radio wave can be detected only because:
 1. it causes the electrons in the antenna to flow.
 2. it causes the antenna itself to move.
 3. it is unmodulated.
 4. the antenna itself is modulated.
 5. the radio wave travels at the speed of light.

Given below are specifications for two different tuners. Evaluate **Tuner A's** performance against **Tuner B** by using the answers below for the next seven questions.

	Tuner A	**Tuner B**
Usable sensitivity	1.5 μV (8.8 dBf) for dB quieting	3 μV (15 dBf) for 30 dB quieting
Selectivity (alternate-channel)	80 dB	55 dB
Capture ratio	1 dB	2 dB
Signal-to-Noise ratio (stereo)	50 dB	80 dB
Frequency response	30 Hz to 15,000 Hz ± 1 dB	30 Hz to 15,000 Hz ± 2 dB
Stereo separation (1 kHz)	20 dB	10 dB
Harmonic distortion (stereo)	0.9%	0.5%

These are the answers to the next seven questions:
 1. tuner A meets the standards of hi-fidelity while B doesn't.
 2. tuner B meets the standards of hi-fidelity while A doesn't.
 3. both A and B meet the standards for hi-fidelity, although A is better than B.
 4. both A and B meet the standards for hi-fidelity, although B is better than A.
 5. Neither A nor B meets the hi-fidelity standards.

45. The usable sensitivity of __2__.
46. The selectivity of __1__.
47. The capture ratio of __3__.
48. The signal-to-noise ratio of __2__.
49. The frequency response of __1__.
50. The stereo separation of __5__.
51. The harmonic distortion of __4__.

52. Which one of the following statements is *not* true about the technical specification called "capture ratio?"
 1. It pertains to FM tuners.
 2. It is measured in dB.
 3. It is another method for measuring selectivity.
 4. The smaller the capture ratio number, the better.
 5. It is an especially important specification if you live in a crowded metropolitan area.

53. Multipath interference can be reduced considerably by purchasing a tuner with a good:
 1. selectivity.
 2. sensitivity.
 3. frequency response.
 4. S/N ratio.
 5. capture ratio.

Chapter 11

ELECTROMAGNETISM

11.1 Introduction

Electricity and magnetism work hand-in-hand within a variety of audio components such as tape decks, speakers, magnetic cartridges, and microphones. The basic principles of electricity were outlined in Chapter 7, and this chapter covers some of the more important concepts in magnetism. One of the salient conclusions to be reached is that electricity and magnetism are not separate concepts. In fact, electric currents actually produce magnetism and conversely, magnetism can be used to generate electric currents. The interplay between electricity and magnetism is so important that scientists refer to the combination as "electromagnetism." Many hi-fi components are designed to operate on the principles of electromagnetism, and much of this chapter will be devoted to showing explicitly how this is accomplished. Before discussing electromagnetism the properties of simple magnets will first be studied.

11.2 Permanent Magnets and Magnetic Fields

Every magnet has two poles, one called the north pole and the other, of course, the south pole. The poles are given their names because the north pole of a compass needle tends to align toward the earth's geographical north. All magnets exert forces on each other, and these magnetic forces obey the following basic law:

> Like magnetic poles repel, and unlike magnetic poles attract.

Figure 11.1 illustrates this law, and, in this respect, the forces between the poles of a magnet are very similar to the forces between electric charges. There are, however, several major differences. First, the magnetic poles are electrically neutral. Although magnetic forces can be either attractive

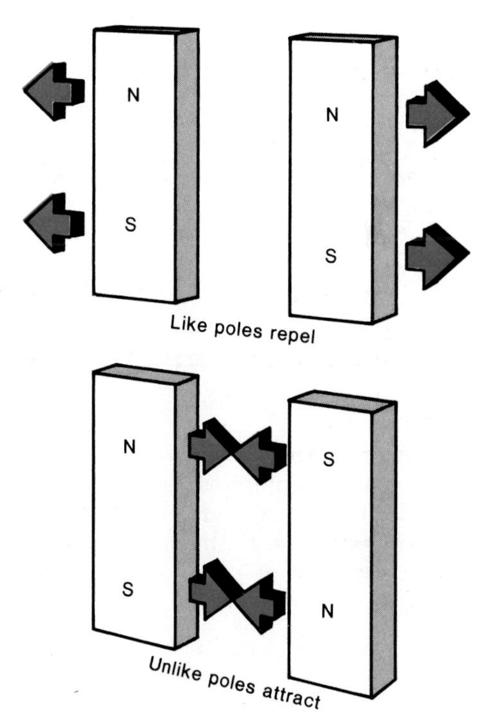

Figure 11.1. The basic force law between magnetic poles; like poles repel and unlike poles attract.

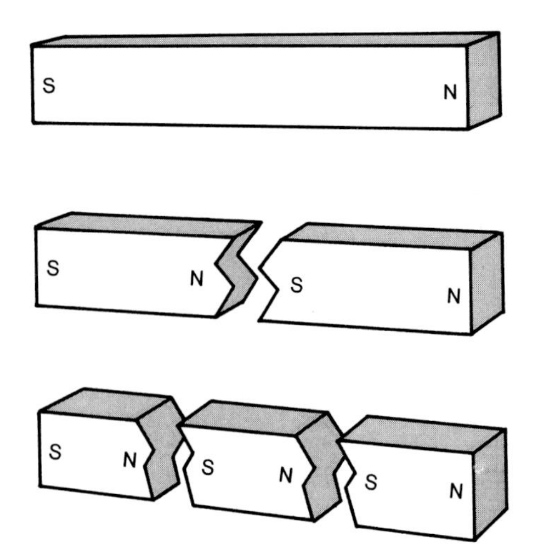

Figure 11.2. Breaking a bar magnet into smaller pieces only results in smaller magnets being formed—each with its own north and south poles.

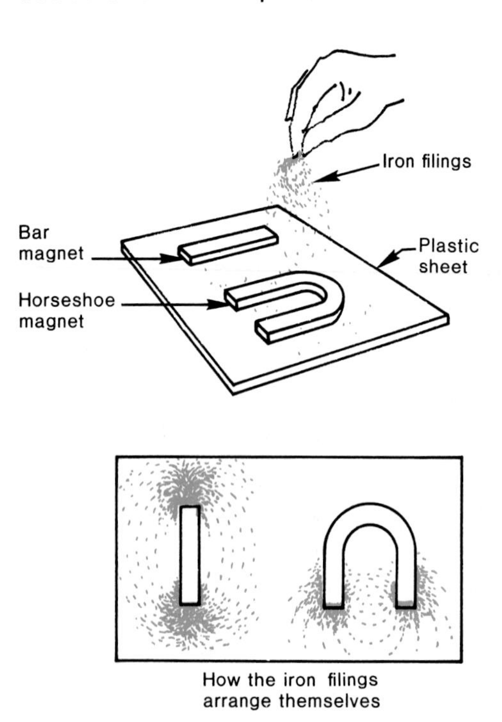

Figure 11.3. Iron filings arrange themselves in definite patterns when sprinkled in the vicinity of magnets. The patterns are a series of "lines" which extend from one pole to another.

or repulsive, they do *not* involve an excess or deficiency of electric charge. Second, magnetic poles always exist in north-south pairs, and it is impossible to isolate one type of pole from the other. Any attempt to break a bar magnet in half simply results in another bar magnet being formed (see Fig. 11.2); magnetic poles, unlike electric charges, cannot be physically separated. Therefore, while electricity and magnetism do share the same type of force law, they appear to be unrelated in most other respects.

Magnetic Fields

Magnetism is an invisible force which can only be seen in terms of the effects which it produces. One can "feel" the force of repulsion between two north poles as they are brought close together. However, there are no visible clues that such an interaction is occurring, because the space between the north poles looks the same whether the poles are there or not. A very simple and useful demonstration can be performed which will, so-to-speak, "develop" the region of space around a magnet, thus making it visible for us to see. Place a sheet of clear plastic over a bar magnet or a horseshoe magnet; then sprinkle some finely-ground iron filings on top of the plastic. If the plastic is gently tapped, the iron filings will form the rather unusual patterns shown in Figure 11.3. It can be seen that the filings arrange themselves into a series of curved lines which begin at one pole of the magnet and end at the other.* Figure 11.4 shows an idealized drawing of these "magnetic lines," where the arrows have been drawn to show, by convention, that the lines are directed from the north pole toward the south pole. The direction of the lines can be determined from the way a compass needle points, as illustrated in the figure. The region of "altered space" which surrounds a magnet is called a *magnetic field.* Figure 11.5 shows the magnetic fields and their associated lines which surround two bar magnets when they are (A) attracting and (B) repelling each other. Notice that the magnetic lines are very concentrated in the region between the two *unlike* poles. Whenever two *like* poles face each other [Fig. 11.5 (B)], the magnetic lines are repulsed from the space between the two poles, thus leaving it as a region of weak magnetic field.

The magnetic field generated by a magnet extends throughout all of space, but it becomes progressively weaker as one moves farther and farther away from the magnet.

*The magnetic lines only appear to begin and end on the poles. Actually they form closed loops which also extend inside the magnet.

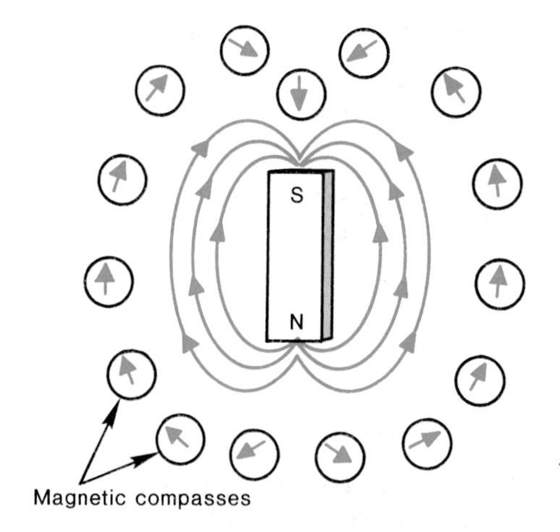

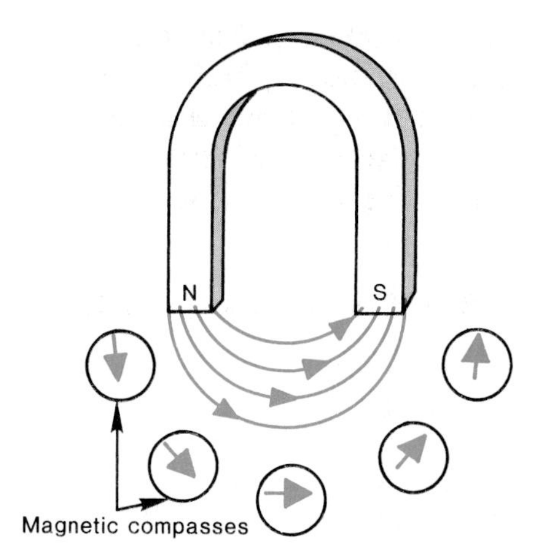

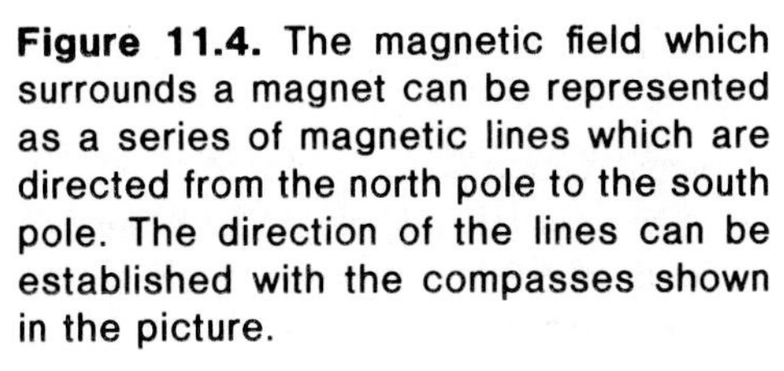

Figure 11.4. The magnetic field which surrounds a magnet can be represented as a series of magnetic lines which are directed from the north pole to the south pole. The direction of the lines can be established with the compasses shown in the picture.

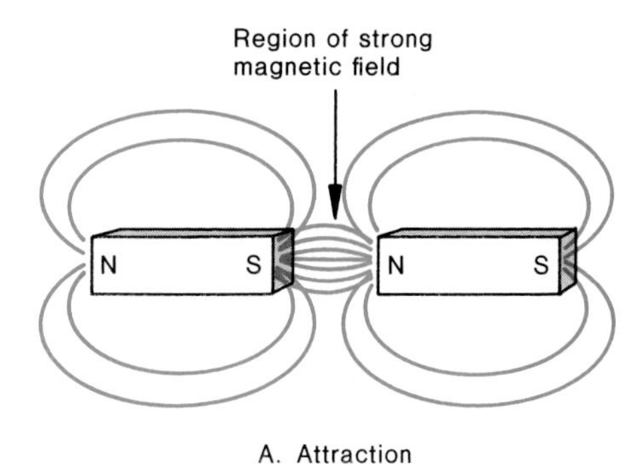

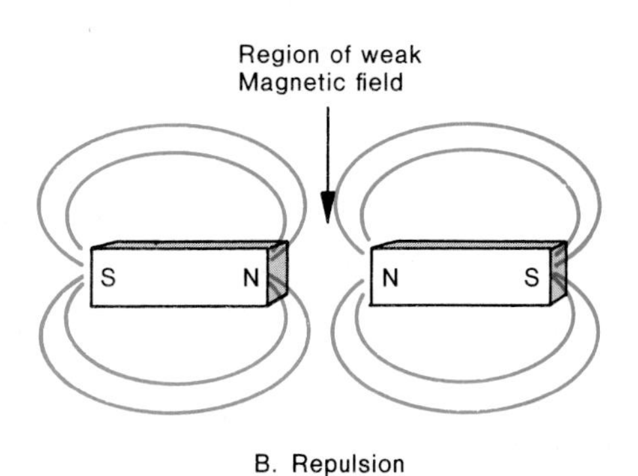

Figure 11.5. A sketch of the magnetic field lines which surround two bar magnets under attractive and repulsive conditions.

The Origin of Magnetism

The property of magnetism may be introduced, or induced, into a piece of iron material which originally was not a magnet. If a piece of unmagnetized soft iron is placed in the magnetic field of a permanent magnet, the iron itself will become a magnet. This process is called *induced magnetism*, and it arises because the magnetic lines of the permanent magnet penetrate the soft iron and cause it to become partially magnetized. The atoms which constitute the iron are themselves tiny magnets. These tiny magnets tend to form regions within the solid called *magnetic domains*. Within a magnetic domain all of the atomic magnets are lined up, so that the domain itself is a magnetized region. Under normal

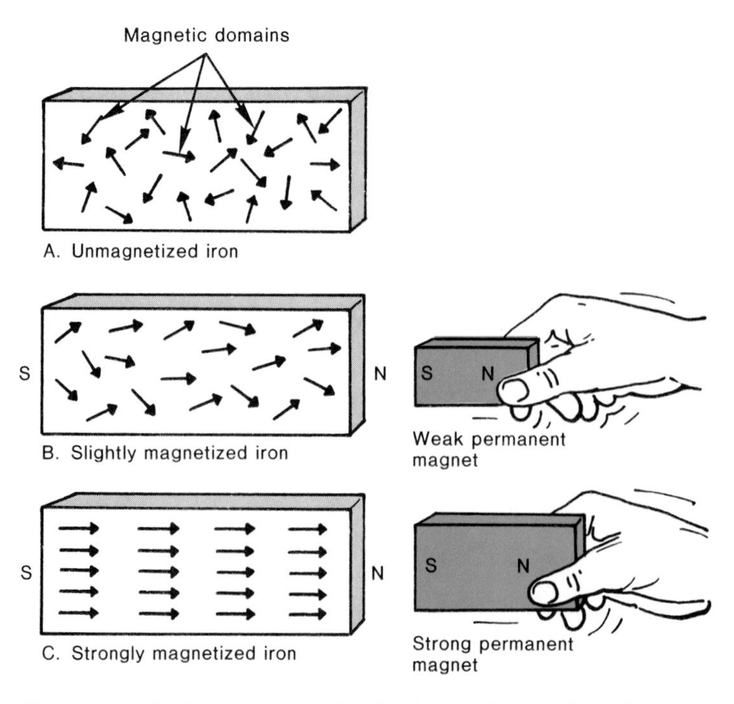

Figure 11.6. An unmagnetized piece of soft iron becomes magnetized when placed in the vicinity of a permanent magnet. Each arrow represents a magnetic domain, which is a region of tiny atomic magnets.

conditions these magnetic domains are randomly aligned, as shown in Figure 11.6(A), so that the iron is initially unmagnetized. When a permanent magnet is brought into the vicinity of the iron, as shown in Figure 11.6 (B), some of the magnetic domains tend to line up such that their north poles are attracted to the south pole of the permanent magnet. Magnetism is thus induced in the iron bar with the polarity as indicated in Figure 11.6 (B). If a stronger permanent magnet is used, more domains are aligned in the iron to produce a stronger induced magnetism, as shown in part (C) of the figure. When the permanent magnet is brought close enough to the iron, the iron will move because of the attraction between the south pole of the permanent magnet and the north pole induced in the iron.

When the permanent magnet is removed, the iron bar will lose some, but not all, of its induced magnetism; the amount which remains is called the *residual magnetism.* Residual magnetism is of prime importance in the operation of magnetic tapes because it actually stores the audio information. This point will be further discussed in Chapter 13.

Since the amount of magnetism depends on the proper alignment of the domains, anything which upsets the alignment reduces the magnetism. Thus, heating a magnet will weaken its magnetism, because heat randomizes the alignment of the domains. Likewise, repeatedly dropping a magnet will weaken it, as the domains are jarred out of alignment.

Summary

Magnets possess north and south poles which allow them to exert magnetic forces on other magnets; like magnetic poles repel each other, while unlike poles are attracted. The altered region of space which surrounds a magnet is called a magnetic field. Magnetic fields can be "seen" by sprinkling iron filings around a magnet; they consist of magnetic lines which are directed, by convention, from the north pole toward the south pole. Materials which can be magnetized are comprised of large numbers of small magnetic regions called domains. Each tiny domain is, in itself, a small magnet which includes many, even smaller, atomic magnets. In a permanent magnet most of the domains are aligned so as to produce a strong amount of magnetism. Conversely, if the domains are randomly aligned, the material does not possess any overall magnetism. Induced magnetism can result when the domains are forced into alignment with the aid of an already magnetized piece of material.

11.3 Magnetic Fields Are Also Produced by Electric Currents

One of the most significant scientific breakthroughs was the discovery that a current-carrying wire also produces a magnetic field. A demonstration of this fact can be performed by sending a current through a vertical copper wire as shown in Figure 11.7. The copper wire is inserted through a hole in a plastic sheet which is held in a horizontal position. When the iron filings are sprinkled on the plastic sheet, they pattern themselves into concentric closed circles about the wire indicating the presence of magnetism. In this case there are no obvious north or south poles as occurs with a permanent magnet. The iron filings can be replaced by magnetic compasses, as shown in Figure 11.8, and the compass needles

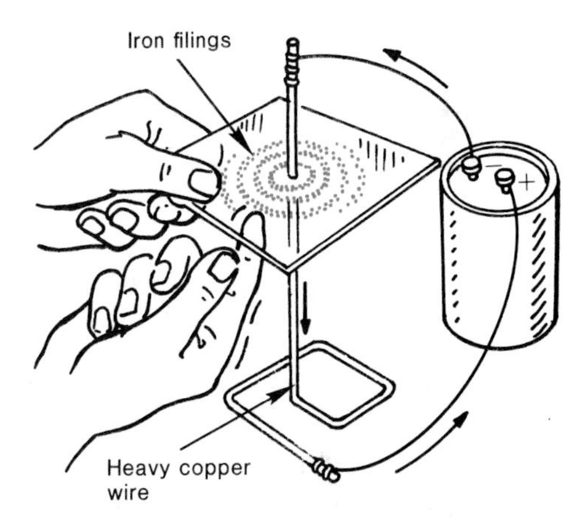

Figure 11.7. A current flowing through a long straight wire produces a magnetic field. The magnetic field lines are concentric circles which are centered about the wire.

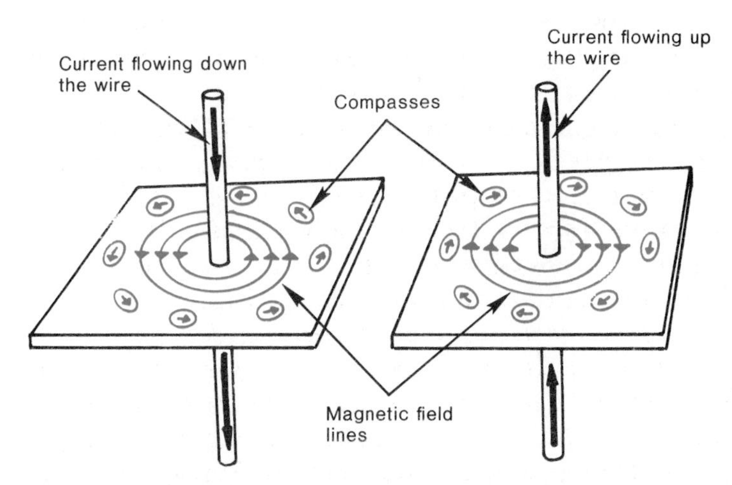

Figure 11.8. The direction of the magnetic field lines around a current-carrying wire reverses when the current reverses. Compasses may be used to establish the direction of the magnetic field lines.

will also line up in a circular pattern. In this case the direction of the magnetic lines is taken to be in the direction of the compasses' north poles. If the direction of current through the conductor is reversed, the compass needles will also point in the opposite direction, indicating that the direction of the magnetic field lines has been reversed. Thus, we see that electric currents create magnetic fields in their vicinity. This phenomenon is called *electromagnetism*, and it is used to produce a very important device in hi-fi called an electromagnet.

Electromagnets

A useful electromagnetic device is produced when the current-carrying wire is bent into the shape of a helix (Fig. 11.9(A)). When iron filings are sprinkled around this configuration, it is found that they form the same pattern of magnetic field lines that exists around the bar magnet. Notice that the magnetic lines of the helix are closed loops which travel both inside and outside of the helix. In a similar fashion the magnetic lines of the permanent magnet also form continuous loops, although the interior lines are not shown in Figure 11.9 (B) because the filings cannot penetrate the magnet. This is quite an interesting observation because, as we shall see, the current-carrying helix possesses most of the magnetic properties of a permanent magnet; such a helix is called an *electromagnet*.

If the iron filings are removed and compasses are placed around the helix, the needles will once again follow the direction of the magnetic field lines as shown in Figure 11.10 (A). The north poles of the compasses will point in the

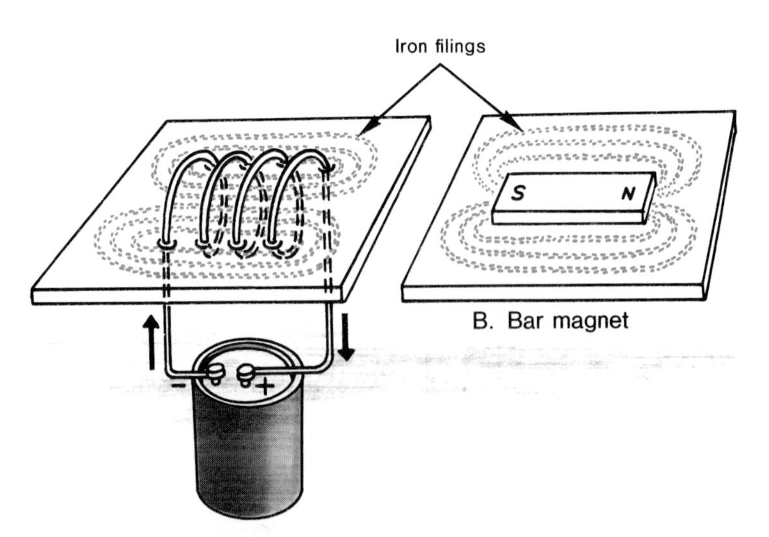

Figure 11.9. The iron filings demonstrate that the magnetic field lines of a current-carrying helix are identical to those of a bar magnet.

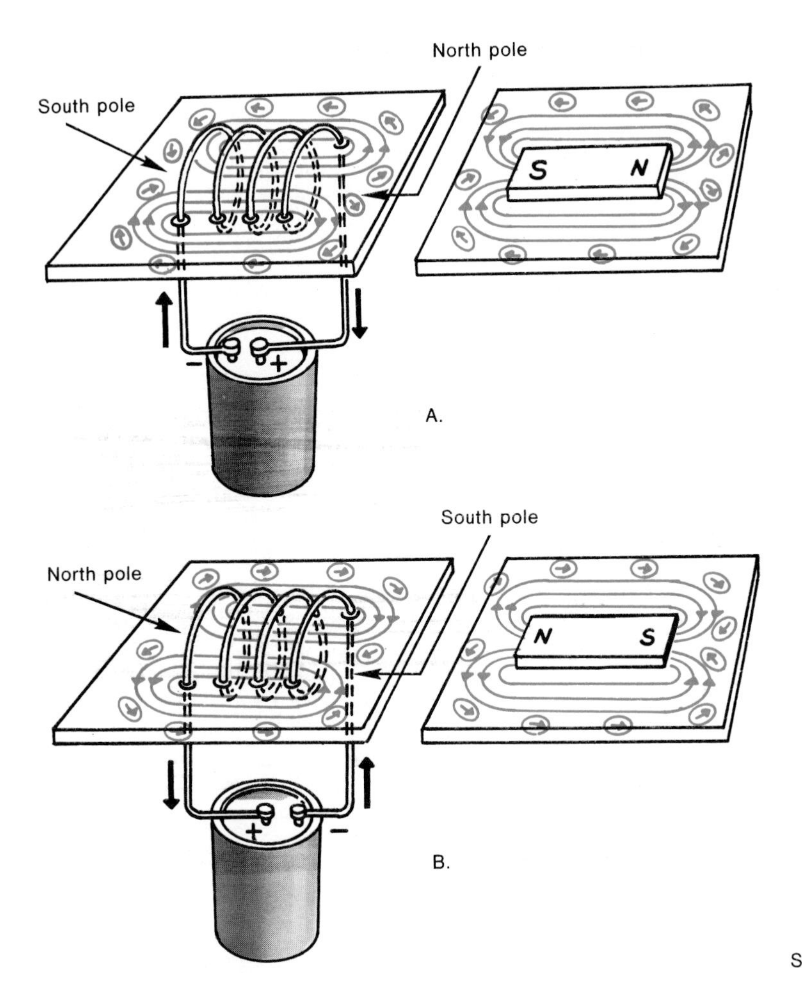

Figure 11.10. (A) Placing compasses around an electromagnet demonstrates that it has a north and a south pole, just like the bar magnet. (B) Reversing the leads to the battery reverses the magnetic poles of the electromagnet.

direction of the magnetic field lines. Notice that the compasses which surround the electromagnet are pointing in the same direction as those which surround the permanent magnet. Therefore, the left end of the helix is a south pole while the right end must be a north pole—exactly the same as the bar magnet! In addition, if the current through the coil is reversed, the compass needles will also reverse their direction indicating that the two poles of the electromagnet have been reversed, as shown in part (B) of the drawing. The left end of the helix now becomes a north pole, while the right end changes to a south pole. The poles of an electromagnet can be switched by simply reversing the direction of the current flow (Fig. 11.11)! The electromagnet and the permanent magnet are identical in the sense that they both possess north and south poles and exhibit the same magnetic field patterns.

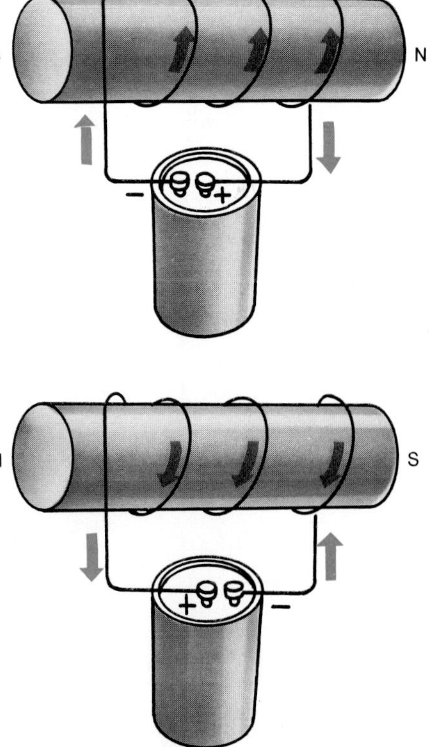

Figure 11.11. Reversing the current flow through an electromagnet reverses its magnetic poles.

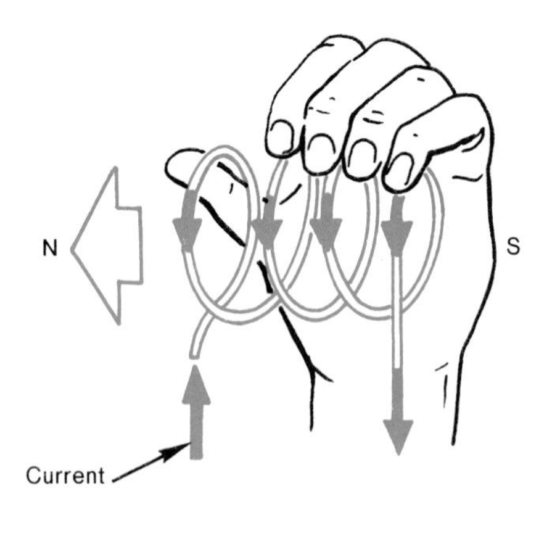

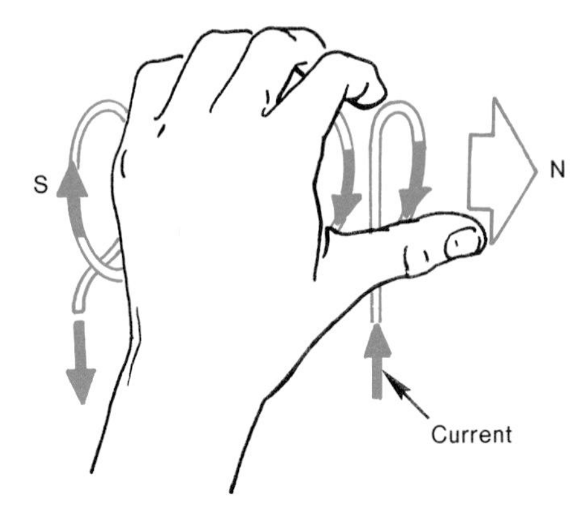

Figure 11.12. A left-hand rule can be used to locate the north pole of an electromagnet. If the fingers of the left hand are wrapped around the coil in the direction of the current flow (shown by the arrows), the thumb will point toward the north pole.

A Left Hand Rule

There is a very nice method, called the *left-hand rule*, which allows one to determine the north pole of an electromagnet. If the fingers of the left hand are wrapped around the helix in the direction of the current flow, the thumb will point toward the north pole of the coil (Fig. 11.12). The other end must, of necessity, be a south pole. The left-hand rule will be of great use when we discuss the operation of magnetic record heads, speakers, and microphones. Because electromagnets possess north and south poles, they can be either attracted to or repelled from other magnets. Some of the possibilities are drawn in Figure 11.13, where the left-hand rule has been used to determine the location of the electromagnet's north pole. The manner in which the helix is wound is unimportant, and the only significant factor is the direction in which the current flows *around* the cylinder. For example, the "A" and "B" helices in Figure 11.13 are wound differently. In addition, the current in "A" flows from left to right, while the current in "B" flows from right to left. Neither of these facts, by themselves, can be used to predict the position of the poles. However, in both cases the currents travel from the top side of the cylinder toward the bottom side. This is the key fact and the left-hand rule, when properly applied, will yield the same result for both helices; the north poles will be on the left and the south poles on the right.

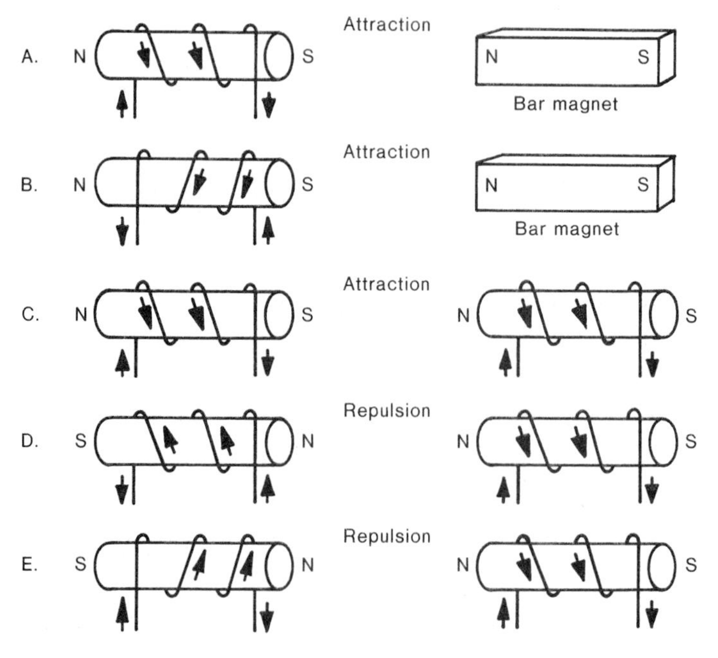

Figure 11.13. An electromagnet can interact with either bar magnets or other electromagnets. The left-hand rule determines the polarity of the electromagnets.

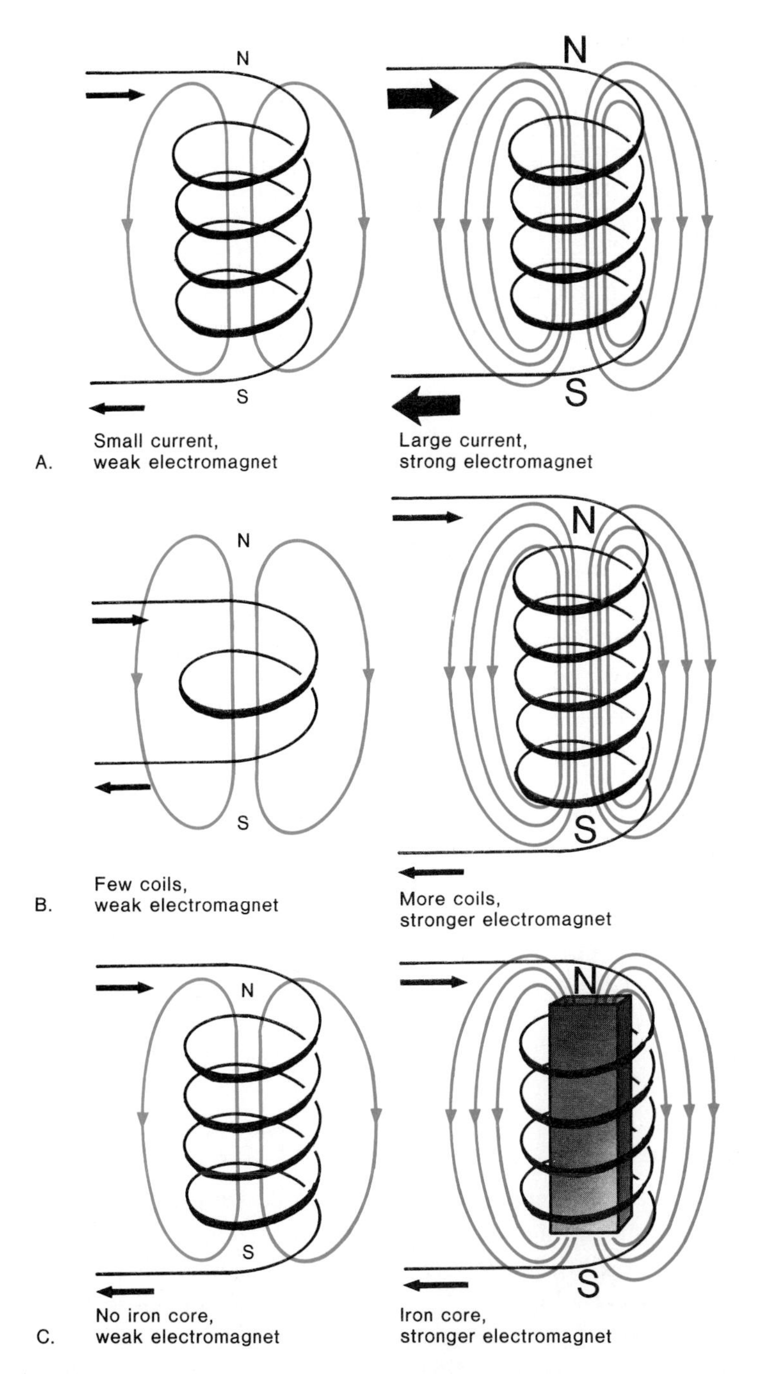

Figure 11.14. There are three methods which will increase the strength of an electromagnet: (A) use more current, (B) add more turns of wire, and (C) add an iron core.

Increasing the Strength of An Electromagnet

There are three methods which can be used to increase the magnetic strength of an electromagnet. First, increasing the current to the helix results in a stronger electromagnet as shown in Figure 11.14 (A). Second, adding more turns of wire to the current-carrying coil also produces a stronger magnet [part (B)]. Therefore, strong electromagnets possess many turns of wire while carrying as large a current as

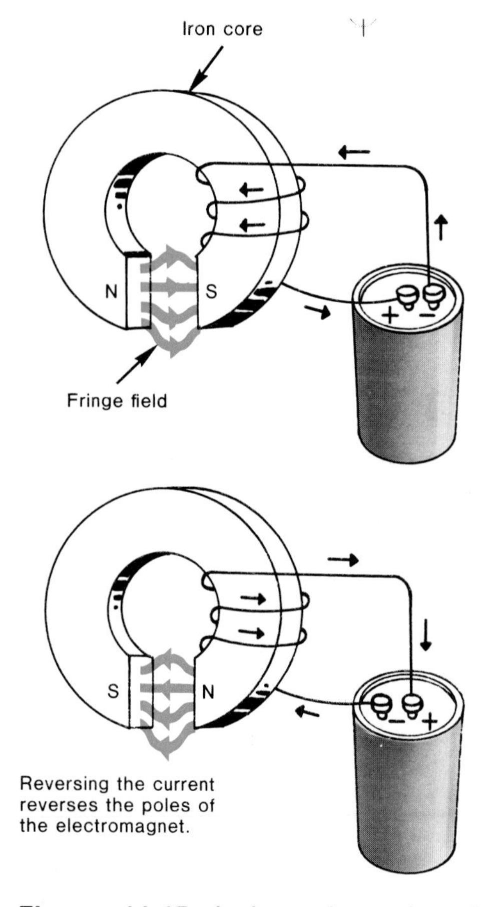

Figure 11.15. A horseshoe shaped electromagnet. The magnetic field lines which bow out from the air gap are called the "fringe field."

possible. Third, adding an iron core inside the helix [Fig. 11.14 (C)] greatly increases the magnetic field strength. The magnetic field produced by the coil alone causes the atomic magnets within the iron core to become aligned, exactly in the same manner as shown in Figure 11.6. The iron core thus becomes a magnet and produces its own magnetic field in addition to that generated by the coil. The combined magnetism of the helix plus the iron core results in a magnetic field which is stronger than either one alone. Electromagnets possess many advantages over permanent magnets. The strength and polarity of an electromagnet can be altered easily by changing the amount and direction of the current; facts which make it useful in hi-fi applications.

An important kind of electromagnet results when the iron core is bent into the form of a horseshoe (Fig. 11.15). The magnetic lines travel across the air gap from the north pole to the south pole. Once inside the iron core the lines travel through the horseshoe and back to the north pole. Hence, the magnetic lines actually form closed loops. It is only *outside* the magnet that the magnetic lines go from the north pole to the south pole. However, *inside* the magnet the lines continuously travel from the south pole toward the north pole in order to maintain the loop continuity. Within the air gap the magnetic lines tend to "bow out," as shown in Figure 11.15. This region is called the "fringe field," and it is of utmost importance in the tape recording process. Horseshoe electromagnets are used for the erase and recording heads in a tape deck. In the next section we shall see how the principles of electromagnetism have been used to produce two important hi-fi devices; speakers and tape recording heads.

Summary

An electromagnet is formed when a current-carrying wire is bent into the shape of a helix. An electromagnet possesses north and south poles, and its magnetic field is identical to that of a permanent bar magnet. The advantage of an electromagnet is that its magnetic poles can be reversed by simply changing the direction of the current, a fact which is used in speaker and tape recording operations. The strength of an electromagnet is governed by the amount of current it carries, the number of windings, and the placement of soft iron in its center region.

11.4 Examples of Electromagnetism in Hi-fi: Speakers and Tape Recording Heads

Speakers

In Chapter 6 it was shown that the cone speaker consisted of, among other things, a permanent magnet and a voice coil. The voice coil is essentially a helix, containing many turns of wire, which is attached to the apex of the cone. As the current from the power amplifier is sent through the voice coil, it becomes an electromagnet whose strength and polarity depend upon the characteristics of the incoming current. This electromagnet works in conjunction with the permanent magnet to produce the vibratory motion of the speaker's cone. Consider Figure 11.16 which is a highly simplified diagram of a cone speaker. The permanent magnet and the voice coil are shown to be physically separated, although in reality the coil actually fits into the permanent magnet structure as shown in Figure 6.2. Consider an instant in time when the current from the amplifier is flowing in the direction indicated by Figure 11.16 (A). According to the left-hand rule, the left end of the voice coil becomes a north pole which is in close proximity to the north pole of the permanent magnet. A mutual repulsion occurs and the voice coil (plus the cone) is magnetically forced to the right. The permanent magnet is rigidly attached to the metal basket so it does not move. Therefore, we see that the incoming current causes the voice coil to become an electromagnet which is repelled from the permanent magnet. The repulsion causes the diaphragm to move forward, thus creating a sound condensation.

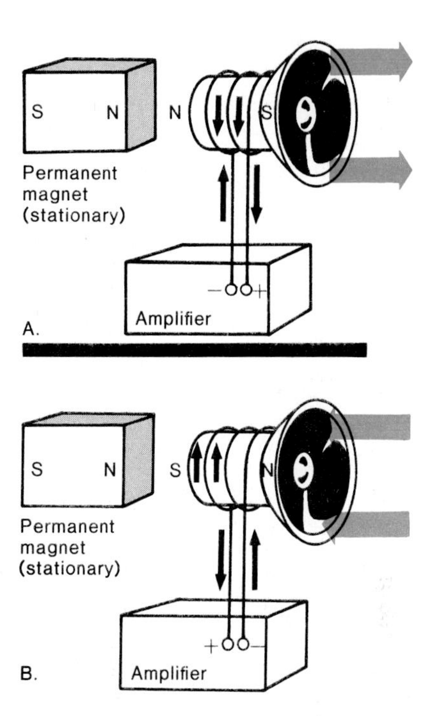

Figure 11.16. The magnetic force between the voice coil and the permanent magnet causes the cone to be forced either (A) outward or (B) inward, depending upon the direction of the AC current flow.

Figure 11.16 (B) illustrates the cone's movement when the current from the amplifier has been reversed. Of course, the poles of the electromagnet also reverse, thus causing an attractive force between itself and the permanent magnet. The attractive force causes the diaphragm to move inward, and a sound rarefaction is created. If, for example, the amplifier is sending a 3 kHz AC signal to the voice coil, then the electromagnet will cycle its poles at the rate of 3,000 times each second. This causes the diaphragm to be "pushed" and "pulled" at the same rate, which then creates a 3 kHz sound wave.

One can also visualize how the speaker creates sounds of different loudness. A large current from the amplifier turns the voice coil into a strong electromagnet. The electromagnet is now repelled from, or attracted to, the permanent magnet with strong magnetic forces. This, in turn, gives rise to greater cone excursions which produce a louder sound. Similarly, small currents produce quiet sounds. In summary, the frequency of the resulting sound is the same as the frequency of the AC current, and the sound loudness depends upon the size, or amplitude, of the current.

Magnetic Recording Heads

The function of a magnetic recording head is to transcribe the electrical signals into magnetic patterns, which are an exact replica of the electric signals. Recording heads operate on the principle of electromagnetism as shown in Figure 11.17. The electrical signal from the tape deck's preamplifier is sent to the recording head where a magnetic field is established in the gap. The magnetic fringe field penetrates the magnetic coating of the tape and causes it to become partially magnetized. The amount of magnetism which is induced into the tape depends upon the strength of the electromagnet. A larger current in the coil produces a larger magnetic fringe field, and the larger fringe field induces more magnetism into the tape. When the tape leaves the vicinity of the recording head, it retains an amount of magnetism which is proportional to the strength of the audio current. If the AC audio current reverses its direction, the fringe field of the recording head also reverses, thus inducing a magnet of opposite polarity into the tape. Therefore, information on both the amplitude and frequency of the audio signal can be stored as magnetic impressions on the tape. The complete recording process will be covered in Chapter 13.

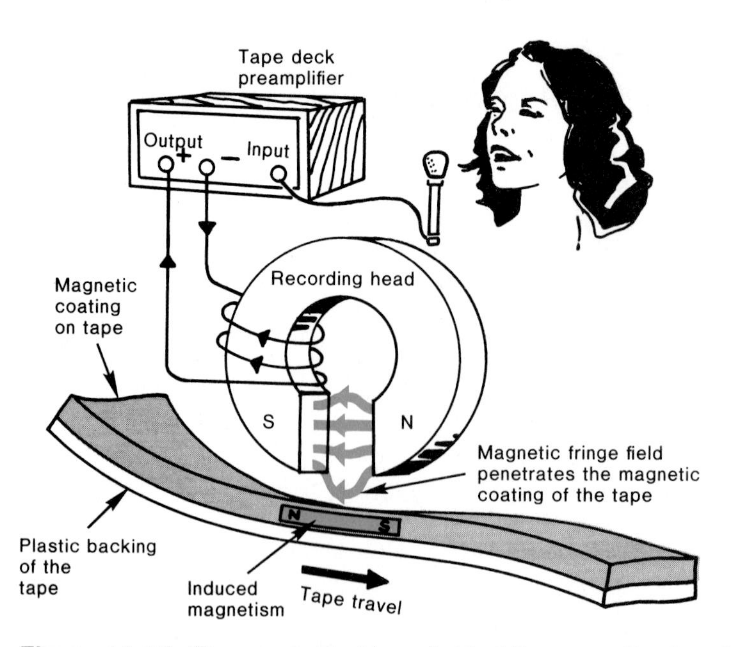

Figure 11.17. The magnetic fringe field of the recording head penetrates the magnetic tape and causes it to become magnetized. Notice that the poles of the induced magnet in the tape are opposite to those of the recording head.

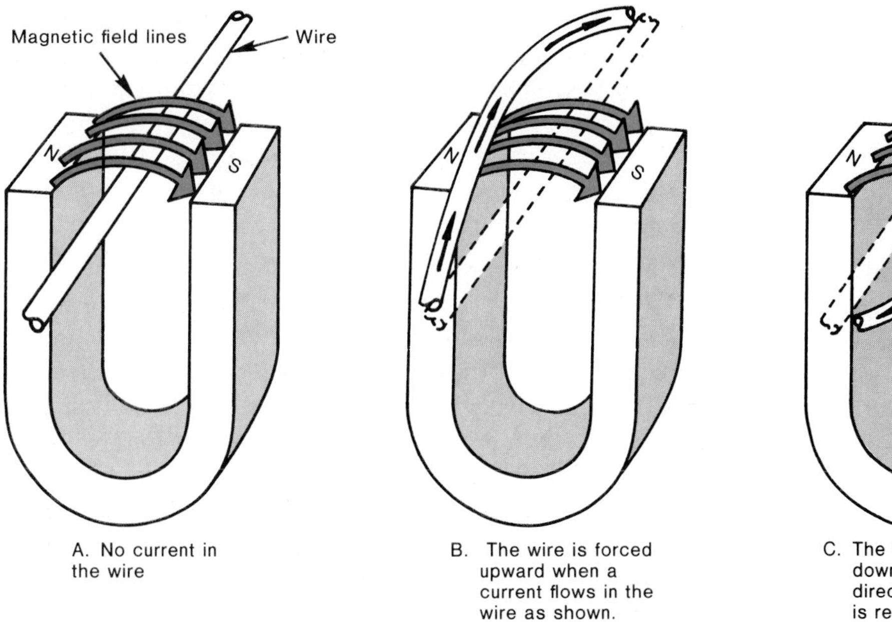

Figure 11.18. A current passing through a magnetic field experiences a deflecting force. The direction of the force depends upon the direction of the current.

11.5 Forces Exerted on Currents by Magnetic Fields

So far, magnetic fields have been shown to exist in the vicinity of both permanent magnets and electromagnets. In addition, magnets exert forces on other magnets—be they permanent magnets or electromagnets. There is another interesting and useful phenomenon in which a current experiences a force when it passes through a magnetic field. Consider Figure 11.18 (A) which shows a straight wire passing through the air gap of a permanent horseshoe magnet. Also drawn in the diagram are the magnetic field lines which extend from the north pole to the south pole. With no current in the wire nothing happens. However, when a current travels through the wire a magnetic force deflects the wire either upward or downward, depending on the direction of the current [see Figure 11.18, (B) and (C)]. In this figure we have chosen the magnetic field lines and the wire to be perpendicular (90°) to each other, because the deflecting force is a maximum under these conditions. On the other hand, if the current and magnetic field lines run *parallel* to each other, there is *no* deflecting force.

A Second Left Hand Rule

It is possible to "predict" the direction of the force on a current by using a "second" left-hand rule which is illustrated in Figure 11.19. Position the thumb and first two fingers of the left hand such that all three form the corner

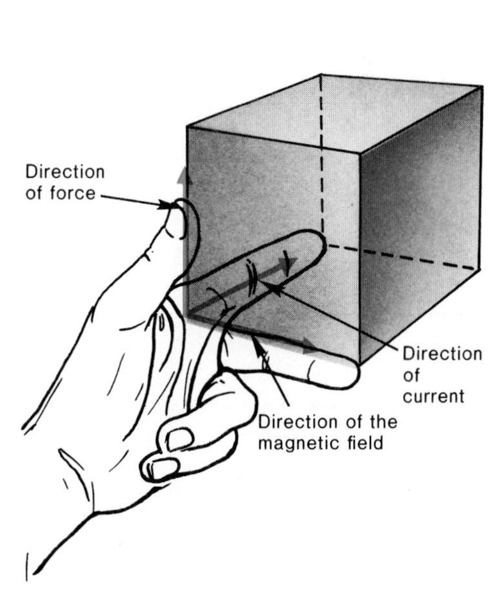

Figure 11.19. A "second" left-hand rule predicts the direction of the force exerted on a current which is traveling through a magnetic field.

of a box. Point the first finger in the direction of the current, the second finger along the magnetic field lines (which always go from north to south), and the thumb will point in the direction of the deflecting force. This rule will allow you to verify that the wire in Figure 11.18 moves as shown. There are two types of speakers which are based on this principle: the air-motion transformer speaker and the magnaplanar speaker.

The Air-motion Transformer Speaker

The air-motion transformer speaker can be constructed by taking a piece of thin paper, folding it like an accordion, and gluing wire along the folds as shown in Figure 11.20. Place a permanent magnet both in front and behind the corrugated sheet, attach the two ends of the wire to the amplifier speaker terminals, and you have a bonafide speaker. The pleats are so designed that at the *front* of the diaphragm all the pleats *close* during the first half-cycle, squeezing the air forward, rather than pushing it as in a conventional diaphragm [Fig. 11.21 (A)]. During the second half-cycle, all the pleats open and the air is "inhaled," creating a pocket of lower air pressure [Fig. 11.21 (B)]. A 1 kHz alternating electrical signal, for example, causes the speaker to "exhale" and "inhale" the air at the same rate, and a 1 kHz sound wave is produced.

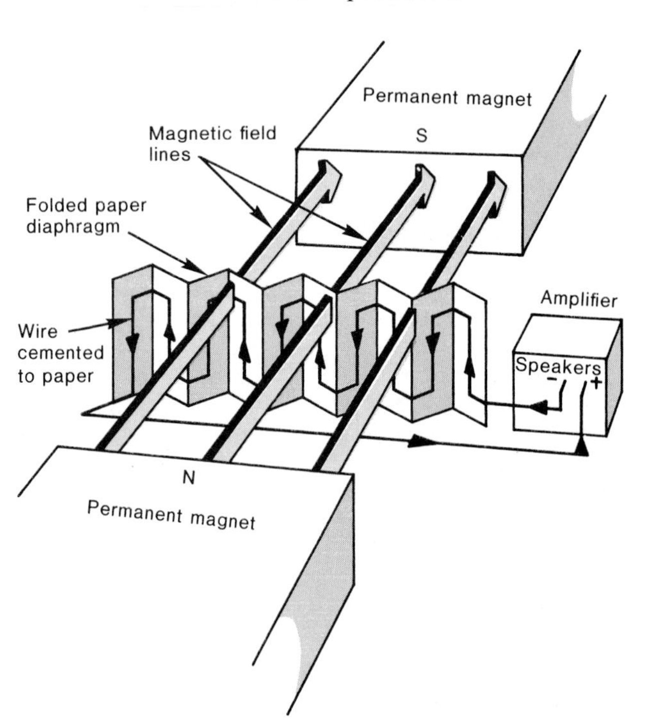

Figure 11.20. An air-motion transformer speaker. The interaction between the magnetic field of the permanent magnet and the current-carrying wire causes the accordion-like diaphragm to squeeze, thus producing sound. The drawing shows the folds to be sharply creased; in practice they are gentle folds.

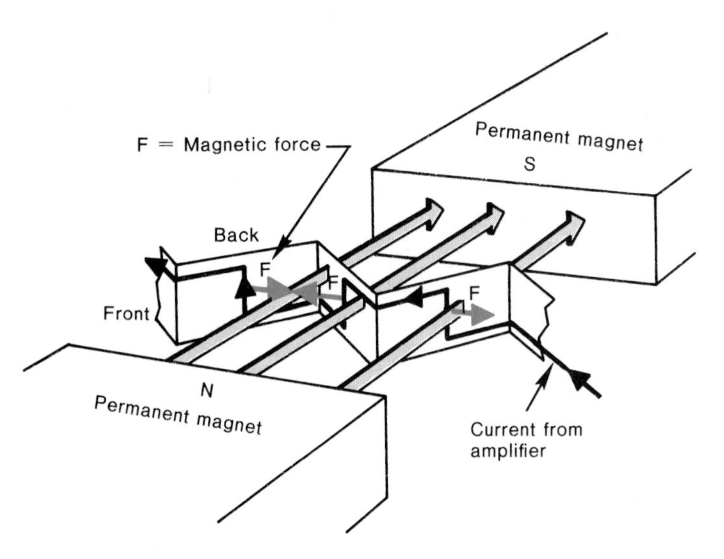

A. First half-cycle of AC current

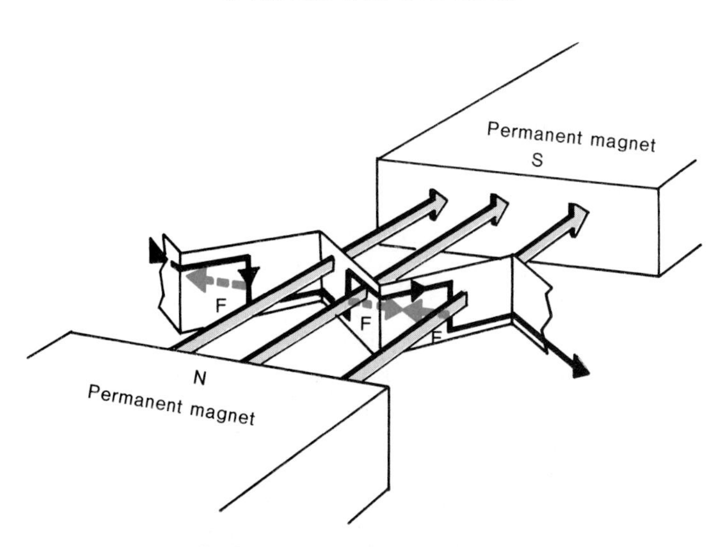

B. Second half-cycle of AC current

Figure 11.21. (A) During the first half-cycle the magnetic forces cause the front pleats to squeeze together, thus forcing air away from the front surface and creating a sound condensation. (B) The magnetic forces reverse their direction during the second half-cycle when the AC current has reversed. This causes the front pleats to open up and "inhale" the air, thus producing a rarefaction.

The action of the air motion transformer speaker is produced by the force exerted on a current which is moving in a magnetic field. Because of the accordion-type geometry, the current travels up one face and down the other face of adjacent pleats. The wire in each pleat experiences a force because the current is moving through the magnetic field lines of the two permanent magnets. Using the "second" left-hand rule, the reader should be able to verify that the current in each pleat is forced in the direction shown in Figure 11.21 (A). Notice that *adjacent* faces of the pleats experience magnetic forces which are in *opposite* directions;

this is because the currents are flowing in opposite directions. There will be a closing up of each pair of conductors *in relation to each other*, and a squeezing of the air takes place at the front of the diaphragm. It should be noted that when the air is squeezed from the front there is a simultaneous inhaling (rarefaction) at the rear surface. In this respect the diaphragm is behaving like an ordinary vibrating cone as discussed in Chapter 6, which simultaneously creates a con-ensation at its front surface and a rarefaction at its rear surface.

When the current from the amplifier reverses during the second half-cycle, the forces on the wires reverse their direction, as shown in Figure 11.21 (B). Now the magnetic forces open the front pleats, and close the rear ones. A rarefaction is now created at the front while the rear surface generates a condensation. Therefore, the air-motion transformer utilizes the alternating magnetic forces, created by the AC signal, to produce the accordion-like motion which creates the sound.

The air-motion transformers have been used extensively by Heil for use as midranges and tweeters. Because the accordion motion does not normally displace a large enough volume of air, it is seldom used for woofer applications.

The Magnaplanar Speaker

The magnaplanar speaker operates on the same principle as does the air-motion transformer speaker. However, the magnaplanar speaker uses a flat diaphragm and a different arrangement of the permanent magnets (see Fig. 11.22). Notice that the north and south poles of the permanent

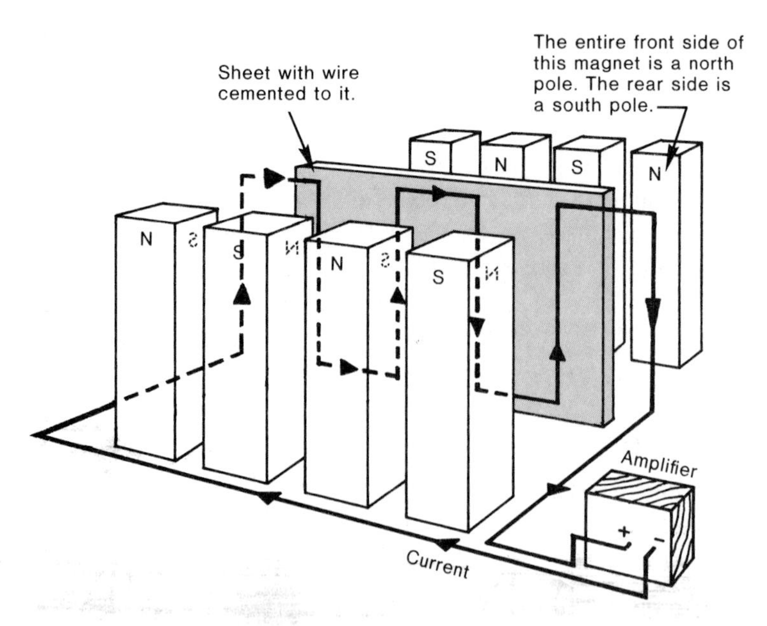

Figure 11.22. A magnaplanar speaker with its permanent magnets and current carrying wire.

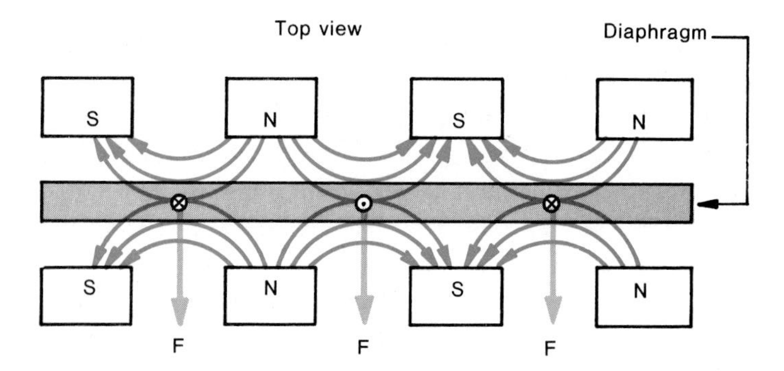

F = magnetic force

Figure 11.23. **The magnaplanar speaker utilizes the force exerted on the current to move the diaphragm. Note that ⊙ indicates a current which is coming out of the page straight at you. The ⊗ indicates a current which is flowing into the page directly away from you.**

magnets are located on either side of the short dimension of the bars, and not on the ends of the long dimension as is customary. The wire is attached to the diaphragm in a "back-and-forth" fashion with its two free ends being connected to the speaker terminals of the power amplifier. Figure 11.23 shows a top view of the speaker with the magnetic field lines, current direction (⊙ = out of page, ⊗ = into page), and magnetic forces. Notice that the magnetic field lines run more-or-less parallel with the surface of the diaphragm. Because of the way in which the permanent magnets are positioned, the magnetic field lines also reverse when moving from one vertical wire to an adjacent one. Although adjacent wires "see" oppositely-directed fields, their currents are also opposite so that the magnetic force is always in the *same direction for all the wires.* (Careful use of the second left-hand rule should convince you that reversing *both* the field and the current directions keeps the force in the *same* direction.) During the first half-cycle Figure 11.23 shows that the entire diaphragm is pushed forward, with a condensation being produced ahead of it. During the second half-cycle the AC current from the amplifier reverses itself, and the diaphragm is pushed backward, as indicated in Figure 11.24. An alternating audio current thus produces a back-and-forth vibratory motion of the diaphragm which produces the sound.

Summary

When an electric current passes through a magnetic field, the current experiences a force. This force is a maximum when the current is perpendicular to the magnetic field, and the force is zero when the current is parallel to the field. The

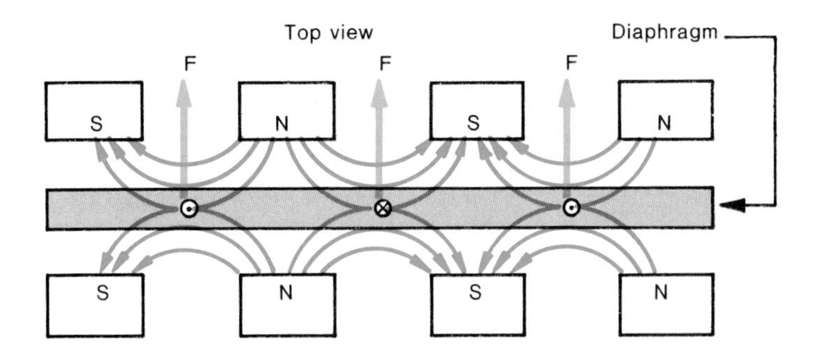

F = magnetic force

Figure 11.24. Reversing the AC current during its second half-cycle reverses the motion of the diaphragm.

magnetic force is perpendicular to both the current and the field, and its direction can be predicted by using a "second left hand rule." Both the air-motion transformer and the magnaplanar speakers utilize the fact that a magnetic field can exert a force on a current-carrying wire.

11.6 Faraday's Law of Induction

Certainly one of the most beautiful and useful of all phenomena is known as Faraday's law of induction. As previously stated, an electric current creates a magnetic field which leads to a device called an electromagnet. The question is this:

> If electric currents create magnetism, can magnetism create electric currents?

The answer, as we shall now discuss, is an emphatic YES. The conditions under which a current (or voltage) can be produced are described by Faraday's law.

> **Faraday's law:** If a magnetic field and a coil of wire are moving relative to each other, a voltage will be created in the coil, thus causing a current to flow.

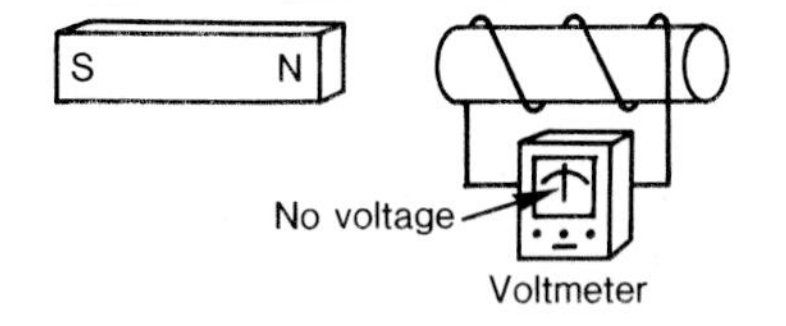

A. No relative motion

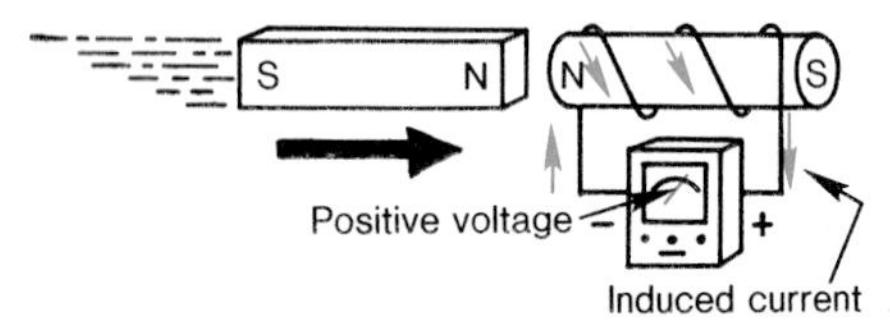

B. Permanent magnet moves toward the helix.

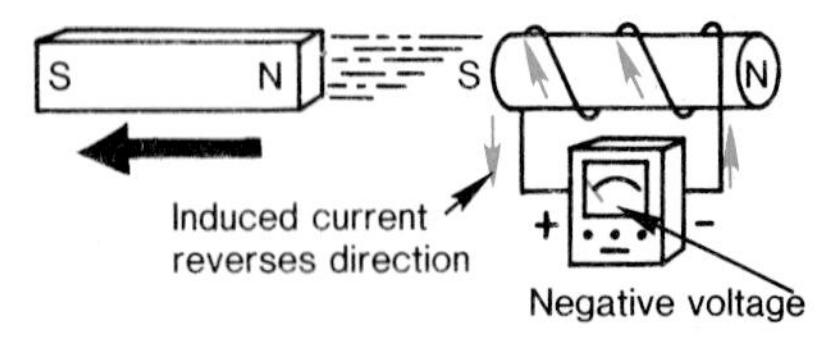

C. Permanent magnet moves away from helix.

Figure 11.25. An illustration of Faraday's law of induction. Whenever there is relative motion between the magnet and the helix, a voltage will be induced (or produced) in the helix.

Consider Figure 11.25 which shows a permanent magnet and a helical coil of wire to which a voltmeter is connected. When there is no relative motion between the permanent magnet and the helix [Fig. 11.25 (A)], the voltmeter registers zero and, consequently, no current flows. The story completely changes as soon as the magnet begins to move either toward or away from the helix. When the magnet moves toward the helix [Fig. 11.25 (B)], a voltage appears, as if by magic, and a current begins to flow! The voltage thus produced is called an "induced voltage," and it is the same as if an imaginary battery had been placed in the circuit. When the magnet moves away from the coil [Fig. 11.25 (C)], an induced voltage is again produced, only with the reverse polarity which causes a reversal in the current.

The process of induction is quite amazing, because it involves the generation of a voltage by the action of a changing magnetic field. Notice that the magnet and coil do not have to actually touch each other! The only requirement is that there be relative motion between the two in order for a voltage to appear in the coil. The magnetic field of the permanent magnet penetrates the coil as shown in Figure 11.26. As the permanent magnet comes closer and closer, the magnetic field which passes through the coil becomes stronger and stronger. Thus, the coil "sees" a *changing* magnetic field due to the approaching magnet, and it is this changing field which induces the voltage into the circuit. The same is true when the magnet is moving away from the helix; only this time the magnetic field is becoming weaker and weaker. Still, the changing field will induce a voltage into the helix. It does not matter whether the magnet approaches, or recedes from, the coil for a voltage to be produced. In either case the coil senses a changing magnetic field and a voltage is induced.

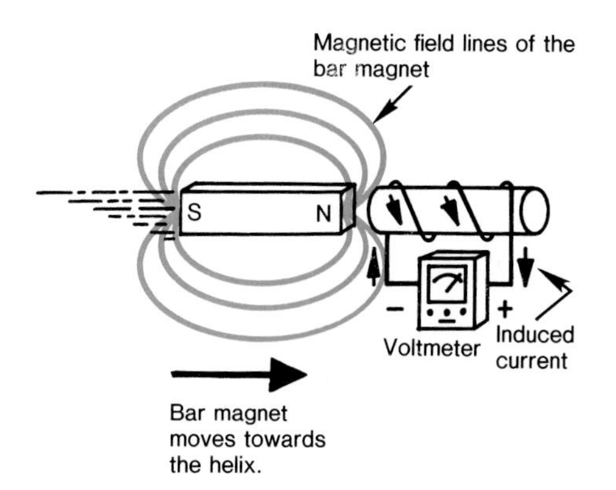

Figure 11.26. A voltage is induced in the helical coil, because it "sees" the changing magnetic field of the approaching permanent magnet. An induced voltage would have also appeared if the magnet had been receding from the coil.

In studying the effect of a moving magnet near a helical wire, it was mentioned that a voltage was produced only when the magnet, and its associated field, were moving relative to the coil. On the other hand, if the permanent magnet were held fixed and the coil allowed to move, a voltage would have also appeared in the coil. Thus induced voltages are produced if either the coil or the magnet is moved; only relative motion between the two is necessary. If there is no relative motion, there will be no induced voltage. It then follows that a continuous source of electricity can be produced only if continuous relative motion of either the magnetic field or the coil can be maintained. If the magnet is vibrated back and forth in a continuous fashion, an alternating voltage will be induced in the coil. The frequency of the AC voltage will precisely match the frequency of the magnet's oscillations. This is how many hi-fi sources, e.g., cartridges, produce their output voltage. Before looking into these applications, however, let us further examine the concept of induced voltage and see what factors influence its generation.

Factors Which Determine the Amount of Voltage Induced in a Circuit

The amount of voltage which is induced in a coil by a changing magnetic field depends upon a number of factors. First, if the relative speed between the magnet and coil increases, then the induced voltage will also increase [Fig. 11.27 (A)]. With all other things being equal, a higher frequency of oscillation means a faster speed which, in turn,

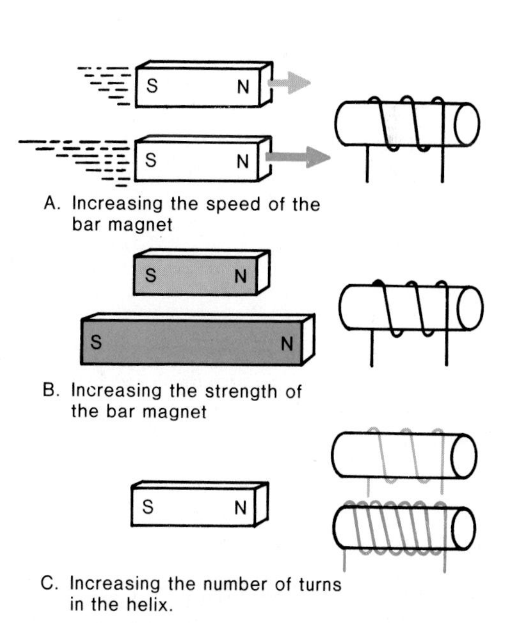

Figure 11.27. Three factors which increase the amount of voltage induced in the helical coil, according to Faraday's law.

creates a larger induced voltage. Second, if the strength of the magnetic field increases, so does the induced voltage, as illustrated in part (B). Third, if the number of turns in the coil is increased, the induced voltage is again increased, as shown in part (C). All three of these methods are used in hi-fi equipment in order to produce source voltages which are as large as possible. Of course, the induced voltage will cause an "induced" current to flow if the ends of the conductor are connected through a closed circuit. According to Ohm's law, the amount of current will be equal to the induced voltage divided by the resistance of the circuit.

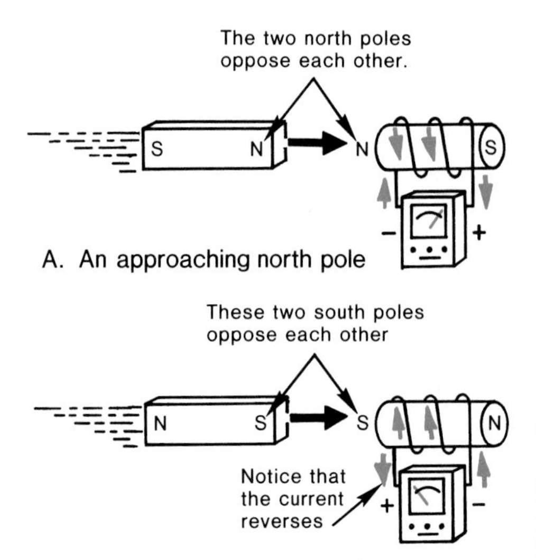

Figure 11.28. According to Lenz' law, the induced current will set up an electromagnet which will oppose the magnetic pole of the approaching magnet.

Lenz' Law—How to Determine the Polarity of the Induced Voltage

Figure 11.25 shows the direction in which the induced current flows when the magnet is both approaching and receding from the coil. Lenz' law tells us about the polarity of the induced voltage, and hence the direction of the induced current which flows in the coil.

Lenz' Law: The polarity of the induced voltage will be in such a direction that the resulting current flow will produce an electromagnet which will oppose the motion which causes the changing magnetic field.

To illustrate Lenz' law consider Figure 11.28 (A) which shows the permanent magnet moving toward the coil. The coil senses an approaching north pole. According to Lenz' law, the coil *opposes* the approaching north pole by establishing a north pole at the end of the coil which is closest to the approaching magnet. Thus, the two north poles will repel each other. By using the left-hand rule (see section 11.3) the induced current must then flow as shown in Figure 11.28 (A). Similarly, an approaching south pole will cause the induced current in the coil to be reversed. In this case the left side of the coil becomes a south pole [Fig. 11.28 (B)] which repels the approaching south pole.

So far we have considered the case when the permanent magnet approaches the coil (or, equivalently, when the coil approaches the magnet). What happens when the magnet and coil are moving *away* from each other? Again, there will be an induced current in the wire, and Lenz' law will tell us the direction of the induced current. Figure 11.29 (A) shows a north pole moving away from the coil. Remember that the coil will attempt to oppose this receding motion. It does so by setting up a *south* pole at its left end in an attempt to attract back the north pole. Similarly, Figure 11.29 (B)

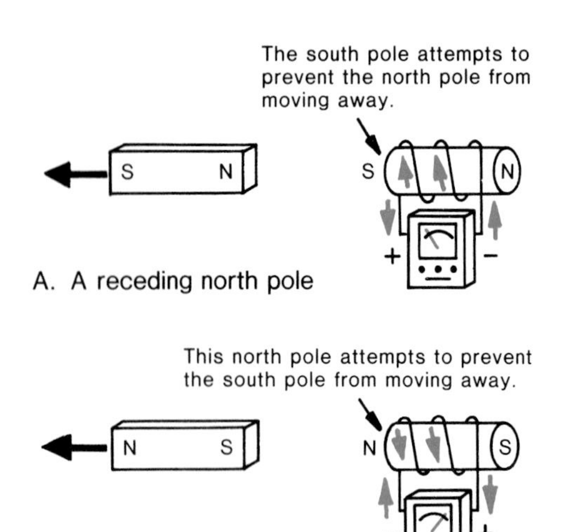

Figure 11.29. According to Lenz' law, if the permanent magnet is moving away from the coil, the coil will oppose this motion by establishing an opposite (attractive) pole.

shows a south pole receding from the coil. The coil, attempting to resist this "loss," establishes a north pole at its left end in order to attract the magnet back. In all cases, once the proper polarity of the induced magnetic poles has been established, the left-hand rule gives the direction of the induced current.

It is important to keep in mind the main feature of Lenz' law. The direction of the induced current depends upon not only the polarity (north or south) of the permanent magnet's closest pole, but also on whether the pole is approaching or receding from the coil! Perhaps you are wondering why the coil sets up an induced magnetic field which opposes, rather than aids, the changing field of the permanent magnet. The answer is tied up with the fact that energy cannot be created or destroyed. Let us assume for the moment that the reverse condition was true—namely, that the induced magnetic pole was established in such a manner that it aided, rather than opposed, the changing field of the moving permanent magnet. Figure 11.30 shows this hypothetical situation, where the left end of the helix is assumed to be a south pole. The induced south pole now provides an *attractive* force to the approaching north pole and, as a consequence, the permanent magnet accelerates and moves even faster towards the coil. The faster moving permanent magnet causes an even greater current to be induced in the coil, with the result that the induced south pole becomes even stronger. It should now be apparent that this situation rapidly becomes self-perpetuating. It leads to an ever-increasing induced electrical current and, hence, an ever-increasing generation of electrical energy. Energy is, therefore, being created which violates the law of energy conservation. In conclusion, the induced current must be established in such a direction around the helix that its magnetic poles will oppose, rather than aid, the changing magnetic field of the permanent magnet. In effect, Lenz' law ensures that the helix acts like a magnetic "brake" which opposes the motion of the moving magnet.

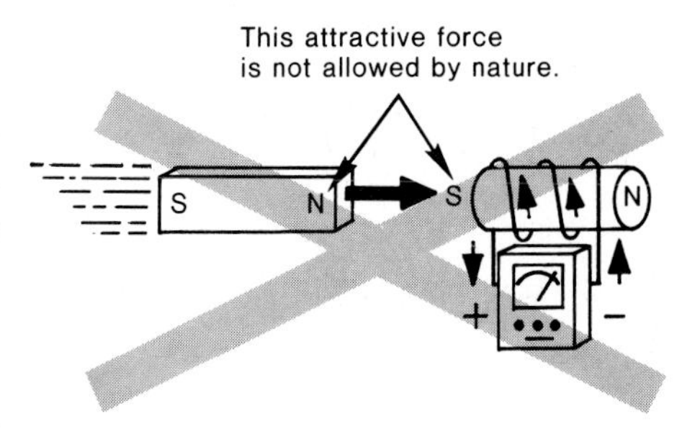

Figure 11.30. The situation depicted can never occur, because it violates the law which states that energy cannot be created. The south pole of the helix would attract the incoming north pole, thus causing it to move faster. The increased speed induces a greater current, which is equivalent to creating electrical energy from nothing.

Summary

Whenever the changing (either increasing or decreasing) magnetic field lines cut the coils of a helix, a voltage is induced in the helix. This fact is known as Faraday's law of induction. The amount of induced voltage depends upon the strength of the magnetic field, how rapidly it is changing, and the number of turns in the helix. An induced voltage will occur when either the magnetic field or the coil is moved—only relative motion is significant. The polarity of the induced voltage is given by Lenz' law which states that the induced current will be established in such a direction so as to oppose, through its magnetism, the motion which causes the changing magnetic field.

11.7 Applications of Induced Voltages in Hi-fi

Phono Cartridges

The basic idea of a phono cartridge is familiar to everyone. The stylus is forced into vibratory motion as it rides in the record groove. This vibratory motion leads to the generation of a small voltage which is then sent to the PHONO input of the preamp/control center for further amplification. The important link between the motion of the stylus and the output voltage is provided by Faraday's law of induction. Figure 11.31 illustrates a simplified version of the so-called "moving magnet" cartridge. The moving magnet cartridge is nothing more than a permanent magnet which is free to vibrate inside a stationary helix. The two ends of the wire helix are then connected to the PHONO input of the preamp/control center. As the record rotates, the groove forces the stylus, and its associated permanent magnet, to vibrate inside the stationary helix. The magnetic field lines "cut" across the coils, and they induce an AC voltage whose polarity changes at the precise rate at which the stylus is forced up and down by the record.* Thus we see that Faraday's law of induction is directly responsible for transforming the motion of the stylus into an electrical voltage. Since only relative motion is important for the generation of induced voltages, it is also possible to move the coil and keep the permanent magnet stationary. Moving coil cartridges work on this principle, with the coil rigidly attached to the vibrating stylus and the magnet held stationary.

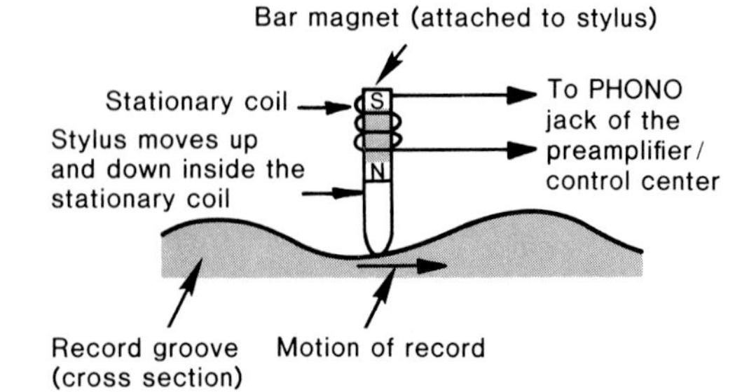

Figure 11.31. The moving magnet phono cartridge utilizes a stylus which vibrates a magnet inside or near a stationary coil. The moving magnet induces an AC voltage in the coil.

Magnetic Microphones

A magnetic microphone, in many respects, is identical to a magnetic phono cartridge. The only difference is that the cartridge utilizes the record groove to push the stylus back and forth, whereas in a microphone the pressure variations of the sound waves provide the force to move a diaphragm. Consider Figure 11.32 which depicts incoming sound waves that are incident upon the microphone's diaphragm. Recall that sound is a series of condensations and rarefactions which represent air pressure that is, respectively, higher and lower than normal. A condensation reaching the diaphragm pushes it inward, and a voltage is induced into the coil because the permanent magnet, attached to the diaphragm, is also moved. A rarefaction reaching the diaphragm causes it to move outward, and creates an induced voltage which is opposite in polarity to that produced by the condensation. A 500 Hz sound wave pushes and pulls on the diaphragm at the rate of 500 times each second. This, in turn, moves the

*This hill-and-dale movement is shown for the sake of illustration. Today, most of the stylus' movement is in the horizontal, rather than the vertical, direction.

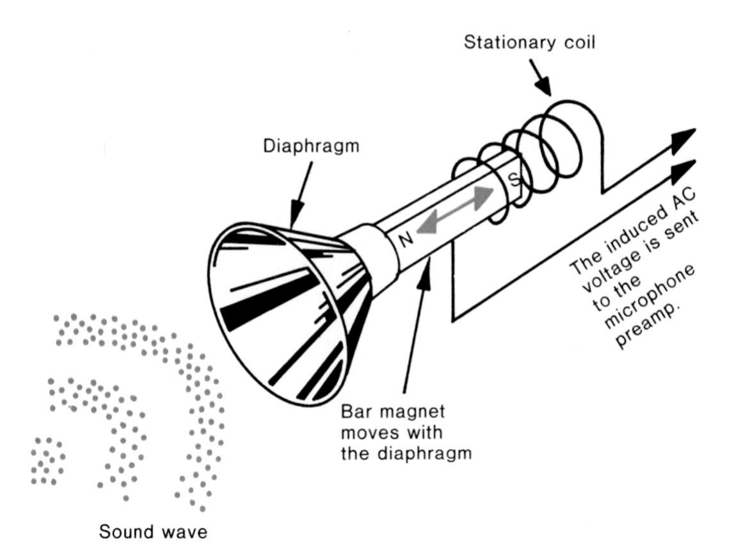

Figure 11.32. A moving magnet (dynamic) microphone utilizes a moving magnet, which is attached to the diaphragm, for inducing an AC voltage in the stationary coil.

permanent magnet at the same rate and induces a 500 Hz AC signal in the coil. A microphone converts pressure variations in the sound wave into mechanical vibration of the diaphragm which, via Faraday's law, produces an alternating voltage.

Like moving coil magnetic phono cartridges, there are also moving coil dynamic microphones. As shown in Figure 11.33, the "voice coil" is attached to the diaphragm while the permanent magnet remains stationary. If you think that this type of microphone looks like a speaker, you're absolutely right! The moving coil microphone is nothing more than a speaker working in reverse! In fact, most intercom

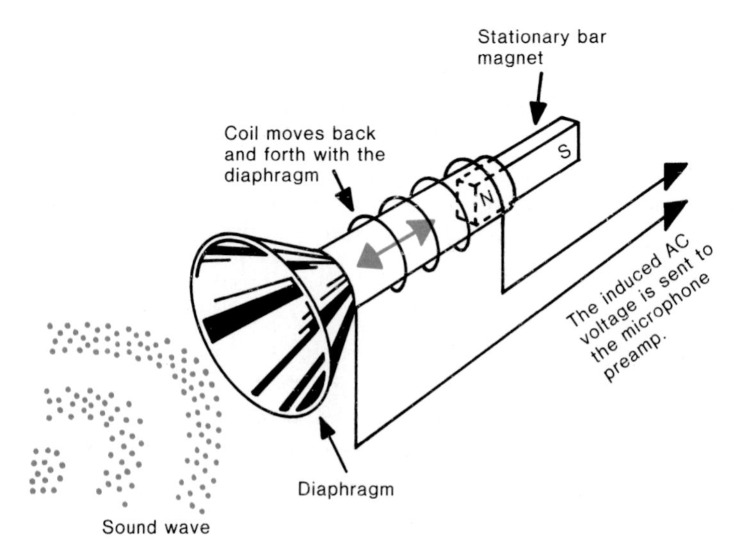

Figure 11.33. A moving coil dynamic microphone. The incident sound waves move the diaphragm and the attached voice coil. The moving voice coil generates an induced voltage which is sent to the microphone preamp for further amplification.

systems used in homes and schools use the same driver for both the speaker and the microphone.

Magnetic Playback Heads

Magnetic playback heads also work on the induced voltage principle. A piece of magnetized tape is like a series of tiny magnets which have been formed on the tape. Of course, each tiny magnet has field lines which extend above the tape from the north pole to the south pole, as shown in Figure 11.34 (A). When the tape is pulled past the playback head, each of the miniature magnets produces magnetic poles in the ends of the soft iron core of the playback head. This causes the magnetic lines to be routed through the iron core rather than through the air [Fig. 11.34 (B)]. As the changing magnetic lines pass through the playback head, a voltage is induced into the coil which is wound around the head. The voltage is then sent to the playback preamp where it is enlarged to approximately 0.5 volt. Faraday's law of induction is thus responsible for the operation of tape decks, where a changing magnetic field, caused by the moving magnets in

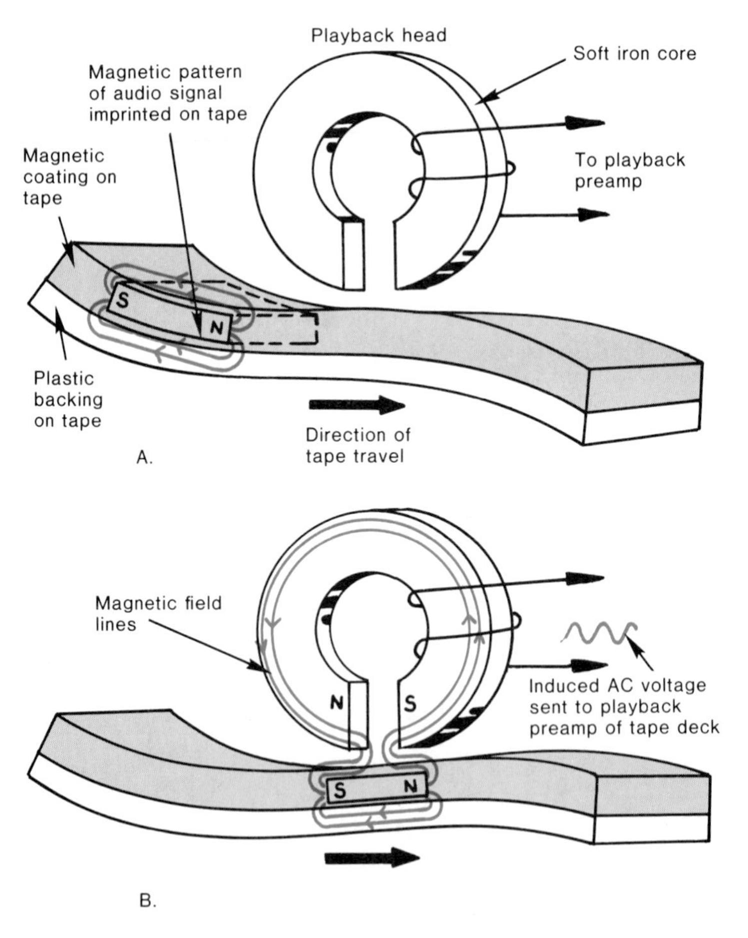

Figure 11.34. (A) The magnetic field lines associated with a piece of magnetized tape. (B) The magnetic lines are rerouted through the soft iron core of the playback head, thus causing an induced voltage to appear in the coil.

the tape, produces the induced voltage in the coil of the playback head.

11.8 Transformers—A Useful Application of Induced Voltages

A transformer is an important device which finds a multitude of uses. The electricity which is sent to your home by the power company is transmitted at very high (several thousand volts) voltage levels because it is cheaper to deliver it in this manner. Before entering your home the high-level voltage is "stepped down" to a more reasonable level of 120 volts (or, perhaps, 220 volts) by the familiar transformer mounted high on the utility pole. In hi-fi systems all amplifiers utilize transformers in their power supply sections to change the level of the 120-volt AC line voltage to smaller values (10 to 40 volts) which are required to operate the transistors properly. Therefore, transformers are useful for changing AC voltage levels to other values which are more appropriate to the needs of a particular device. Transformers can either lower the voltage ("step-down" transformer") or they can raise it ("step-up" transformer); in either case Faraday's law of induction is busy at work.

The transformer consists of two helical coils of wire which are wrapped around a common iron core, as shown in Figure 11.35. The coil to which the input power is applied is called the *primary* coil. The coil from which the output electrical

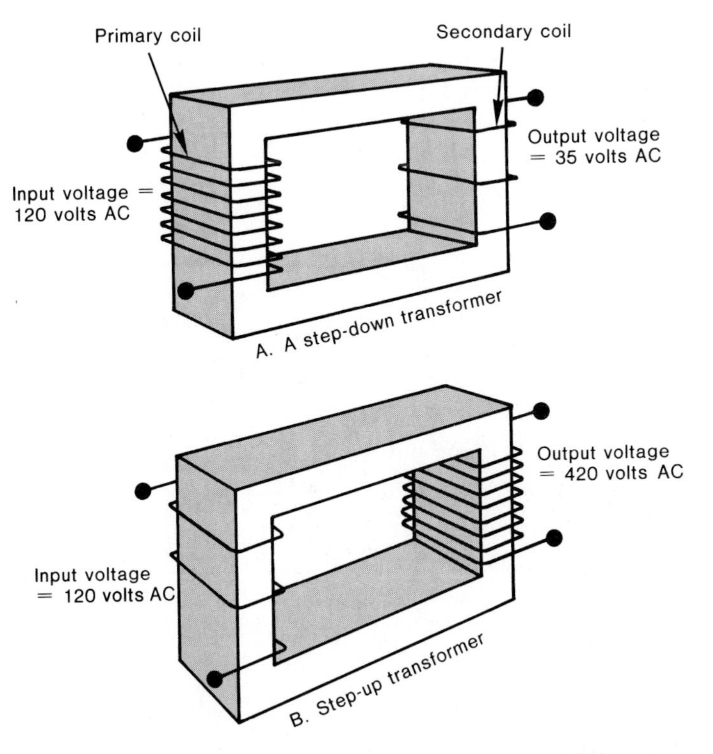

Figure 11.35. (A) A step-down transformer and (B) a step-up transformer.

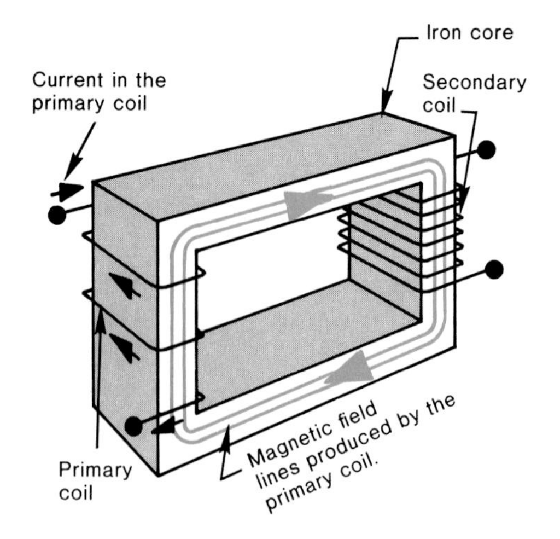

Figure 11.36. The magnetic field lines produced by the primary coil are routed to the secondary coil by the iron core.

power is extracted is called the *secondary* coil. The two coils may, or may not, have the same number of turns, and they are electrically insulated from each other. While the coils are not physically connected to each other, they are magnetically coupled. When an electric current flows in the primary coil, it becomes an electromagnet and establishes its magnetic field lines. These lines are directed to the secondary coil by the soft iron core; in this manner the magnetic coupling is established between the primary and secondary coils of the transformer (see Fig. 11.36). If the current in the primary coil is alternating at the rate of, for example, 60 Hz it will produce an alternating electromagnet whose magnetic field is also changing at the rate of 60 times each second. When the alternating magnetic field generated by the primary cuts through the turns of the secondary coil, a voltage will be induced into the secondary. The action of generating this induced voltage is called "transformer action." In transformer action, the electrical energy is transformed from the primary to the secondary coil by means of an alternating magnetic field. Notice that a voltage will be induced into the secondary any time that it senses a changing magnetic field. It does not matter if the change is brought about by an approaching (or receding) permanent magnet, as shown in Figure 11.25, or by the changing field of the stationary primary coil. All that matters is that a changing field does occur.

Assuming that all the magnetic lines from the primary cut through all the turns of the secondary, the voltage induced in the secondary will depend upon the *ratio* of the number of turns in the secondary to the number of turns in the primary. This relation is expressed by Equation 11.1

$$\text{Secondary voltage} = \left(\frac{\text{Number of secondary turns}}{\text{Number of primary turns}}\right) \times (\text{Primary voltage}). \qquad \text{(Equation 11.1)}$$

For example, if the primary has 50 turns and the secondary 1,000 turns, the voltage induced in the secondary will be 20 times larger than the voltage applied to the primary [Fig. 11.37 (A)]. If the primary voltage is 120 volts, the secondary voltage will be 2,400 volts, since this is a step-up transformer:

$$\text{Secondary voltage} = \frac{1{,}000}{50}(120 \text{ volts}) = 2{,}400 \text{ volts}.$$

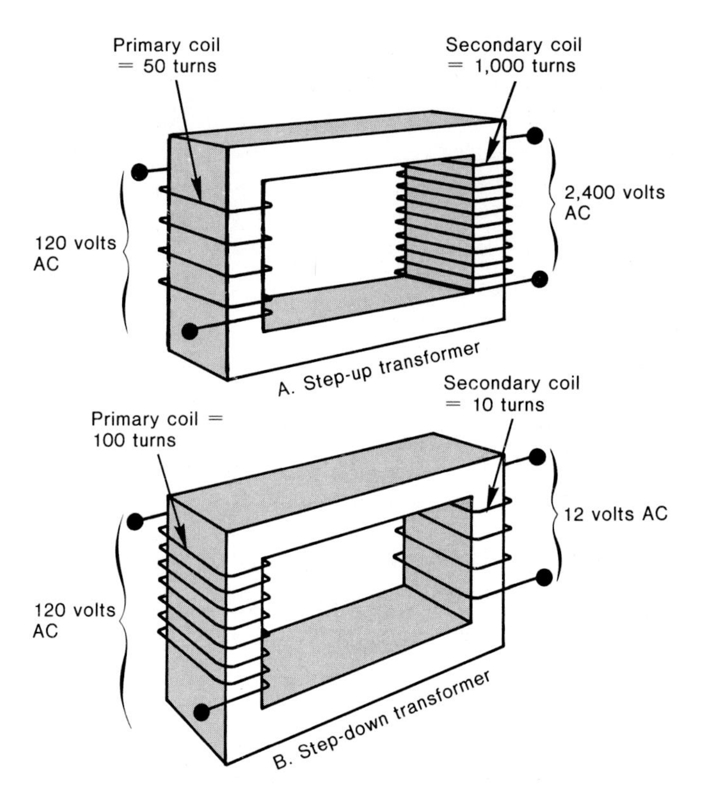

Figure 11.37. The transformer can either step-up or step-down the primary voltage, depending on the ratio of the number of primary turns to the number of secondary turns.

On the other hand, if the secondary coil has 10 turns, compared to 100 for the primary, the secondary voltage will be *less* than the primary voltage (a step-down transformer), as illustrated in part (B) of Figure 11.37.

$$\text{Secondary voltage} = \frac{10}{100}(120 \text{ volts}) = 12 \text{ volts.}$$

Power Transfer

A transformer does not generate electrical power although it can be considered a device which transfers power from the primary circuit to the secondary circuit with minimal loss. Many transformers approach 100% efficiency so that this is a reasonable statement. Since power equals voltage times current (Equation 7.3), then

$$(\text{Voltage} \times \text{Current})_{\text{primary}} = (\text{Voltage} \times \text{Current})_{\text{secondary}} \quad \text{(Equation 11.2)}$$

Figure 11.38 shows what this equation implies. For example, the increased secondary voltage in a step-up transformer is compensated by a *decrease* in the secondary current. Likewise, a step-down transformer reduces the output voltage, but it *increases* the current. The product of voltage and

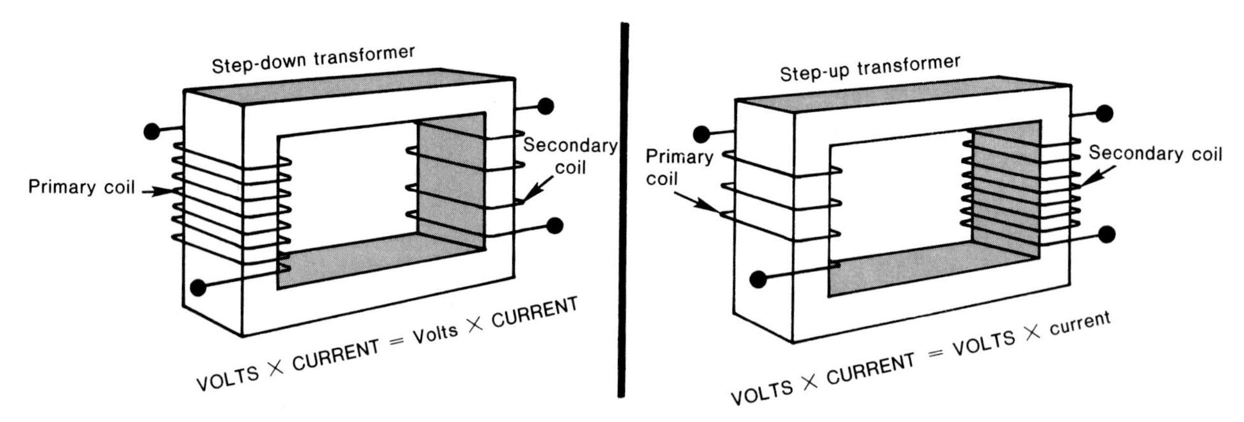

Figure 11.38. The power in the primary coil equals the power transferred to the secondary coil.

current is the same for both primary and secondary coils. Therefore, a transformer allows the voltage to be stepped-up, but only at the expense of stepping-down the current. And vice versa.

Summary

The operation of a transformer is based on Faraday's law of induction. A transformer can either step-up or step-down AC voltages. The ratio of the input voltage across the primary coil to the output voltage across the secondary coil is equal to the ratio of the number of turns in the primary coil to the number of turns in the secondary coil. The electrical power input to the primary coil is equal to the power output from the secondary coil.

Summary of Terms

Air-motion Transformer Speaker—A speaker whose diaphragm has the shape of an accordion. The folds of the diaphragm have straight wire conductors attached to them such that magnetic forces cause a squeezing action of the folds.

Domain—A group of atoms whose miniature atomic magnets are aligned to reinforce one another.

Electromagnet—A helical coil of wire through which an electric current is flowing. The coil becomes a magnet and has a north and a south pole. The strength of the electromagnet can be increased by either adding an iron core, increasing the current, or increasing the number of turns in the coil.

Electromagnetism—A theory which explains the relationship between electric currents and magnetic fields.

Faraday's Law of Induction (Induced Current and Voltage)—A law which states that if a magnetic field and a coil of wire are moving relative to one another, a voltage will be created (induced) in the coil, thus causing a current to flow. The important issue is that the magnetic field lines must pass through the coil and be changing.

Induced Current—(See **Faraday's Law of Induction.**)

Induced Magnetism—The magnetism which results in a magnetizable material like soft iron when it is brought into the vicinity of a magnet. When the magnet is removed, the iron will lose some of this induced magnetism, but a residual amount will remain. In materials like those used in magnetic tapes, much of the induced magnetism is retained.

Induced Voltage—(See **Faraday's Law of Induction.**)

Left-hand Rule—A means for determining the location of an electromagnet's north pole (Rule #1). Also, the left hand can be used to determine the direction of the force acting on a current which is traveling in a magnetic field (Rule #2).

Lenz's Law—A law which allows one to predict the direction in which an induced current will flow in a coil. The polarity of the induced voltage will be in such a direction that the resulting current will produce an electromagnet which opposes the original motion creating the induced voltage.

Magnaplanar Speaker—A speaker with a flat diaphragm which has straight wire conductors attached to it. The magnetic interaction between the current and external magnetic fields causes the diaphragm to vibrate.

Magnetic Field—The region of altered space which exists in the vicinity of either a magnet or an electric current. The field consists of closed magnetic lines whose direction is determined by the way a compass needle points. Outside a magnet the lines appear to originate at a north pole and point toward the south pole.

Magnetic Field Lines—(See **Magnetic Field.**)

Magnetic Pole—The end of a magnet, referred to as either the north pole or south pole. The north pole was so named because a compass needle points to the earth's geographic north pole.

Moving Coil Microphone—A magnetic type of microphone which utilizes a stationary permanent magnet and a moving coil.

Moving Coil Phono Cartridge—A magnetic type of phono cartridge which utilizes a stationary permanent magnet and a moving coil.

Moving Magnet Microphone—A magnetic type of microphone which utilizes a stationary coil and a moving permanent magnet.

Moving Magnet Phono Cartridge—A magnetic type of phono cartridge which utilizes a stationary coil and a moving permanent magnet.

Permanent Magnet—A magnetic material in which the magnetic domains remain permanently aligned.

Playback Head—A device found in tape decks. The playback head responds to the changing magnetic patterns in the recorded tape and generates the electrical audio signal. A playback head is made from a coil of wire wrapped around an iron core, and it uses Faraday's law of induction in its operation. A playback head is not an electromagnet.

Pole—(See **Magnetic Pole.**)

Recording Head—A device found in tape decks. It is an electromagnet which transcribes the audio signals into magnetic patterns on the tape.

Residual Magnetism—(See **Induced Magnetism.**)

Step-down Transformer—(See **Transformer.**)

Step-up Transformer—(See **Transformer.**)

Transformer—A device which consists of two coils of wire wrapped around an iron core, and which utilizes Faraday's law of induction in its operation. The input coil is called the primary coil, or primary, while the output coil is called the secondary coil, or secondary. Transformers are used to increase, or step-up, AC voltages as well as to decrease, or step-down, AC voltages.

Review Questions

1. Explain the nature of the magnetic forces between like and unlike poles.
2. What is a magnetic field and how is the direction of the field lines determined?
3. How does the left-hand rule predict the magnetic polarity of an electromagnet?
4. How does the principle of electromagnetism explain the operation of speakers and tape recording heads?
5. How can a piece of unmagnetized iron become magnetized?
6. A current-carrying wire experiences a force when it is placed in a magnetic field. How can the direction of this force be predicted? What happens when the current runs parallel to the magnetic field lines?
7. Explain the operation of the air-motion transformer speaker and the magnaplanar speaker by using the "second" left-hand rule.
8. What factors influence the amount of voltage which is induced into a coil?
9. Explain Faraday's law of induction and cite some examples of its use in hi-fidelity.
10. How does Lenz' law predict the direction of the current induced in a coil by a changing magnetic field?
11. How does a transformer convert an AC voltage into either a larger or smaller voltage? Upon what law is a transformer based? How is a transformer's output power related to its input power?

Exercises

NOTE: The following questions have up to 5 possible answers. Please select the **one** response which best answers the question.

1. The primary of a transformer has 60 turns and the secondary has 20 turns. If there are 1,200 volts across the primary then there are _______ across the secondary.
   1. 200 volts
   2. 400 volts
   3. 40 volts
   4. 30 volts
   5. 300 volts
2. If 2 amps at 120 volts are entering the primary of a transformer and the voltage at the secondary is 60 volts, the current in the secondary is:
   1. 1 amp.
   2. 2 amps.
   3. 3 amps.
   4. 4 amps.
   5. 5 amps.
3. A step-down transformer is used to:
   1. decrease DC voltage.
   2. decrease AC current.
   3. increase DC current.
   4. decrease AC voltage.
4. A transformer consists of:
   1. a primary coil and a secondary permanent magnet.
   2. a primary permanent magnet and a secondary coil.
   3. two permanent movable magnets; one called the primary, the other called the secondary.
   4. a primary coil and a secondary coil.
   5. None of the above answers is correct.
5. Many transistors inside amplifiers require 6 volts for correct operation. However, electrical outlets always give 120 volts. Therefore, inside each amplifier is a transformer which steps down the 120 volts to 6 volts. If the primary (connected to the 120 volts) has 600 turns, how many turns are in the secondary?
   1. 600.
   2. 60.
   3. 6.
   4. 120.
   5. 30.
6. 120 volts is fed to the primary of a transformer and 30 volts appear at the secondary. What type of a transformer is this, and how many turns are in the primary if the secondary has 100 turns?

Type	Number of Primary Turns
1. step-up	400
2. step-up	25
3. step-down	400
4. step-down	25
5. step-down	100

7. Which answer below determines how much voltage a moving magnet cartridge will produce?
   1. The number of turns in the coil.
   2. The strength of the moving magnet.
   3. How fast the magnet is moving.
   4. 1 and 2 above.
   5. 1, 2, and 3 above.
8. The faster that a magnet moves towards a coil of wire:
   1. the larger the resistance produced in the wire.
   2. the faster the magnetic poles reverse.
   3. the lower the voltage that is induced.
   4. the larger the voltage induced in the wire.
   5. None of the above answers is correct.
9. A voltage can be induced in a coil of wire if:
   1. a magnet moves near the coil.
   2. the coil moves near the magnet.
   3. the coil is in the presence of a changing magnetic field.
   4. All of the above answers are correct.
10. Which of the following may use Faraday's law of induction?
    1. Air-motion transformer speakers.
    2. Cartridges.
    3. Microphones.
    4. Both 2 and 3.
    5. 1, 2, and 3 above.
11. The fact that a voltage can be induced in a coil of wire when a magnet moves near the coil is known as:
    1. Ohm's law.
    2. magnetic poles.
    3. magnetism.
    4. transformer action.
    5. Faraday's law of induction.

12. Below is drawn a permanent magnet and a helical coil of wire.

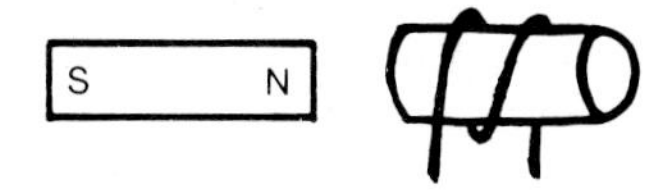

Which one of the answers below is a factor in determining the amount of voltage induced in the coil?
  1. The speed of the coil relative to the magnet.
  2. The strength of the poles.
  3. The speed of the magnet relative to the coil.
  4. The number of turns in the helix.
  5. All of the above are factors which influence the amount of induced voltage.

13. Which direction will the diaphragm move if the current through the coils is as shown?

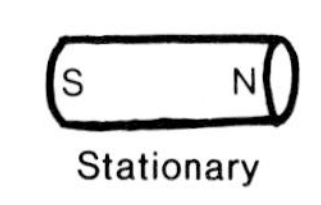

  1. To the left.
  2. Up.
  3. To the right.
  4. To the left, then to the right.
  5. To the right, then to the left.

14. A charged particle experiences a force when:
  1. at rest in a magnetic field.
  2. moving perpendicular to a magnetic field.
  3. moving along the direction of a magnetic field.
  4. None of the above answers is correct.

15. A stationary negative electric charge is shown below between the poles of a stationary horseshoe magnet. The magnetic force on the charge ___ .

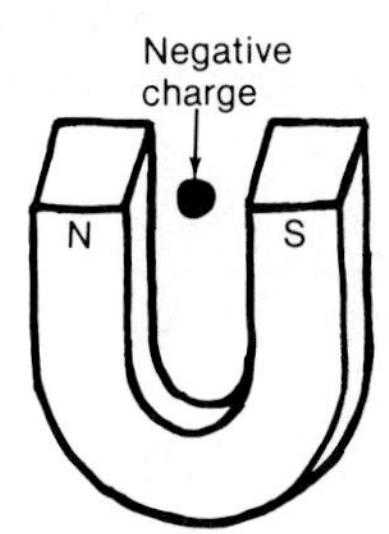

  1. points upward
  2. is zero
  3. points downward
  4. points into the paper
  5. points out of the paper

16. A current is moving through a magnetic field as drawn below. Please indicate the direction of the force exerted on the current.

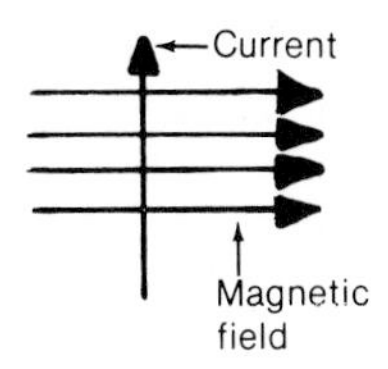

The direction of the force is:
  1. along the direction of the current (from the bottom of the page to the top).
  2. along the direction of the magnetic field (from left to right).
  3. coming straight out of the plane of the paper.
  4. going into the plane of the paper.

17. In order to make an electromagnet stronger, one would:
  1. increase the current through the coil.
  2. add more loops of wire to the coil.
  3. add an iron core.
  4. All of the above answers increase the magnetic field.
  5. None of the above answers is correct.

18. If you double the amount of turns in an electromagnet (keeping the same current) then the magnetic field will:
  1. be just as strong.
  2. be one-half as strong.
  3. be one-fourth as strong.
  4. be 4 times as weak.
  5. be twice as strong.

19. Which part of the helix becomes the north pole?

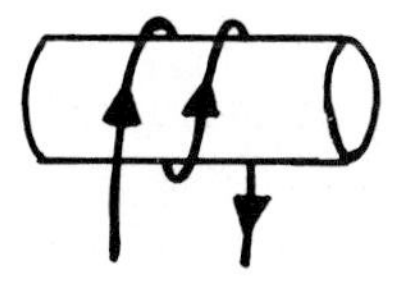

  1. The left end.
  2. The right end.
  3. A north pole will not be produced.
  4. The middle.

20. Below are drawn two electromagnets. Please indicate which one of the 5 answers is correct.

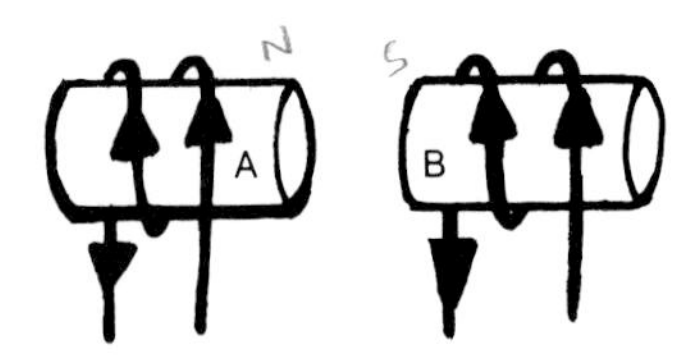

1. They will attract each other because A is north and B is south.
2. They will attract each other because A is south and B is north.
3. They will attract each other because A is north and B is north.
4. They will repel each other because A is north and B is north.
5. They will repel each other because A is south and B is south.

21. If you reverse the direction of the current through an electromagnet:
1. the magnetic field stays the same, but the voltage increases.
2. the magnetic field reverses direction.
3. the magnetic field increases, but the voltage stays the same.
4. the magnetic field reverses direction and then quickly returns to its original direction.
5. the magnetic field disappears.

22. A coil of wire which carries a current and produces a magnetic field is called:
1. a domain.
2. a pole.
3. a Faraday.
4. an electromagnet.
5. zero.

23. A long straight wire carrying a current has a magnetic field which is:
1. in the same direction as the current.
2. in the opposite direction as the current.
3. circular around the wire.
4. from the north pole to the south pole.
5. None of the above answers is correct.

24. The greater the alignment of magnetic domains in a piece of iron, the ___________ the magnetism.
1. weaker
2. stronger
3. Such alignment has nothing to do with magnetism.

25. What will happen to a permanent magnet if it is heated up?
1. Generally speaking, nothing will happen.
2. Its magnetism will increase because the heat tends to align more domains.
3. It will develop several north and south poles.
4. It will change from a permanent magnet to an electromagnet.
5. Its magnetism will be reduced because the heat "agitates" the atoms and causes the domains to become misaligned.

26. Why is a piece of unmagnetized iron attracted by either pole of a permanent magnet?
1. The pole of the magnet closest to the iron causes a like pole to be induced in the iron and attraction results.
2. A permanent magnet cannot attract an initially unmagnetized piece of iron.
3. Electrons in the unmagnetized iron are not moving and thereby can be attracted to the permanent magnet.
4. Atoms in unmagnetized iron can be induced into alignment by the magnetic field of a permanent magnet and cause an opposite pole to be induced in the iron.

27. A region, within a solid, where the magnetic atoms are all aligned is called:
1. magnetic declination.
2. a magnetic field.
3. an electromagnet.
4. a magnetic domain.
5. None of the above answers is correct.

28. Magnetic field lines (exterior to a magnet):
1. extend from the north to the south pole.
2. are generated by electromagnets only.
3. are generated by permanent magnets only.
4. have nothing to do with magnetism.
5. extend from the south pole to the north pole.

29. The region of altered space which surrounds a magnet is called the:
1. pole.
2. domain.
3. magneton.
4. magnetic field.
5. induced voltage.

30. Below are drawn two bar magnets. What will happen if they are free to move?

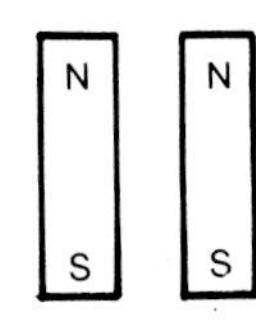

1. They will be attracted towards each other.
2. They will be repelled from each other.
3. Nothing will happen—neither attraction or repulsion.

31. What is the basic law which describes the forces between magnetic poles?
1. Like poles repel, unlike poles attract.
2. Like poles attract, unlike poles repel.
3. Like poles repel, unlike poles repel.
4. Like poles attract, unlike poles attract.
5. None of the above answers is correct.

32. Drawn below are 3 bar magnets and the magnetic field lines associated with the trio. Determine the poles of A, B, and C.

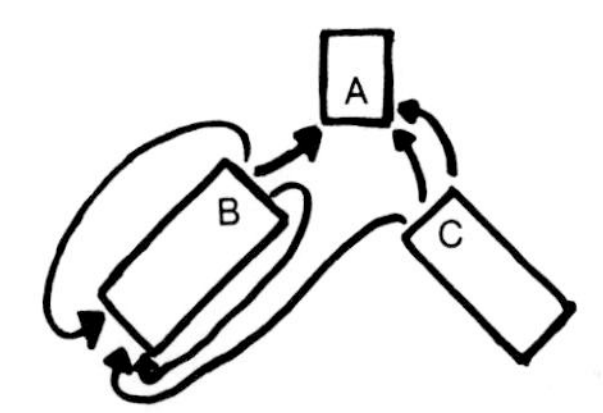

1. A is north, B south, C south.
2. A is north, B north, C south.
3. A is north, B north, C north.
4. A is south, B north, C south.
5. A is south, B north, C north.

33. A magnetic field can be generated by:
1. an electron at rest.
2. a proton at rest.
3. a charged particle in motion.
4. an electric current.
5. Both answers #3 and #4 are correct.

34. Which of the following devices use magnetism in its operation?
1. Speakers.
2. Microphones.
3. Phono cartridges.
4. Tape heads.
5. All of the above devices may utilize magnetism.

35. The role of the *permanent* magnet in a speaker is to:
1. keep any external magnetism from interfering with the speaker's operation.
2. lower the impedance of the speaker.
3. provide a stationary magnetic pole which the moving voice coil can be attracted to or repelled from.
4. There is no permanent magnet in a speaker.

36. The direction of the force experienced by a charged particle moving through a magnetic field is:
1. always perpendicular to both the magnetic field and the velocity of the charged particle.
2. parallel to the magnetic field lines.
3. parallel to the velocity of the particle.
4. parallel to both the velocity of the particle and the magnetic field lines.

37. If the current in a recording head is *tripled*, then the magnetic field that it produces:
1. also triples.
2. does not change.
3. oscillates at 3 Hz.
4. decreases by a factor of 3.
5. oscillates at 6 Hz.

38. A transformer has three times as many turns of wire in the secondary coil as in the primary coil. If the electrical power going into the primary is 300 watts, what is the electrical power coming out of the secondary?
1. 50 watts.
2. 100 watts.
3. 300 watts.
4. 600 watts.
5. 900 watts.

Chapter 12

JUST TO PLAY A RECORD

12.1 Introduction

By and large, the hi-fi industry has evolved around the phonograph record, and playing records continues to be the primary function of any hi-fi system designed for home use. As discussed in Chapter 2, a record player is a complete system which contains the following blocks:

1. a turntable which consists of the platter, motor, tone arm, and the cartridge.
2. the preamp/control center.
3. the power amplifier.
4. the speakers.

Each of these four items can be purchased as separate units or in various combinations. A component hi-fi system usually includes a separate turntable as a source of sound, and most turntables allow the installation of different makes of cartridges. Surprisingly, only a few manufacturers of turntables actually design their own cartridges, and when a turntable is purchased a compatible cartridge must also be selected; often a manufacturer will help by recommending certain models.

At first glance, a turntable appears to have the simple job of rotating the record at a constant speed while holding the cartridge in the correct position to pick up the signal. Yet, they may sell for prices over $900.00, and years of engineering practice have been devoted to their perfection. Good quality turntables can be purchased for less than $100.00, but they obviously will not have the special features of the more expensive units. The turntable's main function is to rotate the record smoothly at a constant speed. Any extraneous mechanical vibrations, no matter how slight, will be picked up by the stylus. Also, small changes in rotational speed will cause variations in the pitch of the sound. Because of this, turntable quality is dependent upon its mechanical design, and the more expensive units are usually engineered to reduce these problems to inaudible levels.

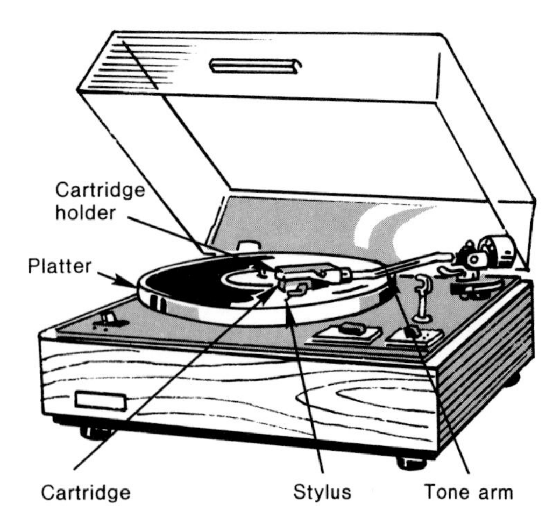

Figure 12.1. A turntable and its various parts.

The pickup, or cartridge, converts the minute wiggles of the record groove into electrical signals. The stylus must be able to follow faithfully these undulations in order to reproduce the recording accurately; as such, it should be light and easily movable. In addition, the cartridge must be guided over the record by a properly designed tone arm which must keep the cartridge positioned at the proper angle to the record groove. Otherwise, a loss occurs in the high frequency response of the cartridge. Therefore, the operation of a good turntable rests upon the successful integration of its individual components. Figure 12.1 shows the various parts of a turntable.

Almost all the important aspects of a turntable's performance can be studied by the use of a few scientific laws which govern its mechanical motion. In this chapter we shall present an overview of all phases of a turntable's operation. In addition, we shall investigate concepts like force, mass, and inertia, which play an all-important role in turntable design. Finally, a discussion of turntable specifications will be given with an eye toward consumer education.

12.2 Types of Turntables

There are three basic types of turntables: the manual, semiautomatic, and fully automatic (commonly called a record changer). All three types of turntables are designed with various styles and accessories to satisfy the needs of almost any customer.

A **manual turntable** holds only one record. The user must select the speed by either changing the belt pulley or by

Model GT35Ap Garrard turntable

using some other control. The tone arm must be manually lowered in order to place the stylus on the record. When the play is complete, the user must lift the tone arm from the record and return it to the rest position. Why would a person want a turntable without a single luxury? The answer is "quality." The manual has always been a mark of quality with its light and freely moving tone arm. Most automatic or semiautomatic turntables have mechanical devices attached to the tone arm for activating the automatic changer. As a general rule, their tracking forces are somewhat greater than necessary so that the tone arm can operate the automatic feature. Increased tracking forces lead to increased record and stylus wear. Also, a very high quality manual turntable can be purchased for relatively little expense. Finally, many people just prefer to change the records themselves.

The **semiautomatic turntable** also holds only one record, but it has certain automatic features. Many semiautomatics initiate the tone arm movement at the beginning of the record play, and then return the tone arm to the rest position after the record is finished. Others only do the latter. Some may provide multiplay features for playing a single record more than once. Cueing devices, which raise and lower the tone arm, are found on most high-quality turntables. The cueing controls should work smoothly, and the tone arm should slowly float down to engage the stylus in the record groove. The semiautomatic is usually priced somewhat higher than a manual, but the extra cost will buy ease in playing records not found on an attention-demanding manual. Their quality today is comparable to the manuals.

A **fully automatic turntable,** commonly called a record changer, automatically plays several records, one after another. Some of these turntables offer a choice of using them as a semiautomatic by replacing the long spindle with a shorter one. The automatic turntables have been considered inferior by audiophiles, because in the past they were not capable of achieving the level of performance possible with the manual units. Differences in performance are extremely small today and many good automatics are available. The biggest objection to automatic turntables is the dropping of records onto other records. Also, as the stack of records grows the tone arm is raised above its optimum tracking height with some degradation in performance. When additional records are placed on the turntable they produce a strain on the drive system which may cause speed drift. Finally, there is always the fear that the changer will "goof" and ruin one of your favorite albums. In spite of these problems, the convenience of not having to change records is very appealing to a large number of consumers. Automatics which are convertible into semiautomatics are very popular

Model ST-8 Harman Kardon turntable

because of their versatility. Favorite albums can be played manually for their protection, and yet the changer can be used when desired.

12.3 Turntable Drive Systems

The heart of a good turntable lies in its drive system. The phrase "drive system" refers to that part of a turntable which rotates the record. Remember that the goal of a perfect turntable is to rotate the disc at a specified, constant speed with no vibration. Any mechanical system, however, generates vibrations and speed inconsistencies. All you have to do is place your hand on an automobile engine and the vibrations become very apparent. Of course, the engine is mounted on the chassis such that these vibrations are generally not transmitted to the driver. Likewise, turntable motors also vibrate and, unless these vibrations are damped, they can cause the platter itself to vibrate. The vibrations transmitted to the platter are picked up by the cartridge which, in turn, translates them into low frequency unwanted noise which is often called *rumble*. Slight variations in the speed of the platter cause changes in the musical pitch. Relatively slow speed changes, less than 6 variations per second, are usually called *wow*, while more rapid changes are known as *flutter*. We shall discuss the options which are available to turntable designers in order to counteract the effects of rumble, wow, and flutter.

Motors

The motor used for a drive system must meet at least four basic requirements:

1. it must revolve at a specified, constant speed, e.g., 33⅓ rpm.
2. it must operate with a minimum amount of vibration.
3. it must provide an adequate rotational force (torque) in order to bring the platter quickly up to its rated speed.
4. it should be magnetically shielded such that the magnetic fields generated by the 60 Hz electrical current within the motor do not leak into the cartridge. Any stray magnetic fields can induce a 60 Hz voltage into the output signal of the cartridge, which can produce an unpleasant "humming" sound from the speakers.

Several types of motors are used in turntables. The lower cost turntables may use *induction motors* which, under normal conditions, are perfectly adequate. However, their speed is subject to fluctuations in either the 120 volt "line" voltage or changes in the platter's load (i.e., stacking several records). The voltage supplied by the wall receptacle may, during the day, drop below its rated value of 120 volts. This is especially true at times when there are heavy power demands. Under such conditions, the induction motor may run slower than normal. The better turntables usually have *synchronous* or *hysteresis synchronous motors.* The speed of these motors depends upon the *frequency* of the 60 Hz line voltage, and they are relatively immune from fluctuations in the size of the voltage. Because power companies carefully regulate the line frequency to be 60 Hz, these motors have excellent constant speed characteristics. The hysteresis synchronous motors produce fewer vibrations and noises than the synchronous motors, and they are used in high quality turntables.

Connecting the Motor to the Platter

The manner in which the motor is connected to the platter is an important part of a turntable's drive system. Three different methods of connection are widely used.

The **rim drive** utilizes an "idler" wheel which is located between the motor shaft and the rim of the platter (see Fig. 12.2). This type of drive has a disadvantage because the idler wheel, in time, may develop flat spots which cause the platter to rotate at an unsteady speed. In addition, the idler wheel is also rigidly pressed against the motor shaft, and any vibrations occurring in the motor are readily transferred to the platter causing unwanted sound. For these reasons the rim drive is the least satisfactory of the three drives, and it is often used in inexpensive units.

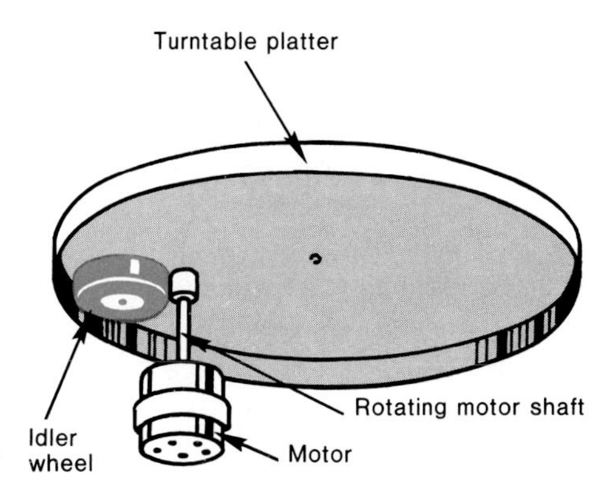

Figure 12.2. The rim drive utilizes an idler wheel to transfer the rotational force from the motor to the platter.

The **belt drive,** as shown in Figure 12.3, consists of a long flexible belt which connects the motor shaft to the platter. It is the most commonly used method of connection, and is often employed in the finest turntables. A salient feature of this drive is that motor vibrations are almost totally absorbed by the belt, with little being transmitted to the platter and ultimately being converted into unwanted sound. Also, the belt is elastic, and it can absorb small changes in the motor's speed, thus reducing speed fluctuations of the platter.

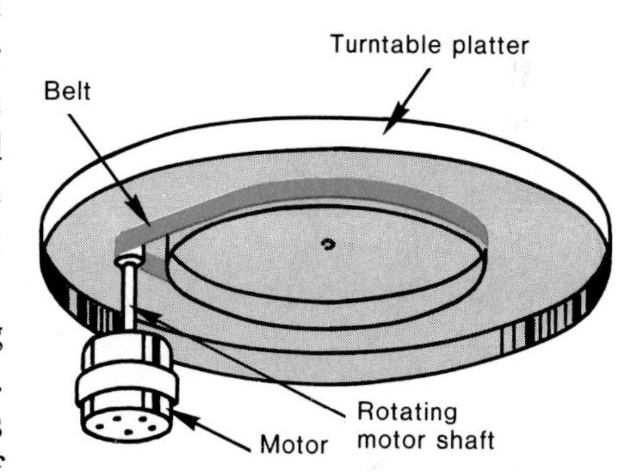

Figure 12.3. The belt drive provides excellent mechanical isolation between the motor and the platter. The belt also absorbs small speed changes of the motor.

The third drive is called the **direct drive** and is found on the more expensive turntables. In this case the platter is mounted directly on the motor shaft as shown in Figure 12.4. Special, slow speed DC motors are required which rotate at the same speed as the platter. The speed of the direct drive motors is accurately maintained by electronic servo-control circuits. These circuits monitor the motor's speed at all times and, if they should detect slight changes, an electronic correction is quickly made to return the motor to its proper speed. The direct drives are expensive, but they are gaining in popularity because their speed control keeps the record rotating at a very constant speed. It goes without saying that the direct drive motors must be extremely quiet and vibration

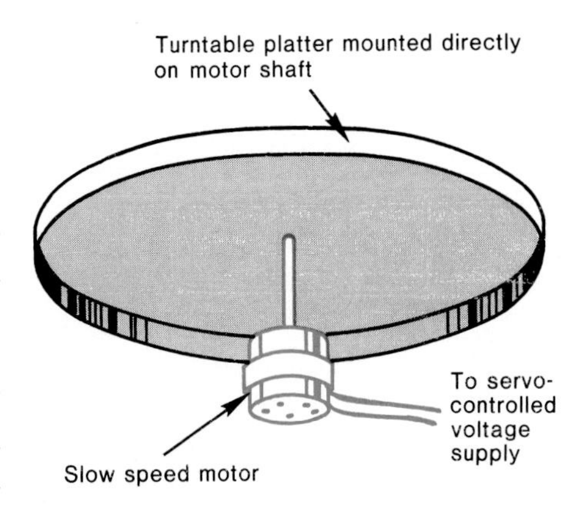

Figure 12.4. In a direct drive turntable the platter is mounted directly on the motor shaft.

free, because they cannot rely upon a damping agent, such as an elastic band, to help reduce the vibrations.

The Speed or Pitch Control

Since the record speed should be as constant as possible, most high quality turntables come equipped with a means for both monitoring the rotational speed and making slight adjustments in its value. This speed control is often called a *pitch control,* as indicated in Figure 12.5, because the rotational speed of the record determines the actual musical pitch that is heard. Typically, a pitch control utilizes a flashing light, called a stroboscope, and precisely spaced markings on the platter. When the platter is rotating at its proper speed, these markings appear to be stationary in the flashing stroboscopic light. A pitch control is important to a person who wants to be certain that the musical pitch is precisely correct.

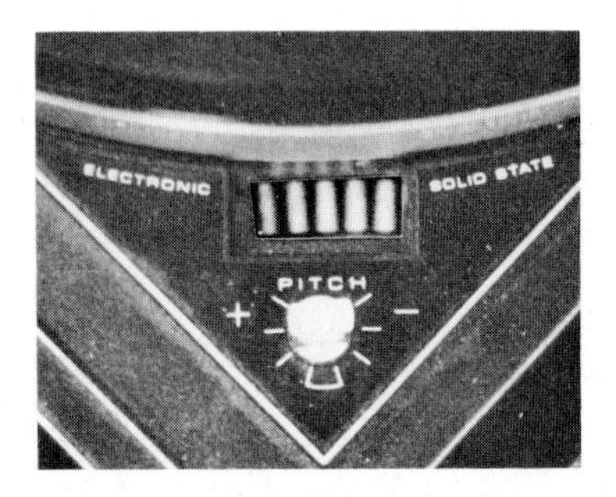

Figure 12.5. A typical pitch control. The control utilizes a stroboscopic effect to ensure that the platter rotates at the correct speed.

The speed of a turntable which does not have a pitch control can still be checked if a cardboard strobe disc is available. These discs can be purchased cheaply at most electronic supply stores. Place the strobe disc on the platter and position a fluorescent lamp near the turntable. A fluorescent lamp does not glow at a steady rate, but it flashes for both the positive and the negative parts of the 60 Hz house current. The markings on the strobe disc are spaced so that they appear to stand still when the light flashes at 120 Hz and the platter is rotating at its correct speed. If the strobe lines appear to move in the direction of the platter's rotation, the platter is turning too fast. The platter is running too slow when the strobe lines appear to turn in the opposite direction.

Platters

The ear is extremely sensitive to small quick changes in the speed of a turntable. This is especially true when, for example, a piano note is sustained for some time. Small speed variations lead to an unsteady, quivering quality of the note as the apparent pitch is raised and lowered about its intended value. A "trained ear" can detect pitch variations when the platter changes its speed by as little as 0.3%. This is very small, and it amounts to a change of only 0.1 rpm in a 33⅓ rpm record. Turntable platters are usually heavy metal disks which serve to act as flywheels (see Fig. 12.6). A flywheel is used in many mechanical systems in order to stabilize the speed. For instance, a car engine uses a flywheel for smoothing out the speed fluctuations which occur between power strokes of the pistons. Motors would "turn over" in a much more erratic fashion without the

A. Top surface

B. Bottom surface

Figure 12.6. The large metal platter serves as a heavy flywheel which tends to keep the speed constant.

smoothing effects of their flywheels. In a similar fashion, the flywheel effect of the platter keeps the record rotating at a constant speed.

Audio engineers describe the flywheel effect in terms of the *platter's moment of inertia.* The larger the platter's moment of inertia, the better it is able to resist small speed changes. Therefore, platters should have a moment of inertia which is as large as possible. The moment of inertia depends upon the mass (weight) of the platter as well as the geometric distribution of the mass. A larger moment of inertia can be obtained if most of the mass is concentrated in a ring around the outer edge of the platter; this can be seen by observing the platter in Figure 12.6 (B). Hence heavy platters, whose mass is distributed toward the perimeter, yield larger moments of inertia which lead to good rotational speed stability. Of course, the motor has difficulty in initially bringing such a platter up to speed; that is why a large torque motor is required. However, once up to the proper speed, slight variations in the motor's power will not readily affect the platter's speed.

If the platter is not properly balanced, it will wobble exactly like an out-of-balance car wheel, and this unwanted vibration would be picked up by the cartridge. Platters are usually die cast from an aluminum alloy and machined for accuracy to ensure proper balance. Magnetic materials, like iron, should not be used for platter construction because stray magnetic fields in the platter will cause induced voltages (noise) in the magnetic cartridge.

One final note about platters. Even the rubber or plastic mat, which rests between the platter and the record, has an important function. The platter has a slight tendency to ring somewhat like a bell in response to extraneous vibrations. If this "ringing" were to reach the cartridge it could be reproduced by the speakers as an undesirable sound. The mat helps to dampen such vibrations much like your fingers do when they touch a ringing bell.

The Suspension

Most turntables are mounted on springs so that an accidental bump on the cabinet will not readily shake the platter (Fig. 12.7). This is one of the biggest mechanical problems in turntable design. The isolation springs must be flexible enough to prevent vibrations from reaching the platter, and they act somewhat like the shock absorbers in a car. At the same time, they must provide adequate support and stability for the heavy platter. Some manufacturers are more successful than others in achieving the proper mountings, although most turntables incur problems under severe conditions. However, if the user places the turntable in an area of the room which is not subject to excessive vibrations (e.g., away from the doors), and removes the speakers from the turntable, external vibrations should cause little trouble in a well-designed unit.

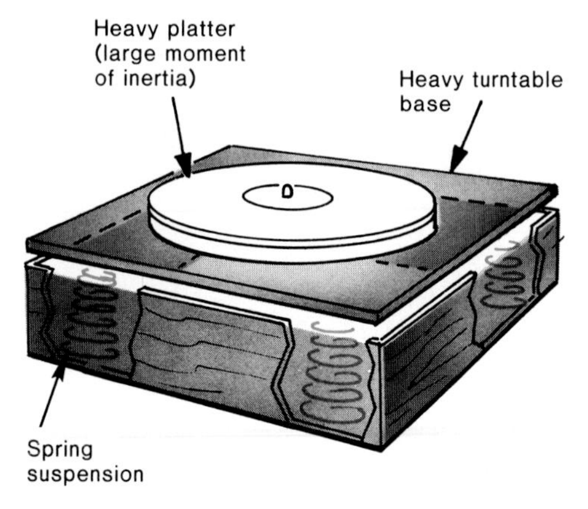

Figure 12.7. The damping of external vibrations relies on the heavy base of the turntable, the spring suspension, and the large inertia of the platter.

12.4 Specifications of Turntable Drive Systems

We have seen that a record must revolve at a steady speed and be unaffected by extraneous vibrations. It is not surprising that the important performance specifications of a turntable drive system are related to these crucial considerations. There are three specifications which indicate how closely a turntable drive system approaches perfection: speed accuracy, rumble, and wow and flutter. Remember, as always, that our guidelines for these specifications are meant to be of help in obtaining a quality turntable at a reasonable price.

Speed Accuracy

The ability of a turntable to maintain an accurate speed over long periods of time (several hours) is an absolute necessity. The specification called "speed accuracy" indicates the extent of this ability, and its definition is as follows.

Speed Accuracy: The maximum percentage change in rotational speed which a turntable exhibits over a period of several hours.

The long-term speed changes are not to be confused with wow and flutter, which are short-term (less than 1 second) changes associated with the motor and drive inaccuracies. A good turntable should be able to keep the long-term speed changes well under 2%. This means that a 33⅓ rpm record will speed up, or slow down, by no more than $(.02 \times 33\frac{1}{3}) = 0.67$ rpm. Turntables equipped with a strobe indicator, as mentioned in section 12.3, can be visually set to an exact speed.

Turntable Buying Guide

Speed Accuracy: Look for a speed accuracy of 2% or less.

Rumble (S/N Ratio)

Low frequency vibrations of the platter, which are picked up by the cartridge, are called rumble. Sometimes rumble can be heard as a low frequency "growling" noise, especially when the musical passage is rather quiet and the volume has been turned up. Often rumble frequencies lie below the audible range, and they are reproduced by systems with subsonic capabilities. An indication of subsonic rumble occurs when a woofer cone can be seen to "breathe" in and out without producing any sound. Excessive rumble, either audible or inaudible, can cause serious distortion by the speakers. The definition of rumble is stated as follows.

Rumble: The number of decibels by which the sound level of extraneous vibrations falls below the sound level of the musical signal.

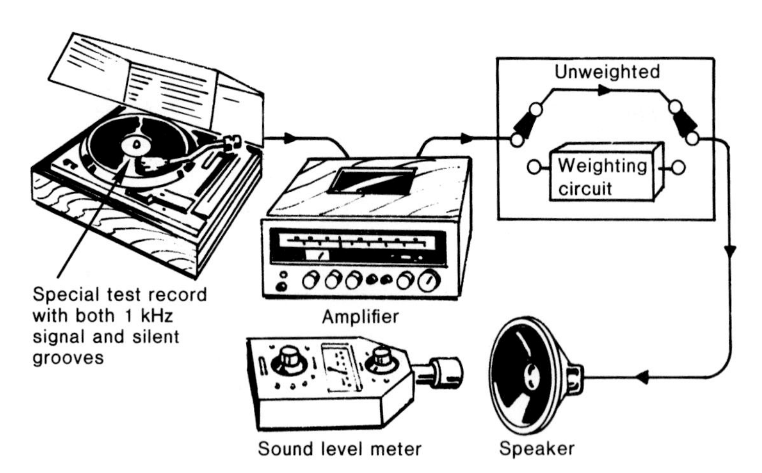

Figure 12.8. Set-up for measuring the unweighted rumble specification of a turntable.

The general method by which rumble is measured is shown in Figure 12.8. First, a standard 1,000 Hz signal is played through the system from a special test record. For the moment ignore the box which is labeled "weighting circuit." The signal is sent to the speaker and a dB meter is used to measure its sound level.* Suppose that the meter registers a level of 90 dB. Now the tone arm is positioned to one of the silent grooves in the record. When rumble is present, the meter will again register an output level, although it will be considerably less than that of the 1 kHz signal; assume that it measures 50 dB. The *rumble* specification is the difference, in dB, between the level of the silent groove and that of the 1 kHz tone. In this case the rumble would be (50—90) = − 40 dB. Notice that the rumble specification is stated exactly like the S/N ratio of an amplifier (see sections 8.6 and 8.7). The only difference is that one type of noise is generated by transistors while the other is produced by a vibrating platter. Also, the rumble specification is usually quoted with a minus sign to emphasize that the rumble sound level is lower than the 1 kHz sound level. This type of rumble measurement is called "unweighted," and it contrasts with the "weighted" measurements which we shall now discuss. The unweighted, or NAB (National Association of Broadcasters), rumble figures for better turntables are -35 dB and lower, e.g., -45 dB.

In section 5.6 we discussed the fact that human hearing becomes less sensitive to frequencies in the bass region. Some manufacturers prefer to specify the rumble by taking into account the *audible* effects of rumble, rather than its absolute value as registered on a dB meter. They argue that it does not make sense to measure the level of rumble for

*In practice this measurement is made by replacing the speaker with an 8 ohm resistor, and connecting the dB meter to measure the electrical, rather than the sound, level.

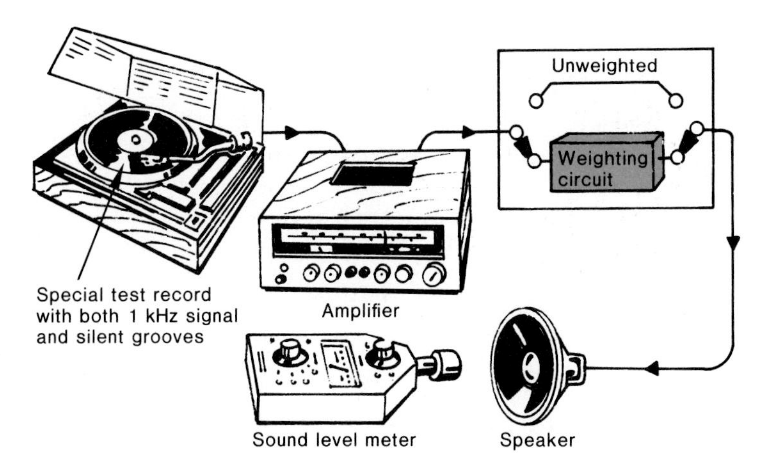

Figure 12.9. Set-up for measuring the weighted rumble specification of a turntable.

frequencies which lie in the deep bass region where the ear becomes insensitive. The process of deemphasizing certain rumble frequencies is called "weighting." In a weighted rumble measurement, the rumble picked up by the cartridge is passed through an electrical weighting network (see Fig. 12.9) which has a frequency response characteristic especially designed to deemphasize certain frequencies. For example, weighting standards developed by German audio manufacturers (and often used by American manufacturers) are known as DIN standards (Deutsche Industrie Normen). The DIN response is shown in Figure 12.10. Notice how the DIN weighting curve has a maximum response for frequencies near 315 Hz, and it falls off sharply for both the lows and highs. Any rumble frequencies below 50 Hz will be completely attentuated by the DIN response curve and, consequently, will not be recorded by the dB meter. As a result, the DIN weighted rumble measurements will always appear better (i.e., larger S/N values) than the corresponding unweighted values. It is not unusual that a DIN rumble figure is -60 dB when the unweighted (NAB) value registers -40 dB.

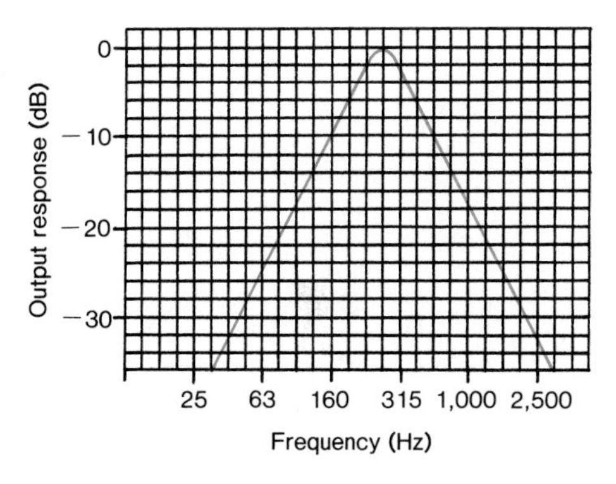

Figure 12.10. A DIN weighting curve used for rumble measurements.

There is a second type of weighting curve which has been developed by CBS laboratories. Known as the ARLL method (audible rumble loudness level), this weighting curve gives values close to the DIN weighted measurements. A typical ARLL rumble figure for a good turntable would be -60 dB or lower. When comparing the rumble specifications of various turntables, it is important that you keep these three different techniques in mind. It is the only way to make fair comparisons among those manufacturers who use different systems of rumble measurement.

Turntable Buying Guide

Rumble

1. **Unweighted (NAB):** Look for an unweighted rumble specification of -35 dB or lower, e.g., -40 dB.
2. **Weighted (ARLL or DIN):** Look for a weighted rumble specification of -60 dB or lower, e.g., -70 dB.

Wow and Flutter

Wow and flutter are short-term changes in a turntable's speed, usually occurring in the subsecond range. Although these two terms are collectively grouped into one specification, wow is considered to represent the slower speed changes of the two. If the motor shaft or the platter bearings should develop a "flat spot," the speed of the record will change each time the flat spot is encountered, or "**W**once-**O**'-**W**round" (**WOW**—get it?) Generally speaking, wow is considered to be any speed changes which occur at a rate between 0.5 and 6 times per second. Flutter, on the other hand, is considered a faster variation (between 6 and 200 times per second) which produces a fluttering sound on long-held notes. Hence the name, flutter. The definition of wow and flutter is given below.

Wow and Flutter: The short term changes in the rotational speed of a turntable platter caused by imperfections in the drive system. Wow and flutter expresses these speed changes as a percentage relative to the desired speed.

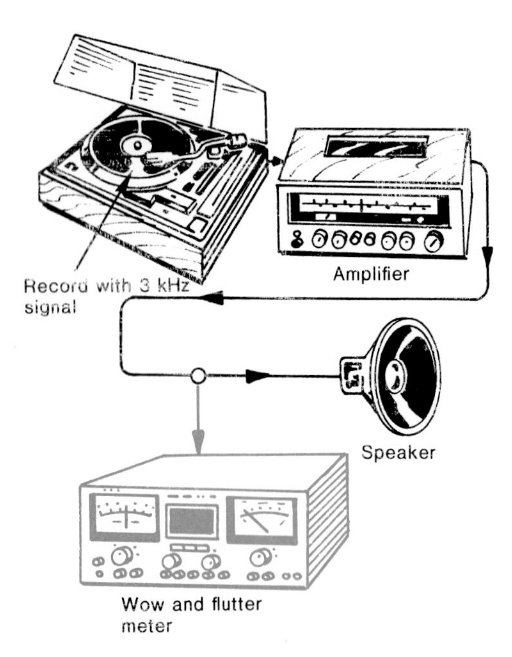

Figure 12.11. Relatively rapid changes in a turntable's speed can be measured with a wow and flutter meter.

Figure 12.11 shows a typical set-up for measuring the wow and flutter. The procedure involves a typical stereo system, except that a special test record is used which has a 3 kHz signal cut into its grooves. The output signal from the amplifier is sent simultaneously to the speakers and to a wow and flutter meter. When the record speeds up, the frequency produced from the record is greater than 3 kHz. Likewise, when the record slows down, the frequency produced is less than 3 kHz. The wow and flutter meter senses these frequency changes, and it gives the wow and flutter specification by dividing the average frequency change by 3 kHz, expressing the ratio as a percentage. For example, a 0.2% wow and flutter measurement means that the frequency of the 3 kHz tone can change by as much as 6 Hz [(.002) × (3,000 Hz) = 6 Hz] due to speed variations in the drive system. Therefore, one can expect a 3 kHz tone to have its reproduced frequency go as low as 2,994 Hz (3,000-6) or as high as 3,006 Hz when played on this turntable.

Most people cannot detect a 0.2% fast-speed change, so this is an acceptable value for high-fidelity performance. The smaller the wow and flutter percentage, the better the turntable.

Wow and flutter values are sometimes "weighted," where the term has the same general meaning it did with the weighted S/N ratios of amplifiers and turntable rumble. The types of weighting curves used for wow and flutter measurements are not the same as those discussed above for turntable rumble. When used with wow and flutter measurements, the term "weighted" means that speed changes which occur at certain rates are counted more heavily than speed changes which occur at other rates. Such preferential treatment is intended to reflect the fact that certain rates of change are more irritating, or have a higher annoyance factor, to the average listener than other rates. For example, many people find that a speed which is fluctuating at the rate of six times per second is far more annoying than a speed which is changing at the rate of once per second. A weighted wow and flutter measurement gives more emphasis to the 6 Hz speed changes than it does to the 1 Hz changes. Without this preferential treatment wow and flutter specifications are said to be "unweighted." Weighted wow and flutter is typically a smaller percentage than the unweighted version.

Turntable Buying Guide

Wow and flutter

1. **Unweighted:** Look for an unweighted wow and flutter specification of 0.2% or smaller.
2. **Weighted:** Look for a weighted wow and flutter specification of 0.1% or smaller.

12.5 Tone Arms and Their Specifications

The tone arm has the job of properly holding the cartridge as it glides across the record. The key word here is "properly," and a good deal of engineering thought goes into a well-designed tone arm. The tracking force, tone arm resonance, tracking angle error, and tone arm pivot friction are the main specifications which pertain to the quality of a tone arm. In addition, there are features found on some of the better turntables, such as cueing and antiskating controls, which may also be of importance to you. We shall now discuss each of these items in turn.

Tracking Force

An important job of any tone arm is to provide the correct amount of force necessary to keep the stylus in contact with the record groove. This force is called the "tracking force," and it is defined as follows.

Tracking Force: The downward force exerted on the stylus in order to keep the stylus in firm contact with the record groove. Turntable manufacturers measure this force in grams.*

The tracking force is a very important specification for the proper operation of a turntable. If it is too large the stylus will exert excessive pressure on the groove, with a substantial increase in both record and stylus wear. Too small a tracking force causes the stylus to lose contact with the groove wall, especially when the groove contains moderate-to-high recorded levels. Such mistracking results in a "shattering" sound, which will almost surely ruin your enjoyment of the music. Less severe mistracking results in a loss of high frequency response and increased distortion. Recommended tracking forces are usually less than 3 grams.**

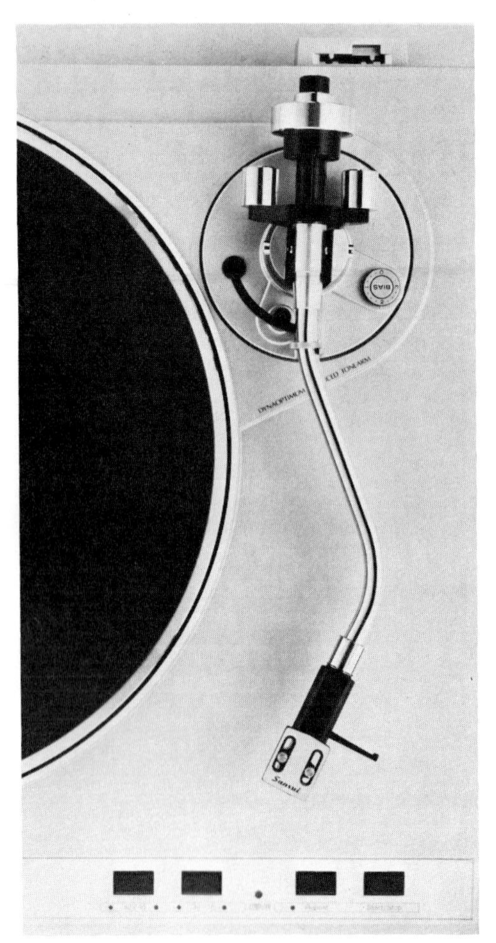
Tone arm on a model FR-Q5 Sansui turntable

The cartridge, being mounted on the end of the tone arm, is also dependent upon the properties of the tone arm for its proper operation. The matching of the cartridge to the tone arm is analogous to the manner in which the performance of a speaker's driver is closely linked to the enclosure; luckily, the correct matching of a cartridge to a tone arm is not nearly as critical. Most tone arm instructions give acceptable values for cartridge weights and other necessary characteristics. When selecting a cartridge/tone arm combination, it is important that these recommendations be carefully followed. A well-designed tone arm has a minimum amount of pivot friction, correct antiskating compensation (discussed below), and a light mass. Such a tone arm will allow the cartridge to track close to its minimum rated force, thus reducing wear. An inferior tone arm requires extra force in order to keep the stylus in contact with the groove walls.

The downward tracking force is provided by the combined action of the cartridge plus the tone arm, and all tone arms have a method for adjusting this force to its proper value. The most widely used method on better quality turntables employs an adjustable counterweight on the end of the tone

*It is not scientifically correct to use the "gram" as a unit of force. The "gram" is a unit of mass. As you will see in sections 12.8 and 14.5, the proper unit of force is called the "dyne," but turntable manufacturers do not use it.

**For a size comparison, 3 grams is approximately the mass of one penny.

arm which is opposite to the cartridge [see Fig. 12.12 (A)]. By moving the weight back and forth, the tone arm can be balanced about its pivot (fulcrum) to yield the correct tracking force; this method is referred to as "static balancing." Another example is a "dynamically balanced" tone arm in which an adjustable spring provides the correct tracking force [see Fig. 12.12 (B)]. This method is widely used on lower priced turntables, but is seldom found on high quality equipment.

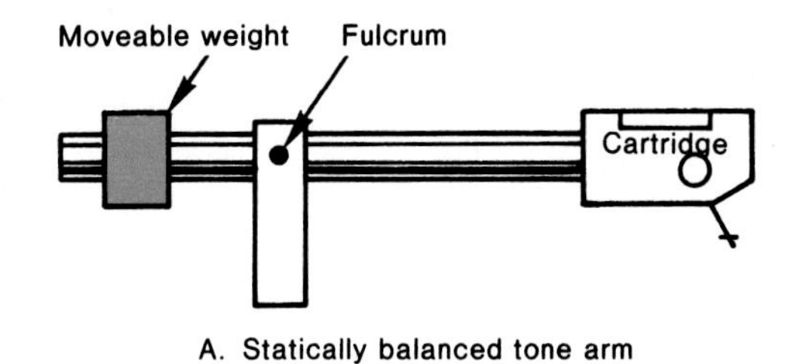

A. Statically balanced tone arm

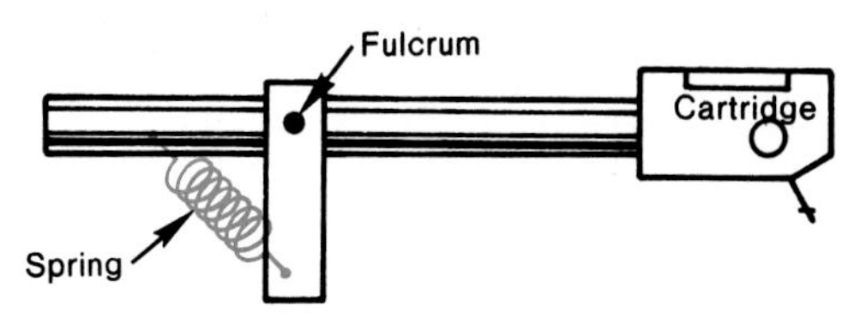

B. Dynamically balanced tone arm

Figure 12.12. Two methods for adjusting the tracking force of a cartridge. (A) The statically balanced tone arm utilizes a counterweight, while (B) the dynamically balanced tone arm employs an adjustable spring.

Tone Arm Buying Guide

Tracking Force: Look for a tone arm which provides for tracking forces up to three grams.

Tone Arm Resonance

As mentioned in section 4.7, every object has one, or more, natural frequencies of vibration at which it "likes" to vibrate. If the object is forced to vibrate at one of its natural frequencies, a condition of resonance occurs. Large amounts of energy can be transferred to the object with potentially disastrous results (remember the Tacoma Narrows bridge!). A tone arm is no exception, for it too has a natural frequency of vibration which is governed by its weight and geometry.

Tone Arm Resonance: The large amplitude vibrations of a tone arm which result when it is forced to vibrate at its natural frequency of vibration.

If the resonant frequency lies within the audible range serious mistracking can result! Suppose, for example, that the tone arm's resonant frequency is 100 Hz. As the stylus tracks the groove, it will almost certainly encounter a 100 Hz signal. This frequency starts the stylus vibrating at 100 Hz which, in turn, causes the tone arm to also vibrate at the same frequency. Since a condition of resonance now exists, the tone arm will begin to oscillate violently with a loss of groove contact and possible groove jumping. To avoid this problem, the tone arm's resonant frequency should fall well below the audible range, perhaps 15 Hz or smaller. On the other hand, if the resonant frequency falls much below 7 Hz there is a potential problem related to record warp. Most record warps occur in a manner which leads to an extra force on the stylus at frequencies between 2 Hz and 6 Hz. If a tone arm resonance frequency occurred in this range, the warped record could force the stylus into mistracking with the loss of reproduction accuracy. In many turntables special damping mechanisms are also used on the tone arm pivot which damp out, or prevent, tone arm resonance.

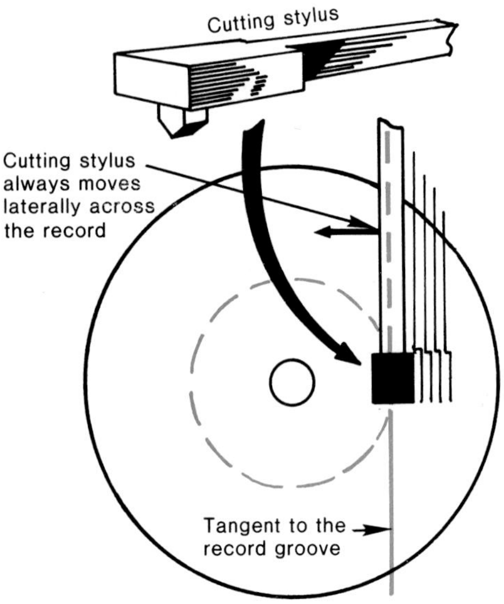

Tone Arm Buying Guide

Tone Arm Resonance: Look for a tone arm resonance between 7 Hz and 15 Hz.

Figure 12.13. The master record is made by moving the cutting stylus laterally across the disc. This ensures that the cutting stylus is always tangent to the record groove.

Tracking Angle Error

When a master record is cut, the cutting head always travels laterally across the record so that the cutting stylus remains tangent to the groove at all times (see Fig. 12.13). However, with pivoted tone arms it is not possible to keep the pickup stylus tangent to the record groove at all positions of the tone arm. Figure 12.14 shows a pivoted arm which has been set tangent to the record groove during the middle of its play (see point "A" in the figure). At the beginning of the record (point "B") the tone arm no longer places the stylus tangent to the groove, and this gives rise to the *tracking angle error* as depicted in the picture. Actually, there may be more than one zero-error point, depending on the tone arm's length and pivot placement. Except at these zero-error points, there is always some tracking angle error in a pivoted tone arm, and it becomes worse as the arm swings away from the zero-error points. In general, as the tracking angle error increases, the stylus loses its ability to pick up the high frequencies. The definition of tracking angle error is stated as follows.

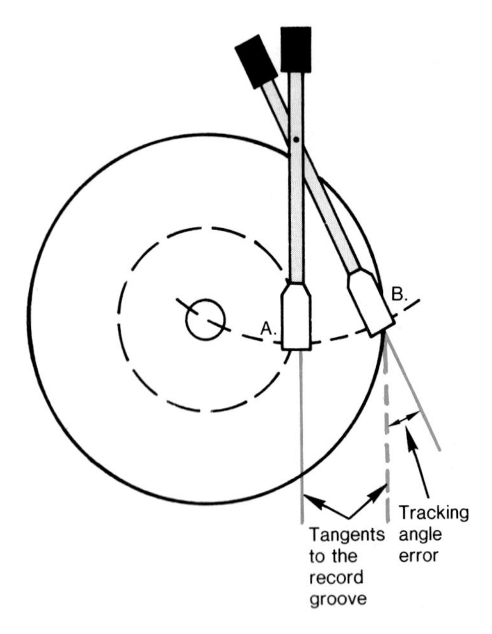

Figure 12.14. A pivoted tone arm leads to tracking angle error, because the stylus is not always positioned tangent to the record groove.

Tracking Angle Error: The largest angular deviation with which the tone arm positions the cartridge away from perfect tangency to the record groove. It is measured in degrees.

For hi-fidelity performance the tracking angle error should be kept below 3° at the worst points on the record.

You may now ask: Why don't turntable manufacturers build laterally moving tone arms which reproduce the same movement as the cutting stylus, and thus eliminate tracking angle errors altogether? Some do, although the lateral, or translational, drives are very difficult and costly to engineer. Figure 12.15 shows one of the excellent translation arm drives now on the market. Pivoted tone arms are far more common because they are relatively inexpensive.

Figure 12.15. A Bang and Olufsen Beogram 4002 translational (lateral) arm turntable.

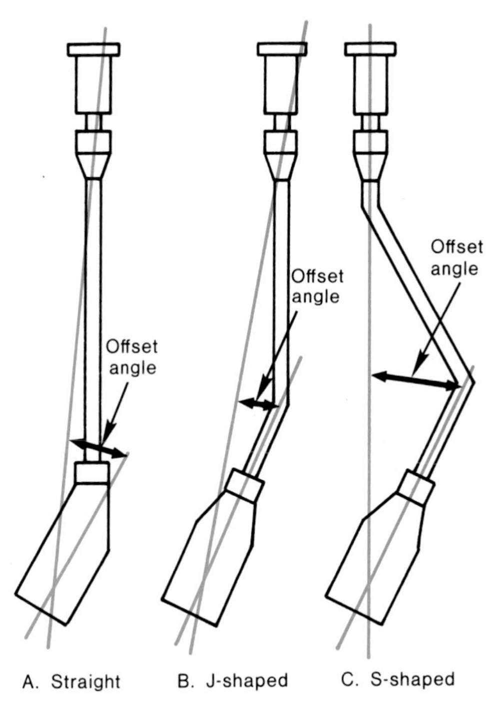

Figure 12.16. The J-shaped, and the S-shaped tone arms help to reduce the tracking angle error.

There are three general shapes of pivoted tone arms, as shown in Figure 12.16: straight types, J types, and S types. In each case the cartridge assembly is positioned at an angle, called the *offset angle*, relative to the tone arm axis. The purpose of the offset angle is to minimize the tracking angle error. An interesting variation has been introduced by Garrard with their Zero-100 turntables—the "zero" meaning that their system introduces no tracking angle error. As shown in Figure 12.17, the Garrard tone arm actually changes the cartridge angle as the tone arm swings inward toward the center of the record. Thus, the stylus is kept tangent to the groove at all times.

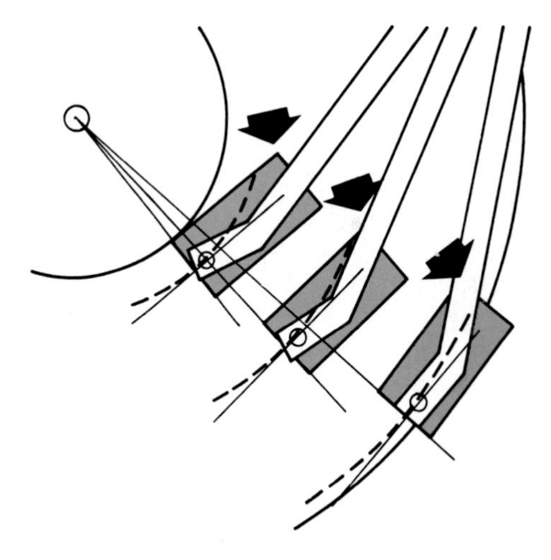

Figure 12.17. The Garrard zero-tracking-error tone arm. The angle between the head and shaft of the tone arm changes continually (articulates) as it moves across the disc, thus providing a constant tangent between stylus and groove. Tracking angle error is negligible.

Tone Arm Buying Guide

Tracking Angle Error: Look for a tracking angle error of 3° or less.

Tone Arm Pivot Friction

A tone arm should move freely as it guides the cartridge across the record. However, there is always some resistance to the motion of the tone arm at the point where the arm is supported on its pivot. This resistance arises because of the friction which is always present when two objects rub against each other. The definition of tone arm pivot friction is as follows.

Tone Arm Pivot Friction: The maximum amount of frictional force encountered by the tone arm as it moves on its pivot. Values are quoted in either grams or milligrams, and separate values may be given for the vertical and horizontal directions.

If the tone arm pivot friction is large, the cartridge carried by the tone arm will experience undue difficulty in following the record groove. Minimum friction is achieved primarily by the use of precision bearings in the pivot assembly.

Tone Arm Buying Guide

Tone Arm Pivot Friction: Look for 50 milligrams or less for both the vertical and horizontal tone arm pivot frictions.

Antiskating Control

There is another important feature provided on many high-quality tone arms; it is called an *"antiskating"* adjustment. In order to understand the need for such a control consider Figure 12.18, which illustrates the frictional force acting on the stylus and tone arm. This frictional force is not the one discussed above as tone arm pivot friction, which occurs at the arm's pivot point. The frictional force shown in Figure 12.18 arises because the surfaces of the stylus and record rub together, thus causing friction. This friction exerts a force on the stylus (and, hence, on the tone arm) in the direction of the record's movement. This force and its direction are illustrated in the figure. You can easily convince yourself that this force is present by replacing the stylus with your fingernail (P.S. Use an old record!). You should be able to "feel" the record pulling your finger in the direction in which it is rotating. Now, once again examine Figure 12.18, and notice the position of this frictional force with respect to the tone arm's fulcrum. Pretend that the tone arm is a door, and the fulcrum represents the hinges. When applying a force to such a door, it should be obvious that the door will swing clockwise toward the center of the record. In a similar manner, the frictional force tends to force the tone arm inward, causing a greater pressure on the inner wall of the groove as compared to the outer wall. That is why a stylus, which accidentally jumps out of the groove, will skate toward the center of the record. The inward skating force causes the

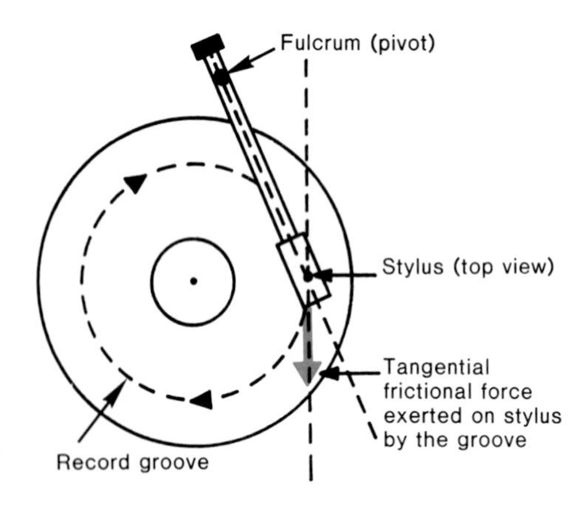

Figure 12.18. The frictional force exerted on the stylus by the groove causes the tone arm to swing inward on its fulcrum toward the center of the record. This force is called a "skating" force.

inner wall to wear faster than the outer wall and, consequently, the frequency response of the inner wall will deteriorate rapidly. The left and right channels of a stereo system are inscribed, respectively, on the inner and outer groove walls. Therefore, the skating force selectively degrades the left channel.

Some turntables provide a counterforce called, appropriately, the *antiskating force*, as illustrated in Figure 12.19. The antiskating force acts on the tone arm to pull it *away* from the inner wall such that the stylus will exert equal pressures on both walls. The antiskating control is usually calibrated in grams, but this does not mean that a setting of two grams gives two grams of antiskating force. Far from it, because the skating force is a much smaller force by comparison. The calibration control actually refers to the size of the vertical tracking force. The antiskating force depends upon the size of the tracking force although it is always much less. If you set the tracking force at two grams, for example, then the antiskate control should be also set on "two"—not because you want two grams of antiskating force—but because "two" gives you the correct amount which is required for the current setting (2 grams) of the tracking force. In some cases the antiskating force is calibrated for the shape of the stylus being used—conical or elliptical—because each type results in a different frictional force between its tip and the groove.

Figure 12.19. An antiskating control for the tone arm. The antiskating force distributes the stylus pressure equally on both walls of the groove by forcing the stylus away from the inner wall.

12.6 Phono Cartridges

One of the truly remarkable feats of modern engineering is the perfection of a phono cartridge which will follow the incredibly tiny (microscopic!) wiggles in the grooves. The purpose of the cartridge is to convert the mechanical movements of the stylus into electrical signals which correspond to the audio information encoded in the grooves. The undulations are so small that it takes a powerful magnifying glass to see even the loudest musical passages. As the stylus sits in the moving groove, it is deflected laterally and vertically when picking up both channels of a stereo program. Considering the fact that each channel may contain frequencies up to 15 kHz, the stylus is forced to move simultaneously up and down, back and forth, at the incredible rate of 15,000 times each second!! It must do so while maintaining contact with the walls at all times without any irregular or erratic motion. In order to achieve this remarkable feat, the mass of the stylus is kept extremely low, because a larger mass means a greater inertia* (the resistance to

*Mass is proportional to the weight of an object. Heavy objects have a large mass and, consequently, they are more difficult to accelerate.

rapid changes in the speed of the stylus). In this respect the masses of the stylus and the turntable platter play opposite roles. The platter is specifically designed with a large rotational inertia which allows it to resist small speed changes. The stylus assembly, on the other hand, must have an extremely small inertia so that it may rapidly follow the quick and sudden undulations of the groove wall. Not only must the stylus possess a low mass, but its mounting must allow it flexibility of movement. This mounting is very important, because it must support the stylus and prevent any irregularities in its motion. A stylus which has both a low mass and a flexible mounting is said to have a *high trackability* and a *large compliance*. These two factors, when combined, allow the cartridge to follow the groove with lower tracking forces which reduce the wear. In general, high quality cartridges possess a high compliance.

Cartridges are commonly divided into two classes: magnetic and non-magnetic. Magnetic cartridges utilize permanent magnets and coils of wire for their operation, and some common types within this class are the moving magnet, moving coil, moving iron, and ribbon cartridges. The ceramic and semiconductor cartridges do not use magnets and, as such, they belong to the non-magnetic class. Generally speaking, most cartridge manufacturers concentrate research and development efforts on one particular type of cartridge for their main product line.

Magnetic Cartridges

The magnetic phono cartridge is used today in virtually all quality hi-fi turntables. As discussed in section 11.7, magnetic cartridges utilize Faraday's law of induction as the basis for their operation. Remember, Faraday's law states that an electrical voltage will be induced in a coil of wire whenever there is relative movement between a coil of wire and a magnetic field.

Figure 12.20 illustrates one possible physical arrangement between the magnet and the coil. Please notice that the magnet and coil arrangement is somewhat different than that illustrated earlier in Figure 11.31. The permanent magnet in Figure 11.31 was shown to vibrate inside the helical coil for the sake of simplicity. However, in virtually all magnetic cartridges the magnet is located near, but not within, the helical coil. Instead, the coil of wire is wrapped around a solid core made from a magnetic material such as iron. The presence of the magnetic core, or "pole piece", increases the amount of voltage induced in the coil. Since only the magnetic field lines (not the magnet itself) need to penetrate the region of the coil, an induced voltage will appear in the

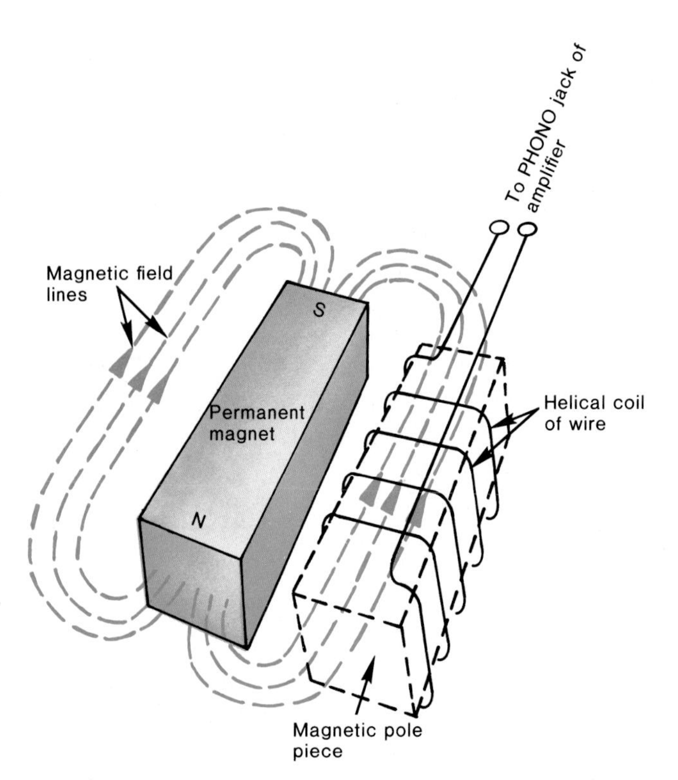

Figure 12.20. A permanent bar magnet in the vicinity of a helical coil of wire. Any relative movement between the magnet and the coil will cause an induced voltage to appear in the coil.

coil of Figure 12.20 when there is any relative motion between the coil and the permanent magnet. Of course, the voltage which has been induced in the coil must then be sent to the PHONO input of an amplifier. Within the amplifier the signal is amplified to the point where it is strong enough to produce sound from the speakers.

We shall now discuss the various types of magnetic cartridges. It will be seen that all of the types depend on Faraday's law of induction for their operation, and they differ only in mechanical design. For simplicity, the following discussions and illustrations are presented for monaural, rather than stereo, cartridges. A discussion on how a stereo cartridge works is presented in section 12.10.

The **moving magnet cartridge** utilizes a small permanent magnet attached to one end of the stylus cantilever, as shown in Figure 12.21 (A). The magnet/cantilever combination is free to rotate about the vertical "shaft" shown in the figure. Located near the magnet, but not attached to it, are situated two helical coils. The coils usually consist of several hundred turns of wire wrapped around an iron or ferrite "pole piece". The pole piece is used to help direct the magnetic field lines through the coils.

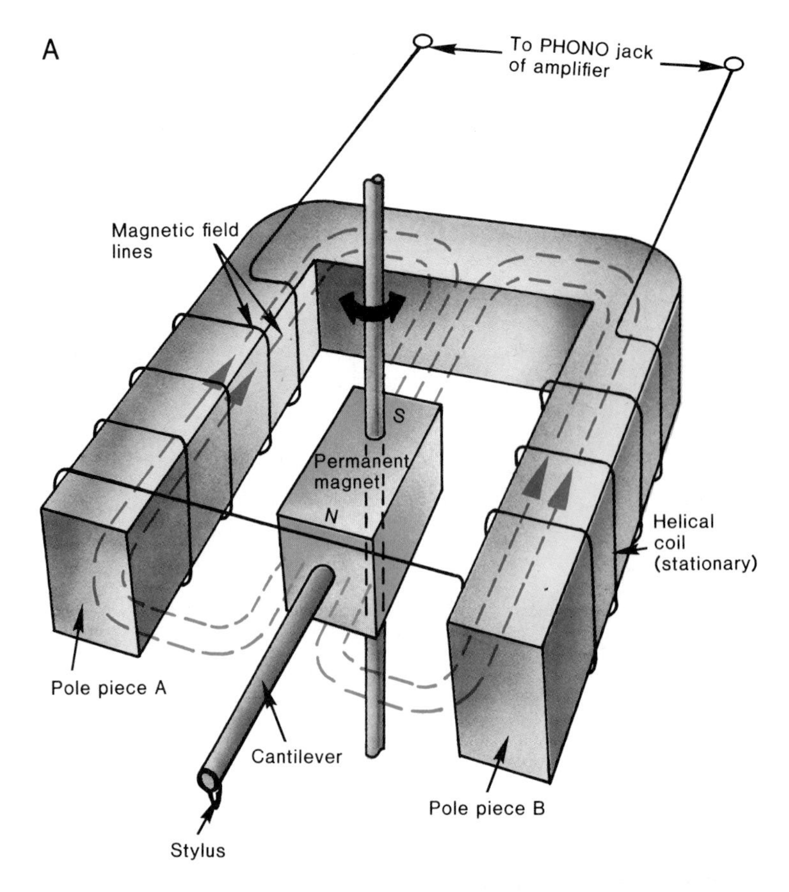

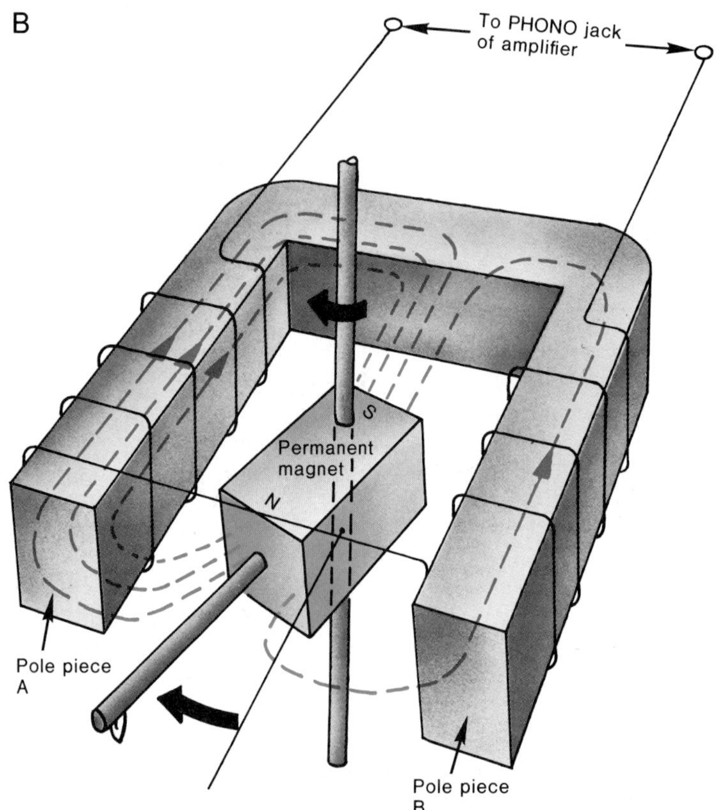

Figure 12.21. A schematic view of a moving magnet cartridge. (A) The stylus and the attached permanent magnet are in their "rest" positions. (B) The stylus and magnet have been forced to the left by the record groove (not shown).

As the stylus moves, so does the magnet, and the distance between the magnet and coils is changed. For example, when the stylus is rotated toward the left, as shown in Figure 12.21 (B), the magnet moves toward coil A and away from coil B; voltages will be induced in both the coils. By connecting the two coils in the proper manner, it is possible to obtain a combined output voltage which is twice that of a single coil. The total output voltage is typically between 2 and 10 millivolts, and most amplifiers have PHONO inputs which are designed to accept this type of cartridge. In addition, the principle of using two coils in this fashion greatly reduces the hum and noise picked up by the cartridge from extraneous signals within the room, such as that due to the 60 Hz line voltage. In fact, the hum and noise reduction is so great that most magnetic cartridges utilize similar humbucking coils.

Moving magnet cartridges have gained vast popularity because they are excellent performers which can be manufactured on an assembly line basis, resulting in lower consumer cost. In addition, the user can easily replace a worn stylus-magnet assembly without having to send the entire cartridge to a repair shop, as illustrated in Figure 12.22.

The **moving coil cartridge** is, in an operational sense, the antithesis of the moving magnet cartridge. The coil is attached to the cantilever and the magnet is stationary, as illustrated in Figure 12.23. As the record groove forces the stylus to vibrate, the attached coil also moves. Since the coil moves relative to the fixed magnetic field, an induced voltage is created.

The moving coil cartridge usually has a much smaller output voltage than does the moving magnet cartridge.

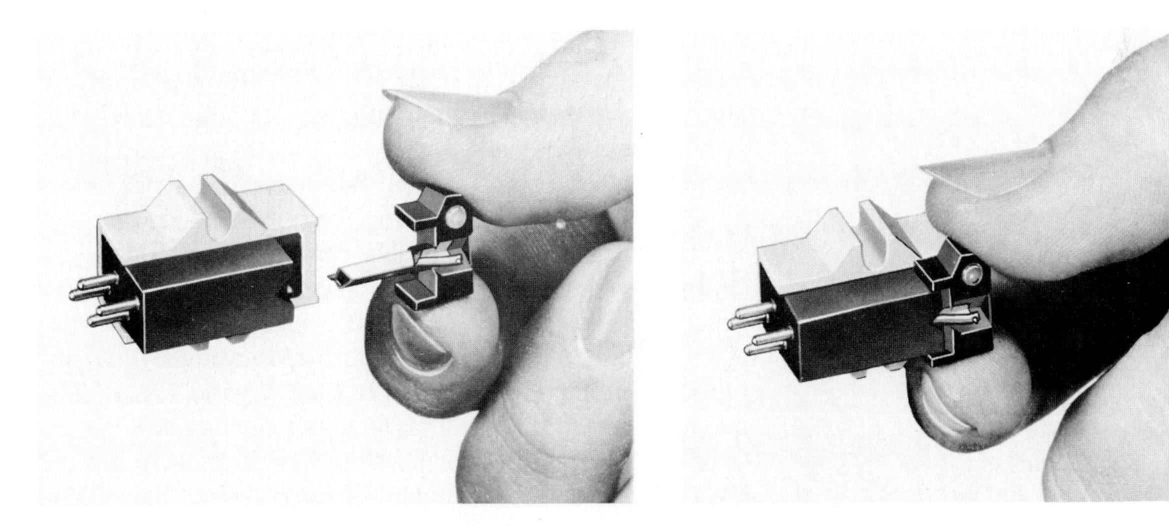

Figure 12.22. A worn stylus-magnet assembly can be easily replaced in a moving magnet cartridge. Photo courtesy of Shure Brothers, Inc.

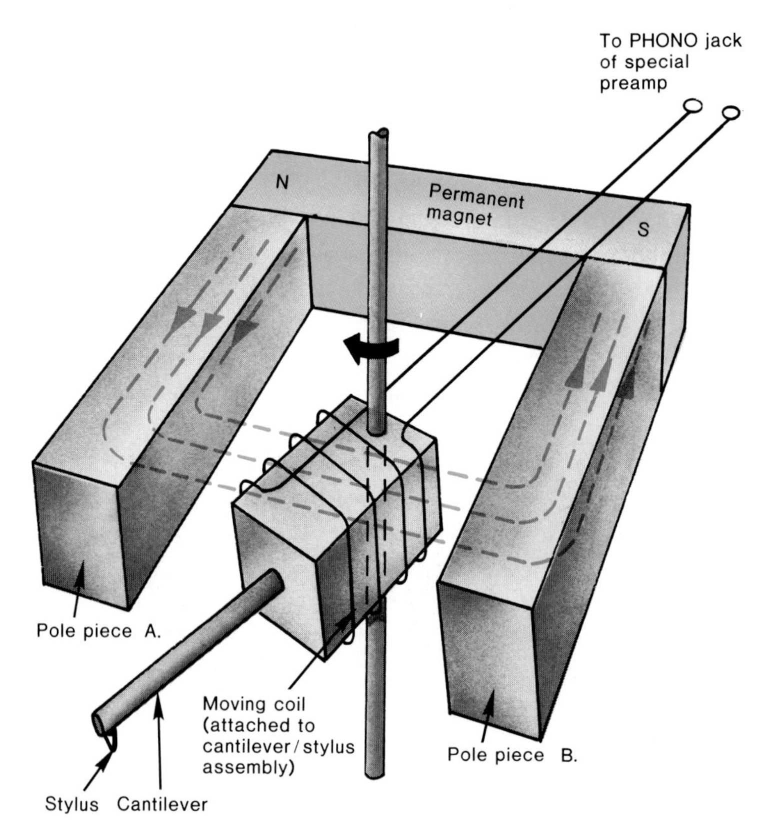

Figure 12.23. A moving coil cartridge. Notice that the coil is attached to the stylus while the permanent magnet remains stationary.

Therefore, the output cable from a moving coil cartridge cannot be connected to the standard PHONO jack found on most amplifiers. Usually a special preamp must be purchased and inserted between the moving coil cartridge and the amplifier. Moving coil cartridges are capable of outstanding performances, and they are considered by many audiophiles to be among the best available. However, they are moderately expensive to manufacture, because some hand work is required. And, to complicate matters, a worn stylus is usually not user replaceable, and the entire cartridge must be returned to the manufacturer for repairs.

The **moving iron** (or **variable reluctance**) **cartridge** has, strangely enough, both a stationary magnet and stationary coils. With this type of cartridge, a piece of magnetic material (usually iron) is attached to the vibrating stylus/cantilever combination, as depicted in Figure 12.24. The magnetic material is not a permanent magnet, but it is used to help direct the magnetic lines through the coils. When the stylus is not vibrating, one-half of the magnetic lines travel through pole piece A and the remaining half travel through pole piece B. As the magnetic material is moved to the left, for example, more of the magnetic lines travel through pole

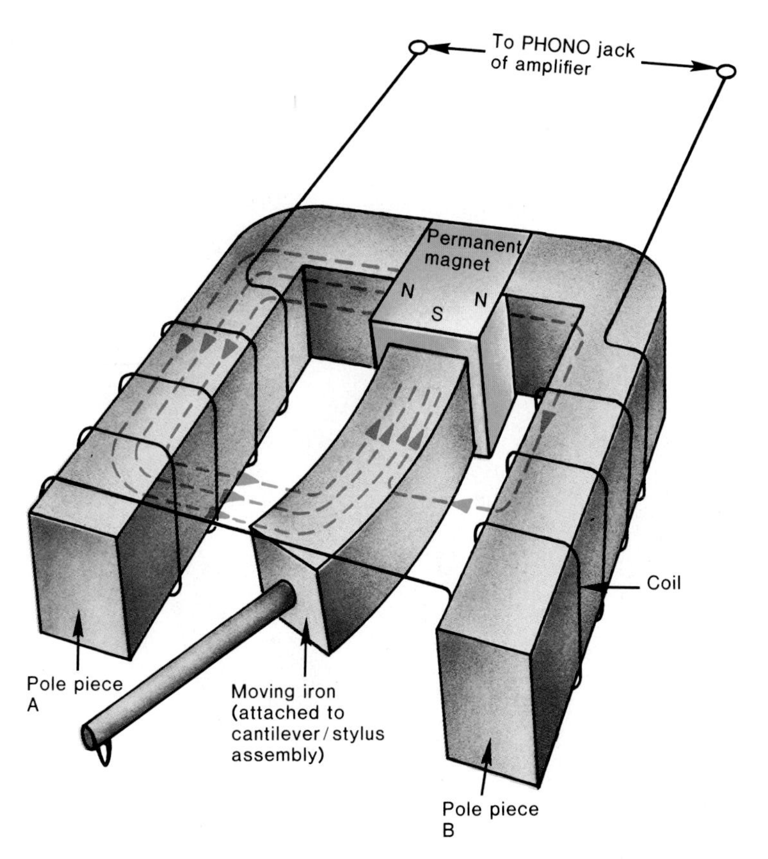

Figure 12.24. A variable reluctance cartridge showing both the stationary magnet and coils.

piece A rather than pole piece B, because the left air gap has been made smaller while the right air gap has been made larger. The net effect is the same as moving a magnet toward one coil and away from the other. Voltages are induced in both coils and, as with the two coils of the moving magnet cartridge, an enhanced, low noise, signal is produced.

The moving iron design allows the use of a large permanent magnet and coils with many turns of wire, both of which are needed to obtain a large output voltage. However, these relatively massive elements do not vibrate and good performance is obtained by keeping the effective mass of the moving iron small. The moving iron cartridge is relatively inexpensive to build, and units can be purchased in all price and quality ranges. The stylus is usually user replaceable.

The **ribbon cartridge** is a relative newcomer in the hi-fi field. As outlined in Figure 12.25, the stylus is attached to an extremely light, thin, and movable piece of metallic foil called the "ribbon". Each end of the ribbon is attached to a fixed wire which leads to the PHONO input of the amplifier. The ribbon and the two wires can be thought of as

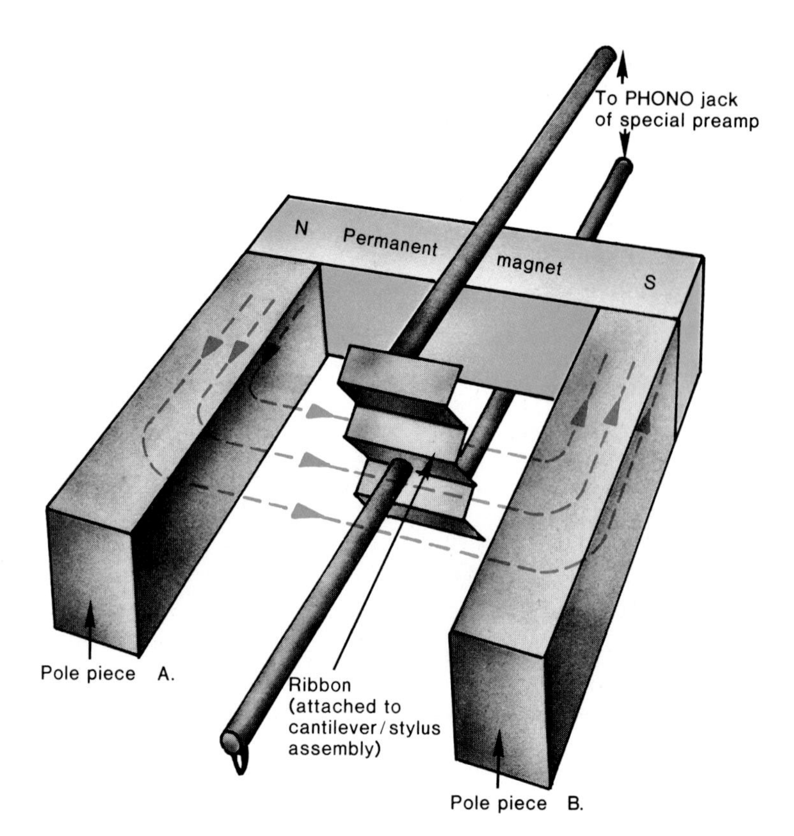

Figure 12.25. A ribbon cartridge.

being a coil with only a single loop. The cantilever is shown to be attached to the ribbon in a manner such that the ribbon's motion will synchronously follow the vibrations of the stylus. As the ribbon vibrates, the area of the single loop is also changed. Since the coil is positioned in a uniform magnetic field, the changing loop area is just like a coil moving relative to a magnet, and a voltage is induced in the coil. The output voltage from the ribbon cartridge is very small and, as with the moving coil cartridge, a special preamp is required. Despite its newness, and expensive cost, this type of cartridge shows promising possibilities in the audio marketplace.

Non-Magnetic Cartridges

The non-magnetic class of cartridges includes all those which do not rely on magnets, coils, and Faraday's law of induction for their operation. Representative of this class are the ceramic and semiconductor cartridges.

Ceramic cartridges are widely used in lower-priced audio systems because of their simple design, large output voltage, and low construction costs. This economy application, along with a lack of serious engineering effort, has led to its reputation as a low-quality performer. Ceramic cartridges do have inherent problems in the reproduction of the low frequencies and they possess a relatively large effective mass.

However, both of these problems could be surmounted with good design, and the performance of ceramic cartridges could be made comparable to the high standards set by magnetic cartridges.

Ceramic cartridges operate on a phenomenon called the piezoelectric (pē-ā′-zō electric) effect. Their operation depends on the ability of certain crystalline materials, such as ceramic, Rochelle salt, and quartz, to produce a voltage when they are bent or twisted. Figure 12.26 illustrates a simplified monaural version of such a cartridge. Two metallic conductors in contact with each side of the ceramic slab detect the alternating voltage which appears as the piezoelectric material is mechanically twisted by the vibrating stylus. The frequency of the alternating voltage will be the same as the vibrational rate of the stylus. Also, the greater the amplitude of vibration the greater will be the output voltage; thus ceramic cartridges can extract both frequency and amplitude information from the groove. It should be noted that ceramic pick-ups do NOT operate on the principle of Faraday's law of induction. No magnet is involved. They are based upon the piezoelectric effect in which certain materials produce a voltage when they are mechanically deformed.

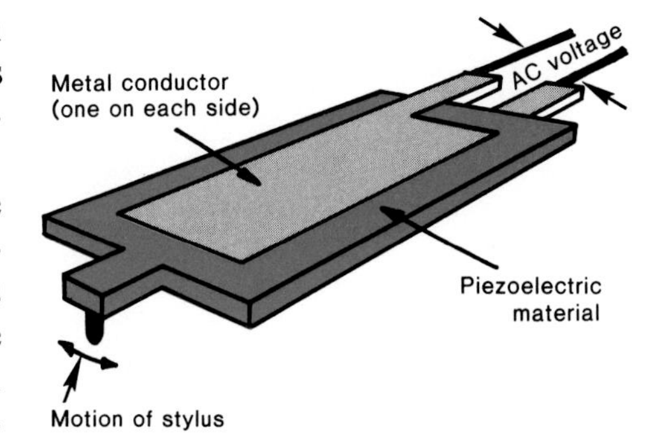

Figure 12.26. An illustration of a monaural ceramic pick-up. When the piezoelectric material is stressed by the vibrating stylus, an AC voltage is produced between the two conducting plates.

Semiconductor cartridges are relatively new, and they show great promise for high fidelity applications. In this type of cartridge, the semiconductor is linked to the stylus, and the vibrating stylus causes a changing pressure to be exerted on the semiconductor. The pressure variations cause the semiconductor to change its electrical resistance. When the semiconductor is attached to a power supply (a source of current), the current passing through the semiconductor will change in accordance with the changing electrical resistance. Through this mechanism, the stylus' motion creates an alternating current which carries all of the audio information encoded in the groove. Because of the need for a special power supply and electronic circuits, the semiconductor cartridges must be used only with the particular turntable system for which they were designed. Although moderately expensive, they possess a very small effective mass and good frequency response.

12.7 Record Grooves and Stylus Shapes

A modern day LP record, like the cartridge, is nothing short of a technological miracle. It must contain within the tiny grooves any possible mixture of complex sounds, loud or soft, bass or treble, and be able to contain at least 20 minutes of playing time. In addition, each stereo groove must

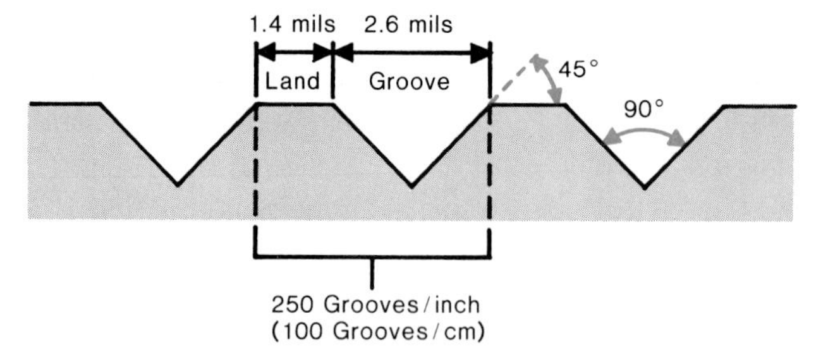

Figure 12.27. A cross-sectional view of a stereo record. The width of a groove is about the thickness of a piece of paper.

encode two separate sound channels and, in the case of quadraphonic records, 4 channels must be contained within the single groove! In order to understand the beauty of a stereophonic record, let us first investigate the nature of the groove itself.

The groove, being extremely small, is often referred to as a "microgroove," and Figure 12.27 illustrates a cross-sectional view. There are approximately 100 grooves per cm, with each groove being a mere four-thousandths of an inch wide* (the thickness of one or two sheets of paper!). The flat surface of the disc is called the "land," and the cutting stylus leaves at least 1.4 mils of land between adjacent grooves. Depending upon the musical content, the actual cut portion of the groove may vary from as small as 1 mil (in order for the stylus tip to fit into the groove) up to a maximum of 2.6 mils. Figure 12.28 shows the action of the cutting stylus as it engraves the audio information into the groove walls. The wiggles in the cut groove are what force the cartridge stylus to vibrate. The undulations are so small that they appear diminutive even when compared to invisible dust particles! An important feature of the stereo groove is that the walls

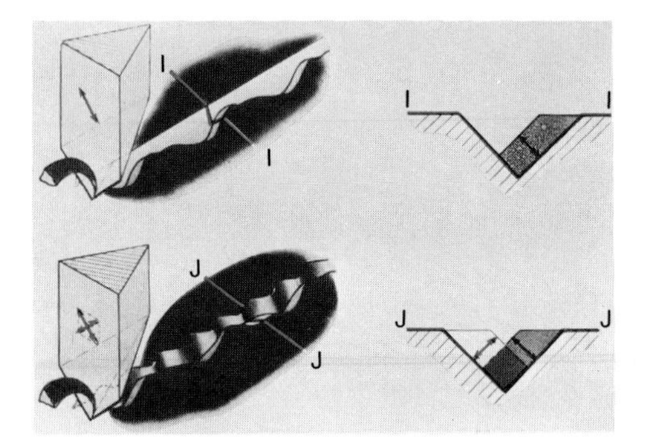

Figure 12.28. An illustration of the cutting stylus as it forms a record groove. The top illustration shows only the right wall being cut. Both walls are cut in the bottom sequence labeled J-J.

Shure V-15 type III super-track "plus" stereo cartridge.

*One thousandth of an inch is often called "1 mil."

are cut at an angle of 45° from the land (see Fig. 12.27), which places an angle of 90° between the two walls. As will be discussed in section 12.10, each wall carries the musical content for one stereo channel.

As much as any other element, the quality of the stylus is closely linked to the final performance of the turntable. The stylus is a carefully shaped and polished piece of diamond which is about the size of a pin point. The advantage of diamond is its hardness; it outlasts all other substances in its ability to resist wear, and a diamond stylus can be expected to last between 600 and 1,000 hours of playing time. The photos in Figure 12.29 illustrate the wearing of a conical shaped stylus in various stages of its useful life. These photos were taken with a special microscope, and they

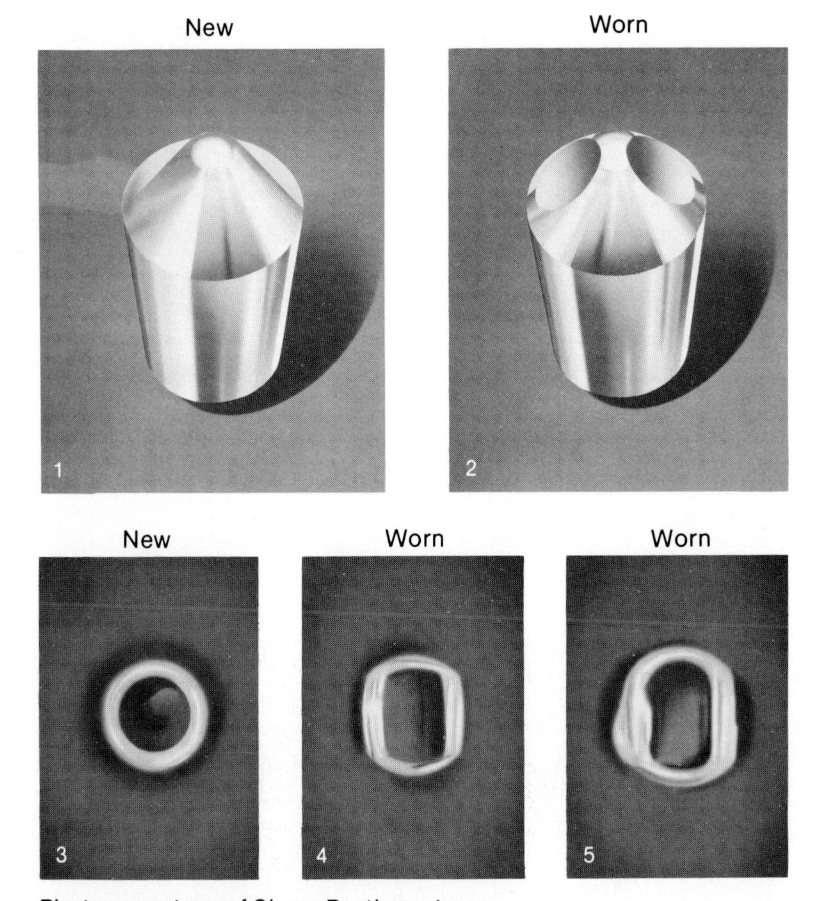

Photos courtesy of Shure Brothers, Inc.

Figure 12.29. The effects of wear are shown for a conically tipped stylus. The bottom three photos show top views of the stylus tip.

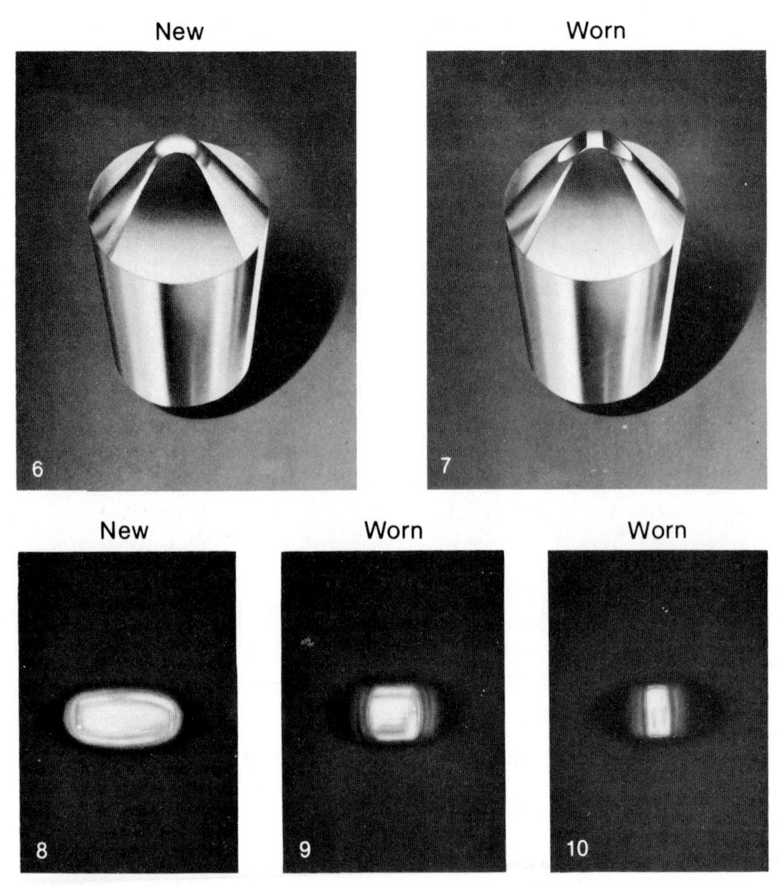

Photos courtesy of Shure Brothers, Inc.

Figure 12.30. The effects of wear are shown for an elliptical stylus. The bottom three photos show top views of the stylus tip.

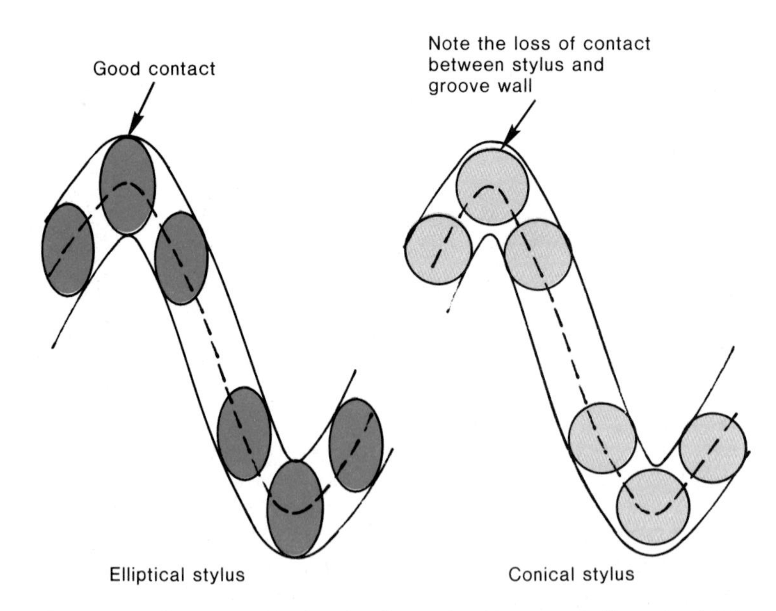

Figure 12.31. An elliptically shaped stylus can follow the high frequency undulations in the groove better than a conical stylus can.

clearly depict the two worn flat spots on the facets of the diamond which were in contact with the groove. Figure 12.30 illustrates the wear on an elliptically-shaped stylus.

Elliptical styli are found on many high quality cartridges, because they are able to follow the groove undulations in a more precise manner than conical styli. Figure 12.31 illustrates that the elliptical stylus is better able to "trace out" a high frequency oscillation of the groove than does the conical stylus. This leads to an improved high frequency response. A three dimensional drawing of an elliptical stylus riding in a stereo grove is shown in Figure 12.32.

In addition to the elliptical and conical styli there are other special shapes called the Shibata, Pramanik, quadrahedral, hyperbolic, etc. These are very high-priced units designed for use with discrete 4-channel records which contain very high (up to 45 kHz!) frequencies. These specially shaped styli have an elongated contact span along the groove wall that distributes the tracking force over a larger area of the groove wall, thereby improving high frequency response and reducing record wear, as depicted in Figure 12.33. Figure 12.34 shows the shape of a multi-radial Pramanik stylus.

Figure 12.32. A three-dimensional representation of an elliptical stylus in a stereo record groove.

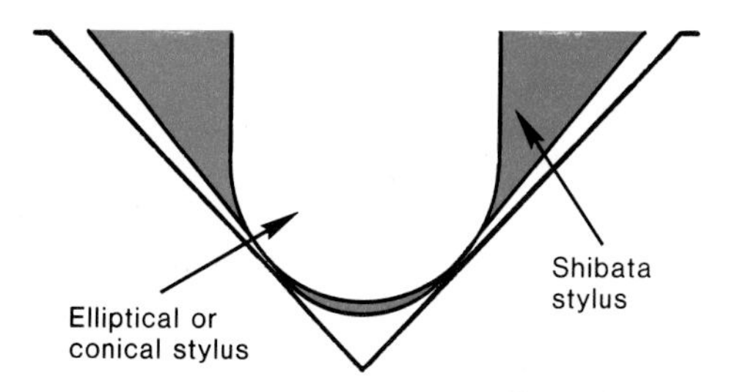

Figure 12.33. The Shibata stylus makes better contact with the groove wall, when compared to either the elliptical or conical stylus. The result is less pressure with a greater record and stylus life, and better high frequency response.

12.8 Cartridge Specifications

There are a number of specifications which are used to describe the quality of a cartridge. It should be noted that a cartridge, like a loudspeaker, is an electromechanical device which involves the interdependence of mechanical motion and electricity. In contrast to purely electronic

Bang & Olufsen a/s

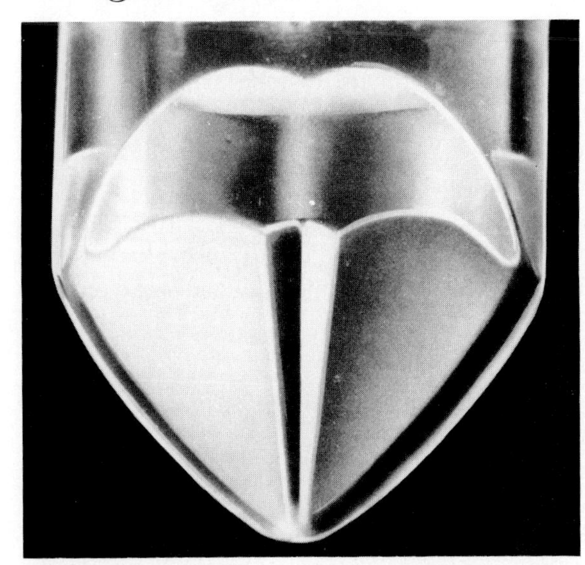

Multi-radial Pramanik stylus as used on the MMC 6000 pickup cartridge

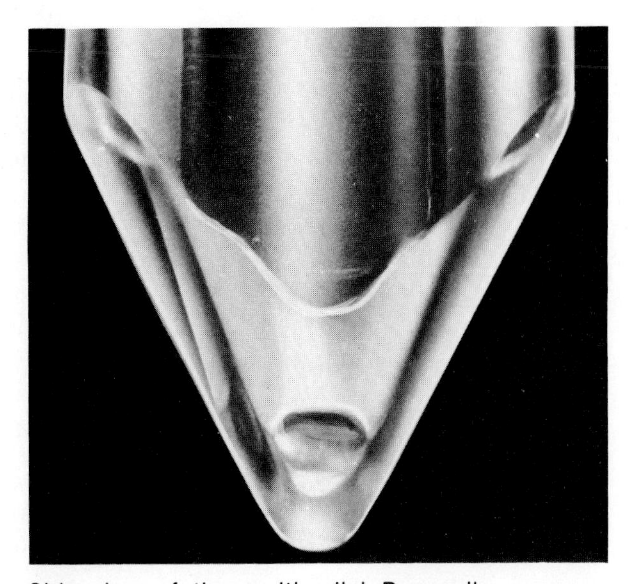

Side view of the multi-radial Pramanik stylus

Figure 12.34. A multi-radial Pramanik stylus.

components, such as tuners and amplifiers, electromechanical units present design problems which are difficult to overcome. Consequently, cartridge performances in such areas as frequency response and stereo separation tend to be somewhat inferior when compared to those of an amplifier or a tuner. So, do not be too alarmed when you discover that some of the following recommended values for cartridge specifications are not as good as the corresponding values for an amplifier or tuner.

As is the case with loudspeakers, the best method of selecting a cartridge is by making listening comparisons with other cartridges. Unfortunately, most hi-fi stores are not equipped to provide the prospective buyer with such comparisons. In addition, cartridge specifications can be confusing because, in some cases, no universally accepted measurement standards exist. For example, two different testing laboratories might quote different values for the frequency response, even though they are both testing the same cartridge. This difference arises because each testing laboratory may employ a different test procedure for measuring the frequency response, and without a universally accepted test procedure such differences in the frequency response are bound to arise. The moral of this story is that the prospective buyer should not be overly concerned with small differences in values for cartridge specifications, especially when units from different manufacturers are being compared. With this qualifying statement, some of the more important cartridge specifications will now be discussed. They include: tracking force, compliance, frequency response, stereo separation, and output voltage.

Tracking Force

The tracking force specification of a cartridge was discussed in section 12.5 in regard to tone arms. A range of acceptable tracking forces is usually given by the cartridge manufacturer. While no one value of the tracking force is a guarantee of sound quality, lower values are typically associated with higher quality cartridges. Cartridges that can be used with lower tracking forces usually provide improved tracking ability, less distortion, and reduced wear on both the stylus and record groove.

Cartridge Buying Guide

Tracking Force: Look for a tracking force of 3 grams or less. Be sure that the value falls within the available range provided by your tone arm.

Compliance

The undulations in a stereo record groove force the cartridge stylus to vibrate at frequencies up to 15,000 Hz and beyond. In order to vibrate this rapidly, the stylus and its mounting must be easily moveable. Compliance is a specification which measures how easily the stylus is able to move, and a highly compliant cartridge has a stylus which is readily moved by the undulations of the groove. The definition of compliance is stated as follows.

Compliance: The distance moved by the stylus in response to a force of 1 dyne. The compliance of a cartridge is quoted in centimeters (cm) per dyne.

The unit of force used in measuring compliance is called a "dyne". A dyne is a very small amount of force, since it requires about 3,000 dynes to equal the weight of one penny. Such a small unit of force is used, because only small amounts are needed if the stylus is easily moveable. Figure 12.35 illustrates the meaning of compliance when the specification is 15×10^{-6} cm/dyne. According to the definition given above, a force of one dyne will move the stylus through a distance of 15 millionths* of a centimeter. Likewise, a force of two dynes will move the stylus twice as far, or 30 millionths of a centimeter. A higher compliance rating of, for example, 25×10^{-6} cm/dyne implies that the stylus moves even further under the influence of the one dyne force.

To a large extent compliance and tracking force go hand-in-hand; a high compliance results in a light tracking force and vice versa. As the stylus rides in the groove, it is forced up and down, as well as sideways, by the two walls. With a highly compliant cartridge the groove needs only to exert a small upward force to displace the stylus. While the upward force does move the stylus/cantilever combination, this force also tends to lift the stylus out of the groove and cause mistracking. The downward tracking force, provided by the unbalanced tone arm and cartridge, is necessary to counteract this upward force, thus ensuring that the stylus is always kept in firm contact with the groove at all times. A high compliance means that only a small upward force is needed to move the stylus, and this, in turn, implies a small counteracting tracking force. Be careful, however. While tracking force and compliance do tend to go hand-in-hand, they are two different specifications.

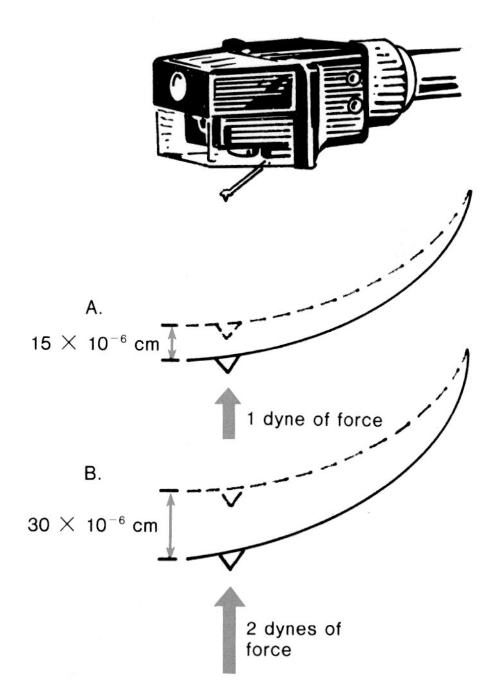

Figure 12.35. A cartridge stylus (compliance = 15×10^{-6} cm/dyne) moving under the influence of (A) 1 dyne of force and (B) 2 dynes of force.

*10^{-6} cm is the scientific notation for one-millionth of a centimeter.

Cartridge Buying Guide

Compliance: Look for a compliance of 10×10^{-6} centimeters/dyne or higher (e.g., 20×10^{-6} centimeters/dyne).

Frequency Response

The frequency response specification for a cartridge has the same general meaning that it did for amplifiers, loudspeakers, and tuners. (See sections 5.8, 6.10, 8.7, and 10.8.) As the stylus rides in the record groove, its vibratory motion causes the cartridge to produce an output AC voltage. The frequency response specification tells how evenly the cartridge reproduces the amplitudes of all the frequencies in the audio range which are contained in the record groove.

Cartridges are not purely electrical devices, for they depend on the mechanical motion of the stylus to generate the output voltage. Because of this mechanical aspect, cartridges do not reproduce the entire audio range as evenly as an all electrical component, such as an amplifier or a tuner. For example, many of today's quality amplifiers routinely amplify all frequencies in the range from 20 Hz to 20,000 Hz to within ±1 dB of perfect evenness or "flatness." However, most cartridges can only reproduce frequencies from 20 Hz to 16,000 Hz, and their output voltage is usually only within ±3 dB of perfect flatness.

Cartridge Buying Guide

Frequency Response: Look for a frequency response of 20 Hz → 16,000 Hz, ±3 dB. The wider the frequency range and/or the smaller the dB figures, the better.

Stereo Separation (Channel Separation)

Stereo separation is the ability of a cartridge to separate channels of information contained in a stereo record, and it has the same general meaning that it did for amplifiers and tuners. (See sections 8.7 and 10.8.)

The stereo separation of a cartridge is typically measured using a 1 kHz audio signal, and is expressed in decibels, as usual. Figures of 20 dB and higher are not uncommon. Stereo separation tends to be lower in value (poorer) for the higher audio frequencies (10 kHz or so) than for the mid-range frequencies near 1 kHz.

Cartridge Buying Guide

Stereo Separation: Look for a stereo separation of 20 dB or higher near 1 kHz, and 10 dB or higher near 10 kHz.

Output Voltage

A magnetic cartridge generates an output AC voltage, as described by Faraday's law of induction, when its stylus follows the movement of the record groove. According to this law, the actual strength of the voltage produced at any given moment depends on the speed of the stylus as it vibrates in the groove. The specification called "output voltage" gives the strength of the voltage produced from a standard record groove according to the following definition.

Output Voltage: The amount of voltage, expressed in millivolts (mV), generated by the cartridge when the stylus is vibrating at a standard frequency (usually 1 kHz) and at a specified vibrational speed (usually between 3.5 and 5 cm/sec.).

The output voltage varies widely among cartridges, and it is not an indication of quality. It is intended to aid the user in properly matching the cartridge to the PHONO input sensitivity of the amplifier. As discussed in section 8.7, the output voltage should be greater than the amplifier's PHONO input sensitivity. Typical values range from fractions of a millivolt to 10 mV for magnetic cartridges.

Cartridge Buying Guide

Output Voltage: Match the output voltage of the cartridge to the PHONO input sensitivity of the amplifier, as described in section 8.7

12.9 The RIAA Phono Equalization Preamplifier

When a turntable is connected into the PHONO input jack of the preamp/control center, the electrical signal is first routed to the RIAA phono preamplifier [see Fig. 8.1 (D)]. This preamp is unique to the phono input, and its job is to boost the signal from approximately 5 mV to 0.5 V as well as to provide an "equalization" of the signal. The equalization can best be visualized as follows. Suppose that it could be arranged such that a record is played through your hi-fi system in a normal fashion, *except* that the so-called "equalization" function was removed from the RIAA equalization preamplifier. What would be heard? The music would not sound normal, because the bass notes would be abnormally quiet and the high frequencies would be abnormally loud. Therefore, a record does *not* contain an exact replica of the music as it was originally produced. To restore the recording to its natural tones, the bass notes must be boosted and the treble notes cut. This process of boosting and cutting is, in general, referred to as *equalization,* and it is performed in the RIAA equalization preamp section of the preamp/control center.

First, we will show why record manufacturers need to use equalization, which will then be followed by a discussion on the method by which it is achieved. It has been discovered that most of the acoustical energy in an orchestral selection lies below 1 kHz and relatively little, perhaps 20%, is found in the region from 1 kHz to 20 kHz. Since the energy of a sound wave is related to its amplitude, the bass notes have extremely large amplitudes while the amplitudes of the treble notes are very small. Because the record groove is a mechanical replica of the sound wave, the electrical signal containing the bass notes causes the cutting stylus to produce large groove widths. By the same token the relatively low energy treble frequencies would give rise to small amplitude, narrow grooves as shown in Figure 12.36. The large excursions of the record groove mean that the grooves would have to be spaced widely apart in order to avoid "overcutting"—a situation in which the large amplitude bass notes cause the cutting stylus to actually overcut one groove and spill over onto an adjacent one. However, large width grooves would severely limit the playing time on a normal 12″ record to values which would be substantially less than 20 minutes. By electronically attenuating the amplitude of the bass notes during the recording of a disc, and then preferentially amplifying them by an equal amount during playback, the groove can be held to a reasonable width which will result in a normal (20–30 minutes) playing time per record side.

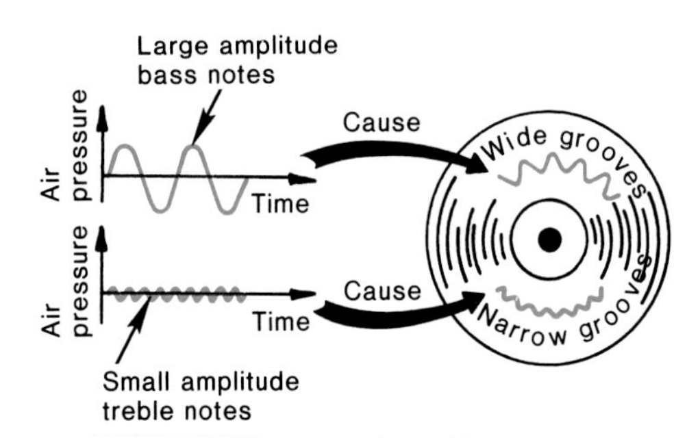

Figure 12.36. **The bass region contains a large amount of acoustical energy, and this gives rise to wide record grooves. The high frequency notes, on the other hand, have much smaller amplitudes.**

There is also a problem with the low amplitude, high frequency notes although it is of a different nature than the bass note problem. If you have ever listened to the few silent grooves either at the beginning or ending of a record, particularly when the VOLUME is turned up, you have noticed a lot of high-frequency "hiss" which is a particularly distracting form of record noise. This hiss, often called "surface noise," is produced when the stylus is forced to follow the imperfections or random irregularities which are present in even the best vinyl discs. If a low amplitude high-frequency record groove were comparable in width to the surface irregularities, the cartridge would pick up both of them in roughly equal amounts. The result would be a noisy recording. To eliminate this problem the amplitudes of the high frequencies are electronically boosted during the recording process. This produces a high-frequency groove width which is substantially larger than the size of imperfections. Of course the process must be reversed during playback in order to restore the recording to its natural sound.

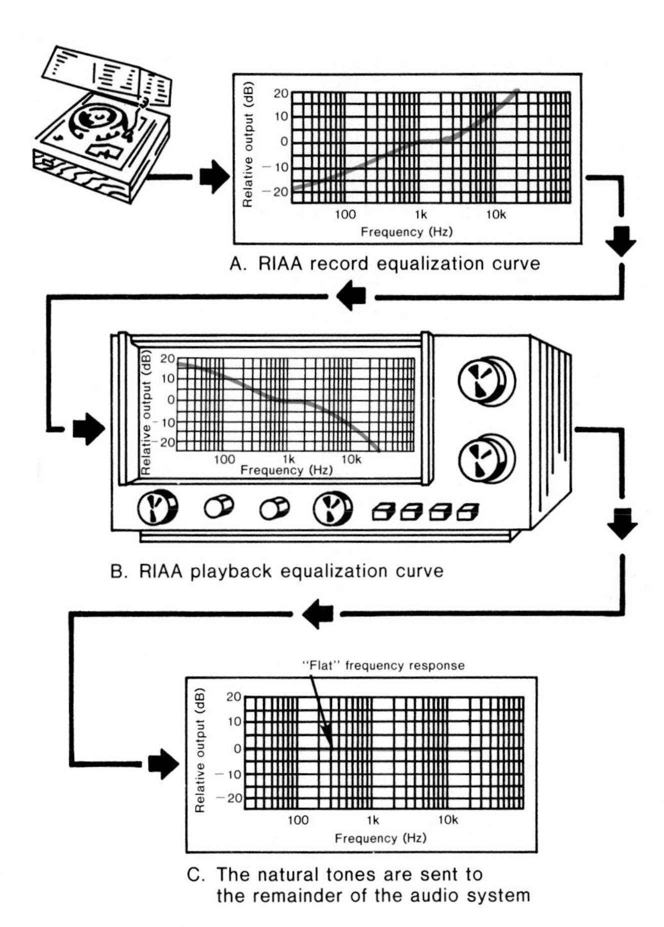

Figure 12.37. The RIAA record and playback equalization curves.

Figure 12.37 (A) shows the frequency response of the audio signal as it leaves the phono cartridge. Notice, as explained previously, how the notes below 1 kHz are attenuated more and more as the frequency is lowered. Again this has been done to keep the grooves at a reasonable width. Above 1 kHz the highs have been boosted in order to make their amplitudes larger than the size of the vinyl imperfections. For example, notice in Figure 12.37 (A) that a 100 Hz note has been attenuated by −13 dB while a 10 kHz note has been boosted by +12 dB.

When this phono signal is fed into the preamp/control center, the RIAA phono preamp must reverse this process by boosting the bass frequencies and attenuating the highs as shown in Figure 12.37 (B). This curve then returns the signal to a flat frequency response, and its natural sound is achieved, as indicated in part (C).

Summary

The signals which are put onto records are *not* exact copies of those produced by the musicians. They have been electronically altered, or equalized, so that the bass tones are deemphasized, while the high frequency tones are boosted. This recording equalization lengthens the playing time and reduces the noise of the records. Upon playback, the RIAA phono equalization preamplifier exactly reverses the alterations so that the natural sound is heard.

12.10 How a Stereo Cartridge Works

One of the great achievements of a stereo cartridge is that a single stylus, moving in a single groove, simultaneously picks up both the left and right stereo channels. It is not at all obvious how this is accomplished, and in this section we shall present the basic ideas behind the stereo cartridge.

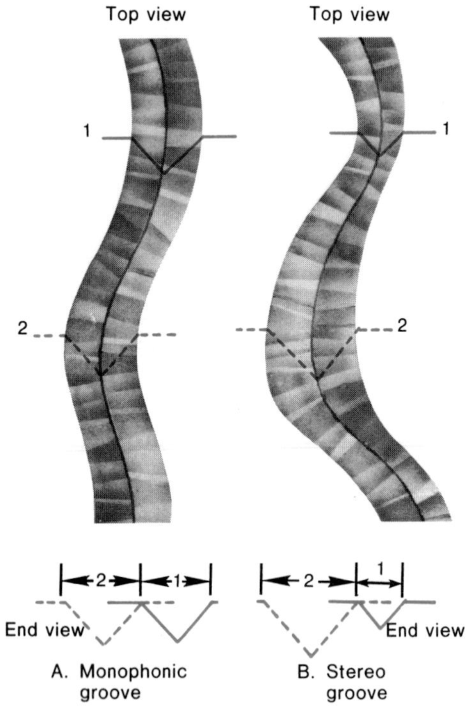

First, let us examine the appearance of a monophonic groove as shown in Figure 12.38 (A). Notice that, when viewed from the top, the groove undulates in the lateral direction with a constant width. The constant width is a result of the constant depth-of-cut as shown in the end-on view in part (A). Contrary to what many people believe, the monophonic groove forces the stylus to vibrate horizontally relative to the record's surface, not vertically. The vertical hill-and-dale method was abandoned years ago because it produced excessive wear on the record and increased mistracking problems. A typical stereo groove is shown in Figure 12.38 (B); it bears some resemblance to the monophonic groove in that it does wiggle in a lateral manner. However, the width and depth of the stereo groove are not constant. This occurs because the two groove walls are cut differently in order for each to contain a separate channel of information.

Figure 12.38. An illustration showing the top and end views of a (A) monophonic and a (B) stereo record groove.

To understand better how the stereo encoding is performed, consider Figure 12.39, which schematically represents a stylus being moved by the left wall alone.* For the sake of clarity, all parts of the cartridge have been removed except the stylus, magnets, and coils. Note especially that the iron cores have been omitted and that the permanent magnets are shown moving with respect to the stationary coils. In addition, the groove and the stylus are drawn very large in comparison to our model "cartridge" in order to illustrate better the stylus movement in relation to the magnets and coils. Opposite the left wall is situated the left-channel coil and the north pole of a permanent magnet. Likewise, the right-channel coil and a south pole are located opposite the right wall.

*The left wall is the inner wall of the groove and modulations of the left wall contain the left channel information. Similarly, the outer wall contains the signal for the right channel.

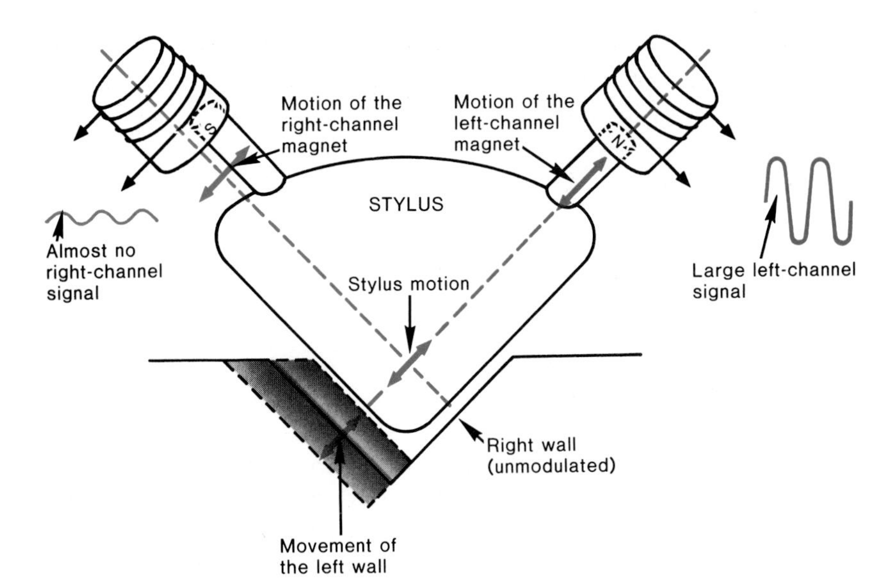

A record groove which contains only left-channel information

Figure 12.39. The stylus is being moved by the left wall alone. A signal is induced in the left-channel coil with (almost) no voltage being produced in the right-channel coil.

In Figure 12.39 undulations in the left wall are forcing the stylus in the direction shown by the arrows. This wall contains only the left channel information so that, under ideal conditions, no sound would be heard coming from the right speaker. As the left wall moves the stylus back and forth, the north pole of the left-channel magnet moves in and out of the left-channel coil. According to Faraday's law of induction, this motion induces an AC voltage in the left-channel coil which is subsequently sent to the PHONO jacks of the preamp for further amplification. Of course the south pole of the right-channel magnet is simultaneously moving in the vicinity of the right-channel coil. However, the motion of the right-channel magnet is perpendicular (90°), rather than parallel, to the center axis of the right-channel coil. This type of perpendicular motion does not produce any significant change in the magnetic lines which cut the right-channel coil, as emphasized in Figure 12.40. As a result, only the left-channel coil has the ability to detect changes in the left wall with almost no voltage being induced in the right-channel coil. Therefore, the perpendicular arrangement of the two coils is of utmost importance in allowing the cartridge to sense which wall contains the undulations. Notice that there is "almost" no voltage induced in the right-channel coil. In practice a small amount does appear and it is responsible for the "stereo separation" of the cartridge. A well-designed cartridge keeps the unwanted signal at least 20 dB below the main channel in the midrange frequencies.

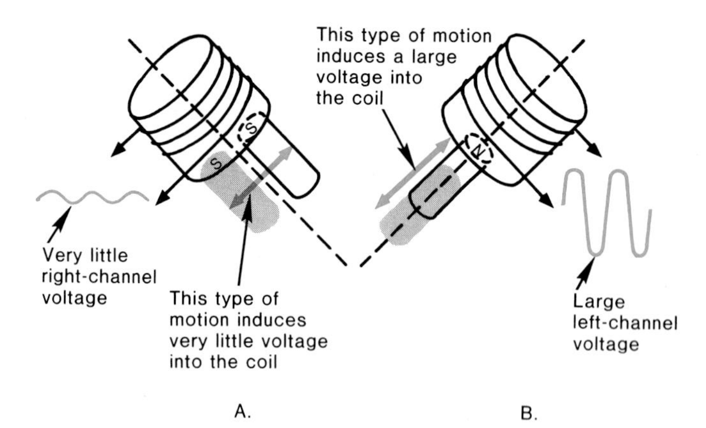

Figure 12.40. (A) A magnet which moves perpendicular (90°) to the coil's axis produces little output voltage. (B) A magnet which moves parallel to the coil's axis generates a large voltage.

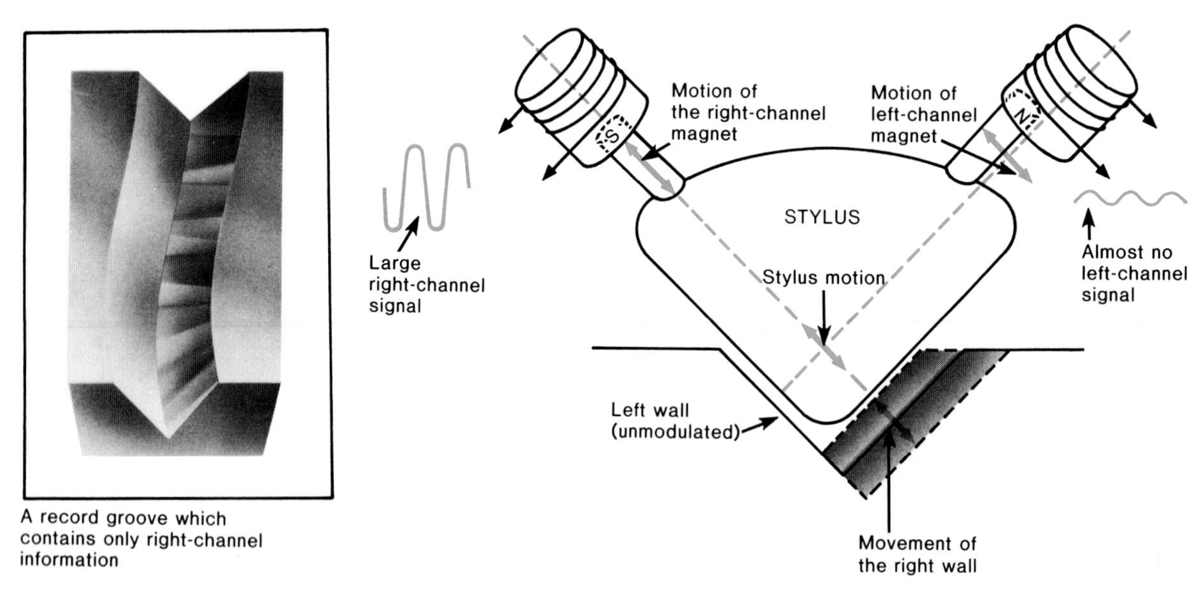

A record groove which contains only right-channel information

Figure 12.41. The stylus is being moved by the right wall only. A signal is induced in the right-channel with (almost) no voltage being produced in the left-channel coil.

Figure 12.41 shows the situation which results when undulations in the right wall are forcing the stylus to move. A voltage is now induced in the right-channel coil according to Faraday's law of induction, with only a small signal appearing in the left-channel coil.

When only one wall of the groove moves the stylus, sound emanates primarily from the corresponding speaker of a stereo system. Similarly, movement along the other wall causes sound to be produced from the other speaker. In this manner different instruments can sound like they are coming from either one speaker or the other when a stereo record is being played. However, it is rare when one groove wall alone moves the stylus, while the other groove wall remains perfectly smooth. After all, instruments are distributed all across the left/right direction of a band, and generally many instruments are playing simultaneously. In fact, instruments which are located precisely in the center of the band should sound like they are situated at the midpoint between stereo speakers. This occurs when each wall of the groove pushes equally on the stylus to make the sound of such instruments come from each speaker with the same loudness. To illustrate the normal appearance of a stereo record groove, Figure 12.42 offers a three dimensional view showing the undulations present on both walls. We will return to this matter of how the signals are cut into the walls of a record groove in Chapter 14, where the use of a mathematical idea called a "vector" will shed additional light on why stereo sounds so realistic!

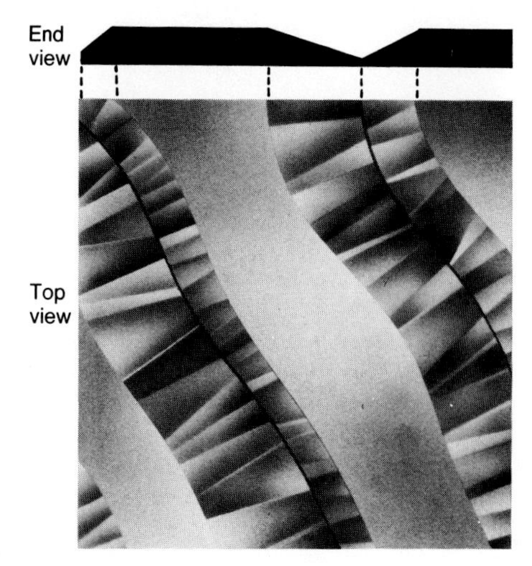

Figure 12.42. A top view of a stereo record groove.

A Summary of Turntable Drive Specifications and Buying Guides

Speed Accuracy

Look for a speed accuracy of 2% or less.

Rumble

1. **Unweighted (NAB):** Look for an unweighted rumble specification of −35 dB or lower, e.g., −40 dB.
2. **Weighted (ARLL or DIN):** Look for a weighted rumble specification of −60 dB or lower, e.g., −70 dB.

Wow and Flutter

1. **Unweighted:** Look for an unweighted wow and flutter specification of 0.2% or smaller.
2. **Weighted:** Look for a weighted wow and flutter specification of 0.1% or smaller.

A Summary of Tone Arm Specifications and Buying Guides

Tracking Force

Look for a tone arm which provides for tracking forces up to 3 grams.

Tone Arm Resonance

Look for a tone arm resonance frequency between 7 Hz and 15 Hz.

Tracking Angle Error

Look for a tracking angle error of 3° or less.

Tone Arm Pivot Friction

Look for 50 milligrams or less for both the vertical and horizontal tone arm pivot frictions.

A Summary of Cartridge Specifications and Buying Guides

Tracking Force

Look for a tracking force of 3 grams or less. Be sure that the value falls within the available range provided by the tone arm.

Compliance

Look for a compliance of 10×10^{-6} centimeters/dyne or higher (e.g., 20×10^{-6} centimeters/dyne).

Frequency Response

Look for a frequency response of 20 Hz → 16,000 Hz, ± 3 dB. The wider the frequency range and/or the smaller the dB figures, the better.

Stereo Separation

Look for a stereo separation of 20 dB or higher near 1 kHz, and 10 dB or higher near 10 kHz.

Output Voltage

Match the output voltage of the cartridge to the PHONO input sensitivity of the amplifier, as discussed in section 8.7.

Summary of Terms

Antiskating Control—A control found on some turntables which applies a slight outward force to the tone arm. This force counteracts the inward-acting skating force.

ARLL (Audible Rumble Loudness Level)—(See **Rumble.**)

Belt Drive—A type of turntable drive system which connects the revolving motor shaft to the platter by means of a flexible belt.

Cartridge—The part of a turntable which generates the electrical signal from the rotating record. There are ceramic cartridges, which utilize the piezoelectric effect in their operation, and magnetic cartridges, which utilize Faraday's law of induction in their operation.

Ceramic Cartridge—(See **Cartridge.**)

Compliance—The distance moved by the cartridge stylus in response to a force of one dyne.

Cutting Stylus—A cutting tool which is used to cut the undulations, representing the sound waves, into the grooves of a master record.

DIN (Deutsche Industrie Normen)—(See **Rumble.**)

Direct Drive—A type of turntable drive system which connects the revolving motor shaft directly to the platter.

Drive System—(See **Turntable Drive System.**)

Dynamic Balancing—A method for adjusting the tracking force using an adjustable spring attached to the tone arm.

Dyne—A unit for measuring force. The weight of one penny is approximately 3,000 dynes.

Flutter—(See **Wow and Flutter.**)

Frequency Response—A specification which indicates how well a cartridge produces one audio frequency compared to another. In order to have full meaning this specification must include both the frequency range and the ± decibel deviation from perfect flatness, e.g., 30 Hz → 15,000 Hz, ± 1 dB.

Friction, Tone Arm Pivot—(See **Tone Arm Pivot Friction.**)

Hysteresis Synchronous Motor—(See **Synchronous Motor.**)

Induction Motor—A motor used in turntables and tape decks. This type of motor is inexpensive and has good short-term speed characteristics. However, it is susceptible to long-term speed changes caused by line voltage changes.

Land—The flat smooth surface of the record located between the grooves.

Magnetic Cartridge—(See **Cartridge.**)

Master Record—The original copy of a recording from which the commercially available records are mass produced.

Maximum Tracking Angle Error—A specification which gives the largest deviation of the stylus (in degrees) away from perfect tangency to the record groove. The error occurs because a pivoted tone arm can not keep the cartridge tangent to the groove at all times.

Mistracking—The loss of contact between the stylus and the groove walls which sometimes occurs during the playing of a record.

NAB (National Association of Broadcasters)—(See **Rumble.**)

Offset Angle—The angle at which the cartridge is held with respect to a line joining the cartridge to the pivot point of the tone arm. Offset angles help to reduce tracking angle errors.

Output Voltage—The amount of voltage generated by a cartridge when the stylus is vibrating at a standard frequency (usually 1 kHz) and at a specified stylus speed (usually in the range from 3.5 cm/sec to 5 cm/sec).

Piezoelectric Effect—The ability of certain materials to produce an electrical voltage when they are bent or twisted.

Pitch Control (Speed Control)—A control used on turntables to correct for slight platter speed errors. This control usually works in connection with a stroboscopic light.

Platter—The rotating heavy metal disk of a turntable, whose function is to support the record as it revolves.

Record Changer—(See **Turntable.**)

Record Groove—The V-shaped trough which is cut into the surface of a record. The sides of the "V" are perpendicular. In a stereo record one wall of the groove contains the undulations which lead to left channel sound, while the other wall contains the undulations which lead to right channel sound.

RIAA Equalization—The specific equalization process used for phonograph records. When a record is cut, the bass frequencies are deemphasized and the high frequencies are boosted relative to the midrange frequencies. On playback, the RIAA (Recording Industry Association of America) phono equalization preamplifier reverses these effects to produce a flat frequency response. The advantages of the RIAA equalization process are that it reduces record noise and allows about 20 minutes or more of audio to be recorded on each side of a record.

Rim Drive—A type of turntable drive system in which the rotating motor shaft is connected to the platter via an idler wheel. The idler wheel presses against the rim of the platter.

Rumble—The low frequency "rumbling" sound which results from extraneous vibrations of the turntable platter. It may be measured in terms of unweighted (NAB) or weighted (DIN, ARLL) decibels.

Skating Force—A force which pushes the stylus toward the center of the record. The skating force is caused by the frictional force between the stylus and the groove.

Speed Control—(See **Pitch Control.**)

Static Balancing—A method for adjusting the tracking force using a movable weight which slides along the back of the tone arm.

Stereo Separation—A specification which tells how much sound has leaked from the channel for which it was intended into the other channel for which it was not intended. Technically it is the ratio of the power of the audio signal, in the cartridge channel for which it is intended, to the power of this signal which has leaked over to the other cartridge channel. This ratio is expressed in decibels.

Stylus—A carefully shaped and polished piece of diamond which moves in the record groove. Cartridges come equipped with stylus shapes of various kinds. The most common kinds are conically or elliptically shaped.

Suspension—The way in which the turntable drive system is supported in its frame. It may rest on four springs, for example. A good suspension helps to isolate the record from extraneous vibrations.

Synchronous Motor and Hysteresis Synchronous Motor—Motors whose speeds depend on the frequency of the power source (i.e., 60 Hz house current), and are more stable than induction motors.

Tone Arm—A rigid rod which holds the cartridge in position while the stylus moves in the record groove. Tone arms can be straight, J-shaped, or S-shaped.

Tone Arm Pivot Friction—The amount of frictional force encountered by the tone arm as it moves on its pivot.

Tone Arm Resonance—The natural frequency of vibration of a tone arm. Its value depends on the cartridge being carried by the tone arm.

Trackability—The ability of a stylus to track the record groove.

Tracking Angle Error—(See **Maximum Tracking Angle Error.**)

Tracking Force—The downward force that the stylus exerts on the record groove.

Turntable—The hi-fi component which is needed to play a record, and includes the motor, the platter, the tone arm, and the cartridge. There are manual, semiautomatic, and fully automatic turntables. The latter kind are sometimes referred to as record changers, since they will automatically play a series of records.

Turntable Drive System—The part of a turntable which makes the record go around. It includes the motor, some means of connecting the motor to the platter, and the platter itself. See **Belt Drive, Direct Drive, and Rim Drive.**

Weighted Measurements—A method of making audio measurements in which certain frequencies are not considered in the measurement as heavily as other frequencies.

Wow and Flutter—Short-term fluctuations in a turntable's speed. Wow is slower changes in the platter's speed, those which occur between 0.5 and 6 times per second. Flutter represents more rapid variations, usually occurring between 6 and 200 times per second.

Review Questions

1. Name and describe the three basic types of turntables. What are the important features of each type?
2. Name three common types of drive motors used in turntables. What are their comparative advantages and disadvantages?
3. What factors are important for reducing the vibrations and speed fluctuations in a turntable drive system? Name three common types of drive mechanisms.
4. What is a pitch control, and why is it important?
5. Why is the turntable mounted on springs?
6. How does speed accuracy differ from wow and flutter?
7. What is the cause of rumble, and how can it be reduced?
8. Why is an adequate tracking force necessary for good reproduction but too much is detrimental?
9. What is tone arm resonance and why must it be kept below the audible range?

10. Why is the tracking angle error a problem with tone arms and how can it be minimized?
11. What is the purpose of the tone arm offset angle?
12. Why does "weighting" yield better values for both the rumble and the wow and flutter specifications?
13. Why is the "skating" force directed toward the center of the record? What is the direction of the antiskating force?
14. Name the different types of phono cartridges and describe how each generates its voltage.
15. Name some various shapes of phono styli, and compare the advantages and disadvantages of each.
16. Describe, briefly, a stereo record groove and how it contains two independent channels of information.
17. How does a highly compliant cartridge increase the frequency response and allow for a smaller tracking force?
18. Describe the purpose of the RIAA equalization process used with all modern records.

Exercises

NOTE: The following questions have up to 5 possible answers. Please select the **one** response which best answers the question.

1. What is the reason that the bass notes are cut according to the RIAA recording curve when a record is made?
   1. To produce less rumble.
   2. To produce less wow and flutter.
   3. To reduce the large groove widths that would be required by the large amplitude bass notes, thus increasing the record's playing time.
   4. To reduce the surface noise produced by the record.
2. If you played a record, **without** using the RIAA playback amplifier, the music would sound like:
   1. too much bass and too much treble.
   2. too much bass and too little treble.
   3. too little bass and too much treble.
   4. too little bass and too little treble.
   5. It would sound perfectly natural.
3. In the RIAA phono equalization, boosting the high frequencies during recording and then attenuating the high frequencies during playback effectively:
   1. reduces the compliance.
   2. reduces record surface noise.
   3. increases the record's playing time.
   4. reduces record wear.
   5. All of the above answers are correct.

The table below contains the specifications for two stereo cartridges. Please answer the next 4 questions using one of the answers given below. (Please note that several questions may have the same answer.)

	A	B
Frequency response	10–20,000 Hz ±1 dB	15–20,000 Hz ±3 dB
Compliance	6×10^{-6} cm/dyne	2×10^{-6} cm/dyne
Stereo separation	30 dB at 1,000 Hz	20 dB at 1,000 Hz
Tracking force	2 grams	6 grams

Here are the answers to be used with the following 4 questions:

1. "A" and "B" **both** meet hi-fi standards although "A" is **better** than "B."
2. "A" and "B" **both** meet hi-fi standards although "A" is **worse** than "B."
3. "A" meets hi-fi standards but "B" does **not.**
4. "A" does **not** meet hi-fi standards although "B" **does.**
5. **Neither** "A" nor "B" meets hi-fi standards.

4. Compare the frequency response of A with that of B.
5. Compare the compliance of A with that of B.
6. Compare the stereo separation of A with that of B.
7. Compare the tracking force of A with that of B.

8. You are considering the purchase of a new phono cartridge and your 81 watt amplifier has a PHONO input sensitivity rating of 0.3 mV. Listed below are the normal output voltages of four cartridges:
   cartridge A: 1 mV.
   cartridge B: 4 mV.
   cartridge C: 7 mV.
   cartridge D: 2 mV.
   Which cartridge(s) would be acceptable for use with your amplifier?
   1. A and B.
   2. Only B.
   3. Only C.
   4. C and D.
   5. B and C.
9. The stereo separation for a stereo cartridge should be (in the midrange):
   1. 5 dB or better.
   2. 10 dB or better.
   3. 15 dB or better.
   4. 20 dB or better.
   5. 60 dB or better.
10. Which cartridge specification tells how much signal 'leaks' from one channel to the other?
    1. Frequency response.
    2. Stereo separation.
    3. Compliance.
    4. THD and IM.
    5. Capture ratio.
11. Which one of the following is the **best** frequency response for a cartridge? (Assuming that money is no object and you can afford the finest):
    1. 20–20,000 Hz ±3 dB.
    2. 20–20,000 Hz ±1 dB.
    3. 15–25,000 Hz ±1 dB.
    4. 15–25,000 Hz ± 3 dB.
    5. 15–25,000 Hz ± 6 dB.
12. The ideal turntable motor would:
    1. produce no vibrations.
    2. turn at a constant speed.
    3. have adequate power.
    4. produce no external magnetic fields.
    5. All of the above answers are correct.
13. Which one of the following is **not** a cartridge specification?
    1. Compliance.
    2. Frequency response.
    3. Tracking force.
    4. Stereo separation.
    5. Tracking angle error.
14. Two specifications of a cartridge which are closely related are:
    1. compliance and tracking angle error.
    2. stereo separation and frequency response.
    3. flutter and rumble.
    4. compliance and tracking force.
    5. cartridge output voltage and antiskating force.
15. The type of phono cartridge in which a magnet is attached to the cantilever is called a:
    1. moving coil cartridge.
    2. variable reluctance cartridge.
    3. moving magnet cartridge.
    4. ceramic cartridge.
    5. moving iron cartridge.
16. A cartridge has a compliance of 20×10^{-6} cm/dyne. If the groove wall exerts a force of 2 dynes on the stylus it will move a distance of:
    1. 10×10^{-6} cm.
    2. 20×10^{-6} cm.
    3. 40×10^{-6} cm.
    4. 80×10^{-6} cm.
    5. 2 cm.
17. The antiskate adjustment is used to:
    1. reduce rumble.
    2. reduce wow.
    3. counteract the force pulling the tone arm toward the rim of the record.
    4. reduce tone arm resonance so the stylus will not bounce out of the record groove.
    5. counteract the force pulling the tone arm toward the center of the record.
18. A tone arm has a resonant frequency of 120 Hz. If a record is playing and the music contains a 120 Hz note then:
    1. the entire tone arm will resonate up and down at 120 Hz.
    2. the tone arm will resonate at 240 Hz (120 × 2).
    3. the tone arm will resonate at 60 Hz (120/2).
    4. nothing abnormal will happen and the tone arm will track the record in a normal fashion.
    5. 120 Hz flutter will be produced.
19. An adjustment which compensates for a force trying to pull the tone arm toward the center of a record is called the:
    1. tracking force adjustment.
    2. tracking angle error adjustment.
    3. center force adjustment.
    4. antiskate adjustment.
    5. pitch control.

20. What generally happens if the tone arm tracking angle error exceeds 3 degrees?
    1. A serious loss of low frequency tones occurs.
    2. A serious loss of high frequency tones occurs.
    3. Groove skipping occurs.
    4. Excessive rumble results.
    5. The wow and flutter increases.
21. Which specification tells how easily a stylus can move in the record groove?
    1. Tracking angle error.
    2. Rumble.
    3. Wow.
    4. Compliance.
    5. Flutter.
22. A phono cartridge should remain tangent to the record groove. The specification which describes the maximum amount that the cartridge deviates from perfect tangency is called the:
    1. antiskate force.
    2. tracking force.
    3. compliance.
    4. tracking angle error.
    5. None of the above answers is correct.
23. Tracking angle error may be eliminated by:
    1. increasing the tracking force.
    2. using a high compliance cartridge.
    3. having the stylus be tangent to the grooves at all times.
    4. reducing the tone arm resonance frequency.
    5. using the antiskating adjustment.
24. What is the tracking force of a typical hi-fidelity tone arm?
    1. 0–0.5 grams.
    2. 0.5–3.0 grams.
    3. 3.0–10 grams.
    4. 10–20 grams.
    5. 20–25 grams.
25. Stylus and record wear, on a good turntable, are **most** related to:
    1. tracking angle error.
    2. the antiskate force.
    3. the tracking force.
    4. tone arm resonance.
    5. wow and flutter.
26. Variations in turntable speed, which produce a fluctuation in musical pitch, is called:
    1. S/N ratio.
    2. rumble.
    3. wow and flutter.
    4. compliance.
    5. resonance.
27. If a constant tone of 500 Hz is being played on a turntable with a wow and flutter rating of 0.2%, what changes in this frequency can be expected when a record is played?
    1. 0.2 Hz.
    2. 1.0 Hz.
    3. 5 Hz.
    4. 10 Hz.
    5. 20 Hz.
28. A turntable is playing a single tone whose frequency, because of wow and flutter, varies between 1,010 and 990 Hz. What is the wow and flutter of the turntable?
    1. 20%.
    2. 10%.
    3. 4%.
    4. 1%.
    5. 0.2%.
29. Suppose that you are listening to a record whose music is being played at a loudness of 95 dB. What would be the **maximum** amount of unweighted rumble (as measured by a dB meter in the room) that could be produced by the turntable and still have it be regarded as hi-fidelity?
    1. 130 dB.
    2. 95 dB.
    3. 80 dB.
    4. 60 dB.
    5. 35 dB.
30. A turntable has an unweighted rumble specification of −40 dB. If a record is being played through a hi-fi system at a loudness of 100 dB, what is the rumble noise level?
    1. 40 dB.
    2. 60 dB.
    3. 140 dB.
    4. 100/40 = 2 1/2 dB.
    5. None of the above answers is correct.
31. The specification which determines the long term ability of a turntable to "hold" the same musical pitch is called:
    1. rumble.
    2. speed accuracy.
    3. S/N ratio.
    4. compliance.
    5. stereo separation.
32. A low frequency noise caused by vibrations of a turntable is called:
    1. rumble.
    2. flutter.
    3. wow.
    4. total harmonic distortion.
    5. dispersion.

33. The better platters are:
    1. made of nonferrous materials.
    2. die-cast and machined.
    3. made with near perfect balance.
    4. All of the above answers are correct.
34. A disadvantage of the rim drive turntable is:
    1. its relatively high cost.
    2. the development of flat spots on the idler wheel.
    3. the requirement of a high torque motor.
    4. low wow and flutter.
    5. low rumble.
35. Which type of turntable utilizes a slow speed motor turning at the actual 33 1/3 rpm record speed?
    1. Rim drive.
    2. Belt drive.
    3. Synchronous.
    4. Direct drive.
    5. Induction.
36. Which one of the following is **not** a type of turntable drive system?
    1. Rim drive.
    2. Pinion drive.
    3. Belt drive.
    4. Direct drive.
37. The type of motor, used in better quality turntables, which has its speed locked to the **frequency** of the AC line voltage (60 Hz), is called:
    1. automatic.
    2. induction.
    3. synchronous.
    4. rim drive.
    5. None of the above answers is correct.

Below is a table which contains the specifications for two tone arms.

	Tone Arm A	Tone Arm B
Tracking angle error	1°	4°
Resonance frequency	25 Hz	10 Hz
Pivot friction	20 milligrams	15 milligrams

Here are the answers to be used with the following 3 questions:

1. "A" and "B" **both** meet hi-fi standards although "A" is **better** than "B."
2. "A" and "B" **both** meet hi-fi standards although "A" is **worse** than "B."
3. "A" meets hi-fi standards but "B" does **not.**
4. "A" does **not** meet hi-fi standards although "B" **does.**
5. **Neither** "A" nor "B" meets hi-fi standards.

38. Compare the tracking angle error of A with that of B. (3)
39. Compare the resonance frequency of A with that of B. (4)
40. Compare the pivot friction of A with that of B. (2)
41. The larger the compliance specification of a cartridge:
    1. the greater is the required tracking force.
    2. the easier it is for the stylus to follow the record groove.
    3. the lower is the tone arm resonance frequency.
    4. the higher is the tone arm resonance frequency.
    5. the harder it is for the stylus to follow the record groove.
42. The weighted rumble specification of a turntable is −65 dB. With a (weighted) dB meter you measure a rumble noise level of 30 dB. The sound level of the music, measured with the same dB meter, would be:
    1. 30 dB.
    2. 65 dB.
    3. 95 dB.
    4. (65–30) = 35 dB.
    5. 65/30 dB.
43. Which answer below determines how much voltage a moving magnet cartridge will produce?
    1. The number of turns in the coil.
    2. The strength of the permanent magnet.
    3. How fast the magnet is moving relative to the coil.
    4. Answers #1, #2, and #3 all determine how much voltage will be produced.
44. Why are the treble notes cut during the recording of a master record?
    1. In order to obtain a long (20 min.) playing time for the record.
    2. To increase the "surface noise" present on all records.
    3. The treble notes are not cut during the recording process.
    4. In order to decrease the amount of rumble produced by the turntable.
    5. In order to decrease the wow and flutter produced by the turntable.

45. If the idler wheel on a turntable develops a severe flat spot, the turntable will:
    1. produce a large amount of rumble.
    2. produce a large amount of wow and flutter.
    3. produce stray magnetic fields.
    4. develop a large tone arm resonance frequency.
    5. develop excessive THD.

    Below is a table which contains the specifications for two turntables.

	Turntable A	**Turntable B**
Speed accuracy	1.0 %	0.5 %
Rumble (NAB)	−20 dB	−40 dB
Rumble (ARLL)	−50 dB	−70 dB
Wow and flutter (unweighted)	0.1 %	0.15 %
Wow and flutter (weighted)	0.04 %	0.06 %

Here are the answers to be used with the following 5 questions:
1. "A" and "B" **both** meet hi-fi standards although "A" is **better** than "B."
2. "A" and "B" **both** meet hi-fi standards although "A" is **worse** than "B."
3. "A" meets hi-fi standards but "B" does **not.**
4. "A" does **not** meet hi-fi standards although "B" **does.**
5. **Neither** "A" nor "B" meets hi-fi standards.

46. Compare the speed accuracy of A with that of B. (2)
47. Compare the unweighted rumble of A with that of B. (4)
48. Compare the weighted rumble of A with that of B. (4)
49. Compare the unweighted wow and flutter of A with that of B. (1)
50. Compare the weighted wow and flutter of A with that of B. (1)

Chapter 13

TAPE DECKS

13.1 A Short History of Magnetic Recording

The magnetic recorder was invented by Valdemar Poulsen in 1898. His "Telegraphone" first used carbon steel piano wire for the "tape." The wire was wound around a brass cylinder in a helical fashion, and the cylinder was rotated while a moving recording head followed the wire. Poulsen used his invention as a dictating machine which recorded for 30 minutes while the wire moved at the astonishing speed of 7 feet per second! In 1900 he took the recorder to the Paris Exposition and, needless to say, it was the sensation of the day. No electronic amplifiers existed at the time, so the record head was driven directly by a carbon microphone. Playback was accomplished by connecting the recording head directly to earphones, and using it as a playback head also. Because of the lack of amplification the sound level was rather low, although it was of very good quality when compared to other audio systems of that time, particularly the disc recordings.

In 1912 DeForest* experimented with Poulsen's Telegraphone by adding an amplifier for the purpose of producing motion picture sound. However, it was not until 1921 when the first commercial recorder with a built-in tube amplifier was introduced. All magnetic recorders used wire or steel bands until BASF perfected the process for coating tape in 1935. In 1937 the first commercial tape recorder was produced in the United States. During the early 1940s significant progress was made, particularly by the Germans, which yielded recordings with frequency ranges out to 10 kHz, good S/N ratios, and relatively little wow and flutter. However, the 30 ips (inches per second) tape speeds on these machines were rather high. In 1947 Ampex began manufacturing tape recorders, and in 1949 Magnecord introduced the first stereo machines.

*Dr. Lee DeForest is best known as the inventor of the triode vacuum tube, the forerunner of the transistor which is widely used in all modern amplifiers.

Throughout the 1950s tape recorders were mainly employed by professionals, and they found very little home use. The cassettes and 8-track cartridges are a result of a search for a tape format which would offer the convenience found with turntables. Envisioned was a system which would allow tapes to be played for hours on end without either requiring attention from the user or dexterity in threading the tape. In 1954 George Eash invented the endless-loop tape cartridge and player. These cartridges were rather large, expensive, and had a short life expectancy; consequently, they did not sell well. From these larger cartridges evolved the smaller and popular "8-track" cartridge (Bill Lear, 1965) and the cassettes (Philips Co., 1964). At about the same time Ray Dolby demonstrated a noise-reduction unit for Decca Records in London. A later (1971) version of his process was to pave the way for high-performance cassettes. Cassettes, because of their extremely slow speeds (1-7/8 ips), must sacrifice signal-to-noise in order to achieve a wide frequency response. The Dolby system significantly improved this situation, and by 1976 the cassette was the favorite tape recording medium for home systems. Today, the traditional open-reel units remain as high-end products for those willing to pay higher prices for a machine with performance and features not found on cassettes or cartridges.

B·I·C Model T-4M cassette deck

13.2 Tape Equipment

Tape recorders are rapidly becoming popular companions to the more established turntables and tuners. This is particularly true in the component systems where spectacular advances in cassette technology have propelled them into a high quality, moderately priced, and convenient medium.

Generally speaking, tape machines may be classified as either tape *decks* or tape *recorders*. The decks consist of a system for moving the tape, a means for recording, playing back, and erasing audio signals, and the required recording and playback preamplifiers. With a deck playback is possible only with the use of headphones or an external preamp/ control center, power amplifier, and speaker system. Often this amplifier combination is replaced by either a receiver or an integrated amplifier. A subclass of the tape deck is the *tape player* (or *player deck*). The tape player is a deck which is minus the recording facilities, and is used solely for the playback of prerecorded tapes. Most of the 8-track cartridge decks are sold as players only, and they utilize the large selection of commercially available prerecorded tapes. A *tape recorder* is similar to a tape deck except, in addition, it is equipped with its own preamp-control center, power amplifier, and loudspeaker(s); the tape recorder is a complete sound system. Tape recorders are found as portable units.

There are three tape formats which are widely used in either decks or recorders: open-reel, cassette, and 8-track cartridge. The open-reel, with its large reels of tape and impressive features, has always been regarded as the tape medium for the highest quality of sound reproduction. Professionals use open-reel decks with tape speeds of either 30 ips or 15 ips. The faster speeds are vital for quality performance because they yield better frequency responses and produce the largest possible signal-to-noise ratios; people who are serious about tape recording almost always prefer the highest available speed. Most nonprofessional decks for home use offer the user either 7-1/2 or 3-3/4 ips speeds, although, occasionally, speeds as high as 15 ips, or as low as 1-7/8 ips, are encountered.*

Open-real tape decks also offer excellent editing, splicing, and sound-mixing capabilities which are either not found or difficult to achieve on cassettes and 8-track cartridges. The open-reel is still the most common, highest performer, most versatile (and most expensive) machine for the serious hi-fi enthusiast.

Cassettes are relative newcomers on the hi-fi scene. Because of their slow tape speed (1-7/8 ips, only) cassettes were initially plagued with poor frequency response and excessive noise. With the development of new precision drive systems, improved heads, better quality tapes, and noise-reduction systems, cassettes have become truly hi-fidelity components. Today cassettes are enjoying the tape spotlight because of their fine performance capabilities, moderate costs ($200-$600, on the average), and ease of handling.

Tandberg model TCD 440A cassette deck

Eight-track cartridges have been used extensively in car stereos and, with a few notable exceptions, they have not achieved the standards of performance of either the open-reels or the cassettes. There are no technological reasons

*Note that all tape speeds are some multiple of the slowest, 1–7/8 ips, speed; for example, 7–1/2 ips is 4 × (1–7/8 ips).

why cartridges could not be designed as quality components since the tape travels at 3-3/4 ips, twice the speed of cassettes. Cartridges, unlike open-reels or cassettes, utilize an endless-loop of tape on *one* reel and, as such, they cannot be reversed (see section 13.3). If you are recording with a cartridge and make a mistake, it is not possible to correct it quickly. Instead the user must "fast forward" the tape in order to reach the problem spot—a process which usually takes from 4 to 10 minutes (in spite of the name "fast" forward). This time delay essentially makes tape editing very cumbersome and, therefore, most cartridge machines are designed only for the playback of prerecorded tapes.

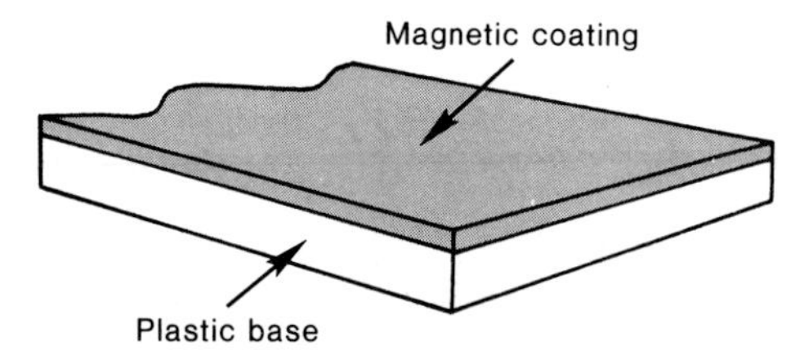

Figure 13.1. A magnetic tape consists of a plastic base with a thin magnetic coating on one side.

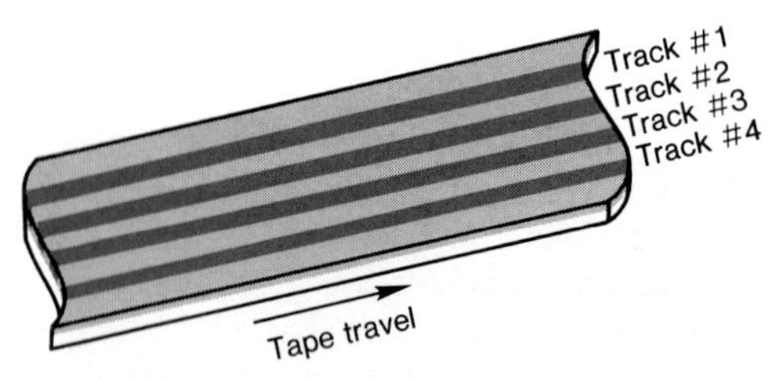

Figure 13.2. All tape is divided into lanes or "tracks" along its length. There are four tracks shown in this example.

13.3 Magnetic Tapes

Magnetic recording tape consists of a thin (5 to 50 micron*) polyester or acetate film coated with a fine layer of magnetic particles (Fig. 13.1). The magnetic particles are usually iron oxide or chromium dioxide, and it is these particles which become magnetized by the recording head and store the audio information.

All tape is divided into multiple lanes called "tracks" which run along the length of the tape, as indicated in Figure 13.2. The first commercial tape recorders built during the

*One micron equals 0.00004 inches, or about 1/100 the thickness of a piece of paper.

1940s used the full width of the tape for recording a monaural program. However, as both tape recording components and the tape itself improved, it became possible to split the tape width into two, four, and then eight recording "tracks," with each track containing one channel of audio information. Multiple tracks provide greater tape economy and the possibility of storing both stereo and quadraphonic programs.

As we shall discuss shortly, the track formats are different for the open-reels, cassettes, and 8-track cartridges. In order to understand these differences better, first consider the magnetic heads illustrated in Figure 13.3. As discussed in sections 11.4 and 11.7, the recording and playback heads consist of a coil of wire wrapped around a horseshoe-like core; the core is usually made from a magnetic material like iron. The left-hand drawing in part (A) shows a magnetic head and the gap between the two ends of the horseshoe. In reality, the gap is extremely narrow (less than the thickness of a piece of paper), and the region of the head which is in contact with the tape is highly polished. Since the moving tape is pressed against the head, the polished surface reduces the friction and abrasion between the two. Usually the magnetic head is enclosed in a metal case, and only the gap is visible from the outside. The gap appears as a tiny slit in the middle of a polished "rectangle," as illustrated by the right-hand drawing in the figure. For stereo, two magnetic heads are enclosed within a single case, as shown in Figure 13.3(B). Of course, a quadraphonic tape deck would have four heads contained within the case.

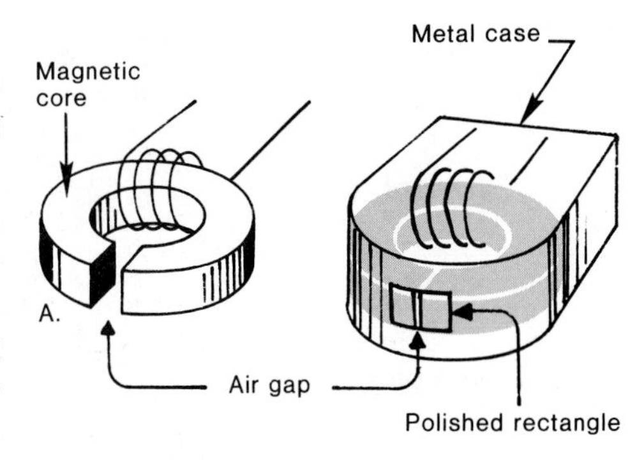

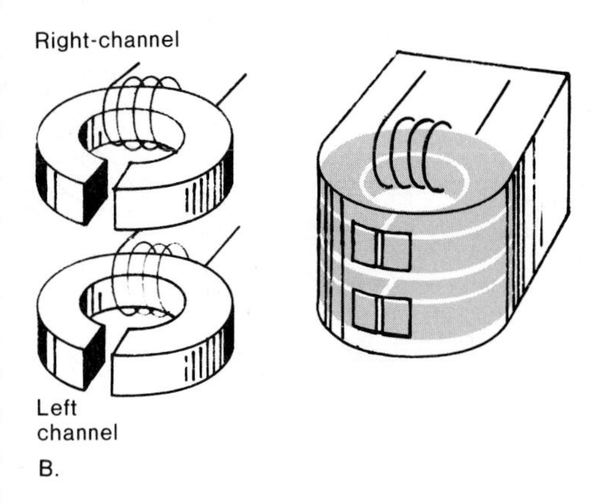

Figure 13.3. (A) A monaural tape head. (B) Stereo tape heads.

Open-Reel Tapes

Most modern open-reel tape decks for home use employ 4 tracks on a 1/4″-wide tape, as illustrated in Figure 13.4. Tracks 1 and 3 contain, respectively, the left and right stereo channels for "side 1." Tracks 2 and 4 contain the right and left stereo channels for "side 2." Notice that sides 1 and 2 are both located on the magnetic side of the tape. To play or record using side 2, the tape is simply rotated clockwise by 180°, so that R2 and L2 are now facing the heads (see Fig. 13.4). This ensures that the magnetic surface is once again in contact with the heads. Figure 13.5 depicts the track arrangements for quadraphonic recording. Of course, one program consumes all 4 tracks, and the total playing time has been cut in half when compared to stereo.

Because of the rather fast tape speeds, open-reel machines must use large reels of 5 or 7 inch diameters; often 10-1/2 inch reels are found in professional decks. Table 13.1 shows the recording times for some common open-reel stereo tapes at 7-1/2 ips recording speed; 15 ips cuts the recording time in half, while 3-3/4 ips provides twice the time compared to those shown in the table.

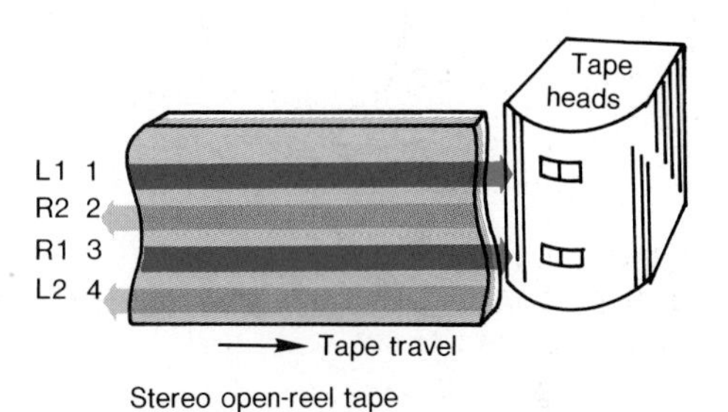

Figure 13.4. The track arrangement for a 4-track open-reel stereo tape. Notice that tracks 1 and 3 carry, respectively, the left and right stereo channels (L1 and R1) for "side 1." Similarly, tracks 2 and 4 contain the stereo program (R2 and L2) for "side 2." Both "side 1" and "side 2" are located on the magnetic side of the tape.

Type	Thickness (microns)	Length (feet)	Recording Time (min.) (side 1 + side 2)
Standard	52	1,200	64
Long play	35	1,800	96
Double play	25	2,400	128

Table 13.1. The recording times for some common types of open-reel stereo tapes at 7-1/2 ips.

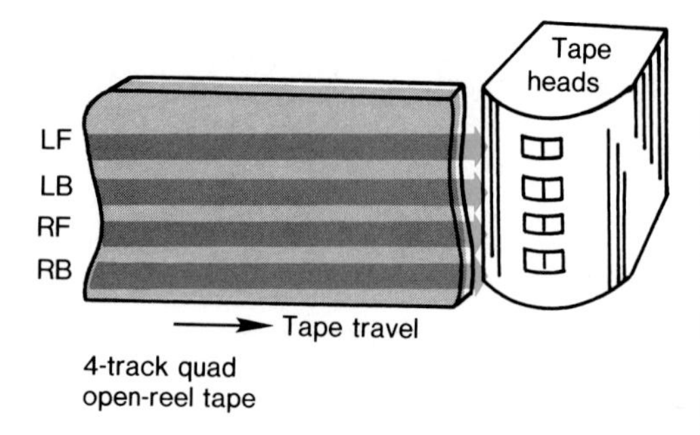

Figure 13.5. A quadraphonic open-reel tape uses all four tracks for the program material. (L = left, R = right, F = front, B = back).

Figure 13.6. The general layout of a cassette case showing the two small tape reels.

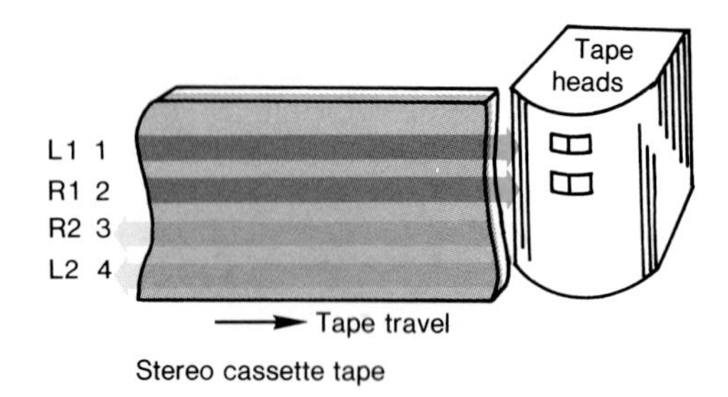

Figure 13.7. The track layout for stereo recording on a cassette tape. (L = left, R = right).

Cassette Tapes

The cassette system is very much like the open-reels. Two tiny reels are contained within the case, and the tape spools from one to the other. Because of the slow speed (1–7/8 ips), cassettes can provide well over an hour's worth of playing time from a small case which measures approximately 4″ × 2–1/2″ × 1/2″ (see Fig. 13.6). Openings at the front edge of the cassette permit the tape to pass across the appropriate erase, record/playback heads. When all the tape has been transferred from one reel to the other, the entire cassette can be turned over to play "side 2." The cassette tape is narrower than that of open-reels (5/32″ vs. 1/4″) and the track configurations are also different. As drawn in Figure 13.7, the cassette utilizes tracks 1 and 2 for the left and right stereo channels of "side 1," while tracks 3 and 4 are reserved for the two stereo channels of "side 2."

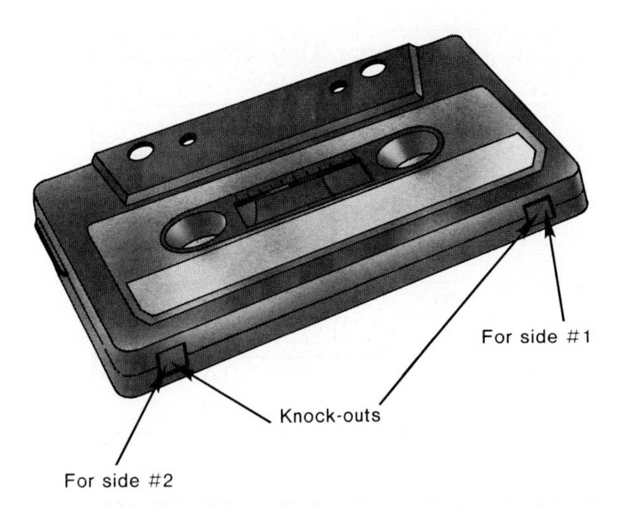

Figure 13.8. The "knock-outs" on a cassette case can be used to prevent accidental erasure of the tape. There is a knock-out for each playing "side" of the tape.

Cassettes have an interesting feature which prevents accidental erasure of previously recorded tapes. This feature consists of two "knock-outs" which are located along the back edge of the cassette (see Figs. 13.6 and 13.8). To activate the erasure-prevention device, the plastic knock-out is removed with, for example, a small screwdriver. **There is an interlock located within the deck which prevents recording when the knock-out has been removed.** This is desirable in order to avoid erasing a tape which contains valuable information; a knock-out is provided for each "side" of the tape. **The erasure-prevention system can later be defeated, if desired, by covering the knock-out holes with adhesive tape.** Neither open-reels nor cartridges have such safety features.

Cassettes, like open-reels, are also available with different playing times, and they are rated as C-30, C-60, C-90, etc. For example, a C-60 tape contains a total of 60 minutes of recording time, 30 minutes for each side. Table 13.2 presents some of the characteristics of various cassette tapes such as thickness, length, and recording times. All of these tapes fit into the standard-sized cassette case. Note that the longer playing time is achieved only by using a greater length of tape. This means that the tape thickness must be reduced in order to fit more tape into the same size cassette case. A phenomenon called "print-through" can occur with the thinnest cassette tapes (i.e., the C-90 and C-120 tapes). Print-through occurs when the tape is wound on the reel, and the magnetic field from one layer of the tape penetrates the adjacent layers and imprints a faint copy of itself on them. Print-through usually occurs during prolonged storage of tightly wound tape.

Type	Thickness (microns)	Length (feet)	Recording Time (min.)	
			Side 1	Side 1 and Side 2 Together
C-30	18	150	15	30
C-60	18	300	30	60
C-90	13	450	45	90
C-120	9	600	60	120

Table 13.2. A comparison of the various cassette playing times.

8-track Cartridge Tapes

The 8-track cartridge was developed primarily for use in automobiles, although it has become popular for home entertainment as well. The cartridge is different from either the cassette or open-reel because it utilizes only one reel. As shown in Figure 13.9, the tape is wound in an "endless-loop" configuration, which is pulled from the inside of the reel and returned to the outside of the reel. Obviously, a "rewind" feature is impossible on an 8-track tape, because the tape cannot be "stuffed" back into the inside of the reel. The "no-rewind" condition makes it awkward to correct errors incurred during the recording process. There are, however, some 8-track machines that have a fast forward speed—usually about twice the normal playing speed.

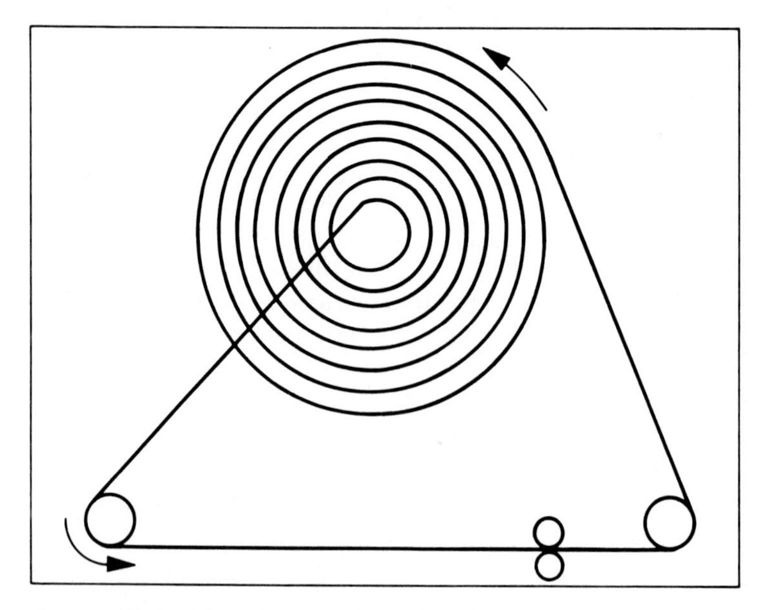

Figure 13.9. The single reel, endless-loop tape of an 8-track cartridge.

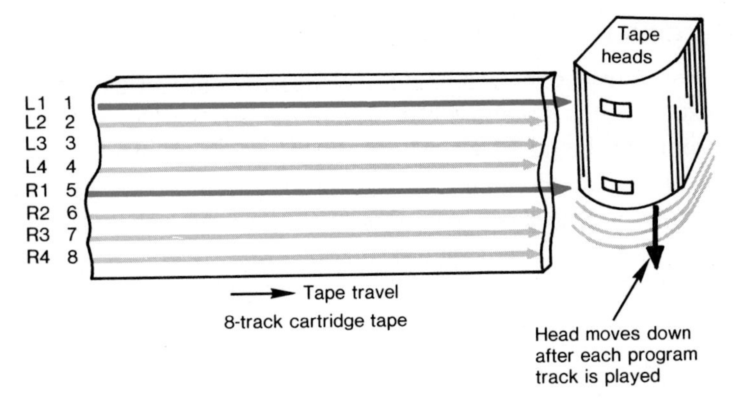

Figure 13.10. An 8-track cartridge tape showing the track assignments for four stereo programs. (L = left, R = right).

All cartridge tapes use a 1/4-inch wide tape, and the width is divided into 8 tracks (Fig. 13.10). The eight tracks allow four separate stereo or two quadraphonic programs to be recorded. Tracks 1 and 5 are used for the first stereo program, 2 and 6 for the second, and so on. Of course, all the tracks play in the same direction since an 8-track cannot be reversed. It is also interesting to note that the selection of various programs is accomplished by keeping the tape fixed and moving the heads across it. A metallic foil strip is attached to the "end" of the tape, and when an electrical sensor detects the strip, it automatically moves the head to play the selection on the following track. In this manner all four stereo programs are played in sequence, and the complete cycle repeats over and over unless an automatic stop is provided.

In theory the higher speed and greater tape width of the 8-track cartridges should make them better performers than the cassettes. However, cartridges were primarily designed for playback of prerecorded music in noisy automobiles and most hi-fidelity manufacturers (with a few exceptions) have not devoted the effort to make these machines comparable in quality with cassette decks. As of now, the 8-track cartridge deck has yet to live up to its full potential.

Low Noise/High Output Tapes

A recent development in magnetic tapes has been the appearance of the so-called "low noise/high output" oxides. These tapes possess a higher magnetic particle density, and they provide an even greater output (6 to 8 dB) than the more conventional iron oxide tapes. The extra sensitivity of the high-output tapes allows a larger sound level range, from a soft whisper to a booming kettle drum, to be recorded onto the tape. It also helps to prevent overload and distortion when the tape is subjected to unexpected and sudden loud sound bursts. Most tape decks must have a special "bias" switch to record properly with these tapes (see section 13.6 for a discussion on bias).

Chromium Dioxide Tapes

For many years iron oxide was the only magnetic coating used on tapes. Recently, however, many tapes are now using chromium dioxide. These tapes yield very high output signals, low distortion, and an excellent frequency response. In order to get the most from this tape, your deck should also have a so-called "chromium dioxide bias" switch. This allows the deck's electronic circuitry to optimize the recording conditions required for this type of tape.

Metal Tapes

The latest generation of magnetic tape is the metal-particle tape. Unlike other magnetic materials used in making tape, the metal particle formulation does not contain the oxygen atoms which are found in the "standard" ferric oxide (Fe_2O_3) and chromium dioxide (CrO_2) tapes. Instead, the metal-alloy tape particles consist of an oxygen free combination of metallic atoms, such as the 70% iron (Fe) and 30% cobalt (Co) used in one brand of tape.

The performance capabilities of the metal-particle tapes are quite impressive when compared to the other types of tape. At high recording levels, where large signals are easily distorted by the tape, the metal tape exhibits a lesser tendency to saturate, particularly for the low-to-middle frequencies. At the high frequencies, 2 kHz and higher, the metal tapes significantly improve the frequency response of the deck, the traditional weak point in cassette perfomance. For example, at 10 kHz and a +10 dB recording level, it is not uncommon for the output from a metal tape to be 15 dB greater than that from a conventional ferric oxide tape.

Currently, the cost of the metal tape is rather high, being about twice as expensive as the premium cassettes. Also, the vastly increased record bias and erase current requirements for the metal-particle tape exceed the capacity of existing cassette decks. The user must thus purchase a new deck which is especially designed to record on to the metal tapes. However, any deck that will handle chromium dioxide tape will *play back* a metal tape.

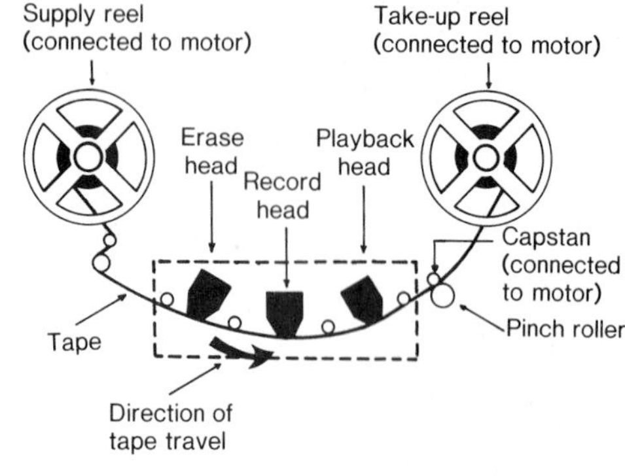

Figure 13.11. The drive system for a 3-head open-reel tape deck.

13.4 The Tape Transport System

The tape transport system performs a similar job as does the drive system of a turntable; namely, pulling the tape past the heads at constant speed. The transport components of primary interest are the capstan and pinch roller, the supply and take-up reels, the drive motors, and the controls. Figure 13.11 shows an open-reel system with the essential transport features. The cassette and cartridge transports are similar to the open-reel, except the cartridge obviously has only one reel. The figure shows a 3-head open-reel in which the tape is pulled from the supply reel on the left, past the erase, record, and playback heads, before it is finally spooled onto the take-up reel. A 2-head tape deck is essentially the same, except that the record and playback features are combined into a single dual-purpose head.

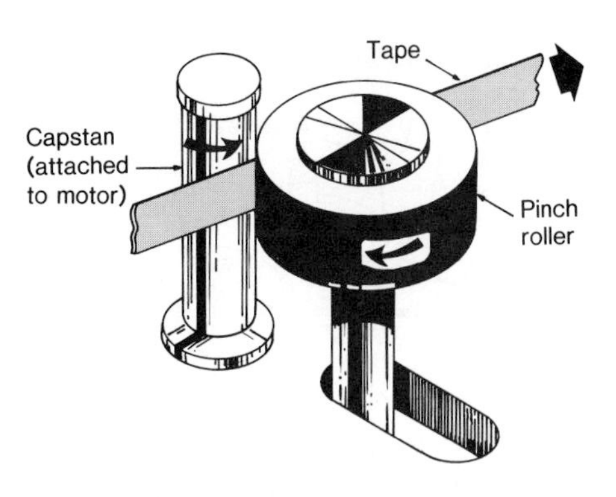

Figure 13.12. The capstan and pinch roller pull the tape past the heads at constant speed. The capstan is rotated by the motor, while the pinch roller presses the tape against the capstan.

The Capstan-Pinch Roller Assembly

The capstan and pinch roller are responsible for pulling the tape across the heads (Fig. 13.12). The capstan is a precision steel shaft which is rotated by the motor drive. The tape is firmly pressed against the capstan by the pinch roller, a passive rubber wheel which is spring loaded for applying the appropriate pressure. The roundness and smoothness of the capstan are very critical in preventing wow and flutter—quick changes in the tape's speed caused by irregularities in the drive system. A flat spot on the capstan would cause the tape to change speed everytime the capstan made one revolution.

Motors

Like the turntable platter, the capstan must be rotated by a motor, of which there are several types available. The most basic and least expensive is the *induction motor*. A conventional induction motor possesses good instantaneous speed regulation, which results in very low wow and flutter. The problem with induction motors is that their speed is, to a certain extent, dependent on the level of the voltage which is applied to them. The ordinary household line voltage can

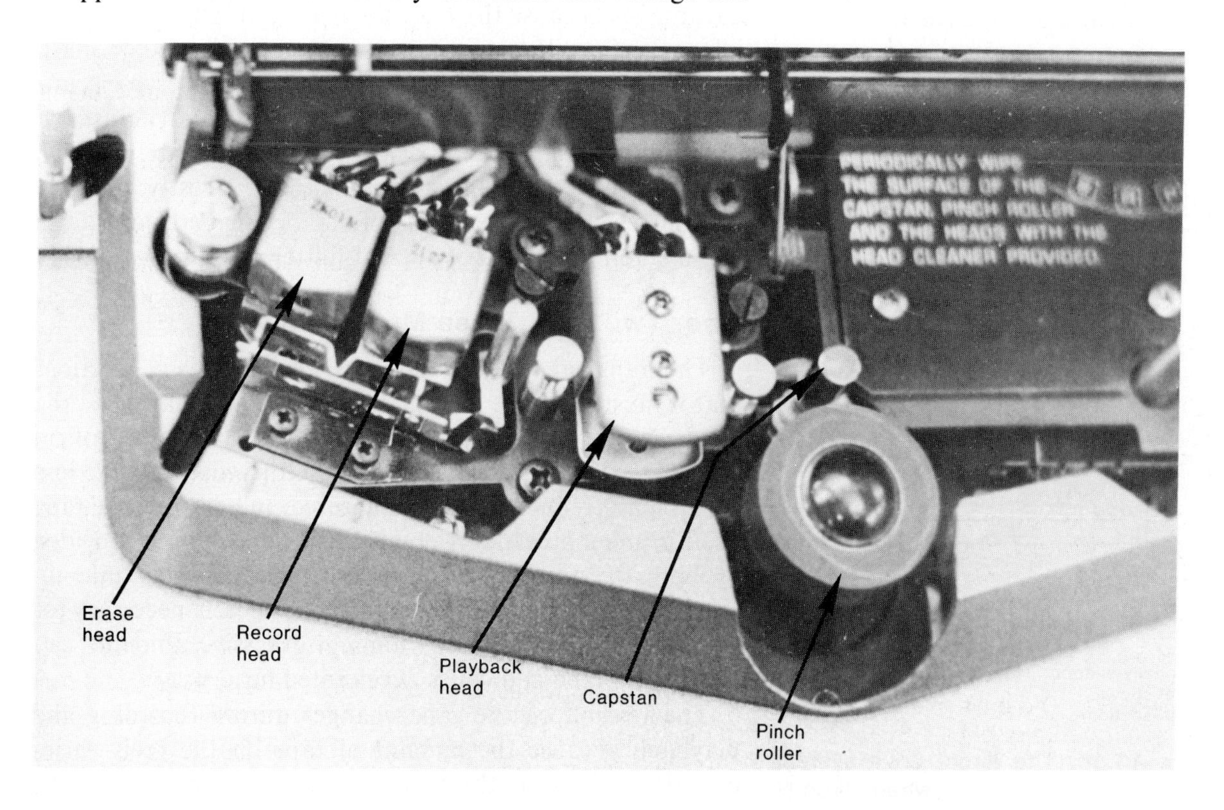

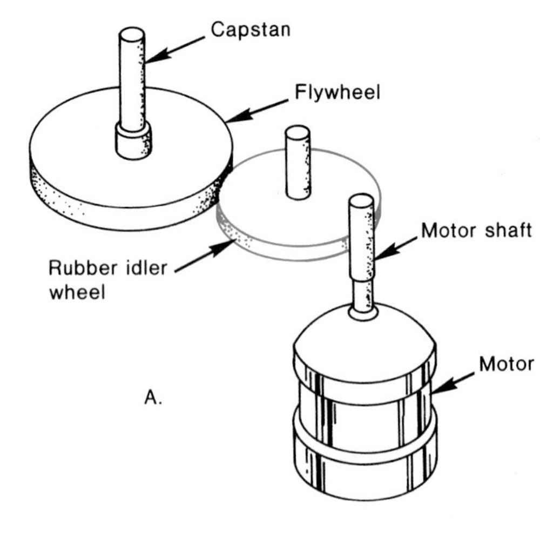

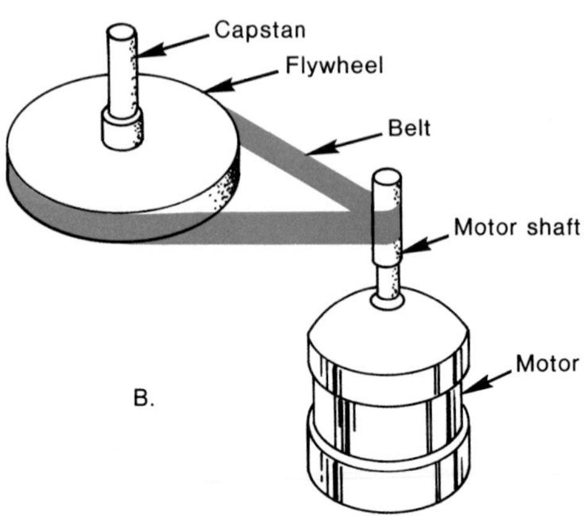

Figure 13.13. Two common drive systems used in tape decks: (A) the idler drive, and (B) the belt drive.

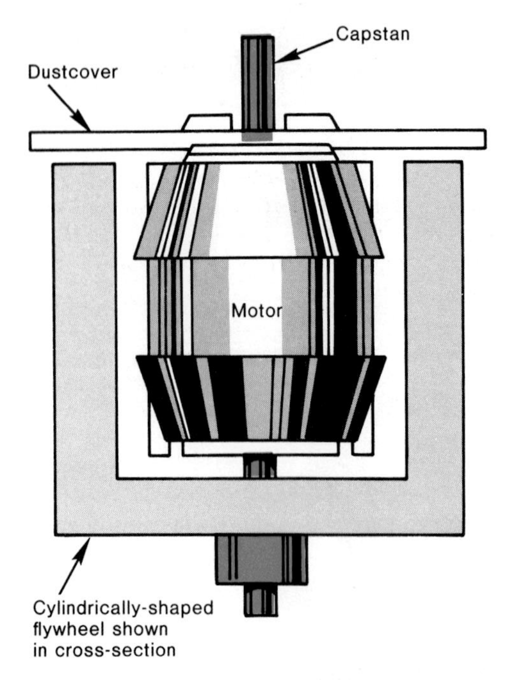

Figure 13.14. The direct drive system mounts the capstan and flywheel directly on the motor shaft.

fluctuate significantly during the day, causing noticeable speed, and, hence, pitch, changes. The speed changes being referred to here are relatively slow compared to those causing wow and flutter. Nonetheless, they are annoying, and for this reason, most high quality tape decks do not use induction motors.

Like the better turntables, the better tape decks use either *synchronous* or *hysteresis synchronous* motors. These motors derive their speed from the 60 Hz line frequency, which is held very constant by the power companies on a long term (24 hour) basis. Thus, they are relatively immune from fluctuations in the line voltage, and have excellent long term speed characteristics.

Connecting the Motor to the Capstan

There are three types of motor drives which are very similar to the systems used with turntables. The *idler wheel* and *belt drives* are illustrated in Figure 13.13. The belt drive is, by far, the most popular method, because the elastic belt helps to keep vibrations and short-term speed changes of the motor from reaching the capstan. Note, also, that a flywheel is attached to the base of the capstan. The large rotational inertia of the flywheel helps to stabilize the capstan's rotation, thereby reducing wow and flutter.

The third type of drive system, found on the most expensive units, is the *direct drive* which employs the motor shaft as the capstan shaft (Fig. 13.14). The flywheel is also directly attached to the shaft. This type of drive has many advantages because there is only one shaft with no need for idler wheels or belts. Since the capstan is the motor shaft, the direct drive motors must employ electronic servo-circuits which sense the motor's rotational speed and adjust for minor fluctuations; this holds the rotation absolutely steady. The servo-controlled direct drive motors are costly, but they can out-perform other types of motors in both long term speed stability and low wow and flutter characteristics.

One, Two, and Three Motor Drives

The transport system in any tape deck has many functions that it must perform. In addition to pulling the tape past the heads at constant speed, it must also place the tape in firm contact with the heads at all times. Otherwise a severe loss of the high frequencies is encountered. In addition, the tape itself is under a certain amount of tension which is provided by both the capstan/pinch-roller and the supply/take-up reels. A certain amount of tension is absolutely necessary for good contact and smooth running; excessive amounts can stretch the tape and cause accelerated head wear.

The tension on the tape changes during recording and playback because the amount of tape on the reels varies considerably during this time. For instance, the tape on the take-up reel goes from a small diameter spool at the beginning to a large diameter at the end. The smaller diameter tape spool exerts a greater tension on the tape than does a full reel. The tape tends to slip by the capstan at a slightly faster rate under heavy tension, thus causing slight pitch changes. Some of the most expensive open-reels employ tension regulators and other devices to protect the tape from excessive tension.

Figure 13.15 illustrates the simplest system which is used in open-reel and cassette machines. The supply and take-up reels, as well as the capstan, are powered from a single motor. This system provides the least amount of control over the tape tension and reel speed. A series of pulleys and springs (not shown in the figure) provide the appropriate reel with the correct tension at the right time. The contact pressure is accomplished by pushing the tape directly against the heads with small felt pads. At slow tape speeds (1–7/8 and 3–3/4 ips) the pressure pads work well, but at higher speeds they are generally not used because of problems associated with head wear, tape scratch, and flutter.

In two-motor machines the tape reels are driven by one motor, and the second motor rotates the capstan (Fig. 13.16). This system, as well as the three-motor drive (Fig. 13.17), are used where the highest precision and rapid winding speeds are desired, and cost is not a significant factor. Needless to say, these types of drives offer much better control of the tape tension and reel speeds, which leads to better performances throughout the entire playing of the tape.

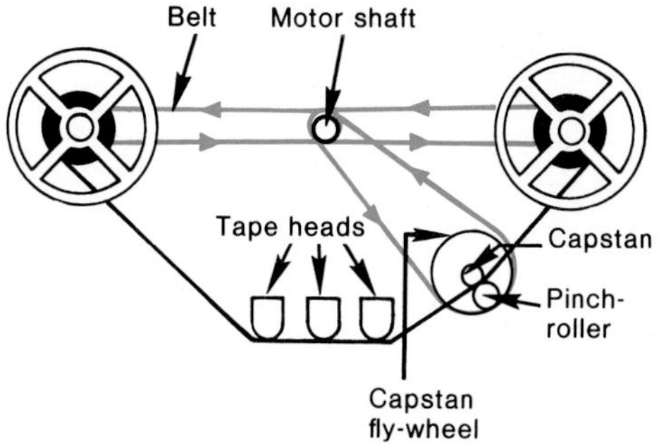

Belt
Motor shaft
Tape heads
Capstan
Pinch-roller
Capstan fly-wheel

Figure 13.15. A one-motor tape drive mechanism.

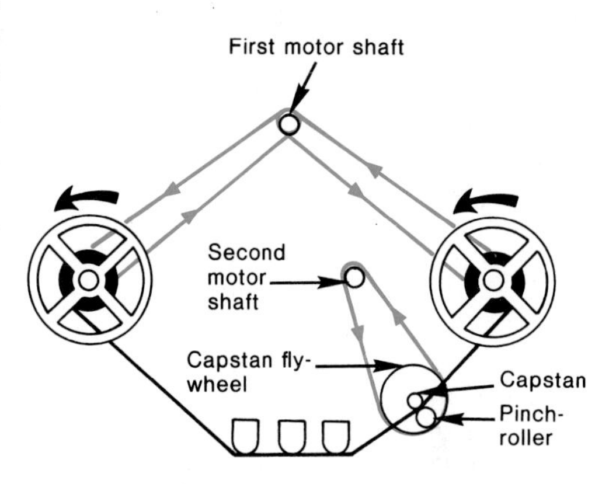

Figure 13.16. A 2-motor tape transport.

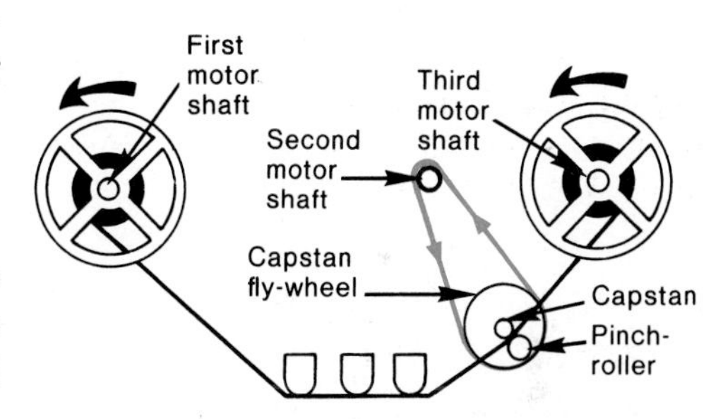

Figure 13.17. A 3-motor tape transport. The supply motor exerts a slight countertorque in order to provide some tension on the tape. The tension ensures good contact between the tape and heads.

The Controls

All tape decks possess a series of controls which allow the user to regulate the movement of the tape. The controls are common to all three types of decks (open-reel, cassette, and cartridge) with the exception of the REWIND feature, which is not found on the 8-track cartridges. These controls are generally referred to as FAST FORWARD, REWIND, PLAY, RECORD, PAUSE, and STOP, as illustrated in Figure 13.18.

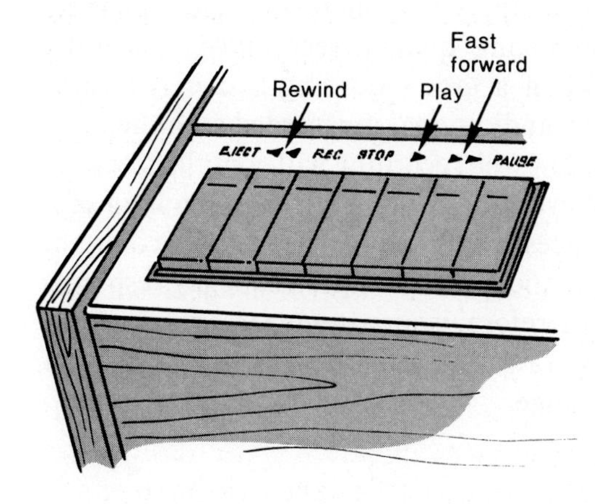

Figure 13.18. Typical controls found on a tape deck.

The FAST FORWARD and REWIND (or FAST REVERSE), found on all open-reels and cassettes, allow the rapid location of any portion of the recording. Usually most tapes can be completely rewound within 40 to 100 seconds, depending upon the machine. Many cartridge decks also have a FAST FORWARD, but it is rather slow due to the intricate nature of the endless-loop reel. The PLAY control allows the playback of a recorded tape at the normal tape speed. To record onto the tape both the PLAY and RECORD controls must be *simultaneously* activated. Presumably, this dual action helps to prevent an accidental erasure which is associated with the recording mode. Suppose that the recording process required only a *single,* rather than two, controls. The user, wishing to playback only, might accidentally depress the wrong switch and enter the recording mode which erases the old program. Having to activate two switches for the recording mode helps to eliminate this problem.

The STOP and PAUSE controls are somewhat similar, although they perform two different functions. The STOP feature both stops the tape movement, and prevents the input signal from reaching the record head. Functionally, it completely halts all aspects of the transport and allows the tape to be removed from the recorder. The PAUSE control is employed when the user wants to momentarily stop the tape, but will shortly resume recording or playback. Depressing the PAUSE control moves the pinch roller away from the capstan, but it usually allows the capstan to maintain rotation for a quick restart. Also, the signal is still sent to the record head so that the user may reset the input level before further recording.

13.5 Aids to a Good Recording: The Meters and the TAPE MONITOR Switch

Meters

All tape decks are equipped with meters which immensely aid the user in producing quality recordings. Generally speaking, the quality of a tape recording is closely tied to the level at which the audio signal is recorded onto the tape. If the level is set too low, the recording will contain excessive noise caused by the ever-present tape "hiss." Too high of a recording level produces tape saturation with a concomitant increase in distortion and a loss of high frequency response. Careful recording, therefore, is an attempt to achieve maximum signal-to-noise ratios, least distortion, and the widest possible frequency range.

The problem in recording music is that it is characterized by two kinds of sound levels. Music is not only characterized by an average sound level, but also by its peaks or "transients" which quickly rise and fall in times measured in fractions of a second. A sudden sharp trumpet blast or the crash of cymbals can easily send the instantaneous sound power soaring 50 to 100 times greater than its average value. Because of their short duration, these transients do not greatly affect the overall average loudness of the music (see Fig. 13.19). However, the transients must be taken into account during the recording process, because they can momentarily cause tape saturation which badly distorts the peaks. The user, in choosing the correct recording levels, must be aware of both the average and transient content of the music.

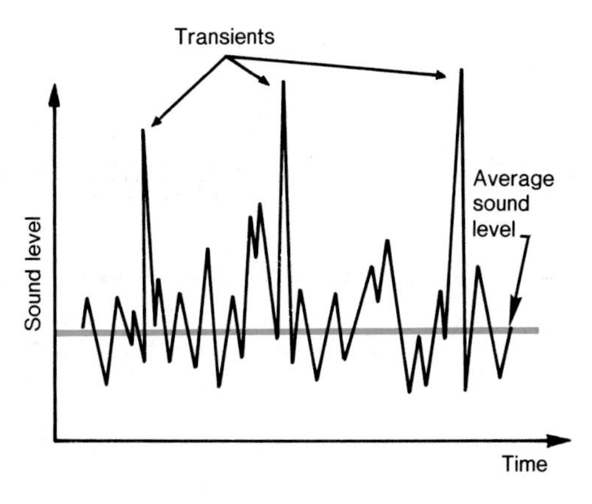

Figure 13.19. Most musical selections are characterized by an average sound level and large intermittent peaks of short duration ("transients"). Even though the peaks may be large, they do not appreciably affect the average sound level, since they last for such short periods of time.

There are two different types of meters which are designed to aid the user in establishing the proper recording levels. Virtually all tape decks contain either one or the other, and some units may have both. The *VU* (volume unit) *meter* is designed to measure the average loudness of a musical passage and respond somewhat, but not fully, to the peaks. The *dB meter,* on the other hand, is very fast-acting, and it responds quite well to the sudden transients. Before discussing how these two meters are employed during the recording process, let us first investigate the method by which they are calibrated.

Both the VU and dB meters are marked with a series of lines which indicate the precise recording level. Each meter has a maximum level, called "0 VU" and "0 dB" respectively, which should not be exceeded during the recording process or distortion may become a problem. As we shall see, these "zero" levels have different meanings. A standard method for adjusting the maximum level on a dB meter ("0 dB") is to increase the level of a 1 kHz tone fed into the deck until the total harmonic distortion of the playback signal reaches 3%. The input level at which this occurs is marked "0 dB" on the meter. The VU meter is calibrated in a slightly different manner. A signal recorded at a level of 0 VU will contain approximately 1% total harmonic distortion (not 3%) in the playback mode. To a certain extent the "zero" levels on these two meters are dependent upon the type and quality of magnetic tape used for calibration. Some tapes will accept a much higher input signal before the prescribed distortion level is reached, others much less; often the manufacturer will state which type of tape was employed. Figure 13.20 illustrates the calibration difference between these two meters. Before we discuss the reasons for this difference, notice from the figure that "0 dB" is approximately 8 dB greater than "0 VU." In other words,

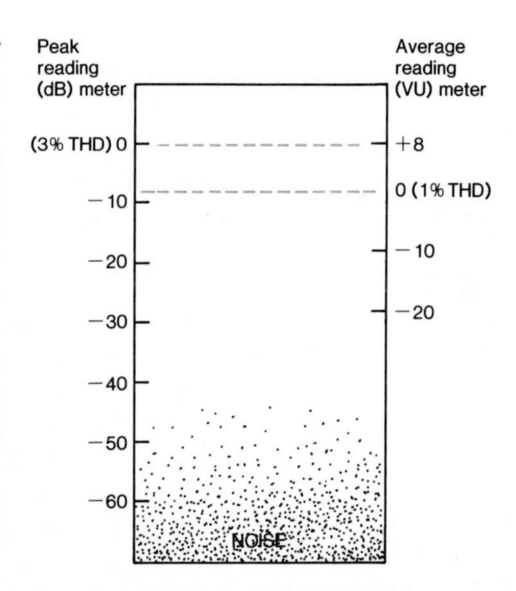

Figure 13.20. A comparison between the recording levels as indicated on a VU and a dB meter. "Zero dB" is the level where a 3% THD occurs in the playback signal, while "zero VU" approximately represents a 1% THD level.

starting with a 1% distorted playback signal (0 VU), it requires approximately 8 dB more input signal to reach the 3% distortion level (0 dB).

Now, consider Figure 13.21 which shows the salient features of both metering systems when applied to the recording process. The comparisons can best be made by using two different musical selections; tune #1 contains many strong peaks while tune #2 is void of any such transients. Even though tune #1 contains transient peaks, they are of sufficiently short duration that the ear perceives both tunes to be of the same relative loudness. As shown in Figure 13.21 (A), both of these tunes would also register the *same level* on the VU meter, because it is, by design, relatively insensitive to the transient peaks. In this sense the VU meter closely approximates the response of the human ear to average loudness levels. If the level of the two tunes is set to read "0 VU," then the music on the tape will be, on the average, 1% distorted. Music containing large transients, such as tune #1, will still have about 8 dB of "headroom" before the peaks become 3% distorted. Therefore, the VU meter allows the user to record different musical selections at the *same average loudness levels,* and yet be somewhat assured that the peaks will not be too badly distorted. The VU meter, being somewhat sluggish, does not indicate their full presence. Of course, it is always possible to record at a lower level, say −10 VU, and allow even more headroom before distortion sets in. However, the signal-to-noise ratio deteriorates by 10 dB, and the recording becomes much noisier.

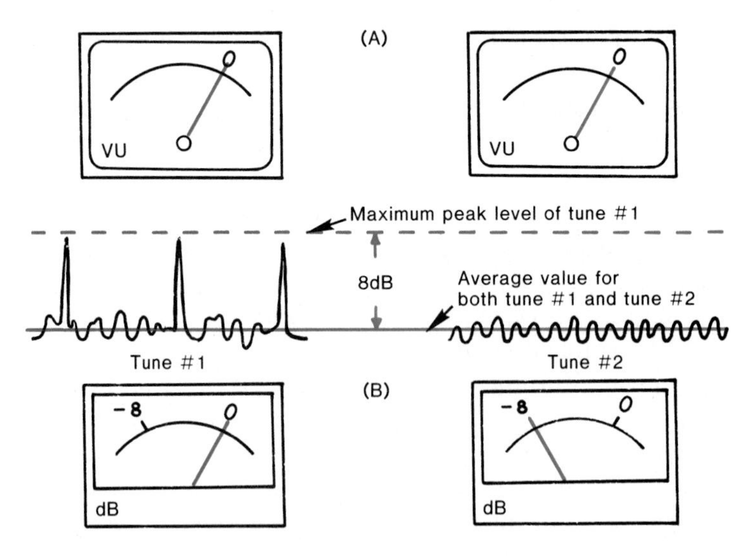

Figure 13.21. A comparison of how the VU and dB metering systems respond to average and transient recording levels. (A) VU meters register O VU for both tunes, since they both have nearly the same average loudness. (B) dB meter registers -8 dB for tune #2, but shows a higher reading for tune #1, since tune #1 contains peaks to which the dB meter responds.

Figure 13.21 (B) illustrates how the two tunes would register on the dB meter, assuming that the maximum level of the many peaks in tune #1 is 8 dB above the average level. Because tune #1 contains many large peaks, the dB meter registers a higher reading than it does for tune #2 (0 dB vs −8 dB in the diagram), even though both tunes appear equally loud to the ear. The dB meter does not give a reliable measure of average loudness, especially when the tunes contain different transient characteristics. Therefore, a user who records all musical selections at the same recording level on the dB meter should not be too surprised when they playback at different average loudness levels.

Some tape decks are equipped with both types of metering systems in order to allow the user a better perspective of the music's average and transient properties. Often the dB meter is replaced with a light emitting diode (LED) which flashes whenever a peak is present that would cause a certain level of distortion (e.g., 3%). In this manner the user is given considerable knowledge concerning the average and peak loudness levels. The recording level should be set as high as possible on the VU meter until an occasional flash on the LED appears. At this point the signal-to-noise ratio has been maximized without encountering any serious distortion of the peaks.

The TAPE MONITOR Switch

While the metering systems can be a great aid in establishing the proper recording levels, the best results are obtained when the user can listen to, or monitor, the recording shortly after it is made. In this manner careful listening becomes the final arbitrator of the recording's quality. Depending upon how quickly the user wishes to hear the recording, either a three-head or a two-head deck is required.

The three-head tape deck consists of separate erase, record, and playback heads. After the audio signal has been placed on the tape by the recording head, it can be monitored a fraction of a second later as the tape moves past the playback head. As shown in Figure 13.22, the tape deck is equipped with a *TAPE MONITOR switch.* This switch, when set to the "tape" position, allows the signals from the playback head, generated by the just-made tape, to be sent on to the rest of the hi-fi system. When the TAPE MONITOR switch is set to "source," the user is listening to the signal *before* it is sent to the record head. By moving the monitor switch back and forth between "tape" and "source," the quality of the taped signal can easily be compared with the original signal. This comparison feature is absolutely essential for consistently producing high-quality recordings, and it is available only if the deck possesses separate record and playback heads.

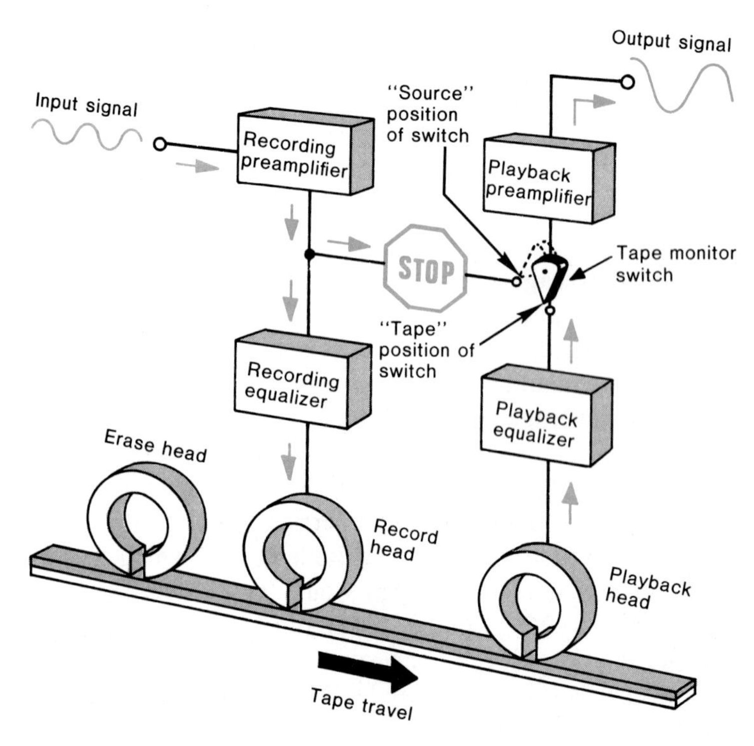

Figure 13.22. A 3-head tape deck allows the tape to be heard from the playback head a fraction of a second after it is recorded, if the monitor switch has been set to "TAPE" rather than "SOURCE." (The recording and playback equalizers shown in the picture will be discussed in section 13.6.)

The monitor feature on a 3-head deck should not be confused with the TAPE MONITOR switch found on most receivers, integrated amplifiers, or preamps. The latter, as explained in section 2.7, determines whether the listener hears the sound from either the source (such as a turntable or a tuner) *before* it is sent to the tape deck, or the sound from the *output jacks* of the tape deck itself. The monitor switch on the tape deck determines what is sent to its output jacks; it can be either the original signal (just before it goes to the record head) or the taped signal from the playback head.

Most, but not all, cassettes and cartridge decks contain only 2 heads, with the record and playback heads being combined into a single unit. At any given time, the dual-purpose head can either record or playback; it *cannot* do both simultaneously. **Therefore, the useful tape monitoring feature which is available with a 3-head system is not found on the 2-head decks.** Figure 13.23 depicts the recording process on a 2-head deck. The user is not able to hear immediately the quality of the recording because the head is being used for recording purposes, and it is *not* available for playback. In this case the output signal from the deck is just the

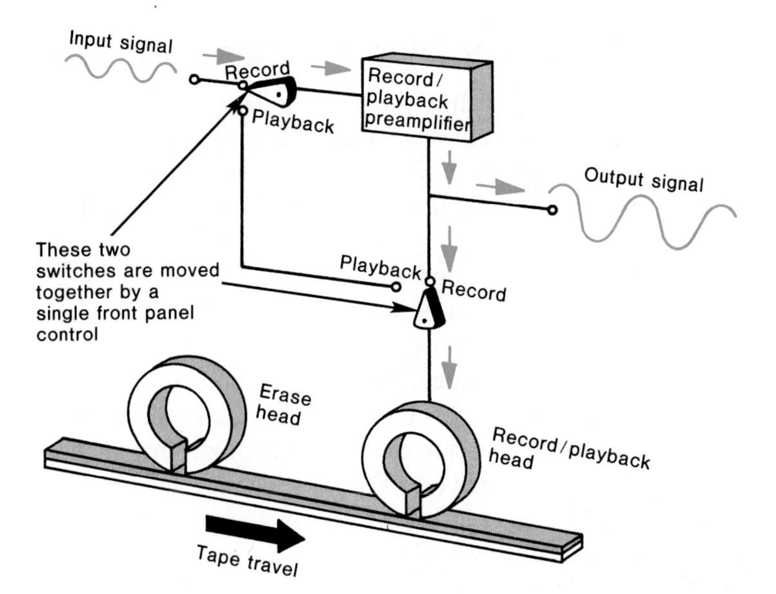

Figure 13.23. Recording with a 2-head deck. Notice, the output signal does not originate from the tape when the switches are set to "RECORD." Instead, it represents the signal just before it is put on tape. The recording and playback equalizers are not shown for the sake of simplicity.

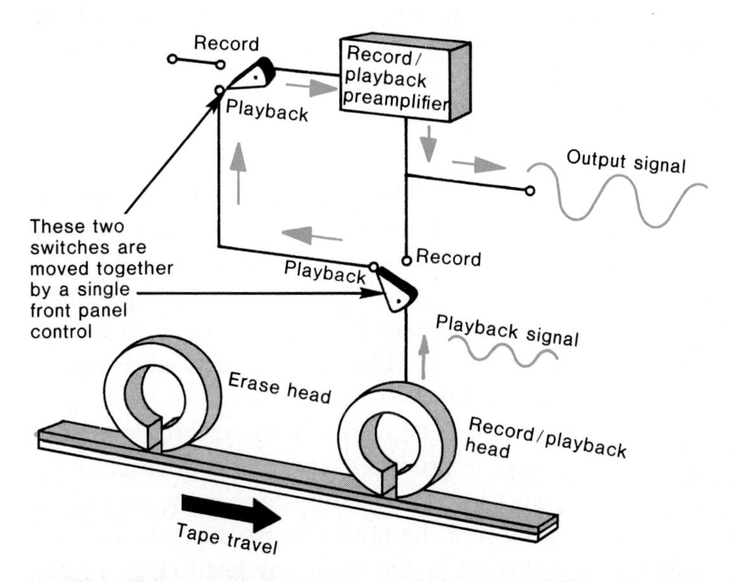

Figure 13.24. Playback from a 2-head deck. When the record preamp also doubles as the playback preamp, the cost of the deck is reduced.

audio signal *before* it is sent to the recording head. Playback of the tape for an evaluation on a 2-head deck involves either a rewind, in the case of cassettes and open-reels, or a fast forward with the cartridges. The deck is then set up for the playback process by appropriate switches as shown in Figure 13.24. In many decks the recording preamp also doubles as the playback preamp, so that a 2-head system is more economical, dollarwise, than the 3-head system.

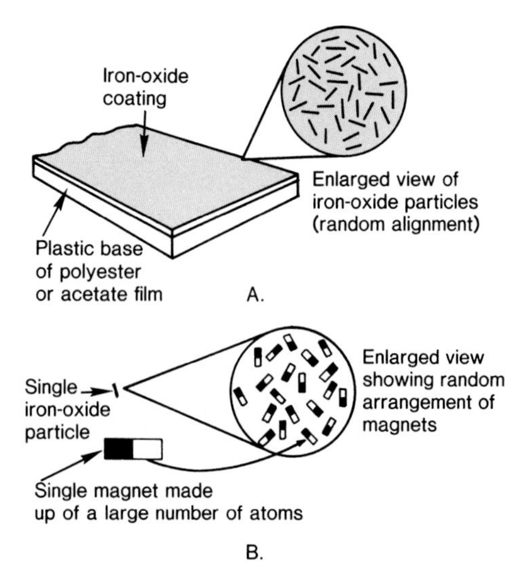

Figure 13.25. (A) An illustration of the iron oxide particles which are coated on one side of the plastic base. (B) An enlarged view showing the collection of tiny magnets within a single iron oxide particle.

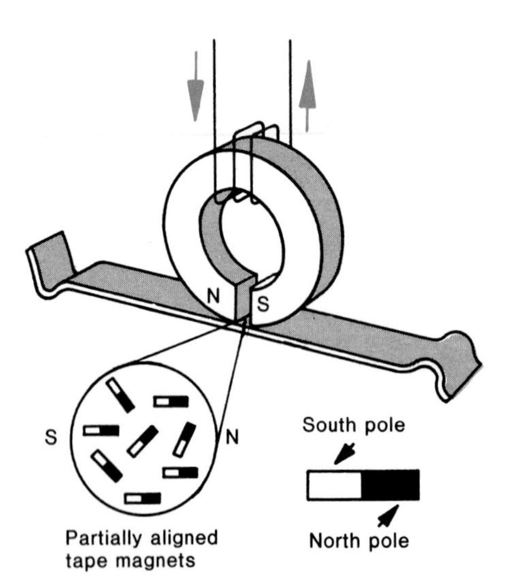

Figure 13.26. The current causes the recording head to become an electromagnet which magnetizes the tape. Notice that the north poles of the tiny magnets are oriented more or less toward the south pole of the recording head because of the attraction between unlike poles.

13.6 Theory of Tape Recording

The magnetic recording process is relatively easy to visualize, and in this section we shall present the ideas which are involved when audio information is both stored on and retrieved from magnetic tape. As mentioned in section 13.3, magnetic tape consists of a plastic base coated on one surface with a thin layer of magnetic particles. The particles are usually iron oxide or chromium dioxide, and in subsequent discussions we shall assume it to be the former. Each "particle" of iron oxide is made up of a large number of extremely small atomic magnets. When the tape is new, these infinitesimal magnets are arranged in a random pattern and the particle is unmagnetized (see Fig. 13.25). It is the job of the recording head to magnetize the tape in accordance with the strength of the audio signal.

The Recording Process

As discussed in section 11.4, the recording head is an electromagnet whose magnetic strength and polarity depend upon the size and direction of the audio AC current. If the magnetic coating on the tape is placed in contact with the gap of the recording head, some of the tiny magnets tend to line up with the magnetic fringe field of the head (Fig. 13.26). Moreover, when the tape leaves the vicinity of the recording head, some of the tiny magnets retain their alignment and the tape acquires a residual amount of magnetism. This is exactly the same process by which a permanent magnet induces magnetism into a piece of soft iron, as discussed in section 11.2. Throughout this process it should be remembered that only the tiny magnets within the iron oxide particles change their positions. The particles themselves remain bonded to the plastic base and do not move.

When more current is delivered to the recording head, a stronger fringe field is produced and proportionately more of the tiny magnets line up. In this manner the amount of magnetism induced into the tape directly depends upon the strength of the current in the recording head (Fig. 13.27). The collection of partially aligned atomic magnets can be thought of as a small magnet which has been placed on the tape. As illustrated in Figure 13.27, the magnetic strength of this "tape-magnet" depends upon the degree to which the atomic magnets are oriented. Greater alignment produces stronger north and south poles. If the current becomes strong enough all the atomic magnets will be lined up, and any further increase in the current will produce no further magnetization of the tape. (See part (D) of the drawing.) The tape is then said to be *saturated*, and severe distortion occurs because the playback signal will no longer be a faithful replica of the input electrical signal.

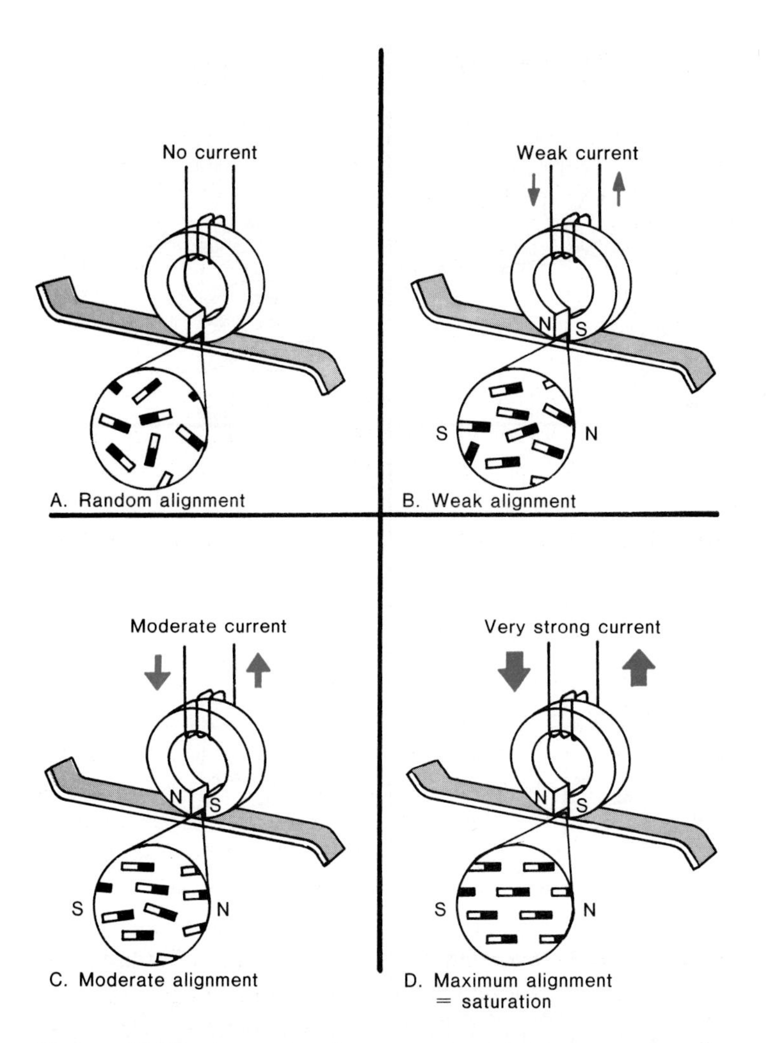

Figure 13.27. **Increasing the current produces a greater alignment of the tiny magnets in the tape. The tape is saturated when all the magnets are completely aligned, as in part D.**

It is important to remember that the electricity coming into the recording head is AC electricity. When the AC current reverses its direction, the tape becomes magnetized again—only with the reverse polarity (Fig. 13.28). Therefore, an alternating current, with its positive and negative half-cycles, produces a series of alternating magnets in the tape. Figure 13.29 shows that one second of a 100 Hz tone encoded on the tape would consist of 100 magnets pointing in one direction alternating with 100 magnets pointing in the other direction. In effect, the iron oxide coating of the tape is broken up into 200 bar magnets, laid end to end, for every second the tape moves across the gap. (Remember, it takes 2 magnets to represent one cycle.) A 15 kHz tone would produce 30,000 such magnets in the same space.

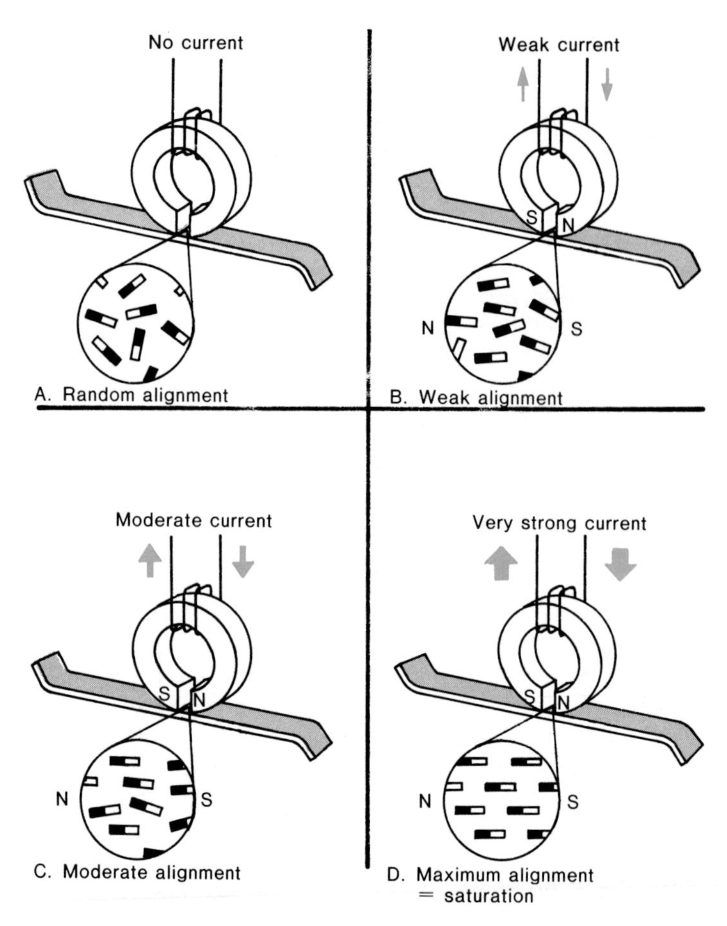

Figure 13.28. **Reversing the current to the recording head also reverses the direction of the tape magnetization. Compare this with the magnetic polarity shown in Figure 13.27.**

Also shown in Figure 13.29 is a graph of the tape magnetization as it changes along the length of the tape. The amplitude of the magnetization curve is, of course, equal to the strength of the magnets induced onto the tape. The strength of the induced magnets, in turn, is proportional to the amplitude of the AC electrical current sent to the recording head. In addition, the wavelength, and hence frequency, of the magnetization curve is determined by the frequency of the AC signal. Thus the tape, through its magnetization, preserves the complete information content (both amplitude and frequency) of the audio signal.

Figure 13.30 shows a photograph of the magnetic patterns on the tape (made "visible" by coating the tape with a special emulsion), and the pictorial arrangement of the effective bar magnets as they are laid end-to-end.

As a summary of the complete recording process, consider the taping of a live performance. First, the complex pressure

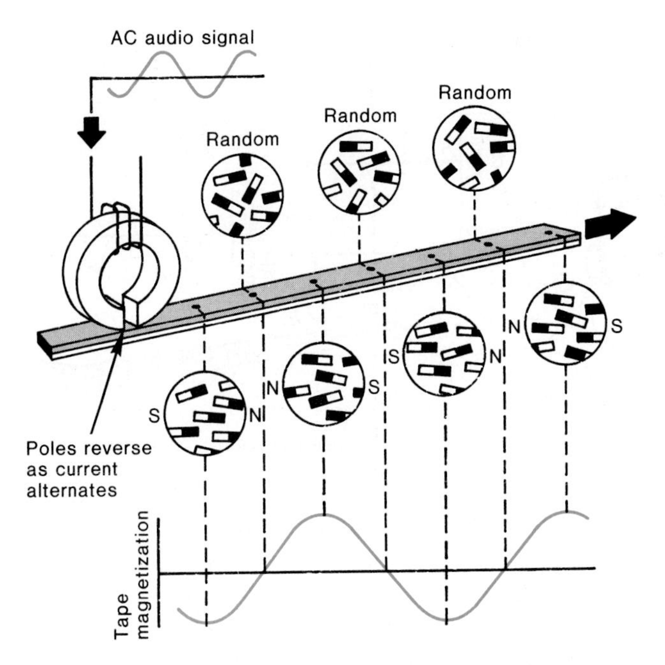

Figure 13.29. When an AC audio signal is applied to the recording head, the tape becomes a series of small bar magnets laid down end-to-end. The tape magnetization reverses whenever the magnets are reversed, in step with the reversal of the AC current direction.

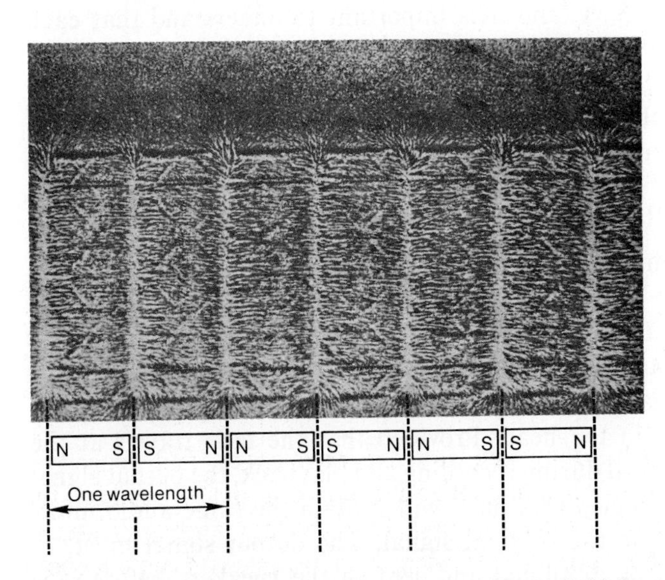

Figure 13.30. A photograph of the magnetic field lines which exist on a piece of tape. Each pattern can be thought of as being produced by a single bar magnet. Two complete magnets represent one cycle of the AC signal.

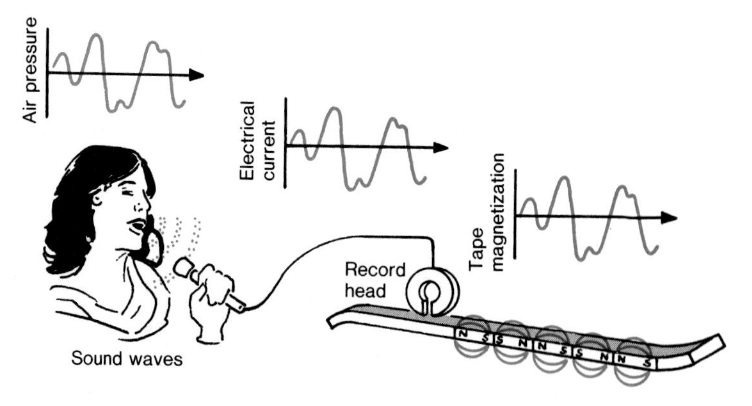

Figure 13.31. The recording process involves the conversion of sound waves (air pressure fluctuations) into AC current which, in turn, creates changes in the magnetization of the tape. Notice that the shape of the waveform must be identical in all three media (sound, electricity, and tape) in order to preserve the audio content.

waveform, which represents the sound from the vocalist, is converted into an exact-replica electrical wave by the microphone. The current is amplified and sent to the recording head which creates a magnetic replica of the current pattern on the tape. This two-step conversion process is depicted in Figure 13.31, and it is important to understand that each step in the process must preserve the exact waveform of the original sound. The versatility of sound systems arises because the waveforms can be transferred to, and stored in, different media.

The Playback Process

In many respects the recording and playback processes are the reverse of each other. During the recording process an electrical current in the coil creates alternating magnets on the tape. In playback, motion of the tape moves the magnets past the playback head, and this induces an electrical voltage in the head. Provided that the tape travels at the same speed during recording and playback, the output signal from the playback head will contain the same audio information as the original signal. The output signal must, of course, be amplified and sent to the speakers before it is finally converted into sound.

The voltage induced in the playback head is a result of Faraday's law of induction, as explained in sections 11.6 and 11.7. However, the locations along the tape where the maximum induced voltages occur are **not** at the same places where the magnetic fields are greatest. Faraday's law states that the maximum induced voltages will occur where the magnetic fields are *changing most rapidly* and *not* where

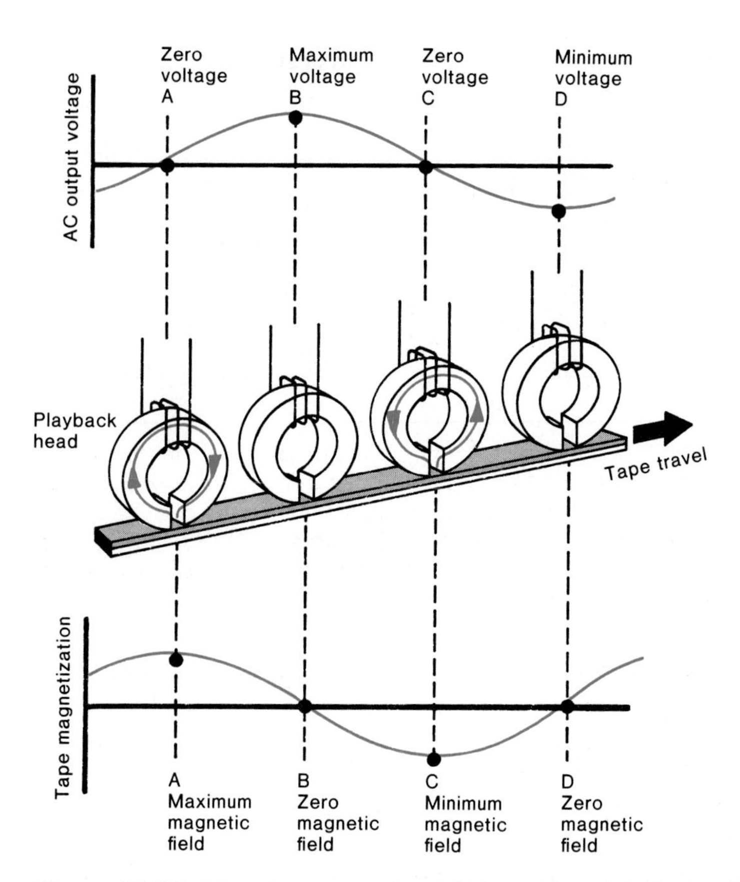

Figure 13.32. The strongest voltage induced into the playback head occurs where the magnetic fields are *changing* by the greatest amount.

the fields themselves are greatest. To see this point, consider Figure 13.32. At point "A" the magnetic field attains its maximum value. Notice, however, that the tape magnetization curve in Figure 13.32 changes very slowly near "A." Thus, the magnetic field in the vicinity of point "A," while strong, is *not changing very much* as the tape moves along, and so no voltage is induced into the head. At the point "B" the magnetic field is zero. Near "B," however, the magnetic field is *changing quite rapidly*, reversing from a north-south to a south-north orientation as the tape moves along. Because of the large change in the field's direction, the greatest voltage will be induced into the playback head at this point. In a similar fashion, Figure 13.32 shows that the induced voltage is absent at the point "C," and present with reversed polarity at the point "D."

Table 13.3 summarizes the various physical laws which are used in the complete tape process from recording to playback.

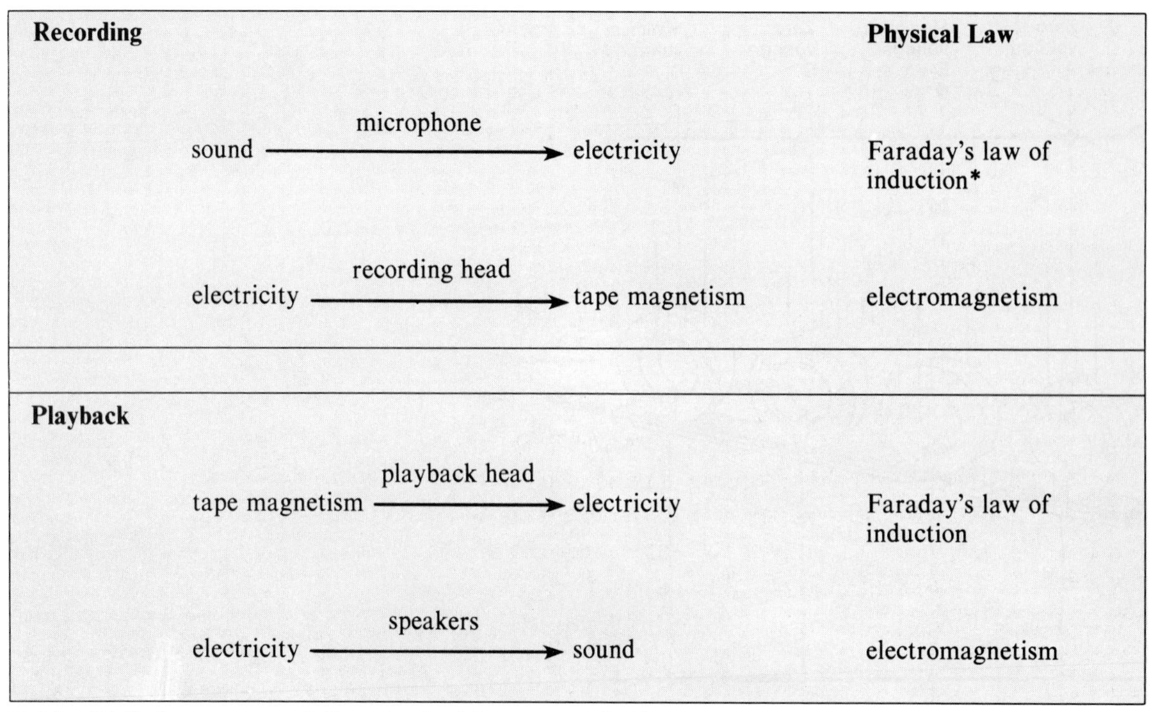

Recording			Physical Law
sound	microphone →	electricity	Faraday's law of induction*
electricity	recording head →	tape magnetism	electromagnetism
Playback			
tape magnetism	playback head →	electricity	Faraday's law of induction
electricity	speakers →	sound	electromagnetism

Table 13.3 A summary of the physical laws used in the recording and playback processes.

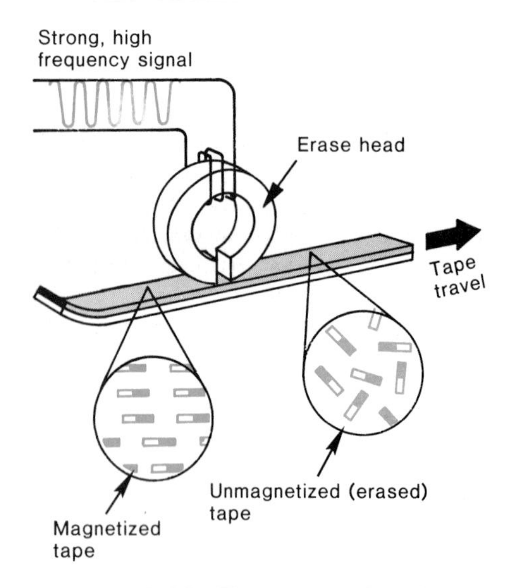

Figure 13.33. The erase head is an electromagnet in which the current is much stronger and of higher frequency than the audio signals. The strong, high-frequency magnetic field of the erase head completely randomizes the tiny magnets in the magnetic tape, thus leaving it in a "new" or blank condition.

How the Tape Is Erased

Erasing the tape consists of scrambling the tiny magnets within each iron-oxide particle back into a completely random condition. All decks have an erase head which, like the record head, is an electromagnet. However the current flowing through the erase head is much stronger and of higher frequency than the audio current. Depending upon the manufacturer, the frequency of the AC erase current is between 60 and 125 kHz. Of course, the magnetic field of the erase head also changes its direction at this rapid rate. As the tape is moved past the erase head (Fig. 13.33), the tiny magnets in each iron-oxide particle are rapidly magnetized in first one and then the other direction. The powerful alternating magnetic field completely scrambles the magnets, and as they move away from the erase head, the magnets finally settle down into the completely random condition characteristic of a new tape.

Tape Bias

The actual recording process is not quite as simple as we have just outlined. Unmagnetized coatings of iron oxide or chromium dioxide tend to remain unmagnetized when only a *small current* is applied to the record head. However, as

*The majority of, but not all, microphones utilize induction to produce an output voltage.

the applied signal becomes stronger and stronger the magnetic field overcomes this tendency and the magnetic particles readily become aligned to the magnetizing field. This situation is analogous to pushing a large heavy box along a flat surface because a gentle shove normally will not start the box moving. The box only begins to move after a rather hefty force is applied to it, although once started, only a nominal force is required to keep it going. The same is true for the tiny magnets which comprise the iron-oxide particles. They need a hefty "shove" in order to start the alignment process, but once partially oriented, any further magnetization occurs in the normal fashion.

The audible effect of this "getting started" problem is the introduction of distortion into the recording. The solution to this problem is to apply a strong, super-high frequency (generally near 100 kHz) to the record head; this signal is called the *bias signal.* The bias signal causes the recording head to generate a strong, high-frequency alternating magnetic field which oscillates the tiny magnets in the tape, thus giving them an initial "shove" to overcome their tendency to remain unaligned. When the audio signal is added to the bias signal, the tape becomes magnetized in a normal fashion.* In conclusion, the introduction of a high-frequency bias signal into the record head, in addition to the normal audio signal, minimizes distortion during the recording process.

There are different types of coatings, such as iron oxide, chromium dioxide, and metal particle, and each requires a different amount of bias current. That is why most quality tape decks have bias switches on their front panel to match the varying bias requirements of the tapes. Figure 13.34 shows a typical bias switch, which provides for three different types of magnetic tape: standard (low bias level), LN or low noise (medium bias), and CrO_2 (high bias). The proper bias setting is usually written on the tape case.

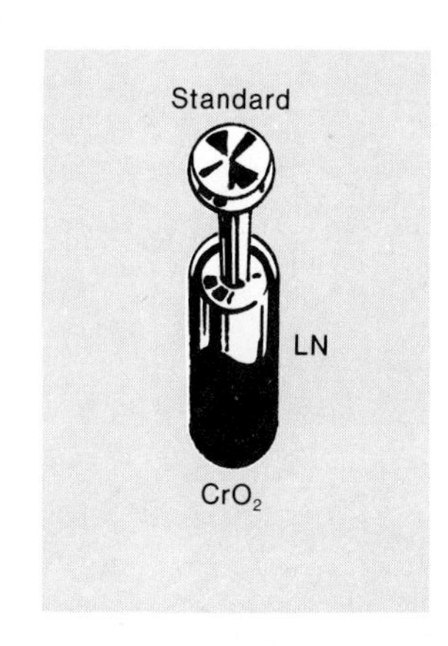

Figure 13.34. A typical bias switch which provides for the proper bias signal for three different kinds of tape. The switch may simultaneously provide for proper equalization, in which case it is referred to as the bias/equalization switch. Some decks provide separate bias and equalization switches.

Tape Equalization

Tapes are very much like records and tuners in that they all require equalization during both recording and playback;** however, the reasons for tape equalization are somewhat different. Equalization arises because of certain problems involved in the recording and playback process. Suffice it to say that when audio signals are recorded with a perfectly flat frequency response, they will *not* be played back with an equally flat response. For a variety of reasons (which are beyond the scope of this book to discuss) the very low and

*In practice, the bias and audio signals are added together, and they are both sent to the record head at the same time.

**Tuner equalization (called pre-emphasis and de-emphasis) is discussed in section 10.6 while the equalization of records is presented in section 12.9.

very high frequencies fall off substantially during playback. Only the midrange frequencies are reproduced at relatively high output levels. To compensate for this,. the low- and high-frequency parts of the signal must be raised such that a flat frequency response is achieved for the entire audio range during playback.

The NAB (National Association of Broadcasters) equalization is an internationally agreed-upon method by which a flat frequency response is restored to the recording. The high-frequency portion is boosted during the recording process, and the low-frequency part is boosted during playback.* In this manner, tape decks are able to achieve a reasonably good frequency response over most of the audio range. Look for an equalization switch on the front panel of a tape deck. It may be in the form of a separate switch, or it may be a combined bias/equalization switch.

13.7 The Dolby** Noise Reduction System

One of the more serious problems encountered in tape recording is that of the ever-present tape "hiss." This form of annoying noise originates in the tape itself, and it increases in severity with the slower tape speeds. Cassettes, because of their slow speed, were particularly susceptible to this form of noise, and for some time it precluded them from reaching their hi-fidelity potential. The Dolby noise reduction system provides a marked improvement in sound quality by significantly reducing the tape hiss, and it is an almost universal feature on quality cassette decks. It can be also used with open-reels and cartridges although their higher speeds make them less dependent on the Dolby process. The Dolby principle takes advantage of a psychoacoustic phenomenon called "masking." Masking is familiar to most of us and it is, simply stated, that our ears tend to conceal soft sounds in the presence of louder sounds. In other words, the listener becomes unaware of the presence of noise when it is accompanied by a much louder musical passage. The noise only becomes objectionable when it is comparable in level to the music.

Figure 13.35 (A) illustrates a large signal, arising from a loud musical passage, being sent to the recording head. During playback, both the large signal and the tape hiss are picked up by the playback head. As long as the musical signal is substantially larger than the tape hiss, the signal will partially mask the presence of the tape hiss and the

*Variations in the NAB equalization method are used for different tape speeds and for the different types of tape decks.

**"Dolby" is a trademark of Dolby Laboratories, Inc.

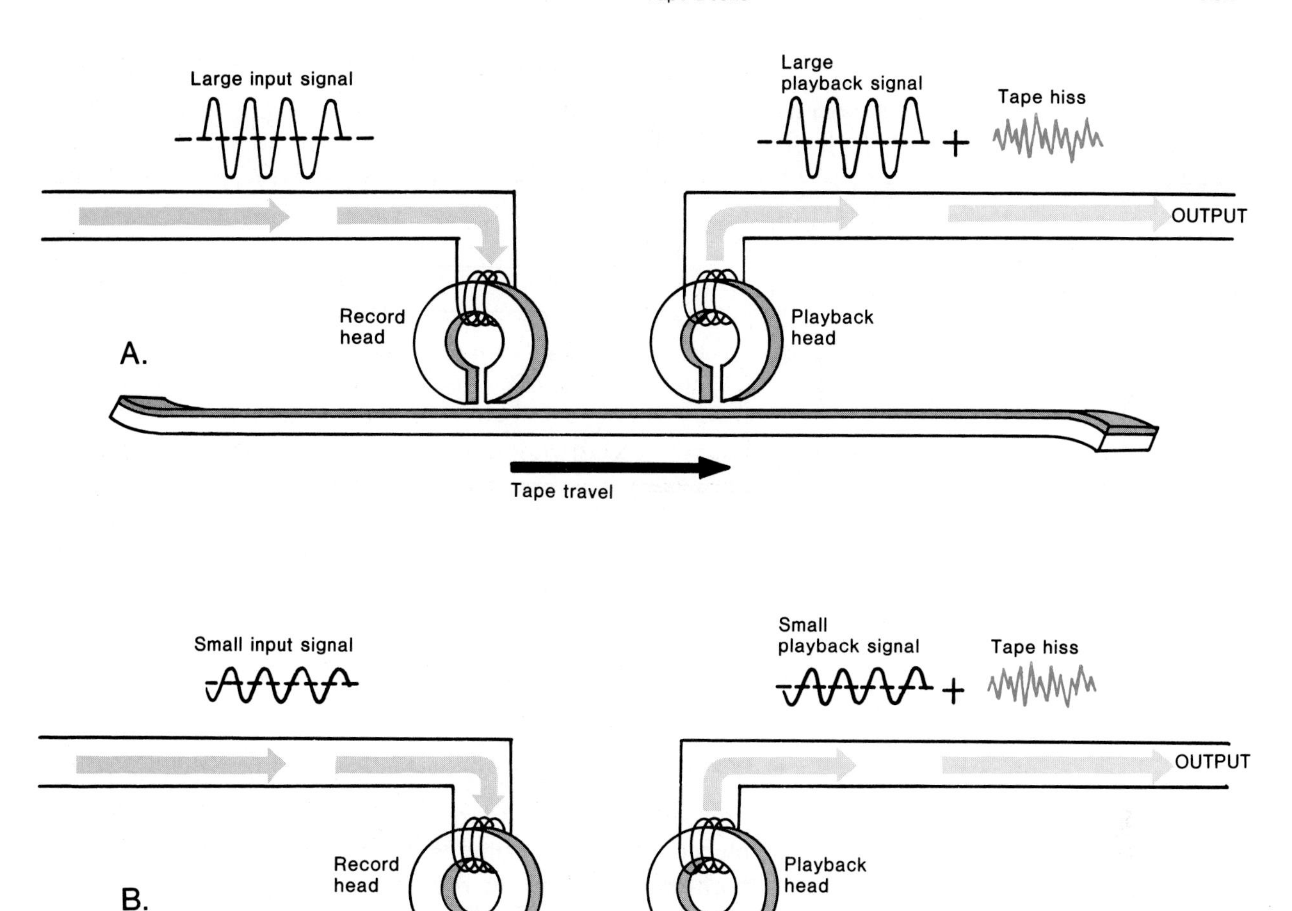

Figure 13.35. (A) The large playback signal will mask the presence of the tape hiss. (B) The small playback signal is comparable in strength to that of the tape hiss.

listener may not be too annoyed by its presence. In part (B) of the figure the situation is somewhat different because a small signal, from a quiet musical passage, is being recorded on the tape. Upon playback, the signal level is comparable to the tape hiss level, and the presence of this noise will be rather obvious to the listener. The Dolby process is a method for reducing the hiss during these soft passages.

The Dolby system is a two-step, symmetrical process that operates on the signal both *before* and *after* recording. In many respects it is analogous to equalization. During the recording process the Dolby circuits "listen" to the *level* of the music. If the music is *loud*, the system allows the sounds to pass *unchanged* on to the record head. Since loud sounds automatically mask the noise, there is no need for Dolby processing. Whenever a *soft passage* comes along, the Dolby circuitry senses its level and automatically *boosts the signal*

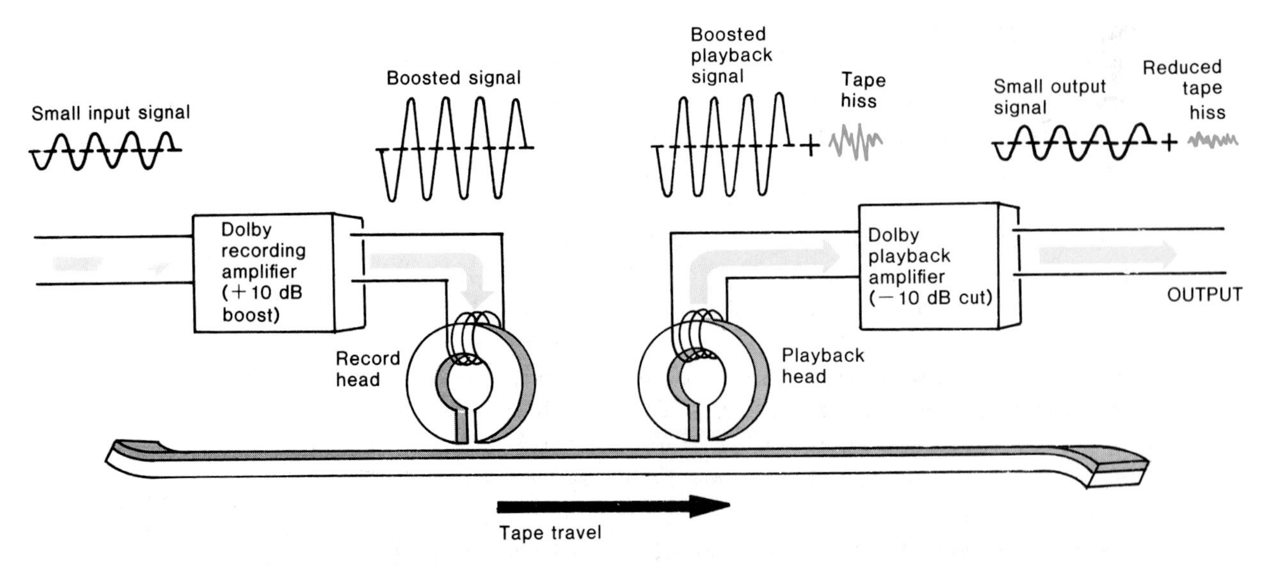

Figure 13.36. The Dolby process reduces tape hiss by approximately 10 dB when the musical signal is small. The signal-to-noise ratio is thus improved by +10 dB.

by a predetermined amount; the softer the signal the more it is raised, with +10 dB being the maximum boost.* Therefore, as shown in Figure 10.36, very low-level signals are placed on the tape some +10 dB above their original level; high-level signals are *not* affected. Once the signals are placed on the tape, tape hiss enters the picture because it is a form of low-level noise which resides on the tape.

During the playback of the Dolbyized tape, the reverse process takes place. The high levels are not affected. However, Figure 13.36 illustrates that the very low level signals, which were boosted by a +10 dB before being put on the tape, are cut by −10 dB on playback in order to restore them to their proper level. In fact, the Dolby playback circuitry cuts *both* the signal and the tape hiss, thus reducing the amount of tape hiss produced by the tape. The significant point to remember is that the tape hiss was *not* originally boosted by +10 dB, since it originated from the tape itself. Therefore, the Dolby system reduces the tape hiss by approximately −10 dB in relation to the low level musical signal, and the result is a significantly quieter recording.** To visualize this last point, compare the output signal and the tape hiss levels in Figures 13.35 (B) and 13.36. In the former the output signal and tape hiss levels were approximately equal, because there was no Dolby processing. In Figure 13.36 the Dolby system reduces considerably the tape hiss relative to the output signal level.

*The boosting only occurs for frequencies greater than 1 kHz where most of the tape hiss frequencies occur.

**Recall from Chapter 5 that a −10 dB cut means that the noise will sound one-half as loud.

The Dolby process, in effect, leaves the overall content of the musical program unaltered (i.e., a flat frequency response), while reducing the tape hiss by -10 dB during the very quiet passages. Notice that the Dolby process does *not* eliminate or even reduce any noise present in the source being taped (like old records). That is because this type of noise is boosted along with any low-level signal during recording, and no relative reduction occurs when the two are cut during playback. Dolby only reduces noise which originates from the tape.

13.8 Tape Deck Specifications

The performance of tape decks, like most other audio components, can be characterized by a number of technical specifications. Familiar concepts such as frequency response, signal-to-noise ratio, wow and flutter, and stereo separation are also used to describe the merits of tape equipment. Tape deck specifications are somewhat more difficult to establish than those for other audio components, because factors such as tape speed, tape type (iron oxide, chromium dioxide, etc.), record and bias levels all have a significant effect on the deck's performance. Since the tape itself is an integral part of the recording process, tape deck manufacturers should include the name and type of tape used for the performance tests; in practice, they rarely do. In short, a deck with good specifications may not perform as well when used with other tapes. In addition, tape deck specifications are not "solid" or "firm" measurements like those of an amplifier. Tape deck design and operation require certain compromises for obtaining the best overall performance. For example, a manufacturer may be able to quote an outstanding frequency response figure by compromising the deck's signal-to-noise ratio. Such a tradeoff will indeed extend the high-end response but at the price of increasing the noise level in the recording.

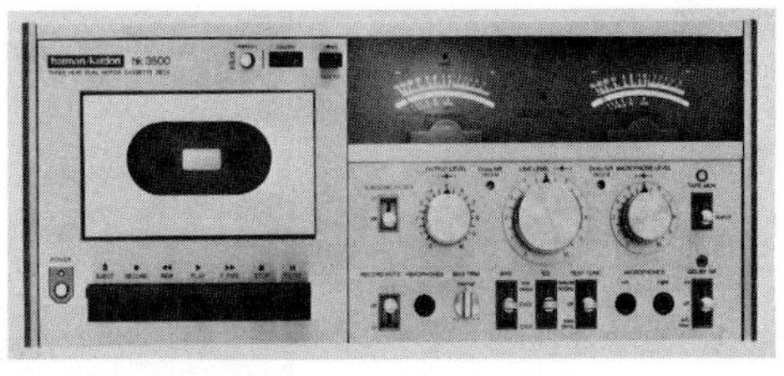

Harman/Kardon model hk 3500 cassette deck

We shall now discuss some of the more important tape deck specifications and indicate, where appropriate, the various performance alternatives which are available. As always, the goal of our recommendations is to help the consumer purchase quality hi-fi equipment in the intermediate price range.

Frequency Response

The frequency response specification of a tape deck conveys the same general information that we discussed in previous chapters for speakers, amplifiers, tuners, and phono cartridges. The frequency response specification indicates how evenly the tape deck handles one audio frequency, as compared to other frequencies, in the overall record/playback process.

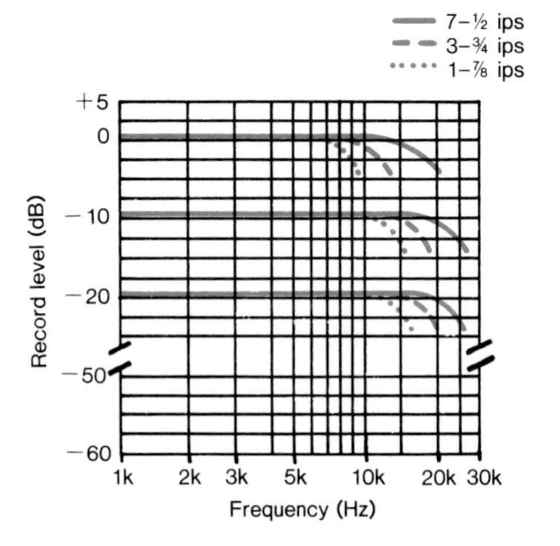

Figure 13.37. The frequency response curves above 1 kHz for three tape speeds (7-1/2, 3-3/4, 1-7/8 ips) measured at different reccording levels.

The frequency response of a tape deck is highly dependent upon both the tape speed and the recording level. The top drawing in Figure 13.37 portrays the frequency responses for the three tape speeds of 7-1/2, 3-3/4, and 1-7/8 ips; in each case, the recording level is set at 0 dB. It is obvious from the curves that the fastest tape speed yields the best frequency response. With a few exceptions, the user has no choice of tape speeds with cassette and cartridge decks, and the user must use 1-7/8 and 3-3/4 ips, respectively. With a variable speed open-reel the fastest speed can be selected in order to achieve the widest possible frequency response.

When the recording level is lowered to either −10 dB or −20 dB in Figure 13.37, the frequency responses for all three speeds improves markedly! Notice that the frequency response of the cassette, recorded at −20 dB, begins to approach that of the 7-1/2 ips open-reel recorded at 0 dB. Thus it is possible, under different recording conditions, for the response of a cassette to approach that of an open-reel; however, the signal-to-noise ratio of the cassette recording will be 20 dB poorer, since the music was recorded with 20 dB less signal. In general, then, most tape recorders cannot achieve a wide frequency response at the maximum recording levels, especially for the slower tape speeds.

Now let us see why it is that high signal levels and slow tape speeds inhibit the frequency response of a tape deck. High signal levels are a problem because of the saturation of the magnetic oxide which occurs when the high frequencies are boosted by the recording equalization. If the level of the input signal to the tape is high to begin with, the additional high frequency boost given to it by the recording equalization can actually cause the resulting signal to saturate the magnetic tape. When saturation occurs, the tape becomes fully magnetized and no further magnetization is possible. Therefore, even though the electrical signal being sent to the recording head is quite large, saturation limits the amount of tape magnetization, with the result that the level of magnetization is less than it should have been. Upon playback, the less-than-natural amount of magnetization causes a reduced output signal to be produced. This reduction appears as a fall-off in the high frequency response of the tape deck.

Slow tape speeds also reduce a deck's frequency response for the reason indicated in Figure 13.38. This figure shows the obvious fact that a greater length of tape passes by the recording head in one second at faster speeds than at slower speeds. When the audio frequency is high, many north/south regions of magnetization must fit onto this length of tape. For example, when the frequency is 15,000 Hz, 30,000 north/south regions must fit onto one second's worth of tape, as discussed in section 13.6. The shorter that one second's

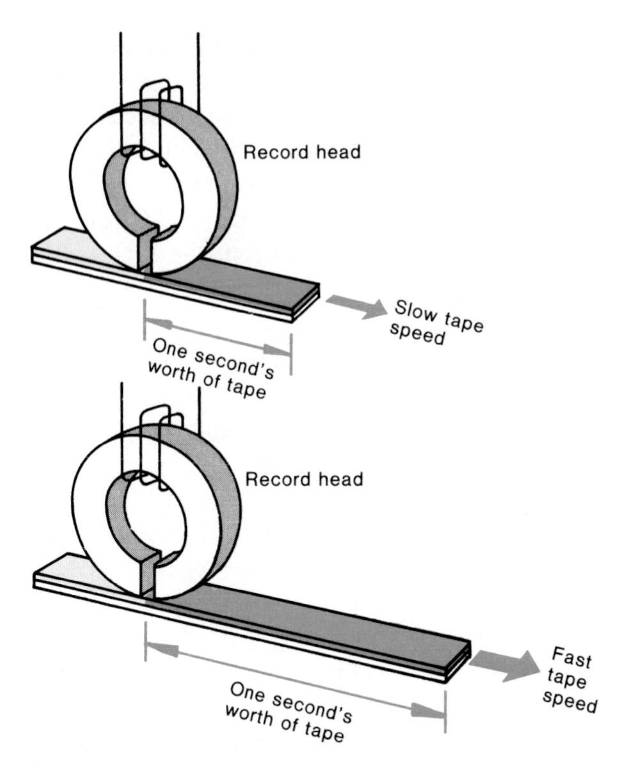

Figure 13.38. At faster tape speeds, more tape passes by the record head than at slower speeds. This provides more room into which the high frequencies can fit at the faster speeds, thus improving the high frequency response of the deck.

worth of tape is, the more closely packed are the 30,000 north/south regions. With very close packing, it becomes difficult for the tape deck to record and playback the 30,000 regions every second, and a relatively poor high frequency response results. Therefore, slow tape speeds always result in a poorer frequency response when compared to the response for faster speeds.

It is customary to measure the frequency response of open-reels around −10 VU and cassettes at −20 VU in order to avoid high frequency fall-off problems. However, there is a price to be paid! A satisfactory frequency response measured at −20 VU means that the signal-to-noise ratio has been reduced by a full 20 dB. Since most manufacturers do not state the recording level when quoting their frequency response measurements, it is probably fair to say that the record levels are less than 0 VU, probably between −10 and −20 VU, because they wish to make their products look as good as possible. Also, the frequency response depends upon the type of tape being used. For example, chromium dioxide tapes played on the same machine usually give a much better high frequency response when compared to the low-noise/high-output tapes. For best results the recording bias should be matched to the tape characteristic, so it is necessary to have a machine with a switch to select the appropriate bias.

The bias and type of tape are important even with open-reel decks, but are absolutely vital with cassette decks. Most manufacturers set the optimum bias on their decks using specific tapes; sometimes these tapes are listed in the instruction manual, but often they are simply left out.

Remember, the frequency response specification is dependent upon the type of tape being used, the tape speed, and the recording level.

Tape Deck Buying Guide

Frequency Response: Look for the following frequency responses for tape decks. The wider the frequency range and the smaller the dB figures, the better.

Deck	Speed	Recording level	Tape	Frequency Response
Open-reel	7-1/2 ips	0 VU	Low-noise/ high-output	50 – 13,000 Hz, ± 3 dB
Open-reel	7-1/2 ips	−10 VU	Low-noise/ high-output	30 – 20,000 Hz, ± 3 dB
Open-reel	3-3/4 ips	0 VU	Low-noise/ high-output	50 – 10,000 Hz, ± 3 dB
Open-reel	3-3/4 ips	−10 VU	Low-noise/ high-output	40 – 14,000 Hz, ± 3 dB
Cassette	1-7/8 ips	0 VU	Chromium dioxide	50 – 7,000 Hz, ± 3 dB
Cassette	1-7/8 ips	−20 VU	Chromium dioxide	30 – 15,000 Hz, ± 3 dB
Cassette	1-7/8 ips	−20 VU	Low-noise/ high-output	30 – 11,000 Hz, ± 3 dB

Signal-to-Noise Ratio (S/N)

In general, the S/N ratio of a tape deck expresses how well it can produce tapes which are free from self-created noise, such as tape hiss. In this sense, the S/N specification conveys the same kind of information for tape decks that it did for amplifiers and tuners. The definition of the S/N ratio is as follows.

Signal-to-Noise Ratio: The ratio of the signal power to the noise power coming from a tape deck, when a maximum signal has been recorded (at approximately 1 kHz). The ratio is expressed in decibels.

Measuring the *signal* is a fairly straightforward procedure. A manufacturer selects one or more types of tape (iron oxide, chromium dioxide, etc.) to be used in the tape deck. For each type, a midrange tone (typically 1 kHz) is recorded onto the tape at the highest possible level before the tape saturates and begins to produce excessive distortion on playback. "Excessive" total harmonic distortion usually means 3% ("0" reading on a dB recording meter), although a few manufacturers choose values as low as 1% or as high as 5%. The "signal" used for the S/N specification is the level of the playback signal which contains the predetermined amount of distortion.

Measuring the noise is somewhat more involved. Leaving all of the deck's controls unchanged from the signal measurement, a recording is made again, but this time with *no* input signal. The "tape hiss" that is heard on playback with this zero-input recording represents the *noise* part of the S/N ratio. The tape hiss usually contains an enormous range of frequencies which extend throughout, and even beyond, the range of human hearing.

There are two methods of measuring noise from a tape deck, unweighted and weighted, and both of these methods are similar to the unweighted and weighted amplifier noise measurements discussed in section 8.7. In an *unweighted* measurement each frequency which contributes to the noise is treated equally in the overall noise measurement. In a *weighted* noise measurement, an electronic weighting circuit is inserted between the tape deck and the meter which measures the noise power. The weighting circuit de-emphasizes the contributions from both the very low and the very high frequency noise components. Therefore, the meter basically measures only the noise at midrange frequencies. Such a "weighted" noise measurement is often used by testing laboratories because they believe that it more closely approximates how our hearing perceives the noise. According to the discussion in section 5.6 concerning the Fletcher-Munson curves, our ears do not hear all sound frequencies with equal loudness. We are most sensitive to the midrange frequencies, and, therefore, the weighted noise measurement gives a greater emphasis to the midrange frequencies present in the noise.

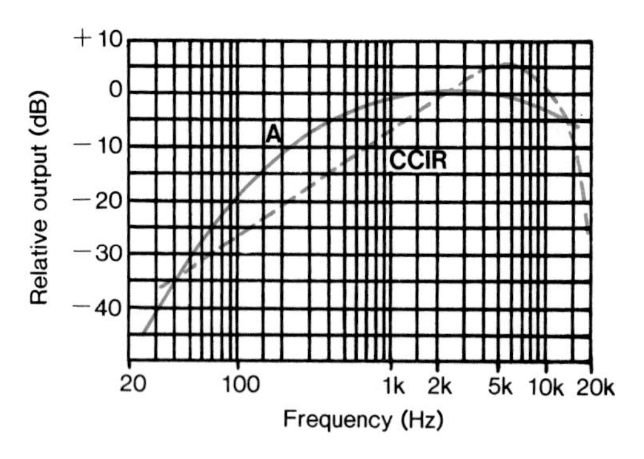

Figure 13.39. Two weighting curves that are often used when measuring the noise from a tape deck.

Figure 13.39 illustrates the frequency response curves for two types of weighting circuits that are widely used today: "A" weighting and "CCIR" weighting. It can be seen from the figure that the CCIR curve gives a higher proportionate "weight" to the noise components in the 2,500 Hz to 12,000 Hz range, and lower significance to the noise in the 45 Hz to 2,500 Hz range than does the A-weighting curve. There is considerable debate as to which weighting curve is better, although A-weighting is so widely used that if you should see a S/N ratio expressed as "65 dB (wtd.)," it is safe to say that the "wtd." involved is A-weighting.

It should be mentioned that weighted S/N ratios are always larger than the corresponding unweighted ratios. This occurs because the weighted noise is less than the unweighted noise, since the filtering action of the weighting circuits removes much of the low and high frequency noise components. Since the noise occurs in the denominator of the S/N ratio, a smaller weighted noise measurement means a *larger* weighted S/N ratio.

Whether or not the user can actually benefit from the full S/N capabilities of the deck depends on the transient characteristics of the program being recorded. Music with few peaks can be recorded at a level fairly close to 0 dB (using a dB meter) and not encounter any serious distortion. In this case the deck's full S/N potential can be reached. Large transients force the user to "back-off" the recording level in order to avoid distortion of the peaks. This reduction in the recording level brings the signal down closer to the noise, and because of this the average S/N of the recording is now less than that specified for the deck. In this respect, the Dolby noise reduction system can help by adding another 5 to 10 dB to the S/N ratio.

Tape Deck Buying Guide

Signal-to-Noise Ratio (S/N): Look for the following S/N ratios for both open-reels and cassettes, using "0" on a dB meter as the reference level (subtract about 8 dB if "0 VU" is used as the reference level):

1. 55 dB or higher, either A- or CCIR-weighted (Dolby off).
2. 60 dB or higher, either A- or CCIR-weighted (Dolby on).

Wow and Flutter

The wow and flutter specification for a tape deck is exactly the same as that used for turntables (see section 12.4). A 0.1% wow and flutter will cause a 3,000 Hz tone to vary in pitch between 2,997 and 3,003 Hz. Higher and lower frequencies will be affected proportionately. Anything below 0.15% is considered acceptable with lower values generally being found with faster tape speeds. There are also weighted wow and flutter measurements which are always less, usually by about 25%–50%, than the unweighted values.

Tape Deck Buying Guide

Wow and Flutter: Look for the following wow and flutter values.

1. 0.15% or less (unweighted).
2. 0.10% or less (weighted).

Stereo Separation and Crosstalk

Stereo separation and crosstalk are different manifestations of the same effect—namely, leakage of the signal from one track of the tape to another. There is a difference between the two terms because adjacent tracks may not represent the two stereo channels for the same program. For example, tracks 1 and 2 in the open-reels are, respectively, the left channel of side 1 and the right channel of side 2, whereas tracks 1 and 2 in cassettes are the left and right channels of side 1 (See section 13.3.) The definitions of these terms are as follows.

Stereo Separation: The ratio of the power of the audio signal, in the tape channel for which it is intended (e.g., left), to the power of this signal which has leaked over to the other channel (e.g., right). This ratio is expressed in decibels.

Crosstalk: The ratio of the power of the audio signal in one track on a tape to the power of this signal which has leaked over to an adjacent track. This ratio is expressed in decibels.

Both stereo separation and crosstalk depend upon the head design, the tape deck's electronics, and the track separation. And both are usually measured at a frequency near 1 kHz, although this frequency is usually not specified. Stereo separation involves recording onto the right stereo channel and measuring the signal which has leaked over onto the left channel, and vice versa. Measured in dB, most cassettes can maintain a stereo separation of 40 dB, while the open-reels achieve 50 dB and higher. The problem of stereo separation is very minimal because any tape recorder is likely to have much better separation than the program being recorded.

Crosstalk is much more serious because it involves the leakage between two adjacent tracks which contain different programs. When the tape is played in the forward direction, the other track is played backwards, and the crosstalk will be in the form of garbled sounds with no relationship to the desired program. Therefore, you should be more demanding when judging the crosstalk specification of a tape deck.

Tape Deck Buying Guide

Stereo Separation and **Crosstalk:** Look for the following specifications for stereo separation and crosstalk.

1. **Stereo Separation**
   Open-reels: 50 dB or higher.
   Cassettes: 40 dB or higher.
2. **Crosstalk**
   Open-reels: 60 dB or higher.
   Cassettes: 60 dB or higher.

A Summary of Tape Deck Specifications and Buying Guides

Frequency Response

Look for the following frequency responses. The wider the frequency range and the smaller the dB figures, the better.

Deck	Speed	Recording level	Tape	Frequency Response
Open-reel	7-1/2 ips	0 VU	Low-noise/ high-output	50 − 13,000 Hz, ± 3 dB
Open-reel	7-1/2 ips	−10 VU	Low-noise/ high-output	30 − 20,000 Hz, ± 3 dB
Open-reel	3-3/4 ips	0 VU	Low-noise/ high-output	50 − 10,000 Hz, ± 3 dB
Open-reel	3-3/4 ips	−10 VU	Low-noise/ high-output	40 − 14,000 Hz, ± 3 dB
Cassette	1-7/8 ips	0 VU	Chromium dioxide	50 − 7,000 Hz, ± 3 dB
Cassette	1-7/8 ips	−20 VU	Chromium dioxide	30 − 15,000 Hz, ± 3 dB
Cassette	1-7/8 ips	−20 VU	Low-noise/ high-output	30 − 11,000 Hz, ± 3 dB

Signal-to-Noise Ratio (S/N)

Look for the following S/N ratios for both open-reels and cassettes, using "0" on a dB meter as the reference level (subtract about 8 dB if "0 VU" is used as the reference level):

1. 55 dB or higher, either A- or CCIR-weighted (Dolby off).
2. 60 dB or higher, either A- or CCIR-weighted (Dolby on).

Wow and Flutter

Look for the following wow and flutter values:

1. 0.15% or less (unweighted).
2. 0.10% or less (weighted).

Stereo Separation

Look for the following stereo separation values:

1. Open-reels: 50 dB or higher.
2. Cassettes: 40 dB or higher.

Crosstalk

Look for the following crosstalk values:

1. Open-reels: 60 dB or higher.
2. Cassettes: 60 dB or higher.

Summary of Terms

Belt Drive—A method for connecting the rotating motor shaft to the capstan by means of a flexible belt.

Bias Signal—A high frequency (above audio) alternating current which is fed to the recording head, along with the audio signal, in order to minimize the distortion of the magnetized signal.

Capstan—A small rotating shaft which pulls the tape across the heads at a constant speed with the aid of the pinch roller.

Cartridge Deck (8-track Tape Deck)—A tape deck which uses cartridge tapes. Cartridges are larger than cassettes and contain a single, endless-loop reel for tape storage. Cartridge tape is 0.64 cm (1/4″) wide, travels at 3-3/4 ips, and contains eight tracks.

Cassette Deck—A tape deck which uses cassettes. A cassette is a small plastic case which contains two reels (supply and take-up) of 0.40 cm (5/32″) wide tape, and openings for the tape heads and the tape drive mechanism. The tape speed is 1-7/8 ips, and the tape contains four tracks.

Chromium Dioxide—A newer type of tape coating which is capable of accepting higher recording levels and offering wider frequency ranges than iron oxide tapes. A special bias and equalization switch is required in order to use these tapes.

Crosstalk—The ratio of the power of the audio signal in one track on a tape to the power of this signal which has leaked over to an adjacent track. Like stereo separation, it is expressed in decibels, but it is not the same thing as stereo separation.

dB Level Meter (Peak Reading Meter)—A meter which is designed to be fast enough to measure the maximum level of the peaks or transients in a signal. It is used as a record or playback level meter. The "0 dB" level corresponds to the level of an audio signal which contains 3% total harmonic distortion during playback.

Direct Drive—A drive system in which the capstan is the rotating shaft of a motor.

Dolby Noise Reduction—A noise reduction technique which reduces tape hiss by approximately

10 dB at low signal levels, when the noise becomes most obvious.

Eight-Track Tape Deck—(See **Cartridge Deck.**)

Equalization—The process of electronically boosting or cutting audio signals in certain frequency ranges relative to those in other frequency ranges.

Erase Head—An electromagnet which uses a strong, high frequency AC signal to completely scramble, or erase, the tiny magnets in the tape.

FAST FORWARD—A control on a tape deck which is used to move the tape in the forward direction more rapidly than its playing speed would allow, in order to facilitate rapid location of any portion of a recording.

FAST REVERSE (REWIND)—A control on a tape deck which is used to move the tape rapidly in the reverse direction. It is absolutely essential for the convenient editing of tape recordings.

Frequency Response—A specification which indicates how uniformly a tape deck reproduces one audio frequency compared to another. In order to have full meaning this specification must include both the frequency range and the ± decibel deviation from perfect flatness, e.g., 30 Hz to 15,000 Hz, ± 3 dB.

Gap—The space between opposite ends of the horseshoe-like iron core of a head, usually measured in microns.

Head—(See **Erase Head, Playback Head, and Record Head.**)

Hysteresis Synchronous Motor—(See **Synchronous Motor.**)

Idler Wheel Drive—A method for connecting the rotating motor shaft to the capstan by means of an idler wheel.

Induction Motor—A type of motor which is inexpensive and has good short-term speed characteristics. However, it is susceptible to long-term speed changes caused by line voltage changes.

Iron Oxide—A type of magnetic coating applied to one surface of the plastic backing during the manufacturing of magnetic tape.

Light Emitting Diode (LED)—A solid state device which emits light when a small current is passed through it. The LED can be used for detecting peaks in the audio signal on record and playback.

Low-Noise/High-Output Tape—A type of magnetic tape that yields a greater (6 to 8 dB) output signal and less noise than does a conventional iron oxide tape. In order to use this tape, the tape deck should have special bias and equalization switches.

Magnetic Tape—A thin polyester or acetate film with a fine layer of magnetic particles coated on one surface. The magnetic particles may be iron oxide, chromium dioxide or other formulations.

Metal Tape—A special type of tape whose magnetic coating does not contain the oxygen atoms found in other types, such as ferric oxide and chromium dioxide. Metal tapes produce significantly higher output levels than other tapes. However, they are very expensive and require specially made tape decks for recording onto them.

Micron—0.00004 inch.

MONITOR Switch (on a Tape Deck)—A switch which allows the user to monitor the signal either before it goes to the record head or after it has been recorded. If the tape deck has separate record and playback heads, the tape can be monitored a fraction of a second after it has been recorded.

Multitrack Tape—Magnetic tape which has been divided along its length into separate lanes or tracks for recording separate signals.

One Motor Drive—(See **Tape Transport.**)

Open-Reel Tape Deck—A tape deck which uses two separate reels for tape storage. Open-reel tapes can travel at 1-7/8, 3-3/4, 7-1/2, 15, and 30 ips. They are 0.64 cm (1/4″) wide and contain four tracks.

PAUSE—A control on a tape deck which can be used to halt the tape, while leaving the capstan rotating and leaving the input signal at the record head (so that it will register on the record level meters). It is very useful for adjusting input levels to produce the highest quality tapes. It is not the same thing as the "STOP" control.

Peak Reading Meter—(See **dB Level Meter.**)

Peaks—(See **Transients.**)

Pinch Roller—A rubber roller which presses the tape against the capstan.

PLAY—A control on a tape deck which allows the tape to move forward at its normal speed for playback. It is also used in conjunction with the "RECORD" control when recording signals onto tape.

Playback Head—A horseshoe-like iron core with a coil of wire wrapped around it. As the magnetized tape moves past the gap between the ends of the iron core, Faraday's law of induction causes a voltage to be produced in the coil. Thus, the playback head detects the audio signals on the tape.

Playback Level Meter—(See **dB Level Meter and VU Level Meter.**)

Player Deck—(See **Tape Player.**)

Print-through—A phenomenon which occurs when a magnetic tape is wound on a spool, and the magnetic fields of one layer penetrate directly through the tape and print faint copies of themselves onto adjacent layers. Print-through is more severe with the thinner tapes.

RECORD—A control on a tape deck which is used in conjunction with the "PLAY" control when recording signals onto the tape.

Record Head—An electromagnet which allows the audio signals to be recorded onto tape.

Record Level Meter—(See **dB Level Meter** and **VU Level Meter**.)

REWIND—(See **FAST REVERSE**.)

RMS—An abbreviation which stands for Root Mean Square, a mathematical process for calculating an average.

Saturation—A condition of a magnetic tape which occurs when all of its miniature atomic magnets are aligned to produce maximum magnetization. In this condition the tape can no longer respond to an input signal and severe audio distortion results.

Signal-To-Noise Ratio—The ratio of the signal power to the noise power, measured at the playback head when a maximum signal has been recorded onto tape. The ratio is expressed in decibels.

Spindle—The post on which a tape reel is held.

Stereo Separation—The ratio of the power of the audio signal, in the tape channel for which it is intended, to the power of this signal which has leaked over to the other channel. This ratio is expressed in decibels. It is not the same thing as crosstalk.

STOP—A control on a tape deck which stops the movement of the tape and prevents the input signal from reaching the record head. It is not the same thing as the "PAUSE" control.

Synchronous Motor and Hysteresis Synchronous Motor—Motors whose speed depends on the frequency of the power source (i.e., 60 Hz house current), and are more stable than that of an induction motor.

Tape Deck—A unit which includes a tape transport, the recording heads (erase, record, and playback), and the recording and playback preamplifiers. Tape decks require a power amplifier in order to drive loudspeakers.

Tape Head—(See **Erase Head, Playback Head**, and **Record Head**.)

Tape Hiss—A constant "hissing" noise which exists on magnetic tapes and can be usually heard during quiet passages.

TAPE MONITOR Switch—(See **MONITOR Switch**.)

Tape Player (Player Deck)—A tape deck or tape recorder which is minus the recording facilities. Thus, it can only play prerecorded tapes.

Tape Recorder—A complete sound system which contains a tape deck and the necessary amplifiers and loudspeakers.

Tape Transport—The drive system for a magnetic tape deck. It can include one motor which rotates both reels and the capstan (one motor drive), a separate motor for the reels and the capstan (two motor drive), or separate motors for each reel and the capstan (three motor drive).

Three Head Deck—A tape deck which includes a separate erase head, a separate playback head, and a separate record head.

Three Motor Drive—(See **Tape Transport**.)

Track—The magnetized strip on a magnetic tape laid down by the recording head. There are typically several tracks on single magnetic tape.

Transients (Peaks)—Sudden increases in sound power which occurs over very short periods of time, such as a fraction of a second. A cymbal crash is a good example of a transient.

Two Head Deck—A tape deck which includes a separate erase head and a single head with the dual purpose of recording and playback.

Two Motor Drive—(See **Tape Transport**.)

VU Level Meter—A meter which responds to the average signal level on recording or playback. It responds only slightly and incompletely to the peaks. The "0 VU" level corresponds to the level of an audio signal which contains about 1% total harmonic distortion on playback.

Weighted Measurements—A method of making audio measurements in which certain frequencies are not considered in the measurement as heavily as other frequencies.

Wow and Flutter—Short-term fluctuations in a magnetic tape's speed. Wow is slower changes in the tape's speed, those which occur between 0.5 and 6 times per second. Flutter represents more rapid variations, usually occurring between 6 and 200 times per second.

Review Questions

1. What are the differences between a tape player, a tape deck, and a tape recorder?
2. How do the magnetic tape containers differ between open-reel, cassette, and cartridge tapes? What tape speeds are available in each format?
3. Describe the track arrangements for an open-reel, a cassette, and a cartridge. Which tracks contain the stereo programs?

4. Describe the construction materials of a magnetic tape. Which side of the tape must be in contact with the heads?
5. What basic elements constitute the tape transport?
6. Describe the function of the capstan and pinch roller.
7. Name the three common types of drive systems, and discuss the merits of each one.
8. What advantages do the two- and three-motor drive systems have over a single-motor drive?
9. Why are the dB and VU meters important in the recording process? To what musical characteristics does each type of meter respond?
10. The monitor feature on a tape deck allows the listener to hear either the source signal or the tape playback signal. Describe how this is accomplished on two- and three-head decks. How are they different?
11. Describe the magnetic recording process of how the magnetic coating on the tape is magnetized by an audio signal.
12. What does tape saturation mean?
13. Where does the maximum playback voltage occur with respect to the "magnets" on the tape?
14. Explain how a tape is erased by a tape deck.
15. What is the purpose of a bias signal?
16. What purpose does tape equalization fulfill in the recording/playback process?
17. Describe how the Dolby noise reduction system reduces tape hiss. Can this process reduce noise from a source such as a noisy record? Why not?
18. Describe the interrelationships between the frequency response, S/N ratio, and tape speed of a tape deck.

Exercises

NOTE: The following questions have up to 5 possible answers. Please select the **one** response which best answers the question.

Please answer the next 5 questions using the data given below in Table 13.4.

Here are the answers for the next 4 questions. An answer may be used more than once.

1. Both "A" and "B" are of high quality although "A" is **better**.
2. Both "A" and "B" are of high quality although "B" is **better**.
3. "A" is of high quality and "B" is not.
4. "A" is **not** of high quality although "B" is high quality.
5. Neither "A" nor "B" is of high quality although "B" is better than "A."

1. Compare the crosstalk. ________
2. Compare the wow and flutter. ________
3. Compare the S/N value. ________
4. Compare the frequency response of the two cassettes. ________
5. How much louder will the noise generated by "A" sound compared to that produced by "B"? The noise from "A" will sound:
   1. slightly louder than "B."
   2. twice as loud as "B."
   3. four times louder than "B."
   4. about 1/2 as loud as that from "B."
   5. about 1/4 as loud as that from "B."
6. The variations in the speed of a turntable or a tape deck, which cause variations in the pitch of the music, are called:
   1. rumble.
   2. wow and flutter.
   3. S/N.
   4. compliance.
   5. crosstalk.

	Cassette "A"	Cassette "B"
Frequency Response (−20 VU, chromium dioxide)	50–10,000 Hz ±3 dB	20–18,000 Hz ±2 dB
Wow and Flutter (weighted)	0.15%	0.1%
S/N (Dolby off), weighted	40 dB	60 dB
Crosstalk	60 dB	50 dB

Table 13.4. Specifications for the two cassette decks used in question 1 through 5.

	Cassette "A"	Cassette "B"
S/N ratio (Dolby off), weighted	60 dB	70 dB
Frequency Response	30–18,000 Hz ± 1 dB	30–15,000 Hz ± 2 dB
Stereo Separation	35 dB	40 dB

Table 13.5. Specification for the two cassette decks used in questions 8, 9, and 10.

7. Which type of motors, used in turntables and tape decks, have their speed locked to the frequency of the electrical outlet voltage (60 Hz)?
   1. Electromagnetic motors.
   2. Direct drive motors.
   3. Electrical outlet motors.
   4. Induction motors.
   5. Synchronous motors.

Note the technical specifications for the two tape decks above in Table 13.5.

Please answer the next 3 questions assuming that both of these decks can be purchased on the market today.

8. If tape deck "A" is playing music through the left channel with a loudness of 90 dB, then the amount which will "leak" over onto the right channel will be:
   1. 90 + 60 = 150 dB.
   2. 90 − 60 = 30 dB.
   3. 90 + 35 = 125 dB.
   4. 90 − 35 = 55 dB.
   5. 90 + 3 = 93 dB.
9. The frequency response (assume −20 VU, chromium dioxide):
   1. of "A" is better than that of "B," even though "B" is of high quality.
   2. of "A" is better than that of "B," although "A" is of poor quality.
   3. of "A" is worse than that of "B," even though "A" is of high quality.
   4. of "A" is worse than that of "B," although "B" is of poor quality.
10. How much **louder** would the noise generated by "A" sound compared to the noise generated by "B"?
    1. About 10 times louder.
    2. About 60 times louder.
    3. About 70 times louder.
    4. About twice as loud.
    5. About one-half as loud.
11. When using a cassette tape deck to record music what advantage is there to setting the record level at −20 dB rather than 0 dB?
    1. Better S/N results.
    2. There is no advantage.
    3. The frequency response becomes better.
    4. Less wow and flutter results.
    5. The tape has a longer lifetime.
12. Suppose that you were shopping around for an open-reel tape deck and found five decks with the frequency responses listed below. Which one has the best specification? (Note: If a spec is **incomplete** you should not even consider it.)
    1. 20,000 Hz, + 2 dB, − 0 dB.
    2. 100 − 15,000 Hz, + 4 dB, − 3 dB.
    3. 50 − 18,000 Hz, + 2 dB.
    4. 30 − 18,000 Hz, + 3 dB, − 1 dB.
    5. 30 − 18,000 Hz, + 1 dB, − 0.5 dB.
13. Suppose that you are listening to an open-reel deck which has the **minimum** allowed value of crosstalk for hi-fidelity performance.
    If the music from the main channel is 80 dB, what will be the level of the crosstalk?
    1. 20 dB.
    2. 30 dB.
    3. 50 dB.
    4. 60 dB.
    5. 80 dB.
14. Which one of the following is **not** a tape deck specification?
    1. S/N ratio.
    2. Wow and flutter.
    3. Frequency response.
    4. Rumble.
    5. All of the above are tape deck specifications.

15. The diagram below shows a piece of magnetic tape being magnetized by a recording head. What will be the direction of the magnetic poles **imprinted upon the tape**?

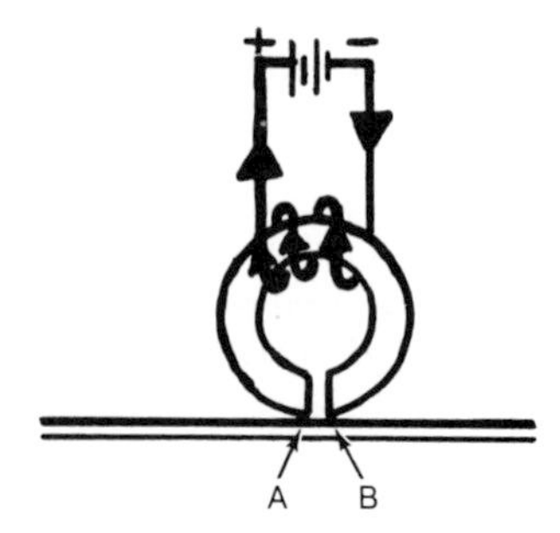

1. Point "A" will become a north pole and "B" a south pole.
2. Point "A" will become a south pole and "B" a north pole.
3. Both "A" and "B" will become north poles.
4. Both "A" and "B" will become south poles.
5. None of the above answers is correct.

16. A piece of magnetic tape is in contact with an electromagnet. The poles of the electromagnet are drawn in the diagram.

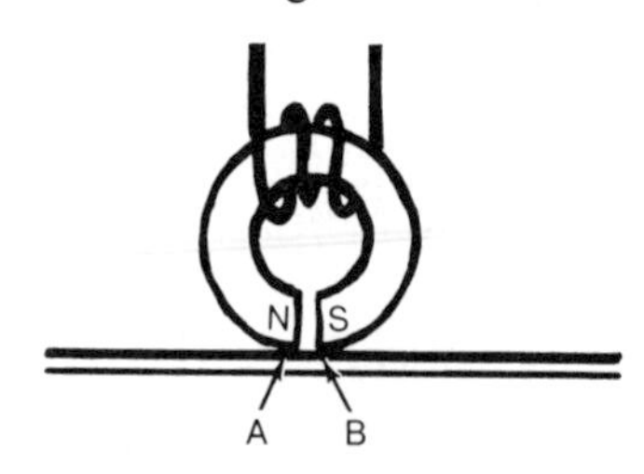

What will be the magnetic poles at the points "A" and "B" **on the tape**?

1. A is north, B is north.
2. A is south, B is south.
3. A is south, B is north.
4. A is north, B is south.

17. What happens to a low level (quiet) audio signal when processed by a Dolby noise reduction system?
   1. It is boosted by +10 dB before being placed on the tape, then reduced by 10 dB upon playback to effectively reduce tape hiss.
   2. It is boosted by +30 dB before being placed on the tape, then reduced by 30 dB upon playback to effectively reduce tape hiss.
   3. It is reduced by 10 dB before being placed on the tape and then boosted by +10 dB upon playback.
   4. Dolby does not process low level signals, only high level loud ones.
   5. None of the above answers is correct.

18. Approximately, how much does the Dolby noise reduction reduce the tape hiss?
   1. 1 or 2 dB.
   2. 10 dB.
   3. 20 dB.
   4. 40 dB.
   5. The Dolby system does not reduce tape hiss.

19. Recording heads are simply:
   1. north poles only.
   2. south poles only.
   3. permanent bar magnets.
   4. permanent horseshoe magnets.
   5. electromagnets.

20. Which button on a tape deck both stops the tape from moving and **prevents** electricity from going to the record head?
   1. STOP.
   2. PAUSE.
   3. PLAY.
   4. RECORD.
   5. No button will do that.

21. The purpose of a **bias** current in a magnetic recording head is to:
   1. erase all previous information on the tape and remagnetize it with new information.
   2. clean the tape of dust and foreign matter.
   3. add a bass boost to the recorded music.
   4. "shake up" the magnetism on the tape so that the tape can be more easily magnetized.
   5. provide equalization.

22. Which button(s) on a cassette deck must be depressed for tape playback?
   1. RECORD.
   2. PLAY.
   3. RECORD and PLAY.
   4. FAST FORWARD.
   5. PAUSE and PLAY.

23. When you play the reverse side of a cassette tape, you physically turn the tape surface over and play the non-magnetic side.
   1. True.
   2. False.

24. What is the primary effect that occurs in tape decks when the capstan and pinch roller are **not** perfectly round?
   1. The wow and flutter specification becomes larger.
   2. The wow and flutter specification becomes smaller.
   3. The crosstalk specification becomes better.
   4. The crosstalk specification becomes worse.
   5. The S/N value becomes better.

25. Which button on a tape deck will stop the tape from moving but still let the electricity from the preamp go to the record head?
    1. No button will do that.
    2. The STOP button.
    3. The PAUSE button.
    4. The PLAY button.
    5. The RECORD button.
26. What part of a tape deck pulls the tape at constant speed?
    1. The take-up reel.
    2. The supply reel.
    3. The playback reel.
    4. The capstan and pinch roller.
    5. The bias.
27. One of the main difficulties in recording with 8-track cartridge tapes (as compared to open-reel or cassettes) is that:
    1. one cannot get more than 4 channels on the width of the cartridge tape.
    2. the tapes cannot record a quadraphonic program.
    3. the tapes move in one direction only and cannot be reversed.
    4. the cartridge tapes are too narrow in width as compared with cassette or open-reel tapes and, therefore, they are of low fidelity.
    5. the recording speed is too slow.
28. Explain the relationship between tape speed and performance.
    1. Faster tape speeds generally produce **better** performance from the tape deck.
    2. Faster tape speeds generally produce **poorer** tape deck performance.
    3. The performance of a tape deck is not related to tape speed.
29. Which tape deck(s) divide the tape into 8 tracks?
    1. Open-reel and cassette.
    2. Open-reel only.
    3. Cartridge and cassette.
    4. Cartridge only.
    5. Cassette only.
30. If the magnetic tape moves from left to right what must be the order of the heads (from left to right) on a three-head tape deck?
    1. Erase, playback, record.
    2. Erase, record, playback.
    3. Record, playback, erase.
    4. Record, erase, playback.
    5. Playback, erase, record.
31. If you bought a tape deck and a pair of loudspeakers (not headphones), would you be able to listen to them?
    1. Yes.
    2. No, because a preamp is needed.
    3. No, because a sound source is needed.
    4. No, because an integrated amp is required.
32. If a weighting filter is used when measuring the S/N of a tape deck, is the resulting S/N higher or lower than if the filter had not been used?
    1. The S/N is larger with the filter.
    2. The S/N is smaller with the filter.
    3. The filter does not change the S/N ratio one way or the other.
33. Which metering system causes a 1% THD to occur when the signal level is set to "0"?
    1. LED meter.
    2. Peak meter.
    3. VU meter.
    4. dB meter.
34. What is the frequency, typically, of tape bias, and what is its purpose?
    1. 20 kHz: Bias reduces the need for recording equalization.
    2. 20 kHz: Bias is used to reduce the S/N value.
    3. 100 kHz: Bias is primarily used for improving the frequency response.
    4. 100 kHz: Bias is used to reduce tape distortion.
    5. 100 kHz: Bias improves the wow and flutter.
35. What is the most critical design parameter of a capstan for reducing wow and flutter?
    1. Its mass.
    2. The material from which it is made.
    3. Its height.
    4. Its roundness.
    5. Its physical size.
36. Approximately, what is the erase frequency used in most tape decks?
    1. 88 MHz.
    2. 20 kHz.
    3. 20 Hz.
    4. 60–125 kHz.
    5. 535 kHz.
37. The "0" level on a VU meter means that a ___ signal, which has been recorded and played back, will contain a THD level of approximately ___.
    1. 5 kHz, 0.1%
    2. 5 kHz, 3.0%
    3. 1 kHz, 0.1%
    4. 1 kHz, 1.0%
    5. 1 kHz, 3.0%

38. If you wanted a meter that measures overall loudness, and responds only somewhat to peaks, you would purchase a:
    1. VU meter.
    2. dB meter.
    3. LED meter.
    4. peak meter.
39. Which type of meter responds well to the sudden peaks, or transients, in the music?
    1. A VU meter.
    2. A dB meter.
    3. A LED meter.
    4. Both answers #1 and #3 are correct.
    5. Both answers #2 and #3 are correct.
40. Suppose that you wanted to record several songs on a tape at the same average overall loudness. You should then record each song at the same level on a ________ meter.
    1. dB
    2. LED
    3. peak reading
    4. VU
    5. Both answers #2 and #4 are correct.
41. Suppose that you are making a tape recording of your favorite record. If you set the recording level too high, the tape recording:
    1. will have a good frequency response.
    2. will have a poor S/N ratio.
    3. will have a poor wow and flutter value.
    4. can have a large amount of distortion.
    5. will have a lower than normal amount of distortion.
42. When the weighted S/N ratio of a tape deck is being measured, which range of audio frequencies is given the most emphasis?
    1. The bass frequencies are given the most emphasis.
    2. Both the bass and midrange frequencies are given approximately the same amount of emphasis.
    3. The midrange frequencies are given the most emphasis.
    4. The treble frequencies are given the most emphasis.
    5. All the frequencies are given an equal amount of emphasis.
43. If you were a tape deck manufacturer and wanted to publish the best possible S/N ratio for your deck, then you would set the signal to ________, and use an ________ noise measurement.
    1. "0" on a dB meter, unweighted
    2. "0" on a dB meter, weighted
    3. "0" on a VU meter, unweighted
    4. "0" on a VU meter, weighted
    5. "−20 dB" on a VU meter, weighted
44. Suppose that you are making a tape recording of your favorite record. If you set the recording level very low, the tape recording will have:
    1. a good S/N ratio.
    2. a poor S/N ratio.
    3. a poor frequency response.
    4. a poor stereo separation.
    5. a good wow and flutter value.
45. If a cassette deck manufacturer wanted to make the S/N ratio as high as possible, would "0" on a VU meter or "0" on a dB meter be used as the reference level for the signal?
    1. "0" on a VU meter, because it would yield a S/N ratio that is approximately 8 dB greater.
    2. "0" on a VU meter, because it would yield a S/N ratio that is approximately 15 dB greater.
    3. "0" on a dB meter, because it would yield a S/N ratio that is approximately 8 dB greater.
    4. "0" on a dB meter, because it would yield a S/N ratio that is approximately 15 dB greater.
    5. Either meter could be used, because they both give the same value for the S/N ratio.

Chapter 14

MECHANICS IN HI-FIDELITY

14.1 Introduction

Mechanics is a field of science which studies the motion of objects. Motion plays a very important role in the operation of a hi-fidelity system, whether it be the rotation of a turntable platter, the vibrations of a speaker diaphragm and a phono stylus, or the movement of a magnetic tape past the recording/playback heads of a tape deck. In all cases, audio engineers are able to design the mechanical features of a hi-fi system because they are familiar with the laws of mechanics. It is the successful synthesis of proper mechanical design, coupled with a thorough knowledge of sound, electricity, magnetism, and heat, that has given us the superb hi-fi components found on the market today. And the future applications of these same laws will bring the ever-improving systems of tomorrow.

In this chapter some of the basic concepts in mechanics will be studied in order to gain a better insight into the operation of turntables, tape decks, and speakers.

14.2 Speed, Velocity, and Acceleration

Speed, velocity, and acceleration are concepts of motion that are familiar to most of us. The speed of an object indicates how fast it is moving. Velocity, like speed, also tells how fast an object is moving; in addition, velocity specifies the direction of the object's motion. Acceleration is different than either speed or velocity, and occurs whenever there is a *change* in the velocity of an object. It is the notion of "change" which is central to the idea of acceleration. If the velocity of an object is increasing, it is said to be "accelerating;" conversely, if the velocity is decreasing, the object is "decelerating." We will now examine each of these concepts in more detail.

Speed

We have previously encountered the concept of speed in Chapter 3 during our discussion of sound waves. The average speed of an object is defined as the distance traveled by the object divided by the time required to make the trip.

$$\text{Average speed} = \frac{\text{Distance traveled}}{\text{Elapsed time}} = \frac{d}{t} \quad \text{(Equation 14.1)}$$

The most common examples of speed in hi-fi are those found in tape decks. The transport system of a tape deck pulls the tape past the heads at a speed which depends on the type of deck being used. Cassettes normally operate at a speed of 1-7/8 ips, while 8-track cartridges run at 3-3/4 ips. Open-reel tape decks usually provide two or three speed choices, ranging from 1-7/8 ips to 30 ips. Of course, tape decks with faster speeds consume more tape in a given amount of time than do slower speed decks. For example, a 15 ips open-reel deck uses eight times more tape per second than does a 1-7/8 ips cassette deck. For this reason, the size of a tape reel is much larger for open-reel decks than for cassette decks.

Question: How long is a cassette tape which can play for 30 minutes (1,800 seconds) on a side?

Answer: Using Equation 14.1, the tape length can be calculated as follows:

Distance traveled = (Average speed) × (Elapsed time)

Distance traveled = (1-7/8 ips) × (1,800 sec.)

Distance traveled = 3,375 inches (85.7 meters).

Therefore, the cassette has 85.7 meters of tape wound on its reels, which is nearly the length of a football field!

The previous definition of speed as the distance traveled divided by the elapsed time is the definition of *average speed. Instantaneous speed*, on the other hand, is the speed an object has at any instant of time. The two speeds may be different, as can be seen by considering a 7-1/2 ips tape deck. The 7-1/2 ips speed is only the average speed of the tape, because at any instant in time the tape may be moving either a little faster or a little slower than its average speed. The fact that a tape's instantaneous speed is usually different from its average speed gives rise to the wow and flutter specification, as discussed in section 13.8. Of course, a *perfect* tape deck would have no wow and flutter, and its instantaneous tape speed would always equal the average speed.

The Units of Speed

An examination of Equation 14.1 shows that the unit in which speed is measured is the unit of distance divided by the unit of time. With tape decks, for example, the distance is usually expressed in "inches" and the time in "seconds," so typical tape speeds are expressed in inches per second (ips). Most car speeds involve "miles" and "hours," and a typical speed is written as 55 miles per hour (mph). However, in the metric system, where distances are measured in kilometers (km), a common unit of speed is kilometers per hour (kph). It is possible to convert from mph to kph, and vice versa, using the following relations:

1 mph = 1.61 kph
1 kph = 0.62 mph.

Very often metric speeds are expressed in meters per second (mps or m/s), rather than in kilometers per hour, and the conversion between these two units is:

1 kph = 0.278 mps (or m/s)
1 mps = 3.60 kph.

Velocity

When describing the motion of an object, both its speed and direction of travel are important. The term "velocity" is used to designate both the speed and direction of a moving object. Figure 14.1 illustrates an example of two cars which have different speeds and different directions of travel. Therefore, the car's velocity in part (A) is 18 m/s east, while the car's velocity in (B) is 30 m/s at 25° south of east.

Another example of velocity is that of a dust particle which is situated on the edge of a 33-1/3 rpm record, as shown in Figure 14.2. Although the speed of the particle remains constant at 0.53 m/s as it rotates, its direction is continuously changing. Since its direction is changing, its velocity is also changing, even though the speed of the particle remains constant. This example illustrates that constant speed is not necessarily the same as constant velocity. An object has a constant velocity only if both its speed *and* direction do not change. Very often "speed" and "velocity" are used interchangeably in hi-fi magazines, although it should be remembered that they are really not the same concepts.

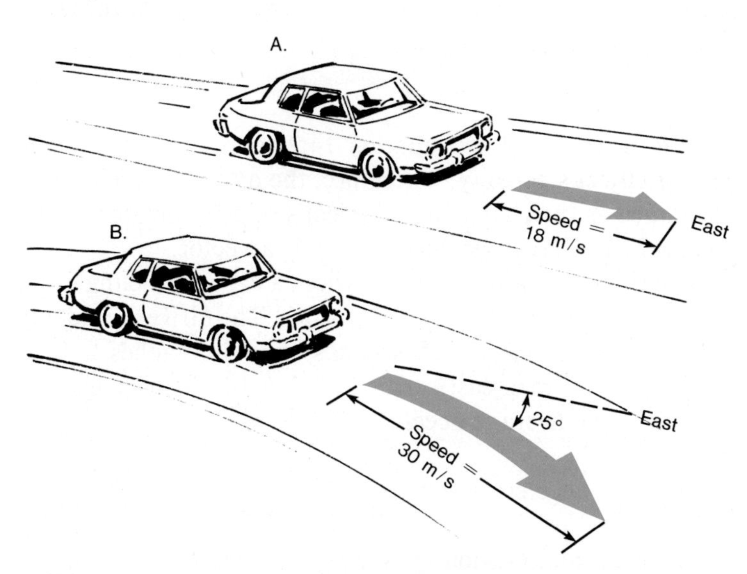

Figure 14.1. (A) The velocity of the car is 18 m/s, eastward. (B) The velocity is 30 m/s at 25° south of east.

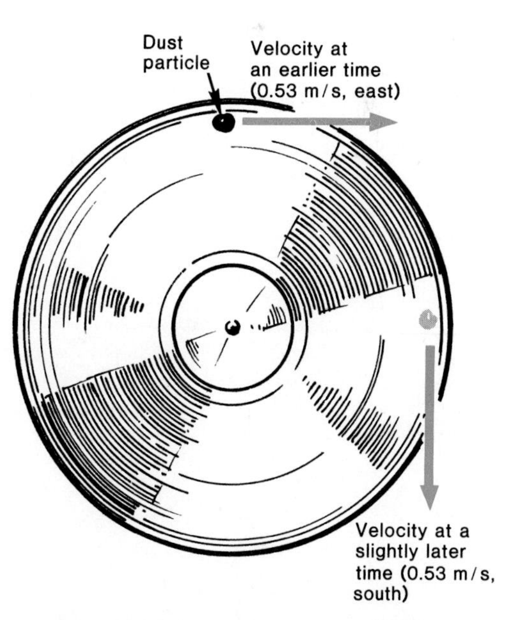

Figure 14.2. As the record rotates, the velocity of the dust particle changes, because the direction of its motion changes. Notice that the speed of the particle remains constant.

Vectors

Quantities, like velocity, which have both a magnitude (size) and a direction are often called *vectors*. In figures and drawings it is customary to represent vectors by using arrows, as indicated in Figures 14.1 and 14.2 for the velocity vector. The arrow is drawn so that it points in the same direction as the vector it represents. The length of the arrow is drawn so that it is proportional to the magnitude of the vector. Thus, in Figure 14.1 the arrow representing the velocity vector in part (B) is longer than the arrow in part (A), because the car's speed is greater in (B). In this example the arrow's length represents the car's speed, and the arrow points in the direction of the car's motion. In Figure 14.2 the two arrows have the same length to indicate that the magnitude of the velocity is constant at 0.53 m/s.

Within text material, it is common to denote a vector by using a boldface letter. For example, the velocity of an object is often written as $\mathbf{v}$. Throughout this chapter we will encounter many examples of vectors, and each one will be symbolized by an appropriate boldface letter.

Acceleration

An object is accelerating whenever its velocity is changing. The average acceleration of an object is defined as follows.

$$\text{Average acceleration} = \frac{\text{Final velocity} - \text{Initial velocity}}{\text{Elapsed time}}$$

or

$$\text{Average acceleration} = \frac{\mathbf{v}_f - \mathbf{v}_i}{t}, \qquad \text{(Equation 14.2)}$$

where $\mathbf{v}_f$ is the final velocity of the object, $\mathbf{v}_i$ is the initial velocity, and t is the elapsed time.

The term $(\mathbf{v}_f - \mathbf{v}_i)$ in Equation 14.2 represents the *change* in the object's velocity. Therefore, the average acceleration is defined as the change in an object's velocity divided by the elapsed time. Average acceleration is a vector.

Consider the familiar scene where a car is traveling at an initial velocity of 10 m/s eastward. The driver suddenly accelerates the car for 5 seconds until it reaches a final velocity of 30 m/s eastward. Equation 14.2 can be used to calculate the average acceleration of the car:

$$\text{Average acceleration} = \left(\frac{30\text{ m/s} - 10\text{ m/s}}{5\text{ sec.}}\right)\text{ eastward,}$$

$$\text{Average acceleration} = 4\text{ m/s/s eastward.}$$

The 4 m/s/s (or 4 m/s^2) term, is read as "4 meters per second per second," and means that the speed of the car is increasing at the rate of 4 meters per second for each second that the car is accelerating. The car, which has an initial speed of 10 m/s, will be traveling at a speed of 14 m/s ($10 + 4 = 14$) at the end of the first second of acceleration. It will be traveling at a speed of 18 m/s ($10 + 4 + 4 = 18$) at the end of the second one, and so on. In this fashion it eventually reaches a velocity of 30 m/s eastward after 5 seconds ($10 + 4 + 4 + 4 + 4 + 4 = 30$).

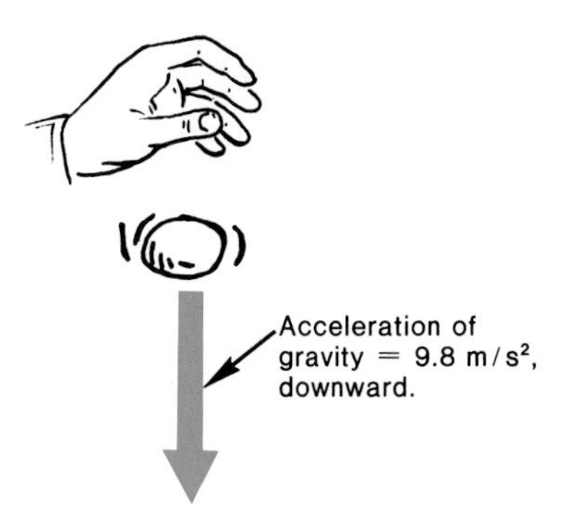

Figure 14.3. In the absence of air friction, all freely falling objects are accelerated downward by gravity at the rate of 9.8 m/s^2.

Another interesting example of acceleration is that due to gravity. In the absence of air friction, *all* freely-falling objects are accelerated downward by gravity at 9.8 m/s^2, as illustrated in Figure 14.3. Often, the gravitational acceleration is represented by the symbol "**g**" so that $\mathbf{g} = 9.8$ m/s^2 downward. Therefore, for each second of free fall, its speed will be 9.8 m/s greater than that of the previous second. Under the action of gravity the object's speed becomes greater and greater until it finally strikes the ground. It is important to note that in the absence of air friction all freely falling objects, large or small, heavy or light, accelerate downward at the same rate of 9.8 m/s^2; a feather falls just as fast as a rock, as illustrated in Figure 14.4 (A). Of course the two objects will fall at *different* rates when air friction is present, as we know from the common situation shown in part (B).

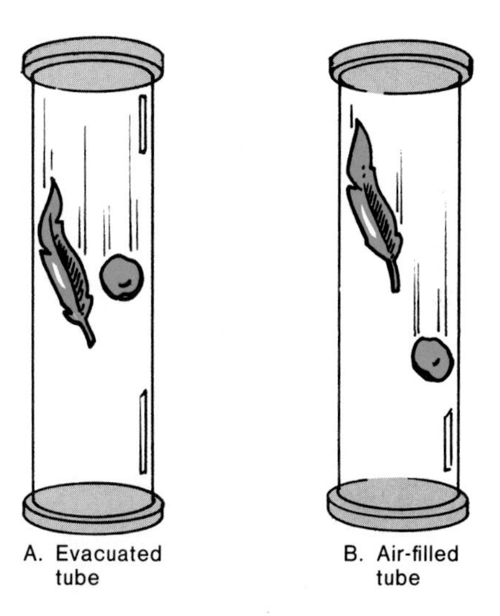

Figure 14.4. (A) In the absence of air friction both the rock and the feather accelerate downward at the same rate. (B) With air friction, the rock accelerates faster than the feather.

When an object is slowing down its velocity decreases, and this decrease is often called a "deceleration" as shown in Figure 14.5. Since the final velocity is *less* than the initial velocity, Equation 14.2 shows that the acceleration is negative. In other words, a deceleration is a negative acceleration.

Figure 14.5. An example of deceleration (negative acceleration).

Summary

The average speed of an object is defined as

$$\text{Average speed} = \frac{\text{Distance traveled}}{\text{Elapsed time}}.$$

The instantaneous speed of an object is its speed at any given instant in time. Velocity is a term used to describe both the speed and direction of an object's motion. Velocity is a vector, and it is graphically represented by attaching an arrow to the object; the length of the arrow represents its speed, and the direction of the arrow represents the direction of the object's motion. Quantities like velocity, which have a size and direction, are called vectors. Whenever an object's velocity is changing, it is accelerating. The average acceleration is a vector, and it is defined as

$$\text{Average acceleration} = \frac{\text{Final velocity} - \text{Initial velocity}}{\text{Elapsed time}}$$

The acceleration due to gravity is $\mathbf{g} = 9.8\ \text{m/s}^2$, downward. In the absence of air friction, all falling objects are accelerated downward by gravity at the same rate of 9.8 m/s^2. When air friction is present, different objects may have different downward accelerations.

14.3 Forces

Force is an important concept in the mechanical operation of hi-fi systems. A force, as most of us have experienced, usually means a push or a pull. If, for example, a tow truck pulls a car, a force is being applied. Or the simple act of opening a door involves a force being applied to the knob. As we have seen in Chapter 11, there are magnetic forces which are exerted between magnets. In fact, it is the magnetic force between the permanent magnet and the voice coil (an electromagnet) of a cone-type speaker that is responsible for its operation. There are also electric forces between charges which give rise to the flow of current in electrical circuits, as discussed in Chapter 7.

In the British system, which is commonly used in the U.S., forces are measured in *pounds*. In the metric system, forces are measured in either *newtons* (abbreviated as N) or *dynes*. The relationships between these three different units of force are:

$$\begin{aligned} 1\ \text{pound} &= 4.45\ \text{newtons}, \\ 1\ \text{pound} &= 4.45 \times 10^5\ \text{dynes}, \\ 1\ \text{newton} &= 1 \times 10^5\ \text{dynes}. \end{aligned}$$

Force, like velocity and acceleration, is also a vector, because both magnitude and direction are required to describe it. The magnitude of the force vector is the strength of the force, and the direction is the direction in which the force is exerted. The force vector is often drawn as an arrow.

Frictional Forces

A frictional force arises from the contact between the surfaces of two objects. There are two types of frictional forces, called *kinetic* and *static*. Kinetic friction occurs when the two objects are in relative motion and rub against each other in the process, such as a car sliding on an icy pavement. Figure 14.6 illustrates a side view of a stylus riding in a moving record groove. As the record rotates, a kinetic frictional force is generated because of the relative motion between the stylus and the groove. Notice that the frictional

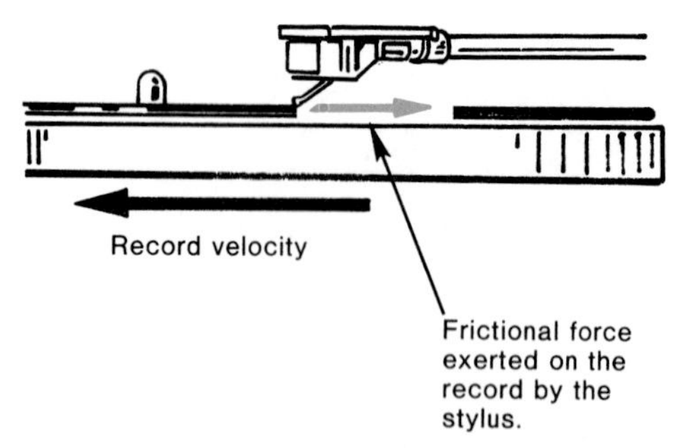

Figure 14.6. The kinetic frictional force, caused by the relative motion between the stylus and the moving record, tends to slow down the record.

force exerted on the record, is opposite in direction to the motion of the record. In other words, kinetic friction always tends to slow down the record, just like the slowing down of a car skidding along the pavement. Another result of the frictional force shown in Figure 14.6 is the skating force, which necessitates the antiskating control found on many turntables (see section 12.5).

Static friction occurs when there is *no relative motion* between objects which are in contact with each other. For example, suppose that you wish to unwind a few meters of magnetic tape from a reel. You simply press the end of the tape between the thumb and first finger and then pull. Since there is no relative motion between the tape and your fingers (remember, you are holding the tape tightly), it is static friction which allows you to pull the tape. In the same manner, the capstan/pinch roller assembly of a tape deck pulls the tape forward, as shown in Figure 14.7. Recall from Chapter 13, that the capstan rotates because it is connected to the drive motor. The pinch roller is a rubber-like, non-skid wheel which is not attached to the motor, but is otherwise free to rotate. When the pinch roller is pressed against the rotating capstan, a static frictional force is produced, which pulls the magnetic tape off the reel. Notice from the figure that there is *no relative motion* between the tape and the capstan/pinch roller assembly. Even though each one is moving, there is no relative motion because the rotating speeds of both the capstan and pinch roller are equal to the tape speed; hence it is static, not kinetic, friction which pulls the tape. Rubber-like material is employed for the pinch roller because it generates a stronger frictional force than would a hard substance like steel.

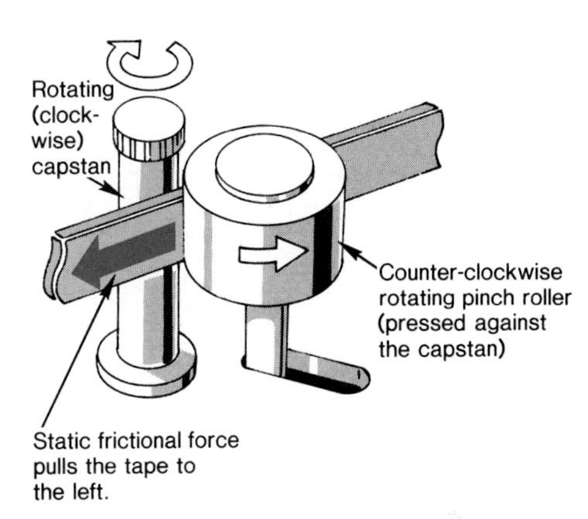

Figure 14.7. The capstan and pinch roller pull the tape by the use of a static frictional force.

Another example of the importance of static friction arises when a record is placed on a rotating turntable platter. The record, as we know, must rotate at precisely the same speed as the platter; there can be no "slipping" between the two. Virtually all quality turntables utilize a rubber-like mat that rests on top of the platter, as shown in Figure 14.8. The mat, with its non-skid surface, provides the necessary static frictional force that prevents the record from slipping relative to the platter.

Gravitational Forces and Weight

Perhaps one of the most common forces is the downward force, or pull, exerted on all objects by gravity. The gravitational force exerted on an object is called its *weight*. If, for example, the gravitational force acting on a person is 120 pounds, the person's "weight" is said to be 120 pounds. On

Figure 14.8. The rubber-like mat found on most turntables keeps the record rotating at the same speed as the platter by the use of a static frictional force.

the moon the gravitational force which pulls on an object is approximately 1/6 of the gravitational pull which would be exerted on earth. The reason for this is that the moon is much smaller than the earth. Therefore, the 120 pound earth weight means only a 20 pound moon weight.

The tracking force which keeps the cartridge stylus in the record groove comes from the gravitational force. Most of the weight of the cartridge is balanced out by the way in which the tone arm is supported. But the small unbalanced portion creates the tracking force (see sections 12.5 and 12.8).

Summary

A force can be thought of as either a "push" or a "pull." Force is a vector that is often represented by an arrow; the length of the arrow is proportional to the strength of the force, while the direction of the arrow is the direction in which the force acts. Magnetic, electrical, frictional, and gravitational forces are important in the operation of hi-fi systems. There are two types of frictional forces: kinetic and static. Kinetic frictional forces arise when there is relative motion between two surfaces which are in contact with each other, i.e., they rub together. Static friction arises when there is no relative motion between the two contacting surfaces. The gravitational force which is exerted on an object is often called its "weight."

14.4 The Addition of Vectors

Very often several forces may be acting on an object at the same time, and it is important to know the combined effect of all such forces. For example, Figure 14.9 illustrates a cross-sectional view of a stylus which is riding in a record groove. As mentioned in section 12.10, a stereo record groove contains two separate channels of information; one channel is encoded in the left wall, and the other channel is encoded in the right wall. To understand better the motion of the stylus, consider Figure 14.9 (A), which shows two positions of the left wall at slightly different times. The movement of the left wall causes a force $\mathbf{F}_L$ to be exerted on the stylus which, in response, moves upward and to the right, as shown in the illustration. Part (B) shows how the stylus moves in response to a force $\mathbf{F}_R$ exerted by the right wall. In general, the force from the left wall will not be equal to the force from the right wall, because the encoded left and right stereo signals are not the same.

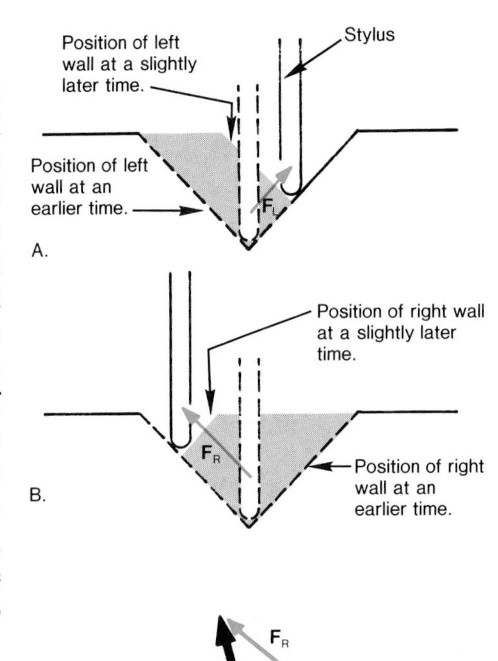

Figure 14.9. (A) The force $\mathbf{F}_L$ exerted on the stylus by the movement of the left wall. (B) The force $\mathbf{F}_R$ exerted on the stylus by the movement of the right wall. (C) The resultant force **F**, which acts on the stylus, is the vector sum of $\mathbf{F}_L$ and $\mathbf{F}_R$.

Of course, during the playing of a stereo record it is rare when either the left wall or the right wall alone exerts a force on the stylus. Usually both walls push on the stylus simultaneously. The combined action of these two forces on the stylus is determined by a method called *vector addition.* Since vectors have direction as well as a magnitude, the sum of two or more vectors cannot be obtained by a simple addition process. Vector addition must always consider the directions of the vectors. Part (C) of the figure illustrates how the $\mathbf{F}_L$ and the $\mathbf{F}_R$ vectors must be "added" together by using the following rules for vector addition:

1. Draw the arrow representing the $\mathbf{F}_L$ vector on a piece of paper, making sure to retain its proper length and orientation.
2. Next, draw in the $\mathbf{F}_R$ arrow such that its tail is at the head of the $\mathbf{F}_L$ arrow. Again, be sure to maintain the proper length and orientation of the $\mathbf{F}_R$ arrow.
3. The vector sum of $\mathbf{F}_L$ and $\mathbf{F}_R$ is the arrow drawn from the tail of $\mathbf{F}_L$ to the tip of $\mathbf{F}_R$, as indicated by the arrow labeled **F** in part (C) of Figure 14.9.

The resultant vector **F** is the net force experienced by the stylus. The length of **F** represents the magnitude, or strength, of the net force that the two walls exert on the stylus. The direction of the **F** arrow shows the direction in which the stylus moves in response to the simultaneous action of $\mathbf{F}_L$ and $\mathbf{F}_R$. Notice that the motion of the stylus is not parallel to either wall; instead it moves in a direction which is determined by the vector sum of the forces exerted on it by the walls.

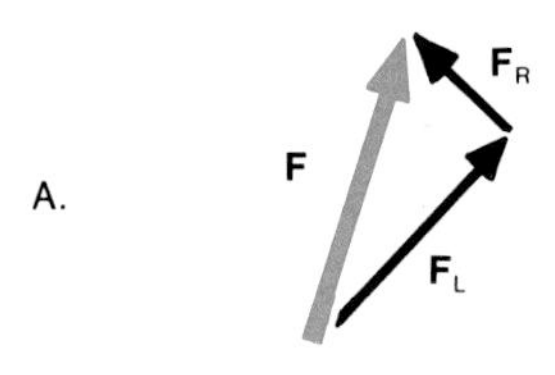

F_L larger than F_R

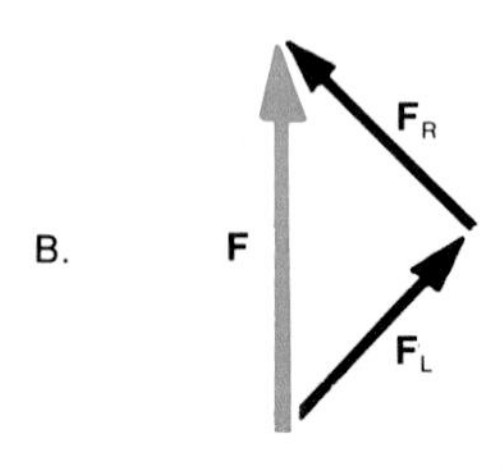

F_L equals F_R

Figure 14.10. Two more examples of vector addition. In both (A) and (B) the vectors $\mathbf{F_R}$ and $\mathbf{F_L}$ are added together to give the resultant vector $\mathbf{F}$.

Figure 14.10 further illustrates the resultant force on the stylus at different times when the record is playing. Part (A) illustrates the resultant force when the left stereo channel is louder than the right channel, i.e., the left wall exerts a larger force than does the right wall. Part (B) shows the case where both stereo channels are equally loud, i.e., when both walls exert equal forces on the stylus. From the previous two figures it is interesting to observe how the resultant force on the stylus changes as the force exerted by each wall changes.

Summary

Velocity, acceleration, and force are vectors because they have both a magnitude and a direction associated with them. Vector quantities are often represented by arrows, with the length of the arrow being proportional to the magnitude of the vector and direction of the arrow indicating the direction of the vector. Two vectors can be added together to obtain the resultant vector by means of vector addition, "head" to "tail" so to speak.

14.5 Newton's First Law

We have seen that the motion of an object was described by its velocity and acceleration. The motion of an object can be changed when various types of forces, such as magnetic, electrical, and frictional forces, act on it. Isaac Newton studied the relationship between these forces and their effect on the motion of an object. In the 17th century he developed three laws, collectively known as "Newton's laws of motion," which form the basis for understanding mechanical systems. These laws play an essential role in the mechanical operation of turntables, phono cartridges, tape decks, and speakers. In this section, and in the next two sections, we will discuss each of Newton's laws and illustrate how they apply to hi-fi systems.

Newton's first law describes what happens to the motion of an object when there is no net force exerted on it. The first law can be stated as follows.

> **Newton's First Law:** Every object continues in a state of rest, or in a state of constant motion along a straight line, unless it is compelled to change that state by forces impressed upon it.

This is a most remarkable law which can be explained best with a few examples. Imagine that you are driving along a long, flat road at a velocity of 60 km/h, eastward. If you shift the car into neutral and turn off the engine, the car will continue to coast for a considerable distance before finally coming to a stop. Of course, it is primarily frictional forces that are responsible for slowing down the car; air friction, friction in the wheel bearings, friction between the tires and the road, etc. If it were not for the forces which act on the car, it would never slow down. The car would travel forever at a constant velocity of 60 km/h, eastward (assuming that the road always remained flat and straight). This conclusion is consistent with Newton's first law; the car will continue in a state of constant motion along a straight line, unless it is compelled to change that state by forces impressed upon it.

Another example of Newton's first law is the spectacular "tablecloth" trick. As illustrated in Figure 14.11, the idea is to pull the tablecloth from under the plates and silverware quickly without (and this is the trick!) disturbing them. Initially, both the objects on the table and the tablecloth are in a "state of rest." As the tablecloth is yanked from under the objects it exerts a force on them due to kinetic friction. However, if the yank is quick enough the frictional force acts for only a *very short time* on the objects; in essence, the frictional force exerts almost no influence on the objects. According to Newton's first law, the objects will continue in their original state of rest, and not be disturbed by the moving tablecloth. Hence, the trick works!

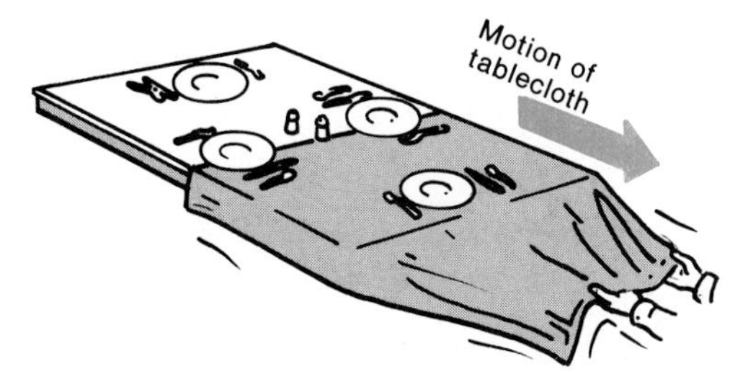

Figure 14.11. The objects remain at rest on the table if the tablecloth is pulled away quickly. This is an example of Newton's first law.

Inertia and Mass

According to Newton's first law, an object will continue in its state of motion unless acted upon by forces. If a net force is applied to an object its state of motion will change. However, it is more difficult to change the motion of some objects than others. The harder it is to change the motion of an object, the more *inertia* it is said to have. Objects such as trucks and freight trains have relatively large amounts of inertia, while baseballs and pencils have relatively little. An object which has a large amount of inertia is difficult to set into motion. Likewise, it is also difficult to slow it down once set into motion. (Have you ever tried stopping a moving freight train?) The inertia of an object can be defined as follows.

> **Inertia**: The property of an object that resists any change in its motion when a force is applied.

The amount of inertia that an object has is quantitatively measured by its *mass*, that is, the "quantity of matter" that

	Mass
One-penny	0.003 kg
Turntable platter	2 kg
Bicycle	20 kg
Medium-sized car	2×10^3 kg
Jet fighter	8×10^3 kg
Saturn V rocket	2.3×10^5 kg
Supertanker	1.5×10^8 kg

Table 14.1. Masses of various objects.

an object possesses. The greater the inertia of an object, the greater its mass, and vice versa. In the metric system the mass of an object is measured either in grams (g) or kilograms (kg). The masses of various objects are listed in Table 14.1

In hi-fi there are some situations where a large mass is desirable, and some where a small mass is desirable. For example, a turntable must be immune from the ever-present vibrations caused by people walking around, doors opening and closing, sound waves, etc. Otherwise, the vibrational forces can set the record into unwanted oscillations which can give rise to very unpleasant sounds. The relatively massive platter, upon which the record rests, helps to minimize this problem. The platter, with its large mass, effectively resists any change in its motion that would be caused by the vibrational forces. Therefore, turntable designers, with a knowledge of Newton's first law, use a platter with a large mass as an aid in reducing extraneous vibrations that could degrade the overall quality of the music.

On the other hand, speaker diaphragms and cartridge styli vibrate at frequencies up to 15,000 Hz and beyond. The back and forth nature of these vibrations means that speakers and cartridges must be able to change their state of motion very quickly. Because of this requirement, Newton's first law implies that both of these units must possess relatively small masses. Audio designers are constantly developing new materials that will permit small mass diaphragms and styli to be designed which have the required mechanical strength.

Mass vs. Weight

The concept of mass is commonly used by many people when they wish to convey a sense of heaviness; a 320 pound football player is said to be rather "massive." Because of this, mass and weight are often used interchangeably. Although more massive objects do weigh more than less massive objects, the two terms are not the same. To illustrate the difference between mass and weight, consider what happens when an astronaut travels from the earth to the moon. Mass, as discussed previously, is a measure of the "quantity of matter" that an object possesses. It does not matter whether the astronaut is on the earth or on the moon, for his mass is the same. The weight of an object is the downward gravitational force exerted on it. The pull of gravity on the moon is only about one-sixth of that on the earth, so a 180 pound astronaut weighs approximately 30 pounds on the moon. Thus, in traveling from the earth to the moon the astronaut changes weight, but not mass.

Summary

Newton's first law states that every object continues in a state of rest, or in a state of constant motion along a straight line, unless it is compelled to change that state by forces impressed upon it. Inertia is a property of an object that resists any change in the object's motion when a force is applied. The amount of inertia that an object has is quantitatively measured by its mass. More massive objects have greater inertia than do less massive objects.

14.6 Newton's Second Law

Newton's first law tells us that if no force is exerted on an object it will continue either in a state of rest or in a state of constant motion along a straight line. But what happens to the object's motion when a force is applied to it? Newton's second law describes how the motion will be affected; in particular, the second law states that the force will cause the object to accelerate.

Newton's Second Law: An object will accelerate when a force is impressed upon it. The acceleration **a** is directly proportional to the net force **F** acting on the object, and inversely proportional to its mass m. The direction of the acceleration is in the direction of the force. Written in symbols, the second law is

$$\mathbf{a} = \frac{\mathbf{F}}{m}, \quad \text{or } \mathbf{F} = m\mathbf{a}. \qquad \text{(Equation 14.3)}$$

Figure 14.12 illustrates the permanent magnet and the voice coil/diaphragm assembly associated with a typical cone speaker. As discussed in section 11.4, the permanent magnet exerts a magnetic force on the voice coil which causes the voice coil and the attached diaphragm to accelerate. If, for example, the force has a value of 20 newtons (20 N) to the right, and the mass of the voice coil/ diaphragm assembly is 0.02 kg, Newton's second law can be used to calculate the acceleration:

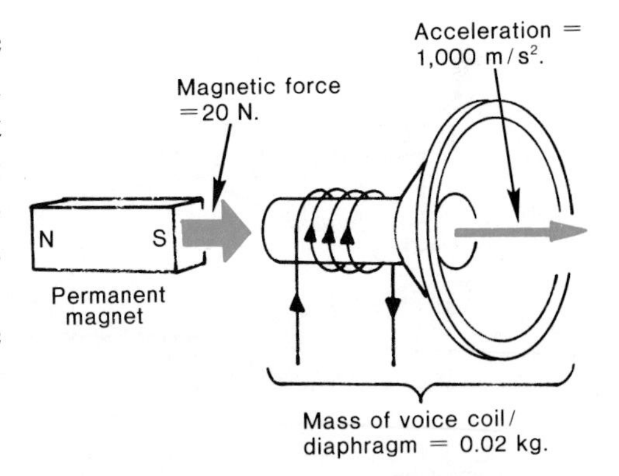

Figure 14.12. The magnetic force causes the voice coil/diaphragm to accelerate to the right. The amount of acceleration is determined from Newton's second law.

$$\mathbf{a} = \frac{20 \text{ N}}{0.02 \text{ kg}} = 1{,}000 \text{ m/s}^2, \text{ to the right.}$$

Therefore, under these conditions the cone's acceleration is an astonishing 1,000 m/s^2, which is more than 100 times the acceleration of gravity ($g = 9.8$ m/s^2)! This acceleration is far greater than that of even the "hottest" drag car racers, which typically have accelerations of 10 m/s^2.

As one might expect, the production of higher frequency sound requires a greater diaphragm acceleration than does the creation of lower frequency sound (assuming that the amplitude of vibration is the same for both frequencies). For the same displacement, a cone that is vibrating at 1,000 Hz travels twice the distance in the same amount of time as does a cone which is vibrating at 500 Hz. In order to cover the greater distance in the same amount of time, the cone producing the 1,000 Hz sound must have a greater acceleration. All other things being equal, the higher the frequency, the greater the acceleration.

Large accelerations of the diaphragm are absolutely necessary if a speaker is to reproduce the sound accurately. Suppose that the amplifier delivers a sudden burst of power to the speakers. This burst could represent a cymbal crash or a sudden "boom" from a kettle drum. In any event, the large burst builds up in a very short period of time, perhaps in less than one-hundredth of a second, and the cone must accelerate quickly in order to follow it. A sluggish speaker will not be able to reproduce the full impact of the music, and, consequently, it will distort the sound.

Using Newton's second law, $\mathbf{a} = \mathbf{F}/m$, it can be seen that there are two ways to increase the acceleration of a speaker diaphragm. First, if the mass of the diaphragm/voice coil assembly is as small as possible, its acceleration will be maximal for a given amount of force. However, the diaphragm must be built sturdily to withstand the large accelerations, and this requirement sets a limit as to how little mass can be used in its construction. Second, a large magnetic force, **F**, can be used to increase the acceleration of the diaphragm. As discussed in section 11.4, the magnetic force is proportional to the strength of the current in the voice coil. Therefore, increasing the acceleration can also be accomplished by increasing the current in the voice coil, which implies the use of a more powerful amplifier. This is one reason why some speakers require more powerful amplifiers to drive them than do other speakers. In addition, the magnetic force can be increased by the use of a more powerful permanent magnet in the construction of a speaker.

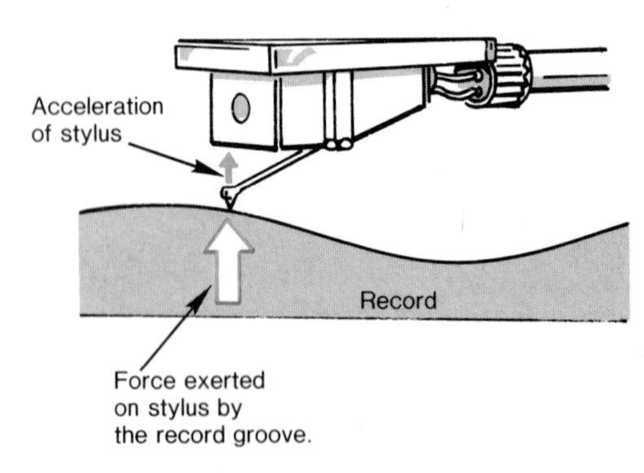

Figure 14.13. The stylus accelerates upward in response to the force exerted on it by the record groove. For large accelerations, the mass of the stylus should be as small as possible.

A phono stylus is another fine example of an object being accelerated by a force. Figure 14.13 illustrates a stylus which is vibrating up and down due to the undulating record groove. A stylus must be able to vibrate at frequencies up to at least 15 kHz in order to pick up the musical content contained in a normal stereo groove. At a frequency of 15 kHz, the stylus is changing its direction of motion 30,000 times each second. According to Newton's second law, such large changes in a stylus' motion can be more easily accomplished if the stylus has a low mass (small inertia).

The acceleration of a stylus can be calculated by using Newton's second law. Assume that a record groove, containing a 1,000 Hz tone of average loudness, exerts an average force of 0.16 N on the stylus. If the mass of the stylus is 0.0005 kg (0.5 g), its average acceleration can be calculated:

$$\mathbf{a} = \frac{\mathbf{F}}{\mathrm{m}} = \frac{0.16\ \mathrm{N}}{0.0005\ \mathrm{kg}} = 320\ \mathrm{m/s^2}.$$

It can be seen that the acceleration of a stylus is quite large, exceeding 30 times the acceleration of gravity (30 g's).

The gravitational force which pulls on an object, often called its "weight," can also be calculated by using Newton's second law. As mentioned in section 14.2, all objects, regardless of their mass, fall toward the earth with an acceleration of approximately $\mathbf{g} = 9.8\ \mathrm{m/s^2}$ (downward) in the absence of friction. The gravitational force, $\mathbf{F}_g$, which produces this acceleration is given by

$$\mathbf{F}_g = \mathrm{m}\mathbf{g} = \mathrm{m}(9.8\ \mathrm{m/s^2}),\ \text{downward.} \qquad \text{(Equation 14.4)}$$

Knowing the mass of an object, the above relation can be used to find its weight. For example, a typical turntable platter has a mass of approximately 2 kg, as indicated in Table 14.1. The weight of the platter is $\mathbf{F}_g = (2\ \mathrm{kg})\ (9.8\ \mathrm{m/s^2}) = 19.6\ \mathrm{N}$, downward, which corresponds to a weight of about 4.4 pounds in the British system.

Since the gravitational force $\mathbf{F}_g$ exerted on an object is often called the weight $\mathbf{W}$ of the object, Equation 14.4 is usually written as

$$\text{Weight} = \mathbf{W} = \mathrm{m}\mathbf{g}.$$

It is not uncommon in hi-fi that forces are stated in terms of just the mass which appears in Equation 14.4. For example, the vertical tracking force of a phono cartridge is often stated as

"Vertical tracking force" = 1.2 grams.

A more proper way to quote the vertical tracking force is to multiply the 1.2 gram (0.0012 kg) mass by the acceleration of gravity ($9.8\ \mathrm{m/s^2}$), as per Equation 14.4:

$$\begin{aligned}\text{Vertical tracking force} &= (0.0012\ \mathrm{kg})(9.8\ \mathrm{m/s^2}) \\ &= 0.012\ \mathrm{N}.\end{aligned}$$

Remember, any time a force is specified in terms of the mass alone, the correct value of the force can be obtained by multiplying the mass by the acceleration of gravity, as necessitated by Newton's second law.

Summary

Newton's second law states that $\mathbf{F} = \mathrm{m}\,\mathbf{a}$, where $\mathbf{F}$ is the force acting on an object, m is its mass, and $\mathbf{a}$ is its acceleration. According to this law, whenever a force acts on an object, the object must accelerate. Conversely, if no force acts on an object, its acceleration is zero, and the object travels with a constant velocity. As a special example of Newton's second law, the gravitational force $\mathbf{F_g}$ which acts on an object is given by $\mathbf{F_g} = m\mathbf{g}$ where $\mathbf{g} = 9.8\ m/s^2$, downward. $\mathbf{F_g}$ is sometimes called the "weight" of an object.

14.7 Newton's Third Law

Newton's third law indicates that forces always occur in equal-but-opposite pairs, and it can be stated as follows.

> **Newton's Third Law:** Whenever one object exerts a force on a second object, the second object exerts an equal, but opposite, force on the first.

The third law is often called the "action-reaction" law, and it is commonly quoted as "For every action (force) there is an equal, but opposite, reaction (force)."

It should be emphasized that the two forces mentioned in the law act on *different* objects; they do *not* act on the same object and, therefore they do *not* cancel out each other. Consider the speaker which was illustrated in Figure 14.12. This figure showed the permanent magnet exerting a 20 N magnetic force on the voice coil/diaphragm. Of course, the force causes the voice coil/diaphragm to accelerate to the right, according to Newton's second law. The third law states that the voice coil/diaphragm exerts an equal, but oppositely-directed, force of 20 N on the permanent magnet.

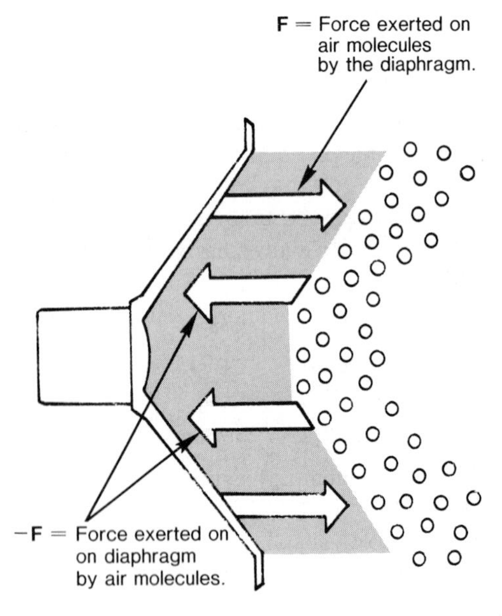

Figure 14.14. The moving diaphragm exerts a force **F** on the air molecules, thus creating a condensation. The air molecules exert an equal, but opposite, force **-F** on the diaphragm, as required by Newton's third law.

Another example of the third law is that of the action-reaction forces which occur when a speaker diaphragm is producing sound. As depicted in Figure 14.14, the forward-moving diaphragm exerts a force on the surrounding air molecules. This force compresses the molecules, thus creating a sound condensation. By Newton's third law, the air molecules exert an equal, but oppositely-directed, force on the diaphragm. This force, being opposite in direction to the motion of the diaphragm, tends to decelerate it. Therefore, when a diaphragm is creating sound its movement is less than it would be if it were placed in a vacuum where no air molecules exist. A power amplifier expends part of its power in overcoming the reaction force created by the air molecules. Notice, as mentioned earlier, that the two forces are exerted on *different* objects; one force is exerted by the diaphragm on the air, while the equal, but opposite, force is exerted by the air on the diaphragm.

The above example is much like throwing a baseball, where the moving hand simulates the diaphragm and the baseball represents an air molecule. When the ball is thrown forward, the hand exerts a force on the ball which causes the ball to accelerate forward. The ball, during this time, also exerts a backward force of equal strength on the hand, which slows its forward movement. Hence, the speed which the hand can attain when an object is being thrown is slower than that which the hand can attain when nothing is being thrown—a situation which is familiar to most of us.

Perhaps the most common situation involving action and reaction forces is during the process of walking or running. Figure 14.15 illustrates the situation. The person, through the use of leg muscles, exerts a frictional force on the earth. The earth accelerates under the application of this force, but being so massive (approximately 6×10^{24} kg) its acceleration is imperceptibly small. According to Newton's third law, the earth exerts an equal force back on the person's leg. It is this reaction force which causes the person to accelerate forward, and, hence, walk. It is important to realize that the person must first exert a force on the earth in order that the earth pushes back on the person. We all know how difficult it is to walk on smooth ice. The ice, with its super-smoothness, does not permit the person to exert a large frictional force on the earth. This, in turn, means that the earth cannot exert a large force on the person. Hence, it is extremely difficult to walk on ice.

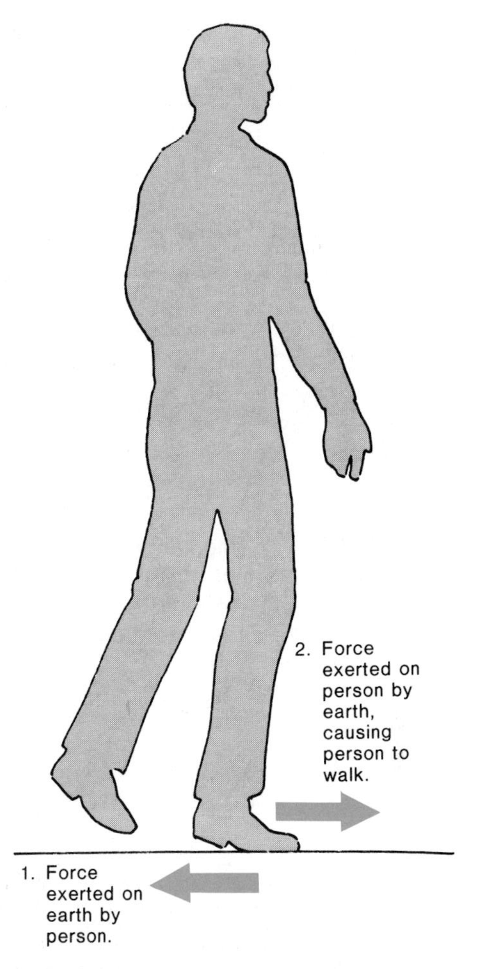

Figure 14.15. In order to walk, a person first exerts a force on the earth. By Newton's third law, the earth exerts an equal, but oppositely directed, force on the person's foot, thus causing the forward movement.

Summary

Newton's third law states that when object #1 exerts a force on object #2, object #2 exerts an equal, but oppositely-directed, force back on object #1. These two forces act on different objects, and therefore they do not cancel out each other.

14.8 Momentum

Momentum is a concept that combines the mass and velocity of an object, and it is defined as follows.

Momentum: The momentum **p** of an object is the product of its mass m times its velocity **v**:

$$\mathbf{p} = m\mathbf{v}. \qquad \text{(Equation 14.5)}$$

Momentum, like velocity, is a vector.

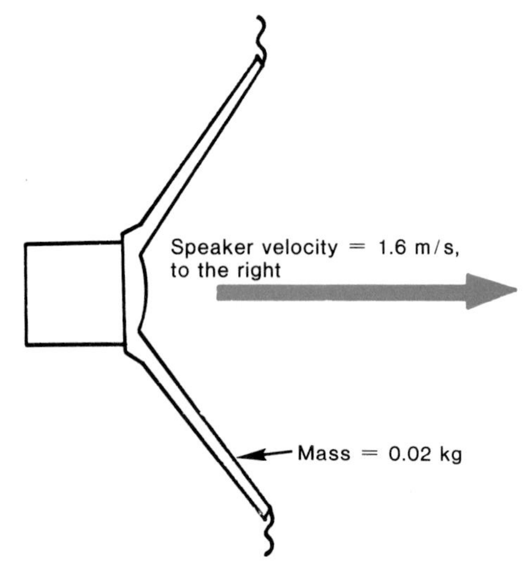

Figure 14.16. The momentum of the speaker is the product of its mass and its velocity. The momentum is a vector which points in the same direction as the speaker's velocity.

Figure 14.16 illustrates a 0.02 kg speaker cone moving with a velocity of 1.6 m/s, to the right. According to Equation 14.5, its momentum at this instant in time is

$$\mathbf{p} = (0.02\ \text{kg})(1.6\ \text{m/s}) = 0.032\ \text{kg}\cdot\text{m/s, to the right.}$$

Momentum is a concept that is often used to convey the fact that a moving object possesses either a large mass or a large velocity, or both. A car moving at 50 m/s has more momentum than when it travels at 20 m/s. And, a car traveling at 20 m/s, being more massive, has more momentum than a bicycle which is moving at the same velocity.

Although the equations for momentum and Newton's second law look somewhat alike, it should be noted that one equation involves the velocity while the other involves the acceleration:

$$\mathbf{p} = \text{m}\mathbf{v}\ \text{(momentum)}$$
$$\mathbf{F} = \text{m}\mathbf{a}\ \text{(2nd law).}$$

Actually, the second law can be (and often is) expressed in terms of the object's momentum, rather than its acceleration. To see how this is accomplished, recall the definition of average acceleration from Equation 14.2:

$$\text{Average acceleration} = \left(\frac{\text{Final velocity} - \text{Initial velocity}}{\text{Elapsed time}}\right)$$

or, in symbols

$$\mathbf{a} = \left(\frac{\mathbf{v}_f - \mathbf{v}_i}{t}\right).$$

Substituting this expression for **a** into Newton's second law yields

$$\mathbf{F} = \text{m}\mathbf{a} = \text{m}\left(\frac{\mathbf{v}_f - \mathbf{v}_i}{t}\right) = \left(\frac{\text{m}\mathbf{v}_f - \text{m}\mathbf{v}_i}{t}\right).$$

Since the quantity $m\mathbf{v}_f$ represents the final momentum $\mathbf{p}_f$ of the object, and $m\mathbf{v}_i$ is its initial momentum $\mathbf{p}_i$, Newton's second law can be written as

$$\mathbf{F} = \left(\frac{\mathbf{p}_f - \mathbf{p}_i}{t}\right). \qquad \text{(Equation 14.6)}$$

This form of the second law is, in every sense, equivalent to the equation **F** = m**a**. Equation 14.6 simply expresses the second law in terms of momentum and time, rather than in terms of mass and acceleration.

Equation 14.6 states, in essence, that the momentum of an object can be changed only when a force is applied to it.

On the other hand, if no resultant force acts on the object ($\mathbf{F} = 0$), its momentum cannot change since

$$\mathbf{F} = 0 = \frac{\mathbf{p_f} - \mathbf{p_i}}{t}, \quad \text{or} \quad \mathbf{p_f} = \mathbf{p_i}.$$

This means that, in the absence of any forces, an object keeps its initial momentum. Essentially, this is a restatement of Newton's first law: every object continues in its state of rest (i.e., zero momentum) or of constant motion (constant momentum), unless it is compelled to change that state (its momentum) by forces impressed upon it. Thus, Newton's first law can be thought of as a special case of the second law when the resultant force which acts on an object is zero.

Summary

The momentum of a moving object is the product of its mass and its velocity. Momentum, like velocity, is a vector. Newton's second law can be expressed in terms of the change in an object's momentum:

$$\mathbf{F} = \frac{\mathbf{p_f} - \mathbf{p_i}}{t},$$

where $\mathbf{p_f}$ is its final momentum, $\mathbf{p_i}$ is its initial momentum, and t is the elapsed time.

14.9 Torque

In hi-fi there are many applications that involve rotational motion rather than straight line motion. A turntable platter, a tape reel, the capstan/pinch roller assembly of a tape deck, and even the rotatable tuning knob found on many tuners, are a few examples of rotation that occur in audio components (see Figure 14.17).

Figure 14.17. Some examples of rotational motion in hi-fi.

In sections 14.5 and 14.6 we discussed how a force causes an object to change its state of motion, i.e., to accelerate. When dealing with a rotating object, such as a turntable platter, it is a *torque* (rhymes with "cork") that causes an object to change its state of rotational motion. As an example of a torque, let us consider the case of a door being opened, as shown in Figure 14.18. In part (A) of the figure a person is exerting a force at the door knob, while in (B) the same force is applied closer to the hinge. Although the forces in the two illustrations are the same, experience tells us that the door in (A) opens easier. Evidentally the ease with which a door opens depends on both the magnitude of the applied force, and its point of application relative to the hinge. The idea of combining a force with a distance leads to the definition of a torque.

Figure 14.18. It is easier to open the door in (A) than in (B), even through the same force is applied in both situations.

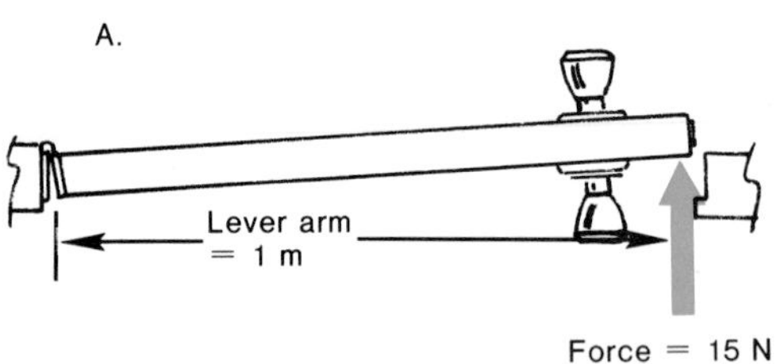

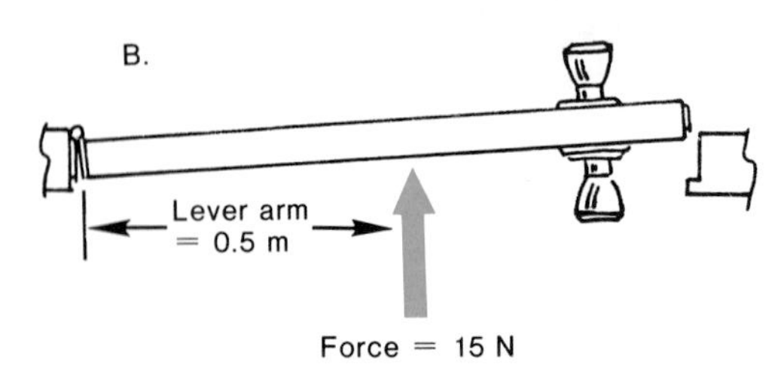

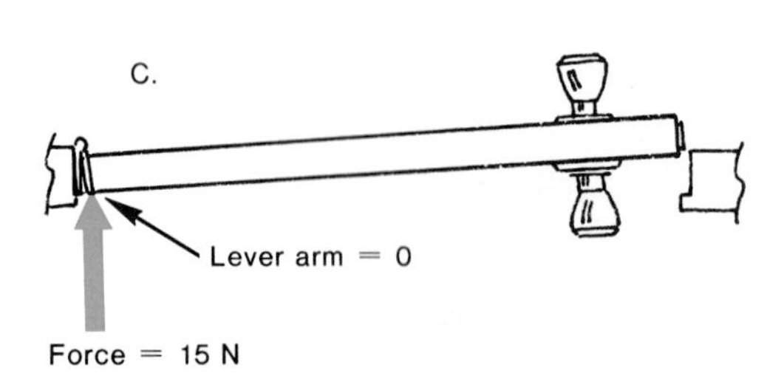

Figure 14.19. A 15 N force is applied to the door at three different positions. In (A) the lever arm is 1 m, in (B) the lever arm is 0.5 m, while in (C) the lever arm is 0.

Torque: The product of the force times the lever arm. The lever arm is the perpendicular distance between the force vector and the pivot point (hinge).

$$\text{Torque} = \text{Force} \times \text{Lever arm}. \qquad \text{(Equation 14.7)}$$

Figure 14.19 illustrates a 15 N force being applied to three different positions along the door. Using the forces and lever arms shown in the figure, the torques exerted in each case can be calculated as follows:

Part (A): Torque = Force × Lever arm
Torque = (15 N) × (1 m) = 15 N·m.

Part (B): Torque = Force × Lever arm
Torque = (15 N) × (0.5 m) = 7.5 N·m.

Part (C): Torque = Force × Lever arm
Torque =(15 N) × (0) = 0.

The largest torque is exerted in part (A) of the figure. In part (C) the force is applied directly to the hinge and, consequently, the lever arm is zero. In this case, no torque is applied to the door, even though a force is exerted on it. Since the door is easiest to open in (A), more difficult to open in (B), and impossible to open in (C), we conclude that the larger the applied torque, the easier it is to rotate the door. In other words, the larger the torque applied to an object, the easier it is to start the object rotating.

Suppose that we wish to turn a stubborn nut with a wrench, and the available force is inadequate to produce the necessary torque to loosen the nut, as shown in Figure 14.20 (A). Part (B) of the figure shows that the nut may be removed with the *same* force if a pipe is inserted over the handle to increase the wrench's lever arm. This increased lever arm produces an increased torque even when the applied force has not been changed.

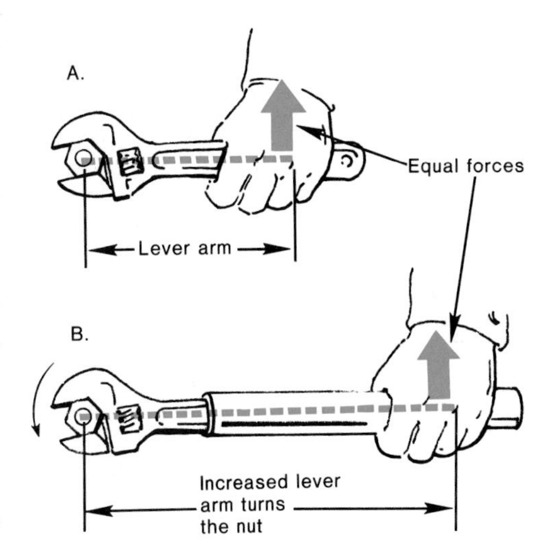

Figure 14.20. (A) The applied torque is insufficient to turn the stubborn nut. (B) Placing a pipe over the end of the wrench increases the lever arm, thus increasing the torque such that the nut can now be loosened.

When a bolt (or nut) is to be tightened on a machine, such as an automobile engine, the manufacturer will often specify the correct tightening torque. Otherwise a mechanic could accidentally over-tighten the bolt and possibly "strip-off" its threads. Often a special torque wrench, which indicates the amount of torque being applied, is used for this purpose.

Many rotating objects in hi-fi are powered by motors. Since the motor provides the necessary torque to start an object rotating, it should not be too surprising to discover that motors are often rated according to the amount of torque that they can produce. Figure 14.21 illustrates two motors which are rated as:

Motor #1
Torque rating = 0.4 N·m (3.54 pound·inches*)
Motor #2
Torque rating = 0.1 N·m (0.88 pound·inches).

Because motor #1 has a greater torque rating, it can bring an identical turntable platter up to speed much faster than motor #2. Most quality turntables have high torque motors which can bring a platter up to full speed within a third of a revolution.

Using the definition of torque in Equation 14.7, it is possible to calculate the amount of force exerted on each platter by the two motors shown in Figure 14.21. As illustrated in the figure, each motor shaft has a radius of 0.005 m (0.5 cm), which is the lever arm distance.

Motor #1: Torque = Force × Lever arm,
0.4 N·m = Force × (0.005 m),
Force = 80 N.

Motor #2: Torque = Force × Lever arm,
0.1 N·m = Force × (0.005 m),
Force = 20 N.

*Torques in the metric system are measured in newton·meters (N·m), while in the British system they are measured in pound·inches. The conversion factor between the two systems is

1 N·m = 8.84 pound·inches.

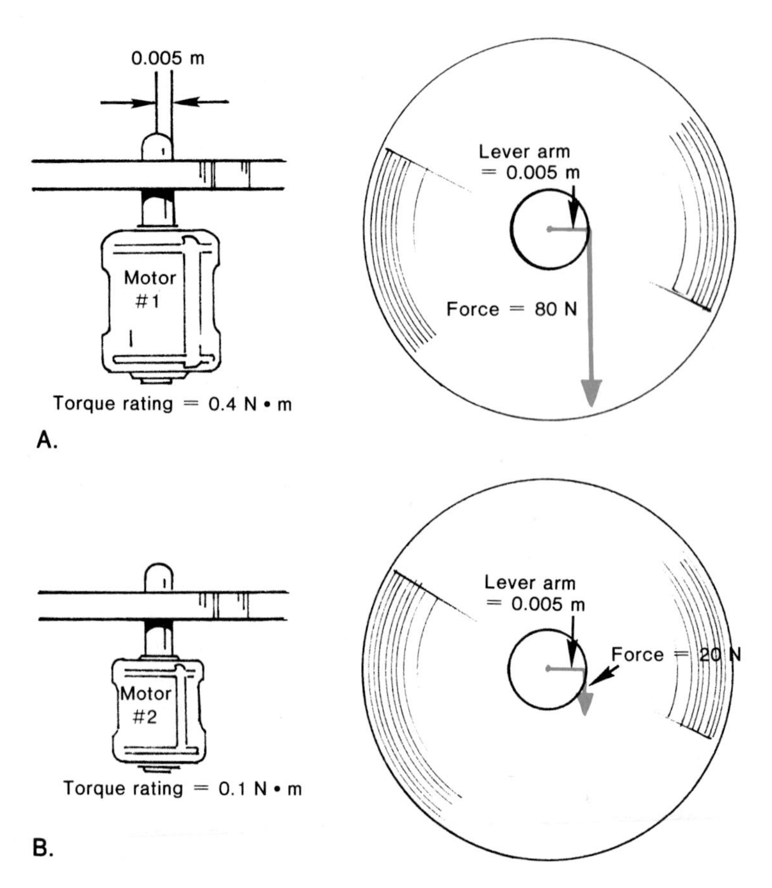

Figure 14.21. (A) The 0.4 N·m torque motor exerts a force of 80 newtons on the inner edge of the record. (B) The smaller, 0.1 N·m torque motor exerts only 20 newtons of force on the inner edge of the record. In both cases the lever arm is 0.005 m.

Hence, motor #1 exerts an 80 N force on the platter while motor #2 only exerts 20 N on its platter.

Summary

Torque is defined as the product of a force times a lever arm. If a net torque is applied to an object at rest, the object will begin to rotate. If the object is rotating, a torque will cause it to either speed up or slow down, depending on how the torque is applied. Motors are often rated according to the amount of torque that they can produce. A high torque motor can bring a turntable platter up to full speed faster than a low torque motor.

14.10 Rotational Inertia

Newton's first law is concerned with the motion of an object which is either at rest or traveling in a straight line. For rotational motion, Newton's first law can be modified as follows.

Newton's First Law for Rotational Motion: Every object continues in a state of rest, or of constant rotational motion, unless it is compelled to change that state by torques impressed upon it.

A good example of this law is that of a turntable platter which is rotating at 33-1/3 rpm. If the motor is suddenly turned off, the platter continues to rotate for some time before finally coming to rest. It is the frictional torques, which arise because of frictional forces, that act on the platter and cause it to slow down. If there were *no* frictional torques, Newton's first law for rotational motion states that the platter would rotate forever at precisely 33-1/3 rpm (even though the motor had been turned off), neither speeding up nor slowing down. When the platter is at rest, Newton's first law for rotational motion also states that the platter would remain at rest until a net torque is impressed upon it. Of course, a motor usually provides the necessary torque to start the platter rotating.

Another example of Newton's first law for rotational motion is that of the earth itself. The earth rotates on its axis at the rate of one revolution per day. Since there are virtually no torques of any magnitude which act on the earth, it continues to rotate at the same rate as it has for the past several million years.

If a torque is applied to an object, the object's state of rotational motion will change. However, it is more difficult to change the rotational motion of some objects than others. The harder it is to change the rotational motion of an object, the more *rotational inertia* it has. The rotational inertia of an object, often called its *moment of inertia*, can be stated as follows.

Rotational Inertia (Moment of Inertia): The property of an object that resists any change in its rotational motion when a torque is applied.

An object which has a large moment of inertia is hard to set into rotational motion. Likewise, once set into motion it is difficult to either speed it up or slow it down. In this respect the moment of inertia of a rotating object plays a role similar to the inertia of an object which is moving along a straight line, as discussed in section 14.5. However, the moment of inertia depends on both the object's mass and shape. To illustrate how an object's mass affects its moment of inertia, consider Figure 14.22 which shows two turntable platters that have different masses but the same shape. The platter

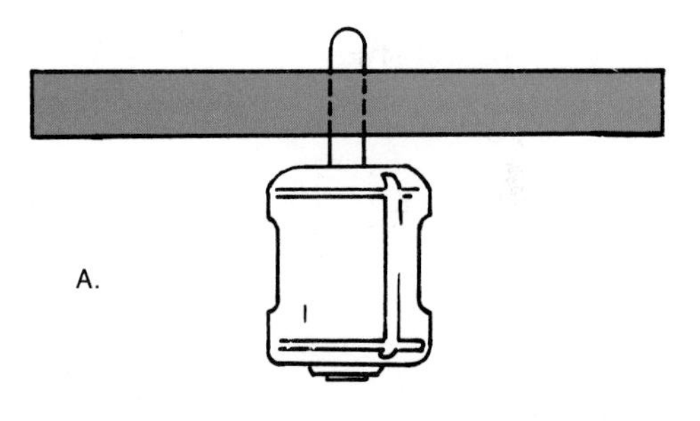

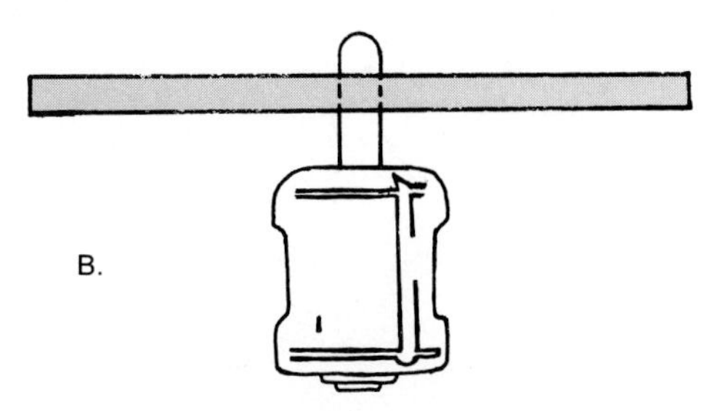

Figure 14.22. The platter in (A) has a larger moment of inertia, due to its larger mass, than does the platter in (B). Both platters have the same shape.

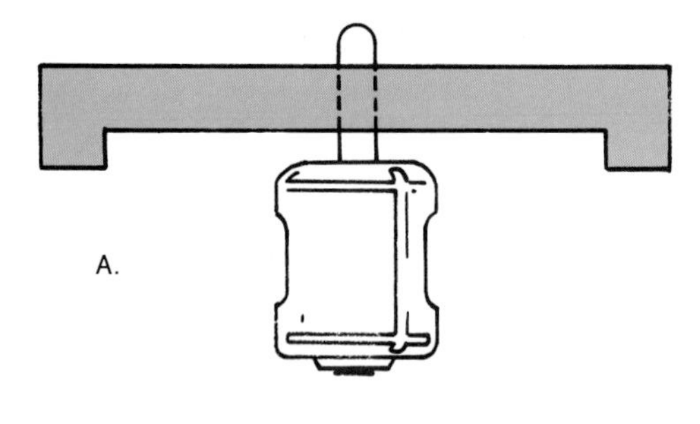

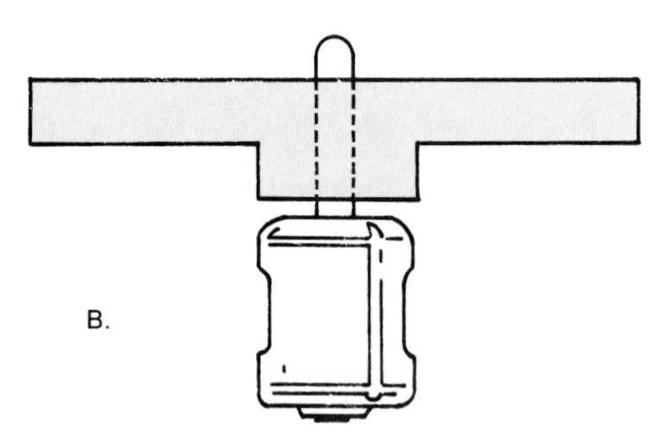

Figure 14.23. The platter in (A) has a greater moment of inertia than the platter in (B), because more of A's mass is distributed toward the outer edge of the platter. Both platters have the same mass.

in part (A) of the figure has a greater mass and, hence, a greater moment of inertia than does the platter in (B). Therefore, the platter in (A), once rotating at the proper speed, will be able to resist better any irregular speed fluctuations caused by frictional torques and imperfections in the drive system. All other things being equal, the more massive platter in (A) will yield a better wow and flutter specification than the platter in (B). Figure 14.23 shows two platters which have identical masses, but different shapes. The platter in part (A) has a concentration of mass located near its outer edge, while the platter in (B) has a concentration of mass located near its center. Without going into all the technical reasons, experiments show that the platter in (A) has a larger moment of inertia than the platter in (B). Therefore, all other things being equal, the platter in (A) is more immune to speed fluctuations, and it will yield a better wow and flutter specification than the platter in (B). In conclusion, Figures 14.22 and 14.23 illustrate that the moment of inertia of an object can be increased by increasing its mass and/or redistributing the mass so that it is farther from the center of rotation. That is why platters found on quality turntables are usually rather massive with a substantial amount of the mass located near the outer rim.

Of course, low wow and flutter specifications are an absolute necessity for quality tape decks, as well as for turntables. In this respect the drive system of a tape deck must also have a large moment of inertia, so as to smooth out any irregular speed fluctuations. Figure 14.24 shows a cylindrical "flywheel" attached to the base of the capstan. The flywheel is designed to have a relatively large moment of inertia, so that the capstan/pinch roller assembly can pull the tape at a speed which is as constant as possible. It is no accident that the flywheel on a capstan serves the same purpose as does the flywheel on an automobile engine; namely, the engine's flywheel helps to smooth out speed fluctuations which arise because of the unevenness of the power strokes.

The tuning knob on a tuner or receiver often utilizes a flywheel to give the knob a "silky smooth" feeling when it is rotated. The flywheel, which is attached to the knob, is usually located within the hi-fi component, as illustrated in Figure 14.25.

Summary

Newton's first law for rotational motion states that every object continues in a state of rest, or of constant rotational motion, unless it is compelled to change that state by torques impressed upon it. The ability of an object to resist changes

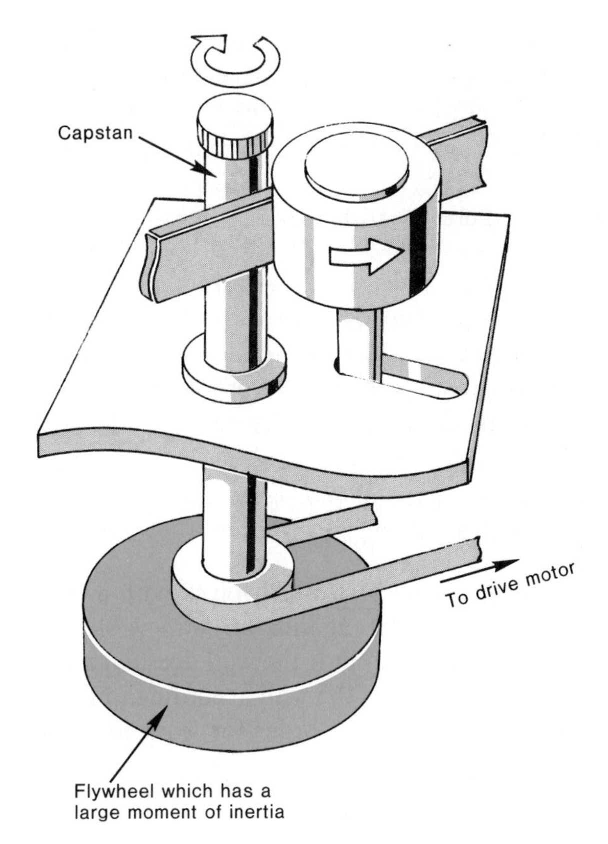

Figure 14.24. The flywheel, which is attached directly to the capstan, helps to reduce speed fluctuations caused by imperfections in the drive system. The flywheel has a large moment of inertia.

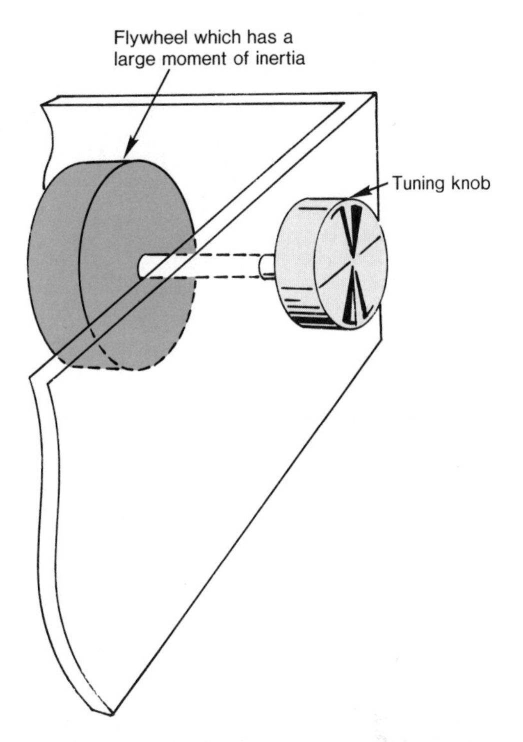

Figure 14.25. The flywheel, which is attached directly to the tuning knob, helps to give the knob a smooth feeling when it is rotated.

in its rotational motion is measured by its moment of inertia (or rotational inertia). The moment of inertia depends on the mass of an object as well as its shape. Turntable platters, capstans, and tuning knobs have large moments of inertia for smooth, constant rotation.

14.11 Work, Power, and Energy

Work, power, and energy are closely related concepts that are often used in describing the mechanical operation of hi-fi components. In this section we will discuss each of these concepts, and present some examples of their usage in audio.

Work

The word "work" is commonly used in our daily lives, and it is usually associated with the performance of some sort of task. Expressions, such as "hard-working" and "working eight hours a day," are familiar to all of us. Whatever its meaning in daily life, work has a very precise meaning in science, and it is defined as follows.

Work: The product of a force that acts on an object and the distance which the object moves under the influence of the force, when the force and the movement are in the same direction.

$$\text{Work}^* = \text{Force} \times \text{Distance}. \qquad \text{(Equation 14.8)}$$

As an example of work, consider the moving speaker diaphragm in Figure 14.16. Suppose that a constant magnetic force of 8.0 N causes the diaphragm to move a distance of 6.4×10^{-3} meters (about ¼ inch). According to Equation 14.8, the work done on the diaphragm by the magnetic force is

$$\text{Work} = (8.0 \text{ newtons}) \times (6.4 \times 10^{-3} \text{ meters}),$$
$$\text{Work} = 5.1 \times 10^{-2} \text{ joules}.$$

Notice that the definition of work involves both a force *and* a distance. One may push against an immovable wall all day and become very tired, but if the wall does *not* move, no work has been performed on the wall according to Equation 14.8. The wall must move in order for work to be accomplished.

Power

Power is the rate at which work is performed, and it is defined as follows.

Power: The work performed divided by the time required to perform the work.

$$\text{Power}^{**} = \frac{\text{Work}}{\text{Time}}. \qquad \text{(Equation 14.9)}$$

Returning to the example of the moving speaker diaphragm, we have seen that 5.1×10^{-2} joules of work were required to move the diaphragm a distance of 6.4×10^{-3} meters. Nothing was said about the time required to move the diaphragm through this distance. If the diaphragm traveled the distance in 5×10^{-3} seconds, Equation 14.9 can be used to calculate the power delivered to the diaphragm:

$$\text{Power} = \frac{5.1 \times 10^{-2} \text{ joules}}{5.0 \times 10^{-3} \text{ seconds}},$$
$$\text{Power} = 10.2 \text{ watts}.$$

*In the metric system the unit of work is newtons times meters. One newton · meter is called a *joule.*

**The unit of power in the metric system is a joule/sec. One joule/sec is called a "watt."

However, if the 5.1×10^{-2} joules of work were delivered in a much shorter time, say 0.5×10^{-3} seconds, the power would then be

$$\text{Power} = \frac{5.1 \times 10^{-2} \text{ joules}}{0.5 \times 10^{-3} \text{ seconds}},$$

$$\text{Power} = 102 \text{ watts.}$$

Therefore, more power is required if a given amount of work is to be performed in a shorter period of time.

Energy

"Energy" is another concept that has also found its way into our lives. Generally speaking, energy can be thought of as "the ability to do work." For example, the gasoline in the tank of a car represents stored chemical energy. This energy can be used by the engine to produce useful work that moves the car. Or, a person who is "full of energy" has the capacity for performing relatively large amounts of work. In science, energy is defined as follows.

Energy: The ability to do work. Like work, the unit of energy is the joule.

Energy takes many forms, and two common forms in mechanics are kinetic energy and gravitational potential energy, which we will now discuss.

Kinetic energy, often called the energy of motion, is the energy that an object has because of its motion. For example, a moving hammer has kinetic energy, because when it strikes a nail the hammer exerts a force which moves the nail through a distance. Thus, the moving hammer does work on the nail, and, by definition, the hammer has kinetic energy. The kinetic energy of an object is given by the following relation.

Kinetic Energy:

$$\text{Kinetic energy} = \frac{1}{2} mv^2, \qquad \text{(Equation 14.10)}$$

where m is the mass of the object and v is its speed.

Notice in Equation 14.10 that the kinetic energy of an object depends on the *square* of its speed. This implies that when the speed of an object is doubled its kinetic energy is quadrupled ($2^2 = 4$); as such, it can do four times more work.

Since a moving object can do work because it possesses kinetic energy, it is of interest to examine how the object obtained its kinetic energy in the first place. To this end, consider an object which is initially at rest. If a constant force of magnitude F is exerted on the object, the object will accelerate from rest to a speed v, while traveling over a distance d. It can be shown* that the kinetic energy acquired by the object is equal to the work done on it by the force; i.e.

$$\underbrace{\frac{1}{2}mv^2}_{\text{Kinetic energy acquired by the object}} = \underbrace{F \times d.}_{\text{Work done on the object by the force F}}$$

In other words, an object possesses kinetic energy because work was initially done on it by a force. If the object is already moving when the force is first applied, the work done by the force goes into *changing* its kinetic energy, as shown below;

$$\frac{1}{2}mv_f^2 - \frac{1}{2}mv_i^2 = F \times d,$$

where m is the mass of the object, and v_f and v_i are its final and initial speeds. Therefore, the work done by the force equals the final kinetic energy minus the initial kinetic energy.

In summary, the relationship between work and kinetic energy operates in two ways. First, if work is done on an object, the kinetic energy of the object increases. Second, if an object has kinetic energy, it can do work on something else.

Gravitational potential energy, often called the energy of position, is the energy that an object possesses because of its position, rather than its motion. Figure 14.26 (A) illustrates a loudspeaker being lifted off the ground. When the loudspeaker is raised it has potential energy because of its position relative to the ground. This potential energy is often called *gravitational potential energy*, because the loudspeaker has potential energy by virture of the gravitational force which pulls it toward the earth. For example, if the loudspeaker is accidentally dropped, as illustrated in part (B), the gravitational force pulls it with increasing speed toward the ground. The loudspeaker's potential energy is being changed into kinetic energy, which allows the loudspeaker to do work when it strikes the ground. Unfortunately, in this case the work that it does is primarily used to destroy itself, as shown in part (C).

*The proof of this statement can be found in any standard physics text.

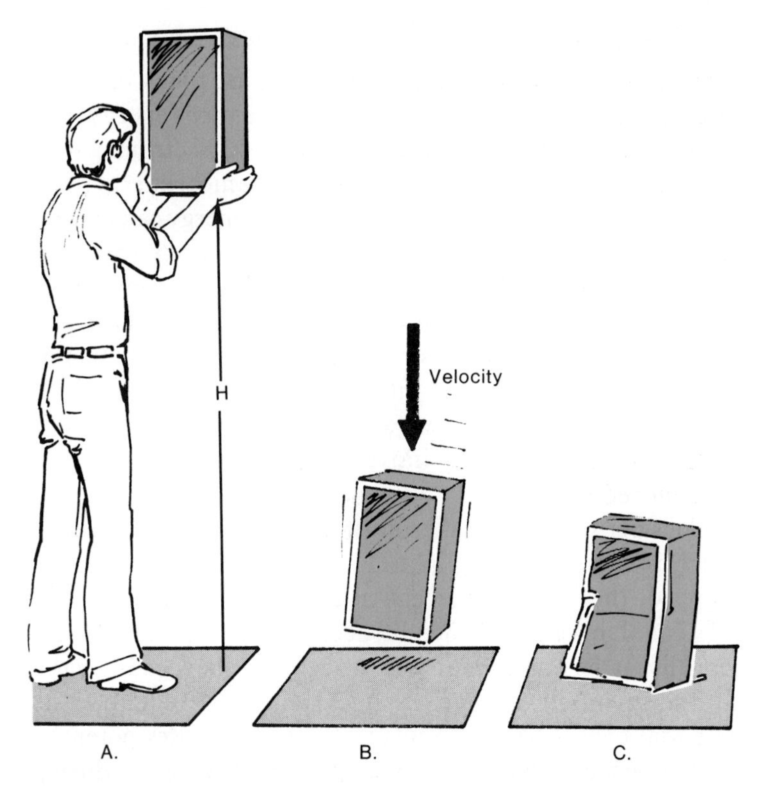

Figure 14.26. (A) The loudspeaker has potential energy because of its height above the floor. (B) The original potential energy of the loudspeaker has been converted into kinetic energy. (C) When the loudspeaker strikes the floor, part of its kinetic energy is used to damage the unit.

The gravitational potential energy of an object is defined as follows.

Gravitational Potential Energy: The product of an object's weight W and its height H above the ground:*

$$\text{Gravitational Potential Energy} = W \times H = (mg) \times H. \qquad \text{(Equation 14.11)}$$

Thus, a heavier object has a greater gravitational potential energy than does a lighter one. Furthermore, the gravitational potential energy can be increased by increasing the height above the ground. In other words, heavier objects falling from greater heights can do more work than lighter objects falling from lesser heights.

*The derivation of this formula can be found in any standard physics text.

Conservation of Mechanical Energy

In section 9.3 we mentioned that there are many different types of energy, such as heat energy, chemical energy, nuclear energy, acoustic energy (sound waves), electromagnetic energy (light and radio waves), and mechanical energy. The Law of Conservation of Energy states that energy can neither be created nor destroyed; it can only be converted from one type to another. In section 9.3 numerous examples were presented which illustrated energy conversions in hi-fi systems. We now wish to illustrate how gravitational potential energy may be converted into kinetic energy. In order to facilitate this discussion, we will call the sum of an object's kinetic energy and its gravitational potential energy its *total mechanical energy*.

Consider the loudspeaker illustrated in Figure 14.26 (A), which is situated at a height H above the floor. Before it is dropped, the loudspeaker has a gravitational potential energy equal to weight × height, according to Equation 14.11. Initially, it has no kinetic energy because its speed is zero. At this point its *total* mechanical energy is equal to its potential energy. As the loudspeaker falls, it loses potential energy because the height is decreasing. However, during this time its kinetic energy is increasing because its speed is increasing. If we examine the moving loudspeaker after it has fallen halfway to the ground, we see that it has both potential energy and kinetic energy. Its potential energy is only one-half its original potential energy, since the height is only one-half of what it was. However, the other half has been transformed into kinetic energy. At this point the loudspeaker has equal amounts of potential and kinetic energy. However, the *total* mechanical energy of the loudspeaker at this point is equal to the total amount of mechanical energy that it originally had at the top. This statement is consistent with the Law of Conservation of Energy; the total mechanical energy (kinetic energy plus gravitational potential energy) of the loudspeaker does not change during its fall. During the fall, though, gravitational potential energy is converted into kinetic energy. This important observation can be formulated by the following equation.

$$\underbrace{mgH}_{\text{Total energy at the top of the fall}} = \underbrace{mgh + \frac{1}{2}mv_{,}^{2}}_{\text{Total energy at a height h above the floor.}}$$

(Equation 14.12)

where m is the mass of the loudspeaker, $g = 9.8\ m/s^2$, H is the original height, h is its height at any point along the fall, and v is its speed at that point. The above equation can be solved to obtain the speed;

$$v = \sqrt{2\,g\,(H - h)}\,.$$

Equation 14.12 shows that the loudspeaker's kinetic energy just before it strikes the floor (h = 0) is exactly equal to the potential energy at the top. The original potential energy has been completely converted into kinetic energy, without any being lost.

Question: What is the speed of the loudspeaker at a height of 2 m above the floor if its original height was 5 m? Using Equation 14.12, the speed is found to be

$$v = \sqrt{2 \times 9.8 \times (5-2)} = 7.7\ m/s.$$

Summary

Work is defined as the product of the net force that acts on an object and the distance which the object moves under the influence of the force, when the force and the movement are in the same direction. Work is measured in units of either foot·pounds or joules. Power is work divided by time, and the unit of power in the metric system is the watt. Energy is the ability to do work. Two common types of mechanical energy are kinetic energy and gravitational potential energy. An object possesses kinetic energy because of its motion. Kinetic energy is given by the equation,

Kinetic energy $= \frac{1}{2}mv^2$, where m is the mass of the object and v is its speed. An object possesses gravitational potential energy because of its height relative to the ground. Gravitational potential energy is given by the formula, Gravitational potential energy = mgH, where m is the mass of the object, $g = 9.8\ m/s^2$, and H is its height above the ground. An object which has either kinetic energy or gravitational potential energy can do work. The total mechanical energy of an object is the sum of its kinetic energy and its gravitational potential energy. When an object falls, gravitational potential energy is converted into kinetic energy; however, its total mechanical energy remains the same during the fall.

Summary of Terms

Acceleration (Average)—(See **Average Acceleration.**)

Average Acceleration—The change in an object's velocity divided by the elapsed time. Average acceleration is a vector.

Average Speed—The distance traveled by an object divided by the time required to make the trip.

Dyne—A unit of force in the metric system.

Energy—The ability of an object to do work. A unit of energy is a joule.

Energy, Gravitational Potential—(See **Potential Energy.**)

Energy, Kinetic—(See **Kinetic Energy.**)

Force—A push or a pull which is exerted on an object. Force is a vector.

Friction—A force which arises when objects are in contact with each other. Kinetic friction occurs then the two objects actually slide against each other. Static friction occurs when there is no relative motion between the two objects.

g—The acceleration of an object due to the earth's gravity. It is $9.8\ m/s^2$, downward.

Gram (gm)—A unit of mass in the metric system.

Inertia—The property of an object that resists any change in its motion when a force is applied. See **Mass.**

Inertia, Rotational—(See **Moment of Inertia.**)

Joule—A unit of energy in the metric system.

Kilogram (kg)—A unit of mass in the metric system.

Kinetic Energy—The energy of motion, described by the relationship: Kinetic energy $= \frac{1}{2} mv^2$.

Kinetic Friction—(See **Friction.**)

Lever Arm—The perpendicular distance between the force vector and the pivot point about which the force causes a rotation. A lever arm is necessary for calculating torque.

Magnitude of a Vector—The "size" of a vector.

Mass—A quantitative measure of an object's inertia. Mass and weight are different. In the metric system mass is measured in kilograms (kg) or grams (gm).

Mechanics—A field of science which studies the motion of objects.

Moment of Inertia—The property of an object that resists any change in its rotational motion when a torque is applied.

Momentum—The product of an object's mass times its velocity, described by the relationship: $\mathbf{p} = m\mathbf{v}$. Momentum is a vector.

Newton—A unit of force in the metric system.

Newton's First Law—Every object continues in a state of rest, or in a state of constant motion along a straight line, unless it is compelled to change that state by forces impressed upon it.

Newton's First Law for Rotational Motion—Every object continues in a state of rest, or of constant rotational motion, unless it is compelled to change that state by torques impressed upon it.

Newton's Second Law—An object will accelerate when a force is impressed upon it. The acceleration **a** is directly proportional to the net force **F** acting on the object, and inversely proportional to its mass m: $\mathbf{a} = \mathbf{F}/m$.

Newton's Third Law—Whenever one object exerts a force on a second object, the second object always exerts an equal, but opposite, force on the first.

Potential Energy (Gravitational)—The energy of an object due to its height above the earth's surface, described by the relationship: Potential energy $= mgh$.

Power—The work performed divided by the time required to perform the work. The unit of power is a watt.

Resultant Vector—A vector which represents the vector sum of two or more vectors.

Speed (Average)—(See **Average Speed.**)

Static Friction—(See **Friction.**)

Torque—The product of the force times the lever arm. A torque causes an object to rotate either faster or slower.

Vector—A quantity that has both a magnitude and a direction. Examples are velocity, acceleration, force, and momentum.

Vector Addition—A method for mathematically adding two or more vectors, which takes into account the directions and magnitudes of the vectors.

Velocity—A measurement which includes both the speed of an object and the direction of its motion. Velocity is a vector.

Watt—A unit of power in the metric system.

Weight—The force exerted on an object by gravity.

Work—The product of a force that acts on an object and the distance which the object moves under the influence of the force, when the force and the movement are in the same direction. The unit of work is a joule.

Review Questions

1. Define the average speed of an object. How does speed differ from velocity?
2. Name three quantities which are vectors.
3. How is acceleration related to the change in an object's velocity? Is it possible for an object to be accelerating even though its speed is constant?
4. Distinguish between constant velocity and constant acceleration. Is is possible for an object's speed to be changing even though it's acceleration is constant?
5. Explain the difference between a kinetic frictional force and a static frictional force.
6. Clearly distinguish between weight and mass. Why is the weight of a person different on the earth than on the moon?
7. Explain how to add two or more vectors by the method of vector addition.
8. What is meant by the "inertia" of an object? Does a 6 kg object have twice as much inertia as a 3 kg object?
9. What kind of motion does an object experience when a net force is exerted on it?

10. Using Newton's second law, what two factors influence the acceleration of an object?
11. Using Newton's third law, explain why a speaker diaphragm tends to slow down when it is creating a sound wave.
12. Do the action-reaction forces mentioned in Newton's third law act on the same object?
13. What two parameters are involved in the concept of momentum? Is momentum a vector?
14. Write down Newton's second law in terms of the change in momentum that an object experiences.
15. Distinguish between a force and a torque. How is it possible to increase the torque without increasing the force?
16. Why do turntable platters and tape deck capstans have relatively large moments of inertia? Name two ways that the moment of inertia of an object can be increased.
17. Explain how work is related to force and distance. How does work differ from power?
18. Distinguish between the concepts of work and energy.
19. What is the difference between kinetic energy and potential energy? Can one type be converted into the other type?
20. State the law of conservation of energy.

Exercises

NOTE: The following questions have up to five possible answers. Please select the **one** response which best answers the question.

1. A person made a 160 km trip in 4 hours. Which one of the following statements is correct?
   1. The instantaneous speed during the trip was 40 kph.
   2. The average speed for the trip was 40 kph.
   3. The average speed for the trip was 160 kph.
   4. The average speed for the trip was 640 kph.
   5. The instantaneous speed during the trip was 640 kph.
2. Speed is a measurement which specifies:
   1. how fast an object is moving.
   2. the direction of an object's motion.
   3. how quickly the object can accelerate.
   4. how fast an object is moving and its direction of motion.
   5. the distance that an object travels.
3. A reel on a tape deck initially had 27,000 inches of tape wound on it. During playback, the tape produced 1,800 seconds of music. How fast was it moving?
   1. 1-7/8 ips.
   2. 3-3/4 ips.
   3. 7-1/2 ips.
   4. 15 ips.
   5. 30 ips.
4. An 8-track cartridge tape deck is playing a prerecorded tape. How much time is required for 400 inches of tape to move past the playback head?
   1. 53 seconds.
   2. 214 seconds.
   3. 1,500 seconds.
   4. 107 seconds.
   5. 750 seconds.
5. What is the difference between speed and velocity?
   1. There is no difference between speed and velocity.
   2. Speed measures only how fast an object is moving while velocity measures only its direction of travel.
   3. Speed measures only the direction of an object's motion, while velocity measures only how fast it is moving.
   4. Speed measures how fast an object is moving, but it does not give its direction of travel. Velocity gives both speed and direction.
   5. Speed can be used to calculate an object's acceleration, and velocity cannot.
6. Suppose that an open-reel tape deck has just been turned on. If it takes 0.5 seconds for the tape to reach its maximum speed of 15 ips, what is the average acceleration of the tape?
   1. 7.5 i/s^2.
   2. 15 i/s^2.
   3. 0.03 i/s^2.
   4. 15.5 i/s^2.
   5. 30 i/s^2.
7. When a car goes around a curve at a constant speed its _________ changes because its direction of travel changes.
   1. instantaneous speed
   2. velocity
   3. average speed
   4. kinetic energy

8. A quantity which specifies both magnitude and direction is:
   1. the average speed.
   2. the instantaneous speed.
   3. a vector quantity.
   4. the elapsed time.
   5. the distance traveled.
9. A 7-1/2 ips open-reel tape deck is turned on, and the tape accelerates at the rate of 10 i/s^2. Calculate the time required to bring the tape up to full speed.
   1. 75 sec.
   2. 10 sec.
   3. 7.5 sec.
   4. 1.3 sec.
   5. 0.75 sec.
10. A speaker is accidentally dropped from the top of a tall ladder. Neglecting air friction, how fast is it traveling at the end of the 2nd second?
    1. 4.9 m/s.
    2. 9.8 m/s.
    3. 19.6 m/s.
    4. 29.4 m/s.
    5. 39.2 m/s.
11. How much speed does an object pick up each second when it has acceleration of 5 m/s^2?
    1. 9.8 m/s.
    2. 2.5 m/s.
    3. 5 m/s.
    4. 10 m/s.
    5. 15 m/s.
12. If an object falls freely from a rest position, what is its *acceleration* at the end of the third second?
    1. 9.8 m/s^2.
    2. 19.6 m/s^2.
    3. 29.4 m/s^2.
    4. 39.2 m/s^2.
    5. 0 m/s^2.
13. A newton is a unit of _______ in the _______ system.
    1. acceleration, metric
    2. velocity, British
    3. velocity, metric
    4. force, metric
    5. force, British
14. Static friction is involved when:
    1. a car skids on a slippery road.
    2. a hockey puck slides across the ice.
    3. (Both answers #1 and #2 are correct.)
    4. the capstan/pinch roller assembly pulls a magnetic tape past the heads of a tape deck.
15. Kinetic friction is involved when:
    1. a car skids on a slippery road.
    2. a hockey puck slides across the ice.
    3. the capstan/pinch roller pulls a magnetic tape past the heads of a tape deck.
    4. a record sits on the rubber-like mat of a turntable.
    5. (Both answers #1 and #2 are correct.)
16. Suppose that your entire hi-fi system weighs 60 pounds. What would it weigh if you lived on the moon?
    1. 10 pounds.
    2. 30 pounds.
    3. 60 pounds.
    4. 180 pounds.
    5. 360 pounds.
17. Suppose that left and right walls of a stereo record groove push on the stylus with equal forces. In response, the stylus will move:
    1. along the left wall.
    2. along the right wall.
    3. vertically.
    4. horizontally.
    5. (The stylus will not move.)
18. A measure of an object's inertia is called its:
    1. velocity.
    2. acceleration.
    3. weight.
    4. mass.
    5. momentum.
19. Suppose that a space ship is traveling through a region of space where no forces are acting on it. Newton's _______ law states that its velocity will _______ .
    1. first law, either increase or decrease
    2. third law, not change
    3. third law, either increase or decrease
    4. first law, not change
    5. second law, decrease
20. A truck has a mass of 3,500 kg, and it is being accelerated by a force of 8,750 N. Calculate the magnitude of the acceleration.
    1. 2.5 m/s^2.
    2. 9.8 m/s^2.
    3. 8,750 m/s^2.
    4. 3,500 m/s^2.
    5. 0.4 m/s^2.
21. A turntable platter has a mass of 2 kg. What is its weight (in newtons)?
    1. 2.0 N.
    2. 9.8 N.
    3. 19.6 N.
    4. 4.9 N.
    5. 0.2 N.

22. Two forces are simultaneously acting on a speaker cone whose mass is 0.06 kg. The first force acts to the right and has a magnitude of 14 N. The second force acts to the left and has a magnitude of 6 N. The acceleration of the cone is:
    1. 8 m/s^2, to the right.
    2. 133 m/s^2, to the right.
    3. 133 m/s^2, to the left.
    4. 0.48 m/s^2, to the right.
    5. 0.48 m/s^2, to the left.
23. A phono cartridge requires 1.8 grams of "vertical tracking force" in order to track properly the record groove. Approximately, what is the value of this force, as expressed in newtons?
    1. 1.8 N.
    2. 9.8 N.
    3. 0.018 N.
    4. 0.00018 N.
    5. 3.6 N.
24. A 0.05 kg speaker cone is being accelerated to the right at the rate of 700 m/s^2. What is the net force exerted on the cone?
    1. 700 N, to the right
    2. 35 N, to the right.
    3. 35 N, to the left.
    4. 14,000 N, to the right.
    5. 14,000 N, to the left.
25. A moving object is under the simultaneous influence of two forces: a 10 N force pulling it to the right, and a 17 N force pulling it down. The object will:
    1. first accelerate to the right and then, at a later time, accelerate down.
    2. first accelerate down and then, at a later time, accelerate to the right.
    3. accelerate along a direction which is parallel to the vector sum of the two forces.
    4. accelerate only to the right.
    5. accelerate only in the downward direction.
26. Suppose that an astronaut is to be transferred from the earth to the moon. Which one of the following statements is true?
    1. The astronaut's mass will change, but not his weight.
    2. The astronaut's weight will change, but not his mass.
    3. Both the weight and mass of the astronaut will change.
    4. Neither the weight nor the mass of the astronaut will change.
27. The downward gravitational force which acts on a person is called the person's:
    1. momentum.
    2. mass.
    3. inertia.
    4. moment of inertia.
    5. weight.
28. The property of an object that resists any change in its motion when a force is applied is called its:
    1. momentum.
    2. acceleration.
    3. velocity.
    4. weight.
    5. inertia.
29. According to Newton's second law, a large acceleration can best be obtained by applying a ______ force to an object which has a ______ mass.
    1. small, large
    2. small, small
    3. large, small
    4. large, large
30. A force of 2.0 N is exerted on a stylus whose mass is 0.005 kg (5 gm). The acceleration of the stylus is:
    1. 0.025 m/s^2.
    2. 40 m/s^2.
    3. 0.075m/s^2.
    4. 400 m/s^2.
    5. 0.01 m/s^2.
31. During the process of walking, a person of 70 kg mass exerts a horizontal force of 40 N on the earth. According to Newton's third law, the earth exerts a ________ force on the foot, thus causing the person to accelerate at the rate of ___
    1. 40 N, 1.75 m/s^2
    2. $(70 \times 9.8) = 686$ N, 9.8 m/s^2
    3. 0 N, 0.0 m/s^2
    4. 20 N, 0.28 m/s^2
    5. 40 N, 0.57 m/s^2
32. A phono stylus exerts a downward force of 0.05 N on a record groove. The record groove exerts ________
    1. an upward force of 0.05 N on the stylus, in accordance with Newton's first law.
    2. an upward force of 0.1 N on the stylus, in accordance with Newton's third law.
    3. an upward force of 0.05 N on the stylus, in accordance with Newton's third law.
    4. a downward force of 0.05 N on the stylus, in accordance with Newton's second law.
    5. a downward force of 0.1 N on the stylus, in accordance with Newton's first law.

33. The product of an object's mass and its velocity is called its:
    1. force.
    2. momentum.
    3. acceleration.
    4. inertia.
    5. weight.
34. A 312 kg object has a velocity of 12 m/s, eastward. Calculate its momentum.
    1. 26 kg·m/s, eastward.
    2. 0.0385 kg·m/s, eastward.
    3. 3,744 kg·m/s, eastward.
    4. 324 kg·m/s, westward.
    5. 300 kg·m/s, westward.
35. A 600 kg object is initially at rest. Calculate the magnitude of the force required to bring its speed up to 40 m/s in 8 seconds.
    1. 0.53 N.
    2. 120 N.
    3. 1.9 N.
    4. 1,800 N.
    5. 3,000 N.
36. At one instant in time a loudspeaker cone has a momentum of 0.5 kg·m/s. At 0.01 seconds later its momentum is 0.73 kg·m/s. Calculate the magnitude of the force exerted on the cone in order to change its momentum.
    1. 73 N.
    2. 0.5 N.
    3. 1.23 N.
    4. 10 N.
    5. 23 N.
37. At one instance in time an object has a momentum of 900 kg·m/s, eastward. If there are no forces acting on the object, what will be its momentum 7 seconds later?
    1. 0 kg·m/s.
    2. $(900 \times 7) = 6{,}300$ kg·m/s, eastward.
    3. $(900 / 7) = 129$ kg·m/s, westward.
    4. 900 kg·m/s, eastward.
    5. 1800 kg·m/s, westward.
38. For tightening purposes, a bolt on an automobile engine has a torque rating of 75 N·m. If a wrench has a radius of 0.15 m, what is the maximum amount of force that should be used to tighten it?
    1. 75 N.
    2. 500 N.
    3. 11.25 N.
    4. 0.002 N.
    5. 0.15 N.
39. A direct drive turntable motor has a torque rating of 0.6 N·m. When the motor is turned on, how much force does it exert on the inner edge of the platter, if the motor shaft has a radius of 0.002 m?
    1. 0.6 N.
    2. 1.7 N.
    3. 32.6 N.
    4. 126 N.
    5. 300 N.
40. The torque exerted on an object, such as a door, can be increased by:
    1. increasing the applied force.
    2. increasing the length of the lever arm.
    3. (Both answers #1 and #2 will cause the torque to increase.)
    4. decreasing the applied force without changing the length of the lever arm.
    5. decreasing the length of the lever arm without changing the magnitude of the applied force.
41. An engine on a motor bike is producing 200 N·m of torque on the rear wheel. If the radius of the rear tire is 0.32 m, what is the magnitude of the force exerted on the surface of the road in order to accelerate the bike?
    1. 200 N.
    2. 625 N.
    3. 0.0016 N.
    4. 64 N.
    5. 0.32 N.
42. The property of an object that resists any change in its rotational motion is called its:
    1. torque.
    2. mass.
    3. moment of inertia.
    4. momentum.
    5. force.
43. The difference between an object's inertia and its moment of inertia is that:
    1. the inertia depends only on the object's mass, while the moment of inertia depends on both its mass and geometrical shape.
    2. the inertia depends on both the object's mass and geometrical shape, while the moment of inertia depends only on its mass.
    3. the inertia measures on object's resistance to any change in its rotational motion, while the moment of inertia measures its resistance to any change in its straight-line motion.
    4. (Both answers #1 and #3 are correct.)
    5. (There is no difference between inertia and moment of inertia.)

44. Turntable platters which have a relatively heavy outer rim can give ____________ wow and flutter figures because of their relatively _________ moment of inertia.
   1. high, large
   2. high, small
   3. low, large
   4. low, small
45. A speaker diaphragm moves 5×10^{-3} m to the right under the influence of a 24 N force. How much work is done on the diaphragm by the force?
   1. 0.005 joules.
   2. 0.024 joules.
   3. 0.082 joules.
   4. 0.120 joules.
   5. 4.8 joules.
46. A loudspeaker weighing 5 N was accidentally dropped from a height of 3 m. How much work was done on it by the gravitational force?
   1. 15 joules.
   2. 5 joules.
   3. 0.6 joules.
   4. 0.1 joules.
   5. 0.02 joules.
47. An amplifier delivers 60 watts of electrical power to a loudspeaker. One percent of this power is used to do work in moving the diaphragm, with the remaining 99% being wasted as heat. How much work is done in moving the diaphragm in 0.003 seconds?
   1. 20,000 joules.
   2. 200 joules.
   3. 0.18 joules.
   4. 1.0 joule.
   5. 0.0018 joules.
48. A stalled car is being pushed by a tow truck. If the tow truck exerts a force of 2,000 N on the car, calculate the car's kinetic energy at the end of 100 meters. (Neglect friction and assume that the car is initially at rest.)
   1. 20 joules.
   2. 2,100 joules.
   3. 2×10^5 joules.
   4. 500 joules.
   5. 12,500 joules.
49. In Problem 48 what will be the car's speed at the end of the 100 meters if it has a mass of 1,200 kg?
   1. 7.6 m/s.
   2. 18.3 m/s.
   3. 22.8 m/s.
   4. 48.9 m/s.
   5. 65.3 m/s.
50. A 700 kg car is traveling at a speed of 30 m/s. How much work was done on the car in order to bring it from rest to 30 m/s?
   1. 700 joules
   2. 7.35×10^6 joules.
   3. 2.1×10^4 joules.
   4. 3.6×10^5 joules.
   5. 3.15×10^5 joules.
51. How much work is done in lifting a 50 N loudspeaker through a vertical distance of 7 meters?
   1. 175 joules.
   2. 350 joules.
   3. 7.1 joules.
   4. 0.14 joules.
   5. 2,450 joules.
52. A rock of weight 10^5 N falls from a high cliff which is 100 m above the ground. How much work can the rock do when it hits the ground?
   1. 1×10^5 joules.
   2. 1×10^2 joules.
   3. 1×10^7 joules.
   4. 1×10^3 joules.
   5. 2.3×10^3 joules.
53. According to the Law of Conservation of Energy, what is the speed of the rock in Problem 52 just before it hits the ground?
   1. 2.70 m/s.
   2. 18.5 m/s.
   3. 31.6 m/s.
   4. 44.3 m/s.
   5. 92.8 m/s.

Appendix

How to Convert Microvolts into dBf

The current standards state that the input RF signal strength, at the tuner's antenna terminals, be specified in "dBf" rather than the older "microvolts." Suppose, for example, that the input signal is 5 microvolts. In scientific notation 5 microvolts is 5×10^{-6} volts, which is a "5" six places to the right of the decimal point (0.000005 volts). There are two steps which are needed in order to convert this voltage into dBf:

1. First, convert voltage into power by using the formula

$$\text{Power} = \frac{(\text{Voltage})^2}{\text{Resistance}}$$

The input impedance (resistance) of most tuners is 300 ohms so that the power which corresponds to 5 microvolts is

$$\text{Power} = \frac{(5 \times 10^{-6}\ \text{volts})^2}{300\ \text{ohms}} = 8.3 \times 10^{-14}\ \text{watts.}$$

2. Second, the conversion from power to dBf is established by using logarithms:

$$\text{dBf} = 10 \log \left(\frac{\text{Power}}{\text{Reference power}}\right).$$

"log" is the logarithm to the base 10, "Power" is the input power supplied to the tuner which was calculated in part #1 above, and "Reference power" is arbitrarily assigned the value of 1×10^{-15} watts. The reference power of 1×10^{-15} watts is called a "femtowatt," and dBf means that the decibel rating has been calculated using 1 femtowatt as the reference power. Using the above numbers one can now calculate the input level as expressed in dBf:

$$\text{dBf} = 10 \log \left(\frac{8.3 \times 10^{-14}\ \text{watts}}{1 \times 10^{-15}\ \text{watts}}\right) = 19.2.$$

This logarithm can be evaluated using a scientific calculator or a table of logarithms. The conversion table shown below lists the input voltage, input power, and dBf rating for some typical tuner input values, assuming a 300 ohm antenna.

Conversion Table

Microvolts	Power (watts)	dBf
1	3.3×10^{-15}	5.2
3	3.0×10^{-14}	14.8
5	8.3×10^{-14}	19.2
10	3.3×10^{-13}	25.2
50	8.3×10^{-12}	39.2
100	3.3×10^{-11}	45.2
500	8.3×10^{-10}	59.2
1,000	3.3×10^{-9}	65.2
5,000	8.3×10^{-8}	79.2

Glossary

A/B Comparison Test—The direct comparison between the sounds of two similar components, such as loudspeakers, which is made by switching back and forth between them.

Acceleration (Average)—The change in an object's velocity divided by the elapsed time. Average acceleration is a vector.

Acoustic Energy—The energy carried by a sound wave.

Acoustic Feedback—An undesirable noise (similar in sound quality to rumble) created when vibrations from the loudspeakers are picked up by the cartridge and amplified by the sound system. The problem of acoustic feedback usually can be greatly reduced by physically separating the loudspeakers from the turntable. In public address systems the microphone/speaker combination also gives rise to acoustic feedback which is often heard as an annoying screech.

Acoustic Image—When sound is reflected from a wall, it appears to originate from a location which is behind the wall. This apparent location is called an acoustic image, and it is similar to a visual image seen in a mirror.

Acoustic Suspension (Air Suspension)—A loudspeaker design which permits good bass reproduction from a relatively small-sized enclosure. The acoustic suspension systems utilize a floppy, highly compliant cone suspension and the trapped air to provide the proper suspension characteristics. They are generally of low-efficiency design, requiring relatively high amplifier powers compared to the vented and horn enclosures.

AFC—(See **Automatic Frequency Control.**)

Air-motion Transformer Speaker—A speaker whose diaphragm has the shape of an accordian. The folds of the diaphragm have straight wire conductors attached to them such that magnetic forces cause a squeezing action of the folds.

Air Suspension—(See **Acoustic Suspension.**)

Alternate-Channel Selectivity—A specification which gives the ability of a tuner to receive only the desired station while rejecting the alternate-channel stations which are not wanted; measured in dB, with larger numbers being better. Technically it is the number of decibels by which the RF signal strength of an undesired station must exceed the RF strength of a tuned in station in order to force through an undesired audio program which is 30 dB below the desired audio program. The word "alternate" means that the undesired station is located on the dial 0.4 MHz away from the desired station.

AM—(See **Amplitude Modulation.**)

Ampere—A measure of electric current. One ampere of current results when one coulomb of charge flows every second.

Amplification—The process of making electrical signals larger.

Amplifier—An electronic device which is used for increasing the size (amplitude) of electrical signals.

Amplitude—The amplitude of a wave is the maximum excursion of a physical quantity from its equilibrium position. The physical quantity can be: the displacement of water for a water wave, the change in pressure for a sound wave, or the displacement of the coils for either a transverse or longitudinal slinky wave.

Amplitude Modulation (AM)—A type of modulation in which the audio information is conveyed by changing the amplitude of the carrier wave. Large audio levels produce large amplitude changes of the carrier wave. AM radio is the most common form of broadcasting, although it does have some defects which limit its use for hi-fidelity listening.

AM Suppression—A specification which gives the ability of an FM tuner to suppress changes in the amplitude of the received signals, thereby improving the S/N ratio by rejecting unwanted noise and interference. It is measured in dB.

Antenna—A structure of metallic rods or wires, with which the radio waves interact to produce radio frequency AC electricity. This electricity is sent as input to the tuner.

Antinodes—Regions along a standing wave where the medium has its maximum amplitude of vibration.

Antiskating Control—A control found on some turntables which applies a slight outward force to the tone arm. This force counteracts the inward-acting skating force.

ARLL (Audible Rumble Loudness Level)—(See **Rumble**.)

Atom—The smallest "particle" of an element. An atom contains atomic units called protons, neutrons, and electrons. The protons are positively charged, and they are located inside the nucleus along with the electrically-neutral neutrons. The negatively charged electrons orbit the nucleus in specific shells or orbits. The protons and neutrons have equal masses and they are about 1,840 times more massive than an electron.

Atomic Number—The number of protons in the nucleus of an atom. Different elements are distinguished by different atomic numbers. It is also the number of electrons in an electrically neutral atom.

Audio Generator—An electronic component, acting as a sound source, which produces a single frequency electrical signal. The frequency can be selected by the user and it may lie anywhere in the audio range. Audio generators are used for testing audio components.

Audio Signal—The wave of AC electricity which is processed by hi-fi components. It is the electrical replica of the sound wave, and, as such, its frequencies lie in the range from 20 Hz to 20 kHz.

Automatic Frequency Control (AFC)—A circuit which automatically keeps the tuner fine-tuned to the selected station. Most hi-fi tuners have an AFC feature, but some provide a front panel switch which allows it to be turned off.

AUX (Auxiliary) Jacks—The source input jacks to a preamp where a tape deck or a tuner may be connected.

A-Weighting—A method commonly used when measuring noise levels which deemphasizes the low and high frequency noise relative to the midrange noise near 1 kHz. An A-weighted noise measurement attempts to account for the annoyance factor of the noise.

Baffle—Any structure which prevents the front and rear surface sound waves of a speaker diaphragm from interfering with each other and causing possible sound cancellation. It is the panel on which the driver is mounted, although the term "baffle" sometimes applies to the entire enclosure that houses the speaker.

BALANCE Control—A control located in the preamp/control center which regulates the relative amount of electrical power in each stereo channel. The BALANCE control is a variable resistor which has been inserted into the amplifier circuits.

Bandwidth—A range of frequencies which may be usable for a particular purpose. For example, the audio bandwidth for FM transmission is from 30 Hz to 15 kHz; the FM broadcast bandwidth is from 88 MHz to 108 MHz; the bandwidth of human hearing is from 20 Hz to 20 kHz, etc.

Basket—The metal frame of a speaker which holds together the cone, spider, voice coil, and magnetic structure.

Bass Reflex (Vented Enclosure, Phase Inverter, or Tuned Port)—A loudspeaker design in which the cone's "back-surface wave" is sent out through a port in the face of the enclosure to reinforce the bass output.

BASS/TREBLE Controls—Continuously adjustable front panel controls on the preamp/control center which change the output sound level in selected frequency ranges. The BASS control can either boost or cut the low frequencies. The TREBLE control performs a similar function for the high frequencies.

Beats—A throbbing sound which is heard when the sound alternately becomes loud, soft, loud, soft, etc. Beats are caused by the interference of two overlapping sound waves which have slightly different frequencies. The interference between the two waves results in regions of alternate reinforcements and cancellations which give rise to the variations in the sound loudness.

Belt Drive—A method for connecting the rotating motor shaft to a turntable's platter or a tape deck's capstan by means of a flexible belt.

Bias Signal—A high frequency (above audio) alternating current which is fed to the recording head of a tape deck, along with the audio signal, in order to minimize the distortion of the magnetized signal.

Bimetallic Strip—Two pieces of metal permanently bonded together to form a single strip, which bends in an arc when heated because the two metals are chosen to have different coefficients of thermal expansion.

Bookshelf Loudspeaker—A loudspeaker which possesses a relatively small enclosure.

Bound Electrons—Electrons which are bound tightly to the nucleus, and are not free to wander about the material of which they are a part.

Building Blocks—The elementary units which comprise a hi-fi system. The four building blocks are: the sound sources (turntable, tape deck, tuner, microphone), the preamplifier/control center, the power amplifier, and the speaker(s).

Capacitor—An electrical component whose impedance decreases as the frequency of the AC current increases. The impedance of a capacitor is measured in ohms.

Capstan—A small rotating shaft in a tape deck which pulls the tape across the heads at a constant speed with the aid of the pinch roller.

Capture Ratio—A specification which gives the ability of a tuner to reject the weaker of two RF signals which have the same frequency; measured in dB, with smaller numbers being better. Technically it is the number of decibels by which a stronger RF signal must exceed a weaker RF signal in order for the audio output from the stronger signal to be 30 dB louder than that from the weaker signal.

Carrier—In hi-fi it is the radio wave which "carries" the audio information from a transmitter. The frequency of the carrier wave is assigned to the station by the FCC, and corresponds to the number of the station "on the dial."

Cartridge—(See **Phono Cartridge**.)

Cartridge Deck (8-track Tape Deck)—A tape deck which uses cartridge tapes. Cartridges are larger than cassettes and contain a single, endless-loop reel for tape storage. Cartridge tape is 0.64 cm (1/4″) wide, travels at 3-3/4 ips, and contains eight tracks.

Cassette Deck—A tape deck which uses cassettes. A cassette is a small plastic case which contains two reels (supply and take-up) of 0.40 cm (5/32″) wide tape, and openings for the tape heads and the tape drive mechanism. The tape speed is 1-7/8 ips, and the tape contains four tracks.

Ceramic Cartridge—(See **Phono Cartridge**.)

Channel—A complete and separate "sound" path through an audio system. A stereophonic system has two channels designated as "left" and "right." A monophonic system has only one channel.

Charges (Electrical)—A positive or negative quantity of electricity. The smallest amount of negative electrical charge is carried by an electron. A proton carries the smallest unit of positive charge and it is equal, but opposite, to that carried by the electron.

Chromium Dioxide—A type of tape coating which is capable of accepting higher recording levels and offering wider frequency ranges than iron oxide tapes. A special bias and equalization switch is required in order to use these tapes.

Circuit Breaker—A device for protecting circuits against excessively large electric currents. Unlike a fuse, it may be reset and used repeatedly.

Clipping—A phenomenon which occurs when the input signal to an amplifier is so large that the amplifier is not able to amplify properly the crests and troughs of the signal. As a result, the crests and troughs of the output electrical wave appear flattened, as if "clipped" off by a pair of scissors. Severe clipping introduces a large amount of distortion into the output wave.

Coefficient of Thermal Expansion—The number of meters by which a one meter length of material will expand when its temperature is increased by one Celsius degree.

Compact—A complete audio system which usually contains all components, except for the speakers, in a single case. The compacts may contain any or all sound sources.

Complete System—A phrase describing an audio system which contains all four building blocks: a sound source, a preamp, a power amp, and speakers.

Complex Wave—A general term which is used to designate any type of wave (sound, water, slinky, etc.) which is composed of more than one sine wave. All voices and musical instruments produce complex sound waves.

Compliance—A measure of a mechanical system's ability to move. A highly compliant phono stylus, or speaker diaphragm, is very flexible and it can move over relatively large distances with only a small applied force. For phono cartridges, compliance is a specification which is quoted in centimeters of movement per dyne of force.

Component—Any element of an audio system which is designed to do a particular job. Typical components include receivers, tuners, speakers, tape decks, turntables, integrated amps, etc.

Compressions—(See **Condensations**.)

Condensations (or Compressions)—Regions along a sound wave where the pressure is largest; also called crests. In a longitudinal slinky, compressions are regions where the coils are "bunched up."

Conduction (Thermal)—A process by which heat energy flows through a material via the activity of its molecules, although the material as a whole does not move.

Conductor—A material which has a low resistance to the flow of electrical current.

Cone—(See **Diaphragm.**)

Conservation of Energy—A very important law which states that energy can neither be created nor destroyed; energy can only be transformed from one form to another.

Console—A complete audio system, including the speakers, contained in a single furniture-type cabinet. Because the turntable and speakers share a single cabinet, consoles usually pose the threat of acoustic feedback at high sound levels.

Constructive Interference—(See **Interference.**)

Continuous Average Power Output—A specification which accurately describes how much power an amplifier can deliver on an uninterrupted basis. In order to have full meaning this specification must give the continuous watts of power per channel, the speaker impedance into which the rated power is delivered, the number of channels which are working simultaneously, the frequency range over which the rated power can be delivered, and the maximum % THD at the rated power.

Convection—A process where heat energy is carried by the mass motion of a medium like the rising of warm air. The molecules of the medium actually move over appreciable distances.

Coulomb—An amount of electric charge that contains 6.24×10^{18} electrons (or protons).

Coulomb's Law—A law which gives the force F between two electric charges Q_1 and Q_2, separated by a distance r. It is $F = kQ_1Q_2/r^2$, where k is a proportionality constant.

Crest/Troughs—The locations along a wave where the displacement of the wave has either its greatest value (crest) or its smallest value (trough).

Crossover Frequency—The frequency at which the crossover network begins to route the signal to a different driver. In a two-way speaker system the crossover frequency is the frequency at which the woofer and high-frequency responses are divided.

Crossover Network—An electronic device used in loudspeakers which is designed to route the appropriate frequencies to the proper drivers; e.g., woofers, midranges, and tweeters.

Crosstalk—The ratio of the power of the audio signal in one track on a tape to the power of this signal which has leaked over to an adjacent track. Like stereo separation, it is expressed in decibels, but it is not the same thing as stereo separation.

Current (AC and DC)—The electric current is the flow of electrons through a wire. In direct current (DC) the electrons move in one direction only. In an alternating current (AC) the electrons simply oscillate back and forth about an equilibrium position. AC and DC are both used extensively in hi-fi systems. Current is measured in amperes.

Cutoff Frequency—The frequency below which a horn-type speaker will no longer produce sound.

Cutting Stylus—A cutting tool which is used to cut the undulations, representing the sound waves, into the grooves of a master record.

Cycle—A term used with waves to designate a crest-trough combination.

Damping Factor—A measure of an amplifier's ability to damp unwanted vibrations of the speaker cone.

dB—(See **Decibel.**)

dBf—A unit for measuring RF signal levels which expresses tuner input levels in terms of decibels rather than microvolts. The "f" in dBf stands for "femtowatt" (10^{-15} watts), and it represents the reference power level used for the 0 dB point. See the Appendix.

dB Level Meter (Peak Reading Meter)—A meter found on tape decks, which is designed to be fast enough to measure the maximum level of the peaks or transients in a signal. It is used as a record or playback level meter. The "0 dB" level corresponds to the level of an audio signal which contains 3% total harmonic distortion during playback.

dB Meter—(See **Sound Level Meter.**)

dBW—A unit for measuring the power of an amplifier, in terms of decibels, relative to a reference level of 1 watt.

Decibel (dB)—A term which compares two powers or two intensities. The decibel is closely correlated with the manner in which the ear perceives loudness changes. One decibel is approximately the smallest change in sound loudness which can be heard.

De-emphasis—(See **Pre-emphasis.**)

Demodulation—(See **Modulation.**)

Destructive Interference—(See **Interference.**)

Diaphragm (or Cone)—The part of a driver which pushes on the air to produce the sound. A cone is a specially shaped diaphragm commonly used in drivers.

Diffraction—The bending of waves around objects or corners.

DIN (Deutsche Industrie Normen)—(See **Rumble.**)

Direct Drive—A type of turntable or tape deck drive system where the revolving motor shaft is connected directly to the platter or capstan.

Dispersion of a Speaker—The ability of a loudspeaker to spread sound into a wide listening area. Dispersion decreases as the frequency increases.

Dolby Noise Reduction—A noise reduction technique which reduces tape hiss by approximately 10 dB at low signal levels, when the noise becomes most obvious.

Domain—A group of atoms whose miniature atomic magnets are aligned to reinforce one another.

Dome Speaker—A speaker whose diaphragm is dome-shaped. Dome speakers increase the sound dispersion.

Doppler Effect—An apparent change in the pitch of sound emitted by a source when either the source or the observer is moving relative to each other. When the source is approaching an observer the pitch sounds higher than normal, and when the source is receding from an observer the pitch appears lower than normal.

Driver—A term that is applied to any sound-producing device which is installed in an enclosure. Sometimes called a speaker.

Drive System—(See **Tape Transport** and **Turntable Drive System**.)

Dynamic Balancing—A method for adjusting the tracking force using an adjustable spring attached to the tone arm.

Dynamic Headroom—A specification which gives the ratio of the power output which an amplifier can deliver for short periods of time to the continuous average power output, expressed in decibels.

Dyne—A unit for measuring force. The weight of one penny is approximately 3,000 dynes.

Efficiency—The ratio, expressed as a percentage, of output signal to input signal; it is often used to estimate the electrical power needed to drive a loudspeaker.

Eight-Track Tape Deck—(See **Cartridge Deck**.)

Electrical Energy—The energy which is given to electrical charges from either a battery or a generator.

Electrical Signal—Electricity which contains the sound information. All audio systems, beginning with the sound sources, process only electrical signals, and the sound itself is not recreated until the signals reach the speakers. In AM and FM broadcasting the radio wave is called the "signal."

Electrolyte—The chemical in a battery which reacts with the plates and separates electric charges onto the positive and negative terminals.

Electromagnet—A helical coil of wire through which an electric current is flowing. The coil becomes a magnet and has a north and a south pole. The strength of the electromagnet can be increased by either adding an iron core, increasing the current, or increasing the number of turns in the coil.

Electromagnetic Waves—A class of waves which travel at the speed of light (3×10^8 meters/sec. or 186,000 miles/sec.) in a vacuum. Members of this class include such well-known types as: AM and FM radio waves, visible light, X-rays, and infrared waves.

Electromagnetism—A theory which explains the relationship between electric currents and magnetic fields.

Electron—(See **Atom** and **Charges**.)

Element—A material whose atoms all have the same atomic number.

Enclosure, Loudspeaker—An acoustically designed cabinet for the driver(s). The enclosure interacts with the driver(s) and it can have a large effect on the sound production, especially at the low frequencies.

Energy—The ability of waves, or any object, to accomplish some type of work. Examples: sound waves forcing your eardrum to vibrate, electrical signals from a power amp which move the speaker diaphragm, etc. The unit of energy is a joule.

Energy, Gravitational Potential—(See **Potential Energy**.)

Energy, Kinetic—(See **Kinetic Energy**.)

Equalization—The process of electronically boosting or cutting the level of audio signals in certain frequency ranges relative to those in other frequency ranges.

Erase Head—An electromagnet which uses a strong, high frequency AC signal to completely scramble, or erase, the tiny magnets in the tape.

Faraday's Law of Induction (Induced Current and Voltage)—A law which states that if a magnetic field and a coil of wire are moving relative to one another, a voltage will be created (induced) in the coil, thus causing a current to flow. The important issue is that the magnetic field lines must pass through the coil and be changing.

FAST FORWARD—A control on a tape deck which is used to move the tape in the forward direction more rapidly than its playing speed would allow, in order to facilitate rapid location of any portion of a recording.

FAST REVERSE (REWIND)—A control on a tape deck which is used to move the tape rapidly in the reverse direction. It is absolutely essential for the convenient editing of tape recordings.

FCC—Federal Communications Commission.

Femtowatt—10^{-15} watts.

Fifty (50) dB Quieting Sensitivity—A specification which indicates how well a tuner can convert weak radio waves into a satisfactory audio signal. Technically, it is the smallest RF signal which causes the tuner to produce a total audio output which is 50 dB greater than the background noise. It is measured in microvolts (μV) or dBf.

Flat Frequency Response—A phrase used to describe a theoretically perfect audio component which produces a uniform, or constant, output signal strength at any frequency within a specified frequency range.

Fletcher-Munson Curves—A series of graphs which characterize the ear's sensitivity to various frequencies at different loudness levels.

Flexible Edge—(See **Suspension.**)

Flutter—(See **Wow and Flutter.**)

FM—(See **Frequency Modulation.**)

FM Band—A range of radio frequencies, from 88 MHz to 108 MHz, set aside for FM broadcasts.

FM MUTE (RUSH, SQUELCH)—A control on a tuner which can be used to suppress the "hissing" noise heard when dialing between stations.

FM Trap—A filtering device which is sometimes installed on TV antenna systems to prevent FM radio waves from interfering with TV reception.

Force—A push or a pull which is exerted on an object. Force is a vector.

Forced Vibrations—The setting up of vibrations within an object, either at the object's natural frequency of vibration, or at any other frequency, by a vibrating force.

Fourier Analysis (or Fourier's Theorem)—A method, discovered by Joseph Fourier, that will resolve any complex wave into a sum of simple sine waves which are harmonically related to each other.

Free (Unbound) Electrons—Electrons which are only loosely bound to the nucleus, and may be easily detached to wander about the material of which they are a part.

Frequency—The number of wave cycles which pass a particular point each second. Frequency is measured in Hertz (Hz). The frequency of a wave is related to its period by: frequency $= \frac{1}{\text{period}}$.

Frequency Modulation (FM)—A type of modulation of radio waves in which the audio information is conveyed by changing the frequency, and not the amplitude, of the radio wave. Large level audio signals cause large changes in the carrier's frequency. FM broadcasting is a hi-fidelity medium used for transmitting both stereo and monophonic programs.

Frequency Range—The range of frequencies which can be reproduced by an audio component. Sometimes confused with frequency response, the frequency range does not give the deviations from perfect flatness, as denoted by a "$\pm$ dB" term.

Frequency Response—A specification which indicates how uniformly an audio component reproduces either electrical signals or sound over a specified range of audio frequencies. In order to have full meaning, this specification must include both the frequency range and the $\pm$ decibel deviation from perfect flatness, e.g., 20 Hz → 20,000 Hz, +0.5 dB, −0.8 dB.

Frequency Response Curve—A graph which shows how uniformly an audio component reproduces either electrical signals or sound over a specified range of audio frequencies.

Friction—A force which arises when objects tend to slide against each other. Kinetic friction occurs then the two objects actually slide against each other. Static friction occurs when there is no relative motion between the two objects.

Friction, Tone Arm Pivot—(See **Tone Arm Pivot Friction.**)

Fundamental (See Harmonics.)—The lowest natural frequency of a vibrating object. In music it is the lowest frequency present in a note.

Fuse—A special thin conductor designed to melt when a predetermined current is passed through it. This stops the current and protects an electrical device from receiving an excess amount of current.

g—The acceleration of an object due to the earth's gravity. It is 9.8 m/s^2 downward.

Gap—The space between opposite ends of the horseshoe-like iron core of a tape head, usually measured in microns.

Generator—Any device which is capable of separating electric charge. If the two terminals of a generator are connected by a wire, a current will flow. Common generators in hi-fi are: phonograph cartridges, tape playback heads, and microphones.

Gram (gm)—A unit of mass in the metric system.

Ground (Electrical)—A term which refers to the third prong on a three prong plug. It is a safety feature which protects the user against being shocked.

Guard Band—A range of frequencies which helps prevent stations (which are adjacent on the dial) from interfering with each other. Each FM station is allowed to change the assigned carrier frequency by as much as ±75 kHz in order to convey the audio information. An additional 25 kHz, called the guard band, is allocated at each extreme.

Harmonics (See Overtones.)—Frequencies which are multiples of the fundamental frequency. The first harmonic is also called the fundamental, the second harmonic has twice the fundamental frequency, etc.

Head—(See **Erase Head, Playback Head, and Recording Head.**)

Heat—(See **Heat Energy.**)

Heat Energy—A form of energy which causes the molecular motion of a substance to increase. A change in an object's heat energy content is detected by a change in its temperature. The amount of heat energy necessary to raise an object's temperature by 1°C depends on its mass and type of material.

Heat Sensitive Resistance Thermometer (Thermistor)—A thermometer constructed from a solid state "semiconductor" whose resistance changes as the temperature changes.

Heat Sink—A metal heat conductor upon which the power transistors are mounted. The heat sink conducts the heat energy away from the transistors.

Heat Transfer—The movement of heat energy from place to place. It can occur via conduction, convection and radiation.

Hertz (Hz)—A unit of frequency which represents one complete wave cycle (crest-trough combination) passing by each second: one Hertz = one cycle/second.

Hi-fi (High-fidelity)—A phrase which is synonymous with "quality performance" in audio components.

HIGH-BLEND—A feature provided on some tuners which mixes some of the high-frequency sounds from both channels, thereby reducing the noise level. The high-blend feature results in some loss of stereo separation.

High Level Inputs—A phrase which is used to refer to the TUNER, TAPE, and AUX inputs of an amplifier, because they can accept higher level input voltages than does the PHONO input.

HIGH (SCRATCH) Filter—A high frequency filter located in the preamp/control center which is used to reduce the noise from records and other sound sources.

Horn Speaker—A type of speaker which contains a "megaphone" (horn) attached to a driver. The megaphone is used to increase the sound level from the driver.

Hysteresis Synchronous Motor—(See **Synchronous Motor.**)

Idler Wheel Drive (Rim Drive)—A type of drive system on a turntable or a tape deck where the motor shaft is connected to the platter or the capstan by means of an idler wheel.

IF Rejection—(See **Image Rejection.**)

IHF Sensitivity—(See **Usable Sensitivity.**)

IM—(See **Intermodulation Distortion.**)

Image Rejection (IF Rejection, Spurious Response Rejection)—One of several specifications which, in general, measure a tuner's ability to reject extraneous signals which occasionally pop up on the FM dial. Two other specifications of this type are IF rejection and spurious response rejection. These specifications are measured in dB, with higher values being better.

Impedance—A general term which characterizes the ability of resistors, capacitors, and inductors to "impede" or to "resist" the flow of current in electrical circuits. Impedance is measured in ohms.

Induced Current—(See **Faraday's Law of Induction.**)

Induced Magnetism—The magnetism which results in a magnetizable material like soft iron when it is brought into the vicinity of a magnet. When the magnet is removed, the iron will lose some of this induced magnetism, but a residual amount will remain. In materials like those used in magnetic tapes much of the induced magnetism is retained.

Induced Voltage—(See **Faraday's Law of Induction.**)

Induction Motor—A motor used in turntables and tape decks. This type of motor is inexpensive and has good short-term speed characteristics. However, it is susceptible to long-term speed changes caused by line voltage changes.

Inductor—An electrical component whose impedance increases as the frequency of the AC current increases. The impedance of an inductor is measured in ohms.

Inertia (See Mass.)—The property of an object that resists any change in its motion when a force is applied.

Inertia, Rotational—(See **Moment of Inertia.**)

Infinite Baffle—A type of loudspeaker design in which the woofer is mounted in either a large enclosure or in a large wall. The enclosure isolates the "back-surface" wave from the "front-surface" wave. The enclosure is so large that it does not appreciably change the resonant frequency of the driver.

Infrasonic Sound—Sound whose frequency lies below the audible range; below 20 Hz.

In-phase—(See **Phase.**)

Input Sensitivity—A specification which gives the input voltage to an amplifier that is required to produce a power output of one watt when the VOLUME control is set to a maximum.

Insulator—A material which has a high resistance to the flow of electricity. On an atomic scale, the atoms of an insulator do not contribute any "free" electrons which are capable of wandering through the material. The electrons of an insulator are all "bound" to their respective atoms.

Integrated Amplifier—A single audio component which combines a preamplifier control/center and a power amplifier. An integrated amplifier does not contain a sound source or speakers.

Intensity—The amount of sound power which passes through a surface divided by the area of the surface, the surface being perpendicular to the direction in which the sound is traveling.

Interference—The adding together of two or more waves, according to the principle of linear superposition, to produce regions of reinforcement and regions of cancellation. Constructive interference results when the crests of one wave meet the crests of another wave, and troughs meet troughs, such that reinforcement of the waves occurs. Destructive interference results when the crests of one wave meet the troughs of another, and vice versa, such that cancellation of the waves occurs.

Intermodulation Distortion (IM)—A specification which gives the distortion of the audio signal that results when two different frequencies are simultaneously passed through an amplifier, or any hi-fi component. IM distortion causes two new frequencies to be added to the two original frequencies. The two new frequencies which are created are the sum and difference of the original frequencies. IM distortion is measured by dividing the combined voltage of the unwanted signals by the combined voltage of the original signals, and expressing the ratio as a percentage.

Iron Oxide—A type of magnetic coating applied to one surface of the plastic backing during the manufacturing of magnetic tape.

Jack—A small, round electrical connector found on all hi-fi units to which the wires, interconnecting the components, are attached.

Joule—A unit of energy in the metric system.

Kilocalorie—A unit for measuring an amount of heat energy. It is the amount of heat needed to raise the temperature of one kilogram of water by one Celsius degree. One kilocalorie equals 4,184 joules.

Kilogram (kg)—A unit of mass in the metric system.

KiloHertz (kHz)—1,000 Hz.

Kilowatt-Hour—A measure of the total amount of electrical energy which is used. It is the product of the power consumed and the total time of consumption. The electric bill is computed using kilowatt-hours.

Kinetic Energy—The energy of motion, described by the relationship: Kinetic energy $= \frac{1}{2} mv^2$.

Kinetic Friction—(See **Friction.**)

Land—The flat smooth surface of the record located between the grooves.

Law of Reflection—(See **Reflection, Law of.**)

Law of Refraction—(See **Refraction, Law of.**)

LED—(See **Light Emitting Diode.**)

Left-hand Rule—A means for determining the location of an electromagnet's north pole (Rule #1). Also the left hand can be used to determine the direction of the force acting on a current which is traveling in a magnetic field (Rule #2).

Lenz's Law—A law which allows one to predict the direction in which an induced current will flow in a coil. The polarity of the induced voltage will be in such a direction that the resulting current will produce an electromagnet which opposes the original motion creating the induced voltage.

Lever Arm—The perpendicular distance between the force vector and the pivot point about which the force causes a rotation. A lever arm is necessary for calculating torque.

Light Emitting Diode (LED)—A solid state device which emits light when a small current is passed through it. The LED can be used for detecting peaks in the audio signal on record and playback.

Limiter—A circuit in an FM tuner that removes any amplitude variations, such as spikes, caused by atmospheric or ignition noise. FM limiters are largely responsible for producing noise free FM reception.

Limits of Human Hearing—The frequency range, from about 20 Hz. → 20 kHz, over which the normal young (25 years or less) person can hear sound.

Listener Fatigue—An auditory exhaustion resulting from loud sound levels together with high distortion. Listener fatigue can be caused by the speakers.

Longitudinal Wave—A wave which causes the particles of the medium to vibrate back and forth parallel to the direction in which the wave travels. Sound waves are longitudinal waves.

Loudness—The subjective perception of the amount of power carried by a sound wave. A more powerful sound wave will produce the sensation of a greater loudness.

LOUDNESS Control—A control which is either on or off according to the setting of a front panel switch, and works in conjunction with the VOLUME control. The LOUDNESS control boosts the low frequencies (and sometimes the highs, too) at low listening levels in order to compensate for the ear's loss of hearing. The LOUDNESS control is found on a preamplifier.

Loudspeaker—The combination of an enclosure plus drivers.

Low-Noise/High-Output Tape—A type of magnetic tape that yields a greater (6 to 8 dB) output signal and less noise than does a conventional iron oxide tape. In order to use this tape, the tape deck should have special bias and equalization switches.

LOW (RUMBLE) Filter—A low frequency filter found on the preamp/control center which usually starts to work below 100 Hz, and is used to reduce turntable rumble.

Magnaplanar Speaker—A speaker with a flat diaphragm which has straight wire conductors attached to it. The magnetic interaction between the current and external magnetic fields causes the diaphragm to vibrate.

Magnetic Cartridge—(See **Phono Cartridge**.)

Magnetic Field—The region of altered space which exists in the vicinity of either a magnet or an electric current. The field consists of closed magnetic lines whose direction is determined by the way a compass needle points. Outside a magnet the lines appear to originate at a north pole and point toward the south pole.

Magnetic Field Lines—(See **Magnetic Field**.)

Magnetic Pole—The end of a magnet, referred to as either the north pole or south pole. The north pole was so named because a compass needle points to the earth's geographic north pole.

Magnetic Structure—The part of a driver which contains a magnet and the necessary iron required to produce a permanent magnetic field around the voice coil.

Magnetic Tape—A thin polyester or acetate film with a fine layer of magnetic particles coated on one surface. The magnetic particles may be iron oxide, chromium dioxide or other formulations.

Magnitude of a Vector—The "size" of a vector.

MAIN AMP IN Jack—An electrical connector on either a receiver, an integrated amplifier or a power amplifier, which can be used to connect an external preamplifier/control center or other accessories to the power amplifier.

Mass—A quantitative measure of an object's inertia. Mass and weight are different. In the metric system mass is measured in kilograms (kg) or grams (gm).

Master Record—The original copy of a recording from which the commercially available records are mass produced.

Maximum Input Signal—A specification which gives the largest input voltage that an amplifier can accept, when its VOLUME control is set low, without producing a clipped output signal.

Maximum Power Rating—A loudspeaker specification which states the maximum electrical power that the loudspeaker can accept before serious damage is incurred.

Maximum Tracking Angle Error—A specification which gives the largest deviation of the stylus (in degrees) away from perfect tangency to the record groove. The error occurs because a pivoted tone arm cannot keep the cartridge tangent to the groove at all times.

Mechanical Energy—Energy which is associated with the motion of a body. For example, a rotating turntable platter and a vibrating stylus have mechanical energy.

Mechanics—A field of science which studies the motion of objects.

Metal Tape—A special type of tape whose magnetic coating does not contain the oxygen atoms found in other types, such as ferric oxide and chromium dioxide. Metal tapes produce significantly higher output levels than other tapes. However, they are very expensive and require specially made tape decks for recording onto them.

Micron—0.00004 inch.

Microphone—A device which converts sound into electrical signals.

Microvolt (μV)—One millionth (10^{-6}) of a volt.

Midrange Speaker—A driver which is used to reproduce the midrange frequencies, usually between 500 and 5,000 Hz.

Millivolt (mV)—One-thousandth of a volt.

Minimum Recommended Power Rating—A loudspeaker specification which states the input electrical power which is required to produce reasonable sound levels in an average room.

Mistracking—The loss of contact between the stylus and the groove walls which sometimes occurs during the playing of a record.

Modulation—A technique by which information is conveyed by a carrier wave in which some property of the wave is allowed to change: e.g., its amplitude or its frequency. The opposite technique is called demodulation, whereby the information is recovered from the carrier wave. Tuners demodulate the modulated carrier waves to recover the audio information.

Moment of Inertia—The property of an object that resists any change in its rotational motion when a torque is applied.

Momentum—The product of an object's mass times its velocity, described by the relationship: $\mathbf{p} = m\mathbf{v}$. Momentum is a vector.

MONITOR Switch (on a Tape Deck)—A switch which allows the user to monitor the signal either before it goes to the record head or after it has been recorded. If the tape deck has separate record and playback heads, the tape can be monitored a fraction of a second after it has been recorded.

Mono (Monophonic)—A word which is used to describe an audio system which has only one "sound" channel.

Moving Coil Microphone—A magnetic type of microphone which utilizes a stationary permanent magnet and a moving coil.

Moving Coil Phono Cartridge—A magnetic type of phono cartridge which utilizes a stationary permanent magnet and a moving coil.

Moving Magnet Microphone—A magnetic type of microphone which utilizes a stationary coil and a moving permanent magnet.

Moving Magnet Phono Cartridge—A magnetic type of phono cartridge which utilizes a stationary coil and a moving permanent magnet.

Multipath Interference—The interference resulting between two radio waves which have identical RF frequencies, and which arrive at the antenna via two or more different paths. One path is a direct path, and the others exist due to reflections of the wave from buildings and other objects.

Multitrack Tape—Magnetic tape which has been divided along its length into separate lanes or tracks for recording separate signals.

MUTE—(See **FM MUTE.**)

NAB (National Association of Broadcasters)—(See **Rumble.**)

Natural Frequency—The frequency at which an object or system naturally vibrates. The natural frequency is determined by the object's weight, shape, and type of material. Also called the resonant frequency.

Negative Feedback—A technique for reducing amplifier distortion, which involves taking a small portion of the output signal, turning it around so that it is exactly out of phase with the input signal, and feeding it back into the input.

Neutron—(See **Atom.**)

Newton—A unit of force in the metric system.

Newton's First Law—Every object continues in a state of rest, or in a state of constant motion along a straight line, unless it is compelled to change that state by forces impressed upon it.

Newton's First Law for Rotational Motion—Every object continues in a state of rest, or of constant rotational motion, unless it is compelled to change that state by torques impressed upon it.

Newton's Second Law—An object will accelerate when a force is impressed upon it. The acceleration $\mathbf{a}$ is directly proportional to the net force $\mathbf{F}$ acting on the object, and inversely proportional to its mass m: $\mathbf{a} = \mathbf{F}/m$.

Newton's Third Law—Whenever one object exerts a force on a second object, the second object always exerts an equal, but opposite, force on the first.

Nodes—Points along a standing wave where the medium does not vibrate.

Normal—The direction perpendicular to a surface.

Note—A musical sound wave which consists of a specific fundamental frequency plus any of its harmonics or overtones. The sound wave corresponding to a note is a complex wave.

Nucleus—(See **Atom.**)

Offset Angle—The angle at which the cartridge is held with respect to a line joining the cartridge to the pivot point of the tone arm. Offset angles help to reduce tracking angle errors.

Ohm—A unit of electrical resistance or impedance.

Ohm's Law—An important law in electricity which states that voltage equals the resistance times the current: Voltage = (Resistance) × (Current).

One Motor Drive—(See **Tape Transport.**)

Open-Reel Tape Deck—A tape deck which uses two separate reels for tape storage. Open-reel tapes can travel at 1–7/8, 3–3/4, 7–1/2, 15, and 30 ips. They are 0.64 cm (1/4″) wide and contain four tracks.

Out-of-phase—(See **Phase**.)

Output Voltage—The amount of voltage generated by a cartridge when the stylus is vibrating at a standard frequency (usually 1 kHz) and at a specified stylus speed (usually in the range from 3.5 cm/sec to 5 cm/sec).

Overtones (See Harmonics.)—Frequencies which are multiples of the lowest or fundamental vibrating frequency. The first overtone has twice the frequency of the fundamental, the second overtone has three times the fundamental frequency, etc.

Parallel Connection—A method of connecting two or more electrical devices to the same source of voltage such that each device receives the full voltage being supplied.

Passive Resonator—A type of loudspeaker design which is similar to the bass reflex design, except that the cabinet is not openly vented. Instead, a second woofer, which is not electrically driven, covers the vent and it is called the passive resonator.

PAUSE—A control on a tape deck which can be used to halt the tape, while leaving the capstan rotating and leaving the input signal at the record head (so that it will register on the record level meters). It is very useful for adjusting input levels to produce the highest quality tapes. It is not the same thing as the "STOP" control.

Peak Reading Meter—(See **dB Level Meter**.)

Peaks—(See **Transients**.)

Period—The time required for one complete wave cycle (crest-trough combination) to pass a particular point. The period of a wave is related to its frequency via: $\text{period} = \dfrac{1}{\text{frequency}}$.

Permanent Magnet—A magnetic material in which the magnetic domains remain permanently aligned.

Phase—A term which describes the relative position of one wave with respect to another. Two identical waves are said to be "in-phase" when crests meet crests and troughs meet troughs; constructive interference results when two waves are in-phase. Two identical waves are "out-of-phase" when the crests of one wave meet the troughs of another, and vice versa; destructive interference results when two waves are out-of-phase.

Phase Inverter—(See **Bass Reflex**.)

Phono Cartridge (or Pick-up)—A small unit which is located on a turntable at the end of the tone arm and holds the stylus (needle). It is the job of the cartridge to transcribe the intricate movements of the stylus, as it vibrates in the groove, into an electrical signal which is sent to the PHONO input of the preamplifier/control center. There are ceramic cartridges which utilize the piezoelectric effect in their operation, and magnetic cartridges, which utilize Faraday's law of induction in their operation.

Phonograph—(See **Record Player**.)

Phono Jacks—The source input jacks to a preamp where a turntable is connected.

Pick-up—(See **Phono Cartridge**.)

Piezoelectric Effect—The ability of certain materials to produce an electrical voltage when they are bent or twisted.

Pinch Roller—A rubber roller which presses the tape against the capstan.

Pitch—The "highness" or "lowness" of a musical tone as perceived by a listener. A high frequency vibrating source produces a sound of high pitch; a low frequency vibrating source produces a sound of low pitch.

Pitch Control (Speed Control)—A control used on turntables to correct for slight platter speed errors. This control usually works in connection with a stroboscopic light.

Platter—A heavy, metal disc which is located on a turntable and which holds and turns the records during playback.

PLAY—A control on a tape deck which allows the tape to move forward at its normal speed for playback. It is also used in conjunction with the "RECORD" control when recording signals onto tape.

Playback Head—A device found in tape decks. The playback head responds to the changing magnetic patterns in the recorded tape and generates the electrical audio signal. A playback head is made from a coil of wire wrapped around an iron core, and it uses Faraday's law of induction in its operation. A playback head is not an electromagnet.

Playback Level Meter—(See **dB Level Meter and VU Level Meter**.)

Player Deck—(See **Tape Player**.)

Pole—(See **Magnetic Pole**.)

Portables—Complete audio systems which are small enough to be conveniently carried.

Potential Energy (Gravitational)—The energy of an object due to its height above the earth's surface, described by the relationship: Potential energy = mgh.

Power—The amount of energy being produced or consumed every second. A common unit of power is called a watt. If the power rating of an amplifier is, for example, 20 watts, then the amplifier will produce and deliver 20 units of electrical energy per second. Likewise, a speaker which is producing 3 watts of acoustical power creates 3 units of sound energy each second.

Power Amplifier—An electronic device which receives the relatively weak electrical signals from the preamplifier/control center and boosts them to such an extent that the signals are capable of driving speakers. The output power of a power amplifier is measured in watts. The power amp may be purchased as a separate audio component, or it may be incorporated into either an integrated amplifier or a receiver.

Power Law—A law which states that power equals voltage times current:

Power = (Voltage) × (Current).

Power Rating—(See **Continuous Average Power Output.**)

Power Supply—An electrical unit found inside amplifiers which provides the source of energy (current) for their operation.

Preamplifier/Control Center (or Preamp)—A collection of electronic circuits which accepts inputs from the various sound sources (turntable, tuner, tape deck, and possibly a microphone), adjusts the signals in level and tonal characteristics according to front panel control settings, and finally passes the signals on to the power amplifier. A preamp may be purchased as a separate audio component or it may be incorporated as part of either an integrated amplifier or a receiver.

PREAMP OUT Jack—An electrical connector from which the preamp output signal may be extracted. This signal can then be used with either another power amplifier or sent to other accessories like an equalizer. This jack may, or may not be present—depends on the manufacturer.

Preassembled Hi-Fi Systems—Complete audio systems in which the four building blocks have been selected and assembled by the manufacturer for the convenience of the consumer.

Pre-emphasis—The boosting of a specified range of input frequencies to a component such as a tuner, turntable, or a tape deck. A reverse process, called de-emphasis, occurs in the output which restores the audio signal to its original level. Any noise introduced between the input and output, which was not pre-emphasized, is reduced by the de-emphasis, thus producing a larger S/N ratio. The process is slightly different for each type of source, and it can be thought of as a type of equalization.

Principle of Linear Superposition—A very important and useful law which states that when two or more waves are present in the same region of space at the same time, the resulting wave is simply the sum of the individual waves.

Print-through—A phenomenon which occurs when a magnetic tape is wound on a spool, and the magnetic fields of one layer penetrate directly through the tape and print faint copies of themselves onto adjacent layers. Print-through is more severe with the thinner tapes.

Proton—(See **Atom.**)

Pure Tone—(See **Note** and **Sine Wave.**) A sound wave which consists only of a single frequency, with no higher harmonics. A pure tone is not the same as a musical note.

Radiation—A process in which energy is transferred by an electromagnetic wave such as infrared or light waves.

Radio—A unit which contains a tuner, a preamplifier/control center, a power amplifier, and speakers. A radio and a tuner are not the same thing.

Radio Frequency (RF) Signal—An electromagnetic wave which can carry information, either by AM or FM, from a transmitter to the tuner. RF signals in the AM broadcast band range from 535 kHz to 1605 kHz, while those in the FM band possess frequencies from 88 MHz to 108 MHz.

Rarefactions—Regions along a sound wave where the pressure is smallest. Also called troughs. In a longitudinal slinky, rarefactions are regions where the coils are "stretched out."

Receiver—A single unit that combines a tuner, preamp, and a power amp. A receiver does not have speakers incorporated into it.

RECORD—A control on a tape deck which is used in conjunction with the "PLAY" control when recording signals onto the tape.

Record Changer—(See **Turntable.**)

Record Groove—The V-shaped trough which is cut into the surface of a record. The sides of the "V" are perpendicular. In a stereo record one wall of the groove contains the undulations which lead to left channel sound, while the other wall contains the undulations which lead to right channel sound.

Recording Head—A device found in tape decks. It is an electromagnet which transcribes the audio signals into magnetic patterns on the tape.

Record Level Meter—(See **dB Level Meter** and **VU Level Meter**.)

Record Player (or Phonograph)—A complete audio system, including a preamp, a power amp, and speaker(s), which has a turntable for its source of sound.

Reflection—The ability of a sound wave to "bounce off" an object.

Reflection, Law of—When a sound wave impinges on a smooth surface, the angle of reflection equals the angle of incidence.

Refraction—The bending of a sound wave caused by differences in the wave's speed as it travels through nonuniform media.

Refraction, Law of—When sound travels from a low-velocity medium into a higher-velocity medium, the refracted sound will bend away from the normal. If the sound travels from a higher-velocity medium into a lower-velocity medium, the refracted sound will bend toward the normal. The amount of bending depends on the sound velocities in the two media and the angle of incidence.

Residual Magnetism—(See **Induced Magnetism**.)

Resistance (Electrical)—The opposition to the flow of electrical current. Resistance is a result of "electrical friction" between the atoms of a material and the flowing electrons. Resistance is measured in ohms.

Resistor—An electronic component which offers resistance to the flow of electrical current.

Resonance—The setting up of vibrations in a body at one of its natural frequencies of vibration by a vibrating force, or wave, which has the same frequency.

Resonance Peak—The large output sound power which is generated when an audio component is vibrating at its natural, or resonant, frequency.

Resonant Frequency—(See **Natural Frequency**.)

Resultant Vector—A vector which represents the vector sum of two or more vectors.

Reverberation—Numerous reflections of sound which arrive at the listener close together and cannot be perceived as individual echoes.

Reverberation Time—The time required for the reverberated sound to decay to one-millionth of its original level after the speakers have been turned off. Typical reverberation times are between 0.5 and 2.2 seconds.

Reverberation Unit—A component that will add reverberation to the sound when connected to an audio sytem.

REWIND—(See **FAST REVERSE**.)

RF—Radio frequency.

RIAA Equalization—The specific equalization process used for phonograph records. When a record is cut, the bass frequencies are deemphasized and the high frequencies are boosted relative to the midrange frequencies. On playback, the RIAA (Recording Industry Association of America) phono equalization preamplifier reverses these effects to produce a flat frequency response. The advantages of the RIAA equalization process are that it reduces record noise and allows about 20 minutes or more of audio to be recorded on each side of a record.

RIAA Phono Equalization Amp—A collection of electronic circuits which boosts and "equalizes" the signal coming from magnetic phono cartridges. The RIAA (Recording Industry Association of America) equalization amplifier is found in the preamplifier/control center section of receivers and integrated amplifiers.

Rim Drive—(See **Idler Wheel Drive**.)

RMS—An abbreviation which stands for "root mean square," a kind of mathematical process for calculating a type of AC voltage or AC current. Sometimes the continuous average power output of an amplifier is referred to (incorrectly) as the RMS power.

Rumble—The low frequency "rumbling" sound which often results from extraneous vibrations of the turntable platter. It may be measured in terms of unweighted (NAB) or weighted (DIN, ARLL) decibels.

RUMBLE (LOW) Filter—(See **LOW Filter**.)

RUSH—(See **FM MUTE**.)

Saturation—A condition of a magnetic tape which occurs when all of its miniature atomic magnets are aligned to produce maximum magnetization. In this condition the tape can no longer respond to an input signal and severe audio distortion results.

SCA—(See **Subsidiary Communications Authorization**.)

SCRATCH Filter—(See **HIGH Filter**.)

Selector Switch—A switch which is located on the front panel of a preamp, integrated amp, or a receiver which directs the electrical signals from a sound source (turntable, tuner, tape deck, microphone) to the remainder of the audio system.

Sensitivity—(See **Fifty (50) dB Quieting Sensitivity** and **Usable Sensitivity**.)

Sensitivity Rating of a Loudspeaker—A specification which gives the level of the sound intensity, measured in dB, that a loudspeaker produces at a distance of (usually) 1 meter when it receives 1 watt of input electrical power.

Series Connection—A method of connecting two or more electrical devices such that all devices have the same current flowing through them.

Signal—(See **Electrical Signal**.)

Signal Strength Meter—A meter which measures the RF level entering the tuner. It is used as an aid when tuning in the desired station.

Signal-to-Noise Ratio (S/N)—A specification which tells how much audio signal there is relative to the noise under specified conditions for a hi-fi component. It is the ratio of the total audio output level to the noise level, expressed in decibels.

Sine Wave—A general term which is used to designate any type of wave (sound, water, slinky, etc.) which consists only of a single frequency. Sine waves are represented by very smooth-looking curves.

Skating Force—A force which pushes the stylus toward the center of the record. The skating force is caused by the frictional force between the stylus and the groove.

Sound—A longitudinal wave consisting of an alternating series of pressure condensations and rarefactions which must have a medium in which to travel.

Sound Intensity—(See **Intensity**.)

Sound Level Meter (dB Meter)—A small portable meter which compares sound levels measured in decibels, relative to a standard level called 0 dB. The scale on a sound level meter usually ranges from 0 dB to 140 dB.

Sound Power—(See **Power**.)

Sound Quality—(See **Timbre**.)

Sound Source—A general phrase used to designate a turntable, tuner, tape deck, or a microphone. Sound sources are not complete audio systems because the sources require additional components (preamp, power amp, and speakers) to produce sound. However, headphones may be used directly with tuners or tape decks.

Source—(See **Sound Source**.)

Source of Sound—(See **Sound Source**.)

Source Preamplifier—A name used in this text to denote a class of amplifiers which equalize and amplify the small source signals to a value of about 0.5 volt. Source preamplifiers are used in tuners, in tape decks, with microphones, and with phono cartridges.

Speaker—An electromechanical device which converts the incoming electrical signals into sound.

Specific Heat—The number of kilocalories of heat needed to raise the temperature of one kilogram of a material by one Celsius degree.

Spectrum Analyzer—An electronic device which will decompose a complex wave into a sum of harmonically related sine waves. The spectrum analyzer will yield both the amplitudes and the frequencies of the constituent sine waves.

Speed (Average)—A measure of how fast an object or a wave is moving. Speed is the distance which is traveled divided by the time required to make the trip; e.g., 344 meters per second. Speed, unlike velocity, does not convey any information on the direction in which the object or wave is moving.

Speed Control—(See **Pitch Control**.)

Spider—A part of the speaker which ensures that the voice coil is properly centered about the magnetic structure, thus preventing rubbing between the two.

Spindle—The post on which a tape reel is held.

Spurious Response Rejection—(See **Image Rejection**.)

SQUELCH—(See **FM MUTE**.)

Standing Waves—Stationary wave patterns, consisting of an alternating sequence of nodes and antinodes, formed in a medium when two identical waves pass through the medium in opposite directions.

Static Balancing—A method for adjusting the tracking force using a movable weight which slides along the back of the tone arm.

Static Friction—(See **Friction**.)

Step-down Transformer—(See **Transformer**.)

Step-up Transformer—(See **Transformer**.)

Stereo—An audio system which has two channels called "left" and "right". "Stereo" does not necessarily imply hi-fidelity, or good quality.

Stereo Separation—A specification which tells how much sound has leaked from the channel for which it was intended into the other channel for which it was not intended. Technically it is the ratio of the power of the audio signal, in the channel for which it is intended, to the power of this signal which has leaked over to the other channel. This ratio is expressed in decibels.

Stereo Threshold—The minimum RF signal level needed at the tuner input for the tuner to be able to produce a stereo program. It is measured in microvolts or dBf.

STOP—A control on a tape deck which stops the movement of the tape and prevents the input signal from reaching the record head. It is not the same thing as the "PAUSE" control.

Stylus—A carefully shaped and polished piece of diamond which moves in the record groove. Cartridges come equipped with stylus shapes of various kinds. The most common kinds are conically or elliptically shaped.

Subcarrier—A 38 kHz electrical signal which is "carried" by the main FM radio wave when a stereo program is being broadcast. This extra signal is a subcarrier in the sense that it carries the extra audio information from which a stereo tuner extracts the second channel of sound. The prefix "sub" is used to distinguish it from the main carrier wave.

Subsidiary Communications Authorization (SCA)—A monaural broadcast (200–5,000 Hz) which is used as background music in restaurants, airline terminals, offices, etc. The SCA program is transmitted by many FM stations right along with their normal stereo programming; however, a special decoder must be added to the tuner in order to receive the SCA program.

SUBSONIC Filter—A low frequency filter, sometimes found on the preamp/control center, which usually starts to work below 20 Hz, and is used to reduce the effects of tone arm resonance and rumble.

Suspension—The way in which the turntable drive system is supported in its frame. It may rest on four springs, for example. A good suspension helps to isolate the record from extraneous vibrations.

Suspension (Flexible Edge)—The elastic material which fastens the edge of the diaphragm to the rim of the basket. The suspension is an important part of a loudspeaker design.

Synchronous Motor and Hysteresis Synchronous Motor—Motors whose speeds depend on the frequency of the power source (i.e., 60 Hz house current), and are more stable than those of induction motors.

Tape Deck—A unit which includes a tape transport, the recording heads (erase, record, and playback), and the recording and playback preamplifiers. Tape decks require a power amplifier in order to drive loudspeakers.

Tape Head—(See **Erase Head, Playback Head,** and **Recording Head**).

Tape Hiss—A constant "hissing" noise which exists on magnetic tapes and can be usually heard during quiet passages.

TAPE IN JACKS—Jacks which are located on the rear panel of a preamp, an integrated amp, or a receiver, to which the outputs from a tape deck, or other audio accessories, are connected.

Tape Monitoring—Listening to a tape recording a fraction of a second after the recording is made. This is possible only if the tape deck has separate record and playback heads. The tape monitoring feature is used in conjunction with the TAPE MONITOR switch found on most preamps.

TAPE MONITOR Switch—A circuit-interrupt switch which is located in the preamp. The TAPE MONITOR switch allows a tape deck to be inserted for either playback-only purposes, or for tape monitoring a recording which has just been made. In addition, the TAPE MONITOR switch also permits a variety of other audio accessories, such as noise reduction units, reverberation units, filters, and equalizers to be connected to the system.

TAPE OUT Jacks—Jacks which are located on the rear panel of a preamp, integrated amp, or a receiver, to which a tape deck is attached for recording purposes. The program material, as selected by the SOURCE switch, is present at the TAPE OUT Jacks.

Tape Player (Player Deck)—A tape deck or tape recorder which is minus the recording facilities. Thus, it can only play prerecorded tapes.

Tape Recorder—A complete sound system which contains a tape deck, preamp, power amp and loudspeakers.

Tape Transport—The drive system for a magnetic tape deck. It can include one motor which rotates both reels and the capstan (one motor drive), a separate motor for the reels and the capstan (two motor drive), or separate motors for each reel and the capstan (three motor drive).

Temperature—A measure of a substance's molecular activity caused by the presence of heat energy. The most common scales on which temperature is measured are the Centigrade, or Celsius, scale and the Fahrenheit scale.

THD—(See **Total Harmonic Distortion.**)

Thermal Conductivity—The ability of a material to conduct heat. In other words, the kilocalories of heat flowing per second through a distance of one meter of material, whose cross-sectional area is one square meter, and between whose ends is maintained a temperature difference of one degree Celsius.

Thermistor—(See **Heat Sensitive Resistance Thermometer**.)

Three Head Deck—A tape deck which includes a separate erase head, a separate playback head, and a separate record head.

Three Motor Drive—(See **Tape Transport**.)

Three-way Speaker System—A loudspeaker which uses three types of drivers to cover the audio range: woofer, midrange, and tweeter.

Threshold of Hearing (TOH)—The smallest sound intensity which can just be heard by a normal ear. At 1 kHz the threshold of hearing is 1×10^{-12} watts/m^2, which is often called the 0 dB level. The threshold of hearing strongly depends on the sound frequency.

Throw—A term which is used to describe the maximum displacement of a speaker diaphragm as it vibrates; e.g., a long throw speaker.

Timbre (or Sound Quality)—A term which describes the characteristic tonal quality of the sounds emitted by a musical instrument. The timbre of a musical note is governed by the number and relative amplitudes of the harmonics or overtones which are present.

Tone—A musical note.

Tone Arm—The arm of a turntable that extends over the record and holds the phono cartridge in place.

Tone Arm Pivot Friction—A specification which gives the amount of frictional force encountered by the tone arm as it moves on its pivot.

Tone Arm Resonance—A specification which gives the natural frequency of vibration of a tone arm. Its value depends on the cartridge being carried by the tone arm.

Tone Control Section—The section of a receiver, integrated amplifier or a preamplifier/contol center which contains the BASS, TREBLE, VOLUME, and FILTER controls.

Torque—The product of the force times the lever arm. A torque causes and object to rotate either faster or slower.

Total Harmonic Distortion (THD)—A specification which gives the undesired harmonic frequencies which are added to a fundamental frequency as it passes through a hi-fi unit. It is expressed as a percentage relative to the total audio output.

Track—The magnetized strip on a magnetic tape laid down by the recording head. There are typically several tracks on a single magnetic tape.

Trackability—The ability of a stylus to track the record groove.

Tracking Angle Error—(See **Maximum Tracking Angle Error**.)

Tracking Force—The downward force that the stylus exerts on the record groove.

Transformer—A device which consists of two coils of wire wrapped around an iron core, and which utilizes Faraday's law of induction in its operation. The input coil is called the primary coil, or primary, while the output coil is called the secondary coil, or secondary. Transformers are used to increase, or step-up, AC voltages as well as to decrease, or step-down, AC voltages.

Transients (Peaks)—Sudden increases in sound power which occur over very short periods of time, such as a fraction of a second. A cymbal crash is a good example of a transient.

Transistor—A solid state device in which the input current controls the amount of current in the output circuit. A transistor is designed such that small changes in the input current lead to large changes in the output current. A transistor is used for amplifying electrical signals.

Transverse Wave—A wave which causes the particles of the medium to vibrate perpendicular (or transverse) to the direction in which the wave travels.

TREBLE Control—(See **BASS/TREBLE Controls**.)

Troughs—(See **Crests/Troughs**.)

Tuned Port—(See **Bass Reflex**.)

Tuner—An electronic component which receives AM and/or FM radio broadcasts and converts them into electrical signals whose frequencies lie in the audio range. A tuner is only a sound source, and a preamp, a power amp, and speakers must be added to complete the audio system. A tuner is not a radio.

Tuning Meter—A meter, sometimes found in a tuner, which can be used in the tuning process. It will indicate the exact center frequency of the FM signal being received, i.e., the exact carrier wave frequency.

Turntable—A sound source which plays records. A turntable usually consists of a platter upon which the record rests, a driving motor, a tone arm, and a cartridge. There are manual, semiautomatic, and fully automatic turntables. The latter are called record changers, since they will automatically play a series of records.

Turntable Drive System (See Belt Drive, Direct Drive, and Idler Wheel Drive.)—The part of a turntable which makes the record go around. It includes the motor, some means of connecting the motor to the platter, and the platter itself.

Tweeter—A driver which is designed to reproduce the high frequency sound, usually between 5 kHz and 20 kHz.

Two Head Deck—A tape deck which includes a separate erase head and a single head with the dual purpose of recording and playback.

Two Motor Drive—(See **Tape Transport**.)

Two-way Speaker System—A loudspeaker which uses two types of drivers to cover the audio range: a woofer and a high frequency driver.

Ultrasonic—Sound whose frequency lies above the audible range; above 20 kHz.

Unbound Electrons—(See **Free Electrons**.)

Usable Sensitivity (IHF Sensitivity)—A specification which indicates how well a tuner can convert weak radio signals into a satisfactory audio signal. Usable sensitivity (or IHF sensitivity) is the smallest RF signal that will produce an audio output signal which is 30 dB above the distortion and noise. It is measured in microvolts or dBf.

Variable Resistor—A device whose resistance can be changed by moving a sliding contact on a resistor. Variable resistors are used for VOLUME and BALANCE controls.

Vector—A quantity that has both a magnitude and a direction. Examples are velocity, acceleration, force, and momentum.

Vector Addition—A method for mathematically adding two or more vectors, which takes into account the directions and magnitudes of the vectors.

Velocity—A measurement which includes both the speed of an object and the direction of its motion. Velocity is a vector.

Vent—An opening or port in a loudspeaker enclosure.

Vented Enclosures—(See **Bass Reflex**.)

Voice Coil—The part of the speaker which carries the electric current.

Volt—(See **Voltage**.)

Voltage—A measure of the amount of electrical energy that a charge possesses. A 1 volt source, such as a battery or generator, gives one joule of electrical energy to each coulomb of charge which passes through it.

VOLUME Control—A control on the preamp/control center which regulates the overall sound level from both stereo channels. The VOLUME control is a variable resistor which has been inserted into the electronic circuits.

VU Level Meter—A meter on a tape deck, which responds to the average signal level on recording or playback. It responds only slightly and incompletely to the peaks. The "0 VU" level corresponds to the level of an audio signal which contains about 1% total harmonic distortion on playback.

Watt—A unit of power. One watt equals one joule of energy per second.

Wave—A process by which energy and information can be transmitted by the periodic motion of a physical quantity. The physical quantity can be: the displacement of water for a water wave, the change in pressure for a sound wave, or the displacement of the coils for either transverse or longitudinal slinky waves.

Wavelength—The distance between successive crests, troughs, or identical parts of a wave, measured in the direction that the wave travels.

Weight—The force exerted on an object by gravity.

Weighted Measurements—A method of making audio measurements in which certain frequencies are not considered in the measurement as heavily as other frequencies.

Woofer—A low frequency driver, usually reproducing sound frequencies between 20 Hz and 500 Hz.

Work—The product of a force that acts on an object and the distance which the object moves under the influence of the force, when the force and the movement are in the same direction. The unit of work is a joule.

Wow and Flutter—Short-term fluctuations in a magnetic tape's speed or a turntable's speed. Wow is slower changes in the tape's speed or the platter's speed, those which occur between 0.5 and 6 times per second. Flutter represents more rapid variations, usually occurring between 6 and 200 times per second.

Index